MAPPING THE PROGRESS OF ALZHEIMER'S AND PARKINSON'S DISEASE

ADVANCES IN BEHAVIORAL BIOLOGY

Recent Volumes in This Series

MAPPING THE PROGRESS OF ALZHEIMER'S AND PARKINSON'S DISEASE

Edited by

Yoshikuni Mizuno
Juntendo University School of Medicine
Tokyo, Japan

Abraham Fisher
Israel Institute for Biological Research
Ness-Ziona, Israel

and

Israel Hanin
Stritch School of Medicine
Loyola University Chicago
Maywood, Illinois

Kluwer Academic/Plenum Publishers
New York, Boston, Dordrecht, London, Moscow

Proceedings of the 5th International Conference on the Progress in Alzheimer's and Parkinson's Disease, held from March 31st to April 5th, 2001, in Kyoto, Japan

ISBN 0-306-46763-1

©2002 Kluwer Academic/Plenum Publishers, New York
233 Spring Street, New York, N.Y. 10013

http://www.wkap.nl/

10 9 8 7 6 5 4 3 2 1

A C.I.P. record for this book is available from the Library of Congress

Printed in the United States of America

SCIENTIFIC ADVISORY BOARD

ACKNOWLEDGEMENTS

The Organizing Committee of the 5th International Conference on Progress in Alzheimer's and Parkinson's diseases gratefully acknowledges the support of the following organizations:

Eli Lily Japan K.K.

Eisai Co., Ltd. & Pfizer Pharmaceuticals Inc.
Pharmacia Corporation
The Japanese Owners of Race Horses Association

Osaka Pharmaceutical Manufactures' Association
The Pharmaceutical Manufacturers' Association of Tokyo

Dr. Tairiku Amakusa
Dr. Banri Amakusa
Fujimoto Pharmaceutical Corp.
Kissei Pharmaceutical Co.
Kyowa Hakko Kogyo Co.
Medtronic Japan Co., Ltd.
National Parkinson Foundation
Nippon Boehringer Ingelheim Co.
Novartis
Sumitomo Pharmaceuticals

Dr. Taro Masaoka
Dr. Masahiro Nomoto
Dr. Masatoshi Morooka
Dr. Mitustoshi Yamamoto
Daiichi Radioisotope Labs., Ltd.
GlaxoSmithKline K.K.
Janssen-Kyowa Co., Ltd.
Meiji Seika Kaisha, Ltd.

Nippon Zoki Pharmaceutical Co., Ltd.
Smoking Research Foundation
Solvay SA
Takeda Chemical Industries, Ltd
Uehara Memorial Life Science Foundation

Electronic Industries Association of Japan
The Federation of Electric Power Companies
The Federation of Steel Manufacturing Association of Japan
Tokyo Bankers Association, INC.
Toyota Motor Corporation

Bearing Industry Association of Japan
Chemical Textile Association of Japan
Communication Device Manufacturing Association of Japan
Electronics Manufacturing Association of Japan
Japan Federation of Construction Contractors
Japan Foreign Trade Council, Inc
Japan Securities Dealers Association
Mining Industry Association of Japan
Paper Manufacturing Industry Association of Japan
Petroleum Association of Japan
Photoindustry Association of Japan
Regional Banks Association of Japan
The Japan Gas Association
The Japan Warehousing Association Inc.
The Japanese Shipowner's Association
The Real Estate Companies Association in Japan
The Life Insurance Association of Japan
Trust and Banking Association of Japan
The second Regional Banking Association of Japan

PREFACE

The 5th International Conference on the Progress in Alzheimer's Disease and Parkinson's Disease took place from March 31st to April 5th, 2001 in Kyoto, Japan. This international conference was organized as a joint Congress with the 9th International Catecholamine Symposium. A total of 1258 clinicians and researchers participated in this joint congress from 38 countries in the world. This book represents the proceedings of the 5th Conference on Alzheimer's and Parkinson's disease.

The International Conference on the Progress in Alzheimer's and Parkinson's disease was first launched by Professor Abraham Fisher of Israel and Professor Israel Hanin of USA. The first conference was held in Eilat, Israel in 1985. The second conference was organized in Kyoto, Japan in 1989; the third one in Chicago, USA, in 1993, and the fourth one in Eilat, Israel in 1997.

The International Catecholamine Symposium (ICS) is an international meeting devoted to the development of basic as well as clinical research on catecholamines. The first Catecholamine Symposium was held in Bethesda, USA in 1958. Since then this symposium has occurred every 5 years. Professor Toshiharu Nagatsu was appointed as the president of the 9th International Catecholamine Symposium, which was to be held in 2001 also in Japan. Therefore, we decided to organize a joint congress of the two meetings, because there is much overlap in research between Alzheimer's disease, Parkinson's disease, and catecholamines. We thank Professor Nagatsu very much for agreeing to organizing this joint congress.

The outcome was greatly successful. This meeting was an extremely good opportunity for investigators and clinicians to meet and discuss with people who were working in somewhat different but closely related fields, and to exchange their scientific knowledge and experiences.

The topics covered in this conference included genetics, molecular biology, biochemistry, pharmacology, pathology, history, epidemiology, clinical phenomenology, diagnosis, imaging, treatment, and future perspectives of Alzheimer's and Parkinson's disease and related disorders. Thus this volume covers most of the important and developing topics related to these disorders.

We chose the city of Kyoto, the beautiful old city and former Capital of Japan for 1,000 years, as the venue for our Conference. More than 1,000 old temples and shrines with many beautiful gardens are scattered in and around Kyoto. In addition, April is the most beautiful season in Kyoto; all the cherry blossoms come into bloom in the beginning of April. Most of the participants enjoyed not only the scientific sessions but also the beautiful landscapes and warm hospitality of the city of Kyoto.

Finally, we thank all the members of the local organizing committee and those of the international advisory board for helping us to prepare and organize this conference. We also want to thank all the participants for their scientific contribution, active discussion, and warm attitude towards colleagues, who joined our meeting in Kyoto. The success largely depended on the participants themselves. Also, we thank the companies, organizations, societies, and individual persons, who supported our conference by their generous financial contributions. Without their support the success of this conference could not have been achieved.

We do hope that this volume is interesting and useful to you to update your knowledge on Alzheimer's and Parkinson's disease.

Yoshikuni Mizuno, M.D.
Abraham Fisher, Ph.D.
Israel Hanin, Ph.D.

CONTENTS

PROGRESS IN ALZHEIMER'S DISEASE

ETIOLOGY AND PATHOGENESIS OF ALZHEIMER'S DISEASE

ANIMAL MODELS OF ALZHEIMER'S DISEASE

GLYCOSAMINOGLYCANS (GAGs) IN ALZHEIMER'S DISEASE

NEW TREATMENT ON THE HORIZON IN ALZHEIMER'S DISEASE

PROGRESS IN PARKINSON'S DISEASE

ETIOLOGY AND PATHOGENESIS

FAMILIAL FRONTOTEMPORAL DEMENTIA AND RELATED DISORDERS

CELL BIOLOGY OF AMYLOIDOGENESIS:

An overview

Dennis J. Selkoe, Weiming Xia, W. Taylor Kimberly, Konstantinos Vekrellis, Dominic Walsh, William Esler and Michael S. Wolfe[*]

INTRODUCTION

Here, we recently summarize recent work from our laboratories on two interrelated aspects of the role of amyloid ß-proteins (Aß) as the initiator of Alzheimer's disease (AD): its production and its degradation. Initial cellular production of these normal, soluble peptides can be followed by degradation inside and/or outside cells, and those peptides which escape degradation can accumulate to form oligomers and polymers (aggregates) that have apparent cytotoxicity.

MATERIALS AND METHODS

All of the methods utilized in the studies summarized in this chapter have been published by us in recent reports (Citron et al., 1997; Esler et al., 2000; Podlisny et al., 1995; Podlisny et al., 1998; Qiu et al., 1998; Qiu et al., 1997; Vekrellis et al., 2000; Walsh et al., 2000; Wolfe et al., 1999; Xia et al., 1998; Xia et al., 1997).

RESULTS

Aß Production

Because several of the APP mutations as well as all of the known mutations in presenilin (PS) 1 and 2 selectively alter the γ-secretase cleavage event to heighten $A\beta_{42}$ production, we have focused in particular on the identity and nature of γ-secretase. A key

[*] Center for Neurologic Diseases, Brigham and Women's Hospital, 77 Avenue Louis Pasteur, Harvard Institutes of Medicine, Room 730, Boston MA 02115

Mapping the Progress of Alzheimer's and Parkinson's Disease
Edited by Mizuno *et al.*, Kluwer Academic/Plenum Publishers, 2002

observation from our perspective was the finding that small amounts of holoAPP could be co-immunoprecipitated with presenilin in lysates and isolated microsomal vesicles from cells expressing transfected or endogenous presenilin (Weidemann et al., 1997; Xia et al., 1997). Although this finding initially generated controversy (Thinakaran et al., 1998), it served as a major impetus for our hypothesizing that presenilin participates intimately as part of the catalytic complex by which γ-secretasemediates the putative intramembranous proteolysis of APP (Selkoe, 1998). An alternative hypothesis for the role of presenilin in the γ-secretase mechanism is that it does not form complexes with APP but rather acts as a mediator of membrane trafficking that brings the components of the γ-secretase reaction together (Naruse et al., 1998; Thinakaran et al., 1998). However, when we examined the maturation of holoAPP through the secretory pathway, i.e. the timing of the acquisition of N- and O-linked sugars, we were unable to detect any difference in this secretory processing between cells that express presenilin 1 and those that entirely lack it (Xia et al., 2000). Likewise, subcellular fractionation on discontinuous iodixanol gradients showed no definable difference in the distribution of the APP C-terminal fragments that are the immediate substrates of γ-secretase (i.e. C99 and C83) between cells that express or lack PS1 (Xia et al., 1998). We extended the original observation of DeStrooper et al (De Strooper et al., 1998) that the absence of PS1 sharply elevates the amount of these fragments in fractionated microsomes, but we observed no change in their subcellular localization. Taken together, these results suggested to us that direct participation of presenilin in the γ-secretase catalytic complex was a more tenable mechanism than an indirect role in the trafficking of the reaction.

A further finding that turned out to be critical in supporting the first over the second hypothesis was the nature of designed peptidomimetic compounds that effectively inhibited γ-secretase (Wolfe et al., 1998; Wolfe et al., 1999). Peptidomimetics based on the $A\beta_{42-43}$ cleavage site that had difluoro alchohol or difluoro ketone moieties installed at the P1 position as non-cleavable transition state mimicking inhibited $A\beta_{40}$ and $A\beta_{42}$. The inhibitory activity of the difluoro alcohol moiety signified that these transition state mimics were acting upon an aspartyl protease rather than one of the other major classes of known proteases. The results obtained with Wolfe's designed inhibitors along with molecular modeling, led to the hypothesis that γ-secretase cleavage involves a helical substrate (the APP transmembrane domain) and an unusual aspartyl protease that cleaves the substrate within the phospholipid bilayer (Wolfe et al., 1999).

These led to close inspection of the presenilin sequence for a possible aspartyl protease motif. The observation of two (and only two) intramembranous aspartyl residues in all presenilins that were predicted to be approximately in the center of TM6 and TM7, flanking the site in the proximal TM6-TM7 loop that undergoes endoproteolysis, led to the hypothesis that presenilin was itself γ-secretase and might also effect the PS endoproteolysis (an autoproteolysis) (Wolfe et al., 1999). Mutation of either TM aspartate in PS1 to alanine or glutamate resulted in a marked decrease in endogenous PS1 endoproteolytic fragments as a result of replacement by the apparently non-cleavable asp-mutant exogenous PS1. Likewise, C83 and C99 levels were markedly elevated, and Aβ and p3 levels were reduced sharply. Thus, mutation of a single TM aspartate residue produced essentially the same biochemical phenotype as deleting the entire PS1 gene (De Strooper et al., 1998).

These results suggested that PS1 was either an unusual diaspartyl cofactor γ-secretase or was itself γ-secretase, an unprecedented intramembranous aspartyl protease activated by autoproteolysis. Because reconstitution of γ-secretase activity, i.e., Aß generation from purified components (presenilin 1 and the C99 substrate) in phospholipid vesicles cannot be achieved without knowledge of the protein cofactors required for the γ-secretase reaction, our laboratories took an alternative approach to confirming presenilin as γ-secretase. Using the aforementioned peptidomimetic transition state analogs, we were able to show that these inhibitors could bind directly to presenilin heterodimers in cell lysates, isolated microsomes and even intact cells (Esler et al., 2000).

When the results summarized above are considered together, there is now strong evidence that presenilin represents the active site of γ-secretase . Although there is no precedent before this work for an intramembranous aspartyl protease, the cleavage of sterol regulatory element binding protein (SREBP) is effected by an unusual multi-transmembrane domain metalloprotease called site 2 protease that appears to have its active site within one of its TM domains (Sakai et al., 1998). To ultimately achieve Aß generation from purified components, we are currently purifying the presenilin/γ-secretase complex and identifying protein partners that may be necessary for proteolytic activity. To date, we have found that only membrane-associated proteins, not soluble PS biding partners such as ß-catenin, are necessary for the γ-secretase activity of PS complexes *in vitro*.

Aß Degradation

Several years ago, our lab conducted an unbiased screen of the conditioned media of several neural and non-neural cell lines for proteolytic activities that could degrade naturally secreted Aß peptides (Qiu et al., 1997). The results of the screen revealed that several cell types released a thiol metalloendopeptidase into their media that actively degraded secreted $Aß_{40}$ and $Aß_{42}$ monomers. Partial purification and analysis of the inhibitory profile indicated that this activity was indistinguishable from that of insulin-degrading enzyme (IDE) (Qiu et al., 1998). We have conducted a series of experiments examining IDE in various neuronal and microglial cell lines, in primary cortical neurons, in human CSF and in human brain samples (both AD and control) (Vekrellis et al., 2000). In both neuronal cultures and total human brain homogenates, the major Aß degrading activity is substantially inhibitable by insulin and 1,10-phenanthroline, consistent with IDE. The use of thiorphan or phosphoramidon, potent inhibitors of neutral endopeptidases such as neprilysin (NEP) (Iwata et al., 2000), produces very little inhibition of Aß degradation in total brain homogenates, from both AD and control subjects. However, when we prepared membrane fractions from these brain homogenates, we observed that ~30% of the Aß-degrading activity was inhibitable by thiorphan, whereas ~70% was inhibitable by insulin. Immunodepletion of IDE from human brain homogenates reduces Aß-degrading activity by approximately 60-80%, and there is a concomitant reduction in the amount of IDE.

These results suggest that IDE is a major Aß-degrading activity in both human brain homogenates and their membrane fractions, consistent with earlier results that both microglia (Qiu et al., 1998) and primary neurons (Vekrellis et al., 2000) can degrade Aß by an insulin-inhibitable protease. We believe that both IDE and neprilysin (and probably other proteases) play a role in Aß degradation in the brain. Our quantitative analyses suggest that IDE contributes the major portion of Aß-degrading activity. It is of related interest that IDE can actively degrade naturally secreted Aß monomers (both $Aß_{40}$

and Aß$_{42}$) but is much less active against low n oligomers of Aß found in conditioned media (Qiu et al., 1997). Therefore, we hypothesize IDE has a principal role in the natural degradation of Aß monomers following their secretion from cells, whereas once the monomers become insoluble, membrane-associated and/or aggregated, neprilysin may play a role as well.

At the time of our initial identification of IDE in our screen for Aß-degrading proteases in cultured cells, we proposed that genetic alterations in IDE might explain some forms of familial AD not currently linked to known genes (Qiu et al., 1998). This suggestion has led to an active collaboration with Rudolph Tanzi and colleagues in which they have performed linkage analysis and allelic association studies using markers on chromosome 10 in the vicinity of the IDE gene. Their work strongly suggests that there is an apparent late-onset FAD locus on chromosome 10q near the IDE gene (Bertram et al., 2000). Two kinds of analyses are now underway: a) further genetic linkage and association studies in late-onset kindreds using polymorphisms in the IDE gene as well as various flanking DNA markers; and b) direct analysis of Aß-degrading activity in lymphoblasts and fibroblasts cultured from AD families linked to chromosome 10.

REFERENCES

Bertram, L., Blacker, D., Mullin, K., Keeney, D., Jones, J., Basu, S., Yhu, S., McInnis, M. G., Go, R. C., Vekrellis, K., Selkoe, D. J., Saunders, A. J., and Tanzi, R. E., 2000, Evidence for genetic linkage of Alzheimer's disease to chromosome 10q, *Science* **290**, 2302-3.

Citron, M., Westaway, D., Xia, W., Carlson, G., Diehl, T., Levesque, G., Johnson-Wood, K., Lee, M., Seubert, P., Davis, A., Kholodenka, D., Motter, R., Sherrington, R., Perry, B., Yao, H., Strome, R., Lieburburg, I., Rommens, J., Kim, S., Schenk, D., Fraser, P., St George-Hyslop, P., and Selkoe, D. J., 1997, Mutant presenilins of Alzheimer's disease increase production of 42-residue amyloid β-protein in both transfected cells and transgenic mice, *Nature Med.* **3**, 67-72.

De Strooper, B., Saftig, P., Craessaerts, K., Vanderstichele, H., Gundula, G., Annaert, W., Von Figura, K., and Van Leuven, F., 1998, Deficiency of presenilin-1 inhibits the normal cleavage of amyloid precursor protein, *Nature* **391**, 387-390.

Esler, W. P., Kimberly, W. T., Ostaszewski, B. L., Diehl, T. s., Moore, C. L., Tsai, J.-Y., Rahmati, T., Xia, W., Selkoe, D. J., and Wolfe, M. S., 2000, Transition-state analogue inhibitors of γ-secretase bind directly to Presenilin-1, *Nat. Cell Biol.* **2**, 428-434.

Iwata, N., Tsubuki, S., Takaki, Y., Watanabe, K., Sekiguchi, M., Hosoki, E., Kawashima-Morishima, M., Lee, H. J., Hama, E., Sekine-Aizawa, Y., and Saido, T. C., 2000, Identification of the major Abeta1-42-degrading catabolic pathway in brain parenchyma: suppression leads to biochemical and pathological deposition [see comments], *Nat Med* **6**, 143-50.

Naruse, S., Thinakaran, G., Luo, J. J., Kusiak, J. W., Tomita, T., Iwatsubo, T., Qian, X., Ginty, D. D., Price, D. L., Borchelt, D. R., Wong, P. C., and Sisodia, S. S., 1998, Effects of PS1 deficiency on membrane protein trafficking in neurons, *Neuron* **21**, 1213-21.

Podlisny, M. B., Ostaszewski, B. L., Squazzo, S. L., Koo, E. H., Rydel, R. E., Teplow, D. B., and Selkoe, D. J., 1995, Aggregation of secreted amyloid β-protein into SDS-stable oligomers in cell culture, *J. Biol. Chem.* **270**, 9564-9570.

Podlisny, M. B., Walsh, D. M., Amarante, P., Ostaszewski, B. L., Stimson, E. R., Maggio, J. E., Teplow, D. B., and Selkoe, D. J., 1998, Oligomerization of endogenous and synthetic amyloid β-protein at nanomolar levels in cell culture and stabilization of monomer by Congo red, *Biochem.* **37**, 3602-3611.

Qiu, W. Q., Walsh, D. M., Ye, Z., Vekrellis, K., Zhang, J., Podlisny, M., Rosner, M. R., Safavi, A., Hersh, L. B., and Selkoe, D. J., 1998, Insulin degrading enzyme regulates the level of monomeric amyloid ß-protein extracellularly via degradation and oligomeriztion, *J. Biol. Chem.* **273**, 32730-32738.

Qiu, W. Q., Ye, Z., Kholodenko, D., Seubert, P., and Selkoe, D. J., 1997, Degradation of amyloid β-protein by a metalloprotease secreted by microglia and other neural and non-neural cells, *J. Biol. Chem.* **272**, 6641-6646.

Sakai, J., Rawson, R. B., Espenshade, P. J., Cheng, D., Seegmiller, A. C., Goldstein, J. L., and Brown, M. S., 1998, Molecular identification of the sterol-regulated luminal protease that cleaves SREBPs and controls lipid composition of animal cells, *Mol Cell* **2**, 505-14.

Selkoe, D. J., 1998, The cell biology of ß-amyloid precursor protein and presenilin in Alzheimer's disease, *Trends in Cell Biol.* **8**, 447-453.

Thinakaran, G., Regard, J. B., Bouton, C. M. L., Harris, C. L., Price, D. L., Borchelt, D. R., and Sisodia, S. S., 1998, Stable association of presenilin derivatives and absence of presenilin interactions with APP, *Neurobiol Disease* **4**, 438-453.

Vekrellis, K., Ye, Z., Qiu, W. Q., Walsh, D., Hartley, D., Chesneau, V., Rosner, M. R., and Selkoe, D. J., 2000, Neurons regulate extracellular levels of amyloid beta-protein via proteolysis by insulin-degrading enzyme, *J Neurosci* **20**, 1657-65.

Walsh, D. M., Tseng, B. P., Rydel, R. E., Podlisny, M. B., and Selkoe, D. J., 2000, Detection of intracellular oligomers of amyloid ß-protein in cells derived from human brain, *Biochemistry* **39**, 10831-10839.

Weidemann, A., Paliga, K., Durrwang, U., Czech, C., Evin, G., Masters, C. L., and Beyreuther, K., 1997, Formation of stable complexes between two Alzheimer's disease gene products: Presenilin-2 and β-amyloid precursor protein, *Nat. Med.* **3**, 328-332.

Wolfe, M. S., Citron, M., Diehl, T. S., Xia, W., Donkor, I. O., and Selkoe, D. J., 1998, A substrate-based difluoro ketone selectively inhibits Alzheimer's gamma-secretase activity, *J Med Chem* **41**, 6-9.

Wolfe, M. S., Xia, W., Moore, C. L., Leatherwood, D. D., Ostaszewski, B. L., Rahmati, T., Donkor, I. O., and Selkoe, D. J., 1999, Peptidomimetic probes and molecular modeling suggest Alzheimer's γ-secretase is an intramembrane-cleaving aspartyl protease., *Biochem.* **38**, 4720-4727.

Wolfe, M. S., Xia, W., Ostaszewski, B. L., Diehl, T. S., Kimberly, W. T., and Selkoe, D. J., 1999, Two transmembrane aspartates in presenilin-1 required for presenilin endoproteolysis and γ-secretase activity, *Nature* **398**, 513-517.

Xia, W., Ostaszewski, B. L., Kimberly, W. T., Rahmati, T., Moore, C. L., Wolfe, M. S., and Selkoe, D. J., 2000, FAD mutations in presenilin-1 or amyloid precursor protein decrease the efficacy of a γ-secretase inhibitor: A direct involvement of PS1 in the γ-secretase cleavage complex, *Neurobiol. Dis.*, in press.

Xia, W., Zhang, J., Ostaszewski, B. L., Kimberly, W. T., Seubert, P., Koo, E. H., Shen, J., and Selkoe, D. J., 1998, Presenilin 1 regulates the processing APP C-terminal fragments and the generation of amyloid ß-protein in ER and Golgi, *Biochem.* **37,** 16465-16471.

Xia, W., Zhang, J., Perez, R., Koo, E. H., and Selkoe, D. J., 1997, Interaction between amyloid precursor protein and presenilins in mammalian cells: Implications for the pathogenesis of Alzheimer's disease, *Proc. Natl. Acad. Sci. USA* **94,** 8208-8213.

ALZHEIMER'S DISEASE AS A PROTEOLYTIC DISORDER
Catabolism and processing of amyloid β-peptide

Nobuhisa Iwata and Takaomi C. Saido[*]

1. INTRODUCTION

The decades-long pathological cascade of Alzheimer's disease (AD) is triggered by the deposition of amyloid β–peptide (Aβ) in the brain[1,2]. Aβ is a physiological peptide, the steady-state levels of which should be determined by the balance between the anabolic and catabolic activities[3-5]. Just a 50% increase in the production of a particular form of Aβ ($A\beta_{1-42}$) caused by familial AD mutations, such as amyloid precursor protein, presenilin (PS) 1 and PS2, leads to aggressive presenile Aβ pathology[1,2], so subtle alterations in metabolic balance over a long period of time are likely to influence not only the pathological progression but also the incidence of the disease. A reduction of catabolic activity of Aβ may also promote the deposition and account for late-onset AD development[6]. In any case, up-regulation of the catabolic activity as well as down-regulation of anabolic activity should prevent or decelerate Aβ deposition (Figure 1).

2. CATABOLIC PATHWAY OF Aβ IN BRAIN

To date, it has been shown that various peptidases can proteolyze synthetic Aβ in the test tube or in cell culture[5]. However, the rate-limiting process of the degradation *in vivo* is still unknown. To address this, we developed a new experimental paradigm in which we injected multiply radio-labeled $^3H/^{14}C\text{-}A\beta_{1-42}$ into rat hippocampus and analyzed the metabolism by tracing the radioactivity[7]. We found that $^3H/^{14}C\text{-}A\beta_{1-42}$ underwent proteolytic degradation in the hippocampus with a half life of 15-20 min, and the degradation was found proceed via $A\beta_{10-37}$ as the major catabolic intermediate. To determine what kind of peptidase is involved in this process, we tested the inhibitory effects of various peptidase inhibitors. Only metalloendopeptidase inhibitors, phosphoramidon and thiorphan, effectively suppressed the degradation of $^3H/^{14}C\text{-}A\beta_{1-42}$

*Nobuhisa Iwata and Takaomi C. Saido, Laboratory for Proteolytic Neuroscience, RIKEN Brain Science Institute, Wako-shi, Saitama 351-0198, Japan.

Mapping the Progress of Alzheimer's and Parkinson's Disease
Edited by Mizuno *et al.*, Kluwer Academic/Plenum Publishers, 2002

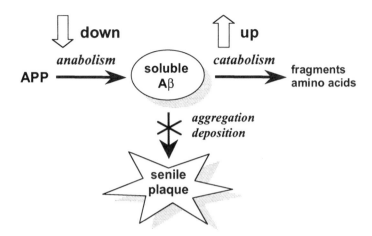

Figure 1. Metabolism of Aβ.   Aβ is a physiological peptide constantly anabolized and caabolized *in vivo*. Either down-regulation of anabolism or up-regulation of catabolism should decelerate deposition by reducing the steady-state levels of Aβ.

and the appearance of the catabolic intermediate. In addition, chronic infusion of thiorphan into rat hippocampus caused accumulation of endogenous $A\beta_{42}$ and formation of deposits. These observations indicate that a neutral endopeptidase sensitive to both phosphoramidon and thiorphan plays a major and rate-limiting role in Aβ catabolism[7]. However, the exact molecular identity of the endopeptidase has remained unclear, because at least four independent-gene-derived endopeptidases, neprilysin, phosphate-regulating gene with homologies to endopeptidases on the X chromosome (PEX), damage-induced neuronal endopeptidase (DINE) and neprilysin-like peptidase α, β, γ, with similar biochemical properties are present in the brain[8]. Despite this apparent molecular redundancy of endopeptidase activity, neprilysin would appear to be the major Aβ-degrading enzyme because this peptidase accounts for the majority of the inhibitor-sensitive endopeptidase activity in the brain and exhibits the most potent Aβ-degrading activity among these isozymes[8,9]. To clarify the contribution of neprilysin, we analyzed degradation of $^3H/^{14}C\text{-}A\beta_{1\text{-}42}$ in neprilysin gene-disrupted mouse brain using the *in vivo* experimental paradigm[10]. Wild-type mice exhibited complete degradation of radiolabeled Aβ in hippocampus 30 min after the injection. This degradation was markedly inhibited by administration of a neprilysin inhibitor, thiorphan, as in the case of rats. In contrast, the majority of the $A\beta_{1\text{-}42}$ peptide injected into the neprilysin-deficient (-/-) mouse brain remained intact even in the absence of thiorphan. Aβ degradation was also decelerated, though to a lesser extent, in the heterozygously deficient (+/-) mice.

3. REDUCTION OF NEPRILYSIN ACTIVITY IN BRAIN CONTRIBUTES TO Aβ DEPOSITION

If neprilysin is the major Aβ-degrading enzyme *in vivo*, neprilysin deficiency should result in the elevation of endogenous Aβ levels in the brain. So, we analyzed endogenous $A\beta_{40}$ and $A\beta_{42}$ contents in mouse brain using an enzyme-linked immunosorbent assay (ELISA) identical to the one employed to examine the effect of PS mutations in transgenic mice[11-13]. As a positive control, we analyzed mice carrying a familial AD-causing PS 1 gene mutation[14] and confirmed a selective 1.5-fold increase in $A\beta_{42}$ content as previously established[1,2,11-14]. The levels of $A\beta_{40}$ and $A\beta_{42}$ were significantly elevated in the neprilysin-deficient mice in a gene dose-dependent manner (Figure 2). The increase in $A\beta_{42}$ caused by the heterozygous deficiency is comparable to that seen in mutant PS transgenic mice[11-14]. In contrast, the met-enkephalin levels in the cortex and brain stem did not increase in the neprilysin-deficient mice[15] despite the fact that neprilysin is a potent enkephalin degrader previously termed 'enkephalinase,' indicating that neprilysin deficiency is compensated for by other peptidases, such as aminopeptidases, in the case of enkephalin metabolism.

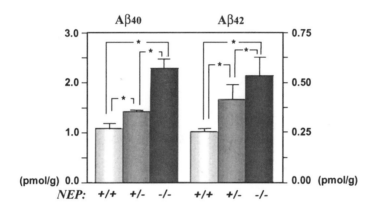

Figure 2. $A\beta_{40}$ and $A\beta_{42(43)}$ levels in wild-type and neprilysin-deficient mice. $A\beta_{40}$ and $A\beta_{42(43)}$ were extracted from mouse brains with guanidine hydrochloride and quantified as described[7]. The antibodies for the ELISA were generously provided by Takeda Chemical Industries, Ltd. Each value represents the average − SE (n = 9 mouse brains). Eight week-old male mice were used. $*p < 0.05$.

Most importantly, the inverse correlation between the neprilysin gene dose and $A\beta$ levels (Figure 2) suggests that even a partial down-regulation of neprilysin activity, which could be caused by aging, will promote $A\beta$ deposition in the brain. This finding is consistent with a recent report describing the selective reduction of neprilysin expression in sporadic AD brains, particularly in high plaque regions such as hippocampus and temporal gyrus[16]. As the expression of two controls, cyclophilin and microtubule-associated protein 2, was not reduced in AD, down-regulation of neprilysin does not seem to be a simple consequence of neurodegeneration. If this change precedes the disease onset, neprilysin down-regulation could have a causal relationship to the $A\beta$ deposition in these AD cases.

Because one feature of $A\beta$ pathology in AD brain is the distinct regional difference in severity[17], we quantified the $A\beta$ levels in various brain regions in wild-type and neprilysin-deficient mice (Figure 3). The regional Aβ levels in the wild-type mouse brain were in the order of hippocampus > cortex > thalamus/striatum > cerebellum, and this tendency was markedly exaggerated in the neprilysin-deficient mice. The order correlates well with the severity of Aβ pathology in AD brains[16,17]. This is consistent with the minimal relative increase of Aβ40 in the cerebellum, a region where mature cored Aβ plaques containing Aβ40 are rarely observed in humans[18].

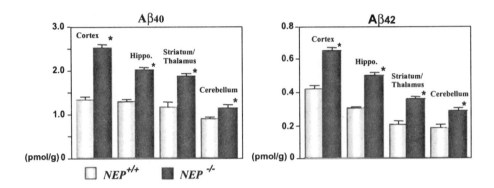

Figure 3. Aβ40 and Aβ42(43) in four brain regions of wild-type and neprilysin-deficient mice. Aβ from the hippocampus, cortex, thalamus/striatum, and cerebellum regions of mouse brains was extracted and quantified as described in Fig. 2. Each bar represents the average − SE (n = 4 mouse brains).

4. PROCESSING OF Aβ

In the pathological condition leading to AD development, the secreted Aβ appears to undergo further processing. The major forms secreted from cells are Aβ$_{1-40/42}$, whereas the Aβ species deposited in senile plaques of human brain are N-terminally truncated or modified[18,19]. A predominantly deposited is Aβ$_{42}$, beginning at pyroglutamate of the third position (Aβ$_{3(pE)-42}$), which is more hydrophobic and insoluble than the secreted forms of Aβ. Therefore, this truncated and modified form may play an important role in senile plaque formation, leading to AD pathogenesis. Transgenic mice overexpressing APP accumulate Aβ$_{1-40/42}$ extensively and predominantly, with only a small amount of Aβ$_{3(pE)-42}$[20]. Aβ forms deposited in senile plaques are clearly distinct in the human case and in transgenic mouse models, suggesting that overexpression of Aβ alone cannot mimic human pathology in a complete manner. We demonstrated that neprilysin-mediated degradation is the major catabolic pathway of Aβ under physiological conditions. If neprilysin activity is reduced with progression of aging or by unknown endogenous or exogenous inhibitory factors, the physiological catabolism of Aβ would break down. Under such aberrant conditions, Aβ$_{1-42}$ may be processed by other pathways, such as sequential truncation at the amino-terminus by aminopeptidases or dipeptidyl peptidase, followed by glutamate cyclization of Aβ$_{3(E)-42}$, affording Aβ$_{3(pE)-42}$, a pathologically deposited form[4] (Figure 4).

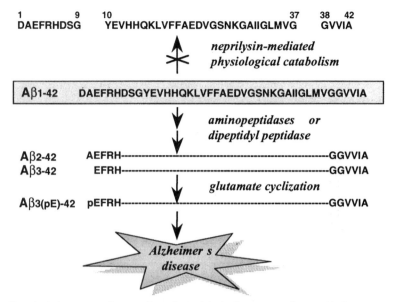

Figure 4. Hypothetical scheme of processing of physiologically secreted forms of Aβ to pathologically deposited forms.   Under aberrant conditions in which neprilysin activity is significantly reduced, a part of Aβ$_{1-42}$ may be processed sequentially at the amino-terminus by aminopeptidases, followed by glutamate cyclization of Aβ$_{3(E)-42}$ to afford Aβ$_{3(pE)-42}$, a pathologically deposited forms. pE, pyroglutamate.

5. MEDICAL APPLICATIONS

Because brain Aβ levels were inversely correlated with the number of neprilysin gene alleles (homozygous neprilysin$^{-/-}$ < heterozygous neprilysin$^{+/-}$ < wild-type neprilysin$^{+/+}$ mice), elevation of neprilysin activity in brains containing high Aβ levels by, for instance, gene therapy or transcriptional up-regulation is likely to effectively lower Aβ levels. This strategy of catabolic up-regulation would be complementary to that of anabolic down-regulation, i.e. secretase inhibition, and should not have side effects on the processing of APP and other secretase substrates[5]. These two strategies may be combined with an Aβ vaccine to achieve a maximum effect in overcoming AD.

6. ACKNOWLEDGMENT

We thank Dr. T. Kudo and Dr. M. Takeda for providing the PS1mutant (I213T) knock-in mice. Our work has been funded by research grants from RIKEN BSI, the Ministry of Education, Special Coordination Funds for promoting Science and Technology of STA, the Ministry of Health and Welfare, and Takeda Chemical Industries.

REFERENCES

1. J. Hardy, Amyloid, the presenilins and Alzheimer's disease, *Trends Neurosci.* **20**(4), 154-159 (1997).
2. D. J. Selkoe, The cell biology of β-amyloid precursor protein and presenilin in Alzheimer's disease, *Trends Cell Biol.* **8**(11), 447-453 (1998).

3. D. J. Selkoe, Physiological production of the β-amyloid protein and the mechanism of Alzheimer's disease, *Trends Neurosci.* **16**(10), 403-409 (1993).

4. T. C. Saido, Alzheimer's disease as proteolytic disorders: anabolism and catabolism of β-amyloid, *Neurobiol. Aging* **19**(Suppl. 1), S69-S75 (1998).

5. T. C. Saido, and N. Iwata, in: Neuroscientific Basis of Dementia, edited by C. Tanaka, Y. Ihara, and P. L. McGeer (Birkhauser Verlag, Basel, 2000), pp. 249-256.

6. T. C. Saido, Degradation of amyloid-beta peptide: a key to Alzheimer pathogenesis, prevention and therapy, *Neurosci. News* **3**(5), 52-62 (2000).

7. N. Iwata, S. Tsubuki, Y. Takaki, K. Watanabe, M. Sekiguchi, E. Hosoki, M. Kawashima-Morishima, H. –J. Lee, E. Hama, Y. Sekine-Aizawa, and T. C. Saido, Identification of the major Aβ$_{1-42}$-degrading catabolic pathway in brain parenchyma: suppression leads to biochemical and pathological deposition, *Nature Med.* **6**(2), 143-150 (2000).

8. K. Shirotani, S. Tsubuki, N. Iwata, Y. Takaki, W. Harigaya, K. Maruyama, S. Kiryu-Seo, H. Kiyama, H. Iwata, T. Tomita, T. Iwatsubo, and T. C. Saido, .Neprilysin degrades both amyloid β peptides 1-40 and 1-42 most rapidly and efficiently among thiorphan- and phosphoramidon-sensitive endopeptidases, *J. Biol. Chem.* **276**(24), 21895-21901 (2001).

9. Y. Takaki, N. Iwata, S. Tsubuki, S. Taniguchi, S. Toyoshima, B. Lu, N. P. Gerard, C. Gerard, H. -J, Lee, K. Shirotani, and T. C. Saido, Biochemical identification of the neutral endopeptidase family member responsible for the catabolism of amyloid β peptide in the brain, *J. Biochem.(Tokyo)* **128**(6), 897-902 (2000).

10. N. Iwata, S. Tsubuki, Y. Takaki, K. Shirotani, B. Lu, N. P. Gerard, C. Gerard, E. Hama, H. –J. Lee, and T. C. Saido, Metabolic regulation of brain Aβ by neprilysin, *Science* **292**(5521), 1550-1552 (2001).

11. K. Duff, C. Eckman, C. Zehr, X. Yu, C. M. Prada, J. Perez-tur, M. Hutton, L. Buee, Y. Harigaya, D. Yager, D. Morgan, M. N. Gordon, L. Holcomb, L. Refolo, B. Zenk, J. Hardy, and S. Younkin, Increased amyloid-β42(43) in brains of mice expressing mutant presenilin 1, *Nature* **383**(6602), 710-713 (1996).

12. D. R. Borchelt, G. Thinakaran, C. B. Eckman, M. K. Lee, F. Davenport, T. Ratovitsky, C. M. Prada, G. Kim, S. Seekins, D. Yager, H. H. Slunt, R. Wang, M. Seeger, A. I. Levey, S. E. Gandy, N. G. Copeland, N. A. Jenkins, D. L. Price, S. G. Younkin, and S. S. Sisodia, Familial Alzheimer's disease-linked presenilin 1 variants elevate Aβ$_{1-42/1-40}$ ratio *in vitro* and *in vivo*. *Neuron* **17**(5), 1005-1013 (1996).

13. N. Sawamura M. Morishima-Kawashima, H. Waki, K. Kobayashi, T. Kuramochi, M. P. Frosch, K. Ding, M. Ito, T. W. Kim, R. E. Tanzi, F. Oyama, T. Tabira, S. Ando, Y. Ihara, Mutant presenilin 2 transgenic mice. A large increase in the levels of Aβ42 is presumably associated with the low density membrane domain that contains decreased levels of blycerophospholipids and sphingomyelin, *J. Biol. Chem.* **275**(36), 27901-27908 (2000).

14. Y. Nakano G. Kondoh, T. Kudo, K. Imaizumi, M. Kato, J. I. Miyazaki, M. Tohyama, and J. Takeda, and M. Takeda, Accumulation of murine amyloidβ42 in a gene-dosage-dependent manner in PS1 'knock-in' mice, *Eur. J. Neurosci.* **11**(7), 2577-2581 (1999).

15. A. Saria, K. F. Hauser, H. H. Traurig, C. S. Turbek, L. Hersh, and C. Gerard, Opioid-related changes in nociceptive threshold and in tissue levels of enkephalins after target disruption of the gene for neutral endopeptidase (EC 3.4.24.11) in mice, *Neurosci. Lett.* **234**(1), 27-30 (1997).

16. K. Yosojima, H. Akiyama, E. G. McGeer, and P. L. McGeer, Reduced neprilysin in high plaque areas of Alzheimer brain: a possible relationship to deficient degradation of β-amyloid peptide, *Neurosci. Lett.*, **297**(2), 97-100 (2001).

17. H. Braak, and E. Braak, Neuropathological stageing of Alzheimer-related changes, *Acta Neuropathol.* **82**(4), 239-259 (1991).

18. T. Iwatsubo, A. Odaka, N. Suzuki, H. Mizusawa, N. Nukina, and Y. Ihara, Visualization of Aβ42(43) and Aβ 40 in senile plaques with end-specific Aβ monoclonals: evidence that an initially deposited species is Aβ 42(43), *Neuron* **13**(7), 45-53 (1994).

19. T. C. Saido, T. Iwatsubo, D. M. Mann, H. Shimada, Y. Ihara, and S. Kawashima, Dominant and differential deposition of distinct β-amyloid peptide species, a N3(pE), in senile plaques, Neuron, **14**(2), 457-466 (1995).

20. T. Kawarabayashi, L. H. Younkin, T. C. Saido, M. Shoji, K. Hsiao Ashe, and S. G. Younkin, Age-dependent changes in brain, CSF, and plasma amyloid β protein in the Tg2576 transgenic mouse model of Alzheimer's disease. *J. Neurosci.* **21**(2), 372-281 (2001).

AMYLOIDOGENESIS AND CHOLESTEROL

Katsuhiko Yanagisawa* and Katsumi Matsuzaki

1. INTRODUCTION

Deposition of amyloid β-protein (Aβ) , a cleaved product of amyloid precursor protein (APP), is a fundamental step in the pathogenesis of Alzheimer's diesease (AD). While Aβ is soluble at its physiological concentrations in biological fluids, including cerebrospinal fluid, blood, and cultured media, it aggregates and forms neurotoxic amyloid fibrils in AD brains. The issue of molecular processes of aggregation of soluble Aβ is one of the crucial subjects in the field of AD research; however, it remains to be elucidated. In the case of familial AD, the expression of the responsible genes is likely to accelerate amyloid fibril formation by increased generation of Aβ; however, no evidence has ever been provided, which confirms that generation and/or clearance of Aβ is altered to lead to an increase in the concentration of soluble Aβ in the brain of cases of sporadic AD, a major form of AD. Thus, it has been hypothesized that aggregation of Aβ is induced by an endogenous seed, which may be formed by binding of Aβ to molecules other than Aβ.

We previously attempted to determine the molecular mechanism underlying the initial process of Aβ aggregation in the AD brain and found a unique Aβ species which was characterized by tight binding of GM1 ganglioside (GM1) in the brains exhibiting early pathologiocal changes of AD[1]. Based on the unique molecular characteristics of GM1-bound Aβ (GM1-Aβ), including its altered immunoreactivity and extremely high aggregation potential, we hypothesized the following: first, Aβ adopts an altered conformation through binding to GM1; second, GM1-Aβ accelerates amyloid fibril formation of soluble Aβ by acting as a seed. Subsequent to our initial report, several investigators performed in vitro studies on GM1-Aβ, and their findings support our hypothesis as follows[2-6]. First, Aβ specifically binds to gangliosides; second, GM1-Aβ significantly enhances the rate of fibril formation of soluble Aβ. As far as we examined,

*Katsuhiko Yanagisawa, Department of Dementia Research, National Institute for Longevity Sciences, Obu, Japan 474-8522. Katsumi Matsuzaki, Graduate School of Biostudies, Kyoto University, Kyoto, Japan 606-8501

Mapping the Progress of Alzheimer's and Parkinson's Disease
Edited by Mizuno *et al.*, Kluwer Academic/Plenum Publishers, 2002

GM1-Aβ has never been detected in normal control brains despite the fact that Aβ is physiologically secreted into the extracellular space and GM1 is expressed in the neuronal membranes. Therefore, the question which remains to be answered is how Aβ binds to GM1 in AD brains.

We previously found that Aβ is more likely to bind to GM1 in cholesterol- and sphingomyelin-rich membranes, the lipid composition of which mimics that of lipid raft[6]. Based on these findings, we further investigated the molecular mechanism underlying the binding of Aβ to GM1.

2. CHOLESTEROL-DEPENDENT BINDING OF Aβ TO GM1 GANGLIOSIDE

To investigate the molecular mechanism underlying the binding of Aβ to GM1, we labeled synthetic Aβ1-40 with the 7-diethylaminocoumarin-3-carbonyl group at its amino terminus (DAC-Aβ), which was custom-synthesized by the Peptide Institute (Minou, Japan) and quantitatively monitored its binding to GM1 in liposomes of various lipid compositions[7]. The compound DAC is as small as a single amino acid, and it was confirmed that DAC-Aβ behaves similarly to the native peptide. Using DAC-Aβ, we first investigated the lipid specificity of Aβ binding and found that DAC-Aβ strongly binds to only liposomes containing GM1 and cholesterol. Significant binding of DAC-Aβ was never observed for the liposomes lacking GM1, such as, liposomes composed of phosphatidylcholine alone, phosphatidylcholine and cholesterol, and phosphatidylserine alone even at their extremely high lipid/peptide ratios. We then investigated how Aβ binding to GM1 was dependent on the concentrations of GM1 and cholesterol, using liposomes composed of GM1, cholesterol and sphingomyelin, which mimic the lipid composition of lipid raft. In the experiment using liposomes containing 20% GM1, it was found that the binding of Aβ to GM1 was enhanced by an increase in the cholesterol concentration in the membranes; whereas the binding of Aβ to GM1 was not dependent on cholesterol concentration in the experiments using liposomes containing 40% GM1, the content being much higher than that in any biological membranes. To clarify the molecular mechanism underlying cholesterol-induced Aβ binding to GM1, we labeled GM1 with a fluorophore, BODIPY, which is known to form an excimer that emits red-shifted fluorescence through the collision of two dye molecules in the excited state. Using BODIPY-labeled GM1, we can determine how GM1 is concentrated in the membrane. In the experiments using BODIPY-labeled GM1, liposomes containing cholesterol exhibited a strong excimer fluorescence intensity; whereas liposomes lacking cholesterol showed a much lower excimer fluorescence intensity. This result suggests that a cluster of GM1 gangliosides is formed in a cholesterol-rich environment.

3. SEEDING ABILITY OF GM1 GANGLIOSIDE-BOUND Aβ

The ability of GM1-Aβ to accelerate fibril formation of soluble Aβ was investigated in detail using liposomes containing cholesterol, sphingomyelin and GM1 at a molar ratio of 2:2:1 (SCG liposomes). The liposomes were incubated with soluble Aβ and the extent of amyloid fibril formation was determined by thioflavin T assay. The fluorescence intensity in the incubation mixture containing Aβ increased immediately following the addition of SCG liposomes without undergoing a lag phase and attained equilibrium hyperbolically. In contrast to the result of the experiment using GM1-containing liposomes, the fluorescence intensity did not increase with the addition of liposomes lacking GM1 ganglioside. To further confirm the amyloid fibril formation in the presence of SCG liposomes, we performed electron microscopic analysis and found that typical amyloid fibrils with 10 nm width and a helical structure were observed. These results strongly suggest that GM1-Aβ is formed on the membrane with high cholesterol concentration and initiates fibril formation of soluble Aβ by acting as a heterologous seed. The seeding ability of GM1-Aβ was previously reported by other group who used liposomes composed of GM1 and phosphatidylcholine[4]. Taken together with the results of our previous study and those of other groups, it is strongly suggested that Aβ specifically binds to GM1, adopts an altered secondary structure, and then accelerates the rate of polymerization of soluble Aβ by acting as a seed.

4. DISCUSSION

Since the discovery that having one of the alleles of apolipoprotein E, a major apolipoprotein in the central nervous system, is a risk factor for the development of AD, cholesterol metabolism has been one of the topics in AD research. Epidemiological studies have suggested that hypercholesterolemia is linked to the susceptibility to AD. Furthermore, it was recently reported, based on the study using animal models of AD, that hypercholesterolemia induced by high-cholesterol diet accelerates the depostion of Aβ in the brain of a transgenic-mouse model of familial AD[8]. In spite of the accumulating evidence indicating the possible association between cholesterol and AD, we are still to clarify how hypercholesterolemia or altered metabolism and/or distribution of cholesterol induces the development of AD, particularly the abnormal deposition of Aβ. In regard to the possibility of the modulation of the amyloid precursor protein (APP) processing by cholesterol, it was first suggested that an increase in the cellular concentration of cholesterol inhibits the generation of the soluble derivative of APP (sAPP)[9, 10]. Alternatively, Simons et al. reported that cholesterol depletion from the cells causes a substantial decrease in the level of secreted Aβ[11]. This finding was followed by a recent observation that cholesterol depletion from the cultured cells inhibited the activity of β-secretase, one of the proteases involved in cleaving Aβ from its precursor[12]. Furthermore, inhibition of cholesterol synthesis in vivo by treating animals with Simvastatin, an inhibitor of a cholesterol synthesis enzyme, HMG-CoA reductase, caused a decrease in Aβ concentration in the cerebrospinal fluid[13].

We previously reported that a unique Aβ species with the ability to accelerate aggregation of soluble Aβ was generated from cultured epithelial cells (MDCK cells) in a cholesterol-dependent manner[14]. Since the unique Aβ species with a seeding ability was secreted from the apical surface of the MDCK cells in an exclusively polarized manner, we hypothesized that this novel Aβ species is generated in lipid rafts, which is rich in cholesterol and transported into the apical surface. Recently, it was also reported that aggregated Aβ42 accumulated in late endosomes in the model cells of Niemann-Pick type C (NPC)[15], which is a genetic disorder in the intracellular trafficking of cholesterol. These lines of evidence strongly suggest that aberrant cholesterol metabolism or altered distribution of cholesterol in neuronal membranes is linked to the abnormal aggregation of Aβ, although it remains to be determined whether the accumulation of aggregated Aβ in NPC cells is induced by the generation of seed Aβ in a cholesterol-dependent manner.

Here, we report that an increase in cholesterol concentration can accelerate the binding of Aβ to GM1, leading to formation of GM1-Aβ, one of the candidate endogenous seed for Aβ aggregation in AD brains. Our current working hypothesis, based on the results of the present and previous studies, is as shown in Fig.1. First, the cholesterol concentration in neuronal membranes increases under certain conditions which are influenced by biological factors, including aging or altered expression of apolipoprotein E. In regard to the possibility of the alteration of the cholesterol concentration in neuronal membranes, it was previously reported that the cholesterol concentration in the exofacial leaflets of synaptic plasma membranes significantly increases with age[16] and/or deficiency of apolipoprotein E[17]. The increase in the cholesterol concentration induces formation of a GM1 cluster, which may be recognized by soluble Aβ. Alternatively, the expression of genes responsible for early familial AD, including *APP* and *presenilins*, -leads to the- enhancement of the generation of Aβ, which may increase the risk of binding of Aβ to the intracellular cholesterol-rich membrane domains with the GM1 cluster, including lipid rafts. In this context, we have to pay attention to the recent study indicating that the lipid rafts accumulate insoluble, but not soluble, Aβ42 in *presenilin 2* mutant mice[18]. Although it was suggested that lipid rafts physiologically contain a substantial amount of Aβ[19,20], it remains to be determined whether the soluble Aβ in the lipid rafts is involved in the pathological formation of insoluble amyloid fibrils. Following its binding to GM1 in cholesterol-rich membrane domains, Aβ adopts an altered structure distinct from that of soluble Aβ, which acts as a seed for soluble Aβ. Amyloid fibril formation is initiated by binding of soluble Aβ to GM1-Aβ and then by consecutive binding to the end of growing fibrils. To the best of our knowledge, GM1 ganglioside is the only molecule, other than fibrillar Aβ used as a nucleus in in vitro study, that can initiate amyloid fibril formation of soluble Aβ. It was previously suggested that amyloid-binding molecules, including acetylcholinesterase, proteoglycan and apolipoprotein E, enhance amyloid fibril formation of Aβ in vitro. However, no evidence has ever been provided, which confirms that these molecules stoichiometrically bind to soluble Aβ and initiate its polymerization through induction of the alteration of the secondary structure of Aβ.

Finally, the present study provides a new insight into the molecular link between cholesterol and Aβ. To date, the results of biochemical analyses of the lipid composition of neuronal membranes of the brain with AD have been controversial. A challenge in future study may be the detailed investigation of the distribution and dynamics of cholesterol in neurons of human brains and/or animal model of AD.

ACKNOWLEDGEMENTS

A. Kakio and H. Hayashi contributed greatly to this study. H. Naiki and K. Hasegawa were helpful in the morphological examination of amyloid formation.

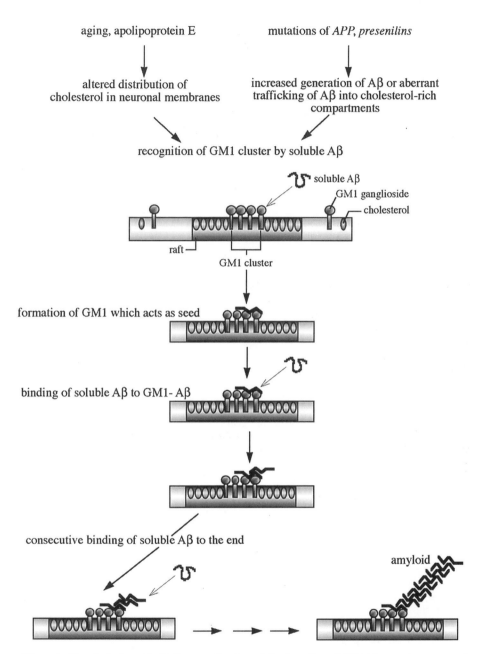

Figure 1. Hypothetical model of Aβ aggregation through the formation of GM1-Aβ, an endogenous seed.

REFERENCES

1. K. Yanagisawa, A. Odaka, N. Suzuki, and Y. Ihara, GM1 ganglioside-bound amyloid β-protein (Aβ): a possible form of preamyloid in Alzheimer's disease. *Nat. Med.,* **1**(10), 1062-1066(1995).
2. J. McLaurin, and A. Chakrabartty, Membrane disruption by Alzheimer β-amyloid peptides mediated through specific binding to either phospholipids or gangliosides. Implications for neurotoxicity. *J. Biol. Chem.,* **271**(43), 26482-26489(1996).
3. L. P. Choo-Smith, and W. K. Surewicz, The interaction between Alzheimer amyloid β(1-40) peptide and ganglioside GM1-containing membranes. *FEBS Lett.,* **402**(2-3), 95-98(1997).
4. L. P. Choo-Smith, W. Garzon-Rodriguez, C. G. Glabe, and W. K. Surewicz, Acceleration of amyloid fibril formation by specific binding of Aβ- (1-40) peptide to ganglioside-containing membrane vesicles. *J. Biol. Chem.,* **272**(37), 22987-22990(1997).
5. J. McLaurin, T. Franklin, P. E. Fraser, and A. Chakrabartty, Structural transitions associated with the interaction of Alzheimer β-amyloid peptides with gangliosides. *J Biol. Chem.,* **273**(8), 4506-4515(1998).
6. K. Matsuzaki, and C. Horikiri, Interactions of amyloid β-peptide (1-40) with ganglioside-containing membranes. *Biochemistry,* **38**(13), 4137-4142(1999).
7. A Kakio, S. Nishimoto, K. Yanagisawa, Y. Kozutsumi, and K. Matsuzaki. Cholesterol-dependent formation of GM1 ganglioside-bound amyloid β-protein, an endogenous seed for Alzheimer amyloid. *J. Biol. Chem.* (in press)
8. M. Refolo, M. A. Pappolla, B. Malester, J. LaFrancois, T. Bryant-Thomas, R. Wang, G. S. Tint, K. Sambamurti, and K. Duff, Hypercholesterolemia accelerates the Alzheimer's amyloid pathology in a transgenic mouse model. *Neurobiol Dis,* **7**(4), 321-331(2000).
9. S. Bodovitz, and W. L. Klein, Cholesterol modulates α-secretase cleavage of amyloid precursor protein. *J. Bio.l Chem.,* **271**(8), 4436-4440(1996).
10. M. Racchi, R. Baetta, N. Salvietti, P. Ianna, G. Franceschini, R. Paoletti, R. Fumagalli, S. Govoni, M. Trabucchi, and M. Soma, Secretory processing of amyloid precursor protein is inhibited by increase in cellular cholesterol content. *Biochem .J.,* **322**(Pt 3), 893-898(1997).
11. M. Simons, P. Keller, B. De Strooper, K. Beyreuther, C. G. Dotti, and K. Simons, Cholesterol depletion inhibits the generation of β-amyloid in hippocampal neurons. *Proc. Natl. Acad. Sci. U S A,* **95**(11), 6460-6464(1998).
12. E. R. Frears, D. J. Stephens, C. E. Walters, H. Davies, and B. M. Austen, The role of cholesterol in the biosynthesis of β-amyloid. *Neuroreport,* **10**(8), 1699-1705 (1999).
13. K. Fassbender, M. Simons, C. Bergmann, M. Stroick, D. Lutjohann, P. Keller, H. Runz, S. Kuhl, T. Bertsch, K. von Bergmann, M. Hennerici, K. Beyreuther, and T. Hartmann, Simvastatin strongly reduces levels of Alzheimer's disease β- amyloid peptides Aβ 42 and Aβ 40 in vitro and in vivo. *Proc. Natl. Acad. Sci. U S A,* **98**(10), 5856-5861(2001).
14. T. Mizuno, M. Nakata, H. Naiki, M. Michikawa, R. Wang, C. Haass, and K. Yanagisawa, Cholesterol-dependent generation of a seeding amyloid β-protein in cell culture. *J. Biol. Chem.,* **274**(21), 15110-15114(1999).
15. T. Yamazaki, T. Y. Chang, C. Haass, and Y. Ihara, Accumulation and aggregation of amyloid {β}-protein in late endosomes of Niemann-Pick type C cells. *J. Biol. Chem.,* **20**, 20(2000).
16. U. Igbavboa, N. A. Avdulov, F. Schroeder, and W. G. Wood, Increasing age alters transbilayer fluidity and cholesterol asymmetry in synaptic plasma membranes of mice. *J. Neurochem.,* **66**(4), 1717-1725(1996).
17. U. Igbavboa, N. A. Avdulov, S. V. Chochina, and W. G. Wood, Transbilayer distribution of cholesterol is modified in brain synaptic plasma membranes of knockout mice deficient in the low-density lipoprotein receptor, apolipoprotein E, or both proteins. *J. Neurochem.,* **69**(4), 1661-1667(1997).
18. N. Sawamura, M. Morishima-Kawashima, H. Waki, K. Kobayashi, T. Kuramochi, M. P. Frosch, K. Ding, M. Ito, T. W. Kim, R. E. Tanzi, F. Oyama, T. Tabira, S. Ando, and Y. Ihara, Mutant presenilin 2 transgenic mice. A large increase in the levels of Aβ 42 is presumably associated with the low density membrane domain that contains decreased levels of blycerophospholipids and sphingomyelin. *J. Biol. Chem.,* **275**(36), 27901-27908.(2000).
19. S. J. Lee, U. Liyanage, P. E. Bickel, W. Xia, P. T. Lansbury, Jr., and K. S. Kosik, A detergent-insoluble membrane compartment contains Aβ in vivo. *Nat. Med.,* **4**(6), 730-734(1998).
20. M. Morishima-Kawashima, and Y. Ihara, The presence of amyloid β-protein in the detergent-insoluble membrane compartment of human neuroblastoma cells. *Biochemistry,* **37**(44), 15247-15253(1998).

TOXICITY OF APP FRAGMENTS

Yoo-Hun Suh[*], Ji-Heui Seo, Yanji Xu, Chaejeong Heo, Najung Kim, Jun Ho Choi, Se Hoon Choi, Jong-Cheol Rah, Keun-A Chang, Won-Hyuk Suh

1. INTRODUCTION

Alzheimer's disease (AD) is a neurodegenerative disorder with progressive decline of cognitive function. The characteristic pathological features of the postmortem brain of AD patients include, among other features, the presence of numerous neurite plaques (NPs) in the brain, which contain a compact deposit of proteinaceous amyloid filaments surrounded by dystrophic neurites, activated microglia, and fibrillary astrocytes[1]. The amyloid filaments are composed of 39-43-amino acid amyloid β-peptides (Aβ), which are derived proteolytically from a larger transmembrane glycoprotein, the amyloid precursor protein (APP)[2]. Several reports suggest that Aβ-related peptides can be secreted constitutively under normal conditions[2]. It is likely that an overproduction, lack of degradation, and / or clearance mechanisms may trigger amyloid deposition and the development of AD pathogenesis. Indeed, several missense mutations in the APP or presenilin (PS1 and PS2) genes, which have been linked to familial AD, result in either excess generation of Aβ peptides or in a shift to the production of longer, more amyloidogenic Aβ peptides[1]. The severity of dementia has been suggested to correlate positively with the Aβ deposition observed in the postmortem brains of AD patients[3]. Furthermore, transgenic mice overexpressing human APP gene display cognitive impairments and / or some pathological features of AD brain, including NPs[4], even though neurotoxic effects of Aβ *in vivo* have been somewhat controversial[5]. These observations, together with several other lines of evidence[6] implicate Aβ deposition as an initiating and / or contributing factor in AD pathogenesis. At present, neither the mechanisms by which Aβ peptides might cause AD pathology nor the role of these peptides in normal brain function is clearly established.

Many studies have shown that Aβ is toxic to neurons *in vitro*[6] and *in vivo*[7]. However, a relatively high concentration (20 μM) of Aβ is needed to induce toxicity and some studies still failed to demonstrate the toxicity of Aβ *in vivo*[5].

[*]Yoo-Hun Suh, Dept. of Pharmacol., Coll. of Med., National Creative Research Initiative Center for Alzheimer's Dementia and Neuroscience Research Institute, MRC, Seoul Nat'l Univ., 110-799 Seoul, Korea

Mapping the Progress of Alzheimer's and Parkinson's Disease
Edited by Mizuno *et al.*, Kluwer Academic/Plenum Publishers, 2002

Table1. Summary of various effects of Aβ and CTF

	Aβ	CTF
Neurotoxicity		
Cultured cells[13-14, 28]	+	+10
in vivo I.C.V.[23-24]	+	++++
Channel Effect		
Xenopus oocyte[15-16]	-	+++
Purkinje Cells[18]	-	+++
Lipid Bilayer[17]	+	++
Microsome Ca^{2+}[20]	-	+++
LTP Hippocampus *in vivo*[19]	+	+++
Learning and memory impairment[23-24]	+	+++
Free radical generation[30]	+++	-
NO generation[25, 27]	+	+++
MAPK Signalling[25-27]	+	+++
NF-κB[25-27]	+	+++
Nuclear Translocation[31]	-	+++
Inflammatory cytokines & chemokines[25-26]	+	+++
Gliosis & Astrocytosis[25, 27]	+	+++

Aβ deposition has been found without accompanying neurodegeneration, and neurodegeneration could occur in areas with no Aβ deposition. Furthermore, it has been reported that under certain conditions in culture Aβ promotes neurite outgrowth[8] instead of exerting toxic action. Thus, Aβ may not be the sole fragment in the neurotoxicity associated with AD

Consequently, the possible effects of other cleaved products of APP need to be explored. Recently, it has been reported that C-terminal fragments of APP are found in media and cytosol of lymphoblastoid cells obtained from patients with early- or late-onset familial AD[9] and Down's syndrome[10]. These C-terminal fragments, which contain the complete Aβ sequence, appear to be toxic to neurons in culture, although the mechanism of action is not fully understood. Also, the C-terminal peptide has been identified in plaques and microvessels, and is possibly a constituent of the neurofibrillary tangles characteristic of AD[11] Furthermore, transgenic mice over-expressing the C-terminal fragment showed extensive neuronal degeneration in the hippocampal area with cognitive impairments[12] and impairment of long-term potentiation (LTP).

Here we reported that a recombinant carboxyl terminal fragment (CTF) itself caused a direct neurotoxicity in PC12 cells and primary cortical neurons[13-14], induced strong non-selective inward currents in *Xenopus oocytes*[15-16] planar lipid bilayers[17], Purkinje cells [18], and blocked the later phase of LTP in rat hippocampus *in vivo*[19]. Recently, we showed the results showing that CTF impaired calcium homeostasis[20-22], and intracerebral injection of CTF was also very toxic to animals accompanying by learning and memory impairment[23-24]. Our results also suggest that CTF may participate in AD pathogenesis by triggering inflammatory reaction through MAPKs- and NF-κB dependent astrocytosis and iNOS induction[25-27]. Collectively these results imply that CTF peptide can damage the neurons both *in vitro* and *in vivo*, and it is thought that both CTF and Aβ may participate in the neuronal degeneration in AD by different mechanisms (Table 1).

2. CARBOXYTERMINAL FRAGMENT

2.1. Neurotoxicity of CTF in Cultured Mammalian Cells

We previously reported CTF-induced cytotoxicity in various neuronal cells including primary neuronal cells[13-14, 28]. In summary, CTF induced a significant LDH release from cultured rat cortical and hippocampal neurons, PC12 cells and SHSY5Y cells in a concentration- and time- dependent manner, but did not affect the viability of U251 cells originating from human glioblastoma. Moreover, when PC12 cells were induced to differentiate into neurons by pretreatment with nerve growth factor (NGF), the cells were much more sensitive against CTF. In contrast to CTF, Aβ increased LDH release only slightly at 50 μM[13-14]. In addition, our studies using C-terminal fragments of APP without Aβ and Transmembrane (TM) or NPTY domain demonstrated that this deletion mutants significantly induced the death of NGF-differentiated PC12 cells and rat cortical neurons. Thus, our findings suggest that C-terminal end of APP without Aβ and TM itself may be involved in the neuronal degeneration, which is associated with AD[28].

2.2. CTF–induced membrane Currents in *Xenopus oocyte* and ion channels formation in planar lipid bilayer

Intracellular injection of CTF into *Xenopus oocyte* resulted in the gradual development of large inward current with a large oscillatory component. CTF also induced strong nonselective inward currents in *Xenopus oocytes* whereas Aβ did not show significant effects[15-16]. Regarding channel formation in planar lipid bilayer, CTF-induced channels are more selective for Ca^{2+} and Na^+ with the permeability sequence $P_{Ca}^{2+}>P_{Na}^+>P_{Ca}^{2+}>P_{Cs}^+>P_{Mg}^{2+}$, whereas Aβ-induced channels are more selective for Cs^+ and Li^+ with the selective sequence $P_{Cs}^+>P_{Li}^+>P_{Ca}^{2+}>P_K^+>P_{Na}^{+}$[17, 29]. Our results showed that CTF increased membrane conductance in a concentration dependent fashion and CTF was more effective than Aβ in forming channels[15-17]. These results suggest that the neuronal ionic homeostasis in AD brain can be more easily disturbed by CTF than by Aβ.

2.3. Effect of CTF on Parallel Fiber-Purkinje Cell Synaptic Transmission

Recently, we reported the effects of CTF on parallel fiber (PF)-Purkinje cell synaptic transmission in the rat cerebellum[18]. Transient inward currents associated with calcium influx were induced by localized applications of CTF to discrete dendritic regions of intact Purkinje cells. Inward currents were also observed following applications of CTF to isolated patches of somatic Purkinje cell membrane. Aβ peptides and CTF induced a great depression of alpha-amino-3-hydroxy-5-methyl-4-isoxazole propionate (AMPA) receptor-mediated synaptic transmission between PF and Purkinje cell, through a combination of pre- and post-synaptic effects. These data indicate that CTF can modulate AMPA-mediated glutamatergic synaptic transmission in the cerebellar cortex. These fragments may, therefore, be considered alternative candidates for some of the neurotoxic effects of AD[18].

2.4. Effects of CTF on the calcium Homeostasis in the Neuronal Cells

The effect of CTF on the microsomal calcium uptake by Mg^{2+}-Ca^{2+} ATPase was analyzed with rat brain microsomes. Pretreatment of CTF for 3 min inhibited calcium uptake into rat brain microsomes significantly after 10 min of incubation. In contrast, Aβ peptides had no effect on the microsomal calcium uptake at 50 µM concentration[20]. Furthermore, CTF inhibited Ca^{2+} uptake in the presence of ouabain and monensin[21]. We reported that the pretreatment with CTF for 24 h at a 10 µM concentration increased intracellular calcium concentration by about twofold in SK-N-SH and PC12 cells, but not in U251 cells[22]. Cholesterol and MK 801 in SK-N-SH and PC12 cells reduced this calcium increase and toxicity induced by CTF, whereas the toxicity of Aβ peptide was attenuated by nifedipine and verapamil. CTF rendered SK-N-SH cells and rat primary cortical neurons more vulnerable to glutamate-induced excitotoxicity. Also, conformational studies using circular dichroism experiments showed that CTF had approximately 15% of beta-sheet content in phosphate buffer and aqueous 2,2,2-trifluoroethanol solutions.

However, the content of beta-sheet conformation in dodecylphosphocholine micelle or in the negatively charged vesicles is increased to 22%-23%[22]. The results of this study showed that CTF disrupted calcium homeostasis and rendered neuronal cells more vulnerable to glutamate-induced excitotoxicity, and that some portion of CTF had partial beta-sheet conformation in various environments, which might be related to the self-aggregation and toxicity. This may be significantly possibly involved in inducing the neurotoxicity characteristic of AD[20-22].

2.5. Learning and memory Impairment and Neuropathological Changes induced by Central Injection of CTF in mice

To elucidate the *in vivo* neurotoxicities of CTF, we examined learning and memory impairments using a passive avoidance task, a Y-maze and water maze task following a single intracerebroventricular injection of CTF to mice[23-24]. CTF caused significant impairments in cued, spatial and working memory performances in a dose dependent manner. The acetylcholine (ACh) level in the cerebral cortex and hippocampus was significantly decreased, accompanying by reduction of mitochondrial pyruvate dehydrogenase (PDH) activity[24]. Also CTF induced reactive gliosis in neocortex and hippocampus and neurodegeneration in neocortex[23]. These results indicate that centrally administered CTF induces behavioral impairment and neuropathologic changes, suggesting a direct toxic effect of CTF per se[23-24].

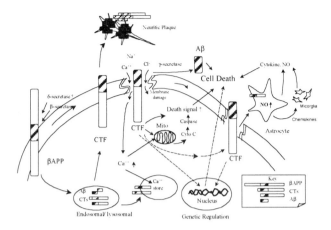

Figure 1. Hypothesis of an etiological role of amyloidogenic CTF of βAPP in AD. βAPP is internalized and processed in the endosomes/lysosomes where CTF and Aβ are produced. In AD and related disorders such as Down's syndrome, excessive production of βAPP and/or reduction of some endosomal/lysosomal activities may induce accumulation of amyloidogenic CTF of βAPP in the neuron and/or near the membrane. Intracellular CTF and Aβ may form ion channels or pores in the cell membrane or puncture holes in Ca^{2+} stores. Both actions could result in a large increase in intracellular Ca^{2+} concentration and cell damage, leading to cell death. CTF may attack mitochondria, which lead to the increased release of cytochrome c and the activation of caspase-3 and may enter nucleus to affect genetic regulation of some genes and finally to death. In addition, CTF may increase the production of NO in astrocytes and microglia, which may induce cell death. CTF may be released from the cell and/or more easily released from the damaged neurons into the extracellular space. Extracellular CTF form de novo ion channels, and this may induce neuronal death from outside the cells.

2.6. CTF-induced inflammatory reaction through MAPKs- and NF-κB dependent astrocytosis and iNOS induction

To explore the direct role of CTF in the inflammatory processes, the effects of the CTF of APP on the production of TNF-α and matix metalloproteinase-9 (MMP-9) were examined in a human monocytic THP-1 cell line. CTF elicited a marked increase in TNF-α and MMP-9 production in the presence of interferon-gamma (IFN-γ) in a dose- and time- dependent manner. Genistein, a specific inhibitor of tyrosine kinase, dramatically diminished both TNF-α secretion and subsequent MMP-9 release in response to CTF through MAPK pathway[26]. Secondly, we investigated the possible role of CTF, as an inducer of astrocytosis. We found that 100 nM CTF induced astrocytosis morphologically and immunologically. CTF exposure resulted in activation of MAPK pathways as well as transcription factor NF-κB. These suggest that CTF may participate in Alzheimer's pathogenesis through MAPKs- and NF-κB-dependent astrocytosis and iNOS induction[27]. Finally, we examined the effects of CTF on the production of cytokines and chemokines. We found that CTF induced IL-1β and TNF-α expression, and the IL-1β induced by CTF up-regulated iNOS gene expression through NF-κB activation in astrocytes and microglial cells. In addition, CTF induced astroglial and microglial chemokines such as

MIP-1α, MCP-1, and RANTES, which play roles in accumulation of microglial cells around amyloid plaques[25]

3. ACKNOWLEDGEMENTS

This study was supported by National Creative Research Initiative Grant (2000-2003) from Ministry of Science & Technology, by BK21 Human Life Sciences, and by the grant (1999-1-213-002-5) from the Basic Research Program of the Korea Science & Engineering Foundation.

REFERENCES

1. M. Mullan and F. Crawford, Genetic and molecular advances in Alzheimer's disease. Trends. Neurosci., **16**(10), 398-403. (1993)
2. C. Haass, M.G. Schlossmacher, A. Y. Hung, C. Vigo-Pelfrey, A. Mellon, B. L. Ostaszewski, I. Lieberburg, E. H. Koo, D. Schenk and D. B. Teplow, Amyloid beta-peptide is produced by cultured cells during normal metabolism Nature, **359**(6393), 322-325 (1992)
3. B. J. Cummings and C. W. Cotman, Image analysis of beta-amyloid load in Alzheimer's disease and relation to dementia severity. Lancet, **346**, 1524-1528 (1995)
4. D. Games, D. Adams, R. Alessandrini, R. Barbour, P. Berthelette, C. Blackwell, T. Carr, J. Clemens, T. Donaldson and F. Gillespie, Alzheimer-type neuropathology in transgenic mice overexpressing V717F beta-amyloid precursor. Nature, **373**(6514), 523-527. (1995)
5. J. A. Clemens and D. T. Stephenson, Implants containing beta-amyloid protein are not neurotoxic to young and old rat brain. Neurobiol. Aging, **13**(5), 581-586. (1992)
6. B. A. Yankner, A. Caceres and L. K. Duffy, Nerve growth factor potentiates the neurotoxicity of beta amyloid. Proc. Natl. Acad. Sci. U S A, **87**(22), 9020-9023. (1990)
7. K. I. Fukuchi, D. D. Kunkel, P. A. Schwartzkroin, K. Kamino, C. E. Ogburn, C.E. Furlong and G. M. Martin, Overexpression of a C-terminal portion of the beta-amyloid precursor protein in mouse brains by transplantation of transformed neuronal cells. Exp. Neurol., **127**(2), 253-264. (1994)
8. B. A. Yankner, L. K. Duffy and D. A. Kirschner, Neurotrophic and neurotoxic effects of amyloid beta protein: reversal by tachykinin neuropeptides. Science, **250**(4978), 279-282. (1990)
9. A. Matsumoto, Altered processing characteristics of beta-amyloid-containing peptides in cytosol and in media of familial Alzheimer's disease cells. Biochem. Biophys. Acta., **1225**(3), 304-310 (1994).
10. F. Kametani, K. Tanaka, T. Tokuda, and S. Ikeda, Secretory cleavage site of Alzheimer's amyloid precursor protein is heterogeneous in Down's syndrome brain. FEBS Lett, **351**(2), 165-167 (1994).
11. T. Dyrks, E. Dyrks, T. Hartmann, C. Masters, and K. Beyreuther, Amyloidogenicity of βA4 and βA4-bearing amyloid protein precursor fragments by metal-catalyzed oxidation. J. Biol. Chem., **267**, 18210-18217 (1992).
12. J. Nalbantoglu, G. Tirado-Santiago, A. Lahsaini, J. Poirier, O. Gonocalves, G. Verge, F. Momoli, S. A. Weiner, G. Massicotte, J. P. Jullien, and M. L. Shapiro, M.L Impaired learning and LTP in mice expressing the carboxy terminus of the Alzheimer amyloid precursor protein. *Nature*, **387**, 500-505 (1997).
13. S. H. Kim and Y. H. Suh, Neurotoxicity of a carboxy terminal fragment of the Alzheimer's amyloid precursor protein. J. Neurochem. **67**, 1172-1182 (1996).
14. Y. H. Suh, An Etiological role of carboxy terminal peptide of Alzheimer's amyloid precursor protein in Alzheimer's disease. J. Neurochem. **68**, 1781-1791 (1997).
15. S. P. Fraser, Y. H. Suh, Y. H. Chong and M. B. A. Djamgoz, Membrane currents induced in Xenopus oocytes by the C-terminal fragment of the β-amyloid precursor protein. J. Neurochem. 66, 2034-40 (1996).
16. S. P. Fraser, Y. H. Suh, and M. B. A. Djamgoz, Ionic effects of the Alzheimer's disease β-amyloid precursor protein and its metabolic fragments. Trends Neurosci. **20**, 67-72 (1997).
17. H. J. Kim, Y. H. Suh, M. H. Lee, and P. D. Ryu, Cation selective channels formed by C-terminal fragment of β-amyloid precursor protein. Neuroreport. **10**(7), 1427-31 (1999).
18. N. A. Hartell, and Y. H. Suh, Peptide Fragments of β-Amyloid Precursor Protein : Effects on Parallel Fiber-Purkinje Cell Synaptic Transmission in Rat Cerebellum. J. Neurochemistry 74(3), 1112-1121 (2000).
19. K. Dullen, Y. H. Suh, R. Anwyl, and M. J. Rowan, Block of LTP in rat hippocampus in vivo by β-amyloid precursor protein fragments. NeuroReport **8**, 3213-3217 (1997).

20. H. S. Kim, C. H. Park, and Y. H. Suh, C-terminal fragment of amyloid precursor protein inhibits calcium uptake into rat brain microsomes by Mg^{2+}-Ca^{2+} ATPase. NeuroReport 9(17), 3875-3879 (1998).
21. H. S. Kim, J. H. Lee, and Y. H. Suh, C-terminal fragment of Alzheimer's amyloid precursor protein inhibits sodium/calcium exchanger activity in SK-N-SH cell. NeuroReport 10, 113-116 (1999).
22. H. S. Kim, C. H. Park, S. H. Cha, J. H. LEE, S. W. Lee, Y. M. Kim, J. C. Rah, S. J. Jeong, and Y. H. Suh, Carboxyl-terminal fragment of Alzheimer's APP destabilizes calcium homeostasis and renders neuronal cells vulnerable to excitotoxicity. FASEB J. 14(11),1508-17 (2000).
23. D. K. Song, M. H. Won, J. S. Jung, J. C. Lee, H. W. Suh, S. O. Huh, S. H. Paek, M. B. Wie, Y. H. Kim, S. H. Kim and Y. H. Suh, Behavioral and Neuropathologic Changes Induced by Central Injection of Carboxyl-Terminal Fragment of β-Amyloid Precursor Protein in Mice. J. Neurochemistry 71(2), 875-878 (1998).
24. S. H. Choi, C. H. Park, J. W. Koo, J. H. Seo, H. S. Kim, S. J. Jeong, J. H. Lee and Y. H. Suh, Memory Impairment and Cholinergic Dysfunction by Centrally Administered Aβ and Carboxyl-terminal Fragment of Alzheimers APP in Mice FASEB J. 15(10):1816-8 (2001)
25. J. C. Rah, H. S. Kim, S. S. Kim, J. H. Bach, Y. S. Kim, C. H. Park, S. J. Jeong, and Y. H. Suh, Effects of Carboxy-Terminal Fragment of Alzheimer's Amyloid Precursor Protein and Amyloid beta Peptide on the Production of Cytokines and Nitric Oxide in Glial Cells. FASEB J. 15(8), 1463-1465 (2001).
26. Y. H. Chong, S. A. Shin, J. H. Sung, J. H. Chung, and Y. H. Suh, Effects of the β-amyloid and C-terminal fragment of Alzheimer's amyloid precursor protein on the production of TNF- and matrix metalloproteinase-9 by human monocytic THP-13. J. Biol. Chem. 276(26):23511-7 (2001)
27. J. H. Bach, H. S. Chae, J. C. Rah, M. W. Lee, S. H. Choi, J. K. Choi, Y. S. Kim, K. Y. Kim, W. B. Lee, Y. H. Suh, and S. S. Kim, C-Terminal Fragment of Amyloid Precursor Protein Induces Astrocytosis. J. Neurochem. 78(1):109-20 (2001)
28. J. P. Lee, K. A. Chang, H. S. Kim, S. S. Kim, S. J. Jeong, and Y. H. Suh, APP Carboxyl-Terminal Fragment Without or With Aβ Domain Equally Induces Cytotoxicity in Differentiated PC12 Cells and
29. Cortical Neurons. J Neurosci. Res. 60 (4), 565-570 (2000).
   N. Arispe, E. Rojas, and H. B. Pollard, Alzheimer's disease amyloid beta protein forms calcium channels in bilayer membranes: blockade by tromethamine and aluminum. Proc Natl Acad Sci USA 90(2), 567-571 (1993)
30. S. Janciauskiene, H. T. Wright, and S. Lindgren, Fibrillar Alzheimer's amyloid peptide Abeta(1-42) stimulates low density lipoprotein binding and cell association, free radical production and cell cytotoxicity in PC12 cells. Neuropeptides 33(6),510-6 (1999).
31. L. A. DeGiorgio, N. DeGiorgio, T. A. Milner, B. Conti, B. T. Volpe, Neurotoxic APP C-terminal and beta-amyloid domains colocalize in the nuclei of substantia nigra pars reticulata neurons undergoing delayed degeneration. Brain Res 874(2),137-46 (2000).

SIGNIFICANCE OF NEUROFIBRILLARY TANGLES IN ALZHEIMER'S DISEASE—LESSONS FROM FRONTOTEMPORAL DEMENTIA AND PARKINSONISM LINKED TO CHROMOSOME 17

Yasuo Ihara[*]

1. INTRODUCTION

Neurofibrillary tangles (NFT), one of the major neuropathological hallmarks of Alzheimer's disease (AD), consist of bundles of unit fibrils, called paired helical filaments (PHF). Because the areas forming NFT match precisely the areas exhibiting neuronal loss, the formation of NFT has been considered to be involved in a pathway leading to neuronal death. This assumption has been strengthened by a recent discovery of a great number of mutations in the tau gene in families afflicted with frontotemporal dementia and parkinsonism linked to chromosome 17 (FTDP-17).[1] This disease entity is characterized by extensive neuronal loss that is predominant in the frontal and temporal cortices, basal ganglia and midbrain, and formation of PHF or PHF-like fibrils in neurons and glial cells.[1] Thus far, ~20 exonic and intronic pathogenic mutations have been identified in the tau gene.[2] The majority of exonic mutations are localized within or close to the microtubule-binding domain, while intronic mutations are clustered in the 5'-splice site of exon 10. Most exonic mutations were claimed to have slightly or significantly decreased ability of promoting microtubule assembly,[3,4] and one mutation (R406W) is associated with a lower extent of phosphorylation within transfected cells.[5]

One speculation is that most exonic mutations reduce the affinity of tau for microtubules, leading to its destabilization, and the resultant cytosolic free tau, preferentially mutant tau, becomes highly phosphorylated and aggregates into PHF-like fibrils, which may in turn exert neurotoxicity. If so, it is important to learn the proportion of wild-type to mutant tau in affected brain which may deviate from the expected ratio (1:1). Thus, we have determined the proportion of mutant to wild-type tau in the soluble

[*] Yasuo Ihara, Department of Neuropathology, Faculty of Medicine, University of Tokyo, Tokyo 113-0033, Japan.

Mapping the Progress of Alzheimer's and Parkinson's Disease
Edited by Mizuno *et al.*, Kluwer Academic/Plenum Publishers, 2002

and insoluble fractions of FTDP-17 brains by immunochemical and protein chemical analyses. To this end, we chose two mutations showing distinct clinical courses: families with P301L which exhibit an aggressive clinical course (early onset and short duration), and families with R406W which exhibit the least aggressive course (late onset and long duration) and abundant NFT in the brain.[6]

2. P301L MUTATION

2.1. Selective Accumulation of Mutant Tau in the Insoluble Fraction and its Selective Decrease in the Soluble Fraction of P301L Brains

To distinguish between wild-type and mutant tau, we synthesized paired peptides, DNIKHVPGGGSVQC and DNIKHVLGGGSVQ, and raised paired antibodies to the peptides which specifically recognize either wild-type or mutant tau. Because the epitopes of AP301 and AL301 are located within the exon 10-encoded second repeat, these antibodies are to label only 4R-tau isoforms.

Using these antibodies, we investigated the tau in the soluble (cytosolic) and insoluble fractions of P301L brains. The Sarkosyl-insoluble fractions were prepared from an AD brain and from various cortices, hippocampi, and cerebella of three P301L patients, and subjected to Western blotting. TM2, (a pan-tau antibody; its epitope is in residues 368-386), labeled two major bands at 64 and 68 kDa and a smear in the Sarkosyl-insoluble fraction of P301L brains. AL301 clearly labeled these two bands and a smear in the same fractions. In contrast, AP301 did not label these two bands in the medial frontal gyrus or the angular gyrus from patient 1 or in medial frontal gyrus, angular gyrus, or the occipital cortex from patient 3. However, weak AP301-labeling was definitely discernible in medial frontal gyrus, angular gyrus and the hippocampus from patient 2 who is likely complicated by AD, and barely seen in the parahippocampus gyrus from patient 1.

Unexpectedly, in the soluble fraction of P301L brains, mutant tau gave consistently weaker signals than wild-type tau, when the signal intensities of the antibodies were normalized. This decrease in the levels of soluble P301L tau was observed across the various, affected regions including the medial frontal gyrus and angular gyrus from all three patients. Similar attenuation of mutant tau was also observed in the occipital cortex and cerebellum, where no Sarkosyl-insoluble tau was detected. Thus, we quantified the mRNA levels for wild-type and mutant tau by RT-PCR, but failed to find any difference in the expression levels between wild-type and mutant tau in both cortices and cerebella of P301L patients.

2.2. Exclusive Incorporation of Mutant Tau into Pretangles in P301L Brain

Using the same antibodies, we analyzed tau deposits in the tissue sections from the P301L brains. As expected, AP301 but not AL301 labeled innumerable NFTs and NTs in the AD brain. In P301L brains, AL301 stained crescent or ring-like perinuclear deposits within the cytoplasm, whereas AP301 did not stain any tau deposits except in patient 2. High magnification with differential interference contrast showed clearly the tau deposits in the P301L brains being pretangles with diffuse, mostly perinuclear, non-fibrillar cytoplasmic staining, which contrasted sharply with typical fibrillar NFTs observed in R406W brains or in AD brain.

3. R406W MUTATION

3.1. Equal Incorporation of Wild-Type and Mutant Tau into PHF-Tau in R406W Brain

To distinguish between wild-type and mutant tau in the fractions of R406W brain, we similarly synthesized paired peptides, SGDTSP<u>R</u>HLSNVSC and SGDTSP<u>W</u>HLSNVSC, and raised paired antibodies, AR406 and AW406, that specifically recognize either wild-type or mutant tau. After preabsorption with the counterpart peptide, AR406 and AW406 reacted exclusively with wild-type and R406W tau, respectively. Using such site-specific antibodies, we examined tau in the TS-soluble and Sarkosyl-insoluble fractions of R406W brains. Both AR406 and AW406 labeled to a similar extent PHF-tau migrating at 60–70 kDa and a smear on a blot of Sarkosyl-insoluble fractions prepared from frontal cortices, temporal cortex, and hippocampus. No specific labeling was found in the fraction from cerebella, which are usually free from NFT. In contrast, only AR406 but not AW406 labeled PHF-tau and a smear in the Sarkosyl-insoluble fraction of an AD brain. Western blotting also showed that similar levels of wild-type and R406W tau were present in the TS-soluble fraction of frontal cortices and cerebella of R406W brains, when the signal intensities of the antibodies were normalized.

3.2. Colocalization of Wild-Type and Mutant Tau in NFT in R406W Brain

We immunostained the tissue sections from various regions of R406W brains using these site-specific antibodies. AR406 and AW406 labeled to a similar extent innumerable intracellular NFT in neuronal perikarya, but not extracellular NFT. Large flame-shaped or globose NFT and fine neuropil threads were abundant in layers II, III, and V in the frontal and temporal cortices. Notably, many NFT were observed in the dentate gyrus, which is barely affected by AD.

While there were some variabilities in AR406 and AW406-signal intensities, the two signals always colocalized in all the NFTs and neuropil threads examined in frontal cortex, hippocampus CA1, and dentate gyrus. These observations, together with the above biochemical result, indicate that wild-type and mutant tau are deposited at an equal proportion in any given affected neuron in R406W brain.

3.3. Aberrant Hyperphosphorylation of R406W Tau in PHF-Tau

In the TS-soluble fraction, mutant tau appeared to be less extensively phosphorylated than wild-type tau, which was shown by the absence of mobility shift of the AW406-immunoreactive bands as compared with an obvious shift of AR406-reactive bands after dephosphorylation.

However, it remains unclear whether R406W tau in NFT in the affected brain is less phosphorylated. In fact, abundant NFT in R406W brain were intensely labeled with C5 and PHF1. Taken together, one might raise the reasonable possibility that wild-type tau rather than mutant tau is preferentially hyperphosphorylated in NFT in the R406W brain. To biochemically assess the phosphorylation of R406W tau *in vivo*, pooled PHF-tau was rechromatographed with a shallow gradient to obtain a better separation of wild-type and mutant tau.[8] Western blotting using AR406 and AW406 showed that early-eluting and late-eluting peaks consist largely of wild-type and R406W tau, respectively. PHF-tau in each peak was labeled intensely with C5 and PHF1, indicating that Ser-396 and -404 of both mutant and wild-type tau are highly phosphorylated in the PHF-tau from R406W brains. Similar results were obtained using other phosphorylation-dependent antibodies including AT8, AT100, M4, and APP422.

4. SIGNIFICANCE OF NFT AND HYPERPHOSPHORYLATION OF TAU

By analogy with polyglutamine disease,[9] the following can be postulated. In FTDP, a certain fragment of tau may have a profound effect on cell viability via certain routes including nuclear translocation, and at the same time it may form cytoplasmic aggregates (PHF-like fibrils). In the neuron, such toxic fragments are constantly generated but efficiently removed or degraded until a certain age in life. After a critical age, when the intracellular degradation machinery becomes slightly defective for an unknown reason, those fragments can stay longer in the affected neurons of FTDP brain, and reach sufficiently high levels to exhibit intracellular toxicity or form PHF. If the neuron succeeds in segregating such toxic tau fragments that have a high aggregation potential, this would lead to PHF formation, and long survival as NFT-bearing neuron.[10] If the neuron fails to segregate such toxic species, this would lead to quick death without leaving behind any trace. Thus, it is quite possible that PHF formation is the protective measures of the neuron. Perhaps, this kind of prevention is less at work in P301L brain, while this is fully at work in R406W brain.

What then is the significance of hyperphosphorylation of tau in PHF or PHF-like fibrils? One exonic mutation, R406W, should provide a key for clarifying the significance of tau hyperphosphorylation in PHF or PHF-like fibrils. R406W within the stably transfected cell is much less phosphorylated,[5] compared to stably transfected wild-type tau and other exonic mutants. In particular, this R406W mutant is much less phosphorylated at Ser-396 and 404, as shown by the immunoreactivities for two monoclonal antibodies, PHF-1 and C5.[5] However, in the Sarkosyl-insoluble fraction of R406W brains, mutant tau is hyperphosphorylated on Ser-396 and 404 as well. One possible explanation (see Figure 1) is that, when R406W tau is bound to tubulin, Trp-406, a bulky residue, prevents access of kinases to an adjacent portion, and Ser-396 and 404 are barely phosphorylated. Once tubulin is from the surroundings lost and free (unbound) mutant tau is left because of its lower susceptibiliy to proteases, this mutant tau takes the same random conformation as wild-type (and other mutant) tau and becomes similarly hyperphosphorylated. If so, the hyperphosphorylation of tau should reflect the absence of tubulin in the surroundings. Further, the hyperphosphorylation may not be the cause of degeneration (loss of tubulin), but may be the effect of degeneration. This assumption is consistent with the observation that PHF or PHF-like fibrils and tubulin scarcely colocalize in the neurites.

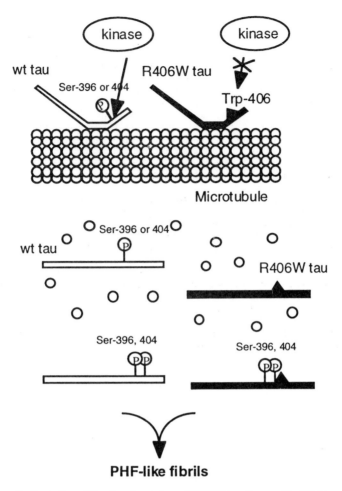

Figure 1. Schematic illustration of the phosphorylation of R406W tau. Because Trp406 (closed triangle) is a bulky residue, a particular protein kinase cannot get access to the Ser-396 and 404 that are close to the microtubule-binding domain, when R406W tau is bound to microtubules or tubulin oligomers. As a result, the mutant tau is much less phosphorylated on these residues. Once tubulin is lost from the surroundings (due to degeneration) and unbound R406W tau (more resistant to proteases) is left, this mutant tau takes the same random conformation as wild-type (and other mutant) tau, and the portion containing Ser-396 and 404 is freely accessible by kinases, and becomes similarly hyperphosphorylated. Thus, the hyperphosphorylation of tau presumably reflects the absence of tubulin in the surroundings.

5. ACKNOWLEDGMENTS

Those P301L (six cases) and R406W brains (two cases) used here are from FTDP-17 patients who were originally found and followed in the Netherlands by Drs. J. van Swieten, P. Heutink, and their colleagues, and provided by Dr. R. Ravid. The present work was done as a part of the doctoral thesis (University of Tokyo) by Dr. T. Miyasaka.

REFERENCES

1. M.G. Spillantini; T.D. Bird; and B. Ghetti, Frontotemporal dementia and Parkinsonism linked to chromosome 17: a new group of tauopathies. *Brain Pathol.* **8**, 387-402 (1998).
2. Alzheimer Research Forum; http://www.alzforum.org/
3. M. Hasegawa; M.J. Smith; and M. Goedert, Tau proteins with FTDP-17 mutations have a reduced ability to promote microtubule assembly. *FEBS Lett.* **437**, 207-210 (1998).
4. M. Hong; V. Zhukareva; V.Vogelsberg-Ragaglia; Z..Wszolek; L. Reed; B.I. Miller; D.H. Geschwind; T.D. Bird; D. McKeel; A. Goate; J.C. Morris; K.C. Wilhelmsen; G.D. Schellenberg; J.Q. Trojanowski; and V.M. Lee, Mutation-specific functional impairments in distinct tau isoforms of hereditary FTDP-17. *Science* **282**, 1914-1917 (1998).
5. N. Matsumura; T. Yamazaki; and Y. Ihara, Stable expression in Chinese hamster ovary cells of mutated tau genes causing frontotemporal dementia and parkinsonism linked to chromosome 17 (FTDP-17). *Am. J. Pathol.* **154**, 1649-1656 (1999)
6. J.C. van Swieten, M. Stevens, S.M. Rosso, P. Rizzu, M. Joosse , I. de Koning, W. Kamphorst, R. Ravid, M.G. Spillantini , Niermeijer, and P. Heutink, Phenotypic variation in hereditary frontotemporal dementia with tau mutations. *Ann. Neurol.* **46**(4), 617-626 (1999).
7. C. Bancher, I. Grundke-Iqbal, K. Iqbal, V.A. Fried, H.T. Smith, and H.M. Wisniewski, Abnormal phosphorylation of tau precedes ubiquitination in neurofibrillary pathology of Alzheimer disease. *Brain Res.* **539**, 11-18 (1991).
8. T. Miyasaka, M. Morishima-Kawashima, R. Ravid, P. Heutink, J.C. van Swieten, K. Nagashima, and Y. Ihara, Molecular analysis of mutant and wild-type tau deposited in the brain affected by the FTDP-17 R406W mutation. *Am. J. Pathol.* **158**(2), 373-379, 2001
9. P. Ferrigno and P.A. Silver, Polyglutamine expansions: proteolysis, chaperones, and the dangers of promiscuity. *Neuron* **26**, 9-12 (2000).
10. R. Morsch; W. Simon; and P.D. Coleman, Neurons may live for decades with neurofibrillary tangles. *J. Neuropathol. Exp. Neurol.* **58**, 188-197(1999).

THE EFFECTS OF APOLIPOPROTEIN E GENOTYPE ON BRAIN LIPID METABOLISM

Alisa Tietz[1] Leah Oron[1], Eliezer Mazliah[2] and Daniel M. Michaelson[1]

1. INTRODUCTION

Apolipoprotein E (apoE), a constituent of several classes of plasma lipoproteins, plays a key role in the transport and metabolism of cholesterol and phospholipids throughout the body.[1] ApoE is the major lipoprotein in the brain and has been shown to be upregulated following brain injury.[2] In view of the marked lipid-related changes which occur in the brain following injury and of the known function of apoE throughout the body, it has been suggested that apoE plays a pivotal role in the metabolism and redistribution of cholesterol and phospholipids during membrane neogenesis associated with synaptic plasticity and repair.[2]

There are three major human apoE isoforms[3] of which apoE3, which contains a cysteine residue at position 112 and an arginine at position 158, is the most common. The other two isoforms, apoE4 and apoE2, contain respectively two arginines and two cysteines at positions 112 and 158. Genetic and epidemiological studies have revealed that the apoE4 genotype is an important risk factor for AD and that the gene dosage of this allele is inversely related to the age of onset of the disease.[4, 5]

In the present study we investigated the possibility that the pathophysiological effects of distinct apoE genotypes are related to isoform-specific effects of apoE on brain lipid metabolism. This was performed by measurements of the effects of apoE genotype on the cholesterol and phospholipid levels of neuronal brain membranes of AD patients and controls and by parallel measurements of the effects of apoE genotype on the brain lipids of human apoE3 and apoE4 transgenic mice.

[1]The Department of Neurobiochemistry, Tel Aviv University, Ramat Aviv 69978, Israel.
[2]The Department of Neuroscience, University of California at San Diego, San Diego, CA, USA.

Mapping the Progress of Alzheimer's and Parkinson's Disease
Edited by Mizuno *et al.*, Kluwer Academic/Plenum Publishers, 2002

33

2. EXPERIMENTAL

2.1 Analysis of Brain Lipids of Alzheimer's Disease Patients and Controls

Brain samples from the frontal cortex of clinically and histologically confirmed AD cases aged 74±5 years and from matched controls aged 68±10 years were obtained at autopsy with a *post mortem* delay of respectively 5±2 and 8±4 hours. The samples were stored at -70°C until used. For preparation of brain membranes, the samples were thawed, homogenized and subjected to subcellular fractionation as previously described.[6] Lipids of the resulting 10,000g x 10 min pellet (fraction P_2) were then chloroform-methanol extracted from the suspensions[7] and separated by TLC on silica gel-G plates employing chloroform:ethanol:triethylamine:water (30:34:30:8, by vol.) as solvent. Commercial (Sigma) phosphatidylcholine (PC), ethanolamine (PE), serine (PS), inositol (PI) and sphingomyelin (SM) were used as standards. The corresponding phospholipids were then eluted and degraded for quantitative analysis. To separate PE and alkenyl 2-acyl-3-glycerylphospho-ethanolamine (plasmalogen-PPE) which have the same mobility on TLC plates, the mixture was hydrolyzed with $MgCl_2$-HCl and then separated by TLC as described above. For quantitative analysis, PC, PE, lysoPE, PE and PI were subjected to alkaline hydrolysis whereas SM was acid hydrolised. The resulting fatty acids were separated and quantitated by GLC on a BPX70 column (SGE, Australia) Cholesterol levels were determined using the total cholesterol kit (Sigma) whereas protein concentration was determined using the BCA reagent (Pierce) and BSA as standard.

2.2 Transgenic Mice

Human apoE3 and apoE4 transgenic mice were generated on an apoE-deficient C57BL/6J background utilizing human apoE3 and apoE4 transgenic constructs as previously reported[8] and were hemizygous for the human apoE transgenes.[9] Six months' old apoE3 and apoE4 male transgenic mice (lines apoE3-453 and apoE4-81, respectively) and matched apoE-deficient (Jackson Labs., USA) and C57BL/6J controls were used in this study. The apoE genotypes of each of the mice used was confirmed by PCR analysis as previously described.[9]

2.3 Analysis of Brain Lipids of the ApoE Transgenic Mice

Brains of apoE3 and apoE4 transgenic mice and of control and apoE-deficient mice (n=3-4 for each group) were removed, homogenized and fractionated as described above except that in addition to the P_2 membrane fraction, microsomal membranes (fraction P3 pelleted at 150,000g x 1 hr) were also prepared and analyzed.

3. RESULTS

3.1 Brain Lipids of Alzheimer's Disease Patients and Controls

Analysis of the lipid composition of brain synaptosomal membranes (fraction P_2) of normal controls homozygous for apoE3 (control 3/3) revealed that they contain similar levels of cholesterol and phospholipids (respectively 927±37 and 925±54 nmole/mg protein; means ± S.E.M.; n=5) and that their phospholipids comprised of PC (33.1± 0.9%), PE (21.5±1.6%), PPE (18.2±1.8%), PS (15.2± 0.3%), PS (4.2±0.3%) and SM (7.6±0.8%). This lipid composition is in accordance with previous reports.[10, 11]

Measurements of the brain lipid composition of AD patients homozygous for apoE3 (AD 3/3) revealed a marginally significant decrease of 10.5±3.5% in cholesterol levels compared to that of controls with the same genotype (p=0.09) but that both groups had similar total phospholipid levels (Figure 1). Comparison of the phospholipid compositions of these groups revealed no differences in the levels of PC, PS, PI and SM. There was however a small difference in the levels of ethanolamine containing phospholipids, such that PE levels were slightly lower in the AD 3/3 group than in the control 3/3 subjects, whereas the opposite was observed with PPE. Further analysis of these results in terms of the percentage of PPE and of the total ethanolamine-containing lipids revealed that it was 45.7 ±4.4% in the control 3/3 samples and 50.7±2% in the AD 3/3 samples. These differences however were not statistically significant. Measurements of the brain lipid composition of AD patients homozygous for apoE4 (AD 4/4) revealed that the cholesterol content of the P_2 membranes was intermediate to those of the apoE3 homozygous AD and controls and that their phospholipid levels were slightly but not significantly lower than those of both groups (Figure 1). Furthermore, there were no differences in the phospholipid compositions of the AD 3/3 and AD 4/4 cases and the slight elevation, relative to control 3/3, in the fraction of PPE from the total ethanolamine-containing lipids was similar in the AD 4/4 and AD 3/3 groups.

Analysis of the fatty acid composition of control 3/3 brain phospholipids revealed that C16:0 which comprises over 40% of the PC was replaced in the other lipids by C18:0 and C18:1 and by considerable levels of polyunsaturated fatty acids (Table 1). Analysis of the brain fatty acid compositions of AD 3/3 and AD 4/4 revealed no differences in those of PC, PE and PI of the three groups (Table 1) and of their SM (not shown). There were, however, small differences in the fatty acid composition of PPE and PS of the three groups, such that

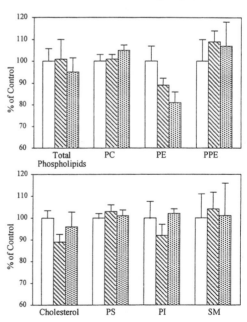

Figure 1. The effects of apoE genotype on the levels of brain phospholipids and cholesterol and of distinct phospholipid classes of Alzheimer's disease patients and controls. Brain synaptosomal membranes (fraction P_2) were prepared and analyzed as described in the text. Results shown (mean ±SEM) were obtained from normal controls homozygous for apoE3 (empty bars, n=5) and from AD patients homozygous for either apoE3 (stapled bars; n=10)) or apoE4 (dotted bars; n=6).

the levels of C22:6 were increased in AD (AD 4/4 > AD 3/3 > control 3/3) and those of C18:1 were decreased (AD 4/4 < AD 3/3 < control 3/3). The difference between the C22:6 levels of PS of the AD 4/4 and the control 3/3 cases were marginally significant (p=0.06), whereas the other differences were not.

3.2 Brain Lipids of ApoE Transgenic Mice

The effects of apoE on brain lipid composition were also studied in an animal model utilizing mice which are apoE-deficient and mice expressing human apoE3 and apoE4 on a null mouse apoE background. Comparison of the brain lipids of control and apoE-deficient mice revealed non significant decreases in the cholesterol, PC, PE and PS levels of the synaptosomal membranes (fraction P$_2$) of the apoE-deficient mice relative to the controls. In contrast, there was a marked decrease in cholesterol (32±3%), PE (31± 4%) and PI (28±3%) in the microsomal membranes (fraction P$_3$) of the apoE-deficient mice relative to the controls (p < 0.05) and a less pronounced decrease in PC and PS levels (Figure 2). These findings are in accordance with results previously obtained with another strain of apoE-deficient mice[6] and suggest that intracellular membranes are affected preferentially by apoE. The extent to which the effects of apoE deficiency on microsomal lipids are reversed in apoE transgenic mice was examined. As shown in Figure 2, the human apoE3 and apoE4 transgenes partially and similarly reversed the decreases in cholesterol PE and PI levels brought about by apoE deficiency. Measurements of the fatty acid composition of the different microsomal phospholipids revealed that they were unaffected either by apoE deficiency or by the human apoE transgenes (not shown).

Table 1. Fatty acid composition of brain P$_2$ membranes of Alzheimer's disease patients homozygous for either apoE3 or apoE4 and of matched controls homozygous for apoE3

Fatty acid Lipid / Group	16:0	18:0	18:1	20:1	20:4	22:4	22:6
PC: Cont 3/3	42.0±1.6	12.1±1.0	31.2±1.4	-	4.2±0.3	1.2±0.2	2.3±0.1
AD 3/3	44.0±1.2	11.1±0.3	31.5±1.0	-	4.5±0.2	0.9±0.1	2.5±0.1
AD 4/4	43.9±1.0	11.0±0.1	30.4±0.5	-	4.6±0.1	0.9±0.1	2.6±0.1
PE: Cont 3/3	5.3±0.5	32.0±1.0	11.0±1.1	-	11.0±0.4	8.5±0.6	24.0±1.0
AD 3/3	5.0±0.5	31.0±0.5	10.6±1.2	-	11.5±0.1	8.3±0.4	26.5±1.5
AD 4/4	6.3±1.0	33.3±1.8	10.2±0.5	-	10.9±0.5	7.3±0.7	24.7±1.9
PPE: Cont 3/3	3.1±0.8	33.0±1.0	20.9±3.5	4.9±0.5	14.2±0.8	25.0±1.0	21.7±1.9
AD 3/3	2.3±0.8	21.0±0.2	18.0±2.0	4.3±0.3	15.2±0.4	24.0±0.6	23.7±1.9
AD 4/4	1.9±0.2	16.0±0.1	15.4±1.3	3.7±0.5	15.4±0.5	25.0±1.3	26.7±2.2
PS: Cont 3/3	1.0±1.2	42.8±0.4	24.6±1.9	2.5±0.4	2.3±0.6	3.9±0.3	16.1±0.8
AD 3/3	1.1±0.2	42.6±0.2	21.4±1.3	1.6±0.4	2.9±0.4	3.5±0.3	18.3±1.2
AD 4/4	1.0±0.1	42.0±0.3	20.7±1.8	2.0±0.3	2.3±0.3	4.0±0.2	20.9±1.2
PI: Cont 3/3	7.1±0.6	35.0±0.6	12.7±1.2	-	26.5±1.6	3.5±0.3	7.6±0.7
AD 3/3	5.4±0.8	37.2±0.5	8.1±0.9	-	31.0±0.7	3.3±0.2	9.0±0.7
AD 4/4	7.7±1.2	33.2±2.4	14.1±3.3	-	26.3±2.4	3.4±0.2	7.4±1.3

The fatty acid contents of brain synaptosomal membranes were determined as described in the text. Results (average ± SEM) were obtained from normal controls homozygous for apoE3 (n=5) and from AD patients homozygous for either apoE3 (n=10) or apoE4 (n=6) and are expressed as % of total fatty acids.

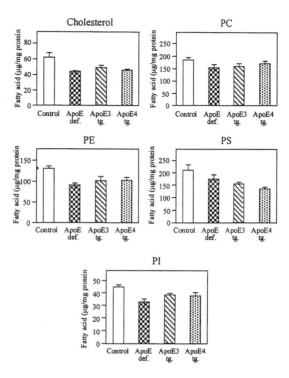

Figure 2. The levels of brain lipids of control, apoE-deficient, apoE3 and apoE4 transgenic mice. Brain microsomal membranes (fraction P₃) from the four mice groups were prepared and their lipids analyzed as described in the text. Results shown are the mean ± SD of 3 - 4 mice in each group.

4. DISCUSSION

Comparison of the lipid composition of brain synaptosomal membranes of AD patients and controls revealed , in accordance with previous reports,[12, 13] a decrease in the levels of PE. This decrease was more pronounced in the AD 4/4 group. The total levels of phospholipids and those of PC, PPE, PS, PI and SM were however not affected either by AD or by the apoE genotype and there was a small decrease in cholesterol levels of the AD 3/3 vs control 3/3 but not in the AD 4/4 group. Measurements of the fatty acid composition of the different phospholipids revealed an increase in the levels of C22:6 in PPE and PS of the AD 4/4 subjects relative to the other groups which was associated with a parallel decrease in the levels of C18:1. The magnitude of these effects was small and the AD 3/3 cases yielded values intermediate to those of the control 3/3 and AD 4/4 groups (Table 1). The small observed effects of the apoE4 genotype may be related to the duration of the disease, which starts earlier in carriers of apoE4 genotype, and not to direct isoform specific effects of apoE4 on the overall lipid composition of brain synaptosomal membranes.

The possibility that other neuronal membranes, such as the microsomal compartment, are affected in AD by the apoE genotype has yet to be examined. The animal model finding that the phospholipid and cholesterol levels of microsomal membranes are affected by apoE deficiency, whereas those of the synaptosomal membranes are not (Figure 2 and ref. 6) supports the assertion that microsomal membranes are preferentially affected by apoE. However, since the levels of microsomal lipids are replenished similarly in the apoE3 and apoE4 transgenic mice (Figure 2), it seems that the effects of apoE on the microsomal lipid composition are not isoform-specific.

In conclusion, the present AD and animal model findings suggest that the pathological effects of the apoE genotype are not associated with changes in the overall lipid composition of brain neuronal membranes. However, the possibility that the lipid composition and trafficking of specialized subcellular compartments are affected isoform-specifically by apoE cannot be excluded.

ACKNOWLEDGEMENTS

This work was supported in part by grants to DMM from the Fund for Basic Research of the Israel Academy of Sciences (grant #43/00-1) and from the Joseph and Inez Eichenbaum Foundation. DMM is the incumbent of the Myriam Lebach Chair in Molecular Neurodegeneration.

REFERENCES

1. R.W. Mahley, Apolipoprotein E: cholesterol transport protein with expanding role in cell biology, *Science* **240**, 622-630 (1988).
2. J. Poirier, Apolipoprotein E in animal models of CNS injury and in Alzheimer's disease, *Trend in Neur. Sci.* **17**, 525-530 (1994).
3. K.H. Weisgraber, Apolipoprotein E: structure-function relationships, *Adv. in Prot. Chem.* **45**, 249-302 (1994).
4. E.H. Corter, A.M. Saunders, W.J. Strittmatter, P.C. Gaskell, *et al.*, Gene dose of apolipoprotein E type 4 allele and the risk of Alzheimer's disease in late onset families, *Science* **261**, 828-829 (1993).
5. A.M. Saunders, W.J. Strittmatter, D. Schmechel, P.H. St. George-Huslop *et al.*, Association of apolipoprotein E allelle E4 with late onset familial and sporadic Alzheimer's diseasem *Neurology* **43**, 1467-1472 (1993).
6. L. Lomnitski, L. Oron, D. Sklan, and D.M. Michaelson, Distinct alterations in phospholipid metabolism in brains of apolipoprotein E-deficient mice, *J. Neurosci. Res.* **58**, 586-592 (1999).
7. J. Folch, M. Lees, and G.H.S. Stanley, A simple method for the isolation and purification of total lipids from animal tissues, *J. Mol. Chem.* **226**, 497-509, 1959.
8. P.-T. Xu, D. Schmechel, T. Rothrock-Christian, D.S. Burkhart, *et al.*, Human apolipoprotein E3 and E4 isoform specific transgenic mice: human like pattern of glial and neuronal immunoreactivity in central nervous system not observed in wild type mice, *Neurol. Dis.* **3**, 229-245 (1996).
9. T. Sabo, L. Lomnitski, A. Nyska, S. Beni, *et al.*, Susceptibility of transgenic mice expressing human apolipoprotein E to closed head injury: The allele E3 is neuroprotective whereas E4 increases fatalities, *Neurosci.* **101**, 279-884 (2000).
10. M. Sodenberg, C. Edlund, K. Kristensson, and G. Dallner, Lipid composition of different regions of the human brain during aging, *J. Neurochem.* **54**, 415-423 (1990).
11. L. Svennerholm, K. Bostrom, C.G. Helander, and B. Jungbjer, Membrane lipids in the aging human brain. *J. Neurochem.* **56**, 2051-2059 (1991).
12. K. Welle, A.A. Farooqui, L. Liss, and L.A. Horrocks, Neuronal membrane phospholipids in Alzheimer's disease. *Neurochem. Res.* **20**, 1329-1333 (1995).
13. M.R. Prasad, M.A. Lovell, M. Yatin, H. Dhillon, and W.R. Markesberg, Regional membrane phospholipid alterations in Alzheimer's disease, *Neurochem Res.* **23**, 81-88 (1998).

APOLIPOPROTEIN E: A NOVEL THERAPEUTIC TARGET FOR THE TREATMENT OF ALZHEIMER'S DISEASE

Judes Poirier[*] and Michel Panisset

1. INTRODUCTION

Alzheimer disease (AD) is associated with neuronal loss, synaptic damage, deposition of beta-amyloid and loss of cholinergic activity in susceptible brain regions. The ladder three pathological markers of AD were shown to be closely associated with the presence of the apolipoprotein ε4 (apoε4) allele in sporadic AD subjects. The apoε4 allele is a well known risk factor for sporadic late onset Alzheimer's disease; patients with two ε4 alleles exhibit an earlier age of onset, higher amyloid plaque counts, cerebrovascular amyloid, marked reductions in choline acetyltransferase and nerve growth factor receptor density as compared to non-ε4 allele subjects. Recent evidence suggest that apoE polymorphism may significantly affect the clinical presentation of the disease as well as the global efficacy of memory enhancer drugs such acetylcholine esterase inhibitors and noradrenergic modulators. Several different hypotheses have been presented to explain the effect of the ε4 allele on the age of onset and clinical progression of the disease. Because of the reported effect of low levels of apoE on synaptic plasticity, reinnervation and lipid homeostasis in apoE knockout mice, it has been proposed a little while ago that the low levels of apoE reported in brain tissues of apoε4 carriers affect lipid homeostasis in such a way that it compromises synaptic plasticity.[1]

This study was designed to examine the effect of Probucol on the clinical course of mild-to-moderate AD subjects in a six month clinical trial. Probucol is a potent inducer of apoE expression that affects both mRNA and protein concentrations in the brain. The mechanism of the reduction in LDL-cholesterol levels is yet to be fully elucidated but it could results from enhanced catabolism of lipoprotein complexes by lipoprotein lipases. There is evidence of an *independent* antioxidant effect. Probucol has an excellent safety profile with very limited side effects.[2] The ability of Probucol to enhance apoE synthesis and secretion was first demonstrated in primary type 1 astrocyte

[*]McGill Centre for Studies in Aging, Douglas Hospital, McGill University, 6825 LaSalle Blvd, Verdun, Québec, Canada, H4H 1R3

Mapping the Progress of Alzheimer's and Parkinson's Disease
Edited by Mizuno *et al.*, Kluwer Academic/Plenum Publishers, 2002

cultures from rodents. The biochemical effect of Probucol on brain apoE metabolism was confirmed in the brain of C57/BL 6J mice after short term intra-peritoneal administration of the active agent. Finally, the drug was used in a 6 month "proof-of-principle" drug trial with patients suffering from mild to moderate Alzheimer's disease.

2. MATERIALS AND METHODS

2.1. Probucol Affects apoE Synthesis and Secretion in Primary Type 1 Astrocyte Cultures and the Brain of C57/Bl 6J Mice

Primary cultures of type 1 astrocytes were derived from the cortex of 1 to 2 day-old Sprague-Dawley rats. Probucol was dissolved is in culture media and added to cell for 24 and 48 hrs. Media was then removed and kept frozen at –80°C until dosage of apoE protein levels by Western blot or apoE mRNA levels by real-time quantitative PCR using a modification of the protocol of Powell and Kroom.[3] C57/BL 6J (Jackson Laboratory, Bar Harbor, USA) mice (n=10) received daily i.p. injections of Probucol dissolved in saline at final concentrations of 0.1, 0.5 and 5 mg/kg. After 10 days of treatment, animals were sacrificed. Tissues samples were kept frozen at -80°C until dosage of apoE.

2.2. Probucol Induces apoE in the Cerebrospinal Fluid and Clinical Outcome in Alzheimer's Disease

Eleven patients with probable AD (NINCD-ADRDA, 1984) were enrolled in a "proof-of -principle" study designed to examine the effect of apoE induction on the clinical course of the disease. All subjects had to have no significant medical problems and not be on any medication that could interfere with their cognitive performance. They had to be between 50 and 80 years of age, to exhibit Mini Mental State Exam (MMSE) score between 10 and 26 and to be fluent in French or English. All patients received Probucol 500 mg twice daily for a period of six months. Lumbar punctures were performed at baseline **and** 1 month after onset of treatment to determine the quantity of CSF apoE, peroxidised lipids, total Tau and beta amyloid 1-40 levels in each subjects. Alzheimer's Disease Assessment Cognitive Scale (ADAS-Cog) and Disability Assessment of Dementia (DAD) scale ratings were performed at baseline, months 3 and 6 to monitor cognitive and global functioning. Blood was obtained at baseline, month 3 and 6 to assay levels of cholesterol and triglycerides.

3. RESULTS

We observed an increased production and secretion of apoE by Probucol in type 1 astrocyte cultures after 24 hours exposure. It returns to near control levels within 48 hours. A concomitant induction of apoE mRNA prevalence was observed in parallel astrocytic cultures at 8, 24 and 48 hrs. Intraperitoneal administration of Probucol was shown to increase steady state levels of apoE in periphery and in the brain. The mid-range dose (0.5 mg/kg/day) was selected to approximate the concentration of Probucol used to lower blood cholesterol in humans. Western blot analyses of apoE levels in mice liver and plasma tissues reveals an inverse bell shape dose response with a peak level at the lowest concentration of Probucol (0.1 mg/kg/day). However, the effect of Probucol is somewhat different in the central nervous system of these animals. A clear dose-dependent increase in the steady state levels of apoE was observed the hippocampus of these animals with the highest induction observed at the 5 mg/Kg/day.

Table 1 summarises the demographic characteristics and CSF biochemical variables of the AD patients at baseline, and one month into treatment. Blinded genotype analysis revealed that four patients exhibited the apoE3/3 genotype whereas the remaining seven subjects belonged to the apoE 4/3 genotype. Plasma cholesterol levels were shown to be significantly reduced after 6 months compared to baseline values (p <0.05).These changes are consistent with the reported cholesterol lowering effect of Probucol in hycholesterolemic subjects.[4] Analysis of apoE levels in the cerebrospinal fluid of AD patients at baseline and month 1 revealed that Probucol was effective in increasing apoE concentration in some, but not in all AD patients.

Induction of apoE concentrations by Probucol varied from 0 to 77% (p <0.05, ANOVA analysis, n=10). The response showed genotype influence: the apoE3/3 subjects exhibiting the strongest induction when compared to E4 carriers (p =0.09). Tau, beta-amyloid 1-40 and lipid peroxides levels remain unchanged in response to the Probucol treatment (Table 1).

Mean clinical responses indicate a relative stabilization of the progression of the disease on the cognitive and the global improvement scales during the 6 month trial. Figure 1 illustrates the results of the correlational analysis that was performed on CSF apoE concentrations as a function of clinical response using the ADAS-Cog scale. It reveals a statistically significant association between the two variables whereby apoE induction above the 35% threshold appears to be associated with clinical improvement (p < 0.05, using Multiple General Linear Model analysis, SPSS Inc.). While two subjects were left out of this particular analysis because of poor compliance in the second portion of the trial, they were included in the group analyses of ADAS-Cog and DAD (Table 1).

Table 1. Demographic characteristics of the AD subjects enrolled in the Probucol clinical drug trial. Biochemical and clinical outcome measures were obtained after 1, 3, and 6 months of treatment

	Clinical Response			
	Baseline +/- SEM	Month 3 +/- SEM	Month 6 +/- SEM	Difference
ADAS-Cog	24.45 +/- 2.35	25.64 +/- 2.51	25.6 +/- 3.27	NS
DAD	78.64 +/- 3.56	85.45 +/- 3.14	82.20 +/- 5.98	NS
MMSE	17.64 +/- 0.96	----	16.00 +/- 1.35	NS

	Biological Markers in the Cerebrospinal fluid			
	Baseline +/- SEM (N=10)	Month 1 +/- SEM (N=10)	Ratio Month 1/Baseline Mean (%)	Difference
Apo E	6.52 +/- 0.90	7.75 +/- 1.10	118,94%	p=0.046 *
ApoE4 Carriers	6.08 +/- 1.24	6.52 +/- 1.30	107,32%	NS
ApoE3/2 Carriers	7.70 +/- 1.83	11.04 +/- 1.99	143,40%	p=0.09 **
Tau	658.20 +/- 112.39	666.25 +/- 114.18	101,22%	NS
Lipid Peroxides	0.71 +/- 0.01	0.7 +/- 0.03	98,35%	NS
Beta Amyloid 1-40	30.31 +/- 2.49	32.32 +/- 2.70	106,63%	NS

	Biological Marker in the Blood			
	Baseline +/- SEM (N=10)	3 Months +/- SEM (N=10)	6 Months +/- SEM (N=10)	Difference
Cholesterol	5.06 +/- 0.37	4.64 +/- 0.33	4.34 +/- 0.33	p<0.05 ***

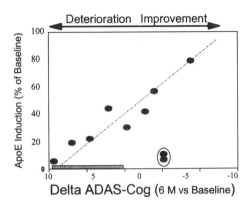

Figure 1. Effect of Probucol-mediated apoE induction in the cerebrospinal fluid of AD subjects at month one as a function of clinical response at month 6, using the ADAS-Cog scale. Shadow area highlights historical placebo range for subjects enrolled in memory enhancer drug trials. Coefficient correlation : 0.51 and p=0.031 (MGLM, SPSS Inc).

4. DISCUSSION

The crucial role played by apoE during normal brain reinnervation and the near complete absence of plasticity in the brain of apoE knockout mice, in apoE4 knockin mice and in apoE4/4 AD subjects point toward lipid delivery as one of the rate limiting steps in neuronal remodelling.[5] The correlation that exists between ADAS-Cog variation and apoE induction highlights the importance of apoE as an important surrogate biological marker to be used in the etiological treatment of Alzheimer's disease. The evidence reported here indicate the pharmacological concentrations of Probucol can promote apoE synthesis in the brain of AD patients without affecting steady state concentrations of Tau, beta amyloid 1-40 or lipid hydroperoxides in the CSF. We cannot exclude the possibility that additional Probucol activities might also be relevant, including modulation of 1) cholesteryl ester transfer protein, 2) other apolipoproteins (apoA-1, apoC-I) and 3) neurotrophin signalling.[2] The exact biological basis for the induction of apoE production in response to Probucol is currently unknown. The concomitant and timely raise of apoE mRNA prevalence and apoE protein secretion in the CNS is certainly consistent with a direct modulatory effect on gene expression *in vivo*. Moreover, circulating levels of cholesterol were found to be significantly reduced by Probucol after 6 months in AD subjects; highlighting an important beneficial side effect to the overall disease stabilization profile of Probucol. This observation is quite consistent with recent reports that the use of cholesterol lowering agents called statins reduced the prevalence of AD by 60 to 72% in a 57,104 elderly subjects cohort from the US.[6] Similar results were reported more recently in the Boston area.[7] The combination of these results with those above clearly suggest that chronic administration of Probucol could be used to activate the beta amyloid scavenging activity of apoE to indirectly reduce amyloid deposition *in vivo* in the CNS.

REFERENCES

1. Poirier J. Apolipoprotein E in animal models of brain injury and in Alzheimer's Disease. *Trends Neurosci.* 12: 525-530 (1994).

2. Davignon J. Probucol. *Handbook Exp. Pharmacol.* 109: 429-469 (1994).
3. Powell E.E. and Kroon PA Measurement of mRNA by quantitative PCR with a nonradioactive label. *J. Lipid Res. 33*, 609-614 (1992).
4. Tedeschi RE, Taylor HL and Martz BL. Safety and effectiveness of probucol as a cholesterol lowering agent. *Artery* 10: 22-34 (1982).
5. Danik M., Champagne D., Petit-Turcotte C., Beffert U. and Poirier J. Brain lipoprotein metabolism and its relation to neurodegenerative disease. *Critical Reviews Neurobiol.* 13: 357-407 (2000).
6. Wolozin B, KellmanW, Rousseau P, Celesia GG and Siegel G. Decreased prevalence of Alzheimer's disease associated with 3-hydroxy-3-methylglutaryl Coensyme A reductase inhibitors. *Arch. Neurol.* 57: 1439-1443 (2000).
7. Jick H, Zornberg GL, Jick SS, Seshadry S. and Drachman DA. Statins and Alzhewimer's disease. Lancet. 356: 1627-1631 (2000).

ANTISENSE INTERVENTION WITH CHOLINERGIC IMPAIRMENTS ASSOCIATED WITH NEURODEGENERATIVE DISEASE

Eran Meshorer and Hermona Soreq[*]

1. INTRODUCTION

Current therapies for Alzheimer's disease are based on suppression of acetylcholine hydrolysis with inhibitors of acetylcholinesterase (AChE) (Figure 1). However, recent data demonstrated that various stressors, including cholinesterase inhibition, promote long-lasting up-regulation of a rare AChE isoform, AChE-R, having isoform-specific non-catalytic morphogenic activities[1,2]. The role of this protein in the etiology of neurodegenerative disease therefore deserves a fresh examination.

2. METHODS

The regulation and biological functions of AChE were studied in patients hypersensitive to anticholinesterases, transfected cells, transgenic mice, and normal mice subjected to psychological stress, anticholinesterase treatment, or closed head injury[3]. AChE gene expression was manipulated in cells and *in vivo*, using partially 2'-*O*-methyl-protected antisense oligonucleotides (ASON)[4].

3. RESULTS AND INTERPRETATION

Several inherited causes are known which confer hypersensitivity to cholinesterase inhibitors[5]. All vertebrates possess two cholinesterase proteins, AChE and

[*] Eran Meshorer and Hermona Soreq, The Hebrew University of Jerusalem, Jerusalem, Israel 91904

Mapping the Progress of Alzheimer's and Parkinson's Disease
Edited by Mizuno *et al.*, Kluwer Academic/Plenum Publishers, 2002

45

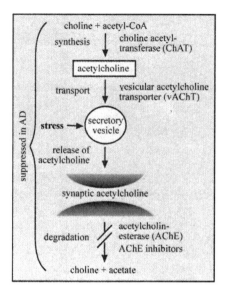

Figure 1. The cholinergic consequences of anti-cholinesterase therapy. Inhibitors of acetylcholinesterase (AChE) are common therapeutic agents employed to treat diseases that involve impaired acetylcholine-mediated neurotransmission. In such diseases (i.e. Alzheimer's Disease or Myasthenia Gravis), the cascade leading from choline and acetyl-CoA to acetylcholine and its subsequent degradation via AChE, is suppressed, leading to a state of hypocholinergic neurotransmission. AChE hydrolyses acetylcholine (ACh) in the synaptic cleft to release choline and acetate. Therefore, its inhibition increases ACh levels in cholinergic synapses, retrieving closer to normal levels of cholinergic communication.

Table 1. Inherited causes for anticholinesterase hypersensitivity

anti-AChE used	Gene (Chr.)	Mutation	Allele frequency	Adverse responses	Reference
Pyridostigmine	BCHE (3q26)	D70G (atypical) *recessive*	1:1000[a]	depression weight loss anxiety	Loewenstein-Lichtenstein et al., 1995
Parathion	BCHE (3q26)	Silent *recessive*	1:100,000[a]	muscle fasciculation leading to asphyxia	Whittaker, 1986 Prody et al., 1989
Pyridostigmine	ACHE (7q22)	Distal enhancer deletion *dominant*	1:50[a]	muscle fasciculations intense headache rhinnorea lacrimation	Shapira et al., 2000
Diazoxon	PON1 (7q21)	Q192R M55L	0.25-0.32[b]	excessive sweating chest tightness nausea muscle twitching	Davies et al., 1996 Haley et al., 1999

[a] Average in the Israeli population
[b] Average in the American population

Table 2. Conditions inducing AChE-R overproduction

Stressor	Overexpressing cell type	Reference
Confined swim	cortical neurons bone marrow cells	Kaufer et al., 1998 Grisaru et al., 2001
DFP	brain muscle retinal neurons	Shapira et al., 2000 Lev-Lehman et al., 2000 Broide et al., 1999
Closed head injury	cortical neurons	Shohami et al., 2000
Glucocorticoids	hematopoietic cells	Grisaru et al., 2001

butyrylcholinesterase (BuChE). While AChE has long been recognized for its catalytic function of hydrolyzing acetylcholine, the function of BuChE, which is non-essential for normal life, has long been a mystery. A suggestion was that BuChE operates as a scavenger, acting to protect the nervous system from anti-cholinesterases (anti-ChEs), thus preventing excessive blockade of the essential AChE. Adverse symptoms were reported for anticholinesterase-exposed carriers of 'atypical' BuChE[6], which is far less sensitive than normal BuChE to inhibition by pyridostigmine and several other carbamate anti-ChEs. Moreover, atypical BuChE demonstrated 1/200th the affinity for tacrine of normal BuChE or AChE. Inherited BuChE mutations may thus explain at least some of the adverse responses to anti-ChE therapies. Another metabolic enzyme that operates as an organophosphate scavenger is paraoxonase (PON1)[7,8]. PON1 mutations, as well as promoter polymorphisms in an upstream enhancer domain of the human ACHE locus, also cause extreme hypersensitivity to anti-ChEs (Table 1). At the post-transcriptional level, alternative splicing associated with overexpression of AChE-R mRNA and its AChE-R protein product were shown following psychological stress, exposure to anti-ChEs or head trauma (Table 2). In addition, unique, non-catalytic, morphogenic properties were attributed to each of the AChE splice variants[3]. Therefore, anticholinesterases used for treating Alzheimer's disease induce AChE-R overproduction and may lead to undesired morphogenic effects.

In principle, most of the currently used drugs are targeted towards proteins with adverse effects. However, thanks to the human genome project, we now know the sequence of many of the mRNA transcripts encoding these proteins. This enables the development of antisense oligonucleotides (ASON) preventing the production of undesired proteins rather than blocking their active sites (Figure 2). Nanomolar doses of ASON mediating selective destruction of stress-induced AChE-R mRNA prevented AChE-R accumulation, as was shown using AChE-R selective hybridization probes[4]. Furthermore, anti-AChE ONs improved memory and behavior in transgenic mice (Cohen et al., unpublished results), protected hippocampal neurons, minimized mortality, and facilitated recovery of transgenic mice following head injury, the highest known risk for non-familial Alzheimer's disease[9].

4. CONCLUSIONS

Inherited promoter polymorphisms, state-of-mind-, drug-, and injury-induced feedback processes all induce accumulation of a previously unknown AChE variant. These

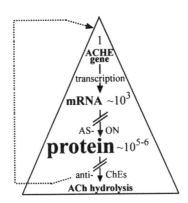

Figure 2. Amplification process from one gene to many protein products. The concentration of AChE mRNA molecules per cell was estimated by Karpel et al. (1994). Given the life span of an AChE mRNA transcript (ca. 4-8 hrs (Chan et al., 1998; Luo et al., 1999 respectively)), and that of the AChE protein (ca. 3.5 day (Wenthold et al., 1974)), together with the amplification at the translation phase, an amplification facto from mRNA to protein of 100-1000 can be estimated. Therefore, ASON drugs can be used in much lowe doses than their counterpart enzyme inhibitors, especially since such inhibitors enhance AChE-R mRNA production (dashed arrow; Soreq and Seidman, 2001).

and the neglected morphogenic properties of AChE should all be considered when contemplating the etiology of cholinergic impairments in neurodegenerative diseases. Antisense technology has emerged as a valuable research tool and a promising direction of novel therapies.

ACKNOWLEDGEMENTS

Supported by the US Army Medical Research and Development Command (DAMD 17-99-1-9547) and by Esther Neuroscience, Ltd. (to H.S.).

REFERENCES

1. D. Kaufer, A. Friedman, S. Seidman, and H. Soreq. Acute stress facilitates long-lasting changes in cholinergic gene expression. *Nature* **393**,373-377(1998).
2. M. Shapira, I. Tur-Kaspa, L. Bosgraaf, N. Livni, A. D. Grant, D. Grisaru, M. Korner, R. P. Ebstein, and H. Soreq. A transcription-activating polymorphism in the ACHE promoter associated with acute sensitivity to anti-acetylcholinesterases. *Hum Mol Genet* **9**, 1273-1281(2000).
3. H. Soreq, and S. Seidman, S. Acetylcholinesterase - new roles for an old actor. *Nat Rev Neurosci* **2**,294-302(2001).
4. N. Galyam, D. Grisaru, M. Grifman, N. Melamed-Book, F. Eckstein, S. Seidman, A. Eldor, and H. Soreq. Complex host cell responses to antisense suppression of ACHE gene expression. *Antisense Nucleic Acid Drug Dev* **11**,51-57(2001).
5. H. Soreq, and D. Glick. Novel roles for cholinesterases in stress and inhibitor responses. In Cholinesterases and Cholinesterase Inhibitors, E. Giacobini, ed. (London: Martin Dunitz), pp. 47-61(2000).
6. Y. Loewenstein-Lichtenstein, M. Schwarz, D. Glick, B. Norgaard Pedersen, H. Zakut, and H. Soreq. Genetic predisposition to adverse consequences of anti-cholinesterases in 'atypical' BCHE carriers. *Nat Med* **1**,1082-1085(1995).
7. H. Davies, R.J. Richter, M. Keifer, C. A. Broomfield, J. Sowalla, C. E. and Furlong. The effect of the human serum paraoxonase polymorphism is reversed with diazoxon, soman and sarin. *Nat Genet* **14**,334-336(1996).
8. R. Haley, S. Billecke, B. N. and La Du, B. N.. Association of low PON1 type Q (type A) arylesterase activity with neurologic symptom complexes in Gulf War veterans. *Toxicol Appl Pharmacol* **157**,227-233(1999).
9. E. Shohami, D. Kaufer, Y. Chen, S. Seidman, O.Cohen, D. Ginzberg, N. Melamed-Book, R. Yirmiya, R., and H. Soreq, Antisense prevention of neuronal damages following head injury in mice. *J. Mol. Med.* **78**,228-236(2000).

NEURONAL SURVIVAL AND DEATH IN ALZHEIMER DISEASE

Arun K. Raina[1], Catherine A. Rottkamp[1], Xiongwei Zhu[1], Osama Ogawa[1], Ayala Hochman[2], Shun Shimohama[3], Atsushi Takeda[4], Akihiko Nunomura[5], George Perry[1], and Mark A. Smith[1]

1. INTRODUCTION

Alzheimer disease (AD) is a prominent neurodegeneration marked clinically, by insidious dementia, and histopathologically, by the appearance of neurofibrillary tangles and senile plaques in affected brain regions. Although almost a hundred years have elapsed since the first description of this debilitating disorder, there is still considerable controversy surrounding the very nature of neuronal cell death. Current views hold that amyloid-β (Selkoe, 2001) oxidative stress (reviewed in Rottkamp et al, 2000a,b), and cell cycle dysregulation (reviewed in Raina et al., 2000), are involved early in the disease process and contribute significantly to the histological phenotype and neuronal cell death that is seen with the progression of the disorder. Here, we review salient aspects of both neuronal death as well as survival in AD.

2. CELL DEATH MECHANISMS

2.1. General Aspects

Despite decades of intense studies in AD, one of the most basic questions remains unanswered, namely the nature of the cell death of vulnerable neurons. In general, mechanisms of cellular death present with either an apoptotic or a necrotic phenotype.

[1] Institute of Pathology, Case Western Reserve University, Cleveland, Ohio 44106
[2] Department of Biochemistry, George S. Wise Faculty of Life Sciences, Tel Aviv University, Tel Aviv, Israel
[3] Department of Neurology, Kyoto University, Kyoto, Japan
[4] Department of Neurology, Tohoku University School of Medicine, Sendai, Japan
[5] Department of Psychiatry and Neurology, Asahikawa Medical College, Asahikawa, Japan

Mapping the Progress of Alzheimer's and Parkinson's Disease
Edited by Mizuno *et al.*, Kluwer Academic/Plenum Publishers, 2002

However, recent evidence seems to point to a far more complex picture, where apoptotic and necrotic phenotypes might present simultaneously (Sperandio et al., 2000). For example, cardiomycytes can demonstrate TUNEL positivity without proceeding to a cell death demonstrating classical apoptotic phenotype (de Boer et al., 2000). The waters are further unnecessarily muddied by the promiscuous use of the same term, namely "apoptosis", to mean different things in different contexts. This confusion arises by conflating a process (a cell death program) with product (the apoptotic phenotype). Perhaps at this early juncture, we should clarify briefly some issues of semantics. Cell death can be classified as *programmed cell death,* that entails a global/extrinsic program of cell death and a *cell death program(s),* that entails a local/intrinsic/cellular death program (Ratel et al., 2001). The latter display a variety of phenotypes ranging from necrosis to apoptosis or as a combination phenotype (Sperandio et al., 2000), while the former is primarily limited to development and presents with an apoptotic phenotype. This classification system allows for a clearer understanding when describing novel phenomena or in situations where there is ambiguity as to the nature of cell death, i.e., in AD.

2.2 Apoptotic Mechanisms and Alzheimer Disease

Research in AD over the last several decades, has focused on defining the precise mechanisms responsible for the demise of vulnerable neurons. Despite these efforts, the nature and time course of neuronal cell death in AD remains controversial (Bancher et al., 1997; Jellinger and Bancher, 1998; Perry et al., 1998a, 1998b). While claims of an apoptotic mechanism abound, there are serious questions as to their plausibility. Indeed, while a cell death program resulting in an apoptotic phenotype has been implicated in the neuronal death in AD, many of the hallmark manifestations characteristic of the terminal phases of apoptotic cell death including detachment, chromatin condensation, nuclear segmentation, blebbing and apoptotic bodies are absent alone or in combination in AD. Moreover, perhaps most striking, however, is the seeming paradox in timing. Apoptosis requires only 16-24 hours for completion and therefore, in a chronic disease like AD with an average duration of almost 10 years, less than one in about 4000 cells should be undergoing apoptosis at any given time, i.e., observation of apoptotic events should be extremely rare (Perry et al., 1998a, 1998b). However, descriptions of apoptosis in AD report numbers of neurons undergoing apoptosis far in excess of this (Su et al., 1997). Indeed, if all the neurons that are reported with having DNA cleavage were undergoing apoptosis the brain would rapidly be void of neurons and the clinical course would be abbreviated to months—this is certainly not the case in AD.

A large array of possible apoptogenic stimuli are established in neurons in AD including reactive oxygen species (Slater et al., 1995), amyloid-β (Yankner, 1996), energy failure (Vander Heiden et al., 2000) and HNE oxidants (Sayre et al., 1997; Mark et al., 1997). These may act alone or synergistically to make neurons vulnerable to apoptotic processes. Additionally, familial AD has been shown to pose an apoptotic risk (Deng et al., 1996). However, while internucleosomal DNA fragmentation (detected by end-labeling techniques) has been used as an absolute index of apoptotic cell death in AD (Anderson et al., 1996), it is now clear that this phenotype, is also characteristic of death due to oxidative stress (Smith et al., 1997a,b; Tsang et al., 1996; Su et al., 1997) and postmortem autolysis (Stadelman et al., 1998). This in combination with the absence of stereotypical endstage signs of an apoptotic cell death program (Perry et al., 1998a), indicate that neurons in AD are not dying via apoptosis. Added to this mechanistic and biochemical ambiguity of neuronal death mechanisms in AD is the already mentioned temporal dichotomy between the acuteness of an apoptotic cell death program and the chronicity of AD (Perry et al., 1998b).

2.3. Ambiguities in the Detection of Apoptosis: Relevance to Alzheimer Disease

Given that the neuronal death in AD may represent a novel death scenario demonstrating aspects of necrosis and apoptosis, the confusion surrounding this issue become more comprehensible. In light of these ambiguities, it becomes essential to revisit the nature of neuronal cell death in AD. Due to the post mortem nature of AD research, consideration of the methodology used in determining cell death is critical. Numerous methods of detecting apoptosis have been developed which are designed to detect the entire apoptotic pathway. These include methods that assay for the presence of death receptors, changes in the plasma membrane structure, activation of the caspase protease cascade, changes in the integrity of mitochondria and fragmentation of nuclear DNA (Raina et al., 2001a). Detection of apoptosis can be conveniently divided into two categories, those that detect changes in cell populations and those that detect changes in individual cells. Each of these have their own sets of advantages and disadvantages, but what is clearly evident is that no single criteria can be used to establish definitively the presence of apoptotic processes because each individually can be indicative of other phenomena. Only when evidence from several of these methods is considered cumulatively is it possible to conclude either the presence or absence of apoptotic cell death.

2.4. Classical Morphology as Evidence of Apoptosis

The earliest criterion established for describing cell death via an apoptotic mechanism was the characteristic cellular phenotype that includes blebbing and nuclear condensation and segmentation. The nuclear features of the apoptotic process include peripheral chromatin condensation and subsequent fragmentation into multiple chromatin bodies surrounded by nuclear envelope remnants. These morphological characteristics continue to be the gold standard for the involvement of apoptosis in the death of any given cell. However, the time course for these changes remains unclear and they are not required for apoptosis given that enucleation does not prevent the process from proceeding in the cytoplasm (Schulze-Ostoff et al., 1994). The fact that such morphological changes, which are considered the most accurate indicators, are absent in AD calls into question the claim that cell death in AD is apoptotic.

Detection of the above mentioned changes by electron microscopy conclusively reveals the presence of apoptosis, although it cannot quantify this since apoptosis is in most cases asynchronous. Other microscopic techniques involve fluorescent microscopy, after staining with propidium iodide or 4,6-diamidino-2-phenolindole to detect nuclear changes or labeling with Annexin V (Zhang et al., 1997), which highlights externalization of phosphatidyl serine. Apoptosis can also be detected by light microscopy using hematoxylin and eosin staining which primarily involves pattern recognition of pyknotic and hyperchromatic nucleui.

2.5. DNA Fragmentation as Evidence of Apoptosis

DNA fragmentation is the most widely used for detection of apoptosis based on the fact that in apoptosis cation-dependent endonucleases cleave genomic DNA between nucleosomes. Therefore, in apoptotic cell death, one expects to see mono- and oligonucleosomal sized DNA fragments as opposed to the high molecular weight species normally seen with genomic DNA. This creates a distinctive electrophoretic banding pattern known as DNA laddering. While this method can be used effectively in combination with other criteria in assessing cells *in vitro* and biopsy tissue, its value in establishing apoptotic processes in the post mortem tissue of AD is limited. DNA fragmentation may in fact represent the greatest source of misleading data used to link AD with apoptosis. Similar mono- and oligonucleosomal patterns can be a result of a variety of conditions including necrosis and oxidative damage, because histones protect DNA from a variety of insults. Another confounding issue in the use of DNA laddering postmortem autolysis can produce a similar pattern (Stadelmann et al., 1998). This is a problem that is unavoidable when studying neuronal death in AD.

In situ DNA fragmentation can also be detected by end-labeling techniques which detect the degradation of DNA by taking advantage of DNA repair enzymes to add labeled nucleotides either nicked strands according to the opposite strand template or addition of nucleotides independent of a template either at nick sites or the blunt end double strand breaks. Respectively, these techniques are called *In Situ* Nick Translation (ISNT) and TUNEL (TdT-mediated X-UTP nick end labeling). Of these techniques, TUNEL is currently the most commonly used. The principle behind terminal end-labeling is that the enzyme terminal deoxyribonucleotidyl transferase (TdT) will catalyze the addition of nucleosides at free 3' hydroxyl groups produced by apoptotic endonucleases.

The advantage of these techniques over DNA laddering is that they allows the investigator to determine the specific subpopulation of cells that is undergoing apoptosis at any given time. It also allows for more careful consideration of the time course of DNA fragmentation and other nuclear morphological changes. Although ISNT and TUNEL have proven to be powerful techniques that have added to our understanding of apoptotic mechanisms, they must be used with a clear understanding of limitations and confounding complications. First, the results of end-labelling techniques are highly variable depending on the manner in which the tissue is preserved and prepared (Tateyama et al., 1998). For example, false negative results can be obtained if the fragmentation is not accessible to the TdT enzyme after permeabilization (Nakamura et al., 1997). The preservation of tissue is of great importance in the study of AD tissue that often is not fixed until many hours postmortem. Additionally, nuclear DNA damage can be caused by other modes of cell death and thereby lead to false positive results in a large number of cells in a given population (Kockx et al., 1998).

2.6. Caspase Cascade Pathways in Alzheimer Disease Neurons

In an effort to clarify the discrepancy between morphological and DNA fragmentation results in AD, our laboratory employed a third mode of detection. We saw the immunocytochemical detection of caspases as an important tool that could be used to resolve some of the confounding issues plaguing our understanding of apoptotic processes in AD. We undertook a systematic study of the caspase cascade proteins in AD by evaluating the key initiator (caspases 8 and 9) and the executioner (caspases 3, 6 and

7) proteins of apoptosis (Raina et al., 2001b). This revealed that while upstream intiator caspases are present in association with the pathological lesions in all cases of AD, downstream executioner caspases including 3, 6 and 7 that signal completion of the apoptosis pathway remain at control levels in vulnerable populations. The presence of caspase 8, which is a downstream component of the Fas > FADD liagand pathway and a specific activator of caspase 3 (Stennicke et al., 1998) argues for the recruitment of the initiator components of the caspase pathway in neurons of AD. This is strengthened by the presence of caspase 9, which leads to caspase 3 activation via a downstream mitochondrial pathway. However, the fact that neither caspase 3 nor 7 figure prominently in the neuropathology of AD (Cohen, 1997) indicates incomplete or effectively absent amplification of the upstream apoptotic signal. Indeed, since such downstream caspases and their proteolytic products are recognized as markers of apoptotic irreversibility (Trucco et al., 1998), their avoidance or sporadic appearance in AD indicates an absence of effective distal propagation of the caspase-mediated apoptotic signal(s). This lack of downstream amplification of signaling via the caspase pathway may well account for the lack of an apoptotic phenotype in AD. Furthermore, the neurons that are present and viable at latter stages in AD must have employed survival compensations to the presumptive pro-apoptotic environment encountered in AD (Raina et al., 2001b).

Although the upregulation of individual caspase does not appear to lead to apoptosis, this does not preclude them from playing a significant role in AD pathogenesis. For example, caspase 6 is found in association with amyloid-β senile plaques and is capable of cleaving amyloid-β protein precursor (AβPP) which resulting in a 6.5 kDa amyloid fragment that is proposed as an alternate AβPP processing pathway (LeBlanc et al., 1999). The lack of caspase 6, a downstream effector caspase, in neuronal pathology, however, argues against a specific role in the apoptotic cascade.

2.7. Cell Cycle

To the surprise of many investigators, the phenotype of pyramidal neurons in AD includes the ectopic presence of various cyclins, cyclin-dependent kinases (Arendt et al., 1996; McShea et al., 1997; Zhu et al., 2000a,b) and cyclin-dependent kinase inhibitors (reviewed in Raina et. al., 2000; Nagy et al., 1998). Indeed, the morphology of neurons that are vulnerable in AD resembles that of cells that are dividing, rather than terminally differentiated, quiescent cells (Vincent et al., 1996; Raina et al., 1999a,b,2000). This evidence, therefore suggests that in AD, susceptible neuronal populations, which are postmitotic and thus restricted in their mitotic competence, actually exit G_0 and re-enter the G_1 phase of the cell cycle during the progression of AD. In addition to the presence of cell-cycle phase dependent kinases, this "re-entrant" phenotype is also supported by activation of selective signal transduction pathways, transcriptional activation that results in cytoskeletal alterations such as τ phosphorylation and increases in mitochondria activity and DNA replication (Perry et al., 1999; Zhu et al., 2000a,b, 2001a-c). In this sense, AD neurons resemble those normally seen only in developmental neurogenesis, in mitotically active cell populations and neoplastic cells.

It has been argued that a number of the cell-cycle related phenomena found ectopically expressed in AD, can also be involved in other cell processes including apoptosis, trophic-deprivation and DNA repair (Elledge, 1996; Stefanis et al., 1996; Park et al., 1997, 1998, 2000; Padmanabhan et al., 1999). Therefore, we looked for unique novel marker of cell cycle progression. The MORF4-related protein family are unique markers of emergence from quiescence and progression into the G_1 phase of the cell cycle. Hence, expression is decreased in quiescent and senescent non-dividing normal human cells and protein levels increase when quiescent cells are stimulated to re-enter the cell cycle (Bertram et al., 1999). We found that a 52kDa protein MORF4-related cell cycle protein is associated with intraneuronal neurofibrillary pathology in select neuronal populations that are vulnerable in AD (Raina et al., 2001c). The fact that a positive transcriptional controlling protein is associated with intraneuronal neurofibrillary pathology in vulnerable neurons of AD is consistent with the changes in gene expression that occur and also lends credence to the proximal role of mitotic mechanisms in AD. The unforseen presence of the cell cycle control elements within the AD intraneuronal environment may in fact indicate early intraneuronal compensations to the primary stressor. The duration of neuronal functionality thus may be a consequence of the degree of success of these compensations. Hence, this return to mitotic competence in AD may have profound implications that should be explored in developing novel strategies for screening and therapeutics.

2.8. Anti-apoptotic Signals in Alzheimer Disease

Apoptosis can be prevented by anti-apoptotic members of the Bcl-2 family, many of which have been reported in AD, including Bcl-xl and Bcl-w (reviewed in Pike, 1999; Zhu et al., unpublished data). Interactions of activated caspases with XIAPs as well as SMAC (DIABLO) provide for post-activation regulation of the death signal (Srinivasula et al., 2001). Additional evidence for the anti-apoptotic nature of the intraneuronal environment in AD includes the expression of GADD 45, a growth arrest DNA damage-inducible protein in select neurons in AD where its early expression is associated with the expression of Bcl-2. Thus, this may in turn confer survival advantages in AD (Torp et al., 1998). Furthermore, the hyperphosphorylated intraneuronal state that exists in AD acts to inhibit downstream substrate proteolysis and thus can promote neuronal survival. Indeed, accumulating evidence points to dephosphorylation being associated with increased capability to cleave PARP (Martins et al., 1998). Additionally, chronic oxidative stress, a major component of early AD pathophysiology (Smith et al., 1996), would further inhibit downstream propagation of caspase-mediated apoptotic signals (Hampton et al., 1998).

3. CONCLUSIONS

The role of apoptosis in the neuronal cell death of AD has been much disputed in the research community. Despite evidence of DNA fragmentation, both DNA laddering and TUNEL, the hallmark features of apoptosis, namely nuclear and cytoplasmic changes in morphology, have remained elusive. However, using detection of activated caspases as another mode of detection to clarify the issue we found that AD a novel form of cell death. Specifically, this represents the first *in vivo* situation reported in which initiation of apoptosis does not directly lead to apoptotic cell death, an interpretation consistent with earlier reports (Lassman et al., 1995; Lucassen et al., 1995, 1997; Sheng et al., 1998a, 1998b). Obviously, this is not a universal process because neurons adjacent those that avoid apoptosis do meet their demise. However, it is unclear whether this degeneration is through successful propogation of the apoptotic cascade or by another pathway such as paratosis. However, in those surviving neurons, it is clear that neuronal viability in AD is, in part, maintained by the lack of distal transmission of the caspase-mediated apoptotic signal(s). This novel phenomenon, which we term **abortosis**, leads to apoptotic avoidance and hence selects for neuronal survival rather than a caspase-induced cell death program that presents as apoptosis.

REFERENCES

Anderson, A. J., Su, J. H., and Cotman, C. W., 1996, DNA damage and apoptosis in Alzheimer's disease: colocalization with c-Jun immunoreactivity, relationship to brain area, and effect of postmortem delay, *J. Neurosci.* **16**:1710-1719.

Arendt, T., Rodel, L., Gartner, U., and Holzer, M., 1996, Expression of the cyclin-dependent kinase inhibitor p16 in Alzheimer's disease, *Neuroreport* **7**:3047-3049.

Bancher, C., Lassmann, H., Breitschopf, H., and Jellinger, K. A., 1997, Mechanisms of cell death in Alzheimer's disease, *J. Neural Transm. Suppl.* **50**:141-152.

Bertram, M. J., Berube, N. G., Hang-Swanson, X., Ran, Q., Leung, J. K., Bryce, S., Spurgers, K., Bick, R. J., Baldini, A., Ning, Y., Clark, L. J., Parkinson, E. K., Barrett, J. C., Smith, J. R., and Pereira-Smith, O. M., 1999, Identification of a gene that reverses the immortal phenotype of a subset of cells and is a member of a novel family of transcription factor-like genes, *Mol. Cell. Biol.* **19**:1479-1485.

Cohen, G. M., 1997, Caspases: the executioners of apoptosis. *Biochem. J.* **326**:1-16.

de Boer, R. A., van Veldhuisen, D. J., van der Wijk, J., Brouwer, R. M., de Jonge, N., Cole, G. M., and Suurmeijer, A. J., 2000, Additional use of immunostaining for active caspase 3 and cleaved actin and PARP fragments to detect apoptosis in patients with chronic heart failure, *J. Card. Fail* **6**:330-337.

Deng, G., Pike, C. J., and Cotman, C. W., 1996, Alzheimer-associated presenilin-2 confers increased sensitivity to apoptosis in PC12 cells, *FEBS Lett.* **397**:50-54.

Elledge, S. J., 1996, Cell cycle checkpoints: preventing an identity crisis, *Science* **274**:1664-1672.

Hampton, M. B., Fadeel, B., and Orrenius, S., 1998, Redox regulation of the caspases during apoptosis, *Ann. NY Acad. Sci.* **854**:328-335.

Jellinger, K. A. and Bancher, C., 1998, Neuropathology of Alzheimer's disease: a critical update, *J. Neural Transm. Suppl.* **54**:77-95.

Kockx, M. M., Muhring, J., Knaapen, M. W. M., and deMeyer, G. R. Y., 1998, RNA synthesis and splicing interferes with DNA *in situ* end labeling techniques used to detect apoptosis, *Am. J. Pathol.* **152**:885-888.

Lassmann, H., Bancher, C., Breitschopf, H., Wegiel, J., Bobinski, M., Jellinger, K., and Wisniewski, H. M., 1995, Cell death in Alzheimer's disease evaluated by DNA fragmentation *in situ*, *Acta Neuropathol.* **89**:35-41.

LeBlanc, A., Liu, H., Goodyer, C., Bergeron, C., and Hammond, J., 1999, Caspase-6 role in apoptosis of human neurons, amyloidogenesis, and Alzheimer's disease, *J. Biol. Chem.* **274**:23426-23436.

Lucassen, P. J., Chung, W. C., Kamphorst, W., and Swaab, D. F., 1997, DNA damage distribution in the human brain as shown by *in situ* end labeling; area-specific differences in aging and Alzheimer disease in the absence of apoptotic morphology, *J. Neuropathol. Exp. Neurol.* **56**:887-900.

Lucassen, P. J., Chung, W. C., Vermeulen, J. P., Van Lookeren Campagne, M., Van Dierendonck, J. H., and Swaab, D. F., 1995, Microwave-enhanced *in situ* end-labeling of fragmented DNA: parametric studies in relation to postmortem delay and fixation of rat and human brain, *J. Histochem. Cytochem.* **43**:1163-1171.

Mark, R. J., Lovell, M. A., Markesbery, W. R., Uchida, K., and Mattson, M. P., 1997 A role for 4-hydroxynonenal, an aldehydic product of lipid peroxidation, in disruption of ion homeostasis and neuronal death induced by amyloid beta-peptide, *J. Neurochem.* **68**:255-264

Martins, L. M., Kottke, T. J., Kaufmann, S. H., and Earnshaw, W. C., 1998, Phosphorylated forms of activated caspases are present in cytosol from HL-60 cells during etoposide-induced apoptosis, *Blood* **92**:3042-3049.

McShea, A., Zelasko, D.A., Gerst, J.L., and Smith, M.A., 1999 Signal transduction abnormalities in Alzheimer's disease: evidence of a pathogenic stimuli, *Brain Res.* **815**:237-242.

Nagy, Z., Esiri, M. M., and Smith, A. D., 1998, The cell division cycle and the pathophysiology of Alzheimer's disease. *Neuroscience* **87**:731-739.

Nakamura, N., Yagi, H., Ishii, T., Kayaba, S., Soga, H., Gotoh, T., Ohsu, S., Ogata, M., and Itoh, T., 1997, DNA fragmentation is not the primary event in glucocorticoid-induced thymocyte death *in vivo*, *Eur. J. Immunol.* **27**:999-1004.

Padmanabhan J., Park D.S., Greene L.A., and Shelanski M.L., 1999, Role of cell cycle regulatory proteins in cerebellar granule neuron apoptosis, *J. Neurosci.* **19**:8747-8756.

Park, D. S., Levine, B., Ferrari, G., and Greene, L. A., 1997, Cyclin dependent kinase inhibitors and dominant negative cyclin dependent kinase 4 and 6 promote survival of NGF-deprived sympathetic neurons. *J. Neurosci.* **17**:8975-8983.

Park, D. S., Morris, E. J., Padmanabhan, J., Shelanski, M. L., Geller, H. M., and Greene, L. A., 1998, Cyclin-dependent kinases participate in death of neurons evoked by DNA-damaging agents, *J. Cell Biol.* **143**:457-467.

Park, D. S., Obeidat, A., Giovanni, A., and Greene, L. A., 2000, Cell cycle regulators in neuronal death evoked by excitotoxic stress: implications for neurodegeneration and its treatment. *Neurobiol. Aging* **21**:771-781.

Perry, G., Nunomura, A., and Smith, M. A., 1998a, A suicide note from Alzheimer disease neurons? (News and Views), *Nat. Med.* **4**:897-898.

Perry, G., Nunomura, A., Lucassen, P., Lassmann, H., and Smith, M. A., 1998b, Apoptosis and Alzheimer's disease (Letter), *Science* **282**:1268-1269.

Perry, G., Roder, H., Nunomura, A., Takeda, A., Friedlich, A. L., Zhu, X., Raina, A. K., Holbrook, N., Siedlak, S. L., Harris, P. L. R., and Smith, M. A., 1999, Activation of neuronal extracellular receptor kinase (ERK) in Alzheimer disease links oxidative stress to abnormal phosphorylation, *Neuroreport* **10**:2411-2415.

Pike, C. J., 1999, Estrogen modulates neuronal Bcl-xL expression and beta-amyloid-induced apoptosis: relevance to Alzheimer's disease, *J. Neurochem.* **72**:1552-1563.

Raina, A. K., Monteiro, M. J., McShea, A., and Smith, M. A., 1999a, The role of cell cycle-mediated events in Alzheimer's disease, *Int. J. Exp. Pathol.* **80**:71-76.

Raina, A. K., Takeda, A., and Smith, M. A., 1999b, Mitotic neurons: a dogma succumbs, *Exp. Neurol.* **159**:248-249.

Raina, A. K., Zhu, X., Rottkamp, C. A., Monteiro, M., Takeda, A., and Smith, M.A., 2000, Cyclin' toward dementia: cell cycle abnormalities and abortive oncogenesis in Alzheimer disease, *J. Neurosci. Res.* **61**, 128-133.

Raina, A. K., Hochman, A., Rottkamp, C. A., Zhu, X., Obrenovich, M. E., Shimohama, S., Nunomura, A., Takeda, A., Sayre, L. M., Perry, G., and Smith, M. A., 2001a, Apoptotic and oxidative indicators in Alzheimer disease, in: *Apoptosis Techniques and Protocols*, A. LeBlanc, ed., Humana Press Inc., New Jersey, in press.

Raina, A. K., Hochman, A., Zhu, X., Rottkamp, C. A., Nunomura, A., Siedlak, S. L., Boux, H., Castellani, R. J., Perry, G., and Smith, M.A., 2001b, Abortive apoptosis in Alzheimer's disease, *Acta Neuropathol* **101**:305-310.

Raina, A. K., Pardo, P., Rottkamp, C. A., Zhu, X., Pereira-Smith, O. M., and Smith, M. A., 2001c, Senescent emergence of neurons in Alzheimer disease. *Mech. Ageing Dev.* **123**:3-9.

Ratel, D., Boisseau, S., Nasser, V., Berger, F., and Wion, D., 2001, Programmed cell death or cell death programme? That is the question, *J. Theor. Biol.* **208**:385-386.

Rottkamp, C. A., Nunomura, A., Raina, A. K., Sayre, L. M., Perry, G., and Smith, M.A., 2000a, Oxidative stress, antioxidants, and Alzheimer disease, *Alzheimer Dis. Assoc. Disord.* **14** (Suppl. 1):S62-S66.

Rottkamp, C. A., Nunomura, A., Hirai, K., Sayre, L. M., Perry, G., and Smith, M.A., 2000b, Will antioxidants fulfill their expectations for the treatment of Alzheimer disease?, *Mech. Ageing Dev.* **116**:169-179.

Sayre, L. M., Zelasko, D. A., Harris, P. L. R., Perry, G., Salomon, R. G., and Smith, M.A., 1997, 4-Hydroxynonenal-derived advanced lipid peroxidation end products are increased in Alzheimer's disease, *J. Neurochem.* **68**:2092-2097.

Schulze-Osthoff, K., Walczak, H., Droge, W., and Krammer, P. H., 1994, Cell nucleus and DNA fragmentation are not required for apoptosis, *J. Cell Biol.* **127**:15-20.

Selkoe, D. J., 2001, Alzheimer's disease results from the cerebral accumulation and cytotoxicity of amyloid β-protein: A reanalysis of a therapeutic hypothesis, *J. Alzheimer's Disease* **3**:75-81.

Sheng, J. G., Mrak, R. E., and Griffin, W. S., 1998a, Progressive neuronal DNA damage associated with neurofibrillary tangle formation in Alzheimer disease, *J. Neuropathol. Exp. Neurol.* **57**:323-328.

Sheng, J. G., Zhou, X. Q., Mrak, R. E., and Griffin, W. S., 1998b, Progressive neuronal injury associated with amyloid plaque formation in Alzheimer disease, *J. Neuropathol. Exp. Neurol.* **57**:714-717.

Slater, A. F., Stefan, C., Nobel, I., van den Dobbelsteen, D. J., and Orrenius, S., 1995, Signalling mechanisms and oxidative stress in apoptosis, *Toxicol. Lett.* **82-83**:149-153.

Smith, M. A., Perry, G., Richey, P. L., Sayre, L. M., Anderson, V. E., Beal, M. F., and Kowall, N., 1996, Oxidative damage in Alzheimer's, *Nature* **382**:120-121.

Smith, M. A., Harris, P. L. R., Sayre, L. M., Beckman, J. S., and Perry, G., 1997a, Widespread peroxynitrite-mediated damage in Alzheimer's disease, *J. Neurosci.* **17**:2653-2657.

Smith, M. A., Harris, P. L. R., Sayre, L. M., and Perry, G., 1997b, Iron accumulation in Alzheimer disease is a source of redox-generated free radicals, *Proc. Natl. Acad. Sci. USA* **94**:9866-9868.

Sperandio, S., de Belle, I., and Bredesen, D. E., 2000, An alternative, nonapoptotic form of programmed cell death, *Proc. Natl. Acad. Sci. USA* **97**:14376-14381.

Srinivasula, S. M., Hegde, R., Saleh, A., Datta, P., Shiozaki, E., Chai, J., Lee, R.-A., Robbins, P. D., Fernandes-Alnemri, T., Shi, Y., and Alnemri, E. S., 2001, A conserved XIAP-interaction motif in caspase-9 and Smac/DIABLO regulates caspase activity and apoptosis, *Nature* **410**:112-116.

Stadelmann, C., Bruck, W., Bancher, C., Jellinger, K., and Lassmann, H., 1998, Alzheimer disease: DNA fragmentation indicates increased neuronal vulnerability, but not apoptosis, *J. Neuropathol. Exp. Neurol.* **57**:456-464.

Stefanis, L., Park, D. S., Yan C. Y., Farinelli S. E., Troy C. M., Shelanski M. L., and Greene L. A., 1996, Induction of CPP32-like activity in PC12 cells by withdrawal of trophic support. Dissociation from apoptosis, *J. Biol. Chem.* **271**:30663-30671.

Stennicke, H. R., Jurgensmeier, J. M., Shin, H., Deveraux, Q., Wolf, B. B., Yang, X., Zhou, Q., Ellerby, H. M., Ellerby, L. M., Bredesen, D., Green, D. R., Reed, J. C., Froelich, C. J., and Salvesen, G. S., 1998, Pro-caspase-3 is a major physiologic target of caspase-8, *J. Biol. Chem.* **273**:27084-27090.

Su, J. H., Deng, G., and Cotman, C. W., 1997, Neuronal DNA damage precedes tangle formation and is associated with up-regulation of nitrotyrosine in Alzheimer's disease brain, *Brain Res.* **774**:193-199.

Tateyama, H., Tada, T., Hattori, H., Murase, T., Li, W.-X., and Eimoto, T., 1998, Effects of prefixation and fixation times on apoptosis detection by *in situ* end-labeling of fragmented DNA, *Arch. Pathol. Lab. Med.* **122**:252-255.

Torp, R., Su, J. H., Deng, G., and Cotman, C. W., 1998, GADD45 is induced in Alzheimer's disease, and protects against apoptosis *in vitro*, *Neurobiol. Disease* **5**:245-252.

Trucco, C., Oliver, F. J., de Murcia, G., and Menissier-de Murcia, J., 1998, DNA repair defect in poly(ADP-ribose) polymerase-deficient cell lines, *Nucleic Acids Res.* **26**:2644-2649.

Tsang, S. Y., Tam, S. C., Bremner, I., and Burkitt, M. J., 1996, Research communication copper-1,10-phenanthroline induces internucleosomal DNA fragmentation in HepG2 cells, resulting from direct oxidation by the hydroxyl radical, *Biochem. J.* **317**:13-16.

Vander Heiden, M. G., Chandel, N. S., Li, X. X., Schumacker, P. T., Colombini, M., and Thompson, C. B., 2000, Outer mitochondrial membrane permeability can regulate coupled respiration and cell survival, *Proc. Natl. Acad. Sci. USA* **97**:4666-4671.

Vincent, I., Rosado, M., and Davies, P., 1996, Mitotic mechanisms in Alzheimer's disease?, *J. Cell Biol.* **132**:413-425.

Yankner, B. A., 1996, New clues to Alzheimer's disease: unraveling the roles of amyloid and tau, *Nat. Med.* **2**:850-852.

Zhang, G., Gurtu, V., Kain, S. R., and Yan, G., 1997, Early detection of apoptosis using a fluorescent conjugate of annexin V, *Biotechniques* **23**:525-531.

Zhu, X., Raina, A.K., Boux, H., Simmons, Z.L., Takeda, A., and Smith, M.A., 2000a, Activation of oncogenic pathways in degenerating neurons in Alzheimer disease, *Int. J. Devl. Neurosci.* **18**:433-437.

Zhu, X., Rottkamp, C. A., Boux, H., Takeda, A., Perry, G., and Smith, M. A. (200b) Activation of p38 kinase links tau phosphorylation, oxidative stress, and cell cycle-related events in Alzheimer disease, *J. Neuropathol. Exp. Neurol.* **59**:880-888

Zhu, X., Rottkamp, C. A., Raina, A. K., Brewer, G. J., Ghanbari, H. A., Boux, H., and Smith, M. A., 2000c, Neuronal Cdk7 in hippocampus is related to aging and Alzheimer disease, *Neurobiol. Aging* **21**:807-813.

Zhu, X., Raina, A. K., Rottkamp, C. A., Aliev, G., Perry, G., Boux, H., and Smith, M. A., 2001a, Activation and redistribution of c-Jun N-terminal kinase/stress activated protein kinase in degenerating neurons in Alzheimer's disease, *J. Neurochem.* **76**:435-441.

Zhu, X., Castellani, R. J., Takeda, A., Nunomura, A., Atwood, C. S., Perry, G., and Smith, M. A., 2001b, Differential activation of neuronal ERK, JNK/SAPK and p38 in Alzheimer disease: the "two hit" hypothesis. *Mech. Ageing Dev.* **123**:39-46.

Zhu, X., Rottkamp, C. A., Hartzler, A., Sun, Z., Takeda, A., Boux, H., Shimohama, S., Perry, G., and Smith, M. A., 2001c, Activation of MKK6, an upstream activator of p38, in Alzheimer's disease. *J. Neurochem.* **79**:311-318.

THE HISTAMINE-CYTOKINE NETWORK IN ALZHEIMER DISEASE: ETIOPATHOGENIC AND PHARMACOGENOMIC IMPLICATIONS

Ramón Cacabelos[*]

1. INTRODUCTION

Alzheimer disease (AD) is a complex disorder characterized by premature neuronal death associated with genetic factors (Cacabelos et al., 1999, 2000a,b). Mendelian genetics, with specific mutations in AD-related genes, accounts for less than 10% of the AD population while allelic associations with potential risk for AD, involving at least 5 different genes (APOE, PS1, PS2, cFOS, A2M) are practically present in all AD cases submitted to genomic screening (Cacabelos et al., 1999, 2000a,b). In addition, in those cases where AD appears associated with only one gene of risk (i.e., APOE-4) the onset of the disease occurs at 65-70 years of age; in contrast, when AD patients exhibit a genotype with more than 2 allelic associations of risk, then the onset of the disease anticipates 10-15 years as compared with cases exhibiting a single gene-related allelic association of potential risk (Cacabelos et al., 1999). At least 10 different genes show some linkage to AD. This fact justifies the great diversity of phenotypic heterogeneity present in AD (Cacabelos, 1996) as well as the lack of homogeneous therapeutic responses to different candidate drugs in current clinical trials (Cacabelos, et al., 2000b). The genomic information available on AD made it possible the construction of the first biochip models for novel drug development (Lombardi et al 1998, 1999a) and the first studies using a polygenic pharmacogenomics approach (Cacabelos, 2000; Cacabelos et al., 2000a) clearly demonstrating these preliminary studies that the therapeutic response in AD is genotype-specific (Cacabelos et al., 2000a,b). These findings suppose two major implications in drug development: (i) it is very unlikely that a single therapeutic strategy be powerful enough to neutralize the pathogenic events present in AD, including β-amyloid deposition in senile plaques and vessels (amyloid angiopathy)(protein conformational disorder), neurofibrillary tangle formation associated with hyperphosphorylation of tau proteins (tauopathic disorder), synaptic loss (neurotransmitter deficit, neurotrophic dysfunction), neuronal apoptosis

[*] Ramón Cacabelos, EuroEspes Biomedical Research Center, Institute for CNS Disorders, 15166-Bergondo, La Coruña, Spain. cacabelos@euroespes.com

Mapping the Progress of Alzheimer's and Parkinson's Disease
Edited by Mizuno *et al.*, Kluwer Academic/Plenum Publishers, 2002

induced by endogenous and/or exogenous factors leading to caspase activation, and other secondary events currently present in the AD brains (inflammatory reactions, free radical formation, glutamate-mediated excitotoxic reactions, alterations in calcium metabolism, cerebrovascular dysfunction)(Fig. 1); and (ii) the polygenic disorder responsible for AD neuropathology in terms of pharmacological treatment requires a multifactorial approach based on the association of several candidate drugs and/or the implementation of novel pharmacogenomics to optimised drug development (efficacy and safety issues) since only approximately 15-20% of the patients show some beneficial effect under conventional treatments (Cacabelos et al., 2000a,b).

Other players in AD neuropathology are those secondary pathogenic events leading to accelerated neuronal death, including cerebrovascular dysfunction with a progressive reduction of blood brain flow and significant changes in brain hemodynamics in parallel to cognitive deterioration, and also the neuroimmune regulatory dysfunction surrounding senile plaques and specific damaged areas of the CNS. Many of these phenomena are also genotype-specific and should be taken into consideration for a better understanding of the etiopathogenesis of AD and for future therapeutic intervention. For instance, in AD patients over 75 years of age, there is an important cerebrovascular component influencing neurodegeneration. The reduction in brain blood perfusion runs in parallel with global deterioration staging (GDS)(Cacabelos et al., 1996). Both degenerative and cerebrovascular changes are accompanied by central and peripheral alterations in the concentration of cytokines and histamine, among many other factors (Cacabelos et al., 1992). Some of these alterations also show a genotype-specific profile. In this paper an overview of the histamine-cytokine network in AD is summarized, emphasizing on basic and clinical data which illustrate the regulation of histamine-cytokine interactions and the changes observed in central and peripheral tissues of AD patients.

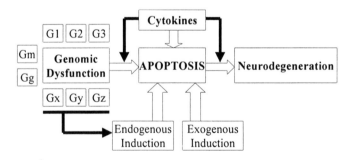

Figure 1. Pathogenic events in Alzheimer disease neurodegeneration

2. HISTAMINE IN ALZHEIMER DISEASE

Histamine (HA) is a neurotransmitter synthesized in the tuberomammillary bodies of the posterior hypothalamus from where ascending and descending pathways innervate the CNS (Watanabe et al., 1984). HA acts on H1, H2, H3 and H4 receptors to regulate inflammation, behavior, neuroendocrine function, learning activities, sleep and arousal, feeding, perception and locomotion (Leurs et al., 1995; Brown et al., 2001; Hough, 2001). Ascending HA pathways to the hippocampus are ipsilateral while HA pathways to the neocortex are bilateral. Brain HA belongs to different compartments (neuronal, mast cell, endothelial, glial) from which approximately 60-80% is neuronal histamine. HA deteriorates cognition, and the administration of both L-histidine and α-fluoromethylhistidine (FMH), a suicide inhibitor of histidine decarboxylase, improve cognition, indicating that brain HA is not essential for mental performance, and that reactive non-neuronal HA responses in brain might be deleterious for neuronal function. HA levels tend to decrease in most regions of the CNS with aging. In contrast, the concentration of HA is markedly elevated in brain tissue (Cacabelos et al., 1989), CSF and blood from AD patients (Fernández-Novoa & Cacabelos, 1995). Elevated concentrations of peripheral HA are not highly modified by GDS, but differ according to specific genotypes, correlating with cerebrovascular changes in both AD and vascular dementia (Fernández-Novoa et al., 1999). FMH-dependent HA depletion enhances ischemic damage in hippocapal CA2 neurons. Neuronal HA might protect against hipoxia-induced apoptosis. Changes in H1 receptors have also been detected by PET in AD (Higuchi et al., 2000). The chronic administration of CDP-choline, an endogenous nucleotide and intermediate factor in membrane phospholipid metabolism, reverses the increase of peripheral HA levels in parallel with improvement in cognition in AD patients, demonstrating that the reduction of non-neuronal HA contributes to alleviate cognitive deterioration and neuronal dysfunction (Fernández-Novoa et al., 1994).

All these data together clearly indicate that HA plays some pathogenic role in AD. However, the basic mechanisms by which HA is altered in AD have never been elucidated. It has been postulated that HA exerts a dual role in the CNS. Neuronal HA seems to display a neurotrophic/neuroprotective effect whereas non-neuronal HA may respond to noxious stimuli associated with cytokine activation producing a neurotoxic effect (Cacabelos et al., 1992).

3. INTERLEUKIN-1β (IL-1) AND TUMOR NECROSIS FACTOR α (TNF)

The presence of a neuroinflammatory reaction in the AD brain has been well-documented in recent times by different authors (Cacabelos et al., 1998). In physiological conditions IL-1 and TNF tend to act synergistically in the regulation of immune responses. In contrast, IL-1 and TNF show an apparent functional dissociation in AD. High levels of IL-1 have been demonstrated in brain tissue, CSF and blood of AD patients (Cacabelos et al., 1991, 1994a). However, TNF levels tend to be lower in AD (Cacabelos et al., 1994b). This antagonic response of IL-1 and TNF in AD might reflect a dysregulation in those neuroimmune mechanisms responsible for neuronal protection in response to toxic, inflammatory and/or neurodegenerative processes. In clinical trials with neuroprotective drugs, cognitive improvement runs in parallel with a tendency to the normalization in the peripheral levels of both IL-1 and TNF (Alvarez et al., 1999).

4. HISTAMINE-CYTOKINE INTERACTIONS

HA, IL-1 and TNF regulate each other in the CNS. HA reduces IL-1 in the hypothalamus in a time- and dose-dependent manner and this effect is not influenced by H1, H2 or H3 receptors since the HA receptor antagonists mepyramine, famotidine and thioperamide, respectively, are not effective in reversing the inhibitory effect that HA exerts on IL-1 (Cacabelos et al., 1993), suggesting that this action might be regulated by H4 receptors. The lack of neuronal HA induced by FMH in turn enhances the levels of hypothalamic IL-1, this clearly indicating the existence of a feedback loop regulating HA-IL1 interactions in the hypothalamus (Fig. 2). HA also regulates TNF response in the CNS in a region-specific manner (Fig. 2). For instance, HA neuronal lesions reduce TNF levels in the posterior hypothalamus and increase TNF levels in the hippocampus (Fig. 2).

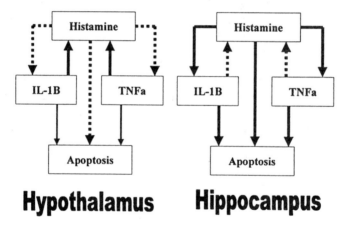

Hypothalamus   Hippocampus

Figure 2. Regional organization of the histamine-cytokine network in the hypothalamus and hippocampus and its potential influence on neuronal apoptosis

Recent evidence suggest that the HA-IL1 network participates in the AD pathogenic cascade according to the following sequential events: (1) β-Amiloid protein (BA4) induces cell damage and neuronal apoptosis; (2) IL-1 and TNF act as mediators of BA4-induced apoptosis; (3) neuronal HA acts as a neurotrophic factor to protect neurons against apoptosis neutralizing IL-1 and TNF activation; (4) a functional dysregulation between IL-1 and TNF in AD contributes to accelerate neuronal death; (5) non-neuronal HA reacts to inhibit IL-1 surge and this effect is deleterious for neurons; and (6) cerebrovascular factors (regional hipoxia, hemodynamic alterations) aggravate neurodegeneration enhancing the activity of the BA4-HA-IL1 cascade (Cacabelos et al., 1989, 1991, 1992, 1994a, 2000b)(Fig. 3)

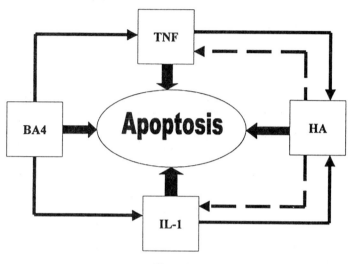

Figure 3.

Most of these events are genotype-specific. For example, serum BA4, ApoE, HA and IL-1 levels clearly show an APOE genotype-related expression. The highest levels of serum BA4 and the lowest levels of serum ApoE are seen in patients with the APOE-4/4 genotype. Microglia activation, IL-1 secretion and MCH-I/II expression in biochip models also show a clear APOE-associated response pattern; and anti-CD3- and Fas-induced lymphocyte apoptosis is also differential according to APOE, PS1 and PS2 genotypes (Cacabelos et al., 2000b; Lombardi et al., 1999a,b).

REFERENCES

Alvarez, X.A., Mouzo, R., Pichel, V., Pérez, P., Laredo, M., Fernández-Novoa, L., Corzo, L., Zas, R., Alcaraz, M., Secades, J.J., Lozano, R., Cacabelos, R., 1999, Effects of citicoline on cognitive performance, brain bioelectrical activity and brain hemodynamics in Alzheimer's disease patients: A double-blind placebo controlled study. *Ann. Psychiat.*, **7**, 332-352.

Brown, R.E., Stevens, D.R., Haas, H.L., 2001, The physiology of brain histamine. *Prog. Neurobiol.*, **63**, 637-672.

Cacabelos, R., 1996, Diagnosis of Alzheimer's disease: defining genetic profiles (genotype vs phenotype). *Acta Neurol. Scand.*, **165**, 72-84.

Cacabelos, R., 2000, Pharmacogenomics in Alzheimer's disease. *Drugs News Perspect.* **13**(4), 252-254.

Cacabelos, R., Yamatodani, A., Niigawa, H., Hariguchi, S., Tada, K., Nishimura, T., Wada, H., Brandeis, L., Pearson, J., 1989, Brain histamine in Alzheimer's disease. *Meth. Find. Exp. Clin. Pharmacol.*, **11**(5), 353-360.

Cacabelos, R., Franco-Maside, A., Alvarez, X.A., 1991, Interleukin-1 in Alzheimer's disease and multi-infarct dementia: Neuropsychological correlations. *Meth. Find. Exp.Clin. Pharmacol.,***13**, 703-708.

Cacabelos, R., Fernández-Novoa, L., Franco-Maside, A., Alvarez, X.A., 1992 Neuroimmune function of brain histamine: implications for neurotrophic activity and neurotoxicity. *Ann. Psychiat.* **3**, 147-200.

Cacabelos, R., Alvarez, X.A., Franco, A., Fernández-Novoa, L., 1993, Dose- and time-dependent effects of histamine on hypothalamic levels of interleukin-1β in rats. *Agents Actions* **38**, C260-C262.

Cacabelos, R., Alvarez, X.A., Fernández-Novoa, L., Franco, A., Mangues, R., Pellicer, A., Nishimura, T., 1994a, Brain interleukin-1β in Alzheimer's disease and vascular dementia. *Meth. Find. Exp. Clin. Pharmacol.,* **16**, 141-151.

Cacabelos, R., Alvarez, X.A., Franco-Maside, A., Fernández-Novoa, L., Caamaño, J., 1994b, Serum tumor necrosis factor in Alzheimer's disease and multiinfarct dementia. *Meth. Find. Exp. Clin. Pharmacol.*, **16**, 29-35.

Cacabelos, R., Caamaño, J., Vinagre, D., Lao, J.I., Beyer, K., Alvarez, A., 1996, Brain mapping and transcranial doppler ultrasonography in Alzheimer disease drug monitoring. In: *Alzheimer Disease: From Molecular Biology to Therapy*, R. Becker, E. Giacobini, ed. Birhäuser, Boston, pp. 469-473.

Cacabelos, R., Alvarez, A., Fernández-Novoa, L., Lombardi, V.R.M., 2000a, A pharmacogenomic approach to Alzheimer's disease. *Acta Neurol. Scand.* **176**, 12-19.

Cacabelos, R., Alvarez, A., Lombardi, V., Fernández-Novoa, L., Corzo, L., Pérez, P., Laredo, M., Pichel, V., Hernández, A., Varela, M., Figueroa, J., Prous, J., Windisch, M., Vigo, C., 2000b, Pharmacological treatment of Alzheimer disease: from psychotropic drugs and cholinesterase inhibitors to pharmacogenomics. *Drugs Today* **36**(7), 415-499.

Cacabelos, R., Beyer, K., Lao, J.I., Mesa, M.D., Fernández-Novoa, L., 1999, Associations of genetic risk factors in Alzheimer's disease and a novel mutation in the predicted TM2 domain of the presenilin-2 gene in late-onset AD. In: *Alzheimer Disease and Related Disorders*. K. Iqbal., D.F. Swaab, B. Winblad, H.M. Wisniewski, ed. John Wiley & Sons, New York, pp. 93-102.

Cacabelos, R., Winblad, B., Eikelenboom, P., ed., 1998, Inflammation and neuroimmunotrophic activity in Alzheimer's disease. *Neurogerontology & Neurogeriatrics*, **2**, Prous Science, Barcelona.

Fernández-Novoa, L., Alvarez, X.A., Franco-Maside, A., Caamaño, J., Cacabelos, R., 1994, CDP-choline-induced blood histamine changes in Alzheimer's disease. *Met. Find. Exp. Clin. Pharmacol.*, **16**, 211-218.

Fernández-Novoa, L., Cacabelos, R., 1995, The histaminergic system in Alzheimer's disease. *Ann. Psychiat.*, **5**, 127-158.

Fernández-Novoa, L., Corzo, L., Zas, R., Laredo, M., Pérez, P., Pichel, V., Mesa, M.D., Cacabelos, R., 1999, Blood histamine levels, brain hemodynamic parameters and risk factors in Alzheimer's disease and vascular dementia. *Ann. Psychiat.*, **7**, 109-117.

Higuchi, M., Yanai, K., Okamura, N., Meguro, K., Arai, H., Itoh, M., Iwata, T., Ido, T., Watanabe, T., Sasaki, H., 2000, Histamine H1 receptors in patients with Alzheimer's disease assessed by positron emisión tomography. *Neuroscience*, **99**(4), 721-726.

Hough, L.B. Genomics meets histamine receptors: new subtypes, new receptors. *Mol. Pharmacol.,***59**(3), 415-419.

Leurs, R., Smit, M.J., Timmerman, H., 1995, Molecular pharmacological aspects of histamine receptors. *Pharmac. Ther.*, **66**(3), 413-463.

Lombardi, V.R.M., García, M., Cacabelos, R., 1998, Microglial activation induced by factor(s) contained in sera from Alzheimer-related APOE genotypes. *J. Neurosci. Res.*, **54**, 539-553.

Lombardi, V.R.M., García, M., Cacabelos, R., 1999a, Genotype-specific microglia response in cell culture systems for the identification of novel therapeutic compounds in Alzheimer's disease and neuroimmune disorders. *Ann. Psychiat.*, **7**, 157-170.

Lombardi, V.R.M., García, M., Rey, L., Cacabelos, R., 1999b, Characterization of cytokine production, screening of lymphocyte subset patterns and in vitro apoptosis in healthy and Alzheimer's disease individuals. *J. Neuroimmunol.*, **97**, 163-171.

Watanabe, T., Taguchi, Y., Shiosaka, S., Tanaka, J., Kubota, H., Terano, Y., Tohyama, M., Wada, H., 1984 Distribution of the histaminergic neuron system in the central nervous system of rats: a fluorescent immunohistochemical análisis with histidine decarboxylase as a marker. *Brain Res.*, **295**, 13-25.

PRESENILIN AND AMYLOIDOGENESIS: A STRUCTURE-FUNCTION RELATIONSHIP STUDY ON PRESENILIN 2

Takeshi Iwatsubo[*], Taisuke Tomita, Tomonari Watabiki, Rie Takikawa, Yuichi Morohashi, and Nobumasa Takasugi

1. INTRODUCTION

Alzheimer's disease (AD) is a progressive dementing neurodegenerative disorder characterized pathologically by the presence of senile plaques and neurofibrillary changes in the brains of affected individuals (1). Senile plaques are composed of amyloid β peptides (Aβ) comprised of ~40 amino acids that are proteolytically produced from β-amyloid precursor protein (βAPP). βAPP is initially cleaved by β-secretase to generate a 99-residue C-terminal fragment (C99) that then is cleaved by γ-secretase to generate Aβ. A subset of AD is inherited as an autosomal dominant trait (familial AD: FAD). Genetic mutations in βAPP genes that cosegregate with the clinical manifestations of FAD increase production of the amyloidogenic Aβ42 species ending at Ala42 (2); Aβ42, which normally comprises only ~10% of total secreted Aβ, aggregates much faster than the predominant Aβ40 species (3), and Aβ42 is the initially and predominantly deposited Aβ species in AD brains (4,5). These data implicated a seminal role of Aβ42 in the pathogenesis of AD.

Mutations in presenilin (PS) 1 and PS2 genes are linked to the majority of early onset FAD. FAD-linked PS mutations affect γ-cleavage of βAPP leading to an increased production of Aβ42 (1). In contrast, ablation of PS1 and PS2 genes in mice completely inhibited production of both Aβ40 and Aβ42, accompanied by accumulation of the βAPP C-terminal stubs that are the direct substrates for γ-secretase (6,7). Furthermore, studies in *Caenorhabditis elegans*, *Drosophila melanogaster* as well as in knockout mice suggested that PS facilitates Notch signaling by activating the ligand-induced intramembranous proteolysis of Notch receptor at site-3 to release their cytoplasmic

[*] Department of Neuropathology and Neuroscience, Graduate School of Pharmaceutical Sciences, University of Tokyo, 7-3-1 Hongo, Bunkyo-ku, Tokyo 113-0033, Japan

Mapping the Progress of Alzheimer's and Parkinson's Disease
Edited by Mizuno *et al.*, Kluwer Academic/Plenum Publishers, 2002

65

domains (NICD) (7,8). This cleavage appears to be very similar to γ-cleavage of βAPP because it occurs close to or within the membrane, is inhibited by inactivation of PS genes, and can be blocked by peptidomimetic γ-secretase inhibitors (8). The physiological role(s) of PS in intramembrane proteolysis has been unclear, but it has been shown that two aspartates within the 6th and 7th transmembrane (TM) domains are required for its γ-secretase activities (9,10). Recently, transition-state analogue γ-secretase inhibitors that inhibit aspartic protease(s) have been shown to directly bind fragment forms of PS, strongly suggesting that PS harbors the catalytic center of γ-secretase (11).

PS1 and PS2 are polytopic integral membrane proteins that undergo endoproteolysis to generate NH_2- and COOH-terminal fragments (NTF and CTF, respectively) (12,13). These fragments are incorporated into the high molecular weight (HMW) protein complexes that are highly stabilized and acquire a long half-life, whereas holoproteins are rapidly degraded (14). Mutagenesis studies showed that the endoproteolysis of PS is neither required for its stabilization, complex formation nor function. In contrast, recent findings indicate that the formation of stabilized HMW complex of PS fragments is closely related to its function. Solubilized membrane fractions containing PS1 fragments exhibit γ-secretase activity (15).

We have found that the integrity of the C terminus of PS is required for its stabilization and abnormal γ-cleavage of βAPP (i.e., Aβ42 overproduction) (16). The C-terminal region of PS is highly conserved. Notably, we found that an amino acid motif comprised of four consecutive amino acids, proline-alanine-leucine-proline, is preserved at the C terminus of all known PS including *C. elegans* SPE-4, which we designated PALP motif. The first proline of PALP motif is replaced with leucine in recessive loss-of-function mutants of *C. elegans spe-4* (17) and *Drosophila Psn* (18). This suggested that the first proline of PALP motif plays an important role in the γ-secretase functions of PS. In contrast, it has been documented that the second proline of PALP motif is replaced with serine (19) or glutamine (20) in some PS1-mutant FAD pedigrees. To elucidate the structural and functional roles PALP motif in the C terminus of PS, we analyzed neuro2a cell lines transfected with PALP-mutated PS and examined their effects on the metabolism of PS proteins as well as on γ-cleavage of βAPP and site-3 cleavage of Notch.

2. HIGH MOLECULAR WEIGHT COMPLEX FORMATION OF PS2

PS proteins are known to form a high molecular weight (HMW) complex with other components (21-23). To examine the capacity of modified PS2 proteins to form HMW complex, we solubilized membrane fractions of N2a cells in 1% CHAPSO, and separated extracted proteins on a linear glycerol velocity gradient (Fig.1). Endoproteolytic fragments derived from wt FL PS2 were predominantly present in the 232~443 kDa HMW range, whereas PS2 holoproteins were fractionated in the 140~232 kDa low molecular weight (LMW) range. PS2/D366A and PS2/ΔloopN, which were stabilized but not cleaved, were present as holoproteins broadly within LMW and HMW ranges. In contrast, unstable PS2/P414L holoproteins were fractionated exclusively in LMW range. Fragments of PS2/P417S, which were stabilized in a similar fashion to FL PS2, were also recovered in HMW fractions, whereas holoproteins were present in LMW range.

To examine the subcellular compartmentalization of PS2 proteins, we separated membrane vesicles by fractionation on discontinuous Iodixanol gradients (21,22). The densest fractions exhibited the strongest immunoreactivity for ER vesicles. Fractions enriched in Golgi/*trans*-Golgi network (TGN) vesicles were identified with antibodies against Adaptin-γ (fractions 6~9). Fraction 9 contained both ER and Golgi/TGN vesicles.

Endoproteolytic fragments derived from FL PS2 were recovered mainly in fractions containing Golgi/TGN vesicles, whereas holoproteins were present in fractions that accumulate ER vesicles (22). PS2/D366A and PS2/ΔloopN were broadly distributed as holoproteins, suggesting that a portion of these mutant holoproteins were transported from ER to Golgi/TGN in a similar manner to endoproteolytic PS fragments. In contrast, PS2/P414L holoproteins remained in fractions rich in ER vesicles. Fragments of PS2/P417S, which were stabilized in a similar fashion to FL PS2, were fractionated mainly in Golgi/TGN fractions, whereas holoproteins were present in ER fractions.

These data suggest that stabilized PS2 proteins participate in the formation of HMW PS2 complexes, which are present in Golgi/TGN. In contrast, unstabilized PS2 proteins form LMW PS2 complexes that remain in ER. These observations support the idea that stabilization and transport to Golgi/TGN of PS proteins require an interaction of PS with unknown cellular cofactor(s), and that the first proline of the PALP motif plays an important role in the formation of PS HMW complex.

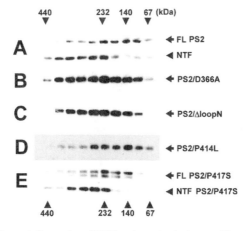

Figure 1. Separation of PS2 by glycerol velocity centrifugation.

3. EFFECTS OF PALP MUTATION ON NOTCH SITE-3 CLEAVAGE

PS is known to serve as a critical component of Notch signaling by executing the site-3 cleavage of Notch (7,8). It has been reported that PS2/D366A abrogates the proteolytic cleavage of mouse Notch-1 in mammalian cells and interferes with the Notch signaling in *C. elegans in vivo* (10), prompting us to speculate that the loss-of-function mutations in invertebrate PS may suppress Notch signaling by inhibiting the proteolytic release of NICD. To examine whether PS2/P414L or PS2/P417S mutations affect the γ-secretase-like site-3 cleavage activity for Notch, we analyzed the proteolytic release of NICD in N2a cells transiently transfected with a cDNA encoding NotchΔE that lacks the extracellular domain (8,10) (Fig.2). Proteolytic generation of NICD from NotchΔE was inhibited in N2a cells overexpressing PS2/D366A (10). However, PS2/P414L mutation did not affect the site-3 cleavage of Notch and normal levels of NICD were generated either on wt or mt basis. In contrast, FAD-linked mt FL PS2 as well as PS2/P417S, the latter harboring an amino acid substitution of the second proline in PALP motif at an analogous position to a FAD-linked PS1 P436S mutation (19), abolished the proteolytic release of NICD.

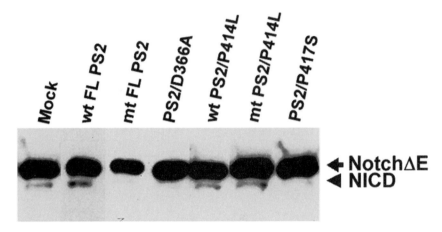

Figure 2. Formation of NICD by PS2.

4. EFFECTS OF PALP MUTATIONS ON Aβ GENERATION

To examine the effects of mutations in PALP motif on γ-cleavage of βAPP, we analyzed N2a cells stably expressing PS2/D366A, PS2/P414L or PS2/P417S (Fig.3). Overexpression of PS2/D366A inhibited γ-cleavage of βAPP, resulting in a marked accumulation of βAPP C-terminal stubs (i.e., C83 and C99), which are the direct precursors for p3 and Aβ, respectively (10). In contrast, cells expressing PS2/P414L or PS2/P417S did not accumulate βAPP CTFs, suggesting that mutations in either of the two prolines of PALP motif in PS2 does not have a dominant negative effect on γ-cleavage of βAPP unlike PS2/D366A (Fig.3A).

To further characterize the effects of PALP mutations of PS2 on Aβ production, we quantitated the levels of secreted Aβ40 and Aβ42 in conditioned media by Aβ C-terminal specific ELISAs (Fig.3B). The total levels of Aβ secreted from cells expressing PS2/P414L or PS2/P417S were comparable to those in cells with wt FL PS2. The percentage of Aβ42 as a fraction of total Aβ (= Aβx-40 + Aβx-42) (%Aβ42) secreted by cells stably expressing mt PS2/P414L was ~10%, and this was similar to the %Aβ42 secreted from cells expressing wt FL PS2 or wt PS2/P414L, whereas the %Aβ42 secreted from cells expressing mt FL PS2 was constantly elevated to ~75% as previously documented (13,16). Moreover, the %Aβ42 in cells expressing PS2/P417S also was elevated to ~30%. These results indicate that the P414L mutation abrogates the overproduction of Aβ42 on a FAD-linked mutant basis, whereas P417S mutation harbors a pathogenic function like FAD mutant PS to increase Aβ42.

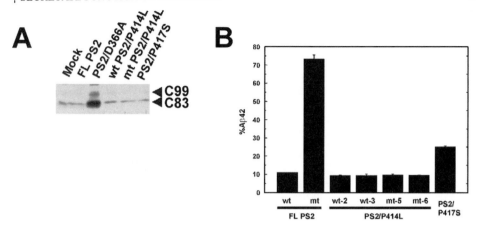

Figure 3. Generation of C-terminal stub (A) and percentage of secreted Aβ42 (B) in N2a cells stably expressing modified PS2.

5. DISCUSSION AND CONCLUDING REMARKS

The PALP motif is completely conserved in all PS species thus far identified. Here we showed that replacement of the first proline of PALP motif with leucine abolished the normal metabolism (i.e., endoproteolysis, stabilization, replacement and HMW complex formation) of PS in mammalian cells. Although subcellular localization of site-3 cleavage activity of Notch remains unknown, Golgi vesicle-enriched fractions have been shown to contain γ-secretase activities for βAPP *in vitro* (21,22). Thus, it is most conceivable that the loss-of-function phenotypes of PS2/P414L for cleavages of βAPP and Notch are due to defects in stable HMW complex formation and proper transportation from ER to Golgi/TGN. Similar mechanisms may underlie the recessive PS-null phenotype exhibited by *C. elegans spe-4 (eb-12)* (17) and *Drosophila Psn (Dps)[46]* (18) alleles *in vivo*. Taken together, it is strongly suggested that the first proline of PALP motif plays an important role in the proper metabolism of PS family proteins to confer normal or abnormal functions.

In contrast, PS2/P417S underwent endoproteolysis to give rise to stabilized PS complexes and %Aβ42 was increased as observed with other FAD-linked PS mutations. Furthermore, Notch site-3 cleavage was partially suppressed both with N141I and P417S PS2 mutants. It has been documented that the proteolytic production of NICD is impaired by some FAD-linked PS1 mutations (23), although NICD is produced normally with other PS1 mutations (48). In this regard, Kulic et al. have recently reported an interesting relationship between Aβ42 overproduction and Notch site-3 cleavage (24): they have introduced arbitrary mutations at position 286 of PS1, and showed that the strength of Aβ42 promoting effects of PS1 mutations is correlated with the extent of reduction in NICD generation. The precise mechanism underlying these changes in γ- and site-3 cleavage remains unknown. However, one could argue that FAD-related PS2 mutations that are more potent in Aβ42 promoting effects compared to those in PS1 (25) might behave like some of the artificial PS1 mutations, leading to inhibition in Notch site-3 cleavage. We hypothesize that this apparent "loss-of-function" in Notch cleavage caused by some of the FAD-linked PS mutations could be mechanistically related to the shift of γ-cleavage from position 40 to 42 of Aβ.

PS2/P414L proteins were fractionated exclusively in LMW ranges, whereas wt PS2 or PS2/P417S were separated in HMW fractions. Moreover, a portion of uncleavable

PS2/D366A or PS2/ΔloopN, that were stabilized as holoproteins, were also fractionated in HMW ranges. These data strongly suggest that the stabilization of PS proteins coincides with the formation of HMW complex, and that the endoproteolysis of PS proteins is not required for these processes. Thus, the difference in molecular size between unstable and stabilized PS complexes may be due to an association with the cofactor(s) including those required for stabilization.

Mutational analysis of the highly conserved PALP motif provided additional evidence that the stabilized HMW complex form of PS represents the active form of γ-secretase. Further attempts to define the molecular mechanism of PS stabilization and to identify the components of PS complexes relevant to intramembranous proteolytic activity will facilitate understanding of the pathogenetic mechanisms of AD.

REFERENCES

1. Selkoe, D. J. (1999) *Nature* **399**, A23-31
2. Suzuki, N., Cheung, T. T., Cai X. D., Odaka, A., Otvos L. Jr., Eckman, C., Golde, T. E. and Younkin, S. G. (1994) *Science* **264**, 1336-1340
3. Jarrett, J. T. and Lansbury, P. T. Jr. (1993) *Cell* **73**, 1055-1058
4. Iwatsubo, T., Odaka, A., Suzuki, N., Mizusawa, H., Nukina, N. and Ihara, Y. (1994) *Neuron* **13**, 45-53
5. Iwatsubo T., Mann, D. M., Odaka, A., Suzuki, N. and Ihara, Y. (1995) *Ann Neurol* **37**, 294-299
6. De Strooper, B., Saftig P., Craessaerts, K., Vanderstichele, H., Guhde, G., Annaert, W., Von Figura, K. and Van Leuven, F. (1998) *Nature* **391**, 387-390
7. Herreman, A., Serneels, L., Annaert, W., Collen, D., Schoonjans, L. and De Strooper, B. (2000) *Nat Cell Biol* **2**, 461-462
8. De Strooper, B., Annaert, W., Cupers, P., Saftig, P., Craessaerts, K., Mumm, J. S., Schroeter, E. H., Schrijvers, V., Wolfe, M. S., Ray, W. J., Goate, A. and Kopan, R. (1999) *Nature* **398**, 518-522
9. Wolfe, M. S., Xia, W., Ostaszewski, B. L., Diehl, T. S., Kimberly, W. T. and Selkoe, D. J. (1999) *Nature* **398**, 513-517
10. Steiner, H., Duff, K., Capell, A., Romig, H., Grim, M. G., Lincoln, S., Hardy, J., Yu, X., Picciano, M., Fechteler, K., Citron, M., Kopan, R., Pesold, B., Keck, S., Baader, M., Tomita, T., Iwatsubo, T., Baumeister, R. and Haass, C. (1999) *J Biol Chem* **274**, 28669-28673
11. Li, Y. M., Xu, M., Lai, M. T., Huang, Q., Castro, J. L., DiMuzio Mower, J., Harrison, T., Lellis, C., Nadin, A., Neduvelil, J. G., Register, R. B., Sardana, M. K., Shearman, M. S., Smith, A. L., Shi, X. P., Yin, K. C., Shafer, J. A. and Gardell, S. J. (2000) *Nature* **405**, 689-694
12. Thinakaran, G., Borchelt, D. R., Lee, M. K., Slunt, H. H., Spitzer, L., Kim, G., Ratovitski, T., Davenport, F., Nordstedt, C., Seeger, M., Hardy, J., Levey, A. I., Gandy, S. E., Jenkins, N. A., Copeland, N. G., Price, D. L. and Sisodia, S. S. (1996) *Neuron* **17**, 181-190
13. Tomita, T., Maruyama, K., Saido, T. C., Kume, H., Shinozaki, K., Tokuhiro, S., Capell, A., Walter, J., Grunberg, J., Haass, C., Iwatsubo, T. and Obata, K. (1997) *Proc Natl Acad Sci U S A* **94**, 2025-2030
14. Ratovitski, T., Slunt, H. H., Thinakaran, G., Price, D. L., Sisodia, S. S. and Borchelt, D. R. (1997) *J Biol Chem* **272**, 24536-24541
15. Li, Y. M., Lai, M. T., Xu, M., Huang, Q., DiMuzio Mower, J., Sardana, M. K., Shi, X. P., Yin, K. C., Shafer, J. A. and Gardell, S. J. (2000) *Proc Natl Acad Sci U S A* **97**, 6138-6143
16. Tomita, T., Takikawa, R., Koyama, A., Morohashi, Y., Takasugi, N., Saido, T. C., Maruyama, K. and Iwatsubo, T. (1999) *J Neurosci* **19**, 10627-10634
17. Arduengo, P. M., Appleberry, O. K., Chuang, P. and L'Hernault, S. W. (1998) *J Cell Sci* **111**, 3645-3654
18. Guo,Y., Livne-Bar, I., Zhou, L. and Boulianne, G. L. (1999) *J Neurosci* **19**, 8435-8442
19. Palmer, M. S., Beck, J. A., Campbell, T. A., Humphries, C. B., Roques, P. K., Fox, N. C., Harvey, R., Rossor, M.N. and Collinge, J. (1999) *Hum Mutat* **13**, 256
20. Taddei, K., Kwok, J. B., Kril, J. J., Halliday, G. M., Creasey, H., Hallupp, M., Fisher, C., Brooks, W. S., Chung, C., Andrews, C., Masters, C.L., Schofield, P. R. and Martins, R. N. (1998) *Neuroreport* **9**, 3335-3339
21. Xia, W., Zhang, J., Ostaszewski, B. L., Kimberly, W. T., Seubert, P., Koo, E. H., Shen, J. and Selkoe, D. J. (1998) *Biochemistry* **37**, 16465-16471

22. Iwata, H., Tomita, T., Maruyama. K. and Iwatsubo, T. (2001) *J Biol Chem* **276**, 21678-21685
23. Song, W., Nadeau, P., Yuan, M., Yang, X., Shen, J. and Yankner, B. A. (1999) *Proc Natl Acad Sci U S A* **96**, 6959-6963
24. Kulic, L., Walter, J., Multhaup, G., Teplow, D. B., Baumeister, R., Romig, H., Capell, A., Steiner, H. and Haass, C. (2000) *Proc Natl Acad Sci U S A* **97**, 5913-5918
25. Citron, M., Westaway, D., Xia, W., Carlson, G., Diehl, T., Levesque, G., Johnson-Wood, K., Lee, M., Seubert, P., Davis, A., Kholodenko, D., Motter, R., Sherrington, R., Perry, B., Yao, H., Strome, R., Lieberburg, I., Rommens, J., Kim, S., Schenk, D., Fraser, P., St George Hyslop, P. and Selkoe, D.J. (1997) Nat. Med. 3, 67-7

PS1 INTERACTS WITH AND FACILITATES β-CATENIN TURNOVER

Edward H. Koo, Salvador Soriano, David E. Kang[•]

1. INTRODUCTION

Mutations in the two related genes, presenilin 1 (PS1) and presenilin 2 (PS2), account for the majority of the cases of familial early onset Alzheimer's disease (AD). The mechanisms by which the mutations cause early onset AD are unknown. The prevailing hypothesis centers on altered γ-secretase activity that generates the C-termini of amyloid β-protein (Aβ) by selectively increasing the amyloidogenic Aβ42 isoform and resulting in accelerated Aβ deposition in brain. Work from a number of laboratories have conclusively demonstrated that presenilins are required for the transmembrane cleavage of both the amyloid precursor protein (APP) and Notch to generate Aβ and the signaling Notch intracellular domain polypeptide, respectively (Annaert and De Strooper 1999). Without doubt, this "regulated intramembrane proteolysis" takes a central role in both AD pathogenesis and development through regulating γ-secretase activity in APP cleavage and Notch signal transduction, respectively (Brown et al., 2000; Capell et al., 2000).

In addition, a number of other molecules have been shown to associate with PS1, but the nature and function of these interactions are unclear. To date, only nicastrin, a recently identified binding partner, has been implicated directly in the amyloidogenic pathway (Yu et al., 2000). Among the other PS1 interacting molecules, the association between PS1 and members of the armadillo family of proteins has been studied in most detail. Armadillo (Arm) was originally identified in *Drosophila* and, like Notch, plays important roles in a number of developmental pathways across multiple species. Arm and its mammalian ortholog β-catenin, bridges cadherins and actin cytoskeleton via α-catenin to form the adherens junction. Therefore, Arm/β-catenin plays an important role in cell-cell adhesion and tissue development. In addition, nuclear translocation of Arm/β-catenin is a central mediator of the Wnt signal transduction pathway (Willert and Nusse 1998). Wnt proteins represent a family of developmentally important signaling molecules, at the

[•] Edward H. Koo, Salvador Soriano, David E. Kang, Department of Neurosciences, University of California, La Jolla, California, 92037.

Mapping the Progress of Alzheimer's and Parkinson's Disease
Edited by Mizuno *et al.*, Kluwer Academic/Plenum Publishers, 2002

73

heart of which is the regulation of the cytosolic pool of β-catenin. At basal conditions without Wnt signals, cytosolic β-catenin is rapidly turned over in the proteasome pathway, where phosphorylation by glycogen synthase kinase-3β (GSK-3β) promotes ubiquitination of β-catenin by binding to the E2/E3 ubiquitin ligase complex via β-transducing repeat-containing protein (β-Trcp) (Polakis 1999). Adenomatous polyposis coli (APC) and axin combine to positively modulate GSK-3β activity and promote β-catenin degradation. Binding of Wnt ligands to cell surface frizzled receptors leads to inactivation of GSK-3β, stabilization of β-catenin and subsequent translocation to the nucleus. In the nucleus, β-catenin binds to the T-cell factor/lymphoid enhancer factor-1 (TCF/LEF abbreviated herein as LEF) family of transcription factors and mediates transcriptional activation of downstream target genes, two of which are c-myc and cyclin D1 (He et al., 1998; Shtutman et al., 1999; Tetsu and McCormick 1999)

This brief review focuses on our work examining the cell biology of PS1 and β-catenin interaction.

2. RESULTS AND DISCUSSION

2.1. Loss of PS1 Decreased β-catenin Turnover

PS1 was initially demonstrated to associate with δ-catenin (also known as NPRAP), a novel member of the armadillo family (Paffenholz and Franke 1997; Zhou et al., 1997). Not surprisingly, other members of the armadillo family, including p0071 and β-catenin since they all contain armadillo repeats, were also shown to associate with PS1 as well (Stahl et al., 1999). Recently, Robakis and colleagues demonstrated that PS1 also binds directly to E-cadherins, an interaction that promotes the binding of β- and γ-catenin and promoting the stability of cadherin junctions (Georgakopoulos et al., 1999). In regards to PS1 and β-catenin, contradictory results have been presented by different laboratories. Wild type PS1 was shown to either stabilize β-catenin by some groups or destabilize β-catenin by other groups (Murayama et al., 1998; Kang et al., 1999; Weihl et al., 1999; Zhang et al., 1999). However, one potential problem associated with the early reports was the use of over-expression of exogenous PS1 or β-catenin to high levels, an approach that may have distorted the true picture. Using PS1 deficient fibroblasts, we initially showed that the turnover rate of endogenous β-catenin in the cytosolic pool was markedly slowed in immortalized PS1 deficient fibroblasts as compared to PS1 +/- control fibroblasts. The turnover rate can be fully restored with re-introduction of wild type PS1 but only partially with a PS1 mutation (ΔX9), suggesting that PS1 mutations may represent partial loss-of-function with regards to β-catenin stability (Kang et al., 1999).

2.2. Loss of PS1 Function Leads to Upregulation of β-catenin/LEF Dependent Signaling

The recent identification of a number of downstream target genes of the β-catenin/LEF family of transcription factors, including c-myc and cyclin D1, presented the opportunity to further test our hypothesis that PS1 is a negative regulator of β-stability and to clarify the confusion surrounding the consequences of PS1/β-catenin interaction. Examination of this downstream pathway did not require manipulation of endogenous β-catenin or the introduction of exogenous β-catenin. These studies were carried out with primary cultured fibroblasts from PS1 deficient animals to exclude potentially confounding changes associated with transformation and over-expression. Accordingly, based on our previous evidence that the turnover of β-catenin is in part regulated by PS1, we predicted comparable alterations in cyclin D1 transcription in primary mouse embryonic fibroblasts deficient in PS1. Indeed, western blotting showed that PS1-/- cells contained higher amounts of β-catenin and cyclin D1 protein. In contrast, levels of cdc2 and cyclin A, two other related genes not subject to β-catenin-mediated transcription, were not elevated in PS1-/- cells (Soriano et al., 2001).

In a previous study, Takashima and colleagues showed that activity of a LEF reporter construct was reduced upon expression of wild type PS1 (Murayama et al., 1998), consistent with our interpretation that PS1 enhanced β-catenin turnover. Therefore, we assayed reporter activity using a cyclin D1 promoter that encompassed positions 0 to -163. At position -81 to -73 is a consensus LEF binding site and has been shown to represent the main contributor to β-catenin transactivation. As predicted, activation of the cyclin D1 promoter was three fold higher in PS1 -/- cells than in PS1 +/- cells. These results demonstrated that higher levels of cytosolic β-catenin in PS1 -/- cells correlate with increased β-catenin/LEF-dependent transcription of cyclin D1, observations consistent with the studies reported by the Takashima laboratory.

The cyclin D1 gene encodes the regulatory subunit of the holoenzyme that phosphorylates and inactivates the Rb tumor suppressor protein. Cyclin D1 is required in G1 phase progression induced by a variety of mitogens. Based on our findings that cyclin D1 transcription is elevated in the absence of PS1, we therefore asked whether cell proliferation, as represented by BrdU incorporation, is altered in cells deficient in PS1. Indeed, BrdU incorporation was approximately two fold higher in PS1-/- than PS1 +/- cells (Soriano et al., 2001). Therefore, PS1 deficiency in mouse embryonic fibroblasts is accompanied by increased levels of β-catenin and cyclin D1 as well as accelerated cell proliferation. Taken together, the studies provided both reporter and functional assays to demonstrate that PS1 facilitates the turnover of β-catenin, in the absence of which cytosolic β-catenin stability is increased, leading to enhanced nuclear signaling and cell proliferation.

2.3. Modulation of β-catenin Signaling by PS1 Requires the Interaction of Both Proteins

The initial report describing the association of PS1 and β-catenin demonstrated that the interaction occurred in the PS1 loop domain (residues 263-407). This domain was narrowed down slightly to residues 299-362 in a yeast two-hybrid screen with p0071, another member of the armadillo family (Stahl et al., 1999). Using GST pull down assay, the region between amino acids 330 to 360 within the loop of PS1 is required for its association with β-catenin (Saura et al., 2000). This result is consistent with the report that following caspase cleavage at position 345, PS1 no longer associated with β-catenin (Tesco et al., 1999). Finally, the requirement for this region within the hydrophilic loop of

PS1 in the β-catenin interaction was confirmed by expressing a PS1deletion mutant lacking residues 330 to 360 (PS1Δcat) in 293 cells. Co-immunoprecipitation results showed that PS1/β-catenin complexes could only be detected with wild type PS1 but not PS1Δcat, indicating that the sequence contained within 330-360 is necessary for PS1/β-catenin interaction.

The PS1Δcat was an important construct in our next analysis of β-catenin stability in PS1-/- primary fibroblasts. The cells were retrovirally infected with constructs containing PS1 wildtype or PS1Δcat cDNAs. As predicted, β-catenin turnover was increased in wildtype PS1-infected PS1-/- cells, but not in PS1Δcat infected PS1 -/- cells. Finally, using the same BrdU incorporation assay, we examined the consequences of PS1Δcat expression in PS1-/- primary fibroblasts. Expression of wildtype PS1 resulted in a reduction of cytosolic β-catenin levels and increased β-catenin turnover by pulse chase experiment, whereas introducing PS1Δcat had no significant effect on either assay. As predicted from our earlier results, the recovery of β-catenin turnover in primary PS1 -/- cells following reintroduction of wildtype PS1 resulted in a concomitant decrease in BrdU incorporation. In marked contrast, entry into S phase in PS1-/- cells was not modified by reintroducing PS1Δcat or vector control thus confirming that the facilitation of β-catenin turnover by PS1 require the hydrophilic loop region encoded by residues 330-360 and the association of PS1 and β-catenin (Soriano et al., 2001).

2.4. PS1 Deficiency Leads to Accumulation of Phosphorylated β-catenin

To begin to understand how PS1 affects β-catenin turnover, we then determined which pool of cytosolic β-catenin was abnormally elevated in the absence of PS1. We reasoned that since phosphorylation and ubiquitination are the events that lead to rapid degradation of β-catenin by the proteosome, we therefore examined the levels of phosphorylated β-catenin in the presence or absence of Wnt3a, a soluble Wnt ligand activates the signaling pathway. Surprisingly, the levels of phosphorylated β-catenin were markedly elevated under basal conditions in PS1 -/- cells as compared to PS1 +/- cells, but were appropriately attenuated after Wnt3a stimulation in both cell types. The rapid reduction in phosphorylated β-catenin levels, attributable to GSK-3β inactivation, indicated that, in response to Wnt3a stimulation, this step of β-catenin dephosphorylation was unaffected by the absence of PS1. However, there were fewer ubiquitinated β-catenin species in PS1 -/- cells after proteosomal inhibition (Soriano et al., 2001). This suggested that PS1 regulates β-catenin turnover at a step between phosphorylation and ubiquitination of β-catenin, just prior to degradation

2.5. Both Wildtype PS1 and PS1Δcat Restore Normal Proteolytic Cleavage of Notch-1 and γ-secretase activity

PS1 plays a central role in the proteolysis of Notch and the modulation of γ-secretase activity. Therefore, while studies described above have provided compelling evidence that one of the functions of PS1 is to regulate cytosolic β-catenin turnover, we cannot exclude the possibility that the cellular consequences we have defined are due to the concomitant impairment of γ-secretase activity or Notch signaling. Therefore, when PSI is deleted of either a large portion of the hydrophilic loop (residues 304-371) or a more limited domain (330-360 in PS1Δcat), neither Notch proteolysis nor Aβ production were impaired by any of these PS1 mutants (Saura et al., 2000; Soriano et al., 2001). However, expression of the PS1 D257A aspartate mutant, which impaired notch and APP proteolysis (Wolfe et al., 1999), had no effect on β-catenin turnover. Thus, the role of PS1 in regulating β-catenin function is clearly separable from its role in γ-secretase processing of Notch and APP.

2.6. Summary

In summary, our recent results provide compelling evidence that PS1 is a negative regulator of β-catenin stability in a model where PS1 normally facilitates ubiquitination of β-catenin. Using functional assays, we showed that in cultured cells, the perturbed stability of β-catenin in PS1 deficient cells is associated with hyper- proliferation of these cells. The results are consistent with the model first proposed for colon cancer wherein abnormal stabilization of β-catenin by APC mutations activates downstream oncogenic signals (reviewed in Barker et al., 2000; Polakis 1999). Thus, our observations add to the cellular machinery that is known to regulate β-catenin degradation. Furthermore, this activity is independent of the known role of PS1 in Notch proteolysis and γ-secretase activity. Whether the role of PS1 in facilitating ubiquitination of β-catenin extends to other molecules that contribute to AD pathology is an interesting concept to test.

3. ACKNOWLEDGEMENTS

We thank Drs. Raphael Kopan, Richard Pestell, Sangram Sisodia, Gopal Thinkaran, and Hui Zheng for helpful discussions, advice, and gift of various reagents.

REFERENCES

1. Annaert W., De Strooper B., 1999, Presenilins: molecular switches between proteolysis and signal transduction. *Trends Neurosci.* **22**:439-43.
2. Barker N., Morin P. J., and Clevers H., 2000, The Yin-Yang of TCF/β-catenin signaling. *Adv.Cancer Res.* **77**:1-24.
3. Brown M. S., Ye J., Rawson R. B., and Goldstein J. L., 2000, Regulated Intramembrane Proteolysis: A Control Mechanism Conserved from Bacteria to Humans. *Cell* **100**:391-8.
4. Capell A., Steiner H., Romig H., Keck S., Baader M., Grim M. G., Baumeister R., and Haass C., 2000, Presenilin-1 differentially facilitates endoproteolysis of the β-amyloid precursor protein and Notch. *Nat.Cell Biol.* **2**:205-11.
5. Georgakopoulos A., Marambaud P., Efthimiopoulos S., Shioi J., Cui W., Li H. C., Schutte M., Gordon R., Holstein G. R., Martinelli G., Mehta P., Friedrich V. L., and Robakis N. K., 1999, Presenilin-1 Forms Complexes with the Cadherin/Catenin Cell-Cell Adhesion System and is Recruited to Intercellular and Synaptic Contacts. *Mol.Cell* **4**:893-902.
6. He T. C., Sparks A. B., Rago C., Hermeking H., Zawel L., da Costa L. T., Morin P. J., Vogelstein B., and Kinzler K. W., 1998, Identification of c-MYC as a target of the APC pathway. *Science* **281**:1509-12.
7. Kang D. E., Soriano S., Frosch M. P., Collins T., Naruse S., Sisodia S. S., Leibowitz G., Levine F., and Koo E. H., 1999, Presenilin 1 facilitates the constitutive turnover of β-catenin: differential activity of Alzheimer's disease-linked PS1 mutants in the β-catenin signaling pathway. *J.Neurosci.* **19**:4229-37.

8. Murayama M., Tanaka S., Palacino J., Murayama O., Honda T., Sun X., Yasutake K., Nihonmatsu N., Wolozin B., and Takashima A., 1998, Direct association of presenilin-1 with β-catenin. *FEBS Lett.* **433**:73-7.

9. Paffenholz R. , and Franke W., 1997, Identification and localization of a neurally expressed member of the plakoglobin/armadillo multigene family. *Differentiation* **61**:293-304.

10. Polakis P., 1999, The oncogenic activation of β-catenin. *Curr.Opin.Genet.Dev.* **9**:15-21.

11. Saura C. A., Tomita T., Soriano S., Takahashi M., Leem J. Y., Honda T., Koo E. H., Iwatsubo T., and Thinakaran G., 2000, The Nonconserved Hydrophilic Loop Domain of Presenilin (PS) is Not Required for Endoproteolysis or Enhanced Aβ42 Production Mediated by Familial Early Onset Alzheimer's Disease-linked PS Variants. *J.Biol.Chem.* **275**:17136-42.

12. Shtutman M., Zhurinsky J., Simcha I., Albanese C., D'Amico M., Pestell R., and Ben Ze'ev A., 1999, The cyclin D1 gene is a target of the β-catenin/LEF-1 pathway. *Proc.Natl.Acad.Sci.,U.S.A* **96**:5522-7.

13. Soriano S., Kang D. E., Fu M., Pestell R., Chevallier N., and Zheng H., 2001, Presenilin 1 Negatively Regulates β-Catenin/T Cell Factor/Lymphoid Enhancer Factor-1 Signaling Independently of β-Amyloid Precursor Protein and Notch Processing. *J.Cell Biol.* **152**:785-94.

14. Stahl B., Diehlmann A., and Sudhof T., 1999, Direct interaction of Alzheimer's disease-related presenilin 1 with armadillo protein p0071. *J.Biol.Chem.* **274**:9141-8.

15. Tesco G., Kim T.-W., Diehlmann A., Beyreuther K., and Tanzi R., 1999, Abrogation of the presenlin 1/β-catenin interaction and preservation of the heterodimeric presenilin 1 complex following caspase activation. *J.Biol.Chem.* **273**:33909-14.

16. Tetsu O., and McCormick F., 1999, β-Catenin regulates expression of cyclin D1 in colon carcinoma cells. *Nature* **398**:422-6.

17. Weihl C. C., Miller R. J., and Roos R. P., 1999, The role of β-catenin stability in mutant PS1 associated apoptosis. *NeuroReport* **10**:2527-32.

18. Willert K., and Nusse R., 1998, β-catenin: a key mediator of Wnt signaling. *Curr.Opin.Gen.Dev.* **8**:95-102.

19. Wolfe M. S., Welming X., Ostaszewski B. L., Diehl T. S., Kimberly W. T., and Selkoe D. J., 1999, Two transmembrane aspartates in presenilin-1 required for presenilin endoproteolysis and γ-secretase activity. *Nature* **398**:513-7.

20. Yu G., Nishimura M., Arawaka S., Levitan D., Zhang L., Tandon A., Song Y. Q., Rogaeva E., Chen F., Kawarai T., Supala A., Levesque L., Yu H., Yang D. S., Holmes E., Milman P., Liang Y., Zhang D. M., Xu D. H., Sato C., Rogaev E., Smith M., Janus C., Zhang Y., Aebersold R., Farrer L. S., Sorbi S., Bruni A., Fraser P., and George-Hyslop P., 2000, Nicastrin modulates presenilin-mediated *notch/glp*-1 signal transduction and β APP processing. *Nature* **407**:48-54.

21. Zhang Z., Hartmann H., Do V. M., Abramowski D., Sturchler-Pierrat C., Staufenbiel M., Sommer B., van de Wetering M., Clevers H., Saftig P., De Strooper B., He X., and Yankner B. A., 1999, Destabilization of β-catenin by mutations in presenilin-1 potentiates neuronal apoptosis. *Nature* **395**:698-702.

22. Zhou J., Liyanage U., Medina M., Ho C., Simmons A. D., Lovett M., and Kosik K. S., 1997, Presenilin 1 interaction in the brain with a novel member of the Armadillo family. *NeuroReport* **8**:1489-94.

β-SECRETASE – A TARGET FOR ALZHEIMER'S DISEASE

Martin Citron[*]

1. INTRODUCTION

The cerebral deposition of amyloid β-peptide (Aβ) is an early and critical feature in both familial and sporadic forms of Alzheimer's disease. Aβ is released from the amyloid precursor protein (APP) by the sequential action of two proteases, β-secretase and γ-secretase, and these proteases are prime targets for therapeutic intervention. We have recently cloned a novel aspartic protease, BACE1, with all the known properties of the elusive β-secretase. This article summarizes the initial identification of BACE1 and describes the analysis of posttranslational modifications. The crystal structure of BACE1 is reviewed and the BACE1 knockout mice that we recently generated are described. The pros and cons of BACE1 inhibition as a therapeutic strategy for familial and sporadic Alzheimer's disease are discussed.

2. CLONING OF β-SECRETASE

Isolating β- and γ-secretases has been a major goal for laboratories in academia and in the pharmaceutical industry since the cloning of APP had revealed that Aβ is located in the middle of a large precursor protein. It took much longer than anyone had expected to accomplish a secretase isolation, primarily because the biochemical purification of the enzymes using peptidic substrates turned out to be exceedingly difficult. Upon homogenization of cells various enzymes capable of cleaving peptidic substrates at the right positions were released and several of these active enzymes performing artifactual cleavages were pursued as β- and γ-secretase candidates over the years. We therefore decided to avoid a biochemical purification strategy and focused on an expression cloning approach to identify

---

[*] Martin Citron, Amgen, Inc., Thousand Oaks, California 91320

Mapping the Progress of Alzheimer's and Parkinson's Disease
Edited by Mizuno *et al.*, Kluwer Academic/Plenum Publishers, 2002

genes that modulate Aβ production. This approach was based on the finding that cultured cells transfected with APP do generate and secrete low levels of Aβ, indicating that they express active secretase enzymes and could therefore be used as both a source of cDNA and as a test system. We specifically hypothesized that overexpression of a secretase in cells transfected with APP would lead to increased Aβ secretion. We constructed a directional cDNA expression library from 293 human embryonic kidney cells in a cytomegalovirus promoter-based expression vector and transiently transfected 8600 pools of 100 clones into a 293 line which overexpresses APP. Changes in the levels of Aβ and other APP metabolites were monitored by ELISA assays. Pools containing a secretase would lead to increased Aβ production, if (i) the secretase activity is limiting for Aβ production and (ii) the secretase consists of only one protein. Using this approach we could ensure that our candidates act on full-length APP in the right subcellular compartment (which most of the protein purification candidates did not do). Our expression cloning approach ultimately led to the identification of BACE1 as the major β-secretase [1]. Subsequently, three other groups independently reported isolation of the same enzyme by different approaches: Yan et al. [2] hypothesized that β-secretase would be an aspartic protease and used a genomics strategy which led to the identifaction of BACE1. Sinha et al. [3] succeeded by biochemical purification using inhibitor affinity chromatography and Hussain et al. [4] also identified β-secretase, but did not reveal in their publication how they selected the clone.

3. PROPERTIES OF β-SECRETASE

The most exciting clone from our primary expression cloning screen encoded a novel protein of 501 amino acids. Based on sequence homology with members of the pepsin subfamily of aspartic proteases it was immediately identified as a protease. The enzyme has an aminoterminal signal peptide of 21 amino acids followed by a proprotein domain spanning amino acids 22 to 45. The lumenal domain of the mature protein extends from residues 46 to 460 and is followed by a transmembrane domain of 17 residues and a short cytosolic tail of 24 amino acids. BACE1 contains two active site motifs at amino acids 93 to 96 and 289 to 292 in the lumenal domain, each containing the highly conserved signature sequence of aspartic proteases D T/S G T/S. Based on the amino acid sequence BACE1 is predicted to be a type I transmembrane protein with the active site on the lumenal side of the membrane where β-secretase cleaves APP. At the amino acid level, BACE1 shows less than 30% sequence identity with human pepsin family members. Immediately after the identification of BACE1 database mining led to the discovery of BACE2 which has 64% amino acid sequence similarity to BACE1 and which also shows a C-terminal transmembrane domain. Together BACE1 and BACE2 define a novel family of aspartic proteases. BACE1 is localized to chromosome 11q23.3, not associated with AD, in contrast, BACE2 localizes to chromosome 21 within the critical Down's syndrome region (Down's syndrome patients develop Alzheimer's disease pathology early in life presumably due to overexpression of APP, also localized on chromosome 21). This has led to the suggestion that BACE2 may also have β-secretase activity and play a role in amyloid formation [5]. BACE1 undergoes a series of posttranslational modifications. All 4 potential N-glycsosylation sites in the ectodomain are occupied by carbohydrate, no O-glycosylation is detected. The six cysteine residues in the ectodomain all form intramolecular disulfide bonds in a pattern which is not conserved in other aspartic proteases. We have thoroughly demonstrated that BACE1 exhibits all the known properties of β-secretase [1] and subsequent publications from other groups have confirmed this analysis. Tissue culture and animal studies indicated that β-secretase is expressed in all tissues, but higher in neurons of the brain. This is exactly what we found for BACE1 mRNA. In contrast, BACE2 is expressed at very low levels in brain neurons [6]. We also demonstrated the presence of BACE1 protein in brain [1] and indeed Sinha et al. [7] arrived at the same enzyme by purification of β-secretase activity from brain. BACE1 has the right topological orientation to attack the β-secretase cleavage site of APP and is localized within acidic intracellular compartments, as expected for β-secretase. Overexpression of BACE1 in cells increases the β-secretase products C99 and APPsβ,while the α-secretase product APPsα is decreased. In cells expressing wild-type APP this directly leads to increased Aβ generation. In contrast, overexpression of BACE2 increases the direct β-secretase cleavage products, but since BACE2 also cleaves within the Aβ region, the overall effect is no increase in Aβ generation [8]. Antisense inhibition of BACE1 decreases β-secretase cleavage and Aβ generation [1,2], while antisense inhibition of BACE2 has no effect on Aβ generation [2]. Purified forms of BACE1 cleave APP substrates in vitro at the correct site and with the same P1 specificity previously described for β-secretase in cell based assays. In pulse chase experiments BACE1 protein (calculated MW 50 kDa) is initially detectable as a 60 kDa immature glycosylated form made in the ER which undergoes rapid maturation to a 70 kDa product which is stable [9]. BACE1 is initially synthesized as a proprotein that is cleaved at residue E46 to form the mature enzyme. This cleavage is not autocatalytic, as in some other aspartic proteases. Instead, furin, or a furin-like protease is likely responsible for the propeptide cleavage, because the furin-specific inhibitor α1-antitrypsin PDX blocks propeptide processing, the furin-deficient Lovo cell line does not process the propeptide and in vitro furin cleaves the propeptide of BACE1 exactly at E46 [10]. In contrast, the propeptide cleavage of BACE2 has been shown to be autocatalytic [11]. The crystal structure of the BACE1 ectodomain complexed with a peptidic inhibitor has recently been solved. The overall structure of the enzyme is very similar to that of other known aspartic proteases. However, there are differences in the details of the active site which is generally more open and less hydrophobic than in other aspartic proteases [12].

4. β-SECRETASE AS A TARGET FOR THE TREATMENT OF ALZHEIMER'S DISEASE

The amyloid hypothesis proposes that Aβ plays an early and critical role in the pathogenesis of Alzheimer's disease [13]. Drugs which shut down Aβ production could therefore be therapeutically useful to block disease progression. However, this strategy has met three major objections. 1. The amyloid hypothesis has not yet been clinically validated. It could be wrong or Aβ production inhibitors could come too late when a patient already exhibits symptoms. This objection is beyond the scope of this article and can only be rejected if a clinical trial of a secretase inhibitor is successful. 2. Aβ may be needed or secretase inhibitors would block the production of other critical APP metabolites. Currently there are no strong data suggesting that Aβ plays a physiological role. A β-secretase inhibitor would reduce only APPsβ and Aβ, but not other APP metabolites like APPsα or the APP C-terminal fragment resulting from γ-secretase cleavage for which a physiological function has been hypothesized. 3. Secretases are widely expressed and are likely to have other substrates in addition to APP. Secretase inhibition could cause unacceptable mechanism based toxicity. We think that this objection is the most significant concern, and we began to address it experimentally by studying the phenotype of knockout mice, as soon as we had identified the β-secretase gene. BACE1 knockout mice that we and others have generated do not make Aβ, providing ultimate in vivo validation of BACE1 as β-secretase [14,15]. We analyzed the phenotype of our BACE1 knockout mice and found them to be healthy and fertile. The knockout mice are normal in terms of gross morphology and anatomy, tissue histology, hematology and clinical chemistry [14]. Obviously, this analysis does not rule out the possibility of more subtle defects in memory, behavior or under stress conditions and in the aged animal. However, it is encouraging that a third group studying young β-secretase knockout mice in various behavioral tests did not detect overt behavioral and neuromuscular defects [16]. This raises the hope that β-secretase inhibition could be safe. Because the knockout mice show no apparent defects, we do not yet know substrates of β-secretase other than APP. In summary, the biology of β-secretase appears relatively straightforward and the knockout mouse data suggest that therapeutic inhibition of β-secretase may be feasible. The big challenge is the identification of highly potent, selective blood brain barrier penetrable β-secretase inhibitors.

REFERENCES

1. Vassar, R. *et al.* β-secretase cleavage of Alzheimer's amyloid precursor protein by the transmembrane aspartic protease BACE. *Science* **286**, 735-741 (1999).
2. Yan, R. *et al.* Membrane-anchored aspartyl protease with Alzheimer's disease β-secretase specificity. *Nature* **402**, 533-537 (1999).
3. Sinha, S. & Lieberburg, I. Cellular mechanisms of β-amyloid production and secretion. *Proc. Natl. Acad. Sci.* **96**, 11049-11053 (1999).
4. Hussain, I. *et al.* Identification of a novel aspartic protease (Asp2) as β-secretase. *Mol. Cell. Neurosci.* **14**, 419-427 (1999).
5. Saunders, A. J. *et al.* BACE maps to chromosome 11 and a BACE homolog, BACE2, resides in the obligate Down syndrome region of chromosome 21. *Science* **286**, 1255a (1999).
6. Bennett, B. D. *et al.* Expression analysis of BACE2 in brain and peripheral tissues. *J. Biol. Chem.* **275**, 20647-20651 (2000).
7. Sinha, S. *et al.* Purification and cloning of amyloid precursor protein β-secretase from human brain. *Nature* **402**, 537-540 (1999).

8. Farzan, M., Schnitzler, C. E., Vasilieva, N., Leung, D. & Choe, H. BACE2, a β-secretase homolog, cleaves at the β-site and within the amyloid-β region of the amyloid-β precursor protein. *Proc. Natl. Acad. Sci.* **97**, 9712-9717 (2000).

9. Haniu, M. *et al.* Characterization of Alzheimer's β-secretase protein BACE - A pepsin family member with unusual properties. *J. Biol. Chem.* **275**, 21099-21106 (2000).

10. Bennett, B. D. *et al.* A furin-like convertase mediates propeptide cleavage of BACE, the Alzheimer's β-secretase. *J. Biol. Chem.* **275**, 37712-37717 (2000).

11. Hussain, I., Christie, G., Schneider, K., Moore, S. & Dingwall, C. Prodomain processing of Asp1 (BACE2) is autocatalytic. *J. Biol. Chem.* **276**, 23322-23328 (2001).

12. Hong, L. *et al.* Structure of the protease domain of memapsin 2 (β-secretase) complexed with inhibitor. *Science* **290**, 150-153 (2000).

13. Hardy, J. & Allsop, D. Amyloid deposition as the central event in the aetiology of Alzheimer's disease. *Trends in Pharmac.* **12**, 383-388 (1991).

14. Luo, Y. *et al.* Mice deficient in BACE1, the Alzheimer's β-secretase, have normal phenotype and abolished β-amyloid generation. *Nat. Neurosci.* **4**, 231-232 (2001).

15. Cai, H. *et al.* BACE1 is the major β-secretase for generation of Aβ peptides by neurons. *Nat. Neurosci.* **4**, 233-234 (2001).

16. Roberds, S. L. *et al.* BACE knockout mice are healthy despite lacking the primary β-secretase activity in brain: implications for Alzheimer's disease therapeutics. *Hum. Mol. Genet.* **10**, 1317-1324 (2001).

CELL BIOLOGY OF PRESENILINS

Seong-Hun Kim and Sangram S. Sisodia[*]

1. INTRODUCTION

Sequential cleavages of APP by β- and γ- secretases generate ~4kD Aβ, which deposits in the brains of Alzheimer's disease (AD) patients (De Strooper and Annaert, 2000). Since Aβ, especially the longer form, is believed to trigger the neuronal damage, reduction of Aβ burden by inhibiting the activity of either enzyme could be useful therapeutic strategies for AD. Extensive controversies have been going on whether presenilins, proteins mutated in the majority of pedigrees of early-onset familial AD, are the γ-secretases themselves. PS1 and PS2 are highly homologous proteins with eight transmembrane (TM) domains. The preponderant PS-related polypeptides that accumulate are ~30kD N-terminal (NTF) and ~20kD C-terminal fragments (CTF) which form 1:1 heterodimers and are detectable as high molecular weight complexes *in vivo* (Capell *et al.*, 1998; Thinakaran *et al.*, 1998; Li *et al.*, 2000a).

Several lines of evidence indicate that the PS1 and PS2 are critically involved in the γ-secretase activity. First, complete deletion of PS1 and PS2 in mice resulted in complete abolition of γ-secretase processing of APP and Notch I (Herreman *et al.*, 2000; Zhang *et al.*, 2000; De Strooper *et al.*, 1999; Struhl and Greenwald, 1999). Interestingly, substitution of either of two conserved aspartate residues within predicted TM 6 and 7 of PS1 (Asp257 and Asp385) or PS2 (Asp263 and Asp366) by other amino acids has a dominant negative effect on the γ-secretase cleavage of APP and Notch I (Wolfe *et al.*, 1999; Steiner *et al.*, 1999; Kimberly *et al.*, 2000). Among these two aspartate residues, the one in the predicted TM7 is less tolerant of mutagenesis compared to the one in the TM6 (Capell *et al.*, 2000). Steiner *et al.* (2000) further demonstrated that presenilins contain a motif near the second aspartate (GA) that is also found in the bacterial type 4 prepilin protease (TFPP) family, leading to the suggestion that the presenilins might be the unusual transmembrane aspartyl proteases. Moreover, Li *et al.* (2000a) reported that detergent solubilized γ-secretase activity coelutes with PS1 at a molecular weight of approximately 2 X 10^6 Da. and PS1 antibody immunoprecipitates γ-secretase activity

---

[*] Department of Neurobiology, Pharmacology and Physiology, The University of Chicago, Chicago, IL60637.

Mapping the Progress of Alzheimer's and Parkinson's Disease
Edited by Mizuno *et al.*, Kluwer Academic/Plenum Publishers, 2002

suggesting that the γ-secretase activity is catalyzed by PS1-containing macromolecular complex. More direct evidence that presenilins are the essential component of γ -secretase activity has recently come from the affinity labeling studies which show that transition state analogue inhibitors of γ-secretase specifically labeled NTF and CTF of PS1 and PS2 (Li *et al.*, 2000b; Esler *et al.*, 2000).

Still several questions remain unanswered. The major problem of the "PS1 and PS2 are γ-secretases" hypothesis is the discrepancy of subcellular distribution of PS and γ-secretase activity. PS1 and PS2 are predominantly localized in the early secretory compartment such as endoplasmic reticulum (ER) or ER to Golgi intermediate compartment (ERGIC) (Capell *et al.*, 1997; Culvenor *et al.*, 1997; Lah *et al.*, 1997; Annaert *et al.*, 1999; Kim *et al.*, 2000) while most γ -secretase activity is present in rather distal secretory pathway, such as Golgi, plasma membrane and endosomes-lysosomes (Koo and Squazzo, 1994; Higaki *et al.*, 1995; Hartman *et al.*, 1997; Perez *et al.*, 1999; Kopan *et al.*, 1996; Schroeter *et al.*, 1998). This "spatial paradox" (Annaert *et al.*, 1999) has led to the suggestion that PS1 may elicit its effects in an indirect way by modulating the trafficking of the "γ-secretase" and/or its substrates. We previously demonstrated that in *PS1*-deficient neurons, the rate of secretion of soluble APP derivatives is quantitatively increased (Naruse *et al.*, 1998). Moreover, maturation and trafficking of other membrane proteins such as APLP1 and TrkB receptor, are also affected with PS1 deletion. Therefore it remains possible that presenilins influence γ -secretase activity indirectly.

In this study, we report that aspartate mutant PS1 not only leads to the accumulation of APP-CTFs but also increases the stability of full length APP. Surprisingly, a fraction of D385A mutant PS1 is mistargeted to the late secretory compartments and partially colocalizes with APP-βCTF. In addition, chemical crosslinking and coimmuno-precipitation studies revealed that aspartate mutant PS1 is associated with APP-CTFs. In contrast, wild type PS1 stays in the ER and is not associated with APP-CTFs.

2. RESULTS & DISCUSSION

2.1. Effects of Aspartate Mutant PS1 on the γ-secretase Processing of APP

We generated mouse neuroblastoma N2a cell lines stably expressing Swedish APP$_{695}$ and either wild type PS1 or PS1 harboring aspartate to alanine substitutions at codons 257 (D257A) or 385 (D385A), singly, or in combination (Double). As previously reported (Wolfe *et al.*, 1999), all the aspartate mutant PS1 failed to undergo endoproteolysis while wild type PS1 generates human derived PS1-CTF. Endogenous mouse PS1-CTF was nearly completely replaced in cell lines expressing aspartate mutant PS1. Treatment with cycloheximde revealed that these aspartate mutant PS1 are as stable as PS1 fragments. Immunoprecipitation of [S^{35}]-methionine labeled lysates with antibody 369 revealed a significant accumulation of APP-CTFs in the cells expressing aspartate mutant PS1. However, while secreted Aβ level was decreased by 60-70% in N2a cells expressing D385A mutant PS1, cells expressing D257A or Double aspartate mutant PS1 failed to show any apparent changes of Aβ secretion. This finding is consistent with previous reports that cells expressing D257A PS1 or naturally occurring splice variant PS1ΔE8 which contains D257A mutation in addition to the deletion of exon 8, did not reduce the Aβ secretion (Capell *et al.*, 2000; Morihara *et al.*, 2000).

2.2. Aspartate Mutant PS1 Stabilizes Full-length APP

To further examine the effects of aspartate mutant PS1 on the processing and secretion of APP, stable N2a pools expressing Swedish variant APP$_{695}$ and either wild type or aspartate mutant PS1 were pulse-labeled for 10 min, and chased for various time periods. The synthetic rate of APP after 10 min pulse is comparable in these pools. During the chase period, immature *N*-glycosylated APP was gradually converted to the mature, *N*- and *O*- glycosylated form of APP and both immature and mature APP were rapidly degraded in wild type PS1 expressing cells. In contrast, full length APP was much more stable and accumulated in cells expressing any aspartate mutant PS1. Even after 3 hr straight metabolic labeling, the level of APP is much higher in these mutant cells. The secretion of APPs into the medium also occurs slightly faster in these aspartate mutant cells although the effect is not as striking as holoprotein. These data suggest that in addition to the inhibition of γ-secretase processing of APP, expression of aspartate mutant PS1 increases the stability of full length APP protein.

2.3. Accumulation of APP-βCTF in Late Secretory Compartments

To characterize the subcellular compartments where APP β-CTF is accumulated in cells expressing aspartate mutant PS1, we performed laser confocal immunofluorescence studies using 3D6 antibody which specifically recognizes the N terminal end of β-secretase cleaved stub of APP. This antibody is quite useful in localizing APP-βCTF because it does not recognize full length APP or other APP-CTFs. In N2a cells expressing D385A PS1, 3D6 immunoreactivity was detected on the plasma membrane as well as in perinuclear vesicular structures and peripheral small vesicles. To identify these compartments, we double labeled the cells with the Golgi marker FITC-Vicia villosa agglutinin (VVA). 3D6 immunoreactivity was overlapped with FITC-VVA in the perinuclear vesicular structures indicating these structures represent the Golgi apparatus. To label endosomes, we incubated cells with FITC- conjugated BSA for 15 min followed by 90 min chase at 37°C. After the chase, internalized FITC-BSA is observed as the dot-like pattern in the cell periphery and 3D6 immunoreactivity is partially overlapped with FITC-BSA. In addition, immunoelectron microscopy using 3D6 antibody revealed that APP-βCTF is not only localized in the cell surface but also in the large vesicular structures which are reminiscent of multivesicular bodies. These results were also biochemically confirmed by cell surface biotinylation experiments, which showed markedly higher accumulation of APP-CTFs on the surface of aspartate mutant PS1 expressing cells. Virtually no APP-CTFs were detected in the streptavidin precipitates in wild type cells while significant fraction of APP-CTFs was surface biotinylated in the aspartate mutant cells. Even considering higher level of APP-CTFs in the straight lysates of aspartate PS1 expressing cells, it is clear that much more proportion of APP-CTFs is present on the cell surface of these cells. Full length APP on the cell surface was also proportionally increased in these cells. Interestingly, surface presentation of APLP2 was also increased in aspartate mutant cells. These results collectively indicate that aspartate mutant PS1 have multiple effects on the membrane trafficking and processing. In this regard, it is noteworthy that trafficking of a limited set of membrane proteins was significantly affected in PS1 deficient neurons (Naruse et al., 1998). In these cells, the rate of secretion of APP$_{s\alpha}$ is quantitatively increased and the rate of acquisition and steady-state levels of complex oligosaccharide modifications of TrkB, is diminished. BDNF-mediated TrkB autophosphorylation is also severely compromised in *PS1*-deficient neurons. Interestingly, a cell adhesion molecule, termed ICAM-5 or telencephalin has been recently found to be dependent on PS1 for proper trafficking to the cell surface in primary hippocampal neurons from PS1-/- embryos (Annaert *et al.*, 2000). These findings suggest that in addition to the intramembraneous γ -secretase cleavage, PS1 is also involved in regulating the trafficking of a limited set of membrane proteins.

2.4. Mistargeting of Aspartate Mutant PS1 and Colocalization with APP-βCTF

Extensive localization studies by have shown that wild type PS1 protein and APP-βCTF are distributed to the different subcellular compartments. PS1 usually resides in early compartments such as ER or ERGIC whereas APP-βCTF is normally accumulated in the rather late secretory compartments. To examine whether subcellular distribution of aspartate PS1 is different from that of wild type PS1, we performed double immunofluorescence of N2a cells expressing either wild type or aspartate mutant with affinity purified αPS1NT and 3D6 antibodies. Surprisingly, D385A mutant PS1 is not only localized in the ER but also detected in Golgi body, endosomes and plasma membrane where it is colocalized with APP-βCTF. In contrast, wild type PS1 is primarily distributed in the reticular network of ER, as previously reported (Kim *et al.*, 2000).

2.5. Association of APP-CTF and Aspartate Mutant PS1

Colocalization of D385A PS1 with APP-CTFs opened the possibility of association of these two proteins. Xia *et al.* (2000a) reported that APP-CTF is coimmunoprecipitated with even wild type PS1 fragments but they used the 1% NP40/0.5% Triton X100 to lyse the cells, which disrupts the association of PS1-NTF and CTF. To examine whether PS1 is associated with APP-CTF, we employed two different experimental paradigms; *in vivo* chemical crosslinking and coimmunoprecipitation with 0.25% *n*-dodecyl α-D-maltoside or 1% CHAPS. Surprisingly, in both experimental paradigms, significant amount of APP-CTFs was coisolated by αPS1NT antibody only in aspartate mutant PS1 expressing cells, not in wild type cells. Even with very long exposure, no APP-CTFs were detectable in the αPS1NT immunoprecipitates in wild type PS1 expressing cells. Interestingly, while more APP-αCTF is present in the straight lysates, αPS1NT antibody pulled down more APP-βCTF suggesting that aspartate mutant PS1 is preferentially associated with APP-βCTF than other CTFs. Full length APP was not detected in the PS1NT immunoprecipitates in both wild type and D385A PS1 expressing cells. Thus, APP-CTFs (preferentially β-stub), not full length APP, is associated with aspartate mutant, not wild type, PS1.

To explain the results, we propose a model that PS1 is critical in folding the γ-secretase in ER. Properly folded γ-secretase is then released from PS1 and transported to the late secretory compartments where it associates with its substrates including APP-CTFs and catalyses their cleavages. In contrast, aspartate mutant PS1 is not able to effectively fold and release the γ-secretase, comigrates with the inactive enzyme to the late secretory compartments and associates with APP-CTFs.

In conclusion, we found that aspartate mutant PS1 has multiple effects on the processing and trafficking of APP and APLP2 suggesting that PS1 could play a much broader role in regulating cellular trafficking of several membrane proteins in addition to promoting intramembrane cleavage of APP and Notch I. Understanding the precise role of PS1 in membrane will be critical for identification of molecular targets of therapeutic measures aimed at reducing Aβ burden.

3. ACKNOWLEDGEMENTS

We'd like to thank Drs. James J. Lah, Allan Levey, Jae-Yoon Leem, Gopal Thinakaran, Hilda H. Slunt and Yuan Zhang for contributions to this work. This study was supported by grants from NIH, NIA and Alzheimer's Association.

REFERENCES

Annaert, W. G., Boeve, C., Esselens, C., Snellings, G., Craessaerts, K., and De Strooper, B., 2000, Presenilin 1 interacts with a neuronal cell adhesion molecule and mediates its intracellular trafficking, *Neurobiol. Aging* **21**:S133.

Annaert, W. G., Levesque, L., Craessaerts, K., Dierinck, I., Snellings, G., Westaway, D., George-Hyslop, P.S., Cordell, B., Fraser, P., and De Strooper, B., 1999, Presenilin 1 controls gamma-secretase processing of amyloid precursor protein in pre-Golgi compartments of hippocampal neurons, *J. Cell. Biol.* **147**:277-294

Capell, A., Grunberg, J., Pesold, B., Diehlmann, A., Citron, M., Nixon, R., Beyreuther, K., Selkoe, D. J., and Haass, C., 1998, The proteolytic fragments of the Alzheimer' s disease associated presenilin-1 form heterodimers and occur as a 100-150-kDa molecular mass complex, *J. Biol. Chem.* **273**: 3205-3211.

Capell, A., Saffrich, R., Olivo, J.C., Meyn, L., Walter, J., Grunberg, J., Mathews, P., Nixon, R., Dotti, C., and Haass, C., 1997, Cellular expression and proteolytic processing of presenilin proteins is developmentally regulated during neuronal differentiation, *J. Neurochem.* **69**: 2432-2440.

Capell, A., Steiner, H., Romig, H., Keck, S., Baader, M., Grim, M. G., Baumeister, R., and Haass, C., 2000, Presenilin-1 differentially facilitates endoproteolysis of the beta-amyloid precursor protein and Notch, *Nat. Cell Biol.* **2**:205-211.

Culvenor, J. G., Maher, F., Evin, G., Malchiodi-Albedi, F., Cappai, R., Underwood, J. R., Davis, J. B., Karran, E. H., Roberts, G., Beyreuther, K., and Masters, C. L., 1997, Alzheimer's disease-associated presenilin-1 in neuronal cells; evidence for localization to the endoplasmic reticulum-Golgi intermediate compartment, *J. Neurosci. Res.* **47**: 719-731.

De Strooper, B. and Annaert, W., 2000, Proteolytic processing and cell biological functions of the amyloid precursor protein, *J. Cell Sci.* **113**:1857-1870

De Strooper, B., Annaert, W., Cupers, P., Saftig, P., Craessaerts, K., Mumm, J. S., Schroeter, E. H., Schrijvers, V., Wolfe, M. S., Ray, W. J., Goate, A., and Kopan, R., 1999, A presenilin-1-dependent gamma-secretase-like protease mediates release of Notch intracellular domain, *Nature* **398**:518-522

Esler, W. P., Kimberly, W. T., Ostaszewski, B. L., Diehl, T. S., Moore, C. L., Tsai, J. Y., Rahmati, T., Xia, W., Selkoe, D. J., and Wolfe, M. S., 2000, Transition-state analogue inhibitors of gamma-secretase bind directly to presenilin-1, *Nat. Cell Biol.* **2**: 428-434.

Hartmann, T., Bieger, S. C., Bruhl, B., Tienari, P. J., Ida, N., Allsop, D., Roberts, G. W., Masters, C. L., Dotti, C. G., Unsicker, K., and Beyreuther, K., 1997, Distinct sites of intracellular production for Alzheimer's disease A beta40/42 amyloid peptides, *Nat. Med.* **3**:1016-1020

Herreman, A., Serneels, L., Annaert, W., Collen, D., Schoonjans, L., and De Strooper, B., 2000, Total inactivation of gamma-secretase activity in presenilin-deficient embryonic stem cells, *Nat Cell Biol.* **2:**461-462.

Higaki, J., Quon, D., Zhong, Z., and Cordell, B., 1995, Inhibition of beta-amyloid formation identifies proteolytic precursors and subcellular site of catabolism, *Neuron* **14:**651-659.

Kim, S. H., Lah, J. J., Thinakaran, G., Levey, A., and Sisodia, S. S., 2000, Subcellular localization of presenilins: association with a unique membrane pool in cultured cells, *Neurobiol. Dis.* **7:**99-117.

Kimberly, W. T., Xia, W., Rahmati, T., Wolfe, M. S., and Selkoe, D. J., 2000, The transmembrane aspartates in presenilin 1 and 2 are obligatory for gamma-secretase activity and amyloid beta-protein generation, *J. Biol. Chem.* **275:**3173-3178

Koo, E. H., and Squazzo, S. L., 1994, Evidence that production and release of amyloid beta-protein involves the endocytic pathway, *J. Biol. Chem.* **269:**17386-17389.

Kopan, R., Schroeter, E. H., Weintraub, H., and Nye, J. S., 1996, Signal transduction by activated mNotch: importance of proteolytic processing and its regulation by the extracellular domain, *Proc. Natl. Acad. Sci. USA.* **93:**1683-1688.

Lah, J. J., Heilman, C. J., Nash, N. R., Rees, H. D., Yi, H., Counts, S. E., and Levey, A. I., 1997, Light and electron microscopic localization of presenilin-1 in primate brain, *J. Neurosci.* **17,** 1971-1980.

Li, Y. M., Lai, M. T., Xu, M., Huang, Q., DiMuzio-Mower, J., Sardana, M. K., Shi, X. P., Yin, K. C., Shafer, J. A., and Gardell, S. J., 2000a, Presenilin 1 is linked with gamma -secretase activity in the detergent solubilized state, *Proc. Natl. Acad. Sci. USA* **97:**6138-6143

Li, Y. M., Xu, M., Lai, M. T., Huang, Q., Castro, J. L., DiMuzio-Mower, J., Harrison, T., Lellis, C., Nadin, A., Neduvelil, J. G., Register, R. B., Sardana, M. K., Shearman, M. S., Smith, A. L., Shi, X. P., Yin, K. C., Shafer, J. A., and Gardell, S. J., 2000b, Photoactivated gamma-secretase inhibitors directed to the active site covalently label presenilin 1, *Nature* **405:**689-694.

Morihara, T., Katayama, T., Sato, N., Yoneda, T., Manabe, T., Hitomi, J., Abe, H., Imaizumi, K., and Tohyama, M., 2000, Absence of endoproteolysis but no effects on amyloid beta production by alternative splicing forms of presenilin-1, which lack exon 8 and replace D257A, *Mol. Brain Res.* **85:**85-90.

Naruse, S., Thinakaran, G., Luo, J. J., Kusiak, J. W., Tomita, T., Iwatsubo, T., Qian, X., Ginty, D. D., Price, D. L., Borchelt, D. R., Wong, P. C., and Sisodia, S. S., 1998, Effects of PS1 deficiency on membrane protein trafficking in neurons, *Neuron* **21,** 1213-1221.

Perez, R. G., Soriano, S., Hayes, J. D., Ostaszewski, B., Xia, W., Selkoe, D. J., Chen, X., Stokin, G. B., and Koo, E. H., 1999, Mutagenesis identifies new signals for beta-amyloid precursor protein endocytosis, turnover, and the generation of secreted fragments, including Abeta42, *J. Biol. Chem.* **274:**18851-18856.

Schroeter, E. H., Kisslinger, J. A., and Kopan, R., 1998, Notch-1 signalling requires ligand-induced proteolytic release of intracellular domain, *Nature* **393:**382-386.

Steiner, H., Duff, K., Capell, A., Romig, H., Grim, M. G., Lincoln, S., Hardy, J., Yu, X., Picciano, M., Fechteler, K., Citron, M., Kopan, R., Pesold, B., Keck, S., Baader, M., Tomita, T., Iwatsubo, T., Baumeister, R., and Haass C., 1999, A loss of function mutation of presenilin-2 interferes with amyloid beta-peptide production and notch signaling, *J. Biol. Chem.* **274:**28669-28673

Steiner, H., Kostka, M., Romig, H., Basset, G., Pesold, B., Hardy, J., Capell, A., Meyn, L., Grim, M. L., Baumeister, R., Fechteler, K., and Haass, C., 2000, Glycine 384 is required for presenilin-1 function and is conserved in bacterial polytopic aspartyl proteases, *Nat. Cell Biol.* **2:**848-851.

Struhl, G., and Greenwald, I., 1999, Presenilin is required for activity and nuclear access of Notch in Drosophila, *Nature* **398:**522-525

Thinakaran, G., Regard, J.B., Bouton, C. M. L., Harris, C. L., Price, D. L., Borchelt, D. R., and Sisodia, S. S., 1998, Stable association of presenilin derivatives and absence of presenilin interactions with APP, *Neurobiol. Dis.* **4:**438-453.

Wolfe, M. S., Xia, W., Ostaszewski, B. L., Diehl, T. S., Kimberly, W. T., and Selkoe, D. J., 1999, Two transmembrane aspartates in presenilin-1 required for presenilin endoproteolysis and γ-secretase activity, *Nature* **398:**513-517.

Xia, W., Ray, W. J., Ostaszewski, B. L., Rahmati, T., Kimberly, W. T., Wolfe, M. S., Zhang, J., Goate, A. M., and Selkoe, D. J., 2000, Presenilin complexes with the C-terminal fragments of amyloid precursor protein at the sites of amyloid beta-protein generation, *Proc. Natl. Acad. Sci. USA* **97:**9299-9304

Zhang, Z., Nadeau, P., Song, W., Donoviel, D., Yuan, M., Bernstein, A., and Yankner, B. A., 2000, Presenilins are required for gamma-secretase cleavage of beta-APP and transmembrane cleavage of Notch-1, *Nat. Cell Biol.* **2:**463-465.

PRESENILIN DEPENDENT γ-SECRETASE PROCESSING OF ß-AMYLOID PRECURSOR PROTEIN AT A SITE CORRESPONDING TO THE S3 CLEAVAGE OF NOTCH

Harald Steiner[1*], Magdalena Sastre[1*], Gerd Multhaup[2], David B. Teplow[3], and Christian Haass[1*]

1. INTRODUCTION

Alzheimer's disease (AD) is the most abundant neurodegenerative disorder worldwide. Senile plaques, composed of Amyloid ß-peptide (Aß) appear to be a major pathological alteration in the brain of AD patients (Selkoe, 1999). Almost all familial AD (FAD) associated mutations affect the generation of Aß by increasing the production of the highly amyloidogenic 42 amino acid variant (Selkoe, 1999). Aß is produced from the ß-Amyloid precursor protein (ßAPP) by endoproteolysis. At least two proteolytic activities are required for Aß generation. ß-Secretase (BACE) mediates the N-terminal cleavage producing a membrane associated C-terminal fragment (CTFß) of ßAPP (Vassar

[*1]Adolf Butenandt-Institute, Dept. of Biochemistry, Laboratory for Alzheimer's and Parkinson's Disease Research, Ludwig-Maximilians-University, 80336 Munich, Germany; [2]Center for Molecular Biology Heidelberg, Im Neuenheimer Feld 282, 69120 Heidelberg, Germany, [3]Department of Neurology, Harvard Medical School, and Center for Neurologic Diseases, Brigham and Women's Hospital, Boston, MA 02115, USA.
[*] These authors contributed equally

Mapping the Progress of Alzheimer's and Parkinson's Disease
Edited by Mizuno *et al.*, Kluwer Academic/Plenum Publishers, 2002

91

and Citron, 2000). The resulting CTFß is the immediate precursor for the intramembraneous γ-secretase cut. This cleavage is facilitated by the presenilins, PS1 and PS2, and there is evidence that presenilins themselves could be unusual aspartyl proteases, which mediate the γ-secretase cut [(Wolfe *et al.*, 1999); for review see (Steiner and Haass, 2000; Wolfe and Haass, 2001)]. γ-Secretase cleavage results in the production of Aß and the generation of the small hydrophobic p7 (Haass and Selkoe, 1993) also called CTFγ. Recent observations indicate that a CTFγ like fragment is indeed generated during ßAPP processing (McLendon *et al.*, 2000; Pinnix *et al.*, 2001). However, it is still unknown if the γ-secretase cut exclusively occurs after position 40 or 42 of Aß. A cleavage further C-terminal terminal would be consistent with the site 3 (S3) cleavage of Notch (Schroeter *et al.*, 1998), which like the ßAPP cleavage occurs by a presenilin dependent proteolytic activity (De Strooper *et al.*, 1999). The identification of the initial cleavage site is of great importance, since γ-secretase inhibition is thought to be a major approach for therapeutic intervention.

2. RESULTS AND DISCUSSION

To prove if CTFγ generation can be observed in living cells, C-terminal fragments of ßAPP were immunoprecipitated from membrane fractions of human embryonic kidney 293 (HEK 293) cells stably transfected with $ßAPP_{695}$ carrying the Swedish mutation (swAPP). This indeed revealed the presence of an approximately 6 kDa C-terminal fragment migrating below the major ßAPP CTFs generated by α-, and ß-secretase. This fragment represents a γ-secretase product, since its generation is blocked by the γ-secretase inhibitor DAPT (Dovey *et al.*, 2001). Moreover, CTFγ generation turned out to be fully dependent on the presence of biological active presenilins as one would expect for a γ-secretase product. This was directly demonstrated by the co-expression of the dominant negative PS1 D385N mutation (Steiner *et al.*, 1999; Wolfe *et al.*, 1999) which fully inhibited CTFγ formation.

In order to identify the precise γ-secretase cleavage sites of ßAPP we have used an *in vitro* assay for CTFγ generation (McLendon *et al.*, 2000; Pinnix *et al.*, 2001). We demonstrate that CTFγ is generated *in vitro* in a PS dependent manner and is inhibited by known γ-secretase inhibitors. Moreover, radiosequencing of the in vitro generated CTFγ revealed that its N-terminus begins at amino acid 50 of the ß-amyloid domain, a site, which corresponds to the S3-cleavage of Notch (Fig. 1).

Our results therefore suggest that γ-secretase mediates at least three different cuts within the transmembrane domain of ßAPP (Fig. 1). The finding of an additional γ-secretase cut close to the predicted border of the transmembrane domain may indicate that the cytoplasmic tail of ßAPP requires "shedding" before/during it undergoes the final γ-secretase cut after positions 40/42 of the ß-amyloid domain. Such a "shedding" event would be very similar to the essential ectodomain shedding of γ-secretase substrates (Struhl and Adachi, 2000). Interestingly, γ-secretase cleavage after amino acid 49 of the

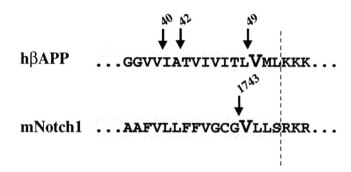

Figure 1: Topologically similar PS dependent γ-secretase/S3 cleavages of ßAPP and Notch. Human ßAPP is cleaved by γ-secretase in a PS dependent manner after position 40, 42 and 49 of the ß-amyloid domain. Mouse Notch1 is cleaved PS dependent at S3 after amino acid 1743 (Schroeter *et al.*, 1998). Note the similar location of the γ-secretase cleavage site at position 49 of the ß-amyloid domain and the S3 cleavage site of Notch. The gray box represents the transmembrane domain; the dashed line the proposed membrane border.

Aß-domain is located at a position, which corresponds to the S3 cleavage of Notch (Schroeter *et al.*, 1998) (Fig. 1). Both cleavages are PS dependent and can be blocked by γ-secretase inhibitors (De Strooper *et al.*, 1999; De Strooper *et al.*, 1998). Thus, it is likely that γ-secretase cleavage at position 49 of ßAPP and S3 cleavage of Notch are mediated by the same PS dependent enzyme. Finally, the identification of CTFγ *in vivo* may also raise the interesting possibility that this fragment similar to the NICD may have a biological function in signal transduction. Based on the striking similarity of the biological mechanisms involved in the generation of NICD and CTFγ as well as potentially similar functions in signal transduction we therefore propose the term AICD for the <u>A</u>myloid precursor protein <u>i</u>ntra<u>c</u>ellular <u>d</u>omain.

REFERENCES

De Strooper, B., Annaert, W., Cupers, P., Saftig, P., Craessaerts, K., Mumm, J.S., Schroeter, E.H., Schrijvers, V., Wolfe, M.S., Ray, W.J., Goate, A. and Kopan, R. (1999) A presenilin-1-dependent γ-secretase-like protease mediates release of Notch intracellular domain. *Nature*, **398**, 518-522.

De Strooper, B., Saftig, P., Craessaerts, K., Vanderstichele, H., Guhde, G., Annaert, W., Von Figura, K. and Van Leuven, F. (1998) Deficiency of presenilin-1 inhibits the normal cleavage of amyloid precursor protein. *Nature*, **391**, 387-390.

Dovey, H.F., John, V., Anderson, J.P., Chen, L.Z., de Saint Andrieu, P., Fang, L.Y., Freedman, S.B., Folmer, B., Goldbach, E., Holsztynska, E.J., Hu, K.L., Johnson-Wood, K.L., Kennedy, S.L., Kholodenko, D., Knops, J.E., Latimer, L.H., Lee, M., Liao, Z., Lieberburg, I.M., Motter, R.N., Mutter, L.C., Nietz, J., Quinn, K.P., Sacchi, K.L., Seubert, P.A., Shopp, G.M., Thorsett, E.D., Tung, J.S., Wu, J., Yang, S., Yin, C.T., Schenk, D.B., May, P.C., Altstiel, L.D., Bender, M.H., Boggs, L.N., Britton, T.C., Clemens, J.C., Czilli, D.L., Dieckman-McGinty, D.K., Droste, J.J., Fuson, K.S., Gitter, B.D., Hyslop, P.A., Johnstone, E.M., Li, W.Y., Little, S.P., Mabry, T.E., Miller, F.D. and Audia, J.E. (2001) Functional γ-secretase inhibitors reduce β-amyloid peptide levels in brain. *J Neurochem*, **76**, 173-181.

Haass, C. and Selkoe, D.J. (1993) Cellular processing of β-amyloid precursor protein and the genesis of amyloid β-peptide. *Cell*, **75**, 1039-1042.

McLendon, C., Xin, T., Ziani-Cherif, C., Murphy, M.P., Findlay, K.A., Lewis, P.A., Pinnix, I., Sambamurti, K., Wang, R., Fauq, A. and Golde, T.E. (2000) Cell-free assays for γ-secretase activity. *Faseb J*, **14**, 2383-2386.

Pinnix, I., Musunuru, U., Tun, H., Sridharan, A., Golde, T., Eckman, C., Ziani-Cherif, C., Onstead, L. and Sambamurti, K. (2001) A novel γ-secretase assay based on detection of the putative C-terminal fragment-γ of amyloid β protein precursor. *J Biol Chem*, **276**, 481-487.

Schroeter, E.H., Kisslinger, J.A. and Kopan, R. (1998) Notch-1 signalling requires ligand-induced proteolytic release of intracellular domain. *Nature*, **393**, 382-386.

Selkoe, D.J. (1999) Translating cell biology into therapeutic advances in Alzheimer's disease. *Nature*, **399**, A23-31.

Steiner, H. and Haass, C. (2000) Intramembrane proteolysis by presenilins. *Nature Reviews Molecular Cell Biology*, **1**, 217-224.

Steiner, H., Romig, H., Pesold, B., Philipp, U., Baader, M., Citron, M., Loetscher, H., Jacobsen, H. and Haass, C. (1999) Amyloidogenic function of the Alzheimer's disease-associated presenilin 1 in the absence of endoproteolysis. *Biochemistry*, **38**, 14600-14605.

Struhl, G. and Adachi, A. (2000) Requirements for presenilin-dependent cleavage of Notch and other transmembrane proteins. *Mol. Cell*, **6**, 625-636.

Vassar, R. and Citron, M. (2000) Aβ-generating enzymes: recent advances in β- and γ-secretase research. *Neuron*, **27**, 419-422.

Wolfe, M.S. and Haass, C. (2001) The role of presenilins in γ-secretase activity. *J Biol Chem*, **276**, 5413-5416.

Wolfe, M.S., Xia, W., Ostaszewski, B.L., Diehl, T.S., Kimberly, W.T. and Selkoe, D.J. (1999) Two transmembrane aspartates in presenilin-1 required for presenilin endoproteolysis and γ-secretase activity. *Nature*, **398**, 513-517.

ENDOPLASMIC RETICULUM STRESS AND ALZHEIMER'S DISEASE

Takashi Kudo,[*] Taiichi Katayama, Kazunori Imaizumi, Yuka Yasuda, Misako Yatera, Masayasu Okochi, Toshihisa Tanaka, Kojin Kamino, Masaya Tohyama, and Masatoshi Takeda

1. INTRODUCTION

The endoplasmic reticulum (ER) performs the synthesis, post-translational modification, and proper folding of proteins. A variety of conditions, for instance the disturbance of calcium homeostasis, nutrient deprivation, or the overexpression of relatively insoluble proteins, can be ER stress, causing the accumulation of unfolding or misfolding proteins in the ER. Because the incompletely folded molecules make threats against living cells, efficient quality control systems have evolved to prevent them from moving along the secretory pathway. Eukaryotic cells have three different mechanisms for dealing with an accumulation of unfolded proteins in the ER known as the unfolded protein response (UPR): transcriptional induction, translational attenuation, and degradation.

Alzheimer's disease is a progressive neurodegenerative disorder, characterized pathologically by cerebral neuritic plaques of amyloid-β peptide (Aβ) and neurofibrillary tangles. Some early-onset cases of autosomal-dominant familial Alzheimer's disease (FAD) are caused by mutations in the amyloid precursor protein (APP) gene located on chromosome 21, *PS1* found on chromosome 14, and *PS2* located on chromosome 1. Of these three loci, mutations in *PS1* are the most prevalent in cases of FAD. However, the mechanism by which mutations in presenilins predispose individuals to FAD has not yet been determined. FAD-linked PS1 variants alter proteolytic processing of APP,[1,2] and mutations in *PS1* increase cellular susceptibility to apoptosis induced by various insults, including withdrawal of trophic factors and exposure to Aβ.[3,4] The mechanism by which *PS1* mutations promote cell death is not known, but cell-culture studies have revealed perturbed calcium homeostasis and increased production of free radicals in affected cells. These apoptotic stimuli provoke the accumulation of unfolded proteins in the ER, and are therefore a type of 'ER stress'. As PS1 is an integral membrane protein and is localized mainly to the ER,[5] it is possible that *PS1* mutations could have a significant role in vulnerability to ER stress.

[*] Takashi Kudo, Department of Clinical Neuroscience, Osaka University Graduate School of Medicine, D3, 2-2 Yamadaoka, Suita, osaka 565-0871, Japan.

Mapping the Progress of Alzheimer's and Parkinson's Disease
Edited by Mizuno *et al.*, Kluwer Academic/Plenum Publishers, 2002

This article focuses on the relationship between ER stress and the pathogenesis of AD, reviewing our recent data mainly.

2. THREE TYPES OF UPR

To date, three cellular responses have been known as UPR. Firstly, in response to the accumulation of unfolded proteins in the ER, eukaryotic cells activate an imtracellular-signaling pathway from the ER to the nucleus, resulting in transcriptional upregulation of molecular chaperones, such as BiP/GRP78 and GRP 94. The chaperones assist or facilitate normal folding of unfolded or misfoled proteins[6]. Continuous delivery of newly synthesized proteins is burden to the ER when proper folding is prevented under ER stress conditions. Therefore the second strategy of cells against ER stress is the generalized suppression of translation mediated by the serine/threonine kinase PERK, which phosphorylates and inactivates the translation initiation factor eIF2 α[7,8]. Thirdly, ER stress activates ER-associated degradation (ERAD) system, by which misfolded proteins are transported out of the ER through the translocon to the cytoplasm and then are ubiquitinated and degraded by the 26S proteasome[9]. By the three protective responses, cells can overcome ER stress, potentially leading to apoptosis. It has been revealed that cells deficient in the UPR are more vulnerable to ER stress-induced apoptosis[10-12].

3. ER-RESIDENT MOLECULES FOR UPR

To date, three ER-resident transmembrane molecules, IRE1, PERK, and ATF6, which sense the accumulation of unfolded proteins to induce UPR, have been identified.

ER stress-induced oligomerization and autophosphorylation of IRE, type I transmembrane proteins in the ER, result in activation of the endonuclease domains that are postulated to initiate splicing of mRNA encoding a putative transcription factor, leading to enhanced transcription of ER chaperone genes, such as BiP/GRP78 and GRP 94[6,13,14].

ATF6 is a type II transmembrane protein localized in the ER and activated by ER stress-induced proteolysis. Upon ER stress, the bZIP-containing N-terminal fragment facing the cytoplasm is liberated from the ER membrane and translocated into the nucleus, resulting in activation of ER chaperone gene transcription, which facilitate protein folding[15].

PERK is type I transmembrane protein kinase localized in the ER. ER stress-induced oligomerization and autophosphorylation of PERK result in phosphorylation of the α subunit of eukaryotic translation initiation factor 2 (eIF2 α) at serine 51. This phosphorylation of eIF2 α leads to inhibition of mRNA with ribosomal 60S and 40S subunits through interference with the formation of a 43S initiation complex, resulting in inhibition of translation initiation[16].

The luminal domains of IRE1 and PERK are so similar that they may use a similar sensing mechanism. Recent studies showed that the luminal domains are indeed functionally interchangeable and that BiP/GRP78 is directly involved in activation of PERK as well as IRE1 by ER stress[17].

4. FAD MUTANTS OF PS1 ATTENUATE UPR INITIATED BY IRE1

Under tunicamycin treatment (3 mg/ml, for 6 hours), which prevents protein glycosylation and causes ER stress, induction of *GRP78* mRNA expression was significantly inhibited in permanent cell lines expressing mutant PS1 (the A246E or the ΔE9 variant) compared with those transfected with wild-type PS1 or empty vector. To confirm that FAD-linked mutants generally cause inhibition of *GRP78* mRNA induction, other FAD-linked PS1 mutants, such as M146V and I213T, were studied with transiently transfected cells. These mutations also suppressed the induction of *GRP78* mRNA. To test this hypothesis directly and to exclude the possibility that the effects were caused by overexpression of PS1 mutants, we studied primary cultured neurons from embryos of PS1 mutant 'knock-in mice, which express the PS1 (I213T) protein at normal physiological levels. Such 'knock-in' mice are thought to provide an animal model that closely reflects FAD in humans[18]. The induction of *GRP78* mRNA by ER stress was slightly decreased in neurons from mice heterozygous for the knock-in *PS1* mutation, and was significantly decreased in those from mice homozygous for the knock-in mutation. In these knock-in mice, the efficiencies of transcription and translation from the mutated *PS1* allele were equivalent to each other, indicating that the reduction of the expression of *GRP78* mRNA was dependent on the expression level of mutated PS1 proteins. Thus, the FAD-linked PS1 mutations appeared to affect the system of the induction.

Immunoprecipitation and immunohistochemical study showed that PS1 interacts with IRE1 on the ER membrane. IRE1 leads to downstream signaling by a process that depends on oligomerization and autophosphorylation of its kinase domain[19-21]. Therefore, we studied the effects of mutations in *PS1* on autophosphorylation of IRE1. SK-N-SH cells stably transfected with wild-type PS1, mutant PS1, or empty vector were stimulated with thapsigargin ($1 \mu M$) for indicated times and western blotting of IRE1 was performed. When cells were treated with $1 \mu M$ thapsigargin, within 15 minutes, the bands of IRE1 were completely shifted (phosphorylated-IRE1) in cells expressing wild PS1. In contrast, no shifts of IRE1-immunoreactive bands were seen in mutant PS1expressing cells that were treated with the same dose of thapsigargin within 30 minutes. These results indicate that FAD-linked PS1 mutants attenuate the autophosphorylation of IRE1 to prevent signaling for GRP78 induction.

To confirm that vulnerability to ER stress in cells expressing PS1 mutants is based on attenuated induction of *GRP78* mRNA by inhibition of IRE1 function, we infected SK-N-SH cells bearing mutant PS1 with recombinant GRP78 using Semliki forest virus (SFV–GRP78 fusion). We then studied responses to ER stress. Sensitivity to ER stress caused by treatment with tunicamycin in neuroblastoma lines expressing PS1 mutants was reversed by infection with recombinant GRP78. These results indicated that expression of GRP78 protects against neuronal death caused by ER stress, and that the reduction in *GRP78* gene expression causes vulnerability to ER stress in cells expressing PS1 mutants.

We examined whether GRP78 protein levels were changed in the brains of Alzheimer's disease patients. Western blotting with anti-KDEL antibody, which detects both GRP78 and GRP94, showed that levels of these proteins were significantly decreased in sporadic Alzheimer's disease (SAD) patients compared with age-matched controls. The levels in the brains of patients with FAD-linked PS1 mutations were decreased to a greater extent. Immunohistochemical analysis revealed that KDEL immunoreactivity was reduced in hippocampal and cortical neurons of both SAD and FAD patients compared with controls. Thus, amounts of ER-resident molecular chaperones such as GRP78 and GRP94 were downregulated in brains of Alzheimer's patients, and these findings are likely to be associated with the pathology of Alzheimer's disease.

5. FAD MUTANTS OF PS1 ATTENUATE UPR INITIATED BY PERK

To investigate whether PS1 interacts with PERK, we studied the subcellular localization of PS1 and PERK in N2a cells stably expressing PS1. Immunofluorescence microscopy showed that PS1 is preferentially distributed in the ER and Golgi. PERK was localized to the ER, as shown by the strict colocalization of PERK immunoreactivity with BiP and GRP94, with are resident in the ER. Double labeling shown that the immunoreactivity of PS1 overlapped with that of PERK in the ER, indicating that PS1 might associate with PERK on the ER membrane. The similar distribution was shown in the cells expressing the mutant PS1.

Activation of PERK during ER stress correlates with the autophosphorylation of those cytoplasmic kinase domains. At first, to examine the effects of PS1 mutants on the activation of PERK, western blotting was perfomed using lysate of N2a cells expressing of wild-type PS1 or mutants PS1 stimulated by 1mM DTT or 1μ M thapsigargin for 5-30minutes. Phosphorylation of PERK retards their mobility on SDS-polyacrylamide gels, and serves as convenient marker for their activation status. In N2a cells expressing mock or wild type PS1, the bands of PERK were completely shifted within 15minutes after treatment with DTT or thapsigargin. In contrast, in cells expressing PS1 mutants, the mobility shifts were not observed 15-30minutes after the treatment. These results indicate that FAD-linked PS1 mutants disturb the autophosphorylation of PERK during ER stress.

Because disturbed function of PERK is known to cause the downregulation of phosphorylation of eIF2 α, we examined the levels of phosphorylated-eIF2 α after ER stress in N2a cells expressing PS1 mutations. Western blotting using anti-phosphorylated eIF2 α antibody showed that phosphorylation of eIF2 α was inhibited in cells expressing mutants PS1. Thus, activation of PERK and resultant phosphorylation of eIF2 α were disturbed by the expression of PS1 mutation.

6. FAD MUTANTS OF PS1 ATTENUATE UPR INITIATED BY ATF6

The translocation of ATF6 fragments into the nucleus was investigated in cell with wild PS1 or mutant PS1 by immunohistochemistry. In wild-typed fibroblasts, the N-terminal fragments of ATF6 were translocated into nucleus quickly. In contrast, in homozygous PS1 knock-in fibroblasts, the translocation was delayed.

The disturbance of ATF6 pathway in homozygous PS1 knock-in fibroblasts was also confirmed by western blotting. In wild-typed fibroblasts, the appearance of 50 kDa ATF6 fragment was detected quickly after ER stress, whereas it was delayed in homozygous fibroblasts. Therefore FAD-linked PS1 mutation also attenuate the signaling pathway though ATF6.

7. DISCUSSION

To date, three ER-resident transmembrane molecules, IRE1, PERK, and ATF6, which sense the accumulation of unfolded proteins to induce UPR, have been identified. Our study showed that FAD-linked mutant PS1 disturbs all molecules. However, it remains unclear why mutant PS1 inhibits the activation of ER stress transducers. Recently, it proposed that activations of the signaling mediated ER stress transducers could be triggered by dissociation of BiP from stress transducers[17]. The dissociation leads to oligomerization of stress transducers inducing its autophosphorylation and the resultant activation of downstream signaling. If PS1 mutants form malfolded structures, BiP may constitutively bind to PS1 molecules to promote its folding. The complex formation of PS1, PERK (or Ire1) and BiP may inhibit the dissociation of BiP from PERK or Ire1 under ER stress conditions.

PS1 mutations alter the processing of APP and cause increased production of the more amyloidogenic Aβ peptide, Aβ (1–42). The ER and ER–Golgi intermediate compartment may be important sites for generation of Aβ (1–42)[22]. Interestingly, a mechanism exists in cells by which unfolded proteins are retrieved to the ER by retrograde transport, and prevented from moving to the cell surface[23]. In view of the alteration in UPR systems in cells expressing PS1 mutants, it is possible that increased levels of Aβ (1–42) in these cells might be the result of retention of the unfolded APP in the ER as a result of the impaired protein-folding system. Amounts of secreted Aβ have been shown to be reduced by transfection with GRP78 [24], a result consistent with our speculation that Aβ production may be associated with the UPR system.

In summary, our results indicate a new mechanism by which PS1 mutations may affect the sensing of ER stress. Experimental manipulation of IRE1, PERK, or eIF2 α phosphorylation or GRP78 expression might allow the development of therapeutic strategies for FAD.

REFERENCES

1. D.R.Borchelt, G. Thinakaran, C. B. Eckman, M. K. Lee, F. Davenport, T. Ratovitsky, C-M. Prada, G. Kim, S. Seekins, D. Yager, H. H. Slunt, R. Wang, M. Seeger, A. I. Leavey, S. E. Gandy, N. G. Copeland, N. A. Jenkins, D. L. Price, S. G. Younkin and S. S. Sisodia, Familial Alzheimer's disease-linked presenilin 1 variants elevate A_1-42/1-40 ratio in vitro and in vivo, *Neuron* 17, 1005–1013 (1996)
2. K.Duff, C. Eckman, C. Zehr, X. Yu, C-M. Prada, J. Perez-tur, M. Hutton, L. Buee, Y. Harigaya, D. Yager, D. Morgan, M. N. Gordon, L. Holcomb, L. Refolo, B. Zenk, J. Hardy and S. Younkin, Increased amyloid-β42(43) in brains of mice expressing mutant presenilin 1, *Nature* 383, 710–713 (1996)
3. Q.Guo, B. L. Sopher, K. Furukawa, D. G. Pham, N. Robinson, G. M. Martin and M. P. Mattson, Alzheimer's presenilin mutation sensitizes neural cells to apoptosis induced by trophic factor withdrawal and amyloid b-peptide: involvement of calcium and oxyradicals, *J. Neurosci.* 17, 4212–4222 (1997)
4. Q.Guo, W. Fu, B. L. Sopher, M. W. Miller, C. B. Ware, G. M. Martin and M. P. Mattson, Increased vulnerability of hippocampal neurons to excitotoxic necrosis in presenilin-1 mutant knock-in mice, *Nature Med.* 5, 101–106 (1999)
5. D. M. Kovacs, H. J. Fausett, K. J. Page, T-W. Kim, R. D. Moir, D. E. Merriam, R. D. Hollister, O. G. Hallmark, R. Mancini, K. M. Felsenstein, B. T. Hyman, R. E. Tanzi and W. Wasco, Alzheimer-associated presenilin 1 and 2: neuronal expression in brain and localization to intracellular membranes in mammalian cells, *Nature Med.* 2, 224–229 (1996)
6. C.Sidrauski, R.Chapman, and P.Walter, The unfolded protein response: an intracellular signalling pathway with many surprising features, *Trends Cell Biol.* 8, 245–249 (1998)
7. Y.Shi, K.M.Vattem, R.Sood, J.An, J.Liang, L.Stramm, and R.C.Wek, Identification and characterization of pancreatic eukaryotic initiation factor 2 α -subunit kinase, PEK, involved in translational control, *Mol.Cell.Biol.* 18, 7499-7509 (1998)

8. H.P.Harding, Y.Zhang, and D.Ron, Protein translation and folding are coupled by an endoplasmic-reticulum-resident kinase, *Nature* 397, 271-274 (1999)

9. J.S.Bonifacino, and A.M.Weissman, Ubiquitin and the control of protein fate in the secretory and endocytic pathways, *Annu.Rev.Cell.Dev.Biol.* 14, 19-57 (1998)

10. H.Liu, R.C.Bowes, B.van de Water, C.Sillence, J.F.Nagelkerke, and J.L.Stevens, Endoplasmic reticulum chaperones GRP78 and calreticulin prevent oxidative stress Ca^{2+} Disturbances, and cell death in renal epithelial cells, *J.Biol.Chem.* 272, 21751-21759 (1997)

11. H.P.Harding, Y.Zhang, A.Bertolotti, H.Zeng, and D.Ron, *Perk* is essential for translational regulation and cell survival during the unfolded protein response, *Mol.Cell* 5, 897-904 (2000)

12. Y.Imai, M.Soda, and R.Takahashi, Parkin suppresses unfolded protein stress-induced cell death through its E3 ubiquitin-protein ligase activity, *J.Biol.Chem.* 275, 35661-35664 (2000)

13. W.Tirasophon, A.A.Welihinda, and R.J.Kaufman, A stress response pathway from the endoplasmic reticulum to the nucleus requires a novel bifunctional protein kinase/endoribonuclease (Ire1p) in mammalian cells, *Genes Dev.*. 12, 1812–1824 (1998)

14. X-Z.Wang, H. P. Harding, Y. Zhang, E. M. Jolicoeur, M. Kuroda and D. Ron, Cloning of mammalian Ire1 reveals diversity in the ER stress responses, *EMBO J.* 17, 5708–5717 (1998)

15. K.Haze, H.Yoshida, H.Yanagi, T.Yura, and K.Mori, Mammalian transcription factor ATF6 is synthesized as a transmembrane protein and activated by proteolysis in response to endoplasmic reticulum stress, *Mol.Biol.Cell* 10, 3787-3799 (1999)

16. H.P.Harding, Y.Zhang, and D.Ron, Protein translation and folding are coupled by an endoplasmic-reticulum-resident kinase, Nature 397, 271-274 (1999)

17. A.Bertolotti, Y.Zhang, L.M.Hendershot, H.P.Harding, and D.Ron, Dynamic interaction of Bip and ER stress transducers in the unfolded-protein response, *Nat.Cell Biol.* 2, 326-332 (2000)

18. Y.Nakano, G. Kondoh, T Kudo, K. Imaizumi, M. Kato, J. Miyazaki, M. Tohyama, J. Takeda and M. Takeda, Accumulation of murine amyloid beta42 with a gene-dosage dependent manner in PS1 'knock-in' mice, *Eur. J. Neurosci.* 11, 2577–2581 (1999)

19. C.Sidrauski, R.Chapman, and P. Walter, The unfolded protein response: an intracellular signalling pathway with many surprising features. *Trends Cell Biol.* 8, 245–249 (1998)

20. W.Tirasophon, A.A. Welihinda, and R.J.Kaufman, A stress response pathway from the endoplasmic reticulum to the nucleus requires a novel bifunctional protein kinase/endoribonuclease (Ire1p) in mammalian cells. *Genes Dev.* 12, 1812–1824 (1998)

21. X.-Z.Wang, H.P.Harding, Y.Zhang, E.M.Jolicoeur, M.Kuroda, and D.Ron, Cloning of mammalian Ire1 reveals diversity in the ER stress responses. *EMBO J.* 17, 5708–5717 (1998).

22. D.G.Cook, M.S.Forman, J.C.Sung, S.Leight, D.L.Kolson, T.Iwatsubo, V.M.-Y. Lee, and R.W.Doms Alzheimer's Aβ (1-42) is generated in the endoplasmic reticulum/intermediate compartment of NT2N cells. *Nature Med.* 3, 1021–1023 (1997)

23. C.Hammond, and Helenius, Quality control in the secretory pathway: retention of a misfolded viral membrane glycoprotein involves cycling between the ER, intermediate compartment, and Golgi apparatus. *J. Cell. Biol.* 126, 41–52 (1994)

24. Y.Yang, , R.S.Turner, and J.R.Gaut, The chaperone Bip/GRP78 binds to amyloid precursor protein and decrease Aβ40 and Aβ42 secretion. *J. Biol. Chem.* 273, 25552–25555 (1998)

PS-1 IS TRANSPORTED FROM THE MOTO-NEURONS TO THEIR AXON TERMINALS

Peter Kasa, Henrietta Papp, and Magdolna Pakaski[*]

1. INTRODUCTION

Missense mutations in the presenilin-1 (PS-1) and presenilin-2 (PS-2) gene are known to play a significant role in early-onset familial Alzheimer's disease (AD). The normal physiological functions of these proteins are still incompletely understood, although evidence is accumulating that they are present mainly in the endoplasmic reticulum and Golgi compartments[1, 2]. It has been suggested that PS-1 can not be found in significant amounts beyond the trans-Golgi network (e.g. in the axons)[3-4]. Others authors, however, have demonstrated the presence of PS-1 in the axons,[5-8] axon terminals[6] and at synaptic contacts[9]. It has been suggested that PS-1 fragments may exit from the cell body and reach the synaptic plasma membranes[10]. The transport of PS-1 fragments in the axons has been postulated, but direct morphological evidence its transport is still lacking.

Our aim in this work was to demonstrate the axonal migration of PS-1 by immunohistochemical means in the motoneurons of the spinal cord, in the sensory ganglion cells of the spinal ganglia and in the sciatic nerve of adult rat.

2. MATERIALS AND METHODS

Adult male Sprague-Dawley rats (4-5 months old, 400 g) were used in the study. The left sciatic nerve of the rats was exposed and a double-ligation procedure was applied. The right sciatic nerve was also exposed and a silk thread was placed loosely around the nerve in the same location in the sham controls. The rats were allowed to recover for 6, 12 or 24 h prior to sacrifice. After the different time intervals, the rats were perfused with

[*] Peter Kasa, Henrietta Papp and Magdolna Pakaski, Alzheimer's Disease Research Centre, Department of Psychiatry, University of Szeged, H-6720 Szeged, Hungary

4% paraformaldehyde solution. The sciatic nerves from the control and ligated animals were cut at a thickness of 15 μm on a cryostat and were mounted on glass slides for immunohistochemical staining. Separate sections were incubated with 0.1 M PBS containing one of the primary antibodies (goat anti-PS-1 polyclonal antibody, AB1575 [Chemicon], or rabbit anti-PS-1 polyclonal antibody, AB5308 [Chemicon]) for 2 days at room temperature. This was followed by incubation in normal serum for 30 min, in biotin-SP-conjugated secondary antibody (1:500) for 90 min, and then in streptavidin-HRP (1:1000) for 90 min. The peroxidase complex was visualized by incubating the sections with 0.05 M Tris-HCl (pH 7.6) containing 0.05% DAB, 0.1% $NiCl_2$ and 0.005% H_2O_2, resulting in a bluish-black colour. In parallel incubations, the primary antibody was omitted, and the specificity of the immunoreactivity was tested. The sections were dehydrated and coverslipped. Photos were taken under normal light field microscopy.

3. RESULTS

PS-1 immunoreactivity was revealed in the cytoplasm of motoneurons (Fig. 1A) and in some axon fibres leaving the grey matter and entering the white matter in the spinal cord of adult rat. Similarly, ganglion cells of different sizes were stained for PS-1 in the spinal ganglia (Fig. 1B). In the control sciatic nerve, PS-1 immunoreactivity was revealed in a discontinuous manner in the axons of the myelinated nerve fibres (Fig. 1C), while no staining appeared in samples where the primary antibodies were omitted (Fig. 1D). After double-ligation, the staining intensity was already increased after 6 h in the segments and reached its maximum at 24 h above the upper ligature (Fig. 1E). Between 6 and 24 h, however, the PS-1 staining gradually increased further only on the proximal side, while the immunoreactivity did not increase further in the distal segment after 6 h (Fig. 1F). Between the two ligatures, the immunostaining was observed only in the cytoplasm of the Schwann cells (not documented) after 24 h.

4. DISCUSSION

In some previous studies on the immunocytochemical localization of PS-1, it was suggested that it is restricted to the cell body and dendrites. It was not found in significant amounts in the axons[3]. Other studies, however, demonstrated the existence of PS-1 in the axons[5-8]. Our results support the data indicating that PS-1 immunostaining may be present not only in the cell bodies and dendrites, but also in the axons of some neurons. All of these results support the suggestion that PS-1 may exit from the cell body and reach the cell surface plasma membrane or the synaptic regions[9, 10] in the neurons.

We used double ligatures, which allow a more accurate assessment of bidirectional transport[11]. Such an experimental set-up clearly reveals that the accumulation of PS-1 is due to bidirectional axonal movement, rather than to accumulation at a site of axonal damage. The presence of PS-1 in the axons, and their accumulation in the segments above the proximal ligature and in the segments below the distal ligature, indicate that this protein may be cycling at all times in the sciatic nerve of the rat.

Our observations demonstrate that there are at least three possible sources of PS-1 in

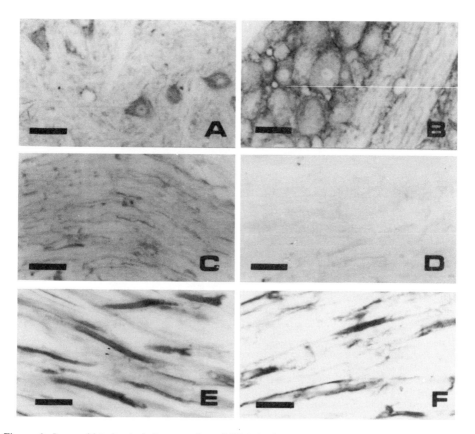

Figure 1. Immunohistochemical demonstration of PS-1 in the various tissue samples, using polyclonal antibodies AB1575 and AB5308. The two antibodies resulted in similar immunoreactivity. PS-1 staining appears in the cytoplasm of motoneurons (A) and in the ganglionic cells of different sizes (B). Immunoreactivity appeared in a discontinuous manner in the axons of the control sciatic nerve bundle (C). No immunoreactivity was detected in the sciatic nerve samples without primary antibody (D). A pronounced accumulation of PS-1 occurred above the proximal ligature after 24 h of double ligation, (E, arrows) and below the lower ligature after 6 h (F). Scale bar = 50 (A, B), 65 μm (C, D), 25 μm (E, F).

the axons of the sciatic nerve: a) in the motoric axons, PS-1 may originate from the ventral horn neurons present in the grey matter of the spinal cord, while in the sensory fibres, PS-1 may be transported from the spinal ganglia, and b) it may be taken up from the PS-1-positive Schwann cells surrounding the axons in the sciatic nerve, and c) from the postsynaptic muscle tissues. Intensive PS-1 immunoreactivity has already been demonstrated in the large neurons of the grey matter of the spinal cord,[5] and in the dorsal root ganglion cells of the mouse. This and our own results lend strong support to our hypothesis that the motoneurons and the sensory ganglion cells are the primary sites of the source of PS-1 in the sensory and motoric axons in the sciatic nerve bundle.

5. SUMMARY

The axonal transport of PS-1 was investigated in the sciatic nerve of rat. The immuno-histochemical results demonstrate a bidirectional transport of this protein. After double ligation of the sciatic nerve, a progressive accumulation of PS-1 immunoreactivity occurred above the upper ligature and to a lesser extent below the lower ligature. It is concluded that PS-1 may leave the trans-Golgi network and be found in the axons and axon terminals of the various neurons.

6. ACKNOWLEDGEMENTS

This work was supported by OTKA (T022683, T030339 and T032458), ETT (T-11/011/2000) and a Széchenyi Professorship to P.K.

REFERENCES

1. D. M. Kovacs, H. J. Fausett, K. J. Page, T. W. Kim, R. D. Moir, D. E. Merriam, R. D. Hollister, O. G. Hallmark, R. Mancini, K. M. Felsenstein, B. T. Hyman, R. E. Tanzi and W. Wasco, Alzheimer-associated presenilins 1 and 2: neuronal expression in brain and localization to intracellular membranes in mammalian cells, *Nature Med.* **2**(2), 224-229 (1996).
2. J. G. Culvenor, F. Maher, G. Evin, F. Malchiodi-Albedi, R. Cappai, J. R. Underwood, J. B. Davis, E. H. Karran, G. W. Roberts, K. Beyreuther and C. L. Masters, Alzheimer's disease-associated presenilin 1 in neuronal cells: Evidence for localization to the endoplasmic reticulum-Golgi intermediate compartment, *J. Neurosci. Res.* **49**(6), 719-731 (1997).
3. W. G. Annaert, L. Levesque, K. Craessaerts, I. Dierinck, G. Snellings, D. Westaway, P. S. George-Hyslop, B. Cordell, P. Fraser and B. De Strooper, Presenilin 1 controls γ-secretase processing of amyloid precursor protein in pre-Golgi compartments of hippocampal neurons, *J. Cell Biol.* **147**(2), 277-294 (1999).
4. D. G. Cook, J. C. Sung, T. E. Golde, K. M. Felsenstein, B. S. Wojczyk, R. E. Tanzi, J. Q. Trojanowski, V. M.-Y. Lee and R. W. Doms, Expression and analysis of presenilin 1 in a human neuronal system: Localization in cell bodies and dendrites, *Proc. Nat. Acad. Sci. (USA)* **93**(17), 9223-9228 (1996).
5. G. A. Elder, N. Tezapsidis, J. Carter, J. Shioi, C. Bouras, H.-C. Li, J. M. Johnston, S. Efthimiopoulos, V. L. Friedrich Jr. and N. K. Robakis, Identification and neuron specific expression of the S182/presenilin I protein in human and rodent brains, *J. Neurosci. Res.* **45**(3), 308-320 (1996).
6. J. Busciglio, H. Hartmann, A. Lorenzo, C. Wong, K. Baumann, B. Sommer, M. Staufenbiel and B. A. Yankner, Neuronal localization of presenilin-1 and association with amyloid plaques and neurofibrillary tangles in Alzheimer's disease, *J. Neurosci.* **17**(13), 5101-5107 (1997).
7. I. Kovacs, I. Torok, J. Zombori and P. Kasa, Immunohistochemical localization of presenilin-1 in hippocampus and entorhinal cortex of human brain samples, Clin. Neurosci. **51**(1), 56-57 (1998).
8. P. E. Fraser, D-S. Yang, G. Yu, L. Lévesque, M. Nishimura, S. Arawaka, L. C. Serpell, E. Rogaeva and P. S. George-Hyslop, Presenilin structure, function and role in Alzheimer disease, *Biochim. Biophys. Acta* **1502**(1), 1-15 (2000).
9. A. Georgakopoulos, P. Marambaud, S. Efthimiopoulos, J. Shioi, W. Cui, H. C. Li, M. Schutte, R. Gordon, G. R. Holstein, G. Martinelli, P. Mehta, V. L. Friedrich Jr. and N. K. Robakis, Presenilin-1 forms complexes with the cadherin/catenin cell-cell adhesion system and is recruited to intercellular and synaptic contacts, *Mol. Cell.* **4**(6), 893-902 (1999).
10. D. Beher, C. Elle, J. Underwood, J. B. Davis, R. Ward, E. Karran, C. L. Masters, K. Beyreuther and G. Multhaup, Proteolytic fragments of Alzheimer's disease-associated presenilin 1 are present in synaptic organelles and growth cone membranes of rat brain, *J. Neurochem.* **72**(4), 1564-1573 (1999).
11. P. Kasa, H. Papp and M. Pakaski, Presenilin-1 and its N-terminal and C-terminal fragments are transported in the sciatic nerve of rat, *Brain Res.* (in press, 2001).

CALSENILIN-PRESENILIN INTERACTION IN ALZHEIMER'S DISEASE

Eun-Kyoung Choi, Joseph D. Buxbaum, Wilma Wasco[*]

1. INTRODUCTION

Most early-onset familial Alzheimer's disease (AD) cases are caused by mutations in the highly related genes presenilin 1 (PS1) and presenilin 2 (PS2). Presenilin mutations produce increases in amyloid β-peptide (Aβ) formation and apoptosis in many experimental systems. A cDNA (ALG-3) encoding the last 103 amino acids of PS2 has been identified as a potent inhibitor of apoptosis[1]. Using this PS2 domain in the yeast two-hybrid system, we have identified a neuronal protein, *calsenilin*, which binds calcium and presenilin[2]. Calcium-binding proteins serve as calcium sensors that transduce calcium signals via specific interactions with intracellular target protein. Here, we will review some of findings from studies of the molecular characterization of calsenilin and its interaction with presenilin.

2. PRESENILIN-INTERACTING PROTEINS

To date, more than 60 mutations in PS1 and at least 2 mutations in PS2 have been genetically linked to early-onset FAD[3-5]. These mutations result in the altered processing of amyloid β precursor protein (APP) and lead to increases in Aβ, which is derived from APP and is the main component of amyloid plaques[6]. The presenilins undergo constitutive endoproteolysis in the brain as well as in cultured cells and generate a stable heterodimers complex composed of the N-and C-terminal fragment (NTF and CTF) that may involved in the biological function of presenilins[7]. The presenilins have also been shown to be substrates for cleavage by caspase-3-like proteases at a site distal to the regulated cleavage site[8, 9]. Notably, FAD mutations in the presenilins are associated with

[*] Eun-Kyoung Choi and Wilma Wasco, Genetics and Aging Unit, Department of Neurology, Massachusetts General Hospital, Harvard Medical School, Charlestown, MA 02129. Joseph D. Buxbaum, Department of Psychiatry and Neurobiology, Mount Sinai School of Medicine, New York, NY 10029.

Mapping the Progress of Alzheimer's and Parkinson's Disease
Edited by Mizuno *et al.*, Kluwer Academic/Plenum Publishers, 2002

increased levels of caspase-derived fragments. Recent findings demonstrate that the phosphorylation of PS2 controls its cleavage by caspase[10], and suggest that cleavage of PS2 is a regulated biological event that occurs physiologically in the absence of FAD mutations.

The normal function of the presenilins is not clear, however roles in membrane trafficking, APP processing, Notch signaling, neuronal plasticity, cell adhesion, the regulation of ER calcium homeostasis, the unfolded protein response and programmed cell death, have all been suggested[5, 6]. In an effort to elucidate the functional role of the presenilins, a number of presenilin-interacting proteins have been identified. It has been reported that presenilins can interact with members of the armadillo family (β-catenin, NPRAP and p0071), GSK-3β, filamins (actin-binding protein 280, ABP280/filamin homologue 1, Fh 1), μ-calpain, calmyrin, APP, Go, Bcl-2 family proteins, QM/Jif1, rab11, sorcin, and calsenilin[2, 5]. Most interactors, including β-catenin, calmyrin, Bcl-X_L, rab11, sorcin, and GSK-3β, bind to the large hydrophilic loop domain of the presenilins, while calsenilin and Go interact with the C-terminal domain. The interaction of the armadillo proteins, GSK-3β and G_O appears to be specific for PS1, whereas μ-calpain, calmyrin and sorcin interact only with PS2. In contrast, only calsenilin, filamins, Bcl-X_L, and APP have the ability to interact with both PS1 and PS2. The characterization of interactors that are specific for either PS1 or PS2 may aid in the elucidation of distinct biological roles for the two proteins, while the characterization of proteins that interact with both presenilins (such as calsenilin) may provide information about common functions in both the normal and diseased brain.

3. IDENTIFICATION OF CALSENILIN

Increasing evidence indicates PS2 is involved in the apoptotic process. ALG-3, a mouse cDNA encoding the last 103 amino acids of PS2 was identified as a potent inhibitor of apoptosis[1]. Overexpression of full-length PS2 in ALG-3 transfected cells reconstitutes sensitivity to apoptosis, indicating that the artificial polypeptide acts as a dominant negative mutant of PS2. It has been argued that the effects of the ALG-3 cDNA product are best explained as reflecting a competition of the ALG-3 cDNA product with full-length PS2 for binding with as yet unidentified protein(s); thus the expression of ALG-3 rescues cells from apoptosis by competing for such as interacting protein.

We have carried out the initial characterization of a protein that might fulfill such a role; a novel calcium-binding protein that interacts with the C-terminus of PS2 which we have termed *calsenilin*[2]. Calsenilin was isolated by using the yeast two-hybrid system and employing the last 40 amino acids of PS2 as the bait. Sequencing of this cDNA clone indicates that it is approximately 3.0 kb in length and contains an open reading frame of ~1.0 kb. Northern blot analysis of RNA derived from a series of human tissues indicates that mRNA for calsenilin is expressed primarily in the brain, and we have used *in situ* hybridization to show that the mRNA is primarily expressed in neuron. Database analysis indicates that the isolated cDNA displays significant homology with members of the recoverin/neuronal calcium sensor (NCS) family of calcium-binding proteins[11]. Like other members of the recoverin/NCS family, calsenilin contains four calcium-binding motifs (EF hands). This family also includes hippocalcin, NCS-1, neurocalcin, frequenin, visinin, visinin-like protein, S-modulin, and guanylyl cyclase activating protein (GCAP)[12]. Although the cellular function of the majority of proteins in this family remain unknown, it is clear that recoverin acts to inhibit rhodopsin kinase and that the GCAPs regulate photoreceptor guanylyl cyclases. In addition, frequenin has been shown to be a positive regulator of neurotransmitter release in *Xenopus* and *Drosophila*, while NCS-1, the mammalian homologue of frequenin, regulates neurosecretion from dense core granules.

Calsenilin appears to be a unique member of the family since it has a novel N-terminal domain that is absent in the other proteins. Since our initial report, calsenilin has also been identified as a calcium-dependent transcriptional repressor that was termed downstream-regulatory-element antagonist modulator (DREAM)[13]. Interestingly, a recent study identifying calsenilin as a A-type voltage-gated potassium (Kv) channel-interacting protein (KChIP3) also reports the identification of two novel proteins (KChIP1 and 2) that are highly homologous to calsenilin, one that contains the novel N-terminal domain and one that does not[14]. These three proteins are encoded by the same unique gene[15]. More recently, splice variants of KChIP2 and calsenilin-like protein (CALP/KChIP4) were discovered (personal communication with Drs. T. Tomita and T. Iwatsubo)[16]. All forms of calsenilin/DREAM/KChIPs have an extended and variable N-terminal domain.

4. MOLECULAR CHARACTERIZATION OF CALSENILIN AND ITS INTERACTION WITH PRESENILIN

Based on our previous data, calsenilin, which runs at ~35-kDa, can interact with PS2 in mammalian cells. Given the high degree of conservation between the last 40 amino acids of PS1 and PS2, we have shown that full-length PS1 also interacts with calsenilin. We have used immunofluorescence analysis to demonstrate that calsenilin is localized in the cytoplasm of transiently transfected cells and that co-expression with PS2 results in a redistribution of the calsenilin expression pattern to resemble the ER/Golgi localization of PS2[2]. The redistribution of calsenilin in the presence of PS2 is further evidence of an *in vivo* interaction and suggests that there might be functional consequences of the interaction. Interestingly, the levels of the caspase-derived 20-kDa PS2 CTF are increased in the presence of calsenilin suggesting that calsenilin either preferentially stabilizes the levels of this fragment or increases its formation.

Although we have successfully detected an endogenous presenilin-calsenilin complex in brain extract[†], it is difficult to detect this complex in cultured cells because endogenous levels of calsenilin levels are relatively low in these cells. To overcome this detection problem, we have used cultured mammalian cells that transiently or stably express calsenilin to extend the characterization of calsenilin and of the calsenilin-PS2 interaction. We have found that calsenilin has the ability to interact with the endogenous 25-kDa CTF that is a product of regulated endoproteolytic cleavage of PS2 and that the presence of the N141I PS2 mutation does not significantly alter the interaction of calsenilin with PS2[17]. Interestingly, when the 25-kDa and the 20-kDa PS2 CTF are both

---

[†] N. F. Zaidi, O. Berezovska, E. K. Choi, H. Chan, C. Lilliehook, B. Hyman, J. D. Buxbaum, and W. Wasco. *Manuscript submitted* (2001).

present, calsenilin preferentially interacts with the 20-kDa CTF. Increases in the 20-kDa fragment are associated with the presence of FAD-associated mutations. However, the finding that the production of the 20-kDa fragment is regulated by the phosphorylation of PS2 suggests that it is a regulated physiological event that also occurs in the absence of the FAD-associated mutations in PS2. Although the molecular mechanism for the preferential interaction between calsenilin and the 20-kDa PS2 CTF remains unclear, this observation may still provide insight into potential effects of calsenilin on the presenilin mutation-associated increases in Aβ.

In addition, we have found that calsenilin itself is a substrate for caspase cleavage and we have used site-directed mutagenesis to demonstrate that this cleavage occurs at a caspase-3 DXXD consensus cleavage sequence, ^{61}DSSD^{64}. Notably, the cleavage of calsenilin by caspase-3 occurs at a site that separates the calcium-binding domain that is conserved in all members of the recoverin/NCS family from the novel N-terminal domain that is specific for calsenilin and one of its recently discovered homologues. Although the function of this N-terminal domain remains unknown, it is tempting to speculate that is involved in the interaction of calsenilin with target proteins such as the presenilins. Caspase cleavage may serve to separate the interacting domain from the calcium responsive domain of the molecule, severing either calcium regulation of the target protein or an interaction with a second molecule that is regulated by the calcium-binding domain.

5. CELLULAR FUNCTIONS OF CALSENILIN/DREAM/KChIPS

Perhaps the most essential physiological roles performed by calcium-binding proteins are to act as calcium sensors that transduce calcium signals via specific interactions with intracellular target protein. Interestingly, our biochemical studies indicate that the association of calsenilin with the presenilins is independent of calcium[17]. Likewise, the interaction of KChIP3 with the potassium channel and the association of NCS-1 with membranes are also calcium-independent[14, 18]. In addition, calcium does appear to be required for the modulatory effects of KChIP3 on the activity of the potassium channel[14]. Whether calcium is similarly required for modulation of presenilin function by calsenilin awaits conclusive elucidation of presenilin function. Although the functional significance of calcium binding to calsenilin remains unclear, we have recently reported that the calsenilin can reverse the potentially pathogenic effects that mutant presenilins have on calcium signals evoked by inositol 1,4,5-triphosphate enhancement[19]. More recently, it has been shown that calsenilin is involved in apoptosis and Aβ formation[20] and that calsenilin regulates calcium signaling and calcium-regulated apoptotic pathway[‡].

6. CONCLUSIONS

Over the years a number of calcium-binding proteins have been identified, reflecting the importance of calcium and its functional role in the cell. Previously, we described the

---

[‡] C. Lilliehook, S. Chan, E. K. Choi, N. F. Zaidi, W. Wasco, M. Mattson, and J. D. Buxbaum. *Manuscript submitted* (2001).

initial identification and characterization of calsenilin, a novel calcium-binding protein that interacts with the C-terminus of the PS1 and PS2 holoproteins. Recent work has focused on the molecular characterization of calsenilin and its interaction with presenilins. Further investigation of the physiological and/or pathological roles of calsenilin may help to understand the functional roles of presenilins in both the normal and diseased brain.

REFERENCES

1. P. Vito, B. Wolozin, J. K. Ganjei, K. Iwasaki, E. Lacana, and L. D'Adamio, Requirement of the familial Alzheimer's disease gene PS2 for apoptosis. Opposing effect of ALG-3. *J. Biol. Chem.* **271**, 31025-31028 (1996).
2. J. D. Buxbaum, E. K. Choi, Y. Luo, C. Lilliehook, A. C. Crowley, D. E. Merriam, and W. Wasco, Calsenilin: a calcium-binding protein that interacts with the presenilins and regulates the levels of a presenilin fragment, *Nat. Med.* **4**, 1177-1181 (1998).
3. J. Hardy, Amyloid, the presenilins and Alzheimer's disease, *Trends Neurosci.* **20**, 154-159 (1997).
4. P. E. Fraser, D. S. Yang, G. Yu, L. Levesque, M. Nishimura, A. Arawaka, L. C. Serpell, E. Rogaeva, and P. St. George-Hyslop, Presenilin structure, function and role in Alzheimer disease, *Biochim. Biophys. Acta* **1502**, 1-15 (2000).
5. G. Van Gassen, W. Annaert, and C. Van Broeckhoven, Binding partners of Alzheimer's disease proteins: are they physiologically relevant?, *Neurobiol. Dis.* **7**, 135-151 (2000).
6. D. J. Selkoe, Alzheimer's disease: genes, proteins, and therapy, *Physiol. Rev.* **81**, 741-766 (2001).
7. G. Thinakaran, J. B. Regard, C. M. Bouton, C. L. Harris, D. L. Price, D. R. Borchelt, and S. S. Sisodia, Stable association of presenilin derivatives and absence of presenilin interactions with APP, *Neurobiol. Dis.* **4**, 438-453 (1998).
8. T. W. Kim, W. H. Pettingell, Y. K. Jung, D. M. Kovacs, and R. E. Tanzi, Alternative cleavage of Alzheimer-associated presenilins during apoptosis by a caspase-3 family protease, *Science*, **277**, 373-376 (1997).
9. H. Loetscher, U. Deuschle, M. Brockhaus, D. Reinhardt, P. Nelboeck, J. Mous, J. Grunberg, C. Haass, and H. Jacobsen, Presenilins are processed by caspase-type proteases, *J. Biol. Chem.* **272**, 20655-20659 (1997).
10. J. Walter, A. Schindzielorz, J. Grunberg, and C. Haass, Phosphorylation of presenilin-2 regulates its cleavage by caspases and retards progression of apoptosis, *Proc. Natl. Acad. Sci.USA.* **96**, 1391-1396 (1999).
11. A. M. Dizhoor, S. Ray, S. Kumar, G. Niemi, M. Spencer, D. Brolley, K. A. Walsh, P. P. Philipov, J. B. Hurley, and L. Stryer, Recoverin: a calcium sensitive activator of retinal rod guanylate cyclase, Science, **251**, 915-918 (1991).
12. R. D. Burgoyne, and J. L. Weiss, The neuronal calcium sensor family of Ca^{2+}-binding proteins, *Biochem. J.* **353**, 1-12 (2001).
13. A. M. Carrion, W. A. Link, F. Ledo, B. Mellstrom, and J. R. Naranjo, DREAM is a Ca^{2+}-regulated transcriptional repressor, *Nature* **398**, 80-84 (1999).
14. W. F. An, M. R. Bowlby, M. Betty, J. Cao, H. P. Ling, G. Mendoza, J. W. Hinson, K. I. Mattsson, B. W. Strassle, J. S. Trimmer, and K. J. Rhodes, Modulation of A-type potassium channels by a family of calcium sensors, *Nature* **403**, 553-556 (2000).
15. F. Spreafico, J. J. Barski, C. Farina, and M. Meyer, Mouse DREAM/Calsenilin/KChIPs: gene structure, coding potential, and expression, *Mol. Cell. Neurosci.* **17**, 1-16 (2001).
16. S. Ohya, Y. Morohashi, K. Muraki, T. Tomita, M. Watanabe, T. Iwatsubo, and Y. Imaizumi, Molecular cloning and expression of the novel splice variants of K^+ channel-interacting protein 2, *Biochem. Biophys. Res. Commun.* **282**, 96-102 (2001).
17. E. K. Choi, N. F. Zaidi, J. S. Miller, A. C. Crowley, D. E. Merriam, C. Lilliehook, J. D. Buxbaum, and W. Wasco, Calsenilin is a substrate for caspase-3 that preferentially interacts with the familial Alzheimer's disease-associated C-terminal fragment of presenilin 2, *J. Biol. Chem.* **276**, 19197-19204 (2001).
18. B. W. McFerran, J. L. Weiss, and R. D. Burgoyne, Neuronal Ca^{2+} sensor 1. Characterization of the myristoylated protein, its cellular effects in permeabilized adrenal chromaffin cells, Ca^{2+}-independent membrane association, and interaction with binding proteins, suggesting a role in rapid Ca^{2+} signal transduction, *J. Biol. Chem.* **274**, 30258-30265 (1999).
19. M. A. Leissring, T. R. Yamasaki, W. Wasco, J. D. Buxbaum, I. Parker, and F. M. LaFerla, Calsenilin reverses presenilin-mediated enhancement of calcium signaling, *Proc. Natl. Acad. Sci. USA.* **97**, 8590-8593 (2000).
20. D. G. Jo, M. J. Kim, Y. H. Choi, I. K. Kim, Y. H. Song, H. N. Woo, C. W. Chung, and Y. K. Jung, Pro-apoptotic function of calsenilin/DREAM/KChIP3, *FASEB J.* **15**, 589-591 (2001).

Acknowledgments—We thank N. F. Zaidi, T. L. Tekirian, C. Lilliehook, T. W. Kim, G. Tesco, R. E. Tanzi, and S. Guénette for helpful discussions; and J. S. Miller and E. J. Choi for technical assistance. This work was supported by grants from the Korea Research Foundation (to E. K. C.), the National Institutes of Health (NS35975 and AG16361 (W.W); AG15801 and AG05138 (J. D. B.), the Alzheimer Association (W. W. and J. D. B.). W. W. is a Pew Biomedical Scholar.

PRESENILIN-1 FUNCTION IN THE ADULT BRAIN

Jie Shen*

Mutations in Presenilin-1 (PS1) are the most common cause of early-onset familial Alzheimer's disease (FAD). Accumulation and deposition of β-amyloid (Aβ) peptides in the cerebral cortex is an early and central process in the pathogenesis of AD. The Aβ peptides are generated from the amyloid precursor protein (APP) as a result of sequential proteolytic cleavages by β- and γ-secretases, which are therefore prime targets for therapeutic intervention. Recent studies showed that Presenilin-1 (PS1) is required for γ-secretase activity and Aβ generation in cultured cells, raising the possibility that targeting PS1 may be an effective therapeutic strategy for AD[1]. Our previous studies of PS1-/- mice, however, revealed that elimination of PS1 expression leads to pleiotropic effects during embryonic development, including impaired neurogenesis and reduced Notch signalling[2, 3]. The perinatal lethality of PS1-/- mice also precluded the investigation of PS1 function in Aβ generation and amyloid plaque formation in the adult brain where these pathological features develop in AD patients.

To investigate the role of PS1 in the adult brain, we employed the Cre/loxP recombination system to generate a floxed PS1 mouse, which was then bred to the CaMKII-Cre transgenic mouse to generate a viable PS1 conditional knockout (cKO) mouse[4, 5]. In the PS1 cKO mouse, PS1 expression is selectively eliminated in the adult forebrain[5]. Western analysis showed that the level of PS1 C-terminal fragment (CTF) in the cortex (neocortex and hippocampus) is slightly reduced at postnatal day 18 (P18) and further reduced at P22. By the ages of 6 weeks and 6 months, very low levels of PS1 CTF were detected in the cortex of cKO mice, while the level of PS1 CTF was unchanged in the cerebellum and brain stem. These results indicate a selective elimination of PS1 expression in the cortex of PS1 cKO mice beginning at P18. The residual amount of PS1 observed in the cortex of cKO mice is

* Jie Shen, Center for Neurological Disease, Brigham and Women's Hospital, Program in Neuroscience, Harvard Medical School, Boston, MA 02115, USA

due to its expression in glial cell types[6] and possibly a small population of neurons lacking Cre expression. In contrast to *PS1-/-* mice, the *PS1* cKO mouse is viable with no obvious phenotypic abnormalities. Nissl staining of the brain sections of cKO mice revealed no gross abnormalities. Despite large numbers of reports that have suggested a role for PS1 in apoptosis using various cell culture systems, we failed to detect an increase in apoptosis in the cerebral cortex of cKO mice. Bisbenzimide staining and TUNEL analysis of the cortex of control and cKO mice showed no significant difference in the low number of apoptotic cells detected for each genotype.

To investigate the role of PS1 in APP processing in the adult cerebral cortex, we used a panel of well-characterized antibodies against APP to examine levels of the APP ectodomains (α–APPs and β–APPs) and membrane bound CTFs (α-CTF and β-CTF) generated by α- and β- secretase cleavage. Western and IP-Western analyses showed that the levels of the full length APP, α–APPs and β–APPs are unchanged in the adult cortex lacking PS1. These results indicate that PS1 is not involved in the regulation of α- and β- secretase cleavage of APP. Western analysis of the APP CTFs showed that the CTFs accumulate in the cortex of cKO mice in an age-dependent manner. At P18, the level of the CTFs is slightly higher in the cKO cortex relative to the littermate control. At P22, levels of the CTFs are further increased, consistent with the reduction in the level of PS1. At the ages of 6 weeks and 6 months, there is a striking progressive accumulation of the CTFs in the cKO cortex. Five distinct CTF species were detected in the cortex of cKO and littermate control mice, as previously reported[7], while only three APP CTF species were detected in CHO cells transfected with the human wild-type *APP* cDNA. The additional APP CTF species in the brain are phosphorylated forms of the α- and β-CTFs, as they disappear following phosphatase treatment. The three CTF species in CHO cells and mouse cortex following phosphatase treatment represent the α-CTF (C83), which is cleaved at +17 (Leu) of Aβ, and the β-CTFs (C99 and C89), which are cleaved at +1 (Asp) and +11 (Glu) of Aβ, respectively[8-10]. The levels of C83 and C89 are increased approximately 30-fold in the cKO cortex, relative to the control, while the increase in the level of C99 is approximately 3-fold. Since elimination of PS1 expression does not affect α- and β-secretase cleavages, these results indicate that PS1 is required for γ-secretase activity, and that lack of PS1 results in accumulation of γ-secretase substrates.

To investigate PS1 function in the generation of Aβ peptides in the adult cerebral cortex, we measured levels of Aβ40 and Aβ42 in the cortex of the cKO and the littermate control mice at the ages of 6 weeks and 6 months by enzyme-linked immunosorbent assay (ELISA). Well-characterized antibodies specific for Aβ40 and Aβ42 - BA27 and BC05, respectively, were used for the assay[11, 12]. We found that levels of Aβ40 are reduced in the cortex of *PS1* cKO mice at both ages examined. Surprisingly, we detected higher levels of Aβ42 in the cortex of cKO mice than in the control, and the increase becomes more substantial by the age of 6 months. Given the pronounced age-dependent accumulation of the CTFs (as much as 30-fold by 6 months) in the cKO cortex, we then examined whether the antibodies used for ELISA detection, BA27 and BC05, crossreact with the CTFs by Western, IP-Western, and IP-ELISA. Immunoprecipitation using either BA27 or BC05 followed by Western analysis, however, revealed that BC05 crossreacts with the APP CTFs, while BA27 does not, suggesting that the levels of Aβ42 detected by BC05 may be artificially elevated in the cKO cortex due to the presence of high levels of the APP CTFs. To measure the levels of Aβ40 and Aβ42 accurately, we removed the APP CTFs from the lysate by immunodepletion, and then determined the level of Aβ40 and Aβ42 in the supernatant by ELISA. Using this method, we found that the level of both Aβ40 and Aβ42 is reduced in the cKO cortex relative to the control. Together, our results showed that elimination of PS1 expression in most neurons of the adult mouse cerebral cortex markedly reduces the production of Aβ peptides, supporting the notion that targeting PS1 is an effective strategy for anti-amyloidogenic therapy in AD.

Our previous studies of *PS1-/-* mice showed that PS1 is involved in the regulation of the Notch signalling pathway during neural development[3, 13]. In the developing brain of both *PS1-/-* mice and *Notch1-/-* mice, the expression of the Notch downstream target gene *Hes5* is reduced and the expression of *Dll1* is increased, while *Hes1* expression is unaffected[3, 14]. PS1 appears to regulate Notch signalling at the level of posttranslational activation, since the level of Notch1 mRNA and protein is unchanged in the absence of PS1[3]. More specifically, PS1 is involved in the proteolytic production of NICD, based on the *in vitro* findings that NICD production is reduced in cultured *PS1-/-* cells transfected with truncated Notch1 constructs[13, 15]. However, a direct *in vivo* assessment of PS1 function in NICD generation has not been possible because endogenous levels of NICD fall below the limits of detection with currently available methods. We therefore investigated PS1 function in Notch signalling in the adult cerebral cortex by examining expression of the Notch downstream target genes, *Hes1*, *Hes5* and *Dll1*. Northern analysis of poly(A)+ RNA derived from the cortex of cKO and littermate control mice at the age of 6 weeks showed similar levels of *Hes1*, *Hes5* and *Dll1* expression. Quantitative comparison of the level of the *Hes1*, *Hes5* and *Dll1* transcripts in the cortex of multiple mice (n=4-5) from multiple experiments (n=3) confirmed normal expression of these Notch downstream target genes in the cKO cortex. These results demonstrate that the regulation of the Notch downstream effector genes differs in the embryonic and adult brain with respect to its dependence on PS1 function.

To investigate whether PS1 is involved in the modulation of synaptic function in the adult brain, we examined the *PS1* cKO mice for deficits in synaptic transmission and plasticity. Long-term potentiation (LTP) and long-term depression (LTD) in the CA1 region of the hippocampus are the best understood models of the synaptic modifications thought to be involved in learning and memory. Previous studies have shown that LTP is enhanced in transgenic mice overexpressing FAD-linked mutant PS1[16, 17]. We investigated the impact of PS1 ablation in the adult forebrain on these forms of synaptic plasticity in the Schaeffer collateral/commissural pathway of acute hippocampal slices. We first evaluated the impact of PS1 inactivation on basal synaptic transmission. There is no significant difference in the input/output curve of the *PS1* cKO and control mice ($p>0.5$, t-test). In addition, the magnitude of the maximal response is similar in the cKO (1.286 ± 1.96 Volts/sec) and in the control (1.376 ± 0.37 Volts/sec). To assess possible effects on the pre-synaptic contribution to synaptic transmission, we next examined paired-pulse facilitation (PPF), a form of short-term plasticity. The cKO and control mice exhibited similar degrees of facilitation at all inter-stimulus intervals tested (F 7,259=0.280; p=0.961). These data indicate that basal synaptic transmission and short-term plasticity are normal in hippocampal area CA1 in the absence of PS1. To determine whether PS1 plays a role in the initiation and maintenance of LTP, we induced LTP with theta burst stimulation (TBS) or a series of high frequency stimulations (HFS, 100 Hz tetanus) in the Schaeffer collateral pathway. The magnitude of LTP induced by 5 TBS (measured 60 min after TBS) was essentially the same in the *PS1* cKO (144.9±10) and in the control (148.5±7.3; p=0.94, *t*-test). A series of HFS (three 100-Hz tetani) produced a larger and longer-lasting form of LTP. The magnitude of LTP, including the late-phase of LTP measured 140 min after the last tetanus was comparable in the *PS1* cKO (167±16) and the control mice (148±15; p=0.26). To investigate further whether PS1 plays a role in the modulation of synaptic plasticity in hippocampal area CA1, we induced LTD with a paired-pulse low frequency stimulation (ppLFS). The magnitude of LTD (measured 60 min after conditioning) was identical in the *PS1* cKO (78±4) and the control (78±4; p=0.94). These results demonstrate that PS1 is not required for the induction and maintenance of LTP and LTD in the Schaeffer collateral pathway of the hippocampus.

To assess the neuronal function of PS1 more globally, we used the Morris water maze task, a hippocampus-dependent paradigm for spatial learning and memory. During the first 7 days of training, both groups of mice improved their performance at a similar rate, as indicated by the decreasing escape latencies and path lengths. However, during the last 3 to 5 days of training, the control mice continued to improve while the performance of the cKO mice plateaued, resulting in significantly longer escape latencies and path lengths for the cKO group. A two-way ANOVA showed a significant interaction effect of group x days (F(9,261) = 2.06; p = 0.03). Subsequent pairwise comparisons (Student's *t*-tests) showed significantly longer latencies and path lengths for the cKO group versus the control group on days 8, 9 and 10 (t(29); p<0.05). Swimming speed and the degree of thigmotaxis (wall hugging) were similar in the cKO and control mice, arguing against non-specific effects due to impaired motor function and/or anxiety. The *PS1* cKO and control mice were further tested in probe trials, in which the platform is removed from the pool, after 1, 5 and 10 days of training. After 1 and 5 days of training, the cKO and control mice showed no preference for the target quadrant and the platform location relative to the remaining quadrants. After 10 days of training, both groups of mice searched preferentially in the target quadrant. The total number of platform crossings by the cKO mice (2.27±0.49), however, was significantly lower than that of the control mice (3.94±0.88; p<0.05), though the radial quadrant occupancy of cKO (0.557±0.058) and control (0.633±0.044) mice was similar (p=0.3). We then moved the hidden platform to a different location in the opposite quadrant of the pool (platform-reversal phase), and repeated the acquisition experiment for 5 more days. The cKO mice again exhibited poorer performance than the controls with longer escape latencies and path lengths. To determine whether the delayed escape latencies of the cKO mice might be caused by deficits in motivation, sensory and/or motor abilities, both groups of mice were also tested in the visible platform version of the task. The escape latencies and path lengths of the cKO and control mice were very low and essentially indistinguishable. Taken together, these results demonstrate a mild but specific impairment in spatial learning and memory in the *PS1* cKO mice.

In summary, our analysis of the *PS1* cKO mice provides direct *in vivo* evidence of the requirement for PS1 in normal APP processing and the generation of amyloid peptides. Surprisingly, the regulation of the Notch downstream genes in the adult brain is independent of PS1, in contrast to the regulation of Notch activity by PS1 during brain development. It is therefore likely that therapeutic γ-secretase inhibitors will be able to achieve reduced production of Aβ peptides in the adult brain without unwanted side effects on the expression of the Notch target genes. PS1 also appears to be required for normal neuronal function in the adult brain, but the observed cognitive deficits in the *PS1* cKO mice are relatively subtle and may be restricted to specific neural circuits. In support of this notion, synaptic transmission and plasticity in hippocampal area CA1 of the cKO mice was entirely normal. It will be important, however, to determine whether mice lacking both presenilins exhibit any additional abnormalities, since therapeutic γ-secretase inhibitors are likely to inhibit both PS1- and PS2-mediated activities. Based on our current study, the benefits of therapeutic γ-secretase inhibitors may outweigh the potentially detrimental effects associated with targeting PS1 function.

REFERENCES

1. B. De Strooper, P. Saftig, K. Craessaerts, H. Vanderstichele, G. Guhde, W. Annaert, K. Figura , and L. FV, Deficiency of presenilin-1 inhibits the normal cleavage of amyloid precursor protein., *Nature* 391, 387-90 (1998)

2. J. Shen, R.T. Bronson, D.F. Chen, W. Xia, D.J. Selkoe , and S. Tonegawa, Skeletal and CNS defects in presnilin-1 deficient mice, *Cell* 89, 629-639 (1997)

3. M. Handler, X. Yang , and J. Shen, Presenilin-1 regulates neuronal differentiation during neurogenesis, *Development* 127, 2593-2606 (2000)

4. H. Yu, J. Kessler , and J. Shen, Heterogeneous populations of Es cells in the generation of a floxed *Presenilin-1* allele, *Genesis* 26, 5-8 (2000)

5. H. Yu, C. Saura, S. Choi, L. Sun, X. Yang, M. Handler, T. Kawarabayashi, L. Younkin, B. Fedeles, M. Wilson, S. Younkin, E. Kandel, A. Kirkwood , and J. Shen, APP processing and synaptic plasticity in the adult cerebral cortex of Presenilin-1 conditional knockout mice, *Neuron in press* (2001)

6. J. J. Lah, C.J. Heilman, N.R. Nash, H.D. Rees, H. Yi, S.E. Counts , and A.I. Levey, Light and electron microscopic localization of presenilin1 in primate brain, *J. Neurosci.* 17, 1971 (1997)

7. J. Buxbaum, G. Thinakaran, V. koliatsos, J. O'Callahan, H. Slunt, D. Price , and S. Sisodia, Alzheimer amyloid protein precursor in the rat hippocampus transport and processing through the perforant path, *J. Neurosci.* 18, 9629-37 (1998)

8. M. Simons, B. de Strooper, G. Multhaup, P.J. Tienari, C.G. Dotti , and K. Beyreuther, Amyloidogenic processing of the human amyloid precursor protein in primary cultures of rat hippocampal neurons, *J Neurosci* 16, 899-908 (1996)

9. Y. Luo, B. Bolon, S. Kahn, B. Bennett, S. Babu-khan, P. Denis, W. Fan, H. Kha, J. Zhang, Y. Gong, L. Martin, J. Louis, Q. Yan, W. Richards, M. Citron , and R. Vassar, Mice deficient in BACE1, the Alzheimer's b-secretase, have normal phenotype and abolished b-amyloid generation., *Nature Neurosci* 4, 231-2 (2001)

10. H. Cai, Y. Wang, D. McCarthy, H. Wen, D. Borchelt, D. Price , and P. Wong, BACE1 is the major b-secretase for generation of Ab peptides by neurons, *Nature Neurosci* 4, 233-4 (2001)

11. N. Suzuki, T. Iwatsubo, A. Odaka, Y. Ishibashi, C. Kitada , and Y. Ihara, High tissue content of soluble ß1-40 is linked to cerebral amyloid angiopathy, *Am. J. Pathol.* 145, 452-460 (1994)

12. K. Duff, C. Eckman, C. Zehr, X. Yu, C.-M. Prada, J. Perez-Tur, M. Hutton, L. Buee, Y. Harigaya, D. Yager, D. Morgan, M.N. Gordon, L. Holcomb, L. Refolo, B. Zenk, J. Hardy , and S. Younkin, Increased amyloid-β42(43) in brains of mice expressing mutant presenilin 1, *Nature* 383, 710-713 (1996)

13. W. Song, P. Nadeau, M. Yuan, X. Yang, J. Shen , and B.A. Yankner, Proteolytic release and nuclear translocation of Notch-1 are induced by presenilin-1 and impaired by pathogenic presenilin-1 mutations, *Proc Natl Acad Sci U S A* 96, 6959-63 (1999)

14. J. L. de la Pompa, A. Wakeham, K.M. Correia, E. Samper, S. Brown, R.J. Aguilera, T. Nakano, T. Honjo, T.W. Mak, J. Rossant , and R.A. Conlon, Conservation of the Notch signalling pathway in mammalian neurogenesis, *Development* 124, 1139-48 (1997)

15. B. De Strooper, W. Annaert, P. Cupers, P. Saftig, K. Craessaerts, J.S. Mumm, E.H. Schroeter, V. Schrijvers, M.S. Wolfe, W.J. Ray, A. Goate , and R. Kopan, A presenilin-1-dependent gamma-secretase-like protease mediates release of Notch intracellular domain, *Nature* 398, 518-22 (1999)

16. A. Parent, D. Linden, S. Sisodia , and D. Borchelt, Synaptic transmission and hippocampal long-term potentiation in transgenic mice expressing FAD-linked presenilin 1, *Neurobiology of Disease* 6, 56-62 (1999)

17. S. Zaman, A. Parent, L. A, M. Lee, D. Borchelt, S. Sisodia , and R. Malinow, Enhanced synaptic potentiation in transgenic mice expressing presenilin-1 familial Alzheimer's disease mutation is normalized with a benzodiazepine, *Neurobiology of Disease* 7, 54-63 (2000)

NEUROPEPTIDE EXPRESSION IN ANIMAL DISEASE MODELS

Tomas Hökfelt, Margarita Diez, Tiejun Shi, Jari Koistinaho, Dora Games,
Karen Kahn, Bernd Sommer, Karl-Heinz Wiederhold and Simone Danner*

1. INTRODUCTION

Lesions can affect the nervous system in many different ways and lead to more or less serious consequences. Our approach has over a period of 15 years been to study the effect of injury on dorsal root ganglion (DRG) neurons projecting into the sciatic nerve and to the dorsal horn of the spinal cord. The most simple approach is to completely transect the sciatic nerve and study the consequences of this injury in the parent cell bodies in the DRGs. This model may have some significance for our understanding of the mechanisms underlying pain that emerges in some patients after various types of surgical procedures, so called neuropathic pain (see 1). It has also been suggested that this model may be relevant for studying neurodegeneration (see refs. in 2). Our work on DRGs has led us to ask questions to what extent the changes we see in DRG neurons occur also in other neuronal systems. In fact, peptide regulation occurs in several systems in the brain in response to nerve injury (see 3). We have then turned our attention to established neurodegenerative diseases. In particular, we have analysed some mouse models for Alzheimer's disease (AD), as will be described below.

2. DORSAL ROOT GANGLION NEURONS

Transection of the sciatic nerve causes a dramatic upregulation of synthesis of the 28 amino acid peptide vasoactive intestinal polypeptide (VIP) in DRGs [4] as well as a number of other peptides such as galanin and neuropeptide Y (NPY) (see 5). Other molecules are

*Tomas Hökfelt and Margarita Diez, Department of Neuroscience, Karolinska Institutet, Stockholm, Sweden, Jari Koistinaho, A.I. Virtanen Institute, University of Kupio, Kupio, Finland, Dora Games and Karen Kahn, Elan Pharmaceuticals, South San Fransisco, CA, USA, Bernd Sommer, Karl-Heinz Wiederhold and Simone Danner, Novartis Pharma Inc., Basel, Switzerland

Mapping the Progress of Alzheimer's and Parkinson's Disease
Edited by Mizuno *et al.*, Kluwer Academic/Plenum Publishers, 2002

117

downregulated, for example the excitatory neuropeptides substance P and calcitonin gene-related peptide (CGRP) (see 5). Also other molecules such as nitric oxide synthase and various peptide and classic neurotransmitter receptors are also regulated (see 5).

Upregulation of molecules may represent a mechanism to promote survival and regeneration of neurons, and downregulation of excitatory molecules as well as upregulation of inhibitory molecules may counteract development of pain. For example, VIP has been shown to exert distinct trophic effects on neurons (see 6), and in galanin null mice regeneration of the sciatic nerve is attenutated, and there are also fewer DRG and cholinergic forebrain neurons, suggesting important roles during development [7, 8]. With regard to pain mechanisms, there is evidence that increased galanin levels after nerve injury may have analgesic actions (see 9).

A much discussed issue has been to what extent sciatic nerve transection in fact causes degeneration and neuron loss. This has been unclear, probably due to several factors, for example the site of the lesion, and in particularly the methodology used to quantitate neuron loss (see refs. in 2). However, recent studies based on unbiased stereological approaches have demonstrated that after transection of the sciatic nerve at mid-thigh level in the rat, there is no loss of neurons during the first four weeks [10]. In contrast, when the rat spinal nerves are transected, that is when the lesion is made closer to the cell body, there is a 20% loss of neurons after 15 days [11]. Moreover, when a transection is made at mid-thigh level in the mouse, there is a significant loss already after one week (down by 25%), and after four weeks more than half of the neurons are lost [2]. Taken together these results suggest that the proximity of the lesion to the cell body is crucial for development of degenerative changes and cell death. These findings thus support the idea that, at least under certain circumstances, axotomy of DRG neurons represents a model for neurodegeneration.

3. ALZHEIMER MOUSE MODELS

It was early recognized that changes in neurotransmitter levels occur in AD. In addition to derangement of cholinergic mechanisms [12, 13], it was reported that the levels of somatostatin are reduced in cortical areas of Alzheimer patients [14, 15]. Subsequently, many other studies showed changes in peptide expression, in most cases decreases but increases were also reported (see 16).

We have had the opportunity to examine the first Alzheimer mouse model generated by Games et al. [17]. In this mouse human beta-amyloid precursor protein with the familial AD mutation V --> F at position 717 is overexpressed under the platelet-derived growth factor-β (PDGFB) promoter. These mice exhibit extensive amyloid plaques in the hippocampus and cortex [18]. We have focused on immunohistochemical analysis of distribution of various peptides [19]. The mice were either 18- or 26-month old. Two types of characteristic staining patterns were seen [19]. First, presence of peptide-containing neurites in amyloid β-peptide positive plaques, and secondly changes in peptide levels in specific subregions and/or neuron populations, especially in the hippocampal formation and ventral cortices. A clear difference was observed between the 18- and 26-month old mice, with only minor changes in the younger ones. Also, there were considerable variations between mice of the same age, and the marked changes reported below were only seen in old mice, but not in all of them. Mostly increases in peptides were observed, but in some cases also decreased levels were encountered. For example, increased peptide levels were observed for galanin, enkephalin and cholecystokinin in stratum lacunosum moleculare, and for galanin, neuropeptide Y, enkephalin and dynorphin in mossy fibers. Many of the peptides also showed elevated levels in ventral cortices. Decreases were observed for cholecystokinin in mossy fibers and substance P in fibers around the granule cells. Finally, the distinct network of galanin-positive fibers, presumably identical to the noradrenaline fibers, could not be detected in the transgenic mice. More recently we have examined a further Alzheimer mouse model which expresses human APP751 under the Thy-1 promoter [20]. Also here changes in peptide expression were observed which were similar but sometimes less pronounced (unpublished observations).

The significance of these changes are unclear, especially since they were detected in very old mice. Nevertheless, it is tempting to make a comparison with the peptide changes in DRG neurons mentioned above, especially the upregulation of peptides, such as galanin and NPY, which are seen in both the DRGs and Alzheimer models. It is possible that these increases in peptide levels serve to improve survival and regeneration, since both galanin [8] and NPY [21] have been shown to have trophic actions and thus may serve to counteract degeneration. On the other hand, it can not be excluded that the increased and decreased peptide levels represent a paraphenomenon without functional significance, or in fact contribute to the symptomatology of these mice. We have previously suggested that lesion-induced upregulation of galanin in cholinergic forebrain neurons may have adverse effects [22, 23], since galanin reduces acetylcholine release from rat hippocampal slices [24]. The same type of 'Alzheimer mice' as studied here has recently been demonstrated to have cognitive defects in the Morris water swim maze [25]. However, these cognitive defects were observed at earlier stages than the changes observed in our study. It would therefore be particularly interesting to analyse with in situ hybridization at what stage the first changes in peptide expression in fact appear, to see if there is any correlation between behaviour and peptide regulation.

4. ACKNOWLEDGEMENTS

This work was supported by Marianne and Marcus Wallenbergs Stiftelse, an Unrestricted Bristol-Myers Squibb Neuroscience Grant and the Swedish MRC (04X-2887).

REFERENCES

1. J. W. Scadding. (1984) Peripheral neuropathies. In *Textbook of Pain.* (eds, P. D. Wall and R. Melzack) pp. 413-425. Churchill Livingstone, Edinburgh.
2. T.-J. S. Shi, T. Tandrup, E. Bergman, Z.-Q. D. Xu, B. Ulfhake and T. Hökfelt, Effect of peripheral nerve injury on dorsal root ganglion neurons in the C57BL/6J mouse: Marked changes both in cell numbers and neuropeptide expression. *Neuroscience* **105**, 249-263 (2001).
3. T. Hökfelt, X. Zhang. T. S. Shi, Y.-G. Tong, H. F. Wang, Z.-Q. D. Xu, M. Landry, C. Broberger, M. Diez, G. Ju, G. Grand and M. Villar. (2000) Phenotypic changes in peripheral and central neurons induced by nerve injury: focus on neuropeptides. In *Challenges for Neuroscience in the 21st Century* (ed, O. Hayaishi) pp. 63-87. Japan Sci. Soc. Press, Karger.
4. S. A. Shehab and M. E. Atkinson, Vasoactive intestinal polypeptide (VIP) increases in the spinal cord after peripheral axotomy of the sciatic nerve originate from primary afferent neurons. *Brain Res.* **372**, 37-44 (1986).
5. T. Hökfelt, X. Zhang and Z. Wiesenfeld-Hallin, Messenger plasticity in primary sensory neurons following axotomy and its functional implications. *TINS* **17**, 22-30 (1994).

6.   I. Gozes and D. E. Brenneman, Neuropeptides as growth and differentiation factors in general and VIP in particular. *J. Mol. Neurosci.* **4**, 1-9 (1993).

7.   F. E. Holmes, S. Mahoney, V. R. King, A. Bacon, N. C. Kerr, V. Pachnis, R. Curtis, J. V. Priestly and D. Wynick, Targeted disruption of the galanin gene reduces the number of sensory neurons and their regenerative capacity. *Proc. Natl. Acad. Sci. USA* **97**, 11563-11568 (2000).

8.   G. O'Meara, U. Coumis, S. Y. Ma, J. Kehr, S. Mahoney, A. Bacon, S. J. Allen, F. Holmes, U. Kahl, F. H. Wang, I. R. Kearns, S.-O. Ögren, D. Dawbarn, E. J. Mufson, C. Davies, G. Dawson and D. Wynick, Galanin regulates the postnatal survival of a subset of basal forebrain cholinergic neurons. *Proc. Natl. Acad. Sci.* **97**, 11569-11574 (2000).

9.   X. J. Xu, T. Hökfelt, T. Bartfai and Z. Wiesenfeld-Hallin, Galanin and spinal nociceptive mechanisms: recent advances and therapeutic implications. *Neuropeptides* **34**, 137-147 (2000).

10. T. Tandrup, C. J. Woolf and R. E. Coggeshall, Delayed loss of small dorsal root ganglion cells after transection of the rat sciatic nerve. *J. Comp. Neurol.* **422**, 172-180 (2000).

11. S. Vestergaard, T. Tandrup and J. Jakobsen, Effect of permanent axotomy on number and volume of dorsal root ganglion cell bodies. *J. Comp. Neurol.* **388**, 307-312 (1997).

12. P. Davies and A. J. Maloney, Selective loss of central cholinergic neurones in Alzheimer's disease. *Lancet* **ii**, 1430 (1976).

13. P. J. Whitehouse, D. L. Price, R. G. Struble, A. W. Clark, J. T. Coyle and M. R. DeLong, Alzheimer's disease and senile dementia: loss of neurons in the basal forebrain. *Science* **215**, 1237-1239 (1982).

14. P. Davies, R. Katzman and R. D. Terry, Reduced somatostatin-like immunoreactivity in cerebral cortex from cases of Alzheimer disease and Alzheimer senile dementa. *Nature* **288**, 279-280 (1980).

15. M. N. Rossor, P. C. Emson, C. Q. Mountjoy, M. Roth and L. L. Iversen, Reduced amounts of immunoreactive somatostatin in the temporal cortex in senile dementia of Alzheimer type. *Neurosci. Lett.* **20**, 373-377 (1980).

16. C.-G. Gottfries, S. O. Frederiksen and M. Heilig, Neuropeptides and Alzheimer's disease. *Eur. Neuropsychopharmacol.* **5**, 491-500 (1995).

17. D. Games, D. Adams, R. Alessandrini, R. Barbour, P. Berthelette, C. Blackwell, T. Carr, J. Clemes, T. Donaldson, F. Gillespie, T. Guido, S. Hagopian, K. Johnson-Wood, K. Khan, M. Lee, P. Leibowitz, I. Lieberburg, S. Little, E. Masliah, L. McConlogue, M. Montoya-Zavala, L. Mucke, L. Paganini, E. Penniman, M. Power, D. Schenk, P. Seubert, B. Snyder, F. Soriano, H. Tan, J. Vitale, S. Wadworth, B. Wolozin and J. Zhao, Alzheimer-type neuropathology in transgenic mice overexpressing V71F B-amyloid precursor protein. *Nature* **373**, 523-527 (1995).

18. E. Masliah, A. Sisk, M. Mallory, L. Mucke, D. Schenk and D. Games, Comparison of neurodegenerative pathology in transgenic mice overexpressing V717F β-amyloid precursor protein and Alzheimer's disease. *J.Neurosci.* **16**, 5795-5811 (1996).

19. M. Diez, J. Koistinaho, K. Kahn, D. Games and T. Hökfelt, Neuropeptides in hippocampus and cortex in transgene mice overexpressing V717F β-amyloid percursor protein - initial observations. *Neuroscience* **100**, 259-286 (2000).

20. C. Sturchler-Pierrat, D. Abramowski, M. Duke, K.-H. Wiederhold, C. Mistl, S. Rothacher, B. Ledermann, K. Bürki, P. Frey, P. A. Paganetti, C. Waridel, M. E. Calhoun, M. Jucker, A. Probst, M. Staufenbiel and B. Sommer, Two amyloid precursor protein transgenic mouse models with Alzheimer disease-like pathology. *Proc. Natl. Acad. Sci. USA* **94**, 13287-13292 (1997).

21. D. E. Hansel, B. A. Eipper and G. V. Ronnett, Neuropeptide Y functions as a neuroproliferative factor. *Nature* **410**, 940-943? (2001).

22. T. Hökfelt, D. Millhorn, K. Seroogy, Y. Tsuruo, S. Ceccatelli, B. Lindh, B. Meister, T. Melander, M. Schalling and L. Terenius, Coexistence of peptides with classical neurotransmitters. *Experientia* **43**, 768-780 (1987).

23. T. Hökfelt, Z.-Q. D. Xu, T.-J. Shi, K. Holmberg and X. Zhang. (1998) Galanin in ascending systems. Focus on coexistence with 5-hydroxytryptamine and noradrenaline. In *Galanin: Basic Research Discoveries and Therapeutic Implications* (eds, T. Hökfelt, T. Bartfai and J. Crawley) pp. 252-263. Ann. New York Acad. Sci., New York.

24. G. Fisone, C. F. Wu, F. Consolo, Ö. Nordström, N. Brynne, T. Bartfai, T. Melander and T. Hökfelt, Galanin inhibits acetylcholine release in the ventral hippocampus of the rat: histochemical, autoradiographic, in vivo and in vitro studies. *Proc. Natl. Acad. Sci. USA* **88**, 7339-7343 (1987).

25. G. Chen, K. S. Chen, J. Knox, J. Inglis, A. Bernard, S. J. Martin, A. Justice, L. McConlogue, D. Games, S. B. Freedman and R. G. M. Morris, A learning deficit related to age and β–amyloid plaques in a mouse model of Alzheimer's disease. *Nature* **408**, 975-979 (2000).

ROLE OF INDUCIBLE NITRIC OXIDE SYNTHASE IN β-AMYLOID-INDUCED BRAIN DYSFUNCTION IN RATS

Kiyofumi Yamada[*], Manh Hung Tran, Hyoung-Chun Kim, and Toshitaka Nabeshima

1. INTRODUCTION

Alzheimer's disease (AD) is histopathologically characterized by the presence of numerous senile plaques, neurofibrillary tangles and marked atrophy in the brain (Katzman and Saitoh, 1991). The senile plaque in its mature form consists of a central core of an extracellular thioflavin S and Congo red-positive fibrous protein (amyloid β-peptide (Aβ)) (Glenner, 1988; Selkoe, 1994). In AD brain tissue two different types of Aβ deposits have been identified, namely, diffuse plaques, which contain nonfibrillar Aβ, and senile plaques, which contain fibrillar Aβ. While the diffuse plaques exert no toxic effect, the senile plaques are associated with degeneration of neurites and neuronal cell bodies. The state of peptide aggregation is correlated with its neurotoxic potency. In AD, neurodegeneration occurs primarily in certain types of neurons, particularly those in the cortex and hippocampus, and the cholinergic dysfunction is believed to be primarily responsible for cognitive deficits (Coyle et al., 1983).

2. BRAIN DYSFUNCTION INDUCED BY CONTINUOUS INFUSION OF AMYLOID β-PEPTIDE INTO LATERAL VENTRICLE IN RATS

A growing number of studies have demonstrated that acute injection or continuous infusion of Aβ into the brain causes brain dysfunction as evidenced by neurodegeneration and an impairment of learning and memory (Yamada et al., 1999a; Yamada and Nabeshima, 2000). An in vivo model for the neurodegenerative effects of

---

[*] K. Yamada, Department of Neuropsychopharmacology and Hospital Pharmacy, Nagoya University Graduate School of Medicine, Nagoya 466-8560, Japan

Mapping the Progress of Alzheimer's and Parkinson's Disease
Edited by Mizuno *et al.*, Kluwer Academic/Plenum Publishers, 2002

Aβ1-40, including neuronal loss and degenerating neurons and neurites with an induction of Alz-50-immunoreactive proteins, was first demonstrated by Kowall et al. (1991).

We have used a technique of continuous intracerebroventricular (i.c.v.) infusion of Aβ_with a mini-osmotic pump. Nitta et al. (1994) demonstrated that a continuous i.cv. infusion of Aβ1-40 at a dose of 300 pmol/day caused a significant impairment of spatial memory formation in a water maze and a deficit of passive avoidance performance, which was accompanied by a slight but significant reduction of choline acetyltransferase (ChAT) activity in the hippocampus. Accumulation of Aβ1-40 in the hippocampus and cerebral cortex was evident immunohistochemically following a 14-day period of infusion (Nitta et al., 1994). Memory impairments in the Aβ-infused rats were recovered 2 weeks after the cessation of the infusion although the reduction of ChAT activity and an increase in glial fibrillary acidic protein (GFAP) immunoreactivity were still present (Nitta et al., 1997). The i.c.v. infusion of Aβ resulted in changes in the ciliary neurotrophic factor protein levels (Yamada et al., 1995), and in the mRNA expression of brain-derived neurotrophic factor (BDNF) in the hippocampus (Tang, et al., 2000). Long-term potentiation (LTP) in the hippocampus following a brief high-frequency stimulation is considered to be a synaptic correlate of memory. The enhanced response after the tetanic stimulation was maintained for more than 45 min in the control hippocampal slices from vehicle-infused rats whereas it declined rapidly to nearly baseline levels in the Aβ1-40-treated group (Itoh et al., 1999). It is also reported that i.c.v. injection of Aβ1-40, Aβ1-42 or C-terminal fragment of APP greatly shortens the hippocampal LTP in vivo (Cullen et al., 1997).

3. IMPAIRMENT OF CHOLINERGIC NEUROTRANSMISSION INDUCED BY AMYLOID β-PEPTIDE

Since the cholinergic hypothesis of AD has been articulated (Coyle et al., 1983), the question of whether Aβ impairs cholinergic neurotransmission is compelling. To assess the cholinergic neurotransmission, we used a microdialysis technique to examine acetylcholine (ACh) release under nicotine stimulation (Itoh et al., 1996; Olariu et al., 2001; Tran et al., 2001). The amplitude of the nicotine-evoked ACh release was not different among naive rats and rats receiving either vehicle or the non-toxic reverse fragment Aβ40-1 whereas continuous infusion of neurotoxic Aβ fragments, including Aβ1-42 and Aβ1-40, at 300 pmol/day for 10 days resulted in a significant decrease in nicotine-stimulated ACh release as compared with control group Aβ40-1 (Itoh et al., 1996; Tran et al., 2001). Aβ25-35 at 3 nmol/day, but not 300 pmol/day, also impaired the nicotine-evoked ACh release when compared with control Aβ40-1 (Olariu et al., 2001). These findings suggest that only the active forms of Aβ impaired nicotine-evoked ACh release. There was no significant difference in the basal level of extracellular ACh among groups, indicating that infusion of Aβ does not affect ACh synthesis, or metabolic processes during this time course. The extracellular choline concentration was not altered, implying that the choline uptake process was not affected either. Hence a lack of precursor for ACh biosynthesis, which leads to a reduced intracellular availability of ACh, was ruled out. Accordingly, the decrease of nicotine-evoked ACh release may be attributable to an impairment of nAChR signal transmission rather than a loss of cholinergic fibers. Because presynaptic nAChRs are involved in modulation of the release of ACh and other neurotransmitters, which appeared to contribute to the symptoms of AD, the impaired nicotine-evoked ACh release may imply a dysfunction in synaptic processes of neurotransmitter release at an early stage caused by Aβ in AD.

A hypofunction of the septo-hippocampal cholinergic system was also found on day 7 or day 21 after acute injection of synthetic Aβ fragments Aβ12-28, Aβ25-35 and Aβ1-40 (Abe et al., 1994). The authors observed an attenuation in K^+-evoked ACh release in Aβ-injected rats and this phenomenon was not associated with decreased ChAT nor inhibition of neuronal firing, therefore an impairment of the ACh release mechanism was undoubtedly involved. Since the impairment of both nicotine- and K^+-evoked ACh and dopamine (DA) release was observed after delays (7-21 days), but not acutely, underlying multiple-step mechanisms involved in a dysfunction of synaptic processes of neurotransmitter release may be responsible, rather than a direct binding to or modulating effect on nAChRs of Aβ.

In contrast to these in vivo findings, Aβ can acutely modulate ACh release in vitro. In rat cortical and hippocampal slices, K^+-evoked ACh release was inhibited by several Aβ fragments including Aβ1-28, Aβ25-35, Aβ1-40, and Aβ1-42 at picomolar to nanomolar concentrations (Kar et al., 1996). The inhibitory effect of Aβ on K^+-evoked ACh release was reversible and tetrodotoxin-insensitive suggesting that Aβ can regulate ACh release by acting on cholinergic terminals. Because the acute toxic effect of Aβ in such experimental conditions was excluded, the authors suggested that endogenous Aβ-related peptides derived from APP act as potent inhibitors of ACh release and may serve as a basis for the functional interrelationship between Aβ toxicity and the vulnerability of some cholinergic populations in AD. Regarding the acute modulatory effect of Aβ on ACh release (Kar et al., 1996) and the high affinity binding of Aβ to α7 nAChRs (Wang et al., 2000), consideration should be given whether the effects of Aβ are mediated via α7 nAChRs and, if so, the extent of the physiological and pathological actions of Aβ.

4. ROLE OF NITRIC OXIDE IN Aβ-INDUCED BRAIN DYSFUNCTION

Evidence that Aβ induces brain dysfunction without apparent neurotoxicity suggests that Aβ disrupts multiple signal transduction systems before exhibiting actual neurotoxicity. Aβ may bind to scavenger receptors and a receptor for the advanced glycation end product on microglia and release nitric oxide (NO) and other reactive oxygen species and, as a consequence, increase oxidative burden (Yamada and Nabeshima, 2000). In fact, several lines of evidence have suggested that oxidative stress is involved in the mechanism of Aβ-induced neurotoxicity (Behl et al., 1994; Yamada et al., 1999b) and pathogenesis of AD (Markesbery, 1997). In AD patients, the expression of inducible NO synthase (iNOS) in a subset of pyramidal neurons of the hippocampus (Lee et al., 1999) and in tangle-bearing neurons (Vodovotz et al., 1996) has been reported.

To investigate whether Aβ can induce this effect in experimental animals in vivo, rats were continuously infused i.c.v. with Aβ1-40 and examined for the expression of iNOS mRNA by RT-PCR and iNOS protein by immunohistochemical staining with specific iNOS antibodies at different time points. Continuous infusion of Aβ1-40 into rat cerebral ventricle induced a time-dependent expression of iNOS mRNA and protein in the hippocampus, especially in the dentate gyrus and to a lesser extent in the CA1 subfield, which was followed by a marked increase in the tissue contents of NO metabolites (nitrite and nitrate). The expression of iNOS protein was found in both microglia and astrocytes, indicating that Aβ activates non-neuronal cells (Tran et al., 2001).

We examined the effects of iNOS inhibitors on cholinergic dysfunction and memory deficit in Aβ-infused rats, to clarify a pathophysiological significance of the Aβ-induced iNOS induction. Inhibition of NO production by iNOS inhibitors, such as aminoguanidine (AG) and S-methylisothiourea (SMT), prevented the Aβ-induced impairment of nicotine-evoked ACh release, as well as spatial learning deficit in a radial arm maze. The protective effects of iNOS inhibitors against Aβ-induced brain dysfunction were specific since: a) both AG and SMT are iNOS-specific inhibitors; b) Co-administration of the iNOS inhibitor AG and the NO precursor L-arginine eliminated the ameliorating effect of AG; and c) AG or SMT treatment in control rats did not alter nicotine-evoked ACh release. In contrast to iNOS inhibitors, the neuronal NOS (nNOS) inhibitor 7-nitroindazole (7-NI) failed to ameliorate the impairment of nicotine-evoked ACh release induced by Aβ. Although only one nNOS inhibitor was used, the result seems to be consistent with the fact that Ca^{2+}-independent NOS activity was specifically increased by Aβ1-40 without any changes in the Ca^{2+}-dependent NOS activity (Tran et al., 2001).

It is worth mentioning that Aβ chronically impaired ACh and DA release in vivo without an apparent association with the decrease of ACh synthesis or cholinergic cell loss (Abe et al., 1994; Itoh et al., 1996; Tran et al., 2001), indicating that an impairment of the neurotransmitter release mechanism may be involved, rather than a direct neurotoxic effect on cholinergic fibers. We speculate that overproduction of NO after Aβ infusion leads to the formation of peroxynitrite and subsequent nitration of synaptic proteins, thus affecting signal transduction pathways of cellular regulation and release. Indeed, immunohistochemical staining with specific nitrotyrosine antibodies to detect peroxynitrite formation showed abundant nitrotyrosine-immunoreactive cells in the hippocampus of Aβ-infused rats (unpublished observation). Additional experiments related to this hypothesis are under way in our laboratory.

5. CONCLUSIONS AND PERSPECTIVE

In conclusion, it is suggested that Aβ induces cholinergic hypofunction via three distinct mechanisms: First, Aβ may exert direct cytotoxic effects on the susceptible cholinergic population, which lead to cholinergic cell death. In general, to induce such a neurotoxicity, the aggregated form of Aβ at micromolar concentrations is necessary, thus in AD, this may occur at a late stage when numerous senile plaques have already formed. Alternatively, Aβ modulates cholinergic function by acting on cholinergic terminals, probably by binding to the nAChR system. More rigorous research needs to be performed since it appears that under physiological conditions, Aβ functions as an endogenous modulator of cholinergic neurotransmission, and it is not clear whether or not the modulating effect of Aβ on cholinergic neurotransmission is mediated via α7nAChRs. Lastly, our findings suggest that Aβ activates non-neuronal cells, such as microglial and/or astroglial cells, which results in an induction of iNOS and the subsequent overproduction of NO. The oxidative stress may lead to dysfunction of multiple signal transduction systems, including impairment of the release of ACh and probably other neurotransmitters, and consequently leading to learning and memory deficits. Interestingly, iNOS inhibitors, as well as antioxidants, ameliorate the cholinergic dysfunction and the memory performance caused by Aβ, thus these drugs may prove a useful tool in pharmacological intervention as a preventive and protective approach to treatment of AD.

6. ACKNOWLEDGMENTS

This study was supported, in part, by Grants-in-Aid for Science Research (No. 99187 and 12670085), and a COE Grant from the Ministry of Education, Culture, Sports, Science and Technology of Japan, and by an SRF Grant for Biomedical Research.

REFERENCES

Abe, E., Casamenti, F., Giovannelli, L., Scali, C. and Pepeu, G., 1994, Administration of amyloid β-peptides into the medium septum of rats decreases acetylcholine release from hippocampus in vivo, *Brain Res.* **636**:162-164.

Behl, C., Davis, J.B., Lesley, R. and Schubert, D., 1994, Hydrogen peroxide mediates amyloid β protein toxicity, *Cell* **77**:817-827.

Coyle, J.T., Price, D.L. and DeLong, M.R., 1983, Alzheimer's disease: a disorder of central cholinergic innervation. *Science* **219**: 1184-1190.

Cullen, W. K., Suh, Y. H., Anwyl, R. and Rown, M. J., 1997, Block of LTP in rat hippocampus in vivo by β-amyloid precursor protein fragment, *NeuroReport* **8**: 3213-3217.

Glenner, G.G., 1988, Alzheimer's disease: its proteins and genes, *Cell* **52**: 307-308.

Itoh, A., Akaike, T., Sokabe, M., Nitta, A., Iida, R., Olariu, A., Yamada, K. and Nabeshima, T., 1999, Impairment of long-term potentiation in hippocampal slices of β-amyloid-infused rats, *Eur. J. Pharmacol.* **382**: 167-175.

Itoh, A., Nitta, A., Nadai, M., Nishimura, K., Hirose, M., Hasegawa, T. and Nabeshima, T., 1996, Dysfunction of cholinergic and dopaminergic neuronal systems in β-amyloid protein-infused rats, *J. Neurochem.* **66**: 1113-1117.

Katzman, R. and Saitoh, T., 1991, Advances in Alzheimer's disease, *FASEB J.* **5**: 278-286.

Kar, S., Seto, D., Gaudreau, P. and Quirion, R., 1996, β-Amyloid-related peptides inhibit potassium-evoked acetylcholine release from rat hippocampal formation, *J. Neurosci.* **16**: 1034-1040.

Kowall, W. N., Beal, F. M., Busciglio, J., Duffy, K. L. and Yankner, B. A., 1991, An in vivo model for the neurodegenerative effects of β-amyloid and protection by substance P. *Proc. Natl. Acad. Sci. USA* **88**: 7247-7251.

Lee, S.C., Zhao, M.L., Hirano, A. and Dickson, D.W., 1999, Inducible nitric oxide synthase immunoreactivity in the Alzheimer disease hippocampus: association with Hirano bodies, neurofibrillary tangles, and senile plaques. *J. Neuropathol. Exp. Neurol.* **58**:1163-1169.

Markesbery, W.R., 1997, Oxidative stress hypothesis in Alzheimer's disease, *Free Radic. Biol. Med.* **23**:134-147.

Nitta, A., Fukuta, T., Hasegawa, T. and Nabeshima, T., 1997, Continuous infusion of β-amyloid protein into cerebral ventricle induces learning impairment and neuronal and morphological degeneration, *Jpn. J. Pharmacol.* **73**: 51-57.

Nitta, A., Itoh, A., Hasegawa, T., Nabeshima T., 1994, β-Amyloid protein-induced Alzheimer's disease animal model, *Neurosci. Lett.* **170**: 63-66.

Olariu, A., Tran, M.H., Yamada, K., Mizuno, M., Hefco, V. and Nabeshima, T., 2001, Memory deficits and increased emotionality induced by β-amyloid (25-35) are correlated with the reduced acetylcholine release and altered phorbol dibutyrate binding in the hippocampus, *J. Neural Transm.* in press.

Selkoe, D.J., 1994, Cell biology of amyloid β-protein precursor and the mechanism of Alzheimer's disease, *Annu. Rev. Cell Biol.* **10**: 373-403.

Tang, Y.P., Yamada, K., Kanou, Y., Miyazaki, T., Xiong, X.-L., Kambe, F., Murata, Y., Seo, H. and Nabeshima, T, 2000, Spatiotemporal expression of BDNF in the hippocampus induced by the continuous intracerebroventricular infusion of β-amyloid in rats, *Mol. Brain Res.* **80**: 188-197.

Tran, M.H., Yamada, K., Olariu, A., Mizuno, M., Ren, X.H. and Nabeshima, T., 2001, Amyloid β-peptide induces nitric oxide production in rat hippocampus: association with cholinergic dysfunction and amelioration by inducible nitric oxide inhibitors, *FASEB J.* in press.

Vodovotz, Y., Lucia, M.S., Flanders, K.C., Chesler, L., Xie, Q.W., Smith, T.W., Weidner, J., Mumford, R., Webber, R., Nathan, C., Roberts, A.B., Lippa, C.F. and Sporn, M.B., 1996, Inducible nitric oxide synthase in tangle-bearing neurons of patients with Alzheimer's disease, *J. Exp. Med.* **184**:1425-1433.

Wang, H.Y., Lee, D.H.S., D'andrea, M.R., Peterson, P.A., Shank, R.P. and Reitz, A.B., 2000, β-Amyloid1-42 binds to α7 nicotinic acetylcholine receptor with high affinity. *J. Biol. Chem.* **275**:5626-5632.

Yamada, K. and Nabeshima, T., 2000, Animal models of Alzheimer's disease and evaluation of anti-dementia drugs, *Pharmacol. Therapeut.* **88**: 93-113.

Yamada, K., Nitta, A., Saito, T., Hu, J. and Nabeshima, T., 1995, Changes in ciliary neurotrophic factor content in the rat brain after continuous intracerebroventricular infusion of β-amyloid (1-40) protein, *Neurosci. Lett.* **201**: 155-158.

Yamada, K., Ren, X. and Nabeshima, T., 1999a, Perspectives of pharmacotherapy in Alzheimer's disease, *Jpn. J. Pharmacol.* **80**: 9-14.

Yamada, K., Tanaka, T., Han, D., Senzaki, K., Kameyama, T. and Nabeshima, T., 1999b, Protective effects of idebenone and α-tocopherol on β-amyloid-(1-42) induced learning and memory deficits in rats: implication of oxidative stress in β-amyloid-induced neurotoxicity in vivo, *Eur. J. Neurosci.* **11**: 83-90.

AMYLOID PATHOLOGY AND CHOLINERGIC NETWORKS IN TRANSGENIC MODELS OF ALZHEIMER'S DISEASE

A. Claudio Cuello[*]

1. INTRODUCTION

The involvement of the forebrain cholinergic system in Alzheimer's disease (AD) pathology is well documented (for review see[1]). Boosting the cholinergic system with the application of acetyl cholinesterase inhibitors remains a valid, current therapeutic approach to modulate the progress of the disease by slowing its characteristic progressive cognitive deficits (see Giacobini and Sugimoto, this volume). However, despite compelling evidence for the involvement of the forebrain cholinergic system in human AD pathology, we do not understand how this compromise of cholinergic neurons comes about. The advancement of mouse transgenic (tg) lines expressing mutated forms of human genes related to familial forms of the disease offers a unique opportunity for such an investigation. We have examined the cholinergic phenotype in three tg lines, as described below. We placed special emphasis on the consequences of the amyloid burden on cortical presynaptic boutons, a synaptic population known to be vulnerable in AD. This brief report summarizes the most salient observations of the ongoing collaborative work between my laboratory and that of Dr. Karen Duff (Nathan Klein Institute, UNY, NY) in these tg animals.

2. METHODS

2.1. Tg Mice and Tissue Samples

Eight-month-old doubly-tg mice, an F_1 hybrid from a cross between an APP mutant ($APP_{K670N, M671L}$, Tg2576,[2]) and a PS1 mutant ($PS1_{M146L}$,[3]), and age-matched non-tg littermates were used in the present study. Brain tissue samples were fixed and processed as previously described.[4] Fifty-μm thick coronal sections were cut with a sledge freezing

[*] A. Claudio Cuello, Depts. of Pharmacology & Therapeutics, Anatomy & Cell Biology, and Medicine, McGill University, 3655 Promenade Sir-William-Osler, Montreal, QC, Canada, H3G 1Y6.

Mapping the Progress of Alzheimer's and Parkinson's Disease
Edited by Mizuno *et al.*, Kluwer Academic/Plenum Publishers, 2002

127

microtome (Leitz) at -20°C between bregma 1.6 mm and bregma -0.6 mm.[5] Free-floating sections were Nissl stained to identify the different cortical laminae and immunohistochemical staining for VAChT, synaptophysin, and Aβ was performed. Some sections were counterstained with Thioflavin-S to determine the boundaries of amyloid plaques. A few LM sections were further processed for VAChT and Aβ immuno-gold staining for electron microscopic analysis following well-established protocols.[6,7]

2.2. Quantification Of Immunoreactive Presynaptic Boutons

Quantification of VAChT-IR bouton and synaptophysin-IR bouton density was performed essentially as previously described.[4,8] Tissue slides were examined with a 100x oil immersion plan achromatic objective and a 10x projection lens. The microscope was equipped with a CCD video camera and was connected to a MCID-M4 image analysis system (Imaging Research Inc., St. Catharines, Ontario, Canada). Immunopositive punctae (VAChT-IR and synaptophysin-IR boutons) were detected by the image analysis system using software designed for silver grain counting. Background staining in all fields from each section was normalized separately by the M3D module of the M4 system. For counting purposes, VAChT-IR and synaptophysin-IR boutons, cell bodies, blood vessels, and cortical areas that were not in focus were excluded.

Boutons were counted both in the neuropile at random and in six successive 300μm² fields (30 μm in height and 10 μm in width) next to the border of amyloid plaque. To study the influence of plaque size on cholinergic and overall bouton density, a comparison of bouton density adjacent to amyloid plaques with non-adjacent boutons of different sizes (<500μm²; 500 to 1000μm²; 1000 to 2000μm²; 2000 to 4000μm²; and 4000μm² or above) was performed. Comparisons of the density and size of pre-synaptic boutons were done using one-way ANOVA, followed by a post-hoc Dunnett's for pair-wise comparisons between the control littermates and tg mice. Other pair-wise comparisons were performed using post-hoc Tukey tests.

3. RESULTS

At the 8-month time point studied, only the doubly tg mice (APP+PS1) displayed an early Aß-IR plaque-like pathology. Many of these plaques are truly amyloid and reactive to Thioflavin-S. At this same time point, the APP (Tg2567) tg line is in a pre-plaque stage, while the PS1 tg line does not show plaque formation at any time point. In these doubly tg animals, we have recently demonstrated[4] a selective loss of cholinergic presynaptic boutons in most areas of the cerebral and hippocampal cortices coincident with the development of the first Aß plaques (Fig 1). In some areas a reduction of the bouton size can also be detected, a situation reminiscent of the recently reported age-related shrinkage of cholinergic boutons after 3D ultrastructural reconstructions.[9] Contrary to what one would expect, we found that, in the APP alone (Tg2567), at the "pre-plaque" stage there was an up-regulation in the number of cortical cholinergic boutons. This would suggest that a sprouting of cholinergic terminals precedes the loss of boutons associated with visible Aß aggregation in the form of plaques. No changes were observed in the PS1 tg mice.

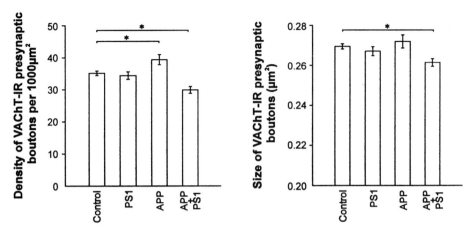

Fig 1. VAChT-IR presynaptic bouton density and size in the frontal cortex of different transgenic mice. A comparison of the global density (weighted mean from all cortical laminae) and the mean size of VAChT boutons between different transgenic mice revealed an elevated number in the $APP_{K670N,M671L}$ transgenic line and a diminished number in the doubly transgenic line accompanied by shrunken bouton size (*$p < 0.05$). Data represent the mean ± SEM. From[4].

When we analyzed the density of synaptophysin-IR boutons (representing the overall pre-synaptic population) in the random neuropile of the neocortex in doubly transgenic mice, no significant changes were found in comparison with littermate controls. However, when the density of the synaptophysin boutons was analyzed in areas adjacent to plaques in the layer V region of the frontal cortex in doubly-tg mice, it was nearly 10 % higher than that measured from the same cortical region in control littermates. This situation is analogous to the cholinergic up-regulation found in the APP tg mice in the random neuropile.

The density of VAChT-IR boutons (representing the cholinergic pre-synaptic population) was shown in the doubly tg mice to be negatively affected by proximity to amyloid plaques. Thus, we observed a loss of cholinergic boutons in the neuropile adjacent to plaques, with a progressive depletion with increasing plaque size. This correlation was shown to be significant by regression analysis (Fig 2).

These investigations also revealed that the immediate perimeter around Thioflavin-S positive and Aß-IR plaques regularly display a corolla of dystrophic synaptophysin-IR boutons. The cholinergic boutons (VAChT-IR) account for a great part of the peri-plaque dystrophic elements. Cholinergic dystrophic terminals are characteristically found in the immediate proximity of amyloid material, a localization confirmed by electron microscopy, while the non-cholinergic (VAChT-negative) dystrophic neurites were found at some distance from the amyloid plaque periphery.

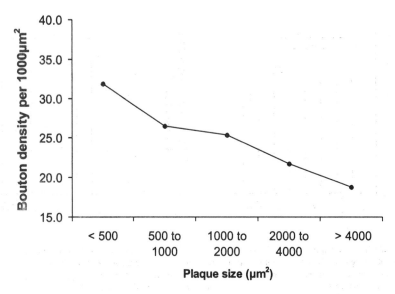

Fig 2. Note that the number of VAChT-IR boutons (density) in the peri-plaque neuropile is affected by the size of plaques. There is a negative correlation between A-beta plaque size and cholinergic bouton density. From Hu et al, in preparation.

4. DISCUSSION AND CONCLUSIONS

These observations point to a selective vulnerability of the cortical cholinergic input to amyloid burden. The cholinergic vulnerability to AD-like pathology in tg mice is similar to the age-related changes observed at light and electron microscopy level in cortical cholinergic synapses.[9] It is possible that the early, pre-plaque stages of the amyloid pathology provoke a (aberrant?) sprouting of cholinergic terminals. If that is the case in the human species, a similar temporary up-regulation of the cholinergic system might precede cholinergic losses in AD, perhaps at pre-clinical stages. Suitable markers and advancements in imaging techniques could test such a scenario and, if confirmed, provide an early biological warning of an on-going neuropathological evolution.

The fact that the number of cholinergic terminals can be affected by presumably non-aggregated Aß peptides present in the extracellular space is not surprising. Aß proteins seem to affect the function of cholinergic terminals in a direct fashion (for review see[10]). Indeed, the steady state number of cortical cholinergic boutons is dependent on minute amounts of endogenous growth factors as the blockade of TrkA receptors induces the retrieval of pre-existent cholinergic synapses in a matter of weeks.[11]

Higher concentrations of extracellular amyloidogenic materials would seemingly affect most neuronal populations as the up-regulation in the number of synaptophysin-IR boutons adjacent to plaques would indicate. The aggregation of the Aß peptide in plaques visible at the light microscopy level provokes grossly distorted neurites. The preferential involvement of cholinergic elements amongst surrounding plaques in tg models have been highlighted by Sturchler-Pierrat and collaborators[12] and ourselves.[4] The highly concentrated milieu containing fibrillar Aß material might have both a tropic and toxic influence on neuritic processes. Thus, in transgenic models, Phinney and collaborators[13] have found that aggregation of Aß material induces the build up of dystrophic axonal elements. Likewise, in human post-mortem material, Knowles and colleagues[14] observed that dendrites crossing through senile plaques radically changed their morphology.

In sum, transgenic mouse models indicate that amyloid pathology is sufficient to have a profound impact on the synaptic organization of the cerebral cortex. It is reasonable to assume that these morphological and functional effects are ultimately responsible for the severe cognitive deficits observed in AD. The observations also point to an early and specific involvement of the cortical cholinergic input as a direct consequence of amyloid burden.

5. ACKNOWLEDGEMENTS

This work has been supported by a grant from the Canadian Institutes of Health Research (MOP-37996) to ACC.

REFERENCES

1. V. Bigl, T. Arendt, S. Fischer, S. Fischer, M. Werner, and A. Arendt, The cholinergic system in aging, *Gerontology* **33**, 172-180 (1987).

2. K. Hsiao, P. Chapman, S. Nilsen, C. Eckman, Y. Harigaya, S. Younkin, Yang, F, and G. Cole, Correlative memory deficits, Aβ elevation, and amyloid plaques in transgenic mice, *Science* **274**, 99-102 (1996).

3. K. Duff, C. Eckman, C. Zehr, X. Yu, C. M. Prada, J. Perez-tur, M. Hutton, L. Buee, Y. Harigaya, D. Yager, D. Morgan, M. N. Gordon, L. Holcomb, L. Refolo, B. Zenk, J. Hardy, and S. Younkin, Increased amyloid-beta42(43) in brains of mice expressing mutant presenilin 1, *Nature* **383**, 710-713 (1996).

4. T. P. Wong, T. Debeir, K. Duff, and A. C. Cuello, Reorganization of cholinergic terminals in the cerebral cortex and hippocampus in transgenic mice carrying mutated presenilin-1 and amyloid precursor protein transgenes, *J.Neurosci.* **19**, 2706-2716 (1999).

5. K. B. J. Franklin and G. Paxinos, *The Mouse Brain in Stereotaxic Coordinates* (Academic Press, San Diego, 1997).

6. A. Ribeiro-da-Silva, J. V. Priestley, and A. C. Cuello, Pre-embedding ultrastructural immunocytochemistry, in: *Immunohistochemistry II*, edited by A. C. Cuello (John Wiley & Sons, Chichester, 1993), pp 181-227.

7. A. Merighi and J. M. Polak, Post-embedding immunogold staining, in: *Immunohistochemistry II*, edited by A. C. Cuello (John Wiley & Sons, Chichester, 1993), pp 229-264.

8. S. Côté, A. Ribeiro-da-Silva, and A. C. Cuello, Current protocols for light microscopy immunocytochemistry, in: *Immunohistochemistry II*, edited by A. C. Cuello (John Wiley & Sons, Chichester, 1993), pp 147-168.

9. P. Turrini, M. A. Casu, T. P. Wong, Y. De Koninck, A. Ribeiro-da-Silva, and A. C. Cuello, Cholinergic nerve terminals establish classical synapses in the rat cerebral cortex: synaptic pattern and age-related atrophy, *Neuroscience* In press (2001).

10. D. S. Auld, S. Kar, and R. Quirion, Beta-amyloid peptides as direct cholinergic neuromodulators: a missing link? *Trends Neurosci.* **21**, 43-49 (1998).

11. T. Debeir, H. U. Saragovi, and A. C. Cuello, A nerve growth factor mimetic TrkA antagonist causes withdrawal of cortical cholinergic boutons in the adult rat, *Proc.Natl.Acad.Sci.USA* **96**, 4067-4072 (1999).

12. C. Sturchler-Pierrat, D. Abramowski, M. Duke, K. H. Wiederhold, C. Mistl, S. Rothacher, B. Ledermann, K.

Bürki, P. Frey, P. A. Paganetti, C. Waridel, M. E. Calhoun, M. Jucker, A. Probst, M. Staufenbiel, and B. Sommer, Two amyloid precursor protein transgenic mouse models with Alzheimer disease-like pathology, *Proc.Natl.Acad.Sci.USA* **94**, 13287-13292 (1997).

13. A. L. Phinney, T. Deller, M. Stalder, M. E. Calhoun, M. Frotscher, B. Sommer, M. Staufenbiel, and M. Jucker, Cerebral amyloid induces aberrant axonal sprouting and ectopic terminal formation in amyloid precursor protein transgenic mice, *J.Neurosci.* **19**, 8552-8559 (1999).

14. R. B. Knowles, C. Wyart, S. V. Buldyrev, L. Cruz, B. Urbanc, M. E. Hasselmo, H. E. Stanley, and B. T. Hyman, Plaque-induced neurite abnormalities: implications for disruption of neural networks in Alzheimer's disease, *Proc.Natl.Acad.Sci.U.S.A* **96**, 5274-5279 (1999).

MECHANISM OF NEURONAL DEATH IN ALZHEIMER'S DISEASE

A transgenic mouse perspective

Michael C. Irizarry

1. INTRODUCTION

Human amyloid precursor protein (hAPP) transgenic mice with cerebral amyloid β-protein (Aβ) deposition have been used to model several of pathophysiological processes implicated in Alzheimer's disease (AD), namely age related amyloid deposition in cortical and limbic regions, amyloid associated gliosis, neuritic changes, and behavioral deficits.[1-5] Two features have not been well represented in hAPP transgenic mice. Despite phospho-tau epitopes present in the periphery of fibrillar amyloid deposits, neurofibrillary tangles do not form. More notable, however, is the absence of the degree of neuronal death seen in AD, where up to 50% of neurons are lost in CA1 and entorhinal cortex.[6-10] This chapter will develop the idea that a subset of amyloid deposits in hAPP transgenic mice is focally toxic to neurons. Since the primary toxic effects in these mice appear to be mediated by amyloid deposition, the second part of the chapter will review how amyloid deposition can be modulated in hAPP transgenic mice. The mice we have studied are detailed in Table 1. [1, 2, 11-16]

2. FOCAL NEURONAL LOSS IN APP TRANSGENIC MICE

Stereologic neuronal counts in PDAPP mice through the age of 18 months, Tg2576 mice through the age of 16 months, and PSAPP double transgenic mice through the age of 12 months do not demonstrate significant neuronal loss in cortical regions or the CA1 subfield of the hippocampus, despite amyloid deposition comparable to or in excess of that seen in human AD.[7, 8, 10] Qualitative observation demonstrates that some amyloid

* Michael C. Irizarry, Alzheimer Disease Research Unit, Massachusetts General Hospital—East, B114-2020, 114 16th Street, Charlestown, MA 02129.

Mapping the Progress of Alzheimer's and Parkinson's Disease
Edited by Mizuno *et al.*, Kluwer Academic/Plenum Publishers, 2002

133

Table 1. Transgenic mice, knockout mice, and crosses reviewed in this chapter

Designation	Transgene	Promoter	Age of Aβ deposition (mo)
hAPP Transgenic Mice			
PDAPP (hAPP$_{V717F}$)[1]	hAPP770 V717F minigene	human PDGF-b	6-9 het, 3 hom
Tg2576 (hAPP$_{Sw}$)[2]	hAPP695 K670N-M671L	hamster PrP	9-11
Other Mice			
hPS1 M146L[11]	hPS1 M146L	human PDGF-b	no deposition
APOE KO[12]			no deposition
MSR KO[13]			no deposition
Crosses			
PSAPP[14]	Tg2576 x hPS1 M146L		3-4
PDAPP hom x APOE KO[15]	PDAPP hom x APOE KO		9
Tg2576 x APOE KO[16]	Tg2576 x APOE KO		12
Tg2576 x MSR KO	Tg2576 x MSR KO		9-11

mo—months; hAPP—human amyloid precursor protein; PDGF-b—platelet derive growth factor b-chain; het—heterozygous; hom—homozygous; Sw—Swedish mutation; hPS1—human presenilin; APOE—apolipoprotein E; MSR—macrophage scavenger receptor.

deposits in hAPP transgenic mice appear to alter the cortical laminar cryoarchitecture. We utilized the radial density function[17, 18]—neuronal density as a function of distance from the center of the Aβ deposit—to determine the local effects of Aβ on neurons in the cortex of 12 month old PSAPP and Tg2576 mice. Non-fibrillar deposits—Aβ immunoreactive deposits that do not stain with thioflavin S (thioS)—had no effect on neuronal density within and outside Aβ deposits. ThioS-stained fibrillar Aβ deposits, however, were associated with a decline in neuronal density of over 50% within plaques, with normal neuronal density immediately outside the edge of the thioS staining. There was no increase in neuronal density at the edge of the thioS plaques to suggest a non-toxic displacement of neurons; and there was no reduction of neuronal density outside the edge of plaques to suggest a toxic "penumbra". These data are compatible with the idea that fibrillar, but not diffuse Aβ deposits, are toxic to neurons. It follows, therefore, that perhaps the best way to affect neuronal toxicity in hAPP transgenic mice is by altering Aβ deposition.

3. MODULATION OF AMYLOID DEPOSITION IN APP TRANSGENIC MICE

Figure 1 presents a simplified model of Aβ metabolism in hAPP transgenic mice. Mutant human APP is processed by β-secretase (BACE) and γ-secretase (PS-1) to produce Aβ, which can deposit as diffuse plaques, and mature to compact plaques.

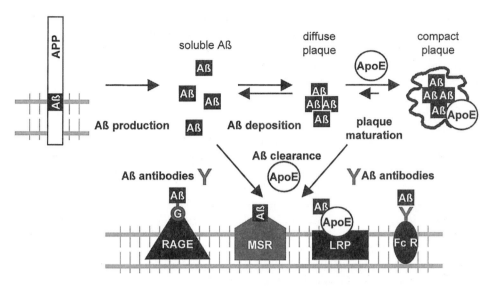

Figure 1. Model of APP and Aβ metabolism in hAPP transgenic mice. Mutant APP is processed by BACE and PS-1 to produce soluble Aβ which can deposit as diffuse or compact (fibrillar) plaques, influenced by chaperone proteins such as apoE. Soluble or plaque associated Aβ may be cleared by receptor mediated processes.

Putative clearance mechanisms for soluble or plaque associated Aβ include the macrophage scavenger receptor (MSR), low density lipoprotein receptor related protein (LRP), the receptor for advanced glycation end-products (RAGE), and antibody-Fc receptor mediated processes.

3.1 Presenilin

The hPS1 M146L transgene accelerates both Aβ and thioS-positive amyloid deposition in the Tg2576 hAPP$_{Sw}$ transgenic mouse.[14] By the age of 12 months, the PSAPP double transgenic mouse has a 20-fold increase of Aβ deposition in cortex and hippocampus relative to the Tg2576 hAPP$_{Sw}$ single transgenic.[10]

3.2 Apolipoprotein E

Crosses of apolipoprotein E (apoE) knockout mice with Tg2576 hAPP$_{Sw}$ mice and PDAPP hAPP$_{V717F}$ mice demonstrate that apoE is required for the development of fibrillar Aβ deposits, neuritic dystrophy, and amyloid angiopathy in hAPP transgenic mice.[15, 16, 19-21] Fibrillar Aβ deposits are absent and diffuse Aβ deposition is reduced in Tg2576 hAPP$_{Sw}$ mice lacking apoE relative to mice containing apoE; nonetheless, the characteristic cortical and limbic pattern of Aβ deposition is maintained.[16] This is in contradistinction to 12-month old homozygous PDAPP hAPP$_{V717F}$ mice lacking apoE, which also lack fibrillar Aβ deposits, but in whom the anatomical localization of diffuse Aβ is altered.[19-21] In the PDAPP hAPP$_{V717F}$ mice containing apoE, focal Aβ deposits occur in all cortical layers, the outer molecular layer of the dentate gyrus, and all layers of the hippocampus. In PDAPP hAPP$_{V717F}$ mice lacking apoE, diffuse Aβ is confined to the deep layers of cortex, and the stratum radiatum and oriens of the CA1/CA3 hippocampal subfields; the dentate gyrus molecular layer and CA1 stratum lacunosum/moleculare is entirely spared. This suggests an additional role for apoE in the anatomical specificity of amyloid deposition in hAPP transgenic mice.

3.3 Macrophage Scavenger Receptor

The macrophage scavenger receptor (MSR) is a receptor for negatively charged macromolecules, such as oxidized-LDL, which has been implicated in microglial migration toward Aβ, and microglial clearance of Aβ.[22-24] One would predict that hAPP transgenic mice lacking MSR would have reduced clearance of Aβ, reduced inflammatory response to plaques, and increased Aβ deposition. However, elimination of MSR in hAPP$_{V717F}$ mice did not affect amyloid deposition or synaptic degeneration.[25] We have extended these studies to Tg2576 hAPP$_{Sw}$ mice. 7 Tg2576 mice lacking MSR and 9 containing MSR were evaluated from 9-16 months of age, and demonstrated the same magnitude and anatomical specificity of Aβ and thioS-positive amyloid deposition, as well as comparable microglial response to Aβ deposits (Figure 2). Thus, MSR does not appear to be the principal active Aβ clearance mechanism or microglial Aβ chemotactic receptor in these mice; there may be redundant processes involving several receptors besides MSR (e.g. RAGE, LRP, macrophage chemoattractant protein receptor, and Fc receptor).

3.4 Antibody Therapy

An Aβ clearance mechanism that has generated considerable enthusiasm over the past two years has been antibody mediated Aβ uptake. PDAPP mice actively and passively immunized against Aβ demonstrate markedly reduced Aβ deposition.[26, 27] Recent studies using in vivo two-photon imaging of living mouse brain via burrhole demonstrate that Aβ deposits can be cleared by topically applied Aβ antibody within 3 days.[28] In addition to demonstrating that Fc receptors can clear antibody-bound Aβ, these studies imply the reversibility of a primary pathological feature of AD in hAPP transgenic mice.

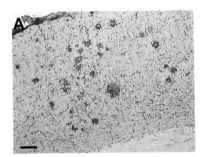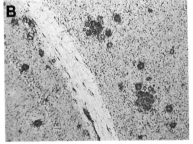

Figure 2. Microglia (stained with biotinylated tomato lectin) associate with amyloid plaques in the cortex of Tg2576 hAPP$_{Sw}$ mice containing MSR (A) and lacking MSR (B). Scale bar 100μm.

4. CONCLUSION

In the Tg2576 and PSAPP mice, the principle neurotoxic effects of Aβ occur within thioS-positive fibrillar Aβ deposits, with focal neuronal loss limited to Aβ deposits in a β-pleated sheet conformation. These toxic amyloid deposits in hAPP transgenic mice are accelerated by mutant PS-1, eliminated in mice lacking apoE, reduced in mice treated with active or passive Aβ immunization, and unaffected in mice lacking MSR. Our data show that fibrillar Aβ toxicity—which can be modulated by PS-1, apoE, and Aβ antibody—is one of the mechanisms of neuronal death and neuritic dystrophy active in hAPP transgenic mice, and by extension, in AD itself.

5. ACKNOWLEGMENTS

All studies were performed in collaboration with B. T. Hyman (MGH). Mice were provided by and studies were performed in collaboration with the following: Tg2576—Karen K. Hsiao-Ashe (U. Minn.); PDAPP—D. Games, D. Schenk (Elan Pharmaceuticals); PSAPP—K. Duff (Nathan Kline Institute, NYU); Tg2576 x *APOE* KO—D. Holtzman (Wash U., St. Louis); PDAPP x *APOE* KO—K. Bales, S. M. Paul (Lilly); *MSR* KO—M. Freeman (MGH). Radial density analysis was performed by L. Cruz-Cruz, B. Urbanc, and G. Stanley (Boston U.) Two-photon studies were performed by B. Bacskai and S. Kajdasz (MGH).

REFERENCES

1. D. Games, D. Adams, R. Alessandrini, *et al.*, Alzheimer-type neuropathology in transgenic mice overexpressing V717F β-amyloid precursor protein, *Nature* **373**, 523-527 (1995).
2. K. Hsiao, P. Chapman, S. Nilsen, C. Eckman, Y. Harigaya, S. Younkin, F. Yang, and G. Cole, Correlative memory deficits, Aβ elevation and amyloid plaques in transgenic mice, *Science* **274**, 99-102 (1996).
3. C. Sturchler-Pierrat, D. Abramowski, M. Duke, K. H. Wiederhold, C. Mistl, S. Rothacher, B. Ledermann, K. Burki, P. Frey, P. A. Paganetti, C. Waridel, M. E. Calhoun, M. Jucker, A. Probst, M. Staufenbiel, and B. Sommer, Two amyloid precursor protein transgenic mouse models with Alzheimer disease-like pathology, *Proc. Natl. Acad. Sci. U.S.A.* **94**, 13287-13292 (1997).
4. D. R. Borchelt, T. Ratovitski, J. van Lare, M. K. Lee, V. Gonzales, N. A. Jenkins, N. G. Copeland, D. L. Price, and S. S. Sisodia, Accelerated amyloid deposition in the brains of transgenic mice coexpressing mutant presenilin 1 and amyloid precursor proteins, *Neuron* **19**, 939-945 (1997).
5. D. Moechars, I. Dewachter, K. Lorent, D. Reverse, V. Baekelandt, A. Naidu, I. Tesseur, K. Spittaels, C. V. Haute, F. Checler, E. Godaux, B. Cordell, and F. Van Leuven, Early phenotypic changes in transgenic mice that overexpress different mutants of amyloid precursor protein in brain, *J. Biol. Chem.* **274**, 6483-6492 (1999).
6. T. Gomez-Isla, J. L. Price, D. W. McKeel, J. C. Morris, J. H. Growdon, and B. T. Hyman, Profound loss of layer II entorhinal cortex neurons occurs in very mild Alzheimer's disease, *J. Neurosci.* **16**, 4491-4500 (1996).
7. M. C. Irizarry, F. Soriano, M. McNamara, K. J. Page, D. Schenk, D. Games, and B. T. Hyman, Aβ deposition is associated with neuropil changes, but not with overt neuronal loss in the PDAPP transgenic mice, *J. Neurosci.* **17**, 7053-7059 (1997).
8. M. C. Irizarry, M. McNamara, K. Fedorchak, K. Hsiao, and B. T. Hyman, APP-Sw transgenic mice develop age related amyloid deposits and neuropil abnormalities, but no neuronal loss in CA1, *J. Neuropath. Exp. Neurol.* **56**, 965-973 (1997).
9. M. E. Calhoun, K. H. Wiederhold, D. Abramowski, A. L. Phinney, A. Probst, C. Sturchler-Pierrat, M. Staufenbiel, B. Sommer, and M. Jucker, Neuron loss in APP transgenic mice, *Nature* **395**, 755-756 (1998).

10. A. Takeuchi, M. C. Irizarry, K. Duff, T. C. Saido, K. Hsiao Ashe, M. Hasegawa, D. M. Mann, B. T. Hyman, and T. Iwatsubo, Age-related amyloid beta deposition in transgenic mice overexpressing both Alzheimer mutant presenilin 1 and amyloid beta precursor protein Swedish mutant is not associated with global neuronal loss, *American Journal of Pathology* **157**, 331-339 (2000).

11. K. Duff, C. Eckman, C. Zehr, X. Yu, C. M. Prada, J. Perez-tur, M. Hutton, L. Buee, Y. Harigaya, D. Yager, D. Morgan, M. N. Gordon, L. Holcomb, L. Refolo, B. Zenk, J. Hardy, and S. Younkin, Increased amyloid-β42(43) in brains of mice expressing mutant presenilin 1, *Nature* **383**, 710-713 (1996).

12. S. H. Zhang, R. L. Reddick, J. A. Piedrahita, and N. Maeda, Spontaneous hypercholesterolemia and arterial lesions in mice lacking apolipoprotein E, *Science* **258**, 468-471 (1992).

13. H. Suzuki, Kurihara, M. Takeya, *et al.*, A role for macrophage scavenger receptors in atherosclerosis and susceptibility to infection, *Nature* **386**, 292-296 (1997).

14. L. Holcomb, M. N. Gordon, E. McGowan, X. Yu, S. Benkovic, P. Jantzen, K. Wright,, I. Saad, R. Mueller, D. Morgan, S. Sanders, C. Zehr, K. O'Campo, J. Hardy, C. M. Prada, C. Eckman, S. Younkin, K. Hsiao, and K. Duff, Accelerated Alzheimer-type phenotype in transgenic mice carrying both mutant amyloid precursor protein and presenilin 1 transgenes, *Nature Medicine* **4**, 97-100 (1998).

15. K. Bales, T. Verina, D. J. Cummins, Y. Du, R. C. Dodel, J. Saura, C. E. Fishman, C. A. DeLong, P. Piccardo, V. Petegnief, B. Ghetti, and S. M. Paul, Lack of apolipoprotein E dramatically reduces amyloid β-protein deposition, *Nature Genetics* **17**, 263-265 (1997).

16. D. M. Holtzman, A. M. Fagan, B. Mackey, T. Tenkova, L. Sartorius, S. M. Paul, K. Bales, K. H. Ashe, M. C. Irizarry, and B. T. Hyman, Apolipoprotein E facilitates neuritic and cerebrovascular plaque formation in an Alzheimer's disease model, *Annals of Neurology* **47**, 739-747 (2000).

17. L. Cruz, B. Urbanc, S. V. Buldyrev, R. Christie, T. Gomez-Isla, S. Havlin, M. McNamara, H. E. Stanley, and B. T. Hyman, Aggregation and disaggregation of senile plaques in Alzheimer's disease, *Proc. Natl. Acad. Sci. U.S.A.* **94**, 7612-7616 (1997).

18. S. V. Buldyrev, L. Cruz, T. Gomez-Isla, E. Gomez-Tortosa, S. Havlin, R. Le, H. E. Stanley, B. Urbanc, and B. T. Hyman, Description of microcolumnar ensembles in association cortex and their disruption in Alzheimer and Lewy body dementias, *Proc. Natl. Acad. Sci. U.S.A.* **97**, 5039-5043 (2000).

19. M. C. Irizarry, G. W. Rebeck, B. Cheung, K. Bales, S. M. Paul, D. Holzman, and B. T. Hyman, Modulation of Aβ deposition in APP trangenic mice by an apolipoprotein E null background, *Ann. N.Y. Acad. Sci.* **920**, 171-178 (2000).

20. K. R. Bales, T. Verina, D. J. Cummins, Y. Du, R. C. Dodel, J. Saura, C. E. Fishman, C. A. DeLong, P. Piccardo, V. Petegnief, B. Ghetti, and S. M. Paul, Apolipoprotein E is essential for amyloid deposition in the APP(V717F) transgenic mouse model of Alzheimer's disease, *Proc. Natl. Acad. Sci. U.S.A.* **96**, 15233-15238 (1999).

21. M. C. Irizarry, B. S. Cheung, G. W. Rebeck, S. M. Paul, K. R. Bales, and B. T. Hyman, Apolipoprotein E affects the amount, form, and anatomical distribution of Aβ deposition in homozygous APPV717F transgenic mice, *Acta Neuropathologica* **100**, 451-458 (2000).

22. J. El Khoury, S. E. Hickman, C. A. Thomas, L. Cao, S. C. Silverstein, and J. D. Loike, Scavenger receptor-mediated adhesion of microglia to beta-amyloid fibrils, *Nature* **382**, 716-719 (1996).

23. D. M. Paresce, R. N. Ghosh, and F. R. Maxfield, Microglial cells internalize aggregates of the Alzheimer's disease amyloid beta-protein via a scavenger receptor, *Neuron* **17**, 553-565 (1996).

24. H. Chung, M. I. Brazil, M. C. Irizarry, B. T. Hyman, and F. R. Maxfield, Uptake of fibrillar β-amyloid by microglia isolated from MSR-A (type I and type II) knockout mice, *NeuroReport* **12**, 1151-4 (2001).

25. F. Huang, M. Buttini, T. Wyss-Coray, L. McConlogue, T. Kodama, R. E. Pitas, and L. Mucke, Elimination of the class A scavenger receptor does not affect amyloid plaque formation or neurodegeneration in transgenic mice expressing human amyloid protein precursors, *American Journal of Pathology* **155**, 1741-1747 (1999).

26. D. Schenk, R. Barbour, W. Dunn, G. Gordon, H. Grajeda, T. Guido, K. Hu, J. Huang, K. Johnson-Wood, K. Khan, D. Kholodenko, M. Lee, Z. Liao, I. Lieberburg, R. Motter, L. Mutter, F. Soriano, G. Shopp, N. Vasquez, C. Vandevert, S. Walker, M. Wogulis, T. Yednock, D. Games, and P. Seubert, Immunization with amyloid-beta attenuates Alzheimer-disease-like pathology in the PDAPP mouse, *Nature* **400**, 173-177 (1999).

27. F. Bard, C. Cannon, R. Barbour, *et al.*, Peripherally administered antibodies against amyloid beta-peptide enter the central nervous system and reduce pathology in a mouse model of Alzheimer disease, *Nature Medicine* **6**, 916-919 (2000).

28. B. Bacskai, S. T. Kajdasz, R. H. Christie, C. Carter, D. Games, P. Seubert, D. Schenk, and B. T. Hyman, Imaging of amyloid-β deposits in brains of living mice permits direct observation of clearance of plaques with immunotherapy, *Nature Medicine* **7**, 369-372 (2001).

INTERACTIONS OF BETA-AMYLOID WITH THE FORMATION OF NEUROFIBRILLARY TANGLES IN TRANSGENIC MICE

Jürgen Götz, Feng Chen and Roger M. Nitsch*

1. INTRODUCTION

Alzheimer's disease (AD) as the most frequent cause of senile dementia is histopathologically characterized by extracellular β-amyloid-containing plaques (that mainly consist of aggregated Aβ peptide derived by proteolysis of the amyloid precursor protein, APP), intracellular neurofibrillary tangles (NFTs), reduced synaptic density, and neuronal losses in selected brain areas. NFTs are composed of hyperphosphorylated microtubule-associated protein tau. They are, in the absence of amyloid plaques, also abundant in additional neurodegenerative diseases including Pick's disease (PiD), progressive supranuclear palsy (PSP), corticobasal degeneration (CBD), argyrophilic grain disease (AgD), and frontotemporal dementia with Parkinsonism linked to chromosome 17 (FTDP-17).

A consistent feature of tau in these tauopathies is its abnormal phosphorylation and relocalization from axonal to somatodendritic compartments where it accumulates in filamentous aggregates that eventually assemble into NFTs (1, 2). In contrast to AD, where tau forms filaments only in neurons, numerous tau filament-containing glial cells are present in a variety of tauopathies including PSP and CBD (3, 4).

The recent discovery of mutations in the tau gene in FTDP-17 has established that dysfunction of tau in itself, in the absence of β-amyloid-containing plaques, can cause neurodegeneration, and lead to dementia (5-7).

---

* University of Zurich, Division of Psychiatry Research, August Forel Strasse 1, CH-8008 Zurich, nitsch@bli.unizh.ch or goetz@bli.unizh.ch

Mapping the Progress of Alzheimer's and Parkinson's Disease
Edited by Mizuno *et al.*, Kluwer Academic/Plenum Publishers, 2002

139

Here, we review recent advances that have been made in the development of various tau transgenic models. These are excellent tools to analyze filament formation and to determine the relationship between the Aβ and the tau pathology in AD. The question how these lesions are pathogenically related to each other, was resolved by two independent experimental approaches: Synthetic Aβ fibrils injected into brains of mutant tau transgenic mice, and doubly transgenic mice expressing both mutant tau and mutant APP demonstrated causal interactions between APP/Aβ and tau that lead to increased NFT formation.

2. TAU, TAU MUTATIONS AND TAUOPATHIES

The microtubule-associated proteins MAP2 and tau are expressed together in most neurons, where they localize to separate subcellular compartments. MAP2 is largely found in dendrites, whereas tau is concentrated in axons. Tau has also been found in astrocytes and oligodendrocytes, although, under physiological conditions, levels are very low. Under physiological conditions, tau is a phosphoprotein; under pathological conditions, it is hyperphosphorylated. Hyperphosphorylation means that tau is phosphorylated to a higher degree at physiological sites and at additional sites. Phosphorylation tends to dissociate tau from its natural partner, the microtubule. Since this increases the soluble pool of tau it might be an important first step in generating protein for the assembly of tau filaments.

Linkage of a familial dementia to chromosome 17q21-22 was reported in 1994, followed by publications that showed linkage of additional inherited dementias with preceding personality changes to the same region of chromosome 17. These diseases were grouped under the heading of "frontotemporal dementia and parkinsonism linked to chromosome 17" (FTDP-17) in a consensus conference. In 1998, the tau gene was sequenced in FTDP-17 families, and missense mutations and intronic mutations in the 5' splice site of the alternatively spliced exon 10 were reported by three groups (5-7). In contrast to AD, where hyperphosphorylated tau forms filaments only in neurons, numerous tau filament-containing glial cells are also present in a variety of tauopathies including PSP and CBD.

Until today, it remains unknown whether tau filaments cause neuronal degeneration or whether filament formation is only a consequence of a more general degenerative process. Studies of the sequence of events that underlie the formation of NFTs may clarify the role of genetic and epigenetic factors in the pathogenesis of this lesion. This is best achieved in experimental animal models that reproduce the neuropathological characteristics of these diseases (8).

3. ANIMAL MODELS

The first human tau transgenic model in mice expressed the longest human four-repeat (4R) tau isoform under control of the human Thy1 promoter (9). Levels of human relative to endogenous murine tau were approximately 10%, as determined by immunoblot analysis of whole brain extracts. Strongly labeled neurons were observed in most brain regions. Their numbers were relatively small, accounting for only a few percent of the total nerve cell population. However, as in AD, transgenic human tau protein was hyperphosphorylated and present in nerve cell bodies, axons and dendrites. Tau filament formation was not observed, and tau staining appeared homogenous or granular, but not fibrillar. A very similar phenotype was reported in mice with neuronal expression of the shortest human tau isoform (10). Together, both mouse strains showed early changes associated with the development of NFTs in AD and related disorders, but they failed to produce NFTs and lacked obvious neurological symptoms.

Subsequently, stronger promoters were chosen by several groups for expression of human tau in transgenic mice. The murine PrP promoter was used to express the shortest human tau isoform (11), and the murine Thy1.2 expression vector was used by two groups to achieve high expression levels of the longest human tau isoform (12, 13). Two different genetic backgrounds (FVB and B6D2F1) were used to generate these latter transgenic strains. Tau hyperphosphorylation (14), somatodendritic localization, and the presence of axonal spheroids were reported for all three models. Evidence for axonal degeneration, including axonal breakdown and segmentation of myelin into ellipsoids ("digestion chambers") was obtained in anterior spinal roots of transgenic mice. To determine filament formation in the transgenic mice, silver impregnation techniques and electron microscopy was applied, however, none of the transgenic models developed tau filaments.

With the identification of the first FTDP-17 associated mutations in tau in 1998, several research groups expressed these mutations in transgenic mice. A human 4R tau isoform lacking the two amino-terminal inserts was expressed together with mutation P301L under control of the murine PrP promoter (15). Line JNPL3 expressed mutant tau at levels comparable to endogenous tau. By 10 months of age, most of the mice developed motor and behavioural disturbances that were more pronounced than reported for wild-type tau transgenic models. NFTs were identified in the diencephalon, brainstem, cerebellum and spinal cord by Congo red, thioflavin-fluorescent microscopy, and Gallyas silver stainings. In a second model (pR5 line) the same mutation was expressed, but the longest human 4R tau isoform was expressed under control of the mThy1.2 promoter (16). NFTs were identified by Gallyas silver stainings and thioflavin S-fluorescent microscopy, and sarcosyl-extracted tau filaments expressed several phosphoepitopes of tau. The FTDP-17 associated mutations V337M and G272V were expressed also in transgenic mice. V337M mutant mice formed neurofibrillary lesions in the hippocampus and neocortex, and showed progressive behavioral changes in the elevated plus maze (A. Takashima, pers. communication). The G272V mutation was expressed in brain by combining a PrP promoter-driven expression system with an autoregulatory, tetracycline-regulatable transactivator loop that resulted in high expression in some neurons and oligodendrocytes. Electron microscopy established filament formation in oligodendrocytes (17). In summary, the FTDP-17 mutant tau transgenic models are suitable to determine which phosphoepitopes of tau are important for filament formation in vivo, which cells are susceptible to filament formation, and how filaments interfere with glial and neuronal functions including axonal transport. Eventually, the components of the pathological signaling cascade in AD and FTDP-17 will be identified.

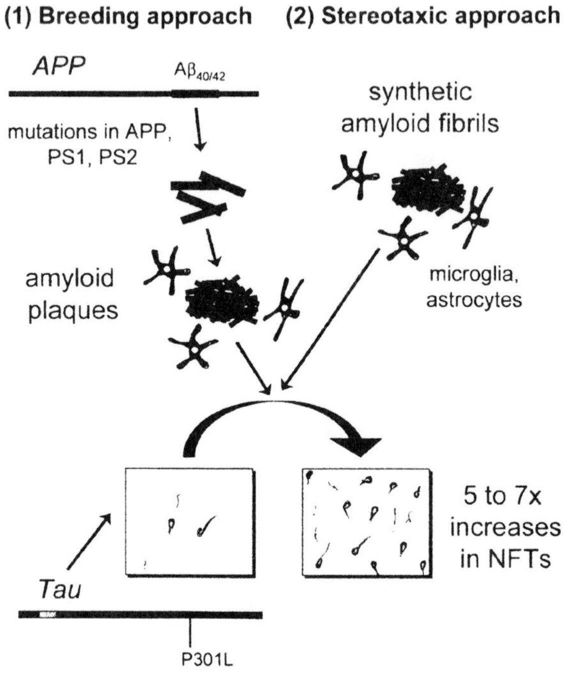

Figure 1. In Alzheimer's disease, mutations in the β-amyloid precursor protein APP itself or in the genes for presenilin 1 and 2 (PS1, PS2) lead to increased production of Aβ peptide (Aβ40, Aβ42). Aβ peptide accumulates and forms amyloid plaques. Microglial cells and astrocytes are likely to be involved in this process. As a consequence of Aβ production, according to the amyloid cascade hypothesis, tau becomes hyperphosphorylated by yet unknown mechanisms and forms filaments which eventually fill up the entire soma of dying cell, and neurofibrillary tangles (NFTs) develop. Transgenic expression of mutant APP lead to amyloid plaque formation but failed to induce NFT formation. In contrast, mice transgenic for FTDP-17 associated mutations formed tau filaments, in the absence of amyloid plaque formation. To determine the relationship between the amyloid and the tau pathology, two alternative approaches were pursued. One involved the intercrossing of APP and Tau mutant mice with plaque and NFT pathology, respectively (breeding approach), the other stereotaxic injection of aggregated, fibrillar Aβ42 peptide into mutant tau transgenic brains (stereotaxic approach). Both approaches resulted in increased NFT formation proving an interaction between APP/Aβ and tau, and thereby confirming the amyloid cascade hypothesis in mice.

4. THE AMYLOID CASCADE HYPOTHESIS: RELATION BETWEEN AMYLOID DEPOSITION AND NEUROFIBRILLARY TANGLE FORMATION

FTDP-17 mutant tau transgenic mice are suitable tools to test the "amyloid cascade" hypothesis that proposes a central role of β-amyloid in the pathogenesis of AD. The precise relation between β-amyloid deposition and NFT formation is unknown. In rhesus monkeys, synthetic A β_{42} fibrils can induce tau-containing NFTs, but this toxicity is highly dependent on species and age: It is more pronounced than in marmoset monkey, and not significant at all in rats or mice, and the toxicity increases with advancing age (18). Currently available transgenic mouse models that co-express human APP and FAD mutations are only partially suited to test this hypothesis, because the development of amyloid plaques in these mice is associated with only marginal neuronal damage, and no NFT formation (19-21).

We addressed the controversy over how the Aβ and the tau pathology are pathogenically related to each other in AD by injecting synthetic $A\beta_{42}$ fibrils stereotaxically into the somatosensory cortex and the hippocampus of five to six month-old P301L tau transgenic pR5 mice and non-transgenic littermates (22). As control peptide, the reversed sequence, Aβ $_{42-1}$, was used (Figure 1). Eighteen days following the injections of $A\beta_{42}$, Gallyas silver impregnation revealed fivefold increases of NFTs, along with neuropil threads and degenerating neurites in the amygdala of P301L, but not wild-type mice. Lewis et al. crossed APP transgenic Aβ-producing mice with tau filament-forming P301L mutant JNPL3 mice (Figure 1) (23). In doubly transgenic mice, the NFT pathology was substantially enhanced in the limbic system and olfactory cortex whereas plaque formation was unaffected by the presence of the tau lesions. Compared to P301L transgenic mice, doubly transgenic mice, at 9-11 moths of age, showed a more than sevenfold increase in the density of NFTs in the olfactory bulb, entorhinal cortex and amygdala.

Together, both experiments demonstrated pathologically meaningful interactions between APP/Aβ and tau that lead to increased NFT formation (Figure 1). Besides their major advantages for understanding the pathophysiology of NFT formation, these models are useful for the development of therapies designed to reduce NFT formation.

5. ACKNOWLEDGEMENTS

This research was supported in parts by grants from the Bayer Pharma AG (BARN; Bayer Alzheimer Research Network), the SNF and by the NCCR "Neuronal plasticity and repair".

REFERENCES

1. M. Goedert, M. G. Spillantini, R. Jakes, R. A. Crowther, E. Vanmechelen, A. Probst, et al., Molecular dissection of the paired helical filament, *Neurobiol Aging* **16**, 325-334 (1995).
2. L. Buee, T. Bussiere, V. Buee-Scherrer, A. Delacourte, and P. R. Hof, Tau protein isoforms, phosphorylation and role in neurodegenerative disorders, *Brain Res Brain Res Rev* **33**, 95-130 (2000).
3. M. B. Delisle, J. R. Murrell, R. Richardson, J. A. Trofatter, O. Rascol, X. Soulages, et al., A mutation at codon 279 (N279K) in exon 10 of the Tau gene causes a tauopathy with dementia and supranuclear palsy, *Acta Neuropathol (Berl)* **98**, 62-77 (1999).

4. T. Komori, Tau-positive glial inclusions in progressive supranuclear palsy, corticobasal degeneration and Pick's disease, *Brain Pathol* **9**, 663-679 (1999).

5. M. Hutton, C. L. Lendon, P. Rizzu, M. Baker, S. Froelich, H. Houlden, et al., Association of missense and 5'-splice-site mutations in tau with the inherited dementia FTDP-17, *Nature* **393**, 702-705 (1998).

6. P. Poorkaj, T. D. Bird, E. Wijsman, E. Nemens, R. M. Garruto, L. Anderson, et al., Tau is a candidate gene for chromosome 17 frontotemporal dementia, *Ann Neurol* **43**, 815-825 (1998).

7. M. G. Spillantini, J. R. Murrell, M. Goedert, M. R. Farlow, A. Klug, and B. Ghetti, Mutation in the tau gene in familial multiple system tauopathy with presenile dementia, *Proc Natl Acad Sci U S A* **95**, 7737-7741 (1998).

8. J. Gotz, Tau and transgenic animal models, *Brain Res Brain Res Rev* **35**, 266-286. (2001).

9. J. Gotz, A. Probst, M. G. Spillantini, T. Schafer, R. Jakes, K. Burki, et al., Somatodendritic localization and hyperphosphorylation of tau protein in transgenic mice expressing the longest human brain tau isoform, *Embo J* **14**, 1304-1313 (1995).

10. J. P. Brion, G. Tremp, and J. N. Octave, Transgenic expression of the shortest human tau affects its compartmentalization and its phosphorylation as in the pretangle stage of Alzheimer's disease, *Am J Pathol* **154**, 255-270 (1999).

11. T. Ishihara, M. Hong, B. Zhang, Y. Nakagawa, M. K. Lee, J. Q. Trojanowski, et al., Age-dependent emergence and progression of a tauopathy in transgenic mice overexpressing the shortest human tau isoform, *Neuron* **24**, 751-762 (1999).

12. K. Spittaels, C. Van den Haute, J. Van Dorpe, K. Bruynseels, K. Vandezande, I. Laenen, et al., Prominent axonopathy in the brain and spinal cord of transgenic mice overexpressing four-repeat human tau protein, *Am J Pathol* **155**, 2153-2165 (1999).

13. A. Probst, J. Gotz, K. H. Wiederhold, M. Tolnay, C. Mistl, A. L. Jaton, et al., Axonopathy and amyotrophy in mice transgenic for human four-repeat tau protein, *Acta Neuropathol (Berl)* **99**, 469-481 (2000).

14. J. Gotz, and R. M. Nitsch, Compartmentalized tau hyperphosphorylation and increased levels of kinases in transgenic mice, *Neuroreport* **12**, 2007-2016. (2001).

15. J. Lewis, E. McGowan, J. Rockwood, H. Melrose, P. Nacharaju, M. Van Slegtenhorst, et al., Neurofibrillary tangles, amyotrophy and progressive motor disturbance in mice expressing mutant (P301L) tau protein, *Nat Genet* **25**, 402-405 (2000).

16. J. Gotz, F. Chen, R. Barmettler, and R. M. Nitsch, Tau Filament Formation in Transgenic Mice Expressing P301L Tau, *J Biol Chem* **276**, 529-534 (2001).

17. J. Gotz, M. Tolnay, R. Barmettler, F. Chen, A. Probst, and R. M. Nitsch, Oligodendroglial tau filament formation in transgenic mice expressing G272V tau, *Eur J Neurosci* **13**, 2131-2140. (2001).

18. C. Geula, C. K. Wu, D. Saroff, A. Lorenzo, M. Yuan, and B. A. Yankner, Aging renders the brain vulnerable to amyloid beta-protein neurotoxicity, *Nat Med* **4**, 827-831 (1998).

19. K. K. Hsiao, D. R. Borchelt, K. Olson, R. Johannsdottir, C. Kitt, W. Yunis, et al., Age-related CNS disorder and early death in transgenic FVB/N mice overexpressing Alzheimer amyloid precursor proteins, *Neuron* **15**, 1203-1218. (1995).

20. D. Games, D. Adams, R. Alessandrini, R. Barbour, P. Berthelette, C. Blackwell, et al., Alzheimer-type neuropathology in transgenic mice overexpressing V717F beta-amyloid precursor protein, *Nature* **373**, 523-527 (1995).

21. C. Sturchler-Pierrat, D. Abramowski, M. Duke, K. H. Wiederhold, C. Mistl, S. Rothacher, et al., Two amyloid precursor protein transgenic mouse models with Alzheimer disease-like pathology, *Proc Natl Acad Sci U S A* **94**, 13287-13292 (1997).

22. J. Gotz, F. Chen, J. van Dorpe, and R. M. Nitsch, Formation of neurofibrillary tangles in P301L tau transgenic mice induced by Abeta42 fibrils, *Science.* in press (2001).

23. J. Lewis, D. W. Dickson, W.-L. Lin, L. Chisholm, A. Corral, G. Jones, et al., Enhanced neurofibrillary degeneration in transgenic mice expressing mutant Tau and APP, *Science.* in press (2001).

HISTORICAL OVERVIEW OF GLYCOSAMINOGLYCANS (GAGs) AND THEIR POTENTIAL VALUE IN THE TREATMENT OF ALZHEIMER'S DISEASE

U. Cornelli, I. Hanin,[1] S. Lorens,[1] J. Fareed, J. Lee, R. Mervis,*
P. Piccolo,** and M. Cornelli***

1. INTRODUCTION

"Dies annorum nostrum sunt septaginta anni aut in valentibus octoginta anni et maior pars eorum labor et dolor...." [1] . The years in our life are seventy, eighty for the more resistant, but most of them are work and pain . This is the translation of Psalm 89 recognizable as one of the oldest documents concerning duration of life; it is attributed to Moses and it was translated 1,300 years later from Hebrew into Latin. Not much has changed for the individual span of life: once a proper environment allows our genoma to express its potentialities, some of us can reach or even exceed the age of one hundred years. In this context senile dementia, and particularly Alzheimer's disease (AD), shortens life and adds a substantial amount of pain to our days; that is why in every part of the world a lot of efforts are devoted to understand all the possible mechanisms that may bring our brain to the loss of life consciousness. The results of these efforts are expressed by a quite variegate number of theories. One of these is related to glycosaminoglycans (GAGs) and proteoglycans (PGs).

2. GAGs AND PGs (GLYCOSAMINOGLYCANS AND PROTEOGLYCANS)

The amount of chemical and biological literature on these molecules is extremely abundant and with appreciable efforts some authors review the most important issues related to PGs and GAGs characteristics [2,3]; here we will report only few information.

Loyola University Stritch School of Medicine, Maywood, Illinois 60153; *Neuro Cognitive Research Laboratories, Columbus, Ohio 43212; **University of Padova, 2nd Division Geriatric Hospital, 35100 Padua, Italy; *** Cornelli Consulting S.a.s. 20129 Milan Italy.

Mapping the Progress of Alzheimer's and Parkinson's Disease
Edited by Mizuno *et al.*, Kluwer Academic/Plenum Publishers, 2002

145

GAGs (HA = hyaluronic acid, HEP = heparin, DS= dermatan sulfate, HS = heparan sulfate, KS = keratan sulfate, CH = chondroitin sulfate) are linear heteropolysaccharides consisting of hexosamine (D-glucosamine [GlcN] or D-galactosamine [GalN]) and either hexuronic acid (D-glucuronic acid [GlcA] or L-iduronic acid [IdoA]) or galactose, disposed in disaccharide units carrying on sulfate substituents.

The most documented activities of GAGs are considered to be dependent from their ability in binding specific proteic susbstrata (e.g. antithrombin or lipoproteinlipase) and the entity of their biologic activity is frequently attributed to four different characteristics: molecular weight, density of charge, degree of sulfation, and type of disaccharide units. These characteristics allow the existence of an infinity of combinations and similarities such that similar GAGs may differ only for potency. A typical example is the affinity for antithrombin (AT III) which determine the anticoagulant activity of GAGs. In terms of AT III HEP > HS > DS [2, 4].

GAGs are not made on a template base and their polymorphism may be related to the necessity of a rapid reaction to the modification of the environment. Most of GAGs present in the human organism derive from PGs and under normal physiological condition they are synthesized on a matrix constituted by a core protein [3].

There are many different proteoglycans (HAPG= hyaluronic acid PG, HEPG= heparin PG, DSPG= dermatansulfate PG, HSPG= heparansulfate PG, KSPG= keratan sulfate PG, CSPG= condroitin sulfate PG) and an attempt has been done to divide them in families according to few common features such as the GAG type or its localization within the biological matrix [2]. PGs can be differentiated on the basis of the core protein and by the GAGs chains which are covalently attached (with the exclusion of HA) via an O-glycoside linkage to a serine residue (or an N-linked asparagine). PGs are homologous if the GAGs chains are formed by the same GAG or not heterologous if the chains are composed by different GAGs. Among PGs the most represented on cell surface are HSPGs (homo or heterologous) which can be found in most animal cells [5]. The same PG may be composed by GAGs of different type [6,7] such as HS, DS, KS or with the same GAG such as perlecan [8]. Some of these PGs such as Syndecans [6] are linked to cells via a transmembrane domain of the core protein, whereas glypicans [7] are anchored through glycosylphosphatidyl inositol (GPI), and perlecans [8] are usually linked to basement membranes. HSPGs are particularly important for AD.

HSPGs in the membranes of neurons and astrocytes of animal brains were found in high concentration [9]. Most of the studies related to HSPGs production and fate were conducted on PC 12 nerve cell lines [10]. However the role of HSPGs is not fully understood, and the importance of the presence of GAGs in these structure is even more intriguing, since many times HSPG is confused with HS. HSPGs (and in general PGs) and the relative GAGs have different roles (Table 1) and interactions with the proteins of the extra cellular matrix (ECM).

Laminin represents the major constituent of the basement membrane and cell adhesion [11]. It has a high affinity for HSPGs, it binds GAGs and collagen IV and it is supposed to have at least three binding sites for HEP. Laminin appears as one of the molecules responsible for neurite outgrowth in different cell cultures; probably HSPGs delay its activity.

While it was shown that HSPGs interact with laminin through a non covalent binding and they form a stable polymerized complex, high concentrations of HEP (opposite to low concentrations) inhibit laminin polymerization [12] .

Vitronectin (or S protein) has an high content of basic aminoacids near the C-terminal [13] and its reaction with GAGs may have a physiological role in the ECM for the regulation of complement system. Moreover vitronectin neutralizes the HEP induced inactivation of ATIII [14] and the interaction of GAGs with vitronectin prevents its binding with the complement protein C9.

TABLE 1. Different activities of HSPGs and GAGs on proteins of ECM

Type of molecule	PGs activity	GAGs activity
Laminin	Increase polymerization	Reduce polymerization
Vitronectin	Undetermined	Inhibitory of complement binding (C 9)
Fibronectin	Synergetic in cell translocation	Inhibitory
Thrombospondin	Receptor binding	Inhibition of binding
FGFs	Undetermined (protective)	Increase of half life

Fibronectin can bind either the core protein of HSPGs [15] or GAGs. GAGs may compete on a common binding site on fibronectin, which does not depend on their polysaccharide nature but it is sensible to a high sulfate content. GAGs induce a conformation change in fibronectin [16] that may have consequences on its function in promoting cell translocation. Conversely the presence of endogenous HSPGs usually promotes the adhesion between Schwann cells and also the neurite outgrowth from dorsal root ganglia cells.

Trombospondin is secreted from platelets (α granules) and in case of injury it becomes incorporated into the ECM, where it plays a role in cell attachment and migration. Trombospondin cell interaction occurs by binding HSPGs on the cell surface probably by its HEP binding region located at the N-terminal [17].

For AD it is important the relationship of Fibroblasts growth factors (FGFs) and PGs or GAGs. Among several peptides collectively known as Heparin-binding growth factors [18], two of them should be mentioned in our contest: acidic and basic fibroblast growth factors (FGFs).

Both were isolated in human brain but they are present in every organ. In vitro they induce differentiation in cell cultures [19] and stimulate neurite outgrowth from neuronal cell lines. FGFs have an HEP binding that probably is a common domain [18] which does not seem to be located in the same region of the receptor binding.

The importance of GAGs in FGF binding to its receptor is pointed out by the observation that cell lines defective of HS biosynthesis are refractory to the mitotic activity of FGF [20]. FGFs bound to the ECM are not mitogenic: that is why it is supposed they are immobilized, driven and protected by HSPGs [21]; these growth factors accelerate the activity either in presence of HEP or heparinase [22].

It is demonstrated that the phosphorilation process activating FGFs becomes easier and more resistant to proteolysis in the presence of GAGs [23]. Neurite outgrowth in the presence of acidic FGF is strongly enhanced by GAGs, probably because they are also able to protect a protease inhibitor. One of this protease inhibitor is the Glial Derived Nexin (GDN) that has many similarities with the serpin family protease inhibitors [24] and is identical to protease nexin 1 (PN-1). GAGs enhance the GDN/PN-1 induced inactivation of thrombin [25] and all this macrocomplex is quickly internalized and degraded by cells.

In this contest it is interesting to underline that in AD brains a dramatic drop of GDN/PN-1 was found [26]. Consequently thrombin or other proteases are free to inhibit neurite outgrowth; moreover GDN/PN-1 increases during axonal regeneration in vivo [45]. Furthermore, fibroblastic receptors for GDN/PN-1 were characterized as containing HSPGs [27].

A general difference between HSPGs and GAGs is emerging for FGFs also: GAGs seem to have a role in acceleration and fluidity of processes of cellular growth and neuronal repairing, whereas HSPGs (homologous or eterologous) keep FGFs in a steady state condition.

There are evidences that cornea endothelial cells do not proliferate until HSPG is able to keep the basic FGF in a sequestered condition; however, during tissue injury, in the presence of cells that secrete heparinase, proteinase, or HEP (mast cell), growth factors can be released from the linkage and increase cells growth [28].

3. AMYLOID, PGs, AND GAGs

The content of GAGs such as CS and HS in amyloid fibrils was defined in the 1960's but not much attention was paid to this findings until 1980. GAGs were detected in every experimental amyloid as well as in all human amyloids [29] including neuritic, vascular plaques and neurofibrillary tangles in AD. Also in brain tissue, isolated from human cases of Gerstman-Straussler syndrome and Creutzfeld-Jakob disease, a significant amount of HSPGs was found [30] thus confirming that they (and also other PGs) are present in other CNS amyloids. However it is not clear if the presence of GAGs is because of PGs or if they were present just as free oligosaccharides.

A high affinity of HSPG -isolated from a mouse sarcoma EHS- for the amyloid precursor protein (βAPPs) was demonstrated and also a reduction of this affinity when GAGs such as HEP are added to the system; moreover a common HEP binding region is present either in βAPP and in the beta amyloid peptide Aβ [31]. These findings demonstrate that HSPG may bind both βAPP and βA, and GAGs inhibit this binding, probably in relation to their molecular weight and charge density.

The amyloid deposit in AD, either in senile plaques (SP) or in brain vessels (vascular amyloid), is composed by a 4.2 kD peptide Aβ. This peptide is heterogeneous being formed predominately of 39/40 residues in vessel amyloid, and of 42-44 residues in SP [32,33]. Aβ is supposed to derive from βAPP composed from a set of 677 to 770 aminoacids. It is well documented that Aβ is released through a cleavage of βAPP by a β secretase (and also γ secretase) but the main metabolic pathway is the cleavage within the Aβ sequence through α secretase that consequently do not allow its formation.

Once 1-40(39) or 1-42(43) Aβ are produced, they have the tendency to aggregate in fibrils to form amyloid. Many substances of different origin were found to trigger the amyloid formation such as Apo E (apolipoprotein E) and also PGs.

At least two different basement-membrane HSPGs were detected [34] either in vascular and non vascular amyloid; it was supposed that −because of the linkage of HSPGs core protein with all the known βAPP within the Aβ area- the α secretase could not be active, and the βAPP cleavage had to be accomplished by other proteases (β for instance) leading to amyloidogenic or neurotoxic fragments. On the other hand HSPGs are able to bind Aβ peptides with the carbohydrate chains also and this was claimed as one of the causes of fibrils formation because it might stabilize fibrils and protect them from proteolysis [35]. The charges on saccharide chains of GAGs may provide sites for lateral and axial aggregation of fibrils [36], as it happens in experimental conditions when synthetic βA peptides are in association with a variety of sulfated GAGs in an acidic medium [37].

In some cases a competition in binding macromolecules was observed, indicating that perhaps GAGs are capable to reduce the polymerization process. Conversely in some in vitro experiences an increase of fibrils was clearly shown indicating that also GAGs may trigger the polymerization.

Following a concomitant infusion of HSPG and a synthetic Aβ1-40 in rat brains [38] it was possible to reproduce an amyloid deposit in the infusion site (similar to the Congo red-stained amyloid of AD patient brains). Fibrils start forming in the osmotic infusion pump and in all the tested solution (Aβ, Aβ+HSPG, Aβ+HS). Surprisingly in those animals infused with HS+Aβ the typical lesion was not detected, whereas in the other two groups of animals it was present with a significant frequency. The explanation of the authors is that HS in vivo may block Aβ fibril formation by competing with the endogenous brain HSPG.

Furthermore it has been shown [39] that small oligosaccharides (from two to eight saccharides) derived from HEP fractionation compete with HEP in binding Aβ.

In conclusion, from the data in the literature it is evident that HSPGs (and also other PGs) may trigger the amyloid fibril formation and GAGs, which do not have a core protein, may have more affinity for Aβ and compete with HSPGs. Finally, small oligoisaccharides have much more affinity for Aβ than GAGs and may form a more clearable amyloid.

4. NEUROFIBRILLARY TANGLES (NFT), PGs, AND GAGs

The cognitive deterioration up to the final stage of dementia is supposed to be positively correlated with the number of NFTs in the association cortex [40] whereas this correlation is not made evident by the number of senile plaques [40,41].

The core of NFTs is constituted by paired helical filaments (PHFs). Immunochemicals and histochemical studies have shown [42,43] that the major component of PHFs is the microtubule-associate protein tau, aggregated in highly phosphorilated state.

The cause of the increase of tau phosphorilation is not defined; one of the possible explanations is the cell regeneration. During cell regeneration or repair, either kinases and phosphatases work as well co-ordinate processes. These enzymes may loose selectivity and coordination when cells are under a chronic stress and the final result of the repair tentative is a sort of abortion.

Higher levels of tau protein were found in CFS of patients suffering from AD or other neurological diseases, therefore tau protein itself can not be considered a proper marker of AD [44] but it can be taken as a marker of neuronal stress condition. Aggregation of tau into PHF can be reconstructed in vitro [45] while the in vivo environment is very different because of the large number of reactive molecules that are detectable within the NFTs might cause aggregation; some of these are: basic FGF [46], HSPG [47], tropomyosin [48], ubiquitin [49], gangliosides [50], AchE [51], βAPP [52], Apo E [53], HEP [54].

The presence of HSPG and HEP in NFTs is interpreted as one of the cause of PHF polymerization [46] because of their capacity to react either with the protein core or the carbohydrate moiety thus creating a bridge within PHF or between PHF and other close macromolecules. In this case the presence of small GAGs competing with HSPG and with HEP can probably reduce the polymerization process.

5. GLYCOSAMINOGLYCAN POLYSULFATE (GAP)

The different activities GAGs were stimulating to test them for the treatment of AD. However, the only available products was HEP or a mixture of GAGs (glycosminoglycan polysulfate or GAP). The choice was in favor of GAP because it was orally bioavailable, not active as anticoagulant, and composed by the most active GAGs (see below).

GAP is composed by HEP (10-20 %), HS (40-60 %), DS(20-35 %) and CS (2-8 %) were admitted for human use in Italy (since 1981) and in few other countries.

The product is extracted from swine mucosa according to a standardized method employed for other GAGs such as HEP. Many studies documented the pharmacological activity of this product [see 55 for review]. GAP has shown to have hypolipemic activity and a protective effect on the development of experimental atherosclerotic lesions, both stimulating the endothelial cell proliferation and reducing the growth of arterial smooth muscle cells.

The plasma lipoproteins rearrangement was tested in rabbits and all the typical parameters like cholesterol, tryglicerdes, phospholipids and proteins of VLDL and LDL were reduced by treatment with GAP [56].

The pharmacokinetic of GAP was studied in animals using markers (tritium for the oral route and fluorescin for intravenous administration). In both cases markers were detected in blood in an amount compatible with a pharmacological activity and a positive correlation was found between the level of markers and lipoproteinlipase activity [57].

An interesting study was conducted by Lorens [58] on aged rats. He showed a significan partial reversal of age-related deficit in rats tested trough a conditioned one-way and two-way avoidance, following GAP administation in drinking water for 12 weeks. A normalization of stress-induced corticosterone secretion was also reported. Furthermore, GAP counteracted the age-related reduction of DOPAC (3,4-dihydroxyphenyl acetic acid) and HVA (homovanillic acid) in nucleus accumbens but it did not counteract the age-related modification of HVA and 5-HIAA (5-hydroxyindoleacetic acid) in neostriatum. Thus the behavioral effect observed may be due to GAP influence on dopamine neurotransmission.

6. GAP IN THE TREATMENT OF PRIMARY DEMENTIA (MID AND AD)

Three different sets of data are available to show the activity of GAP in patients suffering from AD. A pilot study was conducted on 12 patients [59] to check MAO-B activity in platelets and 3-methoxy-4-hydroxy-phenylglicol (MHPG), homovanillic acid (HVA) and 5-hydroxy-indolacetic acid (5-HIAA) levels in CSF. These compounds are main metabolites of norepinephrine, dopamine and serotonin respectively. In AD patients was described that the average levels of MAO-B in platelets are significantly higher than those detected in normal individuals, whereas CSF average levels of MHPG, HVA and 5-HIAA are definitely lower [60,61]. GAP was administered at the dosage of 120 mg (or 1000 lipasemic releasing units LRU) once a day by intramucolar injection for a period of 30 days. The treatment reduced significantly the MOA-B activity in platelets and it increased the levels of 5-HIAA, HVA and MHPG; in other terms the entire scenario of these parameters was reassessed towards normal values. The authors suppose that these activities of GAP are determined by its contribution to the maintenance of endothelial integrity. A general activity on membranes integrity (e.g endothelial cell, platelets) could be a further explanation.

A double blind clinical trial on patients suffering from vascular dementia (MID) was conducted on 30 cases treated for a period of treatment of 8 weeks [62]. Patients over 70 years old were admitted to the trial according to Hachinski (index ≥ 7) and DSM-III criteria. After a wash out period of 2 weeks patients were trated by oral route either with placebo or GAP at the dosage of 24 mg (200 LRU) t.i.d. for a period of 8 weeks. The clinical results were assessed in terms of somatic, psychic and neurological symptoms. Furthermore a battery of psycometric tests was administered to patients before, during and after the treatment. The behavioral and psycometric evaluation consisted of Stuard Hospital Geriatric rating Scale (SHGRS), Gottfreis Rating Scale of Dementia, Digit Repetition from WAIS and Corsi's test. All assessment were made before and after 4 and 8 weeks of treatment. Results showed that GAP was significantly more active than placebo particularly in the SHGRS cluster of parameters related to cooperation, inadequacy and cleanliness. Overall clinical judgements expressed by physicians and by patients' relatives agreed with the results of psychometric tests.

An international multicenter double blind clinical study was conducted on 155 patients suffering from Primary Dementia [63]. Admission criteria were based on DSM-III and HIS to separate diagnosis of AD and MID. Patients with moderate to severe cognitive decline only were included in the study. The cohort of eligible patients consisted of 71 patients suffering from SD and 77 patients suffering from MID.

The patients included in the trial were randomly assigned to one of the treatment regimes which consist of a t.i.d. oral intake of tablets containing either GAP 24 mg (200 LRU) or placebo. The treatment was administered for a period of 12 weeks. The clinical assessment was based on psychiatric, psycogeriatric, psychometric, physical and laboratory analysis.

The treatment with GAP resulted effective in controlling psycopathological symptoms, which are among the earliest detectable manifestations of a dementing process. A significant improvement was also seen on social behavior as reflected in the CGI (Clinical Global Impression scale) and on cognition as reflected by the MMSE (Mini Mental State Examination).

The SCAG scale (Sandoz Clinical Assessment Geriatric) detected improvements of depression, agitation/irritability and cognitive dysfunction. The activity of GAP was similar in AD and MID patients. Treatment emergent symptoms and toxic effects were infrequent and mild.

7. DISCUSSION AND CONCLUSION

"Gutta cavat lapidem"; it is an old Latin sentence expressing the evidence that a day by day continuous, slow and inconsistent stimulus is able to destroy a strong, compact and heavy structure. It is precisely what happens with AD where an undetectable modification undermines the extraordinary architecture of the human brain.

A small damage of the BBB gives rise to a leakage of plasma proteins and also to a difficult discharge of them. Several reports show a reduced glucose transport in microvessels (particularly in the temporo-parietal area) whereas there are no data about the clearance of macromolecules or peptides. This latter event is extremely important because (a in human brains the lymphatic circulation does not exist (b CSF has only a limited capacity to degrade organic substances and (c these substances have to reach CSF first.

The reaction caused by leakage damages segments of the BBB which becomes a sort of semipermeable membrane particularly in those areas where the repairing process is not efficient.

The reaction of the brain in repairing the baseline membrane consists of a production of HSPGs, or other PGs, of GAGs and even of further ECM macromolecules (e.g. fibronectin, spondilin, laminin) usually employed in the basement membrane construction. An analysis of HS structure of AD and normal brains [64] has shown no relevant differences between them. However, the brain HS has peculiar characteristics as compared to HS of other parts of the body; these data indicate that brain needs a specific type of GAGs. Finally, the damage can affect every type of synaptic site. The presence of AchE/ BchE in plaques (cholinergic pathway deficit) and the deficit of other pathways (e.g dopaminergic, serotoninergic) have been demonstrated.

The reaction to the damage involves astrocytes and glyal cells reactivity in the repairing process. Most likely complement proteins and lysosomial enzymes witness the presence of an inflammatory process. It is suggested that [64] single not particularly damaging events may converge creating an unique precipitating event. In general the AD process develops slowly because..... it has time. In all the directions taken by the initial damage HSPGs are directly or indirectly involved. Their presence seems to reflect a tentative of repairing which end up with an abortion because of lack of modulation. This harmonization can be driven by small GAGs which may increase the fluidity of ECM. These small GAGs can reduce the burden of amyloid and NFT, and may increase the efficiency of neuronal repair.

GAP showed to be active giving support to the GAGs theory. Either in AD and in MID the product causes an improvement in those psychopatological symptoms which precede cognitive decline in elderly subjects. However GAP is a polycomponent product, and tentative were made in order to isolate a single component. Among the various compounds tested, one product emerged called C3, which is an HDO (heparin derived oligosaccharide) composed by a narrow range of sulfate saccharides (between 4 and 12 saccharides) which is under study as the candidate for the treatment of AD.

8. SUMMARY

Proteoglycans (PGs) relationship with brain structures is analyzed in view of their role in the pathophysiology of Alzheimer's disease (AD). The difference between Proteoglycans (PGs) and glycosaminoglycans (GAGs) is considered the key for the interpretation of the two opposite roles played by these molecules. PGs are one of the possible determinants of AD whereas GAGs are tied to tentative therapeutical approaches in AD. A particular emphasis is given to the main AD markers: [amyloid substance, neurofibrillary tangles (NFT)] in order to show the background of AD therapy with GAGs. These compounds are capable to mobilize fibroblast growth factors (FGFs) and may harmonize the neuronal repair. They may reduce the formation of fibrils from the amyloid peptide (Aβ) and also the neurofibrillary tangles (NFT) burden. A mixture of GAGs (glycosaminoglycans polysulfate or GAP) was tested in humans. Positive data were obtained in Phase II and III double blind multicenter clinical trials, in which patients suffering from AD were treated with GAP by oral or parenteral route.

REFERENCES

1. Psalm 90 (89): in Nova Vulgata Bibliorum Sacrorum. (Libreria Editrice Vaticana, 1986).
2. Jackson, R.L et al.: Glycosaminoglycans: Molecular Properties, Protein Interaction, and Role in Physiological Processes.Physiol.Rev. 71:481-539 (1991).
3. Lindahl. U.; Lidholt, K.; Spillmann, D.; Kjellén L.: More to "heparin" than anticoagulation. Throm. Res.75: 1-32 (1994).
4. Casu, B.: Structure and biological activity of heparin. In: Tipson R.S.; Horton D.:Advance in Carbohydrate Chemistry and Biochemisrty (Academic Press, Inc.43:51-127 1985).
5. Kraemer, P.M.; Smith,D.A.: High molecular-weigth heparan sulfate from the cell surface. Biochem.Biophys.Res.Commun.56: 423-430 (1974).
6. Bernfield, M:; Kokenyesi, R.; Kato, M.; Hinkes, M.; Spring, J.; Gallo, R.; Lose, E.: Biology of Syndecans: Annu. Rev. Cell. Biol. 8: 333-364 (1992).
7. David, G.; Lories, V.; Decock, B.; Marynen, P.; Cassiman, J.J.; Van Den Berghe, H.: Molecular cloning of a phosphatidilinositol-anchored membrane heparan sulfate proteoglycan from human cell fibroblasts. J. Cell. Biol. 124: 149-160 (1994).
8. Kallunki, P.; Tryggvason, K.: Human basement membrane heparan sulfate proteoglycan core protein: A 467-kD protein containing multiple domains resembling elements of the low density lipoprotein receptor, laminin, neural cell adhesion molecules, and epidermal growth factor. J. Cell. Biol. 116 : 559-571 (1992).
9. Kiang,W.-L.et al.:Glycosaminoglycans and Glycoproteins Associated with Microsomal Subfraction of Brain and Liver.Biochemistry 18: 3841-3848 (1978).
10. Schubert, D.; Schroeder,R.et al.: Amyloid β Protein Precursor Is Possibly a Heparan Sulfate Proteoglycan Core Protein. Science 241: 223-226 (1988).
11. Martin, G.R.; Timpl, R.: Laminin and other basement membrane components. Annu. Rev. Cell.Biol. 3: 57-85 (1987).
12. Kouzi-Koliakos, K.; Koliakos, G:.G.; Tsilibary, E.C.; Furcht,L.T.; Charonis, A.S.: Mapping of three major heparin-bindings sites on laminin and identification of a novel heparin-binding site on the B1 chain. J. Biol.Chem. 264: 17971-17978 (1989).
13. Suzuki, S.; Pierschbaker, M.D.et al.: Domain structure of vitronectin. Alignment of active sites. J.Biol. Chem. 259: 15307-15314 (1984).
14. Peterson, C.B.; Morgon, M.T.; Blackburn, M.N.: Histidine-rich glycoprotein modulation of the anticoagulant activity of heparin. Evidence for mechanism involving competition with both antithrombin and thrombin for heparin binding. J.Biol.Chem. 262: 7567-7574 (1987).
15. Heremans, A.; De Cock, B. et al.: The core protein of the matrix-associated heparan sulfate proteoglycan binds to fibronectin. J. Biol.Chem. 265: 8716-8724 (1990).
16. Khan, M.Y.; Jaikaria, N.S.et al.: Structural changes in the NH_2-terminal domain of fibronectin upon interaction with heparin. J. Biol.Chem. 263: 11314-11318 (1988).
17. Kaesberg, P.R.; Ershler, W.B.; Esko, J.D.; Mosher, D.F.: Chinese hamster ovary cell adhesion to human platelet thrombospondin is dependent on cell surface heparan sulfate proteoglycan. J. Clin. Invest. 83: 994-1001 (1989).
18. Burgess, W.H.; Maciag, T.: The heparin-binding (fifbroblast) growth factor family of proteins. Annu. Rev. Biochem. 58: 575-606 (1989).
19. Morrison, R.S.; Sharma, A.; De Vellis, J.; Bradshaw, R.A.: Basic fibroblast growth factor support the survival of cerebral cortical neurons in primary culture. Proc. Natl. Acad. Sci. USA 83. 7537-7541 (1986).
20. Yayon, A.; Klagsburn, M.; Esko, J.D:; Leder, P.; Ornitz, D.M.: Cell surface, heparin-like molecule are required for binding of basic fibroblast growth factor to its high affinity receptor. Cell. 64: 841-848 (1991).
21. Gordon. P.B.; Choi, H.U.; Conn, G.; Ahmed, A.; Ehrman, B.; Rosenberg, L.; Hatcher, V.B.: Extracellular matrix heparan sulfate proteoglycans modulate the mytogenic capacity of acidic fibroblast growth factor. J. Cell. Physiol. 140: 584-592 (1989).
22. Saksela, O; Moscatelli, D; Sommer, A.; Rifkin, D.B.: Endothelial cell-derived heparan sulfate binds basic fibroblast growth factor and protects it from proteolytic degradation. J.Cell. Biol. 107: 743-751 (1988).
23. Gospodarowicz, D.; Cheng, J: Heparin protects basic and acidic FGF from inactivation. J. Cell. Physiol. 128: 474-484 (1986).
24. Carrel, R.W.; Christey, P.B.; Boswell, D.R.: Serpins: antithrombin and other inhibitors of coagulation and fibrinolysis evidence from amino acid sequences. In: Thrombosys, Haemostasis (Verstraete, M.et al.: Leuven Univ. Press :1-15 1987).
25. Stone, S.R.; Nick, H.; Hofsteenge, J.; Monard, D.: Glial-derived neurite-promoting factor ia a slow-binding inhibitor of tripsin, thrombin, and urokinase. Arch. Biochem.Biophys. 252: 237-244 (1987).
26. Wagner, S.L.; Geddes, J.W.; Cotman,C.W.; Lau, A.L.; Gurwitz, D.; Isakson, P.J.; Cunningham, D.D.: Protease nexin-1, an antithrombin with neurite outgrowth activity, is reduced in Alzheimer disease. Proc. Natl. Acad. Sci. USA 86:8284-8288 (1989).

27. Meier, R.P.; Spreier, P.; Ortmann, A.; Harel, A.; Monard, D.: Induction of glia-derived nexin after lesion of a peripheral nerve. Nature (Lond.) 342: 548-550 (1989).

28. Folkman, J.; Klagsburn, M.; Sasse, J.; Wadzinski, M.; Ingber, D.; Vlodavski, I.: A heparin binding angiogenic protein -basic fibroblast growth factor- is stored within basement membrane. Am. J. Pathol. 130: 393-400 (1988).

29. Snow, A.D.; Willmer, J.; Kisilevski, R.: Sulfated Glycosaminoglycans: A Common Constituent of All Amyloidosis? Lab. Invest. 56: 120-123 (1987).

30. Snow, D.A.; Wight, T.N.; Nochlin, D.; et al.: Immunolocalization of Heparan Sulfate Proteoglycans to the Prion Protein Amyloid Plaques of Gerstmann-Straussler Syndrome, Creutzfeldt-Jakob Disease and Scrapie. Lab. Invest. 63: 601-611 (1990).

31. Narindrasorasak, S.; Lowery, D.; Gonzales-DeWitt, P.; Poorman, R.A.; te al.: High Affinity Interaction between the Alzheimer's β-Amiyloid Precursor Proteins and the Basement Membrane Form of Heparan Sulfate Proteoglycan. J. Biol. Chem: 20: 12878-12883 (1991).

32. Kang, J.; Lemaire, H-G.; Unterbeck, A.; Salbaum, J.M.; Maters, C.L.; Grzeschik, K-H.; Multhaup, G.; Beyreuther, K.; Müller-Hill, B.: The precursor of Alzheimer's disease amyloid A4 protein resembles a cell-surface receptor. Nature 325: 733-736 (1987).

33. Wisniewski, T.; Lalowski, M.; Levy, E.; Marques, M.R.F.; Frangione, B.: The amino acid sequence of neuritic plaque amyloid from a familial Alzeimer's disease patient. Ann. Neurol. 35: 245-246 (1994).

34. Buée, L.; Ding, W.; Anderson, J.P.; et al.: Binding of vascular heparan sulsate proteoglycan to Alzheimer's amyloid precursor protein is mediated in part by the N-terminal region of A4 peptide. Brain Research 627: 199-204 (1993).

35. Fredrickson, R.C.A.; Astroglia in Alzheimer's disease. Neurobiol. Aging 13: 239-253 (1991).

36. Fraser, P.E.; Nguyen, J.T.; Chin, D.T.; Kirschner, D.A.: Effect of sulfate ions on Alzheimer β/A4 peptide assemblies: implication for amyloid fibril-proteoglycan interaction. J. Neurochem. 59: 1531-1540 (1992).

37. Brunden K.R.; Richter-Cook,N.J.; Chaturvedi, N.; Frederikson R:C.A.: pH dependent Binding of Synthetic β-amyloid Peptides to Glycosaminoglycans. J.Neurochem. 61: 2147-2154 (1993).

38. Snow, A.D.; Sekiguchi, R.; Nicholin, D.; et al: An inportant Role of Heparan Sulfate Proteoglycan (Perlecan) in a model System for the depositon and Persistence of Fibrillar Aβ-Amyloid in Rat Brain. Neuron 12: 219-234 (1994).

39. Leveugle, B.; DingW.; Laurence, F.; Dehouck M.P.; Scanameo A.; Cecchelli R. ; Fillit H. Heparin oligosaccharides that pass the blood-brain barrier inhibit beta-amyloid precursor protein secretion and heparin binding to beta-amyloid peptide. J. Neurochem. 70/2: 736-744 (1998).

40. Arriagada, P.V.; Growdon, J.H.; Hedley-Whyte, E.T.; Hyman, B.T.: Neurofibrillary tangles but not senile plaques parallel duration and severity of Alzheimer's disease. Neurology 42: 631-639 (1992).

41. Arnold, S.E.; Hyman, B.T:; Flory, J.; Damasio, A.R.; Van Hoesen, G.W.: The topografical and neuroanatomical distribution of neurofibrillary tangles and neuritic plaques in cerebral cortex of patients with Alzheimer's disease. Cereb. Cortex.1: 103-116 (1991).

42. Grundke-Iqbal, I.; Iqbal, K.; Quinlan, M.,et al.: Microtubule-associated protein tau: a component of Alzheimer paired helical filaments. J. Biol. Chem. 261: 6084-6089 (1986).

43. Trojanowski, J.Q.; Lee, V.M.-Y.: Paired helical filament τ in Alzheimerr's disease: The kinase connection. Am. J. Pathol. 144: 449-453 (1994).

44. Vigo-Pelfrey C.; Seubert, P.; Barbour, R.; et al.: Elevation of microtubule-associated protein tau in the cerebrospinal fluid of patients with Alzheimer's disease. Neurobiology 45: 788-793 (1995).

45. Wille, H.; Drewers,G.; Biernat, J.; Mandelkow, E.-M.; Mandelkow, E.: Alzheimer-like paired helical filaments and antiparallel dimers formed from microtubule-associated protein tau in vitro. J. Cell. Biol. 118: 573-584 (1992).

46. Perry, G.; Siedlak, S.L.; Kawai, M.; Cras,P;.et al.: Neurobibrillary tangles, neuropil threads and seniles plaques all contain abundant binding sites for basic fibroblastic growth factor (β-FGF). J. Neuropathol. Exp. Neurol. 49: 318 (1990).

47. Perry, G.; Sieslak, S.L.; Richey, P.; Kawai, M.; et al.: Association of Heparan Sulfate proteoglycan with Neurofibrillary Tangles of Alzheimer's Disease.J. Neurosci.11: 3679-3683 (1991).

48. Galloway, P.G.; Mulvihll, P.; Siedlak, S.L.; Mijars, M.; et al.:Immunochemical demonstration of tropomyosin in the neurofibrillary pathology of Alzheimer disease. Am. J. Pathol. 137: 291-300 (1990).

49. Tabaton, M.; Perry, G.; Autilio-Gambetti, L.; Manetto, V.; Gambetti, P.: Influence of Neuronal Location on Antigenic properties of Neurofibrillary Tangles. Ann. Neurol. 23: 604-610 (1988).

50. Carolyn, R.E.; Ala, T.A.; frey, W.H.: Ganglioside monoclonal antibody (A2B5) labels Alzheimer's neurofibrillary tangles. Neurobiology 37: 768-772 (1987).

51. Mesulan, M.M.; Moran, M.A.: Cholinesterases within neurofibrillary tangles related to age and Alzheimer's disease. Ann.Neurol. 22: 223-228 (1987).

52. Klier, F.G.; Cole, G.; Stallcup, W.; Schubert, D.: Amyloid β-protein precursor is associated with extracellular matrix. Brain Res. 515: 336-342 (1990).

53. Namba,Y., Tomonaga, M.; Kawasaki, H.; et al.: Apolipoprotein E immunoreactivity in cerebral amyloid deposits and neurofibrillary tangles in Alzheimer's disease and kuru plaque amyloid in Crutzfeld-Jakob disease. Brain Res. 541: 163-166 (1991).

54. Goedert, M.; Jakes, R.; Spillantini, M.G.; Hasegawa, M.; Smith, M.J.; Crowther, R.A.: Assembly of microtubule-associated protein tau into Alzheimer-like filaments induced by sulphated glycosaminoglycans. Nature 383: 550-553 (1996).

55. Prino, G: Pharmacological Profile of Ateroid. In:Ban, T.A.; Lehmann, H.E.: Diagnosis and Treatment of Old Age Dementias (Karger Basel 23: 68-75 1989).

56. Pescador, R.; Mantovani, M.; Niada, R.: Plasma lipoprotein rearramgment in the rabbit induced by mucopolysaccharides from mammalian tissue (Ateroid). Atherogenese 4: suppl IV: 210-216 (1979).

57. Pescador, R.; Madonna, M.: Pharmacokinetics of fluorescin-labelled glycosaminoglycans and of their lipoprotein lipase-inducing activity in the rat. Arzneimittel-Forsh. 32: 819-824 (1982).

58. Lorens, S.; Guschwan, B.S.; Norio, H; Van De Kar, L.: Walenga, J.M.; Fareed, J.: Behavioral, Endocrine, and Neurochemical Effects of Sulfomucopolysaccharide Treatment in the Aged Fisher 344 Male Rat. Semin.Thromb. Haemost. 17 suppl 2: 164-173 (1991).

59. Parnetti, L.; Ban, T.A.; Senin,U.: Glycosaminoglycan Polysulfate in Primary Degenerative Dementia (pilot study of Biologic and Clinical Effects). Neuropsychobiology 31: 76-80 (1995).

60. Brane,G.; Gottfries, C.G.; Blennow, K.; Karlsson, I.; Leckan, A.; Parnetti, L.; Svennerholm L.; Wallin, A.: Monoamine metabolites in cerebrospinal fluid and behavioral rating in patients with early and late onset of Alzheimer dementia. Alzheimer Dis.Ass. Disord. 3: 148-156 (1989).

61. Parnetti, L.; Gottfries, J.; Karlson, I; Langstrom, G, Gottfries, C.G.; Svennerholm, L.:: Monoamines and their metabolites in cerebrospinal fluid of patients with senile dementia of Alzheimer type using high performance liquid chromatography-mass spectrometry. Acta Psychiatr. Scand. 75: 542-548 (1987).

62. Conti, L.; Pacidi, G.F.; Cassano, G.: Ateroid in the Treatment of Dementia: Results of a Clinical Trial. In:Ban, T.A.; Lehmann, H.E.: Diagnosis and Treatment of Old Age Dementias (Karger Basel 23: 76-84 1989).

63. Ban, T.A.; Morey, L.C.; Aguglia, E.; et al.: Glycosaminoglycan polysulphate in th treatment of old age dementias. Prog. Neuro- Psychopharmacol. Biol. Psychiat. 15: 323-342 (1991).

64. Lindahl, B.; Eriksson, L.; Lindahl, U.:Structure of Heparan sulphate from human brain, with special regard to Alzheimer's disease. Biochem. J. 306: 177-184 (1995).

65. Cotman,C.W.: Report of Alzheimer's Disease Working Group A. Neurobiol. Aging 15 Suppl 2: S17-S22 (1994).

GLYCOSAMINOGLYCANS AND GLYCOMIMETICS: POTENTIAL ROLE IN THE MANAGEMENT OF ALZHEIMER'S DISEASE

Qing Ma, Jawed Fareed, Bertalan Dudas, Umberto Cornelli, Stanley Lorens, John Lee and Israel Hanin[*]

As a lead member of the glycosaminoglycan family, unfractionated heparin (UFH) has been used as an anticoagulant and antithrombotic drug for over half a century. Recently, depolymerized heparins commercially known as low molecular weight heparins (LMWHs) have been introduced as potential replacements for UFH in specific indications. These drugs contain significant proportions of oligosaccharides with molecular weights ranging from 800 to 3500 Da and exhibit varying pharmacological profiles including protein and vascular binding and interactions with growth factors and cytokines. It has been demonstrated that some low molecular weight components, e.g. heparin-derived oligosaccharides (HDO), can pass through the blood-brain barrier (BBB), suggesting that these agents may modulate the functionality of the central nervous system. Since vascular deficits, endothelial compromise, and reduction of neuronal repair are associated with Alzheimer's disease (AD), the vascular and anticoagulant effects of these agents may contribute to their potential neuroprotective actions. One of the major vascular and anticoagulant effects of these agents is the endogenous release of tissue factor pathway inhibitor (TFPI), which may be involved in effective neuronal repair. The purpose of this chapter is to provide an updated account on development in the area of heparins, with particular reference to their use in AD and dementia due to vascular deficits.

* Jawed Fareed, John Lee, Departments of Pathology and Pharmacology; Qing Ma, Bertalan Dudas, Umberto Cornelli, Stanley Lorens, Israel Hanin, Department of Pharmacology, Stritch School of Medicine, Loyola University Chicago, Maywood, Illinois 60153, USA.

Mapping the Progress of Alzheimer's and Parkinson's Disease
Edited by Mizuno *et al.*, Kluwer Academic/Plenum Publishers, 2002

157

1. INTRODUCTION

Dementia is characterized by "a decline in intellectual function severe enough to interfere with a person's normal daily activity and social relationships"[1]. Alzheimer's disease (AD) is the most common form of dementing illness in the elderly and remains one of the major causes of death in the US. It is currently estimated that approximately 4 million people in the US have AD. This number is expected to rise to 14 million by the year 2040, as the elderly population continues to increase, with the so-called "baby-boomers" aging. Thus, the potential impact of AD is enormous, and projected costs for health care could rise dramatically if AD cannot be prevented or better managed than it is today. Currently, the national cost of AD is placed at $50-100 billion per year[2].

Although the exact pathogenesis of AD remains to be fully defined, several pharmacological strategies for preventing and treating AD are under active investigation. These strategies include cholinesterase inhibition, M1 agonist administration, administration of nerve growth factor, estrogen replacement therapy, and antioxidants and anti-inflammatory treatment. More recently, drug design has targeted molecular events involved in the pathogenesis of AD including β-amyloid (Aβ) and neurofibrillary tangle formation[3]. LMWHs and their derivatives are promising agents in this category.

UFH, LMWHs and HDO belong to a class of compounds known as glycosaminoglycans (GAGs). Most investigations focus on the interaction between proteoglycans (PGs), the core of senile plaques and GAGs as a potential treatment of AD. It has been demonstrated that the increase of GAGs in brain parenchyma produces a reduction of Aβ cytotoxicity[4]. In addition, GAGs accelerate the activity of several serine protease inhibitors (SERPINs), which inhibit the toxicity of amyloid peptides, suggesting that they may be involved in the anti-AD effects of GAGs[5]. TFPI is one of the SERPINs, localized on the surface of endothelial cells and in glial cells in the brain. Various heparins exhibit different abilities to stimulate TFPI release from vascular endothelial cells[6]. These findings led to examine the possibility that heparins differentially interact with the endothelial cells and/or glial cells to induce TFPI influence on brain vasculature and /or parenchyma.

2. ALZHEIMER'S DISEASE

Three histopathological markers have been characterized in AD: amyloid deposition (senile plaques), neurofibrillary tangles (NFT), and neuronal cell loss in several cortical and subcortical regions[7]. The main component of senile plaques is the β-amyloid peptide. It is derived from a large transmembrane protein, the amyloid precursor protein (APP) and is thought to play an important role in the pathogenesis of AD[8]. Mutations in the APP gene on chromosome 21 are associated with familial AD. The main component of NFT is paired helical filaments, which consist largely of the microtubule associated protein tau in an abnormal state of phosphorylation.

3. HEPARINS

3.1 Chemical Structure

UFH is a linear heteropolysaccharide consisting of alternating hexosamine (D-glucosamine or D-galactosamine) and either hexuronic acid (D-glucuronic acid to L-iduronic acid) or galactose. It forms disaccharides with sulfate substitutes. UFH is synthesized in the rough endoplasmic reticulum of mast and endothelial cells and possibly other cells[9]. A core protein, composed of alternative serine-glycine residues, is formed in which serine residues are capable of attaching polysaccharide chains via linkage regions, i.e. proteoglycans (PGs). The mucopolysaccharide is attached via N-acetylglucosamine and glucuronic acid through the action of two specific glycosyltransferases[10]. This macromolecular UFH is subsequently depolymerized by mast cell endo-glycosidase to give fragments of molecular weight from 6,000 to 25,000 Daltons (Da).

LMWHs and HDO are enzymatically or chemically depolymerized from UFH, resulting in lower molecular weight from 4,000 to 6,000 Da and from 2,000 to 3,000 Da, respectively. The controlled depolymerization results in a narrower molecular weight distribution and much lower polydispersity for LMWHs and HDO compared with UFH. C3, a newly γ-irradiation formed HDO, exhibits a molecular weight of 2.2±0.2 kDa and polydispersity of 1.2-1.3, as measured by gel permeation chromatography.

3.2 Biochemistry and Biological Applications

UFH, LMWHs and HDO are GAGs. They have various affinities to specific proteins, most significantly antithrombin III (AT-III), heparin cofactor II and lipoprotein lipase. Their biological activity is determined by molecular weight, charge density, degree of sulfation, and the type of disaccharide units[10]. UFH and LMWHs are widely used clinically in the prevention and management of deep vein thrombosis (DVT), cardiac surgery, and thrombotic stroke. GAGs combine with a core protein to form PGs.

Recently, heparan sulfate and related PGs have been implicated in the progressive development of polymerized Aβ deposits and other pathophysiological effects in AD[11]. Studies have shown that the association between PGs and amyloid plaques in AD may be due to specific binding of the Aβ sequence and APP to PGs as well as to GAGs[12]. In addition, because of the location of the binding of APP protein in relation to the heparan sulfate PGs, the enzyme (β secretase) that normally cleaves in the middle of the Aβ sequence is blocked, which promotes the aggregation of the Aβ peptide[13]. Therefore, the inhibition of PG binding could theoretically lead to an increase in the amyloid fragments *in vivo* by allowing the beta and gamma secretases to cleave the APP molecule. On the other hand, GAGs, including heparins, which have been shown to block the binding of heparan sulfate PGs to the APP molecule could theoretically be beneficial in preventing plaque deposition with aging.

Moreover, Snow et al.[14] have shown that after Aβ along with heparan sulfate PG was infused into the lateral ventricles of the rat, there was an increased fibrillary Aβ deposition in the neuropil. Interestingly, in the same study when the Aβ and the GAGs heparan sulfate were infused together, no deposition was detected. PGs have also been shown to increase the polymerization of tau proteins into paired-helical filaments, which make up the neurofibrillary tangles of AD[15]. Theoretically, GAGs could also decrease the polymerization of tau and block the formation of neurofibrillary tangles.

In a study by Lorens et al[16] aged rats showed a significant partial reversal of age related behavioral deficits following glycosaminoglycan polysulfate (Ateroid®) administration. Biochemical observations showed that Ateroid® counteracted the age-related reduction in dopamine metabolites and also aided in the normalization of stress induced corticosterone secretion seen in the aged rats. Conti et al.[17] prescribed Ateroid® for primary dementia patients. These patients showed significant improvements in objective performance, psychopathology, and social behavior over placebo groups.

Altogether, GAGs could provide a promising therapy to AD by their inhibitory effects on PGs. In addition, it is hypothesized that GAGs crossing the BBB may interfere directly with endogenous PGs binding with Aβ peptide or APP and, therefore, have potential therapeutic value in AD. Therefore, the ability of GAGs across the BBB becomes crucial for their effects on the brain.

4. BLOOD-BRAIN BARRIER

The blood-brain barrier (BBB) is the result of tight intracellular junctions formed between brain capillary endothelial cells. These tight junctions and the low rate of fluid phase endocytosis of brain endothelial cells limit the transfer of compounds, including therapeutic agents, from the blood to the central nervous system. Despite the BBB, small molecules such as glucose and amino acids enter the brain from the blood stream by active transport mechanisms. In addition, receptors specific for some proteins such as insulin are found on the luminal side (blood) of brain capillary endothelial cells, facilitating their transport [18].

Given that HDO could provide a promising therapy for AD, an understanding of their ability to penetrate the BBB becomes crucial for the determination of their direct CNS effects. In previous studies, an *in vitro* co-culture of brain capillary endothelial cells and astrocytes has been used to mimic the BBB. Leveugle et al.[19] demonstrated the passage of HDO through the BBB and its effects on APP processing using this system. Our studies have also shown that C3 is capable of passing through the un-compromised BBB after intravenous, subcutaneous or oral administration in rats, by measuring the anti-factor Xa activity as a biomarker.

5. TISSUE FACTOR PATHWAY INHIBITOR

TFPI, a member of the family of SERPINs, has a molecular weight of 12-14 kDa. It is one of several physiological anticoagulants, including antithrombin III and activated protein C, which act as potent mediators of the inhibition of the extrinsic pathway of the coagulation cascade. TFPI exerts its anticoagulant properties via direct inhibition of factor Xa and indirect inhibition of factor VIIa. TFPI contains three tandem Kunitz Proteinase Inhibitor (KPI) domains, which mediate its anticoagulant effects[20]. The biosynthesis of TFPI is age-related[21]. In elderly people, a hypercoagulation state due to the low TFPI level has been noted.

Previous studies from our laboratory have demonstrated that UFH and various LMWHs exhibit different potential in TFPI release from endothelial cells, which contributes mostly to their peripheral vascular and anticoagulant effects. However, the mechanism of the interaction of heparins and TFPI is not fully clear.

Although Hollister et al.[22] have recently demonstrated the expression of TFPI in the brain and a significantly elevated amount in frontal cortex from AD patients, the physiological role of TFPI in the central nervous system (CNS) remains unknown. It is still unclear whether the increase in TFPI is due to a compensatory mechanism in which the activity of TFPI decreases, or contributes to the etiology of AD. Moreover, members of the SERPIN family such as AT-III, α_1-antichymotrypsin, and TFPI have been found in association with senile plaques[23]. However, their role in the formation and regulation of plaques is not fully understood. Since TFPI is localized to microglia in both AD and non-AD individuals and is localized to senile plaques, Hollister et al. also suggested that TFPI might play a cell specific role in proteinase regulation in the brain.

Based on these data, we hypothesize that the release of TFPI by heparins would be beneficial in AD because: 1) it may improve the microcirculation in the brain, and 2) the release of functional TFPI into the neuropil perhaps by microglia and/or up-regulation of TFPI expression may be enhanced by heparins to provide a means of clearing proteinase during APP processes in the CNS.

6. POTENTIAL USE OF HEPARIN DERIVED OLIGOSACCHARIDES IN THE TREATMENT OF ALZHEIMER'S DISEASE

As a primary degenerative dementia, AD is not considered to be of vascular origin. Indeed, stroke and severe cardiovascular diseases are generally exclusionary for the clinical diagnosis. However, during recent years both epidemiological and neuropathological studies have suggested an association between AD and several vascular risk factors, such as hypertension, coronary heart disease, atrial fibrillation, diabetes mellitus, ischemic white matter lesions and generalized arteriosclerosis[44].

UFH, LMWHs and HDO are considered as traditional cardiovascular treatments. The first introduction of heparins in the treatment of AD, however, was based on the fact that heparins improved the cerebral microcirculation, as shown in stroke patients[45]. AD may lead to stroke-like lesions in the cerebral microvasculature, as β-amyloid interacting with endothelial cells produces superoxide radicals, which causes endothelial damage; whereas heparins may repair this damage through its vascular effects and/or its neuronal effects after passing through the blood brain barrier. With the advanced understanding of etiology and molecular mechanism, more and more information is being generated to re-consider the role of heparin and its low molecular weight components (HDO) in the pharmacological management of AD.

Table 1. The Characteristics of Heparin Derived Oligosaccharides (HDO) with a Potential in the Management of Alzheimer's Disease (AD)

- New HDO may have direct interactions with PGs and decrease the formation of Aβ.
- New HDO, which have low molecular weight, facilitate their passage through the BBB.
- New HDO may have interactions with SERPINs such as TFPI enhancing its vascular and/or neuronal effects. However, it may only have low anticoagulant effects decreasing the potential of bleeding in aged AD patients.
- New HDO may still have strong anti-inflammatory effects as UFH.

Our preliminary data have demonstrated that an HDO, C3, has a very low molecular weight (2-2.5 kDa) and very low anticoagulant effects (15-20% of UFH). We have also shown that C3 is capable of crossing the BBB. Hanin et al. and Walzer et al. from our group have demonstrated that HDO could decrease immunoreactive tau toxicity in an AD model. Altogether, HDO represent a very promising new category of compounds with potential in the treatment of AD.

REFERENCES

1. P.V. Rabins, Cognition, in: *Oxford Textbook of Geriatric Medicine*, edited by J.G. Evans et al. (Oxford University Press, Oxford and New York, 2000), pp. 917-921.
2. P.D. Meek, K.E. McKeithan, and G.T. Schumock, Economic considerations in Alzheimer's disease. Pharmacotherapy. **18**, 68-73 (1998).
3. R. Kisilevsky, L.J. Lemieux, P.E. Fraser, X. Kong, P.G. Hultin, and W.A. Szarek, Arresting amyloidosis in vivo using small-molecule anionic sulphonates or sulphates: implications for Alzheimer's disease, Nat Med. **1**, 143-8 (1995).
4. S.J. Pollack, I.I. Sadler, S.R. Hawtin, V.J. Tailor, and M.S. Shearman, Sulfated glycosaminoglycans and dyes attenuate the neurotoxic effects of beta-amyloid in rat PC12 cells, Neurosci Lett. **184**, 113-6 (1995).
5. U. Cornelli, The therapeutic approach to Alzheimer's disease, in: *Non-Anticoagulant Action of Glycosaminoglycans*, edited by J. Haremberg and B. Casu (Plenum Press, New York, 1996), pp. 249-79.
6. J. Fareed, W. Jeske, D. Hoppensteadt, R. Clarizio, and J.M. Walenga, Low-molecular-weight heparins: pharmacologic profile and product differentiation. Am J Cardiol. **82**, 3L-10L (1998).
7. D.J. Selkoe, The molecular pathology of Alzheimer's disease, Neuron **6**, 487-498 (1991).
8. J. Kang, H.-G. Lemaire, A. Unterberck, J.M. Salbaum, C.L. Masters, K.-H. Grezschik, G. Multhaup, K. Beyreuther, and B. Muller-Hill, The precursor of Alzheimer's disease amyloid A4 protein resembles a cell-surface receptor, Nature, **325**, 733-736 (1987).
9. J. Fareed, D. Callas, D.A. Hoppensteadt, W. Jeske, J.M. Walenga and R. Pifarre, Current development in anticoagulant and antithrombotic agents, in: *New Anticoagulants for the Cardiovascular Patient*, edited by R. Pifarre (Hanley & Belfus Inc., Philadelphia, 1997), pp. 95-126.
10. B. Casu, Structure of heparin and heparin fragments. Ann NY Acad Sci. **556**,1-17 (1989).
11. A.D. Snow, R.T. Sekiguchi, D. Nochlin, R.N. Kalaria, and K. Kimata. Heparan sulfate proteoglycan in diffuse plaques of hippocampus but not of cerebellum in Alzheimer's disease brain. Am J Pathol. **144**, 337-347 (1994).
12. A.D. Snow, M.G. Kinsella, E. Parks, R.T. Sekiguchi, J.D. Miller, K. Kimata, and T.N. Wight. Differential binding of vascular cell-derived proteoglycans (perlecan, biglycan, decorin and versican) to the beta amyloid protein of Alzheimer's disease. Arch Biochem Biophys. **320**, 84-95 (1995).

13. R. Gupta-Bansal, R.C. Frederickson, and K.R. Brunden, Proteoglycan-mediated inhibition of A beta proteolysis. A potential cause of senile plaque accumulation, J Biol Chem. **270**, 18666-71 (1995).

14. A.D. Snow, R. Sekiguchi, D. Nochlin, P. Fraser, K. Kimata, A. Mizutani, M. Arai, W.A. Schreier, and D.G. Morgan, An important role of heparan sulfate proteoglycan (Perlecan) in a model system for the deposition and persistence of fibrillar A beta-amyloid in rat brain, Neuron. **12**, 219-34. (1994).

15. G. Perry, S.L. Siedlak, P. Richey, M. Kawai, P. Cras, R.N. Kalaria, P.G. Galloway, J.M. Scardina, B. Cordell, B.D. Greenberg, et al. Association of heparan sulfate proteoglycan with the neurofibrillary tangles of Alzheimer's disease. J Neurosci. **11**, 3679-83, (1991)

16. S.A. Lorens, M. Guschwan, N. Hata, L.D. van de Kar, J.M. Walenga, and J. Fareed, Behavioral, endocrine, and neurochemical effects of sulfomucopolysaccharide treatment in the aged Fischer 344 male rat, Sem Thromb Hemost. **17**, S164-73 (1991).

17. L. Conti, G.F. Placidi, and G.B. Cassano, Ateroid in the treatment of dementia: results of a clinical trial, Mod Probl Pharmacopsychiatry. **23**, 76-84 (1989).

18. W.M. Pardridge, Transport of small molecules through the blood-brain barrier: biology and methodology, Advanced Drug Delivery Rev. **15**, 5-36 (1995).

19. B. Leveugle, W. Ding, F. Laurence, M.P. Dehouck, A. Scanameo, R. Cecchelli, and H. Fillit, Heparin oligosaccharides that pass the blood-brain barrier inhibit beta-amyloid precursor protein secretion and heparin binding to beta-amyloid peptide, J Neurochem. **70**, 736-44 (1998).

20. G.J. Broze Jr. The role of tissue factor pathway inhibitor in a revised coagulation cascade. Semin Hematol. **29**, 159-69 (1992).

21. J. Kodama, K. Uchida, H. Kushiro, N. Murakami, and C. Yutani, Hereditary angioneurotic edema and thromboembolic diseases: I: How symptoms of acute attacks change with aging. Intern Med. **37**, 440-3 (1998).

22. R.D. Hollister, W. Kisiel, and B.T. Hyman, Immunohistochemical localization of tissue factor pathway inhibitor-1 (TFPI-1), a Kunitz proteinase inhibitor, in Alzheimer's disease. Brain Res. **728**, 13-9 (1996).

23. V.L. Turgeon and L.J. Houenou, The role of thrombin-like (serine) proteases in the development, plasticity and pathology of the nervous system. Brain Res - Brain Res Rev. **25**, 85-95 (1997).

24. R.N. Kalaria, Cerebral vessels in ageing and Alzheimer's disease. Pharmacol Ther. **72**, 193-214 (1996).

25. R.N. Kalaria, The role of cerebral ischemia in Alzheimer's disease. Neurobiol Aging. **21**, 321-30 (2000).

LOW MOLECULAR WEIGHT GLYCOSAMINOGLYCAN BLOCKADE OF BETA AMYLOID (25-35) INDUCED NEUROPATHOLOGY

M. Walzer, S. Lorens, M. Hejna, J. Fareed, R. Mervis, I. Hanin, U. Cornelli and J. M. Lee*

1. Introduction

Alzheimer's disease (AD) is characterized histopathologically by senile plaques, neurofibrillary tangles (NFTs), reactive astrocytosis and regional specific cell loss. Senile plaques are an extracellular deposition of beta amyloid (Aβ), a 40-42 amino acid peptide, while intracellular NFTs consist of paired helical filaments (PHFs) formed from microtubule associated tau protein in a hyperphosphorylated and/or conformationally altered state[1, 2]. Previous studies have shown that other components, specifically proteoglycans (PGs) [heparan sulfate (HS), chrondroitin sulfate (CS), dermatan sulfate (DS) and keratan sulfate (KS)] and their glycosaminoglycan (GAG) side chains are co-localized with Aβ in senile plaques[3, 4] and hyperphosphorylated tau in NFTs[5] of AD brains. *In vitro* work has shown that PGs and GAGs directly bind Aβ and synthetic Aβ peptides, stimulating fibril nucleation, growth, and stability[6]. These data suggest that PGs and their GAG side chains may act as "seed molecules"[6] for Aβ fibril formation, hence promoting senile plaque formation in AD. Much of the research suggests that both charge and size of these PGs/GAGs influence their ability to interact with Aβ[6].

On the other hand, cell culture studies have shown GAGs (HS and CS) and other synthetic polysulfated compounds attenuate the neurotoxic effects of Aβ[7] by inhibiting Aβ's interaction with cells[8]. In addition, clinical studies in Europe have shown a therapeutic effect of Ateroid (a compound made of both high and low molecular weight glycans) on AD and multi-infarct dementia in elderly patients[9, 10]. Similarly, Ateroid has also been shown to improve both behavioral and neurochemical impairments seen in aged

*M. Walzer, M. Hejna, S. Lorens, I. Hanin, Department of Pharmacology Loyola University Chicago Medical Center. J. Fareed, J. M.Lee, Department of Pathology and Pharmacology Loyola University Chicago Medical Center, Maywood, IL, USA. R. Mervis, Neurocognitive Research Laboratory, Columbus, OH, USA. U. Cornelli, Corcon, Milan, Italy.

Mapping the Progress of Alzheimer's and Parkinson's Disease
Edited by Mizuno *et al.*, Kluwer Academic/Plenum Publishers, 2002

165

F344 rats[11]. Together these studies suggest a role for GAGs in treating not only AD but also other age related dementias.

In the present study we investigated whether smaller low molecular weight (LMW) GAGs can block Aβ aggregation and inhibit Aβ induced AD-like neuropathology. We tested whether subcutaneously administered LMW GAGs could block the effects of Aβ (25-35) in a Fischer 344 rat model of Aβ induced neuropathology[12, 13]. Animals were given twice a day subcutaneous injections of Certoparin (C7; MW 4600), (a LMW heparin clinically used for the prevention and treatment of deep vein thrombosis), or a Certoparin derivative (C6; MW 1600) starting three days before an Aβ (25-35) intra-amygdaloid infusion, and continuing for 32 days post Aβ infusion. Studied endpoints included tau-2 immunoreactivity (IR), reactive astrocytosis measured by increased glial fibrillary acidic protein (GFAP) IR, interleukin-1 beta (IL-1β) IR and Bcl-2 IR in the hippocampus, as well as Congo red staining of Aβ's β-pleated sheet conformation at the injection site in the amygdala. This experimental design allowed us to determine whether LMW GAGs may be beneficial in preventing AD associated neuropathology if treatment begins before Aβ deposition in a characterized rat model of Aβ induced neuropathology.

2. Materials and Methods

The experimental methods have been described previously in detail by Sigurdsson *et al.*[12]

2.1 Drugs

a). Amyloid-β (25-35) (BACHEM, Torrance, CA) is supplied as a TFA salt with the peptide content of 88% (± 3%). The peptide and its respective vehicle (VEH; TFA, sodium salt; Sigma) was dissolved in Nanopure® H_2O (dH_2O) immediately before use and stored at 4°C between infusions.

b). LMW GAGs: Certoparin (Novartis AG, Switzerland) and a Certoparin derivative (C6) were provided by J.F. The molecular weights of these compounds were determined by high-performance size-exclusion chromatography [14] or mass spectrometry, and are approximately 4,600 and 1,600 Da respectively. The LMW heparin Certoparin is mainly composed of hexa-decasaccharide units and is clinically used in Germany for the prophylaxis of postsurgical deep vein thrombosis [15]. Certoparin is produced by deamination cleavage of unfractionated heparin by isoamyl nitrite digestion, leading to the formation of a 2,5-anhydro-D-mannose (5-member ring) at the reducing and non-reducing ends of the heparin chains. Certoparin is heterogeneously composed of 6-10 2-O-SO_3-uronic acid-6-O-SO_3-glucosamine-N-SO_4 disaccharide repeating units with a sulfate/carboxyl ratio of 2.5, equal to that of unfractionated heparin[16]. Further isoamyl nitrite digestion of Certoparin and subsequent isolation of the LMW tetra-hexasaccharides products produces the Certoparin derivative C6. All drugs were provided as a white powder and stored in a desiccator at room temperature. All solutions were made fresh before each injection.

2.2. Data Analysis

Tau-2, Bcl-2 and IL-1β IR cells were counted from 6 coronal sections surrounding the cannula track. The number of IR cells in the hippocampus were analyzed using a One-Way ANOVA (Prism 3.0) followed by a Neuman-Keuls *post hoc* test. Reactive astrocytosis in the ipsilateral hippocampus was rated on a scale of 0-2+.

An animal was considered positive if it exhibited 1-2+ staining. The GFAP staining was analyzed using a Chi-squared test (Prism 3.0) followed by Fisher's exact tests.

3. Results

3.1 Certoparin and C6 Injections Beginning 3 Days Prior to Aβ (25-35) Infusion

In order to determine whether LMW GAGs inhibit Aβ induced neuropathology we started twice a day subcutaneous (2.5 mg/kg) injections of saline, Certoparin or C6 beginning 3 days prior to an intra-amygdaloid Aβ (25-35) infusion. At the 32 day endpoint, we found that the increased tau-2 IR in the right hippocampus of Aβ/ VEH animals was blocked by both Certoparin (Figs. 1A-1C and 2) and C6 (Fig. 2A) injections.

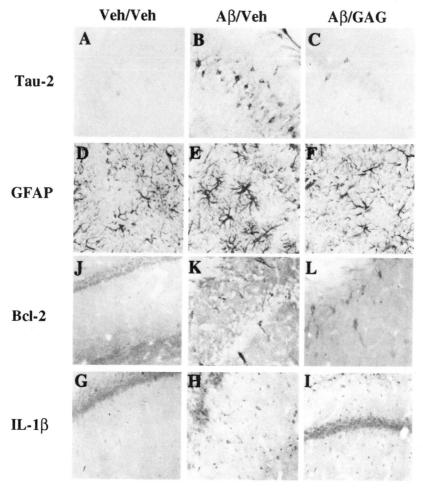

Figure 1. Photomicrograph of Fischer 344 coronal rat brain sections showing: (A-C) tau-2 IR in the CA2 hippocampal region (20X magnification); (D-F) GFAP IR in the CA1 hippocampal region (40X magnification). (G-I) IL-1β IR in the CA2 hippocampal region (20X magnification); and (J-L) Bcl-2 IR in the CA2 hippocampal region (20X magnification).

GFAP IR is one of the earliest and most prominent responses of the CNS to tissue injury and has been found to be increased 8-16 fold in AD brains compared to age matched controls[17]. Increases in GFAP IR found in the right hippocampus of Aβ/VEH animals were also blocked by both Certoparin or C6 injections beginning 3 days prior to Aβ infusion (Fig 2B). Next, we determined whether the Aβ deposit was present in animals given saline or LMW GAG injections. Cresyl violet staining showed no difference between the size of the Aβ deposit found in Aβ/VEH injected animals compared to Aβ/C6 or Aβ/Certoparin injected animals; while all deposits were Congo red positive (i.e. apple-green birefringence under polarized light). In addition, C6 treatment did not decrease Aβ induced Bcl-2 neuronal IR in the right hippocampus (p < 0.05) (Fig. 1H-1J). Also, there was no effect of C6 on Aβ induced IL-1β IR microglia in the right hippocampus (Fig 1K-1M).

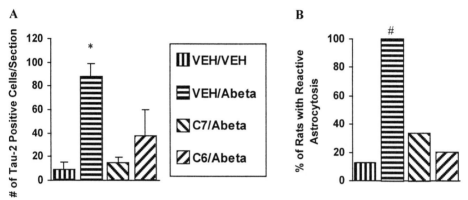

Figure 2. Graphical analysis of (A) Tau-2 IR in the right hippocampus of Fischer 344 male rats. (B) Reactive astrocytosis in the right hippocampus of Fischer 344 male rats. n's = 8-10 animals/group. *p<0.05 compared to VEH/VEH group based on a One-way ANOVA followed by Neuman-Keuls post-hoc analysis. #p<0.05 compared to VEH/VEH group based on a Chi-squared test followed by fischer exact tests.

4. Discussion

Recently we have shown another LMW GAG (C3; MW 1900) is capable of crossing the blood brain barrier (BBB) if given either intravenously, subcutaneously or orally to rats with an uncompromised BBB[18]. Currently it is unknown what molecular weight fraction of these heterogeneous compounds crosses the BBB. However the present data demonstrate that LMW GAGs which range from 1600-4600 Da can block the effects of Aβ (25-35), independent of Aβ fibril formation, in an animal model of Aβ induced neuropathology. Interestingly C6 treatment did not prevent Aβ induced IL-1β microglia or Bcl-2 neuronal IR in this model. Bcl-2 is an anti-apoptotic marker, which is normally increased in neurons suggesting a protective mechanism against a general insult. LMW GAG treatment did not decrease Bcl-2 IR suggesting these neurons maybe still responding to the initial Aβ insult and LMW GAG treatment may not be able to prevent all Aβ induced injury responses. This is further supported by no change in IL-1β IR in the microglia with LMW GAG treatment.

The present data supports at least three hypotheses: (1) LMW GAGs act directly on Aβ inhibiting a neurotoxic mechanism independent of Aβ aggregation; and/or LMW GAGs interact directly with hippocampal neurons, astrocytes and/or their processes, either (2) promoting intracellular protective responses to Aβ and/or (3) effectively blocking Aβ from binding to a membrane bound receptor, thus preventing cellular responses to Aβ.

In support of the second hypothesis, sulfated GAGs, have been shown to bind both fibroblast growth factor (FGF) and the FGF receptor potentiating their intracellular effects[19, 20], specifically neurite outgrowth in cell culture. Hence, LMW GAGs may be acting directly on hippocampal neurons or neuronal processes projecting to the amygdala helping these neurons to recover from Aβ's insult. In addition, recent data suggests that LMW GAG treatment also leads to an increase in both dendritic branching and dendritic spine densities in this animal model[21]. This research supports the idea that LMW GAGs may be working as neurotrophic agents, stimulating cellular processes which allow the cell to better withstand an insult induced by Aβ. However, one may expect to see less GFAP and/or Tau-2 IR compared to baseline in the VEH/LMW GAG group. This study did not allow us to investigate this hypothesis, although we have not seen this effect in similar studies.

Kisilevsky et al.[22] found that small-molecule disulfates (MW 900-1000) could inhibit the in vitro acceleration of Aβ amyloid fibril formation by heparan sulphate. Although these compounds are slightly lower in molecular weight than Certoparin or C6 they do have a similar amount of sulfate residues substantiating the importance of sulfates in inhibiting Aβ induced neuropathology and/or fibril formation. Since there is no measurable change in the Congo red positive deposit after LMW GAG treatment, it is therefore unlikely that LMW GAGs are acting in a similar manner, as Aβ fibrillogenesis inhibitors. Alternatively, in vitro studies have shown GAGs may act by coating Aβ thus preventing Aβ from interacting with neurons[8]. This mechanism would explain why LMW GAG treatment leads to a blockade of Aβ induced neuropathology without altering the Congo Red positive deposit in the present animal model. However the inability to block Bcl-2 or IL-1β IR again suggests a mechanism independent of Aβ or possibly that these proteins are part of the protective effects of LMW GAGs.

The present study provides additional support for the further investigation of LMW GAGs as a potential therapeutic treatment for AD.

Acknowledgments We would like to thank Debra Magnuson, Annamarie Kohn, and Magdalena Zemaitaitis for their work on the Bcl-2 immunocytochemistry and editorial comments. This research was funded by the Retirement Research Foundation grant #98-96 and Illinois Department of Public Health contract # 83880360.

References

1. N. Watanabe, K. Takio, M. Hasegawa, T. Arai, K. Titani and Y. Ihara, Tau 2: a probe for aSer conformation in the amino terminus of tau, *J Neurochem* **58** (3), 960-966 (1992).
2. E. Lang and L. Otvos, Jr., A serine to proline change in the Alzheimer's disease-associated epitope Tau 2 results in altered secondary structure, but phosphorylation overcomes the conformational gap, *Biochem Biophys Res Commun* **188** (1), 162-169 (1992).

3. L. S. Perlmutter, H. C. Chui, D. Saperia and J. Athanikar, Microangiopathy and thecolocalization of heparan sulfate proteoglycan with amyloid in senile plaques of Alzheimer's disease,*Brain Res* **508** (1), 13-19 (1990).
4. A. D. Snow, J. P. Willmer and R. Kisilevsky, Sulfated glycosaminoglycans in Alzheimer's disease, *Hum Pathol* **18** (5), 506-510 (1987).
5. D. A. DeWitt, J. Silver, D. R. Canning and G. Perry, Chondroitin sulfate proteoglycans are associated with the lesions of Alzheimer's disease, *Exp Neurol* **121** (2), 149-152 (1993).
6. G. M. Castillo, C. Ngo, J. Cummings, T. N. Wight and A. D. Snow, Perlecan binds to the beta-amyloid proteins (A beta) of Alzheimer's disease, accelerates A beta fibril formation, and maintains A beta fibril stability, *J Neurochem* **69** (6), 2452-2465 (1997).
7. S. J. Pollack, Sadler, II, S. R. Hawtin, V. J. Tailor and M. S. Shearman, Sulfated glycosaminoglycans and dyes attenuate the neurotoxic effects of beta-amyloid in rat PC12 cells,*Neurosci Lett* **184** (2), 113-116 (1995).
8. Sadler, II, D. W. Smith, M. S. Shearman, C. I. Ragan, V. J. Tailor and S. J. Pollack, Sulphated compounds attenuate beta-amyloid toxicity by inhibiting its association with cells,*Neuroreport* **7** (1), 49-53 (1995).
9. T. A. Ban, L. C. Morey and V. Santini, Clinical investigations with ateroid in old-age dementias, *Semin Thromb Hemost* **17** (Suppl 2), 161-163 (1991).
10. L. Conti, F. Re, F. Lazzerini, L. C. Morey, T. A. Ban, V. Santini, A. Modafferi and A. Postiglione, Glycosaminoglycan polysulfate (Ateroid) in old-age dementias: effects upon depressive symptomatology in geriatric patients,*Prog Neuropsychopharmacol Biol Psychiatry* **13** (6), 977-981 (1989).
11. S. A. Lorens, M. Guschwan, N. Hata, L. D. van de Kar, J. M. Walenga and J. Fareed, Behavioral, endocrine, and neurochemical effects of sulfomucopolysaccharide treatment in the aged Fischer 344 male rat, *Semin Thromb Hemost* **17** (Suppl 2), 164-173 (1991).
12. E. M. Sigurdsson, S. A. Lorens, M. J. Hejna, X. W. Dong and J. M. Lee, Local and distant histopathological effects of unilateral amyloid-beta 25-35 injections into the amygdala of young F344 rats, *Neurobiol Aging* **17** (6), 893-901 (1996).
13. E. M. Sigurdsson, J. M. Lee, X. W. Dong, M. J. Hejna and S. A. Lorens, Laterality in the histological effects of injections of amyloid-beta 25-35 into the amygdala of young Fischer rats,*J Neuropathol Exp Neurol* **56** (6), 714-725 (1997).
14. A. Ahsan, W. Jeske, D. Hoppensteadt, J. C. Lormeau, H. Wolf and J. Fareed, Molecular profiling and weight determination of heparins and depolymerized heparins, *J Pharm Sci* **84** (6), 724-727 (1995).
15. J. Fareed, D. A. Hoppensteadt and R. L. Bick, An update on heparins at the beginning of the new millennium, *Semin Thromb Hemost* **26** (Suppl 1(12)), 5-21. (2000).
16. W. Jeske, H. Wolf, A. Ahsan and J. Fareed, Pharmacologic profile of certoparin, *Exp. Opin. Invest. Drugs* **8** (3), 315-327 (1999).
17. L. Buee, A. Laine, A. Delacourte, S. Flament and K. K. Han, Qualitative and quantitative comparison of brain proteins in Alzheimer's disease, *Biol Chem Hoppe Seyler* **370** (11), 1229-1234 (1989).
18. Q. Ma, Dudas, B., Cornelli, U., Lorens, S. A., Lee., J.M., Mervis, R.F., Fareed, J., Hanin, I., Transport of low molecular weight heparins and related glycosaminoglycans through the blood brain barrier: experimental evidence in a rat model, *FASEB Journal* **14** (8), A1480 (2000).
19. D. H. Damon, P. A. D'Amore and J. A. Wagner, Sulfated glycosaminoglycans modify growth factor-induced neurite outgrowth in PC12 cells, *J Cell Physiol* **135** (2), 293-300 (1988).
20. D. H. Damon, R. R. Lobb, P. A. D'Amore and J. A. Wagner, Heparin potentiates the action of acidic fibroblast growth factor by prolonging its biological half-life,*J Cell Physiol* **138** (2), 221-226 (1989).
21. R. F. Mervis, J. Mckean, S. Zats, A. Gum, R. Reinhart, B. Dudas, U. Cornelli, J. M. Lee, S. A. Lorens, J. Fareed and I. Hanin, Neurotrophic effects of the glycosaminoglycan C3 on dendritic arborization and spines in the adult rat hippocampus: a quantitatve golgi study, *CNS Drug Reviews* **6** (Supp. 1), 44-46 (2000).
22. R. Kisilevsky, L. J. Lemieux, P. E. Fraser, X. Kong, P. G. Hultin and W. A. Szarek, Arresting amyloidosis in vivo using small-molecule anionic sulphonates or sulphates: implications for Alzheimer's disease [see comments], *Nature Medicine* **1** (2), 143-148 (1995).

C3, A PROMISING ULTRA LOW MOLECULAR WEIGHT GLYCOSAMINOGLYCAN FOR THE TREATMENT OF ALZHEIMER'S DISEASE

Israel Hanin, Bertalan Dudas, Ronald F. Mervis, Umberto Cornelli, John M. Lee, Stanley A. Lorens and Jawed Fareed[*]

Based on encouraging reports in the literature both in humans (Perry et al., 1993; Cornelli, 1996; Snow et al., 1995) and in experimental animals (Lorens et al., 1991), it would appear that glycosaminoglycans (GAGs) might be effective as treatment modalities in AD. GAGs (examples include hyaluronic acid, heparin, dermatan sulfate, heparan sulfate, keratan sulfate), are linear heteropolysaccharides consisting of a hexosamine (e.g. hexuronic acid, L-iduronic acid, galactose), disaccharide units, and sulfate substituents. They bind to specific proteins in vivo (e.g. anti-thrombin III and lipoprotein lipase) and their biological activity is determined by their molecular weight, charge density, degree of sulfation, and type of disaccharide units in the molecule. Proteoglycans (PG) provide the matrix for the formation of GAGs in vitro. The core protein is glycosylated, and GAGs then bind covalently to this structure via an O-glycoside linkage to a serine residue, or an N-linkage to asparagine), primarily in the Golgi apparatus. There are many different kinds of PGs, depending on the GAG type attached and their localization within the biological matrix (Kjellen and Lindahl, 1991) (See Figure 1).

PG, such as heparan sulfate PG, have been shown to facilitate the occurrence of classical neuropathological landmarks of AD such as: a) amyloid formation, via an increase in aggregation rate of beta amyloid (Aβ) into β-pleated fibrils (Snow et al., 1994, 1995); and b) increased polymerization of tau into paired helical proteins, which make up the neurofibrillary tangles of AD (Perry et al., 1993). GAGs compete with endogenous PG to protect against these occurrences (Caughey and Raymond, 1993; Snow et al., 1994; Leveugle et al. 1994).

[*] Israel Hanin, Bertalan Dudas, Umberto Cornelli, John M Lee, Stanley A Lorens, Jawed Fareed, Loyola University Chicago, Maywood, Ill 60153. Ronald F Mervis, Neuro-Cognitive Research Laboratories, Columbus, Ohio 43212.

Mapping the Progress of Alzheimer's and Parkinson's Disease
Edited by Mizuno *et al.*, Kluwer Academic/Plenum Publishers, 2002

171

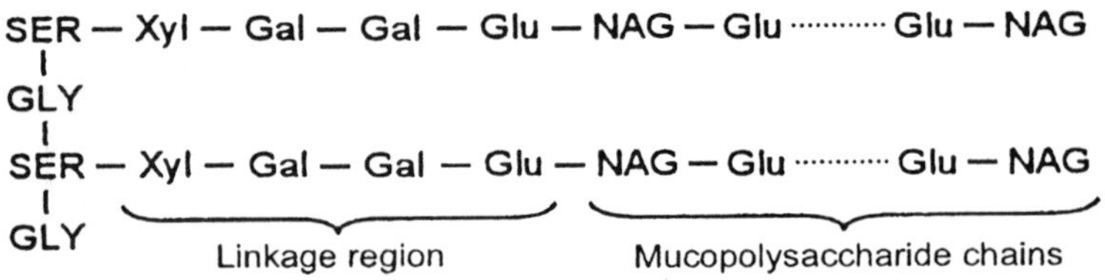

SER — Xyl — Gal — Gal — Glu — NAG — Glu ·········· Glu — NAG
|
GLY
|
SER — Xyl — Gal — Gal — Glu — NAG — Glu ·········· Glu — NAG
|
GLY   Linkage region   Mucopolysaccharide chains

Figure 1. Diagrammatic illustration of the structure of the PG moiety, the origin of endogenously synthesized heparin and other GAGs.

Preliminary studies utilizing a mixture of GAG complexes referred to as Ateroid[R], provide encouraging findings demonstrating improvement in age-related behavioral deficits (rats) and in dementia (man). Ateroid[R] is composed of low molecular weight heparin (10-20%), heparan sulfate (40-60%), dermatan sulfate (20-35%), and chondroitin sulfate (2-8). It has earned a use-patent in several European countries. In aged F344 rats, following 12 weeks of Ateroid[R] administration in the drinking water, Lorens et al. (1991) showed a reversal in age-related deficits in both conditioned one-way and two-way avoidance behaviors. These behavioral observations paralleled a protection against age-related reduction in the dopamine metabolites, DOPAC and HVA, in the nucleus accumbens. Hence, GAGs could conceivably affect factors pertaining to age-related changes in central nervous system (CNS) neurotransmitter function. Ateroid[R] has also been used in man, with positive indications, for the alleviation of various symptoms of primary dementia (Ban et al., 1991; Conti et al., 1993), and for the treatment of symptoms of multiple infarct dementia (Conti et al., 1989). A recent pilot study in AD patients with Ateroid[R] has resulted in a normalization of platelet MAO-B activity (which generally is higher in AD patients than in controls), and of levels of the neurotransmitter metabolites HVA, MHPG, and 5HIAA (which generally are lower than normal in AD patients) in the CSF (Parnetti et al., 1995; Cornelli, 1996).

The Ateroid[R]-induced behavioral and neurochemical effects were obtained with a heterogeneous mixture of compounds, of different sizes and biological properties (see above). A more effective substance would be a homogeneous GAG, which can reproduce the beneficial pharmacological activity reported for the heterogeneous mixture of GAGs (also referred to as GAG-polysulfate, or GAP). We have therefore focused our attention on finding and testing new, homogeneous GAGs, which can reproduce the beneficial pharmacological activity reported for the heterogeneous GAP. Furthermore, because of Ateroid[R]'s central activity, it is reasonable to suspect that the active principle in this mixture is an in vivo metabolite, with a molecular weight (MW) that is low enough to allow permeability of the compound across the blood brain barrier (BBB). Thus, availability of well characterized and reliably reproducible GAGs, with a smaller MW range than that of the parent compounds, would permit more efficient therapeutic application in AD than that which is achieved by a broad mixture of compounds such as GAP.

Compound C3 is the prototype compound in our studies. C3 was selected for testing from among heparins with different MWs. It was our compound of choice because: a) it has an ultra low molecular weight (MW~2100); b) it can be produced in large quantities; and c) the method for its production has already been standardized. It is produced by physical fractionation of heparin via gamma irradiation, followed by gel filtration. The degree of gamma irradiation is modulated in order to obtain an appropriate hydrolysis into fragments within a small range of MWs. Subsequent gel filtration allows the isolation of a mixture of oligosaccharides in the required range (from 4 to 10). Furthermore, C3 exhibits lower anticoagulant, and antiprotease activities, than those of unfractionated heparin, and it can cross the BBB to a limited extent (Ma et al., 2000), thus making it an appealing compound to study as a prospective agent for the treatment of AD. C3 was made available for these studies from Corcon-Intercorcon, Milan, Italy.

We have investigated the effect of C3 administration on tau-2 immunoreactivity in the $A\beta(25-35)$ treated animal model, and have explored its effect on dendritic parameters from Golgi-impregnated hippocampal CA1 neurons. This chapter highlights some of our key findings to date.

Effect on tau-2 immunoreactivity and reactive astrocytosis in an animal model of AD:

We have used in these studies an animal model that has been developed by Drs. Lee and Lorens and their coinvestigators at Loyola University Chicago (Sigurdsson et al., 1996). This model involves the stereotaxic injection of $A\beta(25-35)$ (5.0 nmoles) into the right amygdala of rats, and the measurement of tau immunoreactivity (IR) in several brain areas at different times following this treatment. By day 8 post-injection one can already observe increased tau IR in neuronal cell bodies and processes ipsilaterally in the amygdala and hippocampus, and a progressive Alz-50 IR as well. By 32 days post-injection of $A\beta(25-35)$ reactive astrocytosis was also observed in the ipsilateral hippocampus of the experimental animals.

We demonstrated protection against the tau-2 immunoreactivity and reactive astrocytosis plus cell shrinkage when animals were treated either subcutaneously [2.5 mg/kg, b.i.d., 3 days before, and 32 days after $A\beta(25-35)$ administration (data not shown here)] or orally via gavage [25mg/kg, once daily, 3 days before, and 14 days after $A\beta(25-35)$ administration] (Dudas et al., 2001). This protective effect is dose dependent; we were unable to protect against the neurodegenerative results of $A\beta(25-35)$ administration with a 2.5 mg/kg oral treatment with C3, using the same paradigm as that which was employed with the 25 mg/kg dose (Figure 2A-D).

These data demonstrate a protective effect of C3 against the neurodegenerative consequences of intra-amygdaloid $A\beta(25-35)$ administration. Moreover, the data document one additional important feature of the ultra low molecular weight GAGS. Specifically, they provide further evidence implying that C3 can cross the BBB and exert its protective action in the central nervous system. In addition, the finding that C3 is active following both s.c. and p.o. exposure, illustrates that GAGs, if given to patients, could be provided by conventionally accepted routes of administration.

The injected $A\beta(25-35)$ remains as a deposit at the injection site (Figure 2E-F) and shows Congo red positivity, indicating that the deposit is in the beta pleated form. Sigurdsson et al. (1997) showed that the deposit remains for at least 128 days; C3

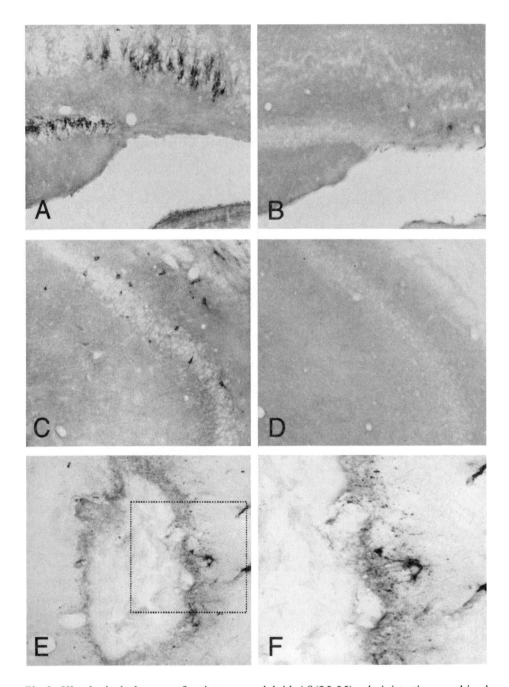

Fig 2. Histological changes after intra-amygdaloid Aβ(25-35) administration combined with 2.5 mg/kg (A, C) and 25 mg/kg (B, D) oral C3 treatment in CA3 (A, B) and CA2 (C, D) zones of the hippocampus. The injected Aβ(25-35) deposit does not show any changes during the C3 treatment (E, F).

administration did not alter the deposit or modify it in any way. Hence, it would appear from these findings that the palliative effect seen with C3 may, at least in part, be due to some neuroprotective and/or neurotrophic mechanism.

Effects of Aβ(25-35) and of C3 on Dendritic Branching: Golgi Analysis:

Neuronal integrity in the animal model, and any subsequent prophylaxis induced by GAGs, may be determined by measuring growth, numbers, and condition of dendritic spines in the cingulate cortex and hippocampus, using Golgi staining methodology. Golgi – impregnated dendritic branches and spines are highly quantifiable parameters. Therefore, it is feasible that one can ascertain fairly precisely not only how much damage a neuronal population may have suffered as a consequence of some brain insult, but – just as importantly – how effective a neuroprotective agent or cognitive enhancer may be in minimizing or reversing such damage (Mervis et al., 2000). A critical factor in this assessment is that – because dendritic branching and spines reflect the integrity of brain circuits – learning and memory are closely correlated with these dendritic parameters (Hill et al., 1993; Vozeh and Myslivecek, 1996).

Neurostructural analysis of dendritic parameters from Golgi-impregnated hippocampal CA1 pyramids showed that C3 administration to control rats (administered s.c. for 32 days) resulted in a dramatic neuroplastic response. Both dendritic branching and dendritic spine density were increased on the basilar tree. It was calculated that there was a 61% increase in total synaptic contacts on the neurons when compared with young-adult controls. The CA1 pyramids from animals injected with the Aβ(25-35) showed no significant branch atrophy but significant spine loss. C3 treatment (for 32 days) to these subjects resulted in a small increase in dendritic branching. There were also indications that longer treatment periods with C3 would be able to elicit a more substantive neuroplastic response from these damaged neurons (Mervis et al., 2000).

Importance of The Experimental Findings:

A number of important observations derived from this research include:
- C3 protects against Aβ(25-35)-induced neurochemical and cytochemical changes in rat brain. This implies that some of this ~2,100 MW mixture can penetrate the brain where it exerts its CNS activity.
- The protective effect of C3 is observed when it is administered either by the s.c., and/or the oral route. Thus, it could be employed using conventional, therapeutically accepted routes of administration.
- C3 appears to have neurotrophic activity, in addition to its already documented ability to prevent amyloidosis and tau-2 immunoreactivity in vivo.
- C3 serves as a useful prototype for other low molecular weight GAGs. Data obtained with this compound will facilitate the design and testing of other, more lipophilic (better penetration of the BBB) and discrete (lower molecular weight range) GAGs.

Thus, C3 (and potentially related GAGs) is an extremely interesting compound with promise of clinical utility in the treatment of Alzheimer's Disease as well as, possibly, other neurodegenerative disorders.

ACKNOWLEDGEMENT

Supported in part by grant #1 R4 AG15740-01

REFERENCES

Ban, T.A., Morey, L.C., Aguglia, E., et al., 1991, Glycosaminoglycan polysulfate in the treatment of old age dementias, Prog. Neuropsychopharmacol. Biol. Psychiat. 15:323-342.

Caughey, B. and Raymond, GJ, 1993, Sulfated polyanion inhibition of scrapie-associated PrP accumulation in cultured cells, J. Virol. 67:643-650.

Conti, L., Placidi, G.F., and Cassano, G, 1989, Ateroid in the treatment of dementia: results of a clinical trial, Mod. Prob. Pharmacopsychiatry 23:76-84.

Conti, L., Cassano, G., Ban, T.A., and Fjetland, O.K., et al., 1993, Treatment of primary degenerative dementia and multi-infarct dementia with glycosaminoglycan polysulfate: a comparison of two diagnoses and two doses, Curr. Ther. Res. 54:44-64.

Cornelli, U., 1996, In: Nonanticoagulant Actions of Glycosaminoglycans, Eds., J. Harenburg and B. Casu, Plenum Press, N.Y., pp. 249-279.

Dudas, B., Cornelli, J., Lee, J.M., Hejna, M., Walzer, S.A., Lorens, S.A., Mervis, R.F., Fareed, J., and Hanin, I., 2001, Oral and subcutaneous administration of the glycosaminoglycan C3 attenuates Aβ(25-35)-induced abnormal tau protein immunoreactivity in rat brain. Neurobiology of Aging, In Press,.

Hill, J.M., Mervis, R.F., Avidor, R., Moody, T.W. and Brenneman, D.E., 1993, HIV envelope protein-induced neuronal damage and retardation of behavioral development in rat neonates, Brain Res. 603:222-233.

Kjellen, L and Lindahl, U. Proteoglycans: structures and interactions, Ann. Rev. Biochem. 60:443-475, 1991.

Leveugle, B., Scanameo, A., Ding, W., and Fillit, H., 1994, Binding of heparan sulfate glycosaminoglycan to beta-amyloid peptide: inhibition by potentially therapeutic polysulfated compounds, Neuroreport 5:1389-1392.

Lorens, S.A., Guschwan, B.S., Hata, N., Van de Kar, L., Walenga, J.M., and Fareed, J., 1991, Behavioral, endocrine, and neurochemical effects of sulfomucopolysaccharide treatment in the aged Fischer 344 male rat, Sem. in. Thromb. Haemost. 17 suppl. 2:164-173.

Ma, Q., Dudas, B., Cornelli, U., Lorens, S.A., Lee, J.M., Mervis, R.F., Fareed, J., and Hanin, I., 2000, Transport of low molecular weight heparins and related glycosaminoglycans through the blood brain barrier: experimental evidence in a rat model, The FASEB J., 969.

Mervis, R.F., McKean, J., Zats, S., Gum, A., Reinhart, R., Dudas B., Cornelli, U., Lee, J.M., Lorens, S.A., Fareed, J., Hanin, I., Neurotrophic effects of the glycosaminoglycan C3 on dendritic arborization and spines in the adult rat hippocampus: a quantitative Golgi study, CNS Drug reviews, 6:44-46.

Parnetti, L., Ban, T.A., and Senin, U., 1995, Glycosaminoglycan polysulfate in primary degenerative dementia (pilot study of biologic and clinical effects), Neuropsychobiology 31:76-80.

Perry, G., Richey, P.L., Siedlak, S.L., Smith, M.A., Mulvihill, P., DeWitt, D.A., Barnett, J., Greenberg, B.D., and Kalaria, R.N., 1993, Immunocytochemical evidence that the beta-protein precursor is an integral component of neurofibrillary tangles of Alzheimer's disease, Am. J. Pathol. 143:1586-1593.

Sigurdsson, E.M., Lorens, S.A., Hejna, M.J., Dong, W.-X., and Lee, J.M., 1996, Local and distant histopathological effects of unilateral amyloid-beta 25-35 injections into the amygdala of young F344 rats, Neurobiol. Aging 17:893-901.

Sigurdsson, E.M., Lee, J.M., Dong, X.W., Hejna, M.J., and Lorens, S.A., 1997, Bilateral injections of amyloid-beta 25-35 into the amygdala of young Fischer rats: behavioral, neurochemical, and time dependent histopathological effects, Neurobiol. Aging 118:591-608.

Snow, A.D., Sekiguchi, R., Nochlin, D., Kalaria, R.N., and Kimata, K., 1994, Heparan sulfate proteoglycan in diffuse plaques of hippocampus but not of cerebellum in Alzheimer's disease brain, Am. J. Pathol. 144:337-347.

Snow, A.D., Kinsella, M.G., Parks, E., Sekiguchi, R.T., Miller, J.D., Kimata, K., and Wight, T.N., 1995, Differential binding of vascular cell-derived proteoglycans (perlecan, biglycan, decorin and versican) to the beta amyloid protein of Alzheimer's disease, Arch. Biochem. Biophys. 320:84-95.

Vozeh, F.J., and Myslivecek, J., 1996, Quantitative changes of dendrites in rat dentate gyrus and basal nucleus of Meynert after passive avoidance training in the neonatal period, Neurosci. Lett. 204:21-24.

MILD COGNITIVE IMPAIRMENT

Vesna Jelic[*] and Bengt Winblad

1. INTRODUCTION

In memory clinics across the world physicians daily meet subjects who complain about their memory loss that can also be evidenced on psychometric testing. However, those individuals usually do not have any deficits in general cognitive functioning and can competently perform daily life activities. Could these individuals be in a preclinical phase of dementia, namely Alzheimer's Disease (AD)? In an attempt to answer this question an operational diagnostic term Mild Cognitive Impairment (MCI) has been designed to focus on the attention of clinicians. MCI refers to subjective and objective cognitive deficits in the elderly which do not cause functional impairment necessary for the clinical diagnosis of dementia (Petersen et al, 1997).

This article will review evolution of the current concept of MCI, various definitions appearing in the literature and sources of evidence that this condition could be considered as a preclinical Alzheimer's disease (AD). Furthermore, biological markers that predict progression towards dementia will be suggested and therapeutical approaches discussed.

2. EVOLUTION OF THE CONCEPT AND DEFINITIONS

On a cognitive continuum between normal aging and dementia MCI represents intermediate part (Figure 1). Therefore, it is a very heterogeneous clinical entity which encompasses subjects with age-related stable cognitive deficits, but also those who are in the preclinical phase of AD.

Numerous definitions of cognitive impairment without dementia in elderly could be found in the literature, starting as early as in 1962 with a Benign Senescent Forgetfulness (Kral, 1962). In regard to their natural course subjects selected according to the Age Associated Memory Impairment (AAMI) (Crook et al, 1986), Aging-Associated Cognitive Decline (AACD) (Levy, 1994) and Age-Related Cognitive Decline (DSM-IV, 1994) criteria represent heterogeneity of normal aging with low progression rate to AD,

[*]V. Jelic, Karolinska Institutet, Department of Geriatric Medicine, Huddinge University Hospital, B-84, 141 86 Huddinge, Sweden

Mapping the Progress of Alzheimer's and Parkinson's Disease
Edited by Mizuno *et al.*, Kluwer Academic/Plenum Publishers, 2002

177

Figure 1. Conversely, Malignant Senescent Forgetfulness (Kral, 1962), Possible Dementia Prodrome (Heyman et al, 1991) and Mild Neurocognitive Disorder (DSM IV, 1994) were all picking up early AD cases. MCI term (Flicker et al, 1991; ICD-10, 1992; Petersen et al, 1999) is considered to represent a high, but still unknown, proportion of future converters to AD.

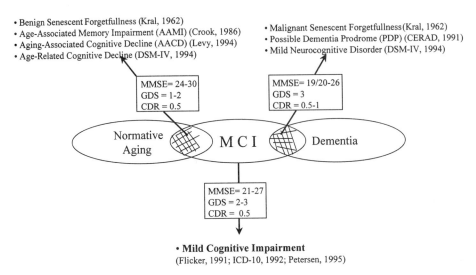

Figure 1. Definitions of cognitive impairment without dementia in elderly. MMSE – Mini Mental State Exam; CDR – Clinical Dementia Rating scale; GDS – Global Deterioration Scale.

Currently, the most popular criteria are those proposed by Petersen et al (1999). They include presence of a memory complain, normal general cognitive function, normal activities of daily living and memory impairment in relation to age and education, but not severe enough to qualify for dementia. Estimated conversion rate to AD is 12–15 % per year as compared to 1% in normal elderly population, Figure 2.

3. CERTAINTY OF MCI PROGNOSIS—BIOLOGICAL MARKERS AS PREDICTORS OF AD

Evidence supporting the MCI concept of preclinical AD is coming from different sources: studies on asymptomatic AD-mutation carriers, prospective clinical and epidemiological studies and autopsy-based studies. Clinicopathological studies have supported the construct validity of MCI. Individuals who prior the death had the Clinical Dementia Rating (CDR) scale scores of 0.5 had senile plaque and neurofibrillary tangle counts in the neocortex sufficient for the diagnosis of AD (Morris et al, 1991) as well as severe neuron loss in layers II and IV of the entorhinal cortex (Gomez-Isla et al, 1996).

Clinical studies are criticized for the selection bias, limited sample sizes and generalizability. In contrast, epidemiological studies show community based prevalence and incidence rates in large cohorts. The most recent ones have suggested that MCI could have differential outcome (Frisoni et al, 1999; Ritchie et al, 2000; Visser et al, 2000). The preliminary results from the Kungsholmen project in Stockholm, Sweden, have shown that after a three years follow-up period 35 % of subjects with cognition impaired, no dementia (CIND) become demented, 29 % died, while 11 % remained stable and 25 % improved (Palmer et al, 2000).

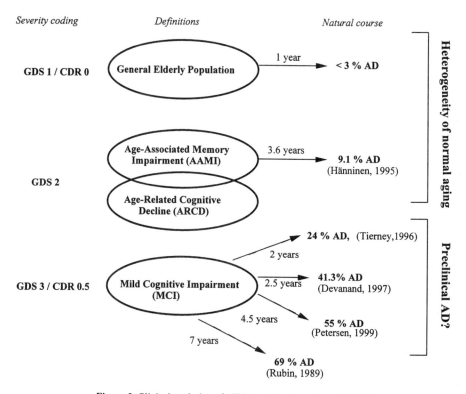

Figure 2. Clinical evolution of Mild Cognitive Impairment (MCI).

To increase the certainty of the MCI prognosis various biological measures have been evaluated as predictors of AD development. Numerous prospective longitudinal studies on neuropsychological predictors of AD in the general population or in MCI subjects at risk agree that deficits in mainly episodic memory precede future development of dementia (Masur et al, 1994; Jacobs et al, 1995; Tierney et al, 1996; Devanand et al, 1997). However, performance on the psychometric tests and consequently their predictive accuracy could be compromised by premorbid intelligence, educational level and occupation. Therefore, additional objective and practice free preclinical markers of the disease are needed. Previous research has provided evidence that biochemical and neuroimaging markers of AD could increase certainty of preclinical diagnosis of the disease in MCI subjects. Longitudinal studies of cerebrospinal fluid (CSF) biochemistry in patients with MCI have shown that at the baseline of the one-year follow-up, 60 % of subjects had high total CSF-tau, 40 % high CSF-phosphorylated-tau and 93 % had low CSF-Aβ42 levels (Andreasen, in press). Reduced volumes of medial temporal lobe structures as measured by Magnetic Resonance Imaging (MRI) were also shown to be strong predictors of the development of dementia (Fox et al, 1996; Jack et al, 1999).

Temporal lobe volume loss was seen in one study to mark the beginning of the disease process within six years prior to dementia onset (Kaye et al, 1997).

Functional imaging also offers possible markers of early AD. Reduced temporo-parietal cerebral glucose metabolism as measured by Positron Emission Tomography (PET) and cerebral blood flow as measured by Single Photon Emission Tomography (SPECT), have been reported to predict progression of clinical symptoms (Herholz et al, 1999; Johnson et al, 1998). It is encouraging that less costly and non-invasive methods such as quantitative EEG could, on the basis of lowered fast and increased slow frequencies, also be used to distinguish MCI subjects who progress to AD from those who remained stable after 2 years follow-up (Jelic et al, 2000).

4. THERAPY OF MCI—TARGETS, OPPORTUNITIES, AND ETHICAL DILEMMAS

Evidence from various sources shows that MCI is a risk factor for the development of AD and may have same underlying pathology as AD. Therefore, intervening in this early phase could have beneficial effects in terms of delaying the disease onset or slowing its progression. Pathophysiological processes leading to AD provide multiple targets for interventions. Aging as a major risk factor for AD sets the stage for a complex cascade of events that compromises calcium and energy homeostasis and together with major genetic risk factors through the oxidative stress leads to synaptic damage. The later results in neuronal death and neuronal loss, consequence of which are neurotransmitter deficits and microglial activation with inflamation. Currently available treatments with acetylcholinesterase inhibitors (AChEI) act at the very end of this chain of processes and it is not surprising that expected effect could be symptomatic with possible delay of the disease onset. This is the aim of major clinical trials that are studying at the moment AChEIs in individuals with MCI.

Antioxidants like E-vitamin and MAO-inhibitor selegiline are targeting oxidative stress and production of free radicals and, therefore, might have disease modifying effects. Although the trial performed on AD patients did not result in improvement of cognition (Sano et al, 1997), intervention in an earlier phase of the disease could have effects and should be further explored. Estrogen replacement therapy in postmenopausal women has generated a debate as to whether it could also have preventive properties. Higher incidence of AD in females encourages this way of thinking but whether benefits outweigh risks should be investigated in larger placebo-controlled trials with control of confounding factors.

The ultimate goal of preventive AD therapy is to ameliorate or prevent pathologic events in the brain, the most important targets being neuritic plaques and tangles. Among these disease-modifying approaches is intervention into the amyloid precursor protein (APP) processing via inhibition of β- and γ-secretases.

The most excitement among these novel approaches has certainly been raised by the recent finding of Schenk et al (1999) that cerebral amyloidosis in transgenic animals could be largely prevented by immunization with APP. That introduces a possibility of a vaccine for AD which should ideally be applied at a preclinical phase when MCI becomes manifest, since hypothesized amyloid deposition starts already in asymptomatic phase.

Less exciting but not less important is treatement of comorbidity, namely hypertension (Guo et al, 1999), and control of other risk factors in MCI subjects. This could have additional disease modifying effects as it has been shown recently that lipid-lowering statins may provide at normal therapeutical doses protection against AD and other forms of dementia (Jick et al, 2000).

Chapter on treatment of MCI on a large scale cannot be closed without mentioning ethical dilemmas raised by tension between the benefits of early diagnosis and interventions on one side and risks of exposing to treatment subpopulation of clinically stable MCI subjects. Is it justified to treat those individuals and raise anxiety in an otherwise healthy population? Should we include criterion of progression in current operational diagnostic criteria for MCI? What will be the consequences of increased demand on public health resources? A consensus is needed as to which are the optimal operational diagnostic criteria and less time consuming neuropsychological assessments which will be useful to define those at highest risk. More efforts should also be directed towards other preclinical manifestations of AD, besides memory impairment. Finally, the most valuable lesson learned from the MCI will be the refinement of currently existing clinical diagnostic criteria for AD.

REFERENCES

Andreasen N, Minthon L, Davidsson P, Vanmechelen E, Vanderstichele H, Winblad B, Blennow K. Evaluation of CSF-tau and CSF-Aβ42 as diagnostic markers for Alzheimer's disease in clinical practice. *Arch Neurol,* in press.

Crook T, Bartus RT, Ferris SH, Whitehouse P, Cohen GD, Gershon S. Age-associated memory impairment: Proposed diagnostic criteria and measures of clinical change. Report of a National Institute of Mental Health Work Group. *Dev Neuropsychology* 1986; **2**:261-276.

Devanand DP, Folz M, Gorlyn M, Moeller JR, Stern Y. Questionable dementia: Clinical course and predictors of outcome. *JAGS* 1997; **45**: 321-328.

Diagnostic and Statistical Manual of Mental Disorders, 4th ed.: DSM-IV. Washington, D.C.: American Psychiatric Association, 1994.

Flicker C, Ferris SH, Reisberg B. Mild cognitive impairment in the elderly: predictors of dementia. *Neurology* 1991; **41**:1006-1009.

Fox MC, Warrington EK, Freeborough PA. Presymptomatic hippocampal atrophy in Alzheimer's disease. A longitudinal MRI study. *Brain* 1996; **119**: 2001-2007.

Frisoni GB, Fratiglioni L, Viitanen M, Winblad B. The natural history of non-demented cognitively impaired subjects: Population data from Kungsholmen project. *Neurology* 1999; **52** [Suppl 2]: A385.

Gomez-Isla T, Price JL, McKeel DW, Morris JC, Growdon JH, Hyman BT. Profound loss of layer II entorhinal cortex neurons occurs in very mild Alzheimer's disease. *The Journal of Neuroscience* 1996; **16**:4491-4500.

Guo Z, Fratiglioni L, Zhu L, Fastbom J, Winblad B, Viitanen M. Occurrence and progression to dementia in a community population aged 75 years and older. *Arch Neurol* 1999; **56**:991-996.

Hänninen T, Hallikainen M, Koivisto K, et al. A follow-up study of age-associated memory impairment: neuropsychological predictors of dementia. *JAGS* 1995; **43**:1007-1015.

Herholz K, Nordberg A, Salmon E, et al. Impairment of neocortical metabolism predicts progression in Alzheimer's disease. *Dement Geriatr Cogn Disord* 1999; **10**: 494-504.

Heyman A, Fillenbaum G, Mirra SS. Consortium to establish a registry for Alzheimer's disease (CERAD): Clinical, neuropsychological, and neuropathological components. *Aging, Clinical and Experimental Research* 1991; **2**: 415-424.

Jack CR, Petersen RC, Xu YC, O'Brien PC, Smith GE, Ivnik RJ, Boeve BF, Waring SC, Tangalos EG, kokmen E. Prediction of AD with MRI-based hippocampal volume in mild cognitive impairment. *Neurology* 1999; **52**:1397-1403.

Jacobs DM, Sano M, Dooneief G, Marder K, Bell KL, Stern Y. Neuropsychological detection and characterization of preclinical Alzheimer disease. *Neurology* 1995; **45**:957-62.

Jelic V, Johansson S-E, Almkvist O, Shigeta M, Julin P, Nordberg A, Winblad B, Wahlund L-O. Quantitative electroencephalography in mild cognitive impairment: longitudinal changes and possible prediction of Alzheimer's disease. *Neurobiol Aging* 2000; **21**: 533-540.

Jick H, Zornberg GL, Jick SS, Seshadri S, Drachman DA. Statins and the risk of dementia. *Lancet* 2000; **356**: 1627-31.

Johnson KA, Jones K, Holman BL, et al. Preclinical prediction of Alzheimer's disease using SPECT. *Neurology* 1998; **50**:1563-1571.

Kaye JA, Swihart T, Howieson D, Dame A, Moore MM, Karnos T, Camicioli R, Ball M, Oken B, Sexton G. Volume loss of the hippocampus and temporal lobe in healthy elderly persons destined to develop dementia. *Neurology* 1997; **48**:1297-1304.

Kral VA. Senescent forgetfulness: Benign and malignant. *The Canadian Medical Association Journal* 1962; **86**: 257-260.

Levy R. New provisional IPA/WHO criteria for age associated cognitive decline (AACD). *Int Psychogeriat Assoc Newsletter* 1994; **10**:3-4.

Masur DM, Sliwinski M, Lipton RB, Blau AD, Crystal HA. Neuropsychological prediction of dementia and the absence of dementia in healthy elderly persons. *Neurology* 1994; **44**:1427-32.

Morris JC, McKeel DW, Storandt M, et al. Very mild Alzheimer's disease: informant based clinical, psychometric, and pathologic distinction from normal aging. *Neurology* 1991; **41**:469-478.

Palmer KM, Fratiglioni L, Frisoni GB, Viitanen M, Winblad B. The evolution of cognitive impairment in the non-demented elderly. Population data from the Kungsholmen project. *Neurobiol Aging* 2000; **21**:S101.

Petersen RC, Smith GE, Waring SC, Ivnik RJ, Kokmen E, Tangelos EG. Aging, memory, and mild cognitive impairment. *International Psychogeriatrics* 1997; **9** (Suppl. 1): 65-69.

Petersen R, Smith G, Waring S, Ivnik R, Tangalos E, Kokmen E. Mild cognitive impairment: clinical characterization and outcome. *Arch Neurol* 1999; **56**: 303-308.

Ritchie K, Ledésert B, Touchon J. Subclinical cognitive impairment: Epidemiology and clinical characteristics. *Comprehensive Psychiatry* 2000; **41** [Suppl 1]: 61-65.

Rubin EH, Morris JC, Grant EA, Vendegna T. Very mild senile dementia of the Alzheimer type. I. Clinical assessment. *Arch Neurol* 1989; **46**:370-382.

Sano M, Ernesto C, Thomas RG, et al. A controlled trial of selegiline, alpha-tocopherol, or both as treatment for Alzheiemr's disease. *N Engl J Med* 1997; **336**:1216-22.

Schenk D, Barbour R, Dunn W, et al. Immunization with b-amyloid attenuates Alzheimer-disease-like pathology in the PDAPP mouse. *Nature* 1999; **400**: 173-177.

The International Classification of Diseases, 10th rev: ICD-10 Geneva: World health Organization, 1992; pp 311-388.

Tierney M, Szalai J, Snow W, et al. Prediction of probable Alzheimer's disease in memory-impaired patients: a prospective longitudinal study. *Neurology* 1996; **46**: 661-665.

Visser PJ, Verhey FR, Ponds RWHM, Cruts M, Van Broeckhoven CL, Jolles J. Course of objective memory impairment in non-demented subjects attending a memory clinic and predictors of outcome. *Int J Geriatr Psychiatry* 2000; **15**:363-372.

FUNCTIONAL ACTIVATION STUDIES IN ALZHEIMER PATIENTS AND STRATEGIES IN DRUG EVALUATION

Agneta Nordberg[*]

1. INTRODUCTION

A great demand is presently put on available functional neuroimaging methods since they may provide a potential to reveal neuronal dysfunction before structural change may appear (Knopman et al. 2001). Positron emission tomography (PET), a non-invasive neuroimaging technique allows quantification and three-dimensional measures of distinct physiological variables such as glucose metabolism, cerebral blood flow, and neurotransmitter and receptor function (Nordberg 1999). Regional deficits in cerebral glucose metabolism (CMRGlu) in the parieto-temporal regions, assessed by [18F] 2-Fluoro-2-deoxy-D-glucose (FDG) as tracer, have consistently been described in patients with Alzheimer's disease (AD). The fact that the metabolic impairment correlates to deficits in neuropsychological domains and increase with progression of the disease suggest that PET also may provide a sensitive marker of disease progression and severity (Nordberg 1993, Mielke et al. 1994).

With the conservative diagnostic criteria for AD the diagnosis may often be given when the clinical symptoms of the disease are quite evident. While the present AD therapies have symptomatic effects with some slowing down of disease progression the drug therapies in the future will hopefully cure/prevent the AD disease (Nordberg and Svensson 1998, Giacobini 2001, Nordberg 2000a, Doody et al. 2001). An early diagnosis at preclinical stages of the disease will be a prerequisite for a successful therapy in the future for AD. The clinical symptoms of AD are probably preceded by a period of unknown duration during which neurophatological alterations may accumulate in the AD brain without detectable changes in cognition (Davis et al. 1999, Goldman et al. 2001). Neuropsychological studies suggest that mild cognitive impairment (MCI) is a condition characterised by subtle cognitive deficits before functional impairments are evident in the patient. MCI might therefore represent an early stage of AD (Almkvist et al. 1998, Petersen et al. 1999, Arnáiz et al. 2000). Selective regional impairments of cortical glucose metabolism are found in MCI patients (Minoshima

[*] A. Nordberg, Karolinska Institutet, NEUROTEC, Molecular Neuropharmacology, Huddinge University Hospital, B-84, S-141 86 Stockholm, Sweden

Mapping the Progress of Alzheimer's and Parkinson's Disease
Edited by Mizuno *et al.*, Kluwer Academic/Plenum Publishers, 2002

et al. 1997, Jelic and Nordberg 2000). It is presently of interest to investigate whether cerebral glucose metabolism or any other biological marker could be used as a predictors for AD predictors and as surrogate markers have an impact on the initiation of drug treatment.

2. PREDICTORS OF ALZHEIMER'S DISEASE

2.1.Mild cognitive impairment

How important is an early detection of MCI patients? How important is it to initiate drug treatment in MCI patients before they may convert to AD? Is there any possibility to predict which MCI patients that
will convert to AD? MCI is considered as an intermediate part between the continuum of normal and dementia. The most common used criteria for MCI include memory complain and memory impairment but normal general cognitive function and daily living (Petersen et al. 1999). In a four year follow-up of 76 subjects selected for MCI the annual conversion rate to AD was 12 % compared to 1 % in normal subjects (Petersen et al. 1999) while in a 2 year follow-up study it was estimated to 25 % (Flicker et al. 1991). It has recently been estimated that among subjects with mild cognitive impairment the rate of progression to dementia or AD is 6-25 % per year (Petersen et al. 2001).

We followed 27 subjects with MCI for 24 months and 26 % developed AD (Jelic and Nordberg 2000). All MCI patients underwent PET studies of brain glucose metabolism before the longitudinal studies was started. Impairment in cerebral glucose metabolism was observed at the start of the follow-up period in the group of MCI patients who progressed to AD. The postero-antero ratio of glucose metabolism (parietal association cortex/ frontal association cortex when corrected for the baseline MMSE scores, correctly predicted the conversion or non-conversion in 93 % of cases (Jelic and Nordberg 2000, Nordberg et al. 2001). Similarly when a group of 20 MCI patients were followed for 36 months 9 patients converted to AD while 11 patients remained clinical stable and were classified as stable MCI. Neuropsychological assessments such as Block Design gave a correct classification of 65 % while the combined measure of Block Design and cerebral glucose metabolism gave a correct classification of 90 % (Arnáiz et al. 2001). It is apparent that early functional metabolic changes can be observed in subjects with MCI and that the degree of functional disturbances in glucose metabolism as measured by PET can be used as a predictive marker for conversion from MCI to AD especially when combined with neuropsychological testing. CSF marker will probably in the future be a valuable complement (Knopman et al. 2001).

2.2. Apolipoprotein E ε4 carriers

Two independent studies have reported that the APOE e4 allele in patients with AD is not associated with specific alterations in glucose metabolism (Corder et al. 1997, Hirono et al. 1998). There is evidence however that it could cause preclinical deficits similar to what can be observed in manifest disease (Small et al. 1995, Reiman et al 1996). It was recently showed that in non-demented APOE ε4 carriers followed longitudinally for two year, memory performance scores did not decline significantly while cortical metabolic ratio did in APOE ε4 carriers (Small et al. 2000). Similar findings were also reported by Reiman et al. (2001) who found a significant decline in cerebral glucose metabolism in the temporal cortex, posterior cingulate cortex, prefrontal cortex, basal forebrain, parahippocampal gyri and thalamus of ε4 heterozygotes in the absence of cognitive alterations during two years. Reiman et al (2001) estimated that probably 50-150 cognitively normal ε4 heterozygotes would be sufficient for a testing the outcome of a preventive AD therapy. Cerebral metabolic rates and genetic factors may provide a mean of preclinical AD detection and importantly assist in the evaluation of the effect of preventive drug treatments in the future.

3. BRAIN ACTIVATION STUDIES IN ALZHEIMER PATIENTS

In order to explore the anatomically activation pattern in brain associated with cognitive process in normal and pathological stages PET and functional magnetic imaging (fMRI) are presently two used in functional imaging studies (Cabeza and Nyberg 2000). By both techniques the changes in cerebral blood flow can be used as sign for neural activity. Since both episodic memory and attention are early affected in AD functional brain imaging has focussed on the regional brain activation pattern for these mental performances in AD patients. Healthy elderly activate the left parietal cortex and left hippocampus during recall of words while AD patients increase brain activation of the left prefrontal cortex and left cerebellum in addition to an increased activation in a number of brain regions common for both controls and AD (Bäckman et al. 1999). Furthermore, studies on sustained attention in healthy young subjects with a Rapid Visual Information Processing Task (RVIP) show an activation of the parietal cortex and right frontal cortex (Coull et al. 1996). When similar task is performed in AD and MCI patients an increased activation was bilaterally observed in the parietal cortex, basal ganglia and a decreased activation is observed in the prefrontal regions bilaterally (Almkvist et al. submitted). Johannsen et al. (1999) observed a deactivation pattern during sustained attention in the frontal cortex of AD patients which was not seen in controls. PET may provides valuable insight into the early dynamic disturbances in brain function related to cognitive function in MCI and AD patients. Since somewhat differences in activation patterns are observed between AD and MCI patients this may reflect compensatory efforts to keep with task requirements and that networks normally used in brain are distressed or impaired by the disease processes in AD. (Almkvist et al. submitted). The possibility of drugs to influence these processes, the underlying mechanisms and to restore the network processes is of high interest to further explore.

Long-term treatment with cholinesterase inhibitors such as tacrine, donepezil, rivastigmine have shown increase in cerebral blood flow and glucose after some months of treatment in AD patients (Nordberg 2000b) and preservation of the cortical glucose metabolism following longer treatment periods with cholinesterase inhibitors (Stefanova et al. submitted). An improvement of nicotinic receptors has been observed following treatment with cholinesterase inhibitors as well as nerve growth factors. (Nordberg 2000b). PET studies have shown a reduced cortical acetylcholinesterase activity during donepezil treatment in AD patients (Kuhl et al. 2000, Shinotoh et al. 2001). The PET findings illustrate the influences of drug interventions on synaptic neuronal markers in relation to changes in cerebral blood flow and glucose metabolism in pathological brain tissue. Functional PET activation studies performed simultaneously with memory tasks will provide further valuable insight into the mechanisms of action of new drugs, how they interact with various brain processes and how they can improve the efficacy of memory processes in the AD patients.

4. ACKNOWLEDGEMENTS

This study was supported by grants from the Swedish Medical Research Council (project no 05817), Stiftelsen för Gamla Tjänarinnor, Stohne's Foundation.

REFERENCES

Almkvist O, Basun H, Bäckman L et al. Mild cognitive impairment-an early stageof Alzheimer'sdisease? J Neural Transm 1999; 54(suppl): 21-29.
Arnáiz E, Blomberg M, Fernaeus SE, Wahlund LO, Winblad B, Almkvist O. Psychometric discrminiation of Alzheimer's disease and mild cognitive impairment. Alzheimer Report 2000:; 2: 97-201.

Arnáiz E, Jelic V, Almkvist O, Wahlund LO, Winblad B, Valind S, Nordberg A. Impaired cerebral metabolism and cognitive functioning predict deterioration in mild cognitive impairment. NeuroReport 2001; 12: 851-855.

Bäckman L, Andersson JLR, Nyberg L, Winblad B, Nordberg A, Almkvist O. Brain regions associated with episodic retrieval in normal aging and Alzheimer's disease, Neurology 1999; 52, 1861-1870.

Cabeza R, Nyberg L. Imaging cognition II. an empirical review of 275 PET and fMRI studies. J Cognitive Neurosci 2000; 12, 1-47.

Coull JT, Frith CD, Frackowiak RSJ, Grasby PM. A fronto-parietal network for rapid visual processing: a PET study of sustained attention and working memory. Neuropsychologia 1996; 34, 1085-1095.

Davis DG, Schmitt FA, Wekstein DR, Markesbery WR, Alzheimer's neuropathologic alterations in aged cognitive normal subjects. J Neuropathol Exp Neurol 1999; 58: 376-388.

Doody RS, Stevens JC, Beck C, Dubinsky RM, Kaye JA, Gwyther L, Mohs RC, Thal L, Whitehouse P, DeKosky ST, Cummings JL . Practive parameter: standard subcommittee of the American Academy of Neurology. Neurology 2001: 56: 1154-1166.

Flicker C, Ferris SH, Reisberg B. Mild cognitive impairment in the elderly: predictors of dementia. Neurology 1991; 41, 1006-1009.

Giacobini E. Do cholinesterase inhibitors have disease-modifying effects in Alzheimer's disease? CNS Drugs 2001; 15: 85-91.

Goldman WP, Price JL, Storandt M, Grant EA, McKee DW, Rubin EH, Morris JC. Absence of cognitive impairment or decline in preclinical Alzheimer's disease. Neurology 2001; 56, 361-367.

Jelic V, Nordberg A. Early Diagnosis of Alzheiemr's disease with postitron emission tomography. Alzheimer Disease and Associated Disorders 2000; 14 (suppl): S109-S113.

Johannsen P, Jacobsen J, Bruhn P, Gjedde A. Cortical responses to sustained and divided attention in Alzheimer's disease. NeuroImage 1999; 10, 269-281.

Knopman DS, DeKosky ST, Cummings JL, Corey-Bloom J, Relkin N, Small GW, Miller B, Stevens JC. Practice parameters: Diagnosis of dementia (an evidence-based review) Report of the quality standard subcommittee of the American Academy of Neurology. Neurology 2001; 56: 1143-1153.

Kuhl DE, Minoshima S, Frey KA, Foster NL, Kilbourn MR, Koeppe RA. Limited donepezil inhibition of acetylcholinesterase measured with positron emission tomography in living Alzheimer cerebral cortex. Ann Neurol 2000; 48; 391-395.

Mielke R, Herholz K, Grind M, Kessler J, Heiss WD. Clinical deterrioation in probable Alzheimer disease correlates with progressive metabolic impairment of association areas. Dementia 1994; 5: 215-234.

Minoshima S, Giordani B, Berent S, Frey KA, Foster NL, Kuhl DE. Metabolic reduction in the posterior cingulate cortex in the very early Alzheimer's disease. Ann Neurol 1997; 42: 85-94.

Nordberg A. Functional brain imaging in Alzheimer's disease. In: Gauthier S,ed. Clinical diagnosis and manage of Alzheimer's disease. 2nd ed. London: Martin Dunitz Ltd, 1999: 117-132.

Nordberg A. Clinical studies in Alzheimer patients with positron emission tomography. Behav Brain Res 1993; 57: 215-234.

Nordberg A. Neuroprotection in Alzheimer's disease-new strategies in treatment. Neurotoxicity Research 2000a; 2, 157-15.

Nordberg A. The effect of cholinesterase inhibitors studied with imaging. In: Giacobini E, ed, Cholinesterases and cholinesterase inhibitors, London: Martin Dunitz, 2000b, 237-247.

Nordberg A, Jelic V, Arnáiz E, Långström B, Almkvist O. In.: Iqbal K, Sisodia S, Winblad B, eds, Alzheimer's disease: advances in etiology, pathogensis and therapeutics, Chichester: John Wiley & Sons, 2001: 153-164.

Nordberg A, Svensson AL. Cholinesterase inhibitors in the treatment of Alzheimer's disease. A comparison of tolerability and pharmacology. Drug Safety 1998; 19: 465-480.

Petersen RC, Smith GE, Waring SC, Ivnik RHJ, Tangalos EG, Krokmen E. Mild cognitive impairment: clinical characterisation and outcome. Arch Neurol 1999; 56: 303-308.

Petersen RC, Stevens JC, Ganguli M, Tangalos EG, Cummings JL, DeKosky ST. Practice parameter: Early detection of dementia: Mild cogntive impairment (an evidence-based review). Report of the quality standard subcommittee of the American Academy of Neurology. Neurology 2001; 56, 1133-1142.

Reiman E, Caselli RJ, Chen K, Alexander GE, Bady D, Frost. Declining brain activity in cognitively normal apolipoprotein e e4 heterozygotes: A foundation for using postitron emission tomography to efficiently test treatments to prevent Alzheimer's disease. PNAS 2001; 98, 3334-3339.

Reiman E, Caselli RJ, Yun LS, Chen K, Bandy D, Minoshima S, Thibodeau SN, Osborne D. Preclinical evidence of Alzheimer's diseas in persons homozygous for the ε4 allele for apolipoprotein E. New Engl J Med 1996; 334, 752-758.

Shinotoh H, Aotsuka A, Fukushi K, Nagatsuka S, Tanaka N, Ota T, Tanada S, Irie T. Effect of donepezil on brain acetylcholinestesterase activity in patients with AD measured by PET. Neurology 2001; 56, 408-410.

Small GW, Ercoli LM, Silverman S, et al. Cerebral metabolic and cognitive decline in persons at genetic risk for Alzheimer's disease. Proc Natl Acad Sci 2000; 97: 6037-6042.

Small GW, Komo S, La Rue A et al. Early detection of Alzheimer's disease by combining apolipoprotein E and neuroimaging. Ann NY Acad Sci 1996; 802: 70-78.

STRUCTURAL AND FUNCTIONAL CORRELATES OF HUMAN ACETYLCHOLINESTERASE MUTANTS FOR EVALUATING ALZHEIMER'S DISEASE TREATMENTS

Functional analysis of E2020 and galanthamine HuAChE complexes

Avigdor Shafferman, Dov Barak, Arie Ordentlich, Naomi Ariel, Chanoch Kronman, Dana Kaplan and Baruch Velan*

1. INTRODUCTION

Inhibitors of AChE constitute presently the only successful approach to symptomatic treatment of senile dementia of the Alzheimer's type, with three such agents (tacrine, E2020 and rivastigmine) in clinical use[1]. The intensive effort to develop more therapeutically efficatious inhibitors is currently aided by the remarkable progress, made during the last decade, in elucidating the structural and functional properties of the enzyme through x-ray crystallography[2, 3] and site directed mutagenesis[4-8]. Investigation of the specific roles for most of the residues in this active-center gorge allowed for identification of several functional subsites including the catalytic triad (S203*(200)***, H447*(440)*, E334*(327)*); the acyl pocket (F295*(288)* and F297*(290)*) and the 'hydrophobic subsite'. The latter accommodates the alcoholic portion of the covalent adduct (tetrahedral intermediate) and may include residues W86*(84)*, Y133*(130)*, Y337*(330)* and F338*(331)*, which operate through nonpolar and/or stacking interactions, depending on the substrate. Stabilization of the charged moieties of substrates and other ligands at the active-centre is mediated by cation - π interactions with the residue at position 86 rather than through true ionic interactions[2, 6, 8].

*Israel Institute for Biological Research Ness-Ziona, 70450, Israel.

** Residue numbers of *Torpedo californica* AChE are listed in italics with the corresponding residues of human AChE in parentheses and in plain text. Residue numbers of human AChE are listed in plain text with the corresponding numbers of Torpedo californica AChE in parentheses and in italics.

Mapping the Progress of Alzheimer's and Parkinson's Disease
Edited by Mizuno *et al.*, Kluwer Academic/Plenum Publishers, 2002

187

In addition, residues Y72*(70)*, Y124*(121)*, W286*(279)* and Y341*(334)*, which are localised at or near the rim of the active-center gorge, together with D74*(72)* constitute the peripheral anionic subsite (PAS) of AChE[4, 5, 7].

Examination of HuAChE interactions with representative potential therapeutic agents like edrophonium, tacrine, huperzine A or the carbamates, pyridostigmine and physostigmine[8], demonstrated that an "aromatic patch" of spatially adjacent aryl moieties of the hydrophobic subsite residues W86, Y133 and Y337 plays dominant role in accommodation of the structurally diverse ligands. Models based on such functional mapping studies suggested also that the structure-function characteristics of HuAChE - inhibitor complexes may not fully correspond to the structural information from x-ray crystallography of the corresponding TcAChE complexes[8]. These discrepancies do not seem to originate from structural differences between the two enzymes, since the recently determined x-ray structure of HuAChE shows close similarity to that of TcAChE[3].

In the present study we examine the complementarity of the HuAChE active center architecture with two of the most extensivly studied second generation AChE inhibitors: E2020 (Aricept, Donepezil) and galanthamine[1]. Our results reveal certain inconsistencies with the crystallographic structures of the respective complexes[9-11] and underline the importance of complementing the structural data with functional characterization of the biological target molecules.

2. RESULTS AND DISCUSSION

2.1. Structural determinants of E2020 selectivity for HuAChE versus HuBChE

One of the reasons for the improved therapeutic profile of E2020, as compared to tacrine, is thought to be its high selectivity for AChE as compared to BChE (Table 1). It has been suggested that inhibition of BChE, which is abundant in human plasma may cause potentiating side effects[12]. Early modelling experiments suggested that the differential specificity of E2020 can be attributed to the structural differences in AChE and BChE at the top of the gorge, at the peripheral anionic site[13]. The recently determined crystallographic structure of the TcAChE- E2020 complex broadly confirms these assignments but points also to the π-cation interaction of the ligand with F*330*(337) as an additional determinant of specificity toward AChE[9] (in BChE the position equivalent to 337 is occupied by alanine). Kinetic studies of inhibition by E2020, of HuAChE enzymes carrying single replacements at the "peripheral anionic site"(residues D74,Y124,W286,Y341), indeed demonstrate that interactions with this site are indeed of major importance to the accommodation of the ligand (e.g. affinity of the triple mutant Y72N/Y124Q/W286A HuAChE is 2500-fold lower than that of the WT). On the other hand, replacement of the third element of the aromatic patch – residue Y337 had no effect on affinity toward E2020 (Table 1), suggesting that this residue does not participate in ligand accommodation neither through hydrophobic interactions nor through the specific cation-π interaction indicated by the crystal structure of the complex (distances between the nitrogen and F*330* ring carbons are 3.9-4.5 Å). Since replacement of Y337 by alanine appears to create a local void in the HuAChE active center rather than affecting its overall functional architecture[8], the present findings indicate that interaction with this residue *does not* contribute to the selectivity of E2020.

According to the crystal structure, the aromatic portion of the benzyl substituent displays a parallel π–π stacking with the indole ring of W*84*(86) (Fig. 1), analogous to the dominant interaction observed in the TcAChE-tacrine complex[14]. This is consistent with the 4-orders of magnitude loss of affinity of E2020 toward the W86A HuAChE compared to the wild type enzyme (Table 1). Thus, as already shown in the past[6, 8], residue W86 is essential for the binding of the HuAChE inhibitors and constitutes a central part in the "aromatic patch" which includes also residue Y133. Indeed replacement of Y133 by alanine resulted in a 325-fold affinity decrease toward E2020, either through loss of a direct interaction with the ligand or through destabilization of the W86 active conformation[15] (Table 1).

Table 1. Interactions of anticholinesterase inhibitors with the hydrophobic site of HuAChE

	Relative K_i (Mutant/WT)				
	Edrophonium	Tacrine	Huperzine A	Galanthamine	E2020
WT	1	1	1	1	1
W86F	55	70	125	8	74
W86A	>64000	17600	>62500	22000	125000
Y133F	26	60	2	220	14
Y133A	1110	110	75	3900	325
Y337A	8	0.1	300	1	0.5
Y337F	1	2	7	1	1
F338A	1	0.5	4	1	13

The actual values of Ki for inhibition of HuAChE and HuBChE are respectively 0.7μM and 50μM for edrophonium; 85nM and 12nM for tacrine; 16nM and 50μM for huperzine A; 90nM and 1.1μM for galanthamine; 2nM and 250 nM for E2020.

Despite the observation that E2020 binds to both the free and the acylated forms of AChE[13] and notwithstanding the crystallographic structure of the complex, which revealed a finger–shaped void at the acyl pocket[9], we examined the effects of replacements at positions 295 and 297 on affinity toward E2020. Opening of the acyl pocket through substitutions of F295 and F297 by alanine shows only a limited effect on the binding capacity of E2020. On the other hand introducing the BChE acyl pocket constituents , leucine in position 295 and in particular valine in position 297, resulted in a dramatic loss of affinity toward E2020 (see Fig. 1). Furthermore, the F295L/F297V HuAChE is 10000-fold less reactive than the wild type enzyme.

We have shown in the past that the functional architectures of HuBChE and of F295L/F297V HuAChE are quite similar with respect to accommodation of noncovalent ligands[6]. Therefore, the surprising result that the F295L/F297V HuAChE is severely impaired in binding E2020, strongly suggests that the *acyl pocket is a major determinant of this ligand specificity toward AChE versus BChE.* Also, in view of this finding it appears that the actual selectivity of E2020 toward HuAChE vs HuBChE (125-fold, see Table 1) is much lower than could have been expected from combination of the effects due to the acyl pocket and the PAS (10000-fold and 500-fold respectively). Yet, binding at the rim of the active site gorge of HuBChE may be quite different from that of the Y72Y/Y124Q/W286A HuAChE, as suggested in the case of the PAS specific ligand propidium[7]. Therefore it is possible that the PAS does not contribute in a major way to E2020 selectivity toward HuAChE and that the acyl pocket structure is actually the main reason for the lower affinity of HuBChE toward this inhibitor.

HuAChE	Relative *Ki* (Mutant/WT) E2020
WT	1
F295A	2
F295L	9
F297A	4
F297V	1350
F295L/F297V	10000

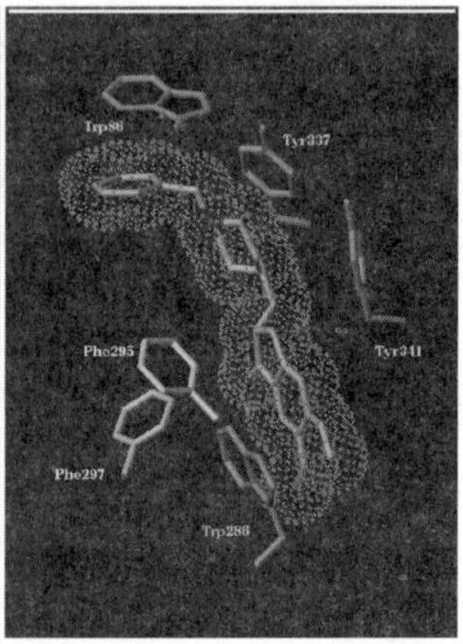

Figure 1. Interactions of E2020 with the acyl-pocket of HuAChE modeled according to the TcAChE complex[9]

2.2. Functional analysis of the HuAChE-galanthamine complex

In the two independently determined crystallographic structures of the TcAChE-galanthamine complex[10, 11] no direct interactions of the protonated amine with either W84 or Y130 could be observed. In fact, the only interaction of the charged group mentioned, in one of these studies, was that with F330 (possibly through water molecule)[10]. Yet, examination of these structures reveals proximity of the nitrogen alkyl substituents and the indole moiety of W84. Since accommodation of all the AChE noncovalent ligands examined to date involves interactions with the "aromatic patch", it was of considerable interest to examine the effects of its structural modification on the affinity of galanthamine. The results show that while replacements of W86 or Y133 by alanine resulted in large decrease in enzyme affinity toward galanthamine (22000-fold and 3900-fold respectively), substitution of Y337 by either phenylalanine or alanine had no effect (Table 1 and Fig. 2). Thus, like for other active center AChE ligands, the protonated cationic moiety of galanthamine appears to be accommodated by the aromatic ring of the residue at position 86. Also, replacement of Y133 by phenylalanine resulted in a 220-fold decrease of affinity toward galanthamine, in agreement with the predicted interaction of the ligands hydroxyl substituent with the tyrosyl oxygen through water bridge[11].

HuAChE	Relative *Ki* (Mutant/WT) Galanthamine	
WT	1
W86F W86A	8 22000
Y337F Y337A	1 1

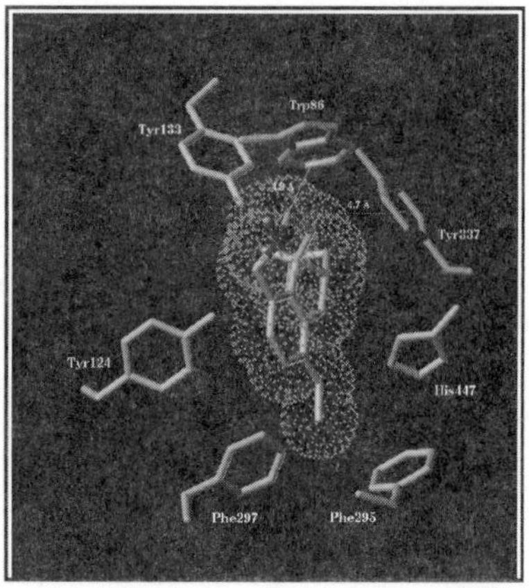

Figure 2. Interactions of galanthamine with the W86 and Y337 of HuAChE modeled according to the TcAChE complex[10, 11]

2.3. Contribution of functional characterization of HuAChE-ligand complexes by site directed mutagenesis to structure based drug design

Both crystallographic data and functional analysis indicate that interaction with residue W86 is the most important single contact of *any* noncovalent inhibitor tested in the AChE binding environment (Table 1). On the other hand, interactions with Y337 are consistently overestimated by crystallographic analysis. In all the TcAChE-inhibitor complexes analyzed the crystallographic orientation of *F330* shows close contacts with the respective ligands and therefore suggests stronger interactions, than are actually observed for the Y337A HuAChE. Even in the case of huperzine A where Y337 was shown to participate in ligand accommodation and stereospecificity, the crystallographic structure of the corresponding TcAChE complex seems to predict a much more pronounced effects. The reason for this pattern, with respect to residue Y337, is not presently clear, yet it may be related to the proposed conformational flexibility of the Y337 side chain[4] which may be more pronounced in solution than in the solid crystallographic structures.

Replacement of Y133 by alanine always results in a decline of enzyme affinity toward the inhibitor, although its direct interactions with the aromatic system have not been observed by x-ray crystallography. This apparent inconsistency has been rationalized by proposing that the main role of Y133 aromatic moiety is in maintaining the active conformation of W86 rather than in direct interactions with the ligands[7, 8]. On the other hand, the H-bond interactions of certain ligands like huperzine-A or galanthamine with the hydroxyl substituent of Y133, which were observed in the respective crystallographic structures, are either supported (galanthamine) or negated (huperzine A) by our analysis.

According to x-ray analysis residue D74 seems to interact directly with certain active center inhibitors like galanthamine and tacrine, yet functional analysis of the corresponding HuAChE-tacrine complex does not seem to support such interaction. All these cases demonstrate that although both methods correctly identify the major interactions in the AChE-inhibitor complexes there are still numerous contacts the significance of which can be assessed only by functional analysis. The effects of the acyl pocket structure on the selectivity of E2020 toward HuAChE vs HuBChE demonstrate this point since they could be anticipated by neither crystallographic analysis nor by the extensive structure-function studies of E2020 analogs[13]. In fact, screening of new inhibitors with our library of single and multiple HuAChE mutants could provide valuable information regarding ligand interactions with the various subsites of the active center. Such information can be converted, through molecular modelling, into specific directions for further optimization of the inhibitor structures. Thus, combination of structural analysis with functional characterisation of the biological targets seems to be the best approach currently available for structure based drug design.

REFERENCES

1. E. Giacobini , in: *Cholinesterases and Cholinesterase Inhibitors*, edited by E. Giacobini (2000).
2. J.L. Sussman, M. Harel, F. Frolow, C. Oefner, A. Goldman, L. Toker and I. Silman, Atomic structure of acetylcholinesterase from Torpedo californica: a prototypic acetylcholine-binding protein, *Science* **253**,872-9 (1991).
3. G. Kryger, M. Harel, K. Giles, L. Toker, B. Velan, A. Lazar, C. Kronman, D. Barak, N. Ariel, A. Shafferman, I. Silman, and J.L. Sussman, Structures of recombinant native and E202Q mutant human acetylcholinesterase complexed with the snake-venom toxin fasciculin-II, *Acta Crystallogr.* **56**(11), 1385-1394 (2000).
4. A. Shafferman, B. Velan, A. Ordentlich, C. Kronman, H. Grosfeld, M. Leitner, Y. Flashner, S. Cohen, D. Barak,. and N. Ariel, Substrate inhibition of acetylcholinesterase : residues affecting signal transduction from the surface to the catalytic center, *EMBO J.* **11**, 3561-3568 (1992).
5. P. Taylor, and Z. Radic, The cholinesterases, *Annu. Rev. Pharmacol. Toxicol.* **34**, 281-320 (1994).
6. A. Ordentlich, D. Barak, C. Kronman, Y. Flashner, M. Leitner, Y. Segall, N. Ariel, S. Cohen, B. Velan, and A. Shafferman, Dissection of the human residues constituting the anionic site, the hydrophobic site, and the acyl pocket, *J.Biol.Chem.* **268**, 17083-17095 (1993).
7. D. Barak, C. Kronman, A. Ordentlich, N. Ariel, A. Bromberg, D. Marcus, A. Lazar, B. Velan, and A. Shafferman, Acetylcholinesterase peripheral anionic site degeneracy conferred by amino acid arrays sharing a common core, *J.Biol.Chem.* **269**, 6296-6305 (1994).
8. N. Ariel, A. Ordentlich, D. Barak, T. Bino, B. Velan, and A. Shafferman, The 'aromatic patch' of three proximal residues in the human acetylcholinesterase active centre allows for versatile interaction modes with inhibitors, *Biochem. J.*, **335**, 95-102 (1998).
9. G. Kryger, I. Silman, and J.L. Sussman, Structure of acetylcholinesterase complexed with E2020 (Aricept): implications for the design of new anti-Alzheimer drugs, *Structure* **7**, 297-307 (1999).
10. C. Bartolucci, E. Perola, C. Pilger, G. Fels., and D. Lamba, Three-dimensional structure of a complex of galanthamine (Nivalin) with acetylcholinesterase from Torpedo californica: implications for the design of new anti-Alzheimer drugs, *Proteins* **42**, 182-191 (2001).
11. H.M. Greenblat, G. Kryger, T. Lewis, I. Silman, and J.L. Sussman, Structure of acetylcholinesterase complexed with (-)-galanthamine at 2.3Å resolution, *FEBS Lett.* **463**, 321-326 (1999).
12. A. Loewenstein, A. Gnatt, L.F. Newille and H. Soreq,Chimeric human cholinesterase:Identification of interaction sites responsible for recognition of acetyl- or butyryl-ChE -specific ligands, *J. Mol Biol.* **234**, 289-296 (1993).
13. A. Inoue, T. Kawai, M. Wakita, Y. Iimura, H. Sugimoto, and Y. Kawakami, The simulated binding of (+/-)-2,3-dihydro-5,6-dimethoxy-2-[[1-(phenylmethyl)-4-piperidinyl]methyl]-1H-inden-1-one hydrochloride (E2020) and related inhibitors to free and acylated acetylcholinesterases and corresponding structure-activity analyses, *J. Med. Chem.* **39**, 4460-4470 (1996).
14. M. Harel, I. Schalk, L. Ehret-Sabatier, F. Bouet, M. Goeldner, C. Hirth, P.H. Axelsen, I. Silman, and J.L. Sussman, Quaternary ligand binding to aromatic residues in the active-site gorge of acetylcholinesterase, *Proc. Natl. Acad. Sci. USA* **90**, 9031-9035 (1993).
15. D. Barak, A. Ordentlich, A. Bromberg, C. Kronman, D. Marcus, A. Lazar, N. Ariel, B. Velan, and A. Shafferman, Allosteric modulation of acetylcholinesterase activity by peripheral ligands involves a conformational transition of the anionic subsite, *Biochemistry* **34**, 15444-15452 (1995).

THE NEW GENERATION OF ACETYLCHOLINESTERASE INHIBITORS

Hachiro Sugimoto[*], Yoichi Iimura, Yoshiharu Yamanishi

1. INTRODUCTION

Acetylcholinesterase inhibitor has been widely recognized as an effective treatment for Alzheimer's Disease (AD). It is known to alleviate the symptoms and slow the progression of the disease. Currently, there are four such kind of drugs in the market. These are *Tacrine*, *Rivastigmine*, *Galantamine* and *Donepezil*. *Donepezil*, developed by my company, Eisai Co., Ltd., was the second drug approved by the U.S. FDA for the treatment of AD.

The loss of presynaptic cholinergic neurons in the cortex and hippocampus is a well-documented aspect of Alzheimer's disease. In the 1970s, Dr. Davies and his group documented both the reduction of choline acetyltransferase and a loss of cholinergic cell bodies in Alzheimer's disease brains. In a subsequent study, Dr. Perry revealed a strong correlation between the extent of damage to the cholinergic system and the degree of dementia. Therefore, enhancement of the activity of cholinergic neurons has been regarded as one of the most promising methods for treating Alzheimer's disease. This theory is referred to as the cholinergic hypothesis.

On the basis of the cholinergic hypothesis, we chose the anti-acetycholinesterase mechanism in our development of anti-Alzheimer's drug. Acetylcholinesterase inhibitors work by preventing the breakdown of acetycholine thereby slowing the development of cognitive impairments in people experiencing the early stages of AD.

2. CONVENTIONAL AChE INHIBITORS

Conventional acetylcholinesterase inhibitors are shown in Figure 1. *Physostigmine* was isolated from eserine bean. The structure was elucidated by Stedman et. al. in 1925. *Galantamine* was isolated from *amaryllidaceae* family in Russia in the 1940s and the

[*]Hachiro Sugimoto, Ph.D., Director, Discovery Research Laboratories I, Eisai Co., Ltd., 5-1-3 Tokodai, Tsukuba-shi, Ibaraki 300-2635, Japan (email: h-sugimoto@hhc.eisai.co.jp)

Mapping the Progress of Alzheimer's and Parkinson's Disease
Edited by Mizuno *et al.*, Kluwer Academic/Plenum Publishers, 2002

structure was elucidated in Japan in the 1950s. These two compounds have been used to treat myasthenia gravis. *Tacrine* was synthesized in 1931 as an agricultural chemical. *Huperzine A* is an alkaloid isolated from the club moss, *huperzia serrata*, which has found use in Chinese herbal medicine. Among these compounds *Tacrine* and *Galantamine* were approved as anti-Alzheimer's disease drugs. The following is a brief description of these compounds.

Tacrine is the first drug approved for the treatment Alzheimer's disease. *Tacrine* inaugurates the first generation of acetylcholinesterase inhibitors. But clinical studies of this drug show some undesirable effects. The most serious is liver toxicity. However, because of the positive effects of *Tacrine* in improving cognitive impairment, a lot of tacrine derivatives studies were conducted. Yet almost all were unsuccessful. Only *Amiridine* is still under clinical trial study in Japan.

Rivastigmine is an acetylcholinesterase inhibitor derived from a first generation compound. This compound was derived from *Physostigmine* or *miotin*. *Rivastigmine* succeeded in overcoming the deficiency of *physostigmine*, which is, the short acting effect in clinical study. These three compounds have carbamate moiety in their molecules.

Galantamine is a very old acetylcholinesterase inhibitor, and like *physostigmine*, it was used for the treatment myasthenia gravis during the 1950s. This compound is a natural product isolated from the *amaryllideceae* family. The compound was subject of a large scale clinical trial study for the treatment of Alzheimer's disease. The results were successful.

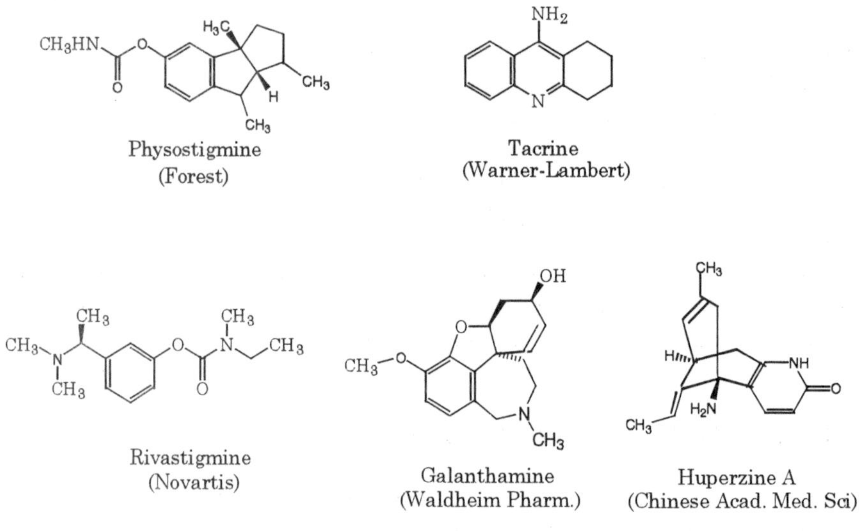

Physostigmine
(Forest)

Tacrine
(Warner-Lambert)

Rivastigmine
(Novartis)

Galanthamine
(Waldheim Pharm.)

Huperzine A
(Chinese Acad. Med. Sci)

Fig. 1 AChE inhibitors

3. DRUG DESIGN

3.1 Seed Compound

Compound 1 is the seed compound of *donepezil hydrochloride*. We found it through random screening. The anti-acetylcholinesterase activity of this compound was 12600nM at IC_{50} in rat brain homogenate. This was not very strong but the compound's novel structure was very promising. I would like to show you our drug design leading to the development of *donepezil*.

3.2 Drug Design - First Stage

When seed compound 1 was replaced with compound 2, the anti-acetylcholinesterase activity increased. Our next plan was to replace the ether moiety with an amide moiety (compound 3). We later discovered that benzylsulfonyl derivative (compound 4) was the most potent AChE inhibitor with an anti-AChE activity 21,000-fold greater compared to the seed compound. However, this compound has a very poor bioavailability rate, and therefore, could not be a candidate for clinical testing.

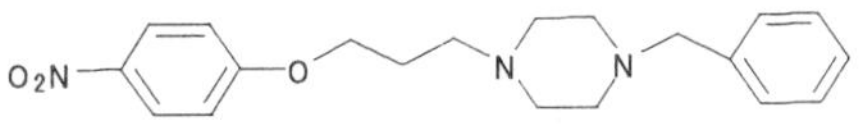

1 IC_{50} = 12600 nM

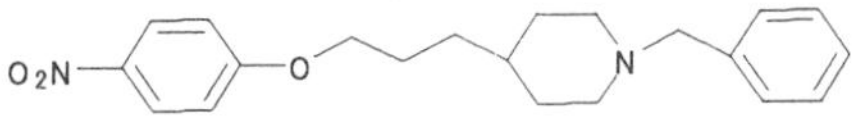

2 IC_{50} = 340 nM

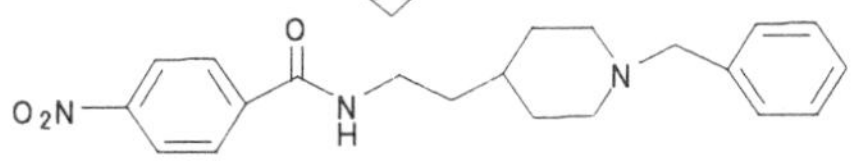

3 IC_{50} = 55 nM

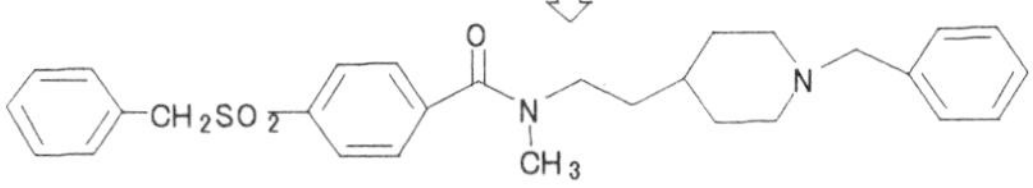

4 IC_{50} = 0.6 nM

Fig. 2 Drug Design 1

3.3 Drug Design - Second Stage

Our next strategy in drug design was the replacement of the amide moiety with ketone moiety (compound 7). This approach maintained the anti-AChE activity of the compound. Furthermore, the cyclic-amide derivative (compound 6) showed enhanced inhibitory action. On the basis of these results, an indanone derivative (compound 8) was designed. The resulting AChE activity was moderate, but we achieved longer duration of action. This new compound was code-named *E2020* with the general name *donepezil hydrochloride*.

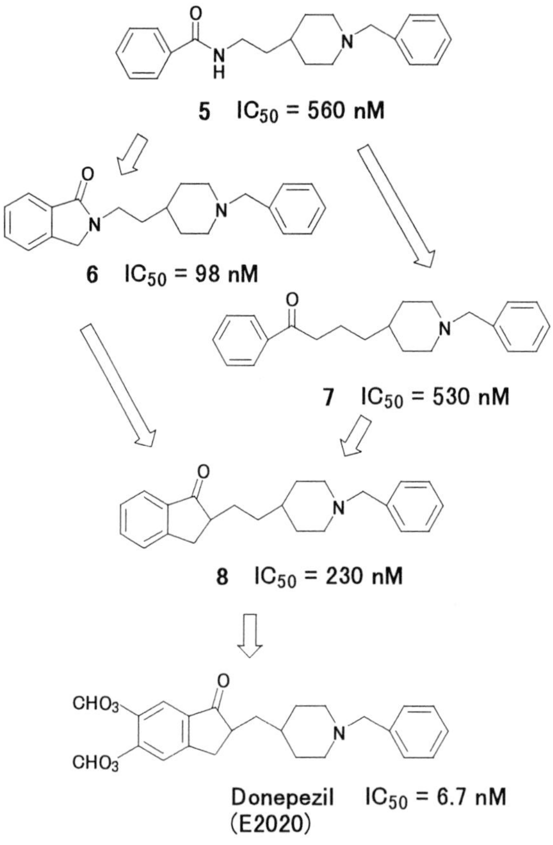

5 IC_{50} = 560 nM

6 IC_{50} = 98 nM

7 IC_{50} = 530 nM

8 IC_{50} = 230 nM

Donepezil IC_{50} = 6.7 nM
(E2020)

Fig. 3 Drug Design 2

4. PHARMACOLOGY

A comparison of the in-vitro inhibitory effects of donepezil and other cholinesterase inhibitors on acetylcholinesterase and butyrylcholinesterase was conducted. In order of inhibitory potency, the ranking are as follows: physostigmine, rivastigmine, donepezil, TAK-147, tacrine, NIK-247 and galantamine. The benzylpiperidine derivatives, donepezil and TAK-147, showed high selectivity for acetylcholinesterase over butyrylcholinesterase. Carbamate derivatives, i.e., physostigmine, rivastigmine and galantamine showed moderate selectivity, while the 4-aminopyridine derivatives, tacrine and NIK-247, showed no selectivity.

The effects of oral administration of donepezil, tacrine, and rivastigmine, on the extracellular acetylcholine concentration in the hippocampus of rats were evaluated by microdialysis technique. These results suggest that oral administration of donepezil, tacrine and rivastigmine increase acetylcholine concentration in the synaptic cleft of the hippocampus. It also suggest that donepezil has a more potent activity than tacrine and a longer-lasting effect than rivastigmine on the central cholinergic system.

Alzheimer's disease is associated with the depletion of cholinergic neurons. To represent this in our experiment, neurotoxin ibotenic acid was injected into the nucleus basalis magnocellularis (NBM) region of the rat brain. Destruction of this region, which contains cholinergic cells that innervate the cerebral cortex, causes a decrease in the concentration of acetylcholine in the cerebral cortex. Treatment with either donepezil or tacrine caused a dose-dependent increase in acetycholine in the cortex. Donepezil appears to be a more potent agent than tacrine in this system.

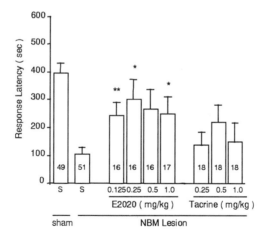

Figure 4:   Effects of E2020 and Tacrine on the Latency of
            Passive Avoidance Response in NBM-Lesioned Rats.
            *, **: $p < 0.05$, $p < 0.01$ (Mann-Whitney's U-test).
            S: Saline, Sham: Sham-operated rats,
            NBM Lesion: NBM lesioned rats.
            Number of animals is given in each column.

An experiment was designed to evaluate the effect of donepezil and tacrine in the passive avoidance task. This is a short-term memory test in the NBM-lesioned animals. NBM-lesioned and sham-operated control animals were placed in a passive avoidance box consisting of light and dark compartments and trained, using electric shock, to avoid entry into the dark compartment. Response latency, that is passive avoidance, was used to measure short-term memory.

Lesioned animals treated with donepezil showed a statistically significant increase in response latency demonstrating that donepezil enhances short-term memory in animals with cholinergic hypofunction. On the other hand, animals treated with tacrine showed no statistically significant improvement in this model.

5. CLINICAL STUDY

We conducted randomized, double-blind, placebo-controlled studies for a period of 24 weeks followed by a 6-week placebo washout period. Patients with mild to moderate Alzheimer's disease were enrolled in the trial. Trial subjects were randomized to treatments of placebo, 5 mg and 10 mg of donepezil administered once daily by the house surgeon. We enrolled about 150 patients per treatment group or a total of about 450 patients.

Using the Alzheimer's Disease Assessment Scale Cognitive Sub-Test or ADAS-cog, we observed statistically significant improvements in patients treated with 5 mg/day and 10 mg/day donepezil for a period of 12 and 18 weeks. Analysis of study endpoint shows that the mean improvement in ADAS-cog at endpoint, adjusted for baseline severity (LS mean) was greater in the 5 mg/day group and 10 mg/day group than in the placebo group.

Donepezil also produced a statistically significant improvement on the Clinician's Interview-Based Impression of Change plus or CIBIC-plus scores of patients compared to those in the placebo group. The CIBIC-plus scores significantly improved in patients receiving both the 5 mg/day and 10 mg/day donepezil.

6. CONCLUSION

The true nature of Alzheimer's Disease is still a much debated issue up to this time. Those of us involved in drug development have still so much to learn from the so-called BAP-tist (those who argue that beta-amyloid plaques are responsible for AD) and the TAU-ist (those who argue that the Tau tangles are responsible for AD). We are hoping that with the abundant genomic information generated everyday by various research groups around the world, we will soon be able to determine exactly the kind of malady that Alzheimer's Disease is. Hopefully we will be able to develop a cure in the very near future.

TV3326, A NOVEL CHOLINESTERASE AND MAO INHIBITOR

For Alzheimer's disease with co-morbidity of Parkinson's disease and depression

Marta Weinstock, Tatyana Poltyrev, Corina Bejar, Yotam Sagi, Moussa B.H.Youdim[*]

1. INTRODUCTION

Alzheimer's (AD) and Parkinson's disease (PD) are progressive neurodegenerative conditions in which a decrease is believed to occur in the activities of mitochondrial enzyme systems (Ghosh et al., 1998; Swerdlow et al., 1998). The cognitive deficits in AD are associated with a loss of cholinergic neurones arising in the basal forebrain (Whitehouse et al., 1981), but also appear in some subjects with PD (Homann et al., 1998). Degeneration of nigrostriatal dopamine neurones, characteristic of PD, is seen to a lesser extent in AD. In both conditions, there are significant reductions in serotoninergic and noradrenergic projections from the dorsal raphe and locus ceruleus nuclei, respectively (Palmer, 1988). These reductions may be responsible for the relatively high incidence of depression that is found as co-morbidity in AD and PD (Newman, 1999; Murray, 1996). In order to delay the rate of progression of the neurodegeneration and ameliorate the cognitive deficits and depressive symptoms we designed a new drug, TV3326, [(N-propargyl-(3R) aminoindan-5-yl)-ethyl methyl carbamate, a selective monoamine oxidase (MAO)-B methyl carbamate], derived

[*] Marta Weinstock (Leon & Mina Deutsch Professor of Pharmacology), Tatyana Poltyrev, and Corina Bejar, Hebrew University, Hadassah School of Medicine, Department of Pharmacology, Jerusalem, Israel. Yotam Sagi and Moussa B.H.Youdim, Technion, Faculty of Medicine, Eve Topf and NPF Centers for Neurodegenerative Diseases, Department of Pharmacology, Haifa, Israel.

from rasagiline, a selective monoamine oxidase (MAO)-B inhibitor (Youdim et al. 2001). The N-propargylamine moiety has been shown to confer neuroprotectiv activity on this and related molecules (Weinstock et al., 2001, Youdim and Weinstock, 2001; Youdim et al., 2001), and reduces the apoptosis and fall in mitochondrial action potential induced by oxidative stress in human neuroblastoma cells (Maruyama et al.. 2000). Moreover, rasagiline was recently shown to delay the progression of motor impairment in PD (Parkinson Study Section, 2000; Rabey et al., 2000). After chronic administration, TV3326 was found to inhibit monoamine oxidase (MAO-A and B) selectively in the brain but not in the intestine and liver (Weinstock et al., 2000). The introduction of the carbamate moiety like that in rivastigmine, a cholinesterase (ChE) inhibitor of proven beneficial effect on cognitive function in AD (Rossler et al., 1999), enabled TV3326 to show significant improvements in memory deficits in rat models of cholinergic hypofunction (Weinstock et al., 2000; 2001). We now describe the effect of this drug on brain 5HT levels, its associated antidepressant-like activity, and its prevention of the loss of dopamine neurons in the mouse 1-methyl-4-phenyl-1,2,3,6-tetrahydropyridine (MPTP) model of PD. MPTP specifically destroys nigrostriatal dopamine neurones resulting in dopamine depletion. The forced swim test (FST) measures the ability of antidepressants to reduce the occurrence of behavioural immobility induced by exposure to a forced swim stress. It is one of the most commonly used tests for antidepressant activity and is sensitive to all major classes of antidepressant drugs that increase 5HT neurotransmission (Borsini and Meli, 1988).

2. METHODS

2.1. Antidepressant activity, measurements of brain MAO activity and 5HT levels.

TV3326, (26 mg/kg) its S-enantiomer, TV3279 (26 mg/kg), moclobemide (20 mg/kg) or amitriptyline (10 chronic or 15 mg/kg acute) or water (1ml/kg) were administered orally to groups of 8 male Sprague-Dawley rats per treatment either acutely, after a 15 min exposure to the FST prior to the 5 min retest the following day (Porsolt et al., 1978), or once daily for two weeks prior to the test. The rats that had received chronic treatment with water, TV3326 or TV3279 were sacrificed within 30 min of the second swim test, the brains removed for measurement of MAO activity (half brain) and 5HT and 5HIAA levels in discrete areas in the remaining half by HPLC with electrochemical detection.

2.2. MPTP model of PD and dopamine levels

Male C57-BL mice were given 26 mg/kg of either TV3326 or TV3279 orally, once daily for two weeks during the last 5 days of which they were also injected with MPTP intraperitoneally (24 mg/kg/day). The mice were killed 3 days later, the brains removed and measurements made of striatal dopamine and its metabolites by HPLC (Ben-Shachar, and

Youdim, 1990) and of tyrosine hydroxylase (TH) activity by the method of Kato et al., (1981).

3. RESULTS

Chronic daily administration for 2 weeks of TV3326 but not TV3279 to rats inhibited brain MAO-A by $66 \pm 2\%$ and MAO-B by $71 \pm 2\%$. Acute administration of TV3326, did not produce any significant inhibition of MAO, neither did it affect the behavior of rats in the FST. However, after chronic administration, it significantly reduced immobility of rats in this test like moclobemide and amitriptyline (Fig. 1).

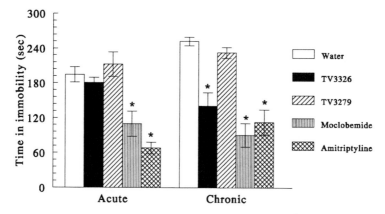

Figure 1. Reduction by chronic treatment with TV3326, moclobemide and amitriptyline in immobility induced in rats in the forced swim test.
TV3326 and TV3279 (26 mg/kg) and moclobemide (20 mg/kg) were given both as acute and chronic treatment, while amitriptyline was given 15 mg/kg in acute treatment and 10 mg/kg daily as chronic treatment. Significantly different from water, * P<0.01.

TV3279, which is a more potent inhibitor of ChE than TV3326, did not inhibit MAO after acute or chronic treatment, neither did it affect the depressive-like behavior in the FST. These findings show that TV3326 has antidepressant activity when given at a dose that caused more than 60% inhibition of brain MAO-A and MAO-B. This resulted in a doubling of 5HT levels in the cortex, hippocampus, hypothalamus and brainstem (Fig. 2).

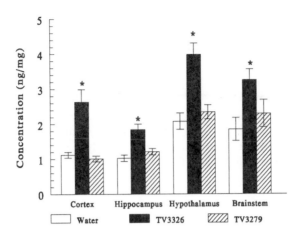

Figure 2. Effect of chronic oral adminisration of TV3326 and TV3279 on concentrations of 5HT in different brain areas in rat.
Significantly different from water, * P<0.01.

When TV3326 was given orally at a dose of 26 mg/kg once a day for 2 weeks to control mice MAO-A was inhibited by 38% and MAO-B by 53% and concentrations of dopamine in the striatum increased by 35%.

MPTP markedly lowered levels of dopamine, DOPAC and HVA and of TH activity in the striatum. Pretreatment with TV3326 almost completely prevented the decline induced by MPTP in the levels of dopamine and its metabolites and of TH (not shown) (Table 1).

Table 1. Effect of TV3326 pretreatment on levels of striatal dopamine and its metabolites in saline and MPTP-treated mice

Group	Dopamine	DOPAC	HVA
Control saline	68.4 ± 9.6	4.39 ± 0.12	34.7 ± 6.3
Control + TV3326 (26mg/kg)	104.6 ± 8.0*	4.54 ± 0.2	50.7 ± 9.9
MPTP	17.7 ± 3.4*	1.15 ± 0.26*	19.7 ± 2.3*
MPTP + TV3326 (26mg/kg)	47.5 ± 5.1§	3.33 ± 1.42	33.1 ± 2.8§

Values represent the mean concentrations (pmol/gm) ± sem. DOPAC= dihydroxyphenylacetic acid; HVA = homovanillic acid
Significantly different from control saline, * P<0.01; significantly different from MPTP, §, P<0.01.

When TV3279 was substituted for TV3326, in this test there was no significant inhibition of MAO or conservation of striatal dopamine levels in MPTP-treated mice. These data show that TV3326, which inhibits MAO-A and B selectively in the brain after chronic treatment (Weinstock et al., 2000), displays similar activity to that of known antidepressants in the rat FST by increasing brain levels of 5HT. It also prevents the loss striatal dopamine in a mouse model of PD by inhibiting MAO-B and the conversion of MPTP to the neurotoxin, MPP^+.

4. DISCUSSION

The complexities in the pathology and concomitant symptomatology in both AD and PD often necessitate the administration of several drugs with different pharmacological activities. Ideally, at an early stage of the disease one should give drugs that can slow its progression by preventing the occurrence of oxidative stress, abnormalities in mitochodrial function and the subsequent neuronal damage. At a later stage, when the characteristic symptoms of each condition appear, drugs should be given to maintain cholinergic activity in the basal forebrain in AD and dopaminergic activity in PD. Antidepressants are often indicated to alleviate depressive symptoms. When PD patients also have cognitive deficits a ChE-inhibitor may be contraindicated, since increasing striatal cholinergic activity can exacerbate the motor dysfunction. Many tricyclic antidepressants, which are more effective in elderly patients than 5HT uptake inhibitors, have significant anticholinergic activity, which exacerbates the cognitive deficits (Edwards, 1995). The reversible MAO-A inhibitor moclobemide not only improves depressive symptoms in PD (Jansen-Steur and Ballering, 1999; Amrein et al., 1999) but also shortens the "off" period when given with L-DOPA (Sieradzan et al., 1995). The selective MAO-B inhibitor, rasagiline has recently been shown to have a beneficial effect in PD both as monotherapy (Parkinson's Study Group, 2000) and as an adjuvant to L-DOPA (Rabbey, 2000). The novel MAO-ChE inhibitor, TV3326, which combines the pharmacological activities of rasagiline, moclobemide and rivastigmine, also shows neuroprotective effects in cell culture and in animal models (Weinstock et al., 2001; Youdim et al., 2001). By elevating striatal dopamine it may be able to counteract any adverse effect induced by a ChE inhibitor in PD and also ameliorate any extrapyramidal symptoms, which may occur in patients with AD. TV3326 may be a potential novel monotherapy for AD, PD with dementia, with concomitant depressive symptoms.

5. ACKNOWLEDGEMENTS

The authors wish to thank the support of Teva Pharmaceuticals Ltd., Israel for funding this research and supplying the test compounds. The support of National Parkinson Foundation, Miami, is gratefully acknowledged.

REFERENCES

Amrein, R., Martin, J.R., and Cameron, A.M., 1999, Moclobemide in patients with dementia and depression. *Adv. Neurolog.* 80: 509-519.

Ben-Shachar, D., and Youdim, M.B.H., 1990, Neuroleptic-induced supersensitivity and brain iron: I. Iron deficiency and neuroleptic-induced dopamine D2 receptor supersensitivity. *J. Neurochem.* 54: 1136-1141.

Borsini, F., and Meli, A., 1988, Is the forced swimming test a suitable model for revealing antidepressant activity? *Psychopharmacol.* 94: 147-160.

Edwards, J.G., 1995, Drug choice in depression. Selective serotonin reuptake inhibitors or tricyclic antidepressants? *CNS Drugs* 4: 141-159.

Ghosh, S.S., Miller, S., and Herrnstadt, C., et al., 1998, Mitochondrial dysfunction in Alzheimer's disease.in: *Progress in Alzheimer's and Parkinson's Diseases*. *Adv. Behav. Biol.* A. Fisher, I. Hanin, M. Yoshida, eds. M. Plenum Press, New York, vol. 49, pp.59-66.

Homann, C.N., Suppan, K., and Polmin, K., et al., 1998, Cognitive impairment in patients with Parkinsonism, in: *Progress in Alzheimer's and Parkinson's Diseases*. *Adv. Behav. Biol.* A. Fisher, I. Hanin, M. Yoshida, eds. M. Plenum Press, New York, vol. 49, pp 337-343.

Jansen-Steur, E.N., and Ballering, L.A., 1999, Combined and selective monoamine oxidase inhibition in the treatment of depression in Parkinson's disease *Adv. Neurol.* 80: 505-508.

Kato, T., Horiuchi, S., Togari, A. and Nagatsu, T., 1981, A sensitive and inexpensive high-performance liquid chromatographic assay for tyrosine hydroxylase. *Experientia* 37: 809-811.

Maruyama, W., Yamamoto, T., Kitani, K., Carrillo, M.C., Youdim, M., and Naoi, M., 2000, Mechanism underlying anti-apoptotic activity of a (-)deprenyl-related propargylamine, rasagiline. *Mech. Aging Dev.* 116:181-191.

Murray, J.B., 1996, Depression in Parkinson's disease. *J. Psychol.* 130: 659-667.

Newman, S.C., 1999, The prevalence of depression in Alzheimer's disease and vascular dementia in a population sample. *J. Affect. Disord.* 52: 169-176.

Palmer, A.M., Stratman, G.C., Procter, A.W., and Bowen, D.M., 1988, Possible neurotransmitter basis of behavioral changes in Alzheimer's disease. *Ann. Neurol.* 23: 616-620.

Parkinson Study Group, 2000, A controlled study with rasagline, a monoamine oxidase B inhibitor in early Parkinson's disease. *Proc. Am. Neurol. Assn. Boston,* p. 137.

Porsolt, R.D., Anton, G., Blavet, N., and Jalfre, M., 1978, Behavioral despair in rats: A new model sensitive to antidepressant treatments. *Eur. J. Pharmacol.* 47: 379-391.

Rabey, J.M,, Sagi,I., and Huberman, M., et al., 2000, Rasagiline mesylate, a new MAO-B inhibitor for the treatment of Parkinson's disease; a double-blind study as adjunctive therapy to levodopa. *Clin. Neuropsychopharmacol.* 23: 324-323.

Rosler, M., Anand, R., and Cicin-Sain, A., et al., 1999, Efficacy and safety of rivastigmine in patients with Alzheimer's disease: international randomised controlled trial. *Br. Med. J.* 318, 633-638.

Sieradzan, K., Channon, S., Ramponi, C., Stern, G.M., Lees, A.J., and Youdim, M.B.H., 1995, The therapeutic potential of moclobemide, a reversible selective monoamine oxidase A inhibitor in Parkinson's disease *J. Clin. Psychopharmacol.* 15 (Suppl 2): 51S-59S.

Swerdlow, S.H., Parks, J.K., and Miller, S.W., et al., 1998, Mitochondrial dysfunction in Parkinson's disease. in: *Progress in Alzheimer's and Parkinson's Diseases*. *Adv. Behav. Biol.* A. Fisher, I. Hanin, M. Yoshida, eds. M. Plenum Press, New York, vol. 49, pp. 67-75.

Weinstock, M., Bejar, C., and Wang, R-H., et al., 2000, TV3326, a novel neuroprotective drug with cholinesterase and monoamine oxidase inhibitory activities for the treatment of Alzheimer's disease. *J. Neural Transm.* (suppl) 60: 157-169.

Weinstock M., Kirschbaum-Slager N., Lazarovici P., Bejar, C., Youdim M.B.H., and Shoham S. 2001, Neuroprotective effects of novel cholinesterase inhibitors derived from rasagiline as potential anti-Alzheimer drugs. *Ann. N Y .Acad. Sci.* 939: 148-161.

Whitehouse, P.J., Price, D.L., and Struble, R.G., et al., 1982, Alzheimer's disease and senile dementia: loss of neurons in the basal forebrain. *Science* 215: 1237-1239.

Youdim, M.B.H., Wadia, A., Tatton, W., and Weinstock, M., 2001, The anti-Parkinson drug rasagiline and its derivatives have neuroprotective activity unrelated to inhibition of monoamine oxidase B. *Ann. N. Y. Acad. Sci.*, 939: 450-458.

Youdim M.B.H., and Weinstock M. 2001, Novel neuroprotective and anti-Alzheimer drugs with cholinesterase and monoamine oxidase inhibitory activities. *J. Pharm. Pharmacol.* in press.

M1 MUSCARINIC AGONISTS AS A THERAPEUTIC STRATEGY IN ALZHEIMER'S DISEASE

M1 muscarinic agonists in Alzheimer's disease

Abraham Fisher, Zipora Pittel, Rachel Haring, Rachel Brandeis, Nira Bar-Ner, Hagar Sonego, Itzhak Marcovitch, Niva Natan, Nadine Mestre-Frances*, and Noelle Bons*

1. INTRODUCTION

M1 muscarinic receptors (M1 mAChR) have an important role in cognitive processing relevant to Alzheimer's disease brains (AD)[1-5]. M1 mAChR are relatively unchanged in AD[1-5] and therefore may serve as a target for an anti-dementia drug treatment. However, some of the tested muscarinic agonists were not highly M1 selective, had major clinical limitations and showed disappointing clinical results in AD[4, 6, 7]. Thus the proof of clinical concept could not be shown.

A cholinergic hypofunction in AD appears to be linked with two other major hallmarks of this disease, β-amyloid (Aβ) and hyperphosphorylated τ proteins[2, 4-7]. Activation of M1 mAChR could beneficially modulate dysfunctions that are associated with AD including - Aβ and τ proteins, Apolipoprotein E (ApoE), as well as some processes involving certain G-proteins and neurotrophins (reviewed in[4, 6]). The property of M1 agonists to alter different aspects of AD pathogenesis could represent their most remarkable, yet unexplored and even ignored clinical value. The compound with minimal adverse effects and a high selectivity for M1 mAChR, in particular *in vivo*, should be selected as a therapeutic strategy designed to influence the progression of AD, and the onset or prevalence of various target populations at risk to develop AD. Select compounds from the *AF series* [*e.g.* AF102B, AF150(S) and AF267B, Fig 1] may fulfil such rigorous acceptance criteria as described in the text below. Comparison is made, when possible, with cholinesterase inhibitors (AChE-Is) used in AD treatment.

Correspondence: Abraham Fisher, Ph.D. Israel Institute for Biological Research; POBOX 19, Ness Ziona 74100, Israel, Tel:+972-8-9381603; Fax: +972-8-981615; fisher_a@netvision.net.il,
* Univ Montpellier II, Montpellier, France

Mapping the Progress of Alzheimer's and Parkinson's Disease
Edited by Mizuno *et al.*, Kluwer Academic/Plenum Publishers, 2002

2. RESULTS AND DISCUSSION

2.1 M1 agonists and β-amyloid (Aβ) processing

Stimulation of M1 mAChRs can increase cleavage of amyloid precursor proteins (APP) in the middle of its Aβ region[8] (reviewed also in[2, 4, 6, 7]; Fig 2). This produces the secreted, neurotrophic and neuroprotective APP (α-APPs), preventing the formation of Aβ (Fig 2). M1 agonists may alter APP processing in cortex[9] and hippocampus where M1 mAChRs are abundant[4, 6]. Notably, an increased secretion of α-APPs in various *in vitro* systems results in decreased synthesis of Aβ following treatment with muscarinic agonists[10, 11].

Studies *in vivo* support the relation between the cholinergic system and Aβ metabolism and strengthen *in vitro* reports about a tight linkage between APP processing and acetylcholine. Thus chronic cholinergic hypofunction in rats induced by AF64A[6] or the immunotoxin 192IgG-saporin[12] decrease α-APPs, indicating a reduction in α-secretase activity. In rabbits, where the sequence of Aβ(1-42) is similar to humans, a chronic cholinergic hypofunction elevates Aβ in the cortex and hippocampus[13]. The immunotoxin-induced effects in rats on α-APPs were attenuated by RS-86 (a muscarinic agonist)[12]. Furthermore, in rabbits, AF102B decreased Aβ(1-40), while AF267B and AF150(S) reduced levels of both Aβ(1-42) and Aβ(1-40)[14] in the cerebrospinal fluid (CSF). This is the first study that showed such an effect *in vivo*. So, in principle, we could design M1 agonists that have select activities on Aβ(1-42) and Aβ(1-40), respectively.

Two clinical studies revealed that chronic treatment with the M1 agonists AF102B and talsaclidine significantly decreased CSF Aβ in AD patients[15, 16]. These may indicate that M1 agonists have an important role in affecting Aβ processing probably by reducing Aβ burden in brains of AD patients. No other compounds were yet reported with such a unique profile in AD. Moreover, physostigmine (an AChE-I) and hydroxychloroquine (an anti-inflammatory drug) did not show a significant effect on CSF Aβ levels when tested in AD patients in the same study as AF102B[15].

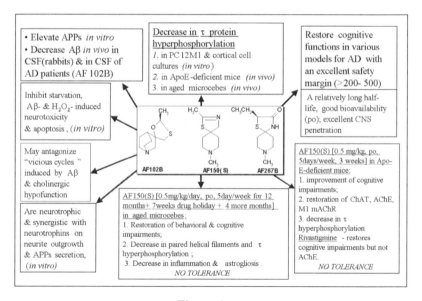

Figure 1

Inhibition of β and/or γ-secretases, would lead to accumulation of APP or its Aβ-containing fragments, if the inactivated α-secretase remains unmodified. These fragments can be potentially neurotoxic. In this context, the ability of M1 agonists to stimulate APPs secretion and Aβ reduction (*via* activation of α-secretase) might be a relatively direct route to decrease Aβ level in the brain. Interestingly, the effects observed with AF102B on CSF Aβ in AD patients, and in rabbits with the M1 agonists, resemble results obtained with γ-secretase inhibitors as reported by Bristol Myers Squibb. An indirect inhibitory effect on γ-secretase *via* M1 mAChR- mediated phosphorylation of presenilin-1 was proposed to explain the findings regarding AF102B in AD patients[15].

2.2. The linkage between starvation-, oxidative stress-, Aβ-mediated neurotoxicity and the M1 mAChR

Coupling of M1 mAChR with G-proteins in hippocampus is impaired following a septal-hippocampal cholinergic lesion[17]. In AD, this disruption in coupling is observed in hippocampal areas most affected by Aβ plaques, but is intact in less affected areas[2, 3]. Notably, sub-toxic concentrations of Aβ disrupt coupling of M1 mAChR with G-proteins linked to phosphoinositide(s) (PI) hydrolysis without producing neuronal cell injury or death[6, 18] (Fig 2). Presently, it is unclear if this effect relates to Aβ-induced cell death. In fact, Aβ-induced M1 mAChR-G protein uncoupling may represent an unexplored role for Aβ (*e.g.* a negative feedback to a prolonged activation of M1 mAChR). However, in AD such uncoupling may lead to a chronic impaired signaling, a decrease of α–APPs and generation of more Aβ, aggravating further the cholinergic deficiency. Some of these "vicious cycles" (Fig 2) may be blocked by M1 selective agonists, providing a rational therapeutic strategy in AD[4, 6].

In addition to the above, Aβ- and oxidative stress-induced apoptosis is mediated by reactive oxidative species (ROS). This effect is attenuated by antioxidants and M1 agonists [AF150(S) and AF267B][6]. The attenuation of Aβ- and oxidative stress-induced apoptosis *via* M1 mAChR activation is a novel finding. This adds further support to AF150(S) and AF267B as disease modifying agents in AD.

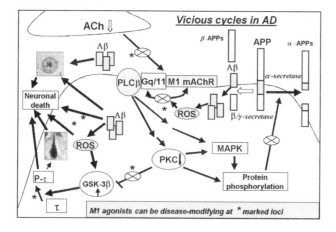

Figure 2

2.3. M1 mAChR-dephosphorylation of τ proteins

Activation of M1 mAChR decreases τ protein phosphorylation (Figs 1, 2). This was shown in PC12 cell transfected with M1 mAChR studies[19], confirmed *in vitro* [20] as well as *in vivo*, in ApoE-deficient mice[21] and in aged microcebes (this paper). A decreased hyperphosphorylation of τ *via* M1 mAChR activation may indicate a linkage between the muscarinic signal transduction system(s) and the neuronal cytoskeleton, *via* regulation of phosphorylation of τ[2, 4, 6, 19-21]. Activation of M1 mAChRs might provide a novel treatment strategy for AD by modifying τ processing in the brain. Moreover this may indicate a beneficial role for M1 agonists - as inhibitors of "vicious cycles" induced by Aβ and over-activation of certain kinases (Fig 2) and/or down-regulation of phosphatases, respectively. Activation of M1 mAChR decreases τ hyperphosphorylation *via* activation of PKC and inhibition of glycogen synthase kinase-3β (GSK-3β)[20] (see Fig 2). Notably, GSK-3β has been postulated to mediate AD τ hyperphosphorylation, Aβ-induced neurotoxicity and presenilin-1 mutation pathogenic effects[22]. Taken together, this may add unique value to M1 agonists since inhibition of GSK-3β is presently a major therapeutic target at several big pharma companies.

Interestingly, unlike M1 muscarinic agonists[19-21], nicotinic agonists and AChE-Is increased both τ phosphorylated and dephosphorylated levels in cell cultures[23]. In case increased phosphorylation of τ induced by nicotinic agonists and AChE-Is will also be observed *in vivo,* this may be detrimental in AD [*e.g.* paired helical filament (PHF) formation and aggregation[6, 22]]. Notably, AF150(S) decreased PHF-τ formation and aggregation in the aged microcebes (see below and Fig 1).

2.4 Studies in animal models for AD

AF102B, AF150(S) and AF267B restored memory and learning deficits in animal models, that mimic cholinergic and/or other deficits reported in AD, without producing adverse central and peripheral side-effects at effective doses and showing a relatively wide safety margin (>200-500 fold)[4, 6]. Two examples are described herein briefly:

2.4.1 Studies in ApoE-deficient mice

ApoE-deficient mice have memory deficits, synaptic loss of basal forebrain cholinergic projections and hyperphosphorylation of distinct epitopes of τ. In this model AF150(S) restored cognitive impairments and reduced choline acetyltransferase, AChE and elevated M1 mAChR brain levels, respectively to normal and decreased τ hyperphosphorylation on select epitopes[20] (Fig 1; reviewed in[4, 6]). These findings indicate that our M1 agonists may have neurotrophic effects *in vivo*. Notably, rivastigmine, an FDA-approved AChE-Is for AD treatment, restored only cognitive impairments, but not AChE levels[6].

2.4.2 Studies in the aged primate Microcebus murinus

The primate *Microcebus murinus* show with ageing similar cognitive deficits and cerebral lesions to those observed in AD[24-26]. In this model AF150(S) - i) improved cognitive and behavioral impairments; ii) decreased hyperphosphorylated τ & the number of neurons containing aggregated τ (*e.g.* indicative of diseased brains) and the number of PHF; and iii) decreased astrogliosis and inflammation (Fig 1). This supports AF150(S) as a candidate drug in AD that is not producing tolerance following a prolonged treatment, and the aged microcebes as a relevant model of the disease.

3. FUTURE PERSPECTIVES

An M1 muscarinic treatment of AD may also enlarge their use in other neurological, neuropsychiatric diseases and autoimmune diseases with a documented cholinergic deficiency. Notably, AF102B (EVOXAC^{TM}, Cevimeline) was recently approved in the USA and Japan for the treatment of dry mouth in Sjogren's syndrome, an autoimmune disease with harsh symptoms of dry mouth and dry eyes. AF102B might prove to be effective in AD, either alone or combined with other available treatments.

ACKNOWLEDGEMENTS

Supported in part by the Institute for the Study of Aging, New York, NY, USA.

REFERENCES

1. J.A. Court JA and E.K. Perry, Dementia: the neurochemical basis of putative transmitter orientated therapy, *Pharmac. Ther.* 52, 423-443 (1991).
2. C.J. Ladner and J.M Lee, Pharmacological drug treatment of Alzheimer disease: the cholinergic system revisited, *J. Neuropathol. Exptl. Neurol.* 57, 719-731 (1998).
3. E. Perry, J. Court, R. Goodchild, M. Griffiths, E. Jaros, M. Johnson, S. Lloyd, M. Piggott, D. Spurden, C. Ballard, I. McKeith and R. Perry, Clinical neurochemistry: developments in dementia research based on brain bank material, *J Neural Transm* 105, 915-933 (1998).
4. A. Fisher, Muscarinic receptor agonists in Alzheimer's disease: More than just symptomatic treatment. *CNS Drugs* 12, 197-214 (1999).
5. A. L. Svensson, I. Alafuzoff and A. Nordberg, Characterization of muscarinic receptor subtypes in Alzheimer and control brain cortices by selective muscarinic antagonists. *Brain Res.* 596, 142-148 (1992)
6. A. Fisher, Therapeutic strategies in Alzheimer's disease: M1 muscarinic agonists, *Jap. J. Pharmacol.* 84, 101-112 (2000).
7. J. Pavia, M. de Ceballos, and F.S. de la Cuesta, Alzheimer's disease: relationship between muscarinic cholinergic receptors, β-amyloid and τ proteins. *Fundam. Clin. Pharmacol.* 12, 473-481 (1998).
8. R.M. Nitsch, B.E. Slack, R.J. Wurtman, and J.H. Growdon, Release of Alzheimer amyloid precursor derivatives stimulated by activation of muscarinic acetylcholine receptors, *Science* 58, 304-307 (1992).
9. Z. Pittel, E. Heldman, J. Barg, R. Haring, and A. Fisher, Muscarinic control of amyloid precursor protein secretion in rat cerebral cortex and cerebellum. *Brain Res.* 742, 299-304 (1996).
10. A.Y. Hung, C. Haass, R.N. Nitsch, W.Q. Qiu, M. Citron, R.J. Wurtman, J.H. Growdon, and D.J. Selkoe, Activation of protein kinase C inhibits cellular production of the amyloid β-protein. *J. Biol. Chem.* 268, 22959-22962 (1993).
11. B.A. Wolf, A.M. Wertkin, Y.C. Jolly, R.O. Yasuda, B.B. Wolfe, R.J. Konrad, D. Manning, S. Ravi, J.R. Williamson, and V. M-Y Lee, Muscarinic regulation of Alzheimer's disease amyloid precursor protein secretion and amyloid β-protein production in human neuronal NT2N cells. *J. Biol. Chem.* 270, 4916-4922 (1995).
12. L. Lin, B. Georgievska, A. Mattsson, O. and Isacson, Cognitive changes and modified processing of amyloid precursor protein in the cortical and hippocampal system after cholinergic synapse loss and muscarinic receptor activation. *Proc Natl Acad Sci USA* 96, 12108-12113 (1999).

13. T.G. Beach, P.E. Potter, Y. Kuo, M.R. Emmerling, R.A. Durham, S.D. Webster, D.G. Walker, L.I. Sue, S. Scott, K.J. Layne, A.E. and Roher, Cholinergic deafferentation of the rabbit cortex: a new animal model of Aβ deposition, *Neurosci. Lett.* **31**, 9-12 (2000).

14. T.G. Beach, D.G. Walker, P.E. Potter, and A. Fisher, Reduction of cerebrospinal fluid amyloid beta after systemic administration of M1 muscarinic agonists, *Brain Res.* **905**, 220-223 (2001).

15. R.M. Nitsch, M. Deng, M. Tennis, D. Schoenfield, and J.H. Growdon, The selective muscarinic M1 agonist AF102B decreases level of total Aβ in CSF of patients with Alzheimer's disease, *Arch. Neurol.* **48**, 913-918 (2000).

16. C. Hock, A. Maddalena, M. Deng, J.M. Growdon, and R.M. Nitsch, Treatment with talsaclidine decreases cerebrospinal fluid levels of total amyloid β-peptide in patients with Alzheimer's disease. In: *9th Meeting of the Internat. Study Group on the Pharmacol. of Memory Disorders Associated with Aging.* Zurich, Switzerland, Feb 18-20 (2000).

17. P.E. Potter, C. Gaughan, and Y. Assouline, Lesion of septal-hippocampal neurons with 192IgG-saporin alters function of M1 muscarinic receptors, *Neuropharmacol.* **38**, 579-586 (1999).

18. J.F. Kelly, K. Furukawa, S.W. Barger, M.R. Rengen, R.J. Mark, E.M. Blanc, G.S. Roth and M.P. Matson, Amyloid β-peptide disrupts carbachol-induced muscarinic cholinergic signal transduction in cortical neurons, *Proc. Natl. Acad. Sci.* **96**, 6753-6758 (1996).

19. E. Sadot, D. Gurwitz, J. Barg, L. Behar, I. Ginzburg, and A. Fisher, Activation of m1-muscarinic acetylcholine receptor regulates *tau* phosphorylation in transfected PC12 cells, *J. Neurochem.* **66**, 877-880 (1996).

20. O.V. Forlenza, J.M. Spink, R. Dayanandan, B.H. Anderton, O.F. Olensen, and S. Lovestone, Muscarinic agonists reduce τ phosphorylation in non-neuronal cells via GSK-3β inhibition and in neurons, *J. Neural. Transm.* **107**, 1201-1212 (2000).

21. I. Genis, A. Fisher, and D.M. Michaelson, Site-specific dephosphorylation of τ in apolipoprotein E-deficient and control mice by M1 muscarinic agonist treatment, *J. Neurochem.* **12**, 206-213 (1999).

22. J. Lucas, F. Hernandez, P. Gomez-Ramos, M.A. Moran, R. Hen, and J. Avila, Decreased nuclear β–catenin, τ hyperphosphorylation and neurodegeneration in GSK-3β conditional transgenic mice, *EMBO J.* **20**, 27-39 (2001).

23. E. Hellstrom-Lindahl, H. Moore H, and A. Nordberg, Increased levels of τ protein in SH-SY5Y cells after treatment with cholinesterase inhibitors and nicotinic agonists, *J Neurochem.* **74**, 777-784 (2000).

24. N. Bons N, V. Jallageas, N. Maestre-Frances, S. Silhol, A. Petter, and A. Delacourte, *Microcebus murinus*, a convenient laboratory animal for the study of Alzheimer's disease, *Alzheimer's Res.* **1**, 83-87 (1995).

25. S. Silhol, A. Calenda, V. Jallageas, N. Mestre-Frances, M. Bellis, and N. Bons, Beta-amyloid protein precursor in *Microcebus murinus*: genotyping and brain localization, *Neurobiol. Dis.* **3**, 169-182 (1996).

26. P. Giannakopoulos, S. Silhol, V. Jallageas, J. Mallet, N. Bons, C. Bouras, and P. Delaere, Quantitative analysis of tau protein-immunoreactive accumulations and beta-amyloid protein deposits in the cerebral cortex of the mouse lemur, *Microcebus murinus*, *Acta Neuropathol.* **94**, 131-139 (1997).

PHENSERINE REGULATES TRANSLATION OF β–AMYLOID PRECURSOR PROTEIN MESSAGE

A Novel Drug Target for Alzheimer's Disease

D.K. Lahiri,* T. Utsuki, K.T.Y. Shaw, Y.-W. Ge, K. Sambamurti, P.S. Eder, J. T. Rogers, M.R. Farlow, T. Giordino, N.H. Greig[1]

1. INTRODUCTION

One of the major hallmarks of Alzheimer's disease (AD) is the appearance of senile plaques primarily composed of amyloid β-peptide (Aβ), which is derived from a larger protein called β-amyloid precursor protein, βAPP[1]. βAPP is cleaved by three enzymes, α-, β- and γ-secretases, to different protein fragments, including the toxic Aβ and other C-terminal fragments implicated in the pathogenesis of AD.[1] Our goal is to study agents that reduce βAPP expression, as this is the precursor to all the toxic fragments. Recently, we have synthesized a family of novel cholinesterase inhibitors (ChEIs), phenserine and analogues.[2] Phenserine improves cognitive performance in rodents.[3] Studies of rats with forebrain cholinergic lesions, known to increase βAPP in cholinergic projection areas, have shown that phenserine can prevent this rise and, additionally, reduce βAPP production in naive animals.[4] As both βAPP processing and cholinesterase activity are affected in the AD brain, our current studies have focused on the molecular changes induced by ChEIs.[5-7] We have reported that tacrine, a first generation of ChEIs for treatment of AD, could reduce the levels of secreted βAPP and Aβ.[7] We report herein, the mechanism through which phenserine interacts with the cellular processing of βAPP.

---

*D.K. Lahiri (corresponding author), Y.-W. Ge, M.R. Farlow: Indiana University School of Medicine, 791 Union Drive, Indianapolis, IN; K. Sambamurti: Mayo Clinic, Jacksonville, FL; J. T. Rogers: Genetics & Aging Unit, Harvard Medical School, Boston, MA; P.S.Eder, T. Giordino: Message Pharmaceuticals, Malvern, PA; T. Utsuki, K.T.Y. Shaw, N.H. Greig:: Drug Design & Development, NIA/NIH, Baltimore, MD, USA.

Mapping the Progress of Alzheimer's and Parkinson's Disease
Edited by Mizuno *et al.*, Kluwer Academic/Plenum Publishers, 2002

2. METHODS

2.1. Cell culture, Drug treatment and LDH assay: Human neuroblastona (SK-N-SH, SH-SY-5Y) and astrocytoma (U373) cell lines were cultured in regular media until they reached 70% confluence as described[6]. To start the experiment, cells were fed with fresh media with low serum (0.5%) containing 0, 5 or 50 μM phenserine. Levels of lactate dehydrogenase (LDH) in conditioned medium were measured as a marker of cell viability and integrity, as described previously[7].

2.2. Lysate preparation and Western blot analysis: At each time point, the conditioned medium was collected for the analysis of secretory βAPP levels and the cell lysate prepared to measure total cellular βAPP levels[7]. Fifteen μg of protein from each sample was analyzed by Western immunoblotting using 22C11 antibody. The samples were detected by chemiluminiscence. Densitometric quantification of blots was undertaken by using a CD camera and NIH-IMAGE (version 4.1).

2.3. Aß Assay: Total Aß peptide levels were assayed in the conditioned media samples using a sandwich ELISA method as previously described[7].

2.4. Northern Blotting: Total RNA (10μg) was prepared from treated astrocytoma cells and subjected to northern blotting using the βAPP-cDNA probe. Equal loading of samples was verified by rehybridizing the filter using a β actin-cDNA probe.

2.5. APP synthesis rate: The effect of phenserine treatment on the rate of βAPP synthesis was determined as described previously[7].

2.6. Statistical analysis: The statistical analysis was carried out by ANOVA/ Post-hoc.

3. RESULTS

3.1. Phenserine decreases βAPP levels irrespective of its anticholinesterase activity:
βAPP protein levels were measured after treatment of the SH-SY-5Y cells with both 5 μM (+)-phenserine and (-)-phenserine for 0.5, 1, 2, and 4 h. These stereoisomeric forms of the drug have opposite affects on ChEI activity: (+)-phenserine exhibits no anti-cholinesterase activity whereas (-)-phenserine has potent enzymatic activity[2]. In both experiments, the βAPP levels in the cell lysates slowly decreased at each time point with the most dramatic decline observed after 4 h (data not shown).

3.2. Phenserine decreases both βAPP and Aβ levels in neuroblastoma cells: Following phenserine treatment of SK-N-SH cells for 16 h, βAPP levels were reduced in a time- and concentration-dependent manner in both conditioned media and cell lysates. This was not associated with cellular dysfunction, as determined by measurement of LDH levels versus untreated controls. Quantification of levels of total Aβ was undertaken at 8 and 16 h, and results demonstrate a phenserine induced reduction of 14% and 34% ($p<0.002$), respectively, versus untreated controls (Table 1).

Table 1. Effects of Phenserine on levels of different derivatives of βAPP, Aβ and cellular toxicity at 16 hour of treatment in human neuroblastoma (SK-N-SH) cells[a].

Treatment	Dose	SAPP (%)	βAPP (%)	Aβ (pM)	LDH (A490 nm)
Control	-	100	100	11.8±0.23	0.099±0.001
Phenserine	5 µM	51.6±5.5*	84.4±5.7	Not done	Not done
Phenserine	50 µM	33.8±2.26 *	59.7±0.4*	8.1±0.43*	0.092±0.004

[a]* Value differs significantly from corresponding value for control (n=3).

3.3. Phenserine decreases βAPP levels in astrocytoma cell line: Following an extended period of phenserine treatment, U373 cells exhibited a similar pattern of decreased βAPP protein synthesis as in the human neuroblastoma cells (data not shown).

3.4. Phenserine does not affect βAPP mRNA levels in U373 cells: In contrast to its effect on protein levels, phenserine treatment of U373 cells did not affect βAPP mRNA levels. Northern of U373 treated with phenserine for 48 h showed consistent levels of βAPP mRNA without any major fluctuations in densitometry readings after standardization of each sample to actin mRNA expression (data not shown). Clearly, phenserine's action on βAPP protein is at the level of translation, as northern blot analysis of untransfected and transfected cells show little differences in mRNA levels.

3.5. Phenserine decreases βAPP synthesis. To assess the effects of phenserine on the translational efficiency of ßAPP mRNA, a short pulse-labeling experiment was performed, which shows that phenserine significantly decreased ßAPP synthesis (50% reduction) without changing TCA precipitable counts under the condition when the drug significantly reduced newly translated βAPP in these cells[9].

4. DISCUSSION

The present study characterizes the regulation of βAPP processing and gene expression by a novel ChEI, phenserine, currently in AD clinical trials. Phenserine improves learning in rats[3], and protects rats with forebrain cholinergic lesions from associated elevations in βAPP protein synthesis[4]. Here we examined the molecular mechanisms of βAPP regulation by phenserine using a cell culture model. Neuroblastoma cell lines have been considered as an accepted model of neuronal differentiation. In its undifferentiated state, it is an ideal candidate for this study as these cells express adrenergic neurotransmitter enzymes, exhibit muscarinic membrane receptors and display a basal level of βAPP expression. Phenserine treatment of SH-SY-5Y cells rapidly reduced levels of endogenous βAPP, under the dose and conditions described. This action was not due to the ChEI activity of phenserine as the stereoisomeric (+) phenserine, that lacks ChEI activity[2], reduced the level of βAPP protein to a similar magnitude as its ChEI potent (-)-enantiomer. For phenserine-treated cells, the reductions in βAPP also translated into reduced secreted Aβ levels. As a lipophilic compound, phenserine is able to diffuse across membrane barriers with rapid ease. Pharmacological reports of its ability to cross the blood-brain barrier, reaching brain/plasma ratios of 10:1[2], substantiate its ability to penetrate and exert its action within neuroblastoma cells. These actions were replicated in two other cell lines of CNS origin; SK-N-SH neuroblastoma and U373 astrocytoma cells. In contrast, other anti-cholinesterase drugs, such as physostigmine, are unable to alter βAPP levels in neuronal cell lines[6].

Recently, we have shown that phenserine's action is neither mediated through classical receptor signaling pathways involving transcription factor, ERK or PI 3 kinase activation, nor is it associated with the anticholinesterase activity of the drug[5]. Rogers et al[8] identified an important IL-1 responsive element localized to the 5' untranslated region (5'UTR) of βAPP mRNA that conferred translational control of βAPP protein synthesis in U373 cells. Phenserine's action on βAPP 5'UTR mRNA resulted in a reduced CAT reporter expression in the transfected astrocytoma cells[5], although βAPP 3'UTR sequences will also have to be tested for their involvement. In parallel experiments, phenserine treated transfected cells showed decreased βAPP protein levels and concurrent maintenance of mRNA steady-state levels in these cells, suggesting that the drug exerts its effect through translational modifications[5]. However, it is not clear whether the drug's effect at the post transcriptional level via 5'-UTR element is specific for a certain cell line or a general phenomenon. Other possibilities for phenserine's effects on translation include action through a yet undefined regulatory site in the βAPP mRNA or through modulation of an RNA binding protein that regulates βAPP translation.

5. CONCLUSION

Our cell culture studies support the ability of phenserine to reduce βAPP as demonstrated *in vivo*[4], and lower secreted levels of Aβ. Phenserine action on βAPP likely occurs via a translational mechanism. This mechanism appears to be independent of cholinergic activity and may involve direct interactions with other proteins. In light of this, phenserine action on βAPP, identifying a novel drug target for AD therapy, is a phenomenon that warrants further study.

6. ACKNOWLEDGMENTS

This work was supported in part by grants from the Alzheimer's Association, Axonyx and National Institutes of Health.

REFERENCES

1. Selkoe DJ. The origins of Alzheimer disease: A is for amyloid. *JAMA* **283**, 1615-1617 (2000).
2. Greig NH, et al. Phenserine and ring-C hetero-analogues: drug candidates for the treatment of Alzheimer's disease. *Med Chem Rev* **15**, 3-31 (1995).
3. Patel N, et al. Phenserine, a novel acetylcholinesterase inhibitor, attenuates impaired learning of rats in a 14-unit T-maze induced by blockade of the NMDA receptor. *Neuroreport* **9**, 171-176 (1998).
4. Haroutunian V, et al. Pharmacological modulation of Alzheimer's beta-amyloid precursor protein levels in the CSF of rats with forebrain cholinergic system lesions. *Mol Brain Res* **46**, 161-168 (1997).
5. Shaw, K.T.Y., et al. Phenserine regulates translation of β-amyloid precursor mRNA by a putative interleukin-1 responsive element, a new target for drug development. *PNAS* **98**, 7605-7610 (2001).
6. Lahiri D, et al. Effects of cholinesterase inhibitors on the secretion of beta-amyloid precursor protein in cell cultures. *Ann N Y Acad Sci* **826**, 416-421 (1997).
7. Lahiri D, et al. The secretion of amyloid beta-peptide is inhibited in tacrine-treated human neuroblastoma cells. *Mol. Brain Res.* **62**, 131-140 (1998).
8. Rogers J, et al. Translation of the Alzheimer amyloid precursor protein mRNA is up-regulated by interleukin 1 through 5'-untranslated region sequences. *J Biol Chem* **274**, 6421-6431 (1999).
9. Utsuki, T., et al. Regulation of ß-amyloid (Aß) precursor protein (ßAPP) translation: a new target for Alzheimer disease (AD) drug development. *SFN Abstract*, **27**, 91.3 2001

THE FORKHEAD FAMILY OF TRANSCRIPTION FACTORS ARE TARGETS OF AKT IN IGF-1 MEDIATED SURVIVAL IN NEURONAL CELLS

Wen-Hua Zheng, Satyabrata Kar and Remi Quirion[*]

1. INTRODUCTION

Apoptosis is part of normal mammalian development. About half of all neurons of the neuroaxis and 99% of the total number of cells generated during the human lifetime die through apoptosis (Datta et al., 1999; Vaux and Korsmeyer 1999). The process of apoptosis is tightly regulated and suppressed by survival factors such as the insulin-like growth factor-1 (IGF-1).

IGF-1 is a polypeptide growth factor essential for the normal development and maintenance of cellular integrity including cells of the central nervous system (de Pablo and de la Rosa, 1995). Recent studies have shown that IGF-1 also plays an important role in the adult, especially during neurodegenerative conditions (Loddick and Rothwell, 1999). The expression of IGF-1 and its receptors in the central nervous system are upregulated in response to brain injuries and this most likely reflects a protective effect as exogenous IGF-1 protects the brain from hypoxic and ischemic injuries *in vivo* (Beilharz et al., 1998; Gluckman et al., 1998; Loddick et al., 1998). IGF-1 also promotes the survival of a broad range of cells from various death insults *in vitro* (Cheng et al., 1992; Dore et al., 1997; Galli et al., 1995; Sortino et al., 1996, Zheng et al., 2000a). For example, our laboratory has shown that the treatment of cultured hippocampal neurons with IGF-1 prevents and rescues these neurons from β-amyloid toxicity (Dore et al., 1997).

The biological effects of IGF-1 are mediated by its receptors, the type I IGF receptors (Butler et al., 1998). Binding IGF-1 to its receptor activates a receptor tyrosine kinase, resulting in the tyrosine phosphorylation of the receptor and its substrates such as IRS1/2 proteins or Shc (LeRoith et al., 1995), leading to the activation of multiple signaling pathways such as mitogen-activated protein kinase (MAPK) and phosphatidylinositol 3-kinase (PI3K)/Akt pathways and cell survival (Butler et al., 1998).

[*] Douglas Hospital Research Center, Departments of Psychiatry, and Pharmacology and Therapeutics, McGill University, Montreal, Quebec, Canada H4H 1R3

Mapping the Progress of Alzheimer's and Parkinson's Disease
Edited by Mizuno *et al.*, Kluwer Academic/Plenum Publishers, 2002

217

2. AKT KINASE, ITS PHOSPHORYLATION AND ACTIVATION

Akt kinase is a serine/threonine kinase and the downstream target of PI3 kinase playing an important role in cell survival (Datta et al., 1999; Butler et al., 1998, Zheng et al., 2000a; Zheng et al., 2000b). The activation of Akt is mediated by survival factors such as IGF-1 and the neurotrophins. These survival factors activate PI3 kinase which increases intracellular phospholipids in target cells, leading to the recruitment of Akt to the plasma membrane where it is phosphorylated by its upstream kinase PI (3.4.6)P3-dependent kinase 1(PDK1) and PDK 2 at Thr308 and Ser473 (Datta et al., 1999; Alessi et al., 1997). We have recently shown that IGF-1 stimulates the activation of Akt in neuronal cells by the PI3 kinase pathway (Zheng et al., 2000b). The phosphorylation of these two residues in Akt is required for its full activation (Alessi et al., 1996a; Datta et al., 1999). PDK1 phosphorylates Thr308 and has been cloned (Alessi et al., 1997) while PDK2 remains to be characterized. A recent report suggested that PDK1 forms complex with the protein kinase C-related kinase 2 which is capable of phosphorylating both Thr308 and Ser473 residues of Akt (Balendran et al., 1999). Other reports have indicated that the integrin-linked kinase (ILK) may be the Ser473 kinase (Delcommenne et al., 1998), while MAPKAPK-2, a downstream target of p38 MAP kinase, and PDK1-dependent autophosphorylation may also be involved in the phosphorylation of Ser473 of Akt (Alessi et al., 1996a; Rane et al., 2000; Toker and Newton, 2000).

The mechanisms by which Akt mediates the survival effect of IGF-1 in neuronal cells is not clear. Activated Akt phosphorylates proapoptotic proteins such as the Bcl-2 family member Bad (del Peso et al., 1997), caspase-9 (Cardone et al., 1998) and most recently the Forkhead transcription factor FKHRL1(Brunet et al., 1999; Zheng et al., 2000c), FKHR (Guo et al., 1999; Tang et al., 1999; Nakae et al., 1999) blocking their proapoptotic properties and leading to cell survival. Among these targets, the role of Bad is limited in neurons as Bad knockout mice do not show alternations in neuronal apoptosis (Miller et al., 1997; Kaplan and Miller, 2000). Similarly, caspase-9 may not be essential in neurons on the basis of the lack of conservation of the Ser196 Akt phosphorylation site in nonhuman procaspase-9 and the reports of Akt induced phosphorylation of endogenous procaspase-9 in neurons (Fujita et al., 1999, Kaplan and Miller, 2000). Thus, the possible role of forkhead proteins in IGF-1/Akt-mediated survival in neurons deserves to be fully investigated.

3. FORKHEAD FAMILY OF TRANSCRIPTION FACTORS ARE DOWNSTREAM TARGETS OF AKT

The Forkhead transcription factors are a family of transcription factors characterized by the presence of a highly conserved forkhead domain having a winged-helix motif and DNA binding activity (Lai et al., 1993). The forkhead family of transcription factors contains about 90 members that have been described in a wide range of organisms from yeast to humans (Kaufmann and Knochel, 1996). Among them, FKHR, FKHRL1 and AFX are the closest mammalian homologs of DAF-16 and belong to a subgroup of forkhead proteins known as FKHR proteins (Anderson et al., 1998). FKHR proteins contain three putative Akt phosphorylation sites RXRXXS/T (Alessi et al., 1996b) and are downstream targets of Akt (Kops and Burgering, 1999).

4. AKT KINASE PHOSPHORYLATES FORKHEAD FAMILY TRANSCRIPTION FACTORS IN VITRO AND IN VIVO

In vitro, various active forms of Akt are able to phosphorylate recombinant human FKHR proteins or immunoprecipitated ectopically expressed Forkhead proteins including the nematode DAF-16, human and rat FKHRL1, human and mouse FKHR and human AFX (Datta et al., 1999). Brunet et al., (1999) have shown that constitutively active but not kinase dead Akt immunoprecipitated from HEK293T cells effectively phosphorylated wild-type GSK-FKHRL1 in an *in vitro* kinase assay. Site-directed mutagenesis demonstrated that Thr32, Ser253 and Ser315 residues are the three residues phosphorylated by Akt in FKHRL1 (Brunet et al., 1999). Consistent with these findings, we have shown that pure, active Akt directly phosphorylated FKHRL1 imunoprecipitated from neuronal cells at Thr32 and Ser253 residues (Zheng et al., 2000c). Similar results were obtained for HKHR and AFX, which were phosphorylated at Thr24/Ser256/Ser319 and Thr28/Ser193/Ser258, respectively (Rena at al., 1999, Tang et al., 1999; Nakal et al., 1999).

In vivo, Akt phosphorylates all three Forkhead isoforms at all three sites in various cell types in response to different stimuli via a PI3K/Akt pathway (Kops and Burgering, 1999). IGF-1 induced the phosphorylation of transfected FKHRL1 on Thr32 and Ser253 in HEK293T cells and fibroblast cell line CCL39, and these actions were blocked by LY294002, an inhibitor of PI3K. Moreover, co-transfection of constitutively active Akt and HA-tagged FKHRL1 induced the phosphorylation of HA-FKHRL1 at Thr32 and Ser253, and Ser315 in CCL39. IGF-1 and insulin can also induce the phosphorylation of endogenous FKHRL1 on Thr32 and Ser253 residues in CCL39 cells (Brunet et al., 1999).

5. IGF-1 INDUCED THE PHOSPHORYLATION OF ENDOGENOUS FKHRL1 IN NEURONAL CELLS BY THE PI3 KINASE PATHWAY

However, whether endogenous forkhead proteins are also signaling components of PI3K/Akt pathway in neuronal cells is not known. Accordingly, we have investigated the effect of IGF-1 on the phosphorylation of FKHRL1 and its regulation in PC12 cells and primary cultured neurons. Figure 1A and 1B showed that IGF-1 time- and concentration-dependently induced the phosphorylation of FKHRL1 in PC12 cells. The phosphorylation of FKHRL1 induced by IGF-1 was inhibited by the PI3K inhibitors wortmannin and LY294002 (not shown) but not by the MEK inhibitor PD98059 or the p70 S6 kinase pathway inhibitor rapamycin suggesting that the action of IGF-1 is mediated by the PI3K/Akt pathway in these cells (Fig. 1C and Zheng, et al., 2000c). This hypothesis was confirmed by data from *in vitro* kinase assay with pure recombinant active Akt and the transient expression of constitutively active Akt or kinase dead Akt in PC12 cells (Zheng et al., 2000c). As FKHRL1 associated with Akt in PC12 cells, these data taken together demonstrate that endogenous FKHRL1 is a substrate of Akt in IGF-1 receptor-mediated survival in neuronal cells (Zheng et al., 2000c). In addition to IGF-1, insulin, EGF, FGF and neurotrophins can also induce the phosphorylation of endogenous FKHRL1, FKHR and AFX in neuronal cells (Zheng et al., unpublished data).

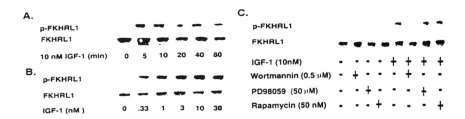

Figure 1. IGF-1 induced the phosphorylation of FKHRL1 in PC12 cells via the PI3K pathway. A. time course; B. concentration dependency and C. putative pathways studied using inhibitors

6. THE PHOSPHORYLATION OF FORKHEAD TRANSCRIPTION FACTORS BLOCKS THEIR NUCLEAR TRANSLOCATION AND LEADS TO CELL SURVIVAL

The functional relevance of the phosphorylation of endogenous forkhead proteins induced by IGF-1 in neuronal cells remained to be fully elucidated. Forkhead proteins have recently been suggested to act as proapoptotic proteins as the expression of unphosphorylated FKHRL1 or FKHR induced cell death in various cells (Brunet et al., 1999; Tang et al., 1999). FKHRL1 contains a core domain of 100 amino acids that mediates its interaction with DNA consensus sequence. Under conditions during which Akt is inactive, FKHRL1 is not phosphorylated and is localized to the nucleus where it binds to DNA consensus sites of the Fas ligand gene inducing its transcription and leading to apoptosis (Brunet et al., 1999). In contrast, exposure to IGF-1 induces the activation of Akt kinase which phosphorylates FKHRL1 on Thr32 and Ser253, leading to the association of FKHRL1 with the cytoplasmic protein 14-3-3 and hence its retention in the cytoplasm. This in turn leads to the inhibition of the Fas ligand gene transcription and cell survival (Brunet et al., 1999). In PC12 cells and primary cultured neurons, IGF-1 induced the phosphorylation of FKHRL1 and promotes cell survival (Zheng et al., 2000c and unpublished data). These findings suggest that the phosphorylation of FKHRL1 may play a role in IGF-1 mediated survival of neuronal cells (Fig. 2). Hence, the survival effect of IGF-1 in PC12 cells and primary cultured neurons is likely related to their effects on Akt kinase and the phosphorylation of death inducing factors such as the Forkhead transcription factors.

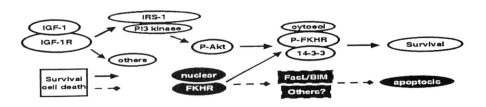

Figure 2. Survival pathway of PI3K/Akt/FKHRL1 activated by IGF-1 in neuronal cells.

7. SUMMARY AND FUTURE STUDIES

The Forkhead family of transcription factors are proapoptotic proteins which are able to induce cell death in the absence of survival factors. In the presence of survival factors such as IGF-1, these transcription factors are phosphorylated by Akt kinases and their proapoptotic property is hence blocked, leading to cell survival (Fig. 2). These findings provide a new mechanism to explain the survival and degeneration of neuronal cells which may have important implications for the treatment of neurodegenerative conditions. It will now be important to investigate the physiological relevance of endogenous forkhead proteins using knockout animal models. Moreover, it will be of interest to uncover the target genes of the forkhead proteins to obtain new insights into the functions and specificity of these unique transcription factors.

REFERENCES

Alessi, D. R., Andjelkovic, M., Caudwell, B., Cron , P., Morrice, N., Cohen, P., and Hemmings, B. A. 1996a, Mechanism of activation of protein kinase B by insulin and IGF-1, *EMBO J.* **15**: 6541-6551.

Alessi, D. R., Caudwell, F. B., Andjelkovic, M., Hemmings, B. A., and Cohen P. 1996b, Molecular basis for the substrate specificity of protein kinase B; comparison with MAPKAP kinase-1 and p70 S6 kinase, *FEBS Lett.* **399**: 333-338.

Alessi, D. R., James, S. R., Downes, C. P., Holmes, A. B., Gaffney, P. R., Reese, C. B., and Cohen, P. 1997, Characterization of a 3-phosphoinositide-dependent protein kinase which phosphorylates and activates protein kinase Balpha, *Curr Biol.* **7**: 261-9.

Anderson, M. J., Viars, C. S., Czekay, S., Cavenee, W. K., and Arden, K. C. 1998, Cloning and characterization of three human forkhead genes that comprise an FKHR-like gene subfamily, *Genomics.* **47**: 187-199.

Balendran, A., A. Casamayor, M. Deak, A. Paterson, P. Gaffney, R. Currie, C. P. Downes, and Alessi, D. R. 1999, PDK1 acquires PDK2 activity in the presence of a synthetic peptide derived from the carboxyl terminus of PRK2, *Curr. Biol.* **9**: 393-404.

Beilharz, E. J., Russo, V. C., Butler, G., Baker, N. L., Connor, B., Sirimanne, E. S., Dragunow, M., Werther, G. A., Gluckman, P. D., Williams, C. E., et al., 1998, Co-ordinated and cellular specific induction of the components of the IGF/IGFBP axis in the rat brain following hypoxic-ischemic injury. *Brain Res. Mol. Brain Res.* **59**:119-134.

Brunet, A., Bonni, A., Zigmond, M. J., Lin, M. Z., Juo, P., Hu, L. S., Anderson, M. J., Arden, K. C., Blenis, J., and Greenberg, M. E. 1999, Akt promotes cell survival by phosphorylating and inhibiting a Forkhead transcription factor, *Cell* **96**: 857-868.

Butler, A. A., Yakar, S., Gewolb, I. H., Karas, M., Okubo, Y., and LeRoith, D. 1998, Insulin-like growth factor-I receptor signal transduction: at the interface between physiology and cell biology, *Comp Biochem.Physiol B Biochem.Mol.Biol.* **121**: 19-26.

Cardone, M. H., Roy, N., Stennicke, H. R., Salvesen, G. S., Franke, T. F., Stanbridge, E., Frisch, S., and Reed, J. C. 1998, Regulation of cell death protease caspase-9 by phosphorylation, *Science* **282**: 1318-1321.

Cheng, B., and Mattson, M. P., 1992, IGF-I and IGF-II protect cultured hippocampal and septal neurons against calcium-mediated hypoglycemic damage, *J. Neurosci.* **12**:1558-1566.

Datta S. R., Brunet, A., and Greenberg, M. E., 1999, Cellular survival: a play in three Akts, *Genes Dev.* **13**:2905-2927.

Delcommenne, M., C. Tan, V. Gray, L. Rue, J. Woodgett, and Dedhar, D. 1998, Phosphoinositide-3-OH kinase-dependent regulation of glycogen synthase kinase 3 and protein kinase B/AKT by the integrin-linked kinase, *Proc. Natl. Acad. Sci. USA.* **95**: 11211-11216.

del Peso, L., M. Gonzalez-Garcia, C. Page, R. Herrera, and Nunez G. 1997, Interleukin-3-induced phosphorylation of BAD through the protein kinase Akt, *Science.* **278**: 687-689.

de Pablo, F., and de la Rosa, E. J., 1995, The developing CNS: a scenario for the action of proinsulin, insulin and insulin-like growth factors, *Trends Neurosci.* **18**:143-150.

Dore, S., Kar, S., and Quirion, R., 1997, Insulin-like growth factor I protects and rescues hippocampal neurons against -amyloid- and human amylin-induced toxicity, *Proc. Natl. Acad. Sci. USA.* **94**: 4772-4777.

Galli, C., Meucci, O., Scorziello, A., Werge, T. M., Calissano, P. and Schettini, G., 1995, Apoptosis in cerebellar granule cells is blocked by high KCl, forskolin, and IGF-1 through distinct mechanisms of action: the involvement of intracellular calcium and RNA synthesis, *J. Neurosci.* **15**: 1172-1179.

Gluckman, P. D., Guan, J., Williams, C., Scheepens, A., Zhang, R., Bennet, L., and Gunn, A., 1998, Asphyxial brain injury--the role of the IGF system, *Mol. Cell. Endocrinol.* **140**: 95-99.

Fujita, E., Jinbo, A., Matuzaki, H., Konishi, H., Kikkawa, U., and Momoi, T. 1999, Akt phosphorylation site found in human caspase-9 is absent in mouse caspase-9, *Biochem.Biophys.Res.Commun.* **264**: 550-555.

Guo, S., Rena, G., Cichy, S., He, X., Cohen, P., and Unterman, T. 1999, Phosphorylation of serine 256 by protein kinase B disrupts transactivation by FKHR and mediates effects of insulin on insulin-like growth factor-binding protein-1 promoter activity through a conserved insulin response sequence, *J.Biol.Chem.* **274**: 17184-17192.

Kaplan, D. R. and Miller, F. D. 2000, Neurotrophin signal transduction in the nervous system., *Curr.Opin.Neurobiol.* **10**: 381-391.

Kaufmann, E. and Knochel, W. 1996, Five years on the wings of fork head, *Mech.Dev.* **57**: 3-20.

Kops, G. J., and Burgering, B. M. 1999, Forkhead transcription factors: new insights into protein kinase B (c-akt) signaling, *J. Mol. Med.* **77**: 656-665.

Lai, E., Clark, K. L., Burley, S. K., and Darnell, J. E., Jr. 1993, Hepatocyte nuclear factor 3/fork head or "winged helix" proteins: a family of transcription factors of diverse biologic function, *Proc.Natl.Acad.Sci.U.S.A.* **90**: 10421-10423.

LeRoith, D., Werner, H., Beitner-Johnson, D., and Roberts, C. T. 1995, Molecular and cellular aspects of the insulin-like growth factor I receptor, *Endocr.Rev.* **16**: 143-163.

Loddick, S. A., Liu, X.-J., Lu, Z.-X., Liu, C., Behan, D. P., Chalmers, D. C., Foster, A. C., Vale, W. W., Ling, N., and De Souza, E. B., 1998, Displacement of insulin-like growth factors from their binding proteins as a potential treatment for stroke, *Proc. Natl. Acad. Sci. USA.* **95**: 1894-1898.

Loddick, S. A., and Rothwell, N. J., 1999, Mechanisms of TNF alpha action on neurodegeneration: interaction with insulin-like growth factor-1. *Proc Natl Acad Sci U S A..* **96**: 9449-9451.

Miller, T. M., Moulder, K. L., Knudson, C. M., Creedon, D. J., Deshmukh, M., Korsmeyer, S. J., and Johnson, E. M., Jr. 1997, Bax deletion further orders the cell death pathway in cerebellar granule cells and suggests a caspase-independent pathway to cell death, *J.Cell Biol.* **139**: 205-217.

Nakae, J., Park, B.C., and Accili, D. 1999, Insulin stimulates phosphorylation of the forkhead transcription factor FKHR on serine 253 through a Wortmannin-sensitive pathway, *J.Biol.Chem.* **274**: 15982-15985.

Rena, G., S. Guo, S.C. Cichy, T.G. Unterman, and Cohen, P. 1999, Phosphorylation of the transcription factor forkhead family member FKHR by protein kinase B, *J. Biol. Chem.* **274**:17179-17183.

Rane, M. J., Coxon, P. Y., Powell, D. W., Webster, R., Klein, J. B., Ping, P., Pierce, W., and McLeish, K. R. 2000, P38 kinase-dependent MAPKAPK-2 activation functions as PDK2 for Akt in human neutrophils, *J.Biol.Chem.* In press.

Sortino, M. A., and Canonico, P. L., 1996, Neuroprotective effect of insulin-like growth factor I in immortalized hypothalamic cells. *Endocrinology.* **137**: 1418-1422.

Tang, E. D., Nunez, G., Barr, F. G., and Guan, K. L. 1999, Negative regulation of the forkhead transcription factor FKHR by Akt, *J.Biol.Chem.* **274**: 16741-16746.

Toker A. and Newton A.C. 2000, Akt/protein kinase B is regulated by autophosphorylation at the hypothetical PDK-2 site. *J.Biol.Chem.* **275**: 8271-8274.

Vaux, D. L., and Korsmeyer, S. J. 1999, Cell death in development, *Cell.* **22**: 245-254.

Zheng, W.-H., Kar, S., Dore, S., and Quirion, R. 2000a, Insulin-like growth factor-1 (IGF-1): a neuroprotective trophic factor acting via the Akt kinase pathway, *J.Neural Transm. Suppl*, 261-272.

Zheng, W.-H., Kar, S., Dore, S., and Quirion, R. 2000b, Stimulation of protein kinase C modulates insulin-like growth factor-1-induced akt activation in PC12 cells, *J Biol Chem.* **275**: 13377-13385.

Zheng, W.-H., Kar, S., and Quirion, R. 2000c, Insulin-like growth factor-1-induced phosphorylation of the forkhead family transcription factor FKHRL1 is mediated by akt kinase in PC12 cells, *J.Biol.Chem.* **275**: 39152-39158.

DEVELOPMENT OF ANTI-DEMENTIA DRUGS FOR ALZHEIMER'S DISEASE: PRESENT AND FUTURE

Toshitaka Nabeshima[*] and Kiyofumi Yamada

1. INTRODUCTION

Alzheimer's disease (AD) is a neurodegenerative disorder that is neuropathologically characterized by the presence of numerous senile plaques and neurofibrillary tangles accompanied by neuronal loss. The extracellular senile plaques are composed of amyloid β-peptides (Aβ), 40-42 amino acid peptide fragments of the β-amyloid precursor protein (APP), whereas the intracellular neurofibrillary tangles are composed of highly phosphorylated tau proteins (Selkoe, 1994). Clinical manifestations of AD are primarily the progressive loss of memory and language. With disease progression, patients may have psychiatric and behavioral disturbances. In this article, we reviewed the recent progress in pharmacotherapy for AD, as well as possible future therapeutic strategies.

2. CHOLINERGIC THERAPY

Cholinergic neurons in the nucleus basalis of Meynert are reduced early in the course of the disease, and the dysfunction of cholinergic neurons is believed to be primarily responsible for cognitive deficits in AD (Coyle et al., 1983). Improvement of cognitive function in AD has been demonstrated with acetylcholinesterase (AChE) inhibitors. Tacrine was the first AChE inhibitor approved for the treatment of AD, and three other AChE inhibitors, including donepezil, rivastigmine and galantamine, have since been approved. Compared with tacrine, other AChE inhibitors produced mild adverse side effects and very low hepatotoxicity.

Direct activation of muscarinic and nicotinic cholinergic receptors has also been considered a therapeutic strategy in AD. Specific M1 receptor agonists could potentially

[*] T. Nabeshima, Department of Neuropsychopharmacology and Hospital Pharmacy, Nagoya University Graduate School of Medicine, Nagoya 466-8560, Japan

Mapping the Progress of Alzheimer's and Parkinson's Disease
Edited by Mizuno *et al.*, Kluwer Academic/Plenum Publishers, 2002

223

ameliorate cognition impairment without the peripheral side effects associated with stimulation of M2 and M3 receptors. Activation of nicotinic receptors with selective agonists such as GTS-21 may also provide some benefits in AD since it shows cognition-enhancing effects as well as neuroprotective effects in animal studies (Nabeshima and Yamada, 2001).

3. ANTI-INFLAMMATORY THERAPY

Inflammatory mechanisms contribute to the neurodestructive process occuring in the senile plaques in the AD brains (Aisen and Davis, 1994). The acute phase proteins, cytokines and other proteins associated with inflammation are found in brains of AD patients but not in brains from age-matched controls. The presence of complement components suggests that complement-mediated neuronal lysis occurs in brains of AD patients. Activated microglial cells which are associated with senile plaques may be the source of these mediators. Epidemiological studies demonstrated that previous use of corticosteroid or nonsteroidal anti-inflammatory drugs (NSAID) reduces the risk of AD. McGeer and Rogers (1992) have proposed anti-inflammatory therapy as a therapeutic approach to AD. In a clinical trial, indomethacin has been shown to protect mild to moderately impaired AD patients (Rogers et al., 1993). Recent studies suggested inhibitors for cyclooxygenase-2 (COX-2) and inducible nitric oxide synthase (iNOS) as potential drugs for AD (Tran et al., 2001).

4. ESTROGEN REPLACEMENT THERAPY

Epidemiological studies have indicated that the prevalence of AD after age 65 is two to three times higher in women than men, suggesting gender difference as a risk of AD. Studies also indicated that replacement therapy with estrogens in postmenopausal women delays the onset and decreases the risk of AD (Tang et al., 1996). It is demonstrated that estrogens modulate cholinergic neuronal activity, monoamine metabolism and the expression of brain-derived neurotrophic factor (BDNF) mRNA in the brain. Estrogens have also been shown to attenuate excitotoxicity, oxidative injury and $A\beta$ toxicity, and to reduce the neuronal degeneration of $A\beta$ (Simpkins et al., 1997)

5. ANTI-OXIDANT THERAPY

Several lines of evidence suggest that oxidative stress is important in the pathogenesis of AD and that antioxidants may protect neurons from $A\beta$-induced toxicity (Yamada et al., 1999; Yamada and Nabeshima, 2000) In clinical trials, a potent antioxidant, idebenone, showed a clear dose-related antidementia activity in AD (Gutzmann et al., 1997). Clinical trials with selegiline (L-deprenyl), a selective monoamine oxidase (MAO) B inhibitor, in AD found significant effects on cognition. A recent controlled trial showed that treatment with selegiline and/or α-tocopherol in patients with moderately severe impairment from AD slows the progression of disease (Sano et al., 1997).

6. ANTI-AMYLOID THERAPY

Accumulated evidence supports the idea that the formation and deposition of Aβ in the brain, especially the more hydrophobic/fibrillogenic 1-42 form of Aβ (Aβ1-42), are early and key pathological events in the development of AD. Mutations in all three genes known to cause familial AD (APP, presenilin 1 (PS1) and PS2) results in an increase in Aβ1-42 production relative to Aβ1-40 or the total Aβ levels. Transgenic mice with mutant forms of APP develop fibrillar plaques that contain predominantly Aβ1-42 and show many of the pathological characterizations associated with AD. Diffuse plaques, the earliest forms of senile plaques, consist exclusively of Aβ1-42. These findings support the amyloid cascade hypothesis of AD (Hardy, 1997), and the anti-amyloid therapy, which refers to intervention to stop the deposition of amyloid in the brain, may be effective for the treatment of AD. Recent progress in understanding the molecular mechanisms of cytotoxic Aβ formation offers a number of potential targets for novel therapeutic strategies in AD.

Firstly, inhibition of β- (BACE/Asp2; Hussain et al., 1999; Sinha et al., 1999; Vassar et al., 1999; Yan et al., 1999) and/or γ-secretases (PS1; Wolf et al., 1999) could decrease Aβ production. Alternatively, diversion of APP metabolism from an amyloidogenic to non-amyloidogenic processing pathway may be possible by using muscarnic agonists that increase the percentage of APP molecules cleaved by α-secretase. Thirdly, inhibition of the aggregation of Aβ, which leads to the formation of a cytotoxic form of Aβ, would make for a novel pharmacotherapy. Finally, activation of the enzyme activities, which are capable of degrading Aβ, may be a potential therapeutic strategy for preventing or slowing the pathogenetic mechanism of AD (Qiu et al., 1997; Iwata et al., 2000).

7. ANTI-TAU THERAPY

Neurofibrillary tangles consist of paired helical filaments (PHF) and related straight filaments. The major constituent of PHF is hyperphosphorylated tau. Aβ fibril induces tau phosphorylation, resulting in the loss of microtubule binding capacity and somatodendritic accumulation (Busciglio et al., 1995). The discovery of mutations in the tau gene in familial frontotemporal dementia with Parkinsonism linked to chromosome 17 (FTDP-17) has firmly established the relevance of tau pathology to the neurodegenerative dementing disorders (Spillantini and Goedert, 1998). Recently it was demonstrated that the prolyl isomerase Pin 1 can restore the ability of phosphorylated tau to bind microtubules and promote microtubule assembly in vitro, providing a new insight into the pathogenesis of AD (Lu et al., 1999)

8. NEUROTROPHIN THERAPY

Neurotrophins, such as nerve growth factor (NGF) and BDNF, promote the maintenance and survival of numerous peripheral and CNS neurons. NGF provides a trophic support on cholinergic neurons of the basal forebrain. Although it is not clear whether memory dysfunction and neuronal loss in AD are caused by a lack of NGF, NGF has been shown to prevent the loss of basal forebrain neurons following axotomy in experimental animals. Clinically, three AD patients were treated with NGF administered continuously into the lateral cerebral ventricle. In this clinical research, no clear cognitive amelioration was demonstrated in any patient. An increase in nicotine binding was found in some brain areas in these patients 3 months after the end of NGF treatment. The amount of slow-wave cortical activity was reduced in the first 2 patients but not the third one. Two negative side effects, a dull constant back pain and a marked weight loss, accompanied NGF treatment. It was concluded that long-term i.c.v. NGF administration may cause certain potentially beneficial effects, but the i.c.v. route of administration is also associated with negative side effects that appear to outweigh the positive effects (Jönhagen et al., 1998). Several means of deliverying neurotrophins into the brain, as well as other alternative ways to provide neurotrophic support, have been proposed (Nabeshima and Yamada, 2000).

9. VACCINE THERAPY

A novel and exciting result of vaccine therapy in AD was reported by Schenk et al (1999). PDAPP tansgenic mice, which overexpress mutant human APP, were immunized with Aβ either before the onset of AD-like neuropathologies or at an older age when Aβ deposition and other neuropathological changes were established. Aβ immunization of the young animals, which produced high titres of serum antibody, almost completely prevented the development of Aβ deposition, dystrophic neurites, astrocytosis and microgliosis. The immunization of the older animals also markedly reduced the extent and progression of the AD-like brain lesions. The authors also showed that peripherally administered antibodies against Aβ1-42 enter the central nervous system and reduce pathology in a mouse model of AD (Bard et al., 2000). The vaccine therapy also prevents memory impairment in a mouse model of AD (Morgan et al., 2000).

10. GENE THERAPY

Gene therapy can deliver a single or several well-defined molecules in a highly specific spatial and temporal fashion. It includes ex vivo and in vivo approaches. In ex vivo gene transfer, cells genetically modified to secrete a certain protein are grafted into a discrete area of the brain. The in vivo approach is a direct gene transfer into the brain by the delivery of transgenes to the host cells. So far, studies have focused on the ex vivo approach because of its ability to generate various types of modified cells that can be characterized in vitro before use. Transplantation of fibroblasts, modified to secrete NGF, into the basal forebrain protected cholinergic neurons and promoted functional recovery

(Chen and Gage, 1995). Similarly, when primary fibroblasts, modified to express choline acetyltransferase (ChAT), the synthetic enzyme of ACh, were transplanted into the cerebral cortex, the learning and memory impairment caused by the basal forebrain lesion was ameliorated (Winkler et al., 1995).

11. ACKNOWLEDGMENTS

This study was supported, in part, by Grants-in-Aid for Science Research (No. 99187 and 12670085), and a COE Grant from the Ministry of Education, Culture, Sports, Science and Technology of Japan, and by an SRF Grant for Biomedical Research.

REFERENCES

Aisen, P.S. and Davis, K.L., 1994, Inflammatory mechanisms in Alzheimer's disease: Implications for therapy, *Am. J. Psychiat.* **151**: 1105-1113.

Bard, F., Cannon,C., Barbour, R., Burke, R.-L., Games, D., Grajeda, H., Guido, T., Hu, K., Huang, J., Johnson-Wood, K., Khan, K., Kholodenko, D., Lee, M., Lieberburg, I., Motter, R., Nguyen, M., Soriano, F., Vasquez, N., Weiss, K., Welch, B., Seubert, P., Schenk, D. and Yednock, T., 2000, Peripherally administered antibodies against amyloid β-peptide enter the central nervous system and reduce pathology in a mouse model of Alzheimer disease, *Nat. Med.* **6**: 916-919.

Busciglio, J., Lorenzo, A., Yeh, J. and Yankner, B.A., 1995, β-Amyloid fibrils induce tau phosphorylation and loss of microtubule binding, *Neuron* **14**: 879-888.

Chen, K.S., and Gage, F.H., 1995, Somatic gene transfer of NGF to the aged brain: behavioral and morphological amelioration, *J. Neurosci.* **15**: 2819-2825.

Coyle, J.T., Price, D.L., and DeLong, M.R., 1983, Alzheimer's disease: a disorder of central cholinergic innervation, *Science* **219**: 1184-1190.

Gutzmann, H., Hadler, D. and Erzigkeit, H., 1997, Long-term treatment of Alzheimer's disease with idebenone, in: Alzheimer's Disease: *Biology, Diagnosis and Therapeutics*, K. Iqbal, B. Winblad, T. Nishimura, M. Takeda and H.M. Wisniewski, ed., John Wiley & Sons, Chichester, pp. 687-705.

Hardy, H. 1997, Amyloid, the presenilins and Alzheimer's disease, *Trends Neurosci.* **20**: 154-159.

Hussain, I., Powell, D., Howlett, D. R., Tew, D. G., Meek, T. D., Chapman, C., Gloger, I. S., Murphy, K. E., Southan, C. D., Ryan, D. M., Smith, T. S., Simmons, D. I., Walsh, F. S., Dingwall, C. and Christie, G., 1999, Identification of a novel aspartic protease (Asp 2) as β-secretase, *Mol. Cell. Neurosci.* **14**: 419-427.

Iwata, N., Tsubuki, S., Takaki, Y., Watanabe, K., Skiguchi, M., Hosoki, E., Kawashima-Morishita, M., Lee, H.J., Hama, E., Sekine-Aizawa, Y. and Saido, T.C., 2000, Identification of the major Aβ1-42-degrading catabolic pathway in brain parenchyma: suppression leads to biochemical and pathological deposition, *Nat. Med.*, **6**: 718-719.

Jönhagen, M.E., Nordberg, A., Amberla, K., Bäckman, L., Ebendal, T., Meyerson, B., Olson, L., Seiger, A., Shigeta, M., Theodorsson, E., Viitanen, M., Winblad, B., and Wahlund, L.-O., 1998, Intracerebroventricular infusion of nerve growth factor in three patients with Alzheimer's disease, *Dement. Geriatr. Cogn. Disord.* **9**: 246-257.

Lu, P.-J., Wulf, G., Zhou, X.Z., Davies, P. and Lu, K.P. 1999, The prolyl isomerase Pin 1 restores the function of Alzheimer-associated phosphorylated tau protein, *Nature* **399**: 784-788.

McGeer, P.L. and Rogers, J., 1992, Anti-inflammatory agents as a therapeutic approach to Alzheimer's disease, *Neurology* **42**: 447-449.

Morgan, D., Diamond, D.M., Gottschall, P.E., Ugen, K.E., Dickey, C., Hardy, J., Duff, K., Jantzen, P., DiCarlo, G., Wilcock, D., Connor, K., Hatcher, J., Hope, C., Gordon, M. and Arendash, G.W., 2000, A beta peptide vaccination prevents memory loss in an animal model of Alzheimer's disease, *Nature* **408**: 982-985.

Nabeshima, T. and Yamada, K., 2000, Neurotrophic factor strategies for the treatment of Alzheimer's disease, *Alz. Dis. Assoc. Disor.* **14** (Suppl 1): S39-S46.

Nabeshima, T. and Yamada, K., 2001, New therapeutic approaches to Alzheimer's disease, in: *Neuroscientific Basis of Dementia*, C. Tanaka, P.L. McGeer and Y. Ihara, ed., Birkhäuser Verlag, Basel, pp. 287-294.

Qiu, W.Q., Ye, Z., Kholodenko, D., Seubert, P. and Selkoe, D.J., 1997, Degradation of amyloid β-protein by a metalloprotease secreted by microglia and other neural and non-neural cells, *J. Biol. Chem.* **272**: 6641-6646

Rogers, J., Kirby, L.C., and Hempelman, S.R., 1993, Clinical trial of indomethacin in Alzheimer's disease, *Neurology* **43**: 1609-1611.

Sano, M., Ernesto, C., Thomas, R.G., Klauber, M.R., Schafer, K., Grundman, M., Woodbury, P., Growdon, J., Cotman, C.W., Pfeiffer, E., Schneider, L.S. and Thal, L.J., 1997, A controlled trial of selegiline, alpha-tocopherol, or both as treatment for Alzheimer's disease, *New Eng. J. Med.* **336**: 1216-1222.

Selkoe, D.J., 1994, Cell biology of amyloid β-protein precursor and the mechanism of Alzheimer's disease, *Annu. Rev. Cell Biol.* **10**: 373-403.

Schenk, D., Barbour, R., Dunn, W., Gordon, G., Grajeda, H., Guido, T., Hu, K., Huang, J,. Jonson-Wood, K., Khan, K., Kholodenko, D., Lee, M., Liao, Z., Lieburg, I., Motter, R., Mutter, L., Soriano, F., Shopp, G., Vasquez, N., Vandervert, C., Walker, S., Wogulis, M., Yednock, T, Games, D. and Seubert, P., 1999, Immunization with amyloid-β attenuates Alzheimer-disease-like pathology in the PDAPP mouse, *Nature* **400**: 173-177.

Simpkins, J.W., Green, P., Gridley, K.E., Singh, M., de Fiebre, N.C. and Rajakumar, G., 1997, Role of estrogen replacement therapy in memory enhancement and prevention of neuronal loss associated with Alzheimer's disease, *Am. J. Med.* **103** (3A): 19S-25S.

Sinha, S., Anderson, J. P., Barbour, R., Basi, G. S., Caccavello, R., Davis, D., Doan, M., Dovey, H. F., Frigon, N., Hong, J., Jacobson-Croak, K., Jewett, N., Keim, P., Knops, J., Lieburg, I., Power, M., Tan, H., Tatsuno, G., Tung, J., Schenk, D., Seubert, P., Suomensaari, S. M., Wang, S., Walker, D., Zhao, J., McConlogue, L. and John, V., 1999, Purification and cloning of amyloid precursor protein β-secretase from human brain, *Nature* **402**: 537-540.

Spillantini, M.G. and Goedert, M., 1998, Tau protein pathology in neurodegenerative diseases, *Trends Neurosci.* **21**: 428-433.

Tang, M.-X., Jacobs, D., Stern, Y., Marder, K., Schofield, P., Gurland, B., Andrews, H. and Mayeux, R., 1996, Effects of oestrogen during menopause on risk and age at onset of Alzheimer's disease, *Lancet* **348**: 429-432.

Tran, M.H., Yamada, K., Olariu, A., Mizuno, M., Ren, X.H. and Nabeshima, T., 2001, Amyloid β-peptide induces nitric oxide production in rat hippocampus: association with cholinergic dysfunction and amelioration by inducible nitric oxide inhibitors, *FASEB J.* in press.

Vassar, R., Bennett, B. D., Babu-Khan, S., Kahn, S., Mendiaz, E. A., Denis, P., Teplow, D. B., Ross, S., Amarante, P., Loeloff, R., Luo, Y., Fisher, S., Fuller, J., Edenson, S., Lile, J., Jarosinski, M. A., Biere, A. L., Curran, E., Burgess, T., Louis, J. -C., Collins, F., Treanor, J., Rogers, G. and Citron, M., 1999, β-Secretase cleavage of Alzheimer's amyloid precursor protein by the transmembrane aspartic protease BACE, *Science* **286**: 735-741.

Winkler, J., Suhr, S.T., Gage, F.H., Thal, L.J. and Fisher, L.J., 1995, Essential role of neocortical acetylcholine in spatial memory, *Nature* **375**: 484-487.

Wolf, M. S., Xia W., Ostaszewski, B. L., Diehl, T. S., Kimberly, W. T. and Selkoe, D. J., 1999, Two transmembrane aspartates in presenilin-1 required for presenilin endoproteolysis and γ-secretase activity, *Nature* **398**: 513-517.

Yamada, K., Tanaka, T., Han, D., Senzaki, K., Kameyama, T. and Nabeshima, T., 1999, Protective effects of idebenone and α-tocopherol on β-amyloid-(1-42) induced learning and memory deficits in rats: implication of oxidative stress in β-amyloid-induced neurotoxicity in vivo. *Eur. J. Neurosci.* **11**: 83-90.

Yamada, K. and Nabeshima, T., 2000, Animal models of Alzheimer's disease and evaluation of anti-dementia drugs, *Pharmacol. Therapeut.* **88**: 93-113.

Yan, R., Bienkowski, M. J., Shuck, M. E., Miao, H., Tory, M. C., Pauley, A. M., Brashler, J. R., Stratman, N. C., Mathews, W. R., Buhl, A. E., Carter, D. B., Tomaselli, A. G., Parodi, L. A., Heinrikson, R. L. and Gurney, M. E., 1999, Membrane-anchored aspartyl protease with Alzheimer's disease β-secretase activity, *Nature* **402**: 533-537.

HUMAN NEURAL STEM CELL TRANSPLANTATION IMPROVES COGNITIVE FUNCTION OF AGED BRAIN

Possibility of Neuroreplacement Therapy

Kiminobu Sugaya, Tingyu Qu, Hojoong M. Kim and Christopher L. Brannen

1. INTRODUCTION

The discovery of multipotent neural stem cells (NSCs) in the adult brain[1,7,8,11] has brought revolutionary changes in the theory on neurogenesis, a theory that now suggests that regeneration of neurons can occur throughout life. To further this revolution, we have recently shown that Human NSCs (HNSCs) differentiated and survived more than 3 weeks *in vitro* without the addition of any supplements or exogenous factors[5]. This result suggests that HNSCs are capable of producing endogenous factors necessary for their own survival and neuronal differentiation. Together, these recent findings stimulated us to investigate the transplantation of HNSCs to determine whether or not the aged brain will provide the necessary environment needed for a successful HNSCs transplantation. HNSCs not only expanded more than a year *in vitro* but survived 30 days after xenotransplantation, retaining both multipotency and migratory capacity, but more remarkably, the HNSCs transplantation improved the cognitive function in 24 months old rats[13].

2. TRANSPLANTATION OF HNSCS AND MORRIS WATER MAZE TEST

The HNSCs expanded without differentiation under the influence of mitogenic factors in supplemented serum free media[5]. To differentiate between host and transplanted cells, the nuclei of the HNSCs were labeled by the incorporation of bromodeoxyuridine (BrdU) into the DNA. These labeled cells were subsequently injected unilaterally into the lateral ventricle of matured (6 month old) and aged (24 month old) rats. The cognitive function of the animal was assessed by the Morris water maze before and 4 weeks after the transplantation of HNSCs. Before the HNSCs transplantation, some

Department of Psychiatry, University of Illinois at Chicago, The Psychiatric Institute, Room #232, 1601 West Taylor Street, Chicago, IL 60612

Mapping the Progress of Alzheimer's and Parkinson's Disease
Edited by Mizuno *et al.*, Kluwer Academic/Plenum Publishers, 2002

of the aged animal (aged memory unimpaired animal) had cognitive function in the range of the matured animal, while others (aged memory impaired animal) had cognitive function entirely out of the range of the matured animals. After the HNSCs transplantation, most of the aged animals had cognitive function in the range of the matured animals. Strikingly, one of the aged memory impaired animal displayed behavior that was dramatically better than the level of the matured animals (Fig. 1a). Statistical analysis showed that cognitive function significantly improved in both matured ($p < 0.001$, n=8) and aged memory impaired animals ($p < 0.001$, n=6). In contrast, no improvement in cognitive function was observed in vehicle injected control animals (n=6), or aged memory unimpaired animals (n=7) after the HNSCs transplantation (Fig. 1b). Three (3) out of the 13 aged animals showed deterioration of performance in the water maze after the HNSCs transplantation. This fact needs to be further analyzed, but this may be due to the deterioration of the physical strength of these animals during the experimental period. There are two possible mechanisms to explain the beneficial effects of the HNSCs transplantation to cognitive function of the host brain. One is replacement or augmentation of neuronal circuit by the HNSCs derived neurons and other is the neurotrophic action of factors released from the transplanted HNSCs. Although the following morphological study shows extensive incorporation of HNSCs and massive growth of neuronal fibers in the host brain area related to spatial memory task, HNSCs may still migrate toward the damaged neurons and rescue them by the production of neurotrophic factors. Therefore, a synergism between these two mechanisms may exist that allowed for the successful transplantation.

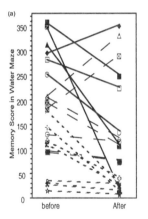
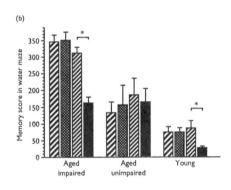

Fig. 1. Effect of HNSCs transplantation on memory score in the water maze. (a) Individual memory score before and after the transplantation shows improvement in the majority of the animals.——— : Aged memory impaired animals,——: Aged memory unimpaired animals, ---: Matured animals. (b) Mean of memory score in each animal group before (▨) and after ▦) HNSCs transplantation shows a significant improvement in aged memory impaired and young animals. While the animal received vehicle injection do not show significant difference in memory scores between before (▨) and after (■) the injection[13].

3. POSTMORTEM IMMUNOHISTOCHEMICAL ANALYSIS

To investigate the morphology and population of differentiated HNSCs, we further analyzed brain sections taken after the second water maze task by immunohistochemistry with cell specific markers. The transplanted HNSCs, with BrdU immunopositive nuclei, were stained for human βIII-tubulin and human glial fibrillary acidic protein (GFAP). The presence of these cell specific antigens indicates that the transplanted HNSCs successfully differentiated into neurons and astrocytes, respectively.

3.1. Neuronal differentiation of the transplanted HNSCs

Immunohistochemical analysis of brain sections revealed that the cells intensely and extensively positive for human βIII-tubulin staining had BrdU positive nuclei, indicting neuronal differentiation of the transplanted HNSCs (Fig. 2). Specifically, these cells are located primarily in the bilateral singular and parietal cortexes (layer II, IV and V) and hippocampus, brain areas known to be related to spatial memory [6,10,16]. Interestingly, the animals' spatial memory also improved as assessed by the Morris water maze. Thus, it is arguable that the HNSCs incorporation into these brain areas allowed for an improvement in spatial memory.

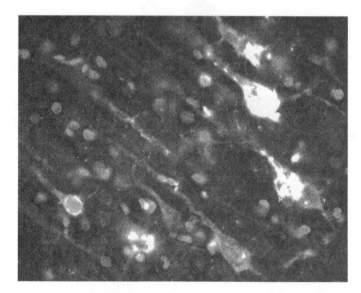

Fig. 2
βIII-tubulin immunopositive cells in the cortex shows pyramidal neuron like morphology. The size of the transplanted HNSCs is noticeably larger than the host (rat) cells.

Although physiological and   electromicroscopic investigation may required to determine the functional incorporation of these HNSCs derived neurons, the morphological observation indicates that functional association of these cells to the host brain occurred. Further histochemical analysis revealed that the βIII-tubulin-positive donor-derived cells found in the cerebral cortex are characterized by having dendrites pointing to the edge of the cortex whereas in the hippocampus, donor-derived neurons exhibited morphologies with multiple processes and branches. These differential morphologies of the transplanted HNSCs in different brain regions indicate that site-specific differentiation of HNSCs occurs according to various factors expressed in each brain region.

3.2. Astroglial differentiation of the transplanted HNSCs

In addition, the transplanted HNSCs also differentiated into GFAP- and BrdU-immunopositive staining cells localized near the area where neuronal cells were found. Further analysis with double immunostaining revealed that donor-derived astrocytes co-localized with the neuronal fibers in the cortex layer III and CA2 region of hippocampus.

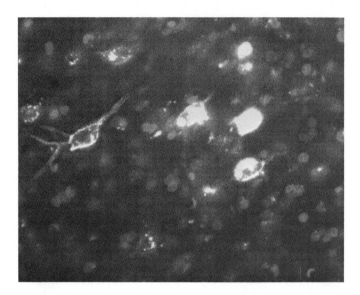

Fig. 3
GFAP immuno-positive cells form a layer in the cortex. The size of the transplanted HNSCs is noticeably larger than the host (rat) cells.

These donor-derived astrocytes were large compared to host glia, having cell bodies 8-10 μm in diameter with thick processes and BrdU immunopositive nuclei. We did not detect the above-mentioned morphologies and distribution of GFAP positive cells in the control rats that received no HNSC transplantation. The regions rich in astrocyte staining are also the same regions where the extensively stained neuronal fibers were identified. During development, glial cells have many complex functions such as neuronal and axonal guidance, and production of trophic factors[12]. It has been suggested that following transplantation, migrating glial cells guide and support the growth and extension of neuronal fibers[9]. However, other studies have argued that glial cells may be detrimental by forming an extensive interface between the host and graft[2,19]. Although the mechanism(s) of glia-neuron interaction in the HNSCs-transplanted host brain is not well understood, this overlapping distribution of glial and neuronal fibers strongly suggests that this interaction plays a pivotal role in the survival, migration, and differentiation of transplanted HNSCs.

4. OTHER CONSIDERATIONS

The most significant difference in our experimental procedure compare to the previous studies is the lateral intra ventricular transplantation of undifferentiated HNSCs in the form of neuro-spheroids. While many studies were done with intra-tissue injection of dissociated and partially differentiated NSCs[3,4,14,15,17], we employed spheroid injection because the dissociation of neuro-spheroids is known to cause immediate senescence of NSCs and increase the vulnerability of NSCs in culture[18]. Another added benefit of intraventricular injection is that since there is less tissue destruction, it may induce less recruitment of immune cells by the host. This is evidenced by the lack of ventricular distortion, and increased host astrocyte staining. The mechanism behind transplanted cell migration into the brain through the ventricle as of yet is still unknown. However, our results indicate that the mechanism may lie behind a direct integration into the host brain. Specifically, immunohistochemical analysis revealed that some of the BrdU cells were found to be situated along the lateral ventricular wall while a few appeared to have integrated directly into the cells lining the ventricle. Similar observations were reported in a variety of studies using neuronal transplantation to the lateral ventricle of animals. The intra-ventricular transplantation used in this study may provide an alternative route to the site-specific injection by which the grafted cells may gain access to various structures by the flow of CSF.

Following immunohistochemistry, a symmetrical distribution of neurons and astrocytes at both sides of the host brain was observed, indicating that the progeny of these HNSCs have a great potential for migration. This may be due to the fact that we transplanted undifferentiated HNSCs and such an immature stage for both glia and neuron possesses the potential to migrate over long distances. The extensive incorporation of neuronal and glial cells found in the cortex and other sub-regions of the hippocampus may be interpreted as evidence for the significance of local cues or signals within these regions which enable these grafted NSCs to migrate. It remains to be demonstrated, however, to what extent these newly formed neurons can undergo complete maturation with physiologically functional connections to the host brain. In a previous study, we observed that initial glial differentiation of HNSCs was followed by neuronal differentiation in a serum-free culture media without any additional factor[5]. This finding suggests that glial differentiation caused by the serum deprivation produced factors that allowed neurons to differentiate. Since we observed an association of astrocytes and neurons derived from HNSCs in this study, we may have to consider the possibility that the donor astrocytes may direct the neuronal differentiation.

5. CONCLUSION

In order to facilitate therapeutic HNSC application to the general adverse consequences of aging and neurodegenerative diseases, it is important to understand these environmental factors which direct the differentiation fate of these HNSCs to diverse lineage *in vivo*. While future studies are needed to elucidate these environmental factors, we have nonetheless demonstrated that HNSCs transplantation into the brains of aged memory impaired rats significantly improved their cognitive function. Moreover, not only did the HNSCs successfully differentiate into neurons and astrocytes, but more importantly, both neurons and astrocytes migrated to the cortex and hippocampus in a symmetrical, well-defined and organized pattern in the adult brain.

6. ACKNOWLEDGEMENT

The author would like to thank Andrew K. Sugaya for his technical assistance.

REFERENCES

[1]   Alvarez-Buylla, A. and Kirn, J.R., Birth, migration, incorporation, and death of vocal control neurons in adult songbirds, *J Neurobiol*, **33** (1997) 585-601.
[2]   Azmitia, E.C. and Whitaker, P.M., Formation of a glial scar following microinjection of fetal neurons into the hippocampus or midbrain of the adult rat: an immunocytochemical study, *Neurosci Lett*, **38** (1983) 145-150.
[3]   Benninger, Y., Marino, S., Hardegger, R., Weissmann, C., Aguzzi, A. and Brandner, S., Differentiation and histological analysis of embryonic stem cell- derived neural transplants in mice, *Brain Pathol*, **10** (2000) 330-341.
[4]   Blakemore, W.F. and Franklin, R.J., Transplantation options for therapeutic central nervous system remyelination, *Cell Transplant*, **9** (2000) 289-294.
[5]   Brannen, C.L. and Sugaya, K., In vitro differentiation of multipotent human neural progenitors in serum-free medium, *Neuroreport*, **11** (2000) 1123-1128.
[6]   Cassel, J.C., Duconseille, E., Jeltsch, H. and Will, B., The fimbria-fornix/cingular bundle pathways: a review of neurochemical and behavioural approaches using lesions and transplantation techniques, *Prog Neurobiol*, **51** (1997) 663-716.
[7]   Goldman, S.A. and Nottebohm, F., Neuronal production, migration, and differentiation in a vocal control nucleus of the adult female canary brain, *Proc Natl Acad Sci U S A*, **80** (1983) 2390-2394.
[8]   Gould, E., Reeves, A.J., Graziano, M.S. and Gross, C.G., Neurogenesis in the neocortex of adult primates [see comments], *Science*, **286** (1999) 548-552.
[9]   Isacson, O., Deacon, T.W., Pakzaban, P., Galpern, W.R., Dinsmore, J. and Burns, L.H., Transplanted xenogeneic neural cells in neurodegenerative disease models exhibit remarkable axonal target specificity and distinct growth patterns of glial and axonal fibres, *Nat Med*, **1** (1995) 1189-1194.
[10] Kesner, R.P., Behavioral analysis of the contribution of the hippocampus and parietal cortex to the processing of information: interactions and dissociations [In Process Citation], *Hippocampus*, **10** (2000) 483-490.
[11] Paton, J.A. and Nottebohm, F.N., Neurons generated in the adult brain are recruited into functional circuits, *Science*, **225** (1984) 1046-1048.
[12] Pundt, L.L., Kondoh, T. and Low, W.C., The fate of human glial cells following transplantation in normal rodents and rodent models of neurodegenerative disease, *Brain Res*, **695** (1995) 25-36.
[13] Qu, T., Brannen, C.L., Kim, H.M. and Sugaya, K., Human neural stem cells improve cognitive function of aged brain, *Neuroreport*, **12** (2001) 1127-1132.
[14] Rosser, A.E., Tyers, P. and Dunnett, S.B., The morphological development of neurons derived from EGF- and FGF-2- driven human CNS precursors depends on their site of integration in the neonatal rat brain, *Eur J Neurosci*, **12** (2000) 2405-2413.
[15] Rubio, F.J., Bueno, C., Villa, A., Navarro, B. and Martinez-Serrano, A., Genetically perpetuated human neural stem cells engraft and differentiate into the adult mammalian brain, *Mol Cell Neurosci*, **16** (2000) 1-13.
[16] Sugaya, K., Greene, R., Personett, D., Robbins, M., Kent, C., Bryan, D., Skiba, E., Gallagher, M. and McKinney, M., Septo-hippocampal cholinergic and neurotrophin markers in age-induced cognitive decline, *Neurobiol Aging*, **19** (1998) 351-361.
[17] Svendsen, C.N., Caldwell, M.A., Shen, J., ter Borg, M.G., Rosser, A.E., Tyers, P., Karmiol, S. and Dunnett, S.B., Long-term survival of human central nervous system progenitor cells transplanted into a rat model of Parkinson's disease, *Exp Neurol*, **148** (1997) 135-146.
[18] Svendsen, C.N., ter Borg, M.G., Armstrong, R.J., Rosser, A.E., Chandran, S., Ostenfeld, T. and Caldwell, M.A., A new method for the rapid and long term growth of human neural precursor cells, *J Neurosci Methods*, **85** (1998) 141-152.
[19] Zhou, F.C., Buchwald, N., Hull, C. and Towle, A., Neuronal and glial elements of fetal neostriatal grafts in the adult neostriatum, *Neuroscience*, **30** (1989) 19-31.

BiP/GRP78-INDUCED MICROGLIAL ACTIVATION AND INCREASE OF Aβ CLEARANCE

Yoshihisa Kitamura, Jun-ichi Kakimura, Takashi Taniguchi,
Shun Shimohama, and Peter J. Gebicke-Haerter*

1. INTRODUCTION

One hallmark of Alzheimer's disease (AD) is the accumulation of fibrillar amyloid-β peptides (Aβ), these aggregations and formation of plaques in the brain (Selkoe, 1999). Therefore, it is important to understand the balance of production and clearance of Aβ. However, the regulation of Aβ clearance is not fully understood. Missense mutations in the presenilin-1 (PS1) gene, that link to some early-onset cases of the familial Alzheimer's disease (FAD), increase the production of Aβ(1-42) (Duff et al., 1996), the vulnerability to neuronal apoptosis through increase of Ca^{2+} release from the endoplasmic reticulum (ER) (Mattson et al., 2000), and the downregulation in the signaling of unfolded protein response (UPR) (Katayama et al., 1999). These observations suggest that dysfunction of the ER may participate in the pathogenesis in AD brains.

The immunoglobulin heavy chain-binding protein (BiP, also known as a 78-kDa of glucose-regulated protein: GRP78) is induced in the ER by stressful conditions that activate the pathway of UPR signaling involving ER-resident transmembrane kinases, IRE1 and PERK (Kaufman, 1999). Recent studies suggest ER stress-induced apoptosis is mediated by activation of caspase-12 that is localized in the ER (Nakagawa et al., 2000) or of c-Jun N-terminal kinases that is associated with IRE1 and TRAF2 (Urano et al., 2000). BiP/GRP78 is expressed by the UPR and may act to alleviate the ER stress and/or to prevent apoptosis (Kaufman, 1999). Therefore, BiP/GRP78 is considered to play a molecular chaperone in the ER for naive, aberrantly folded or mutated proteins, and also cytoprotection against stressful conditions (Kaufman, 1999). However, other functions of BiP/GRP78 are unknown. Therefore, we examined effects of exogenous BiP/GRP78 on cytokine production and Aβ clearance in the microglia (Kakimura et al., 2001).

* Y. Kitamura, J. Kakimura, T. Taniguchi: *Department of Neurobiology, Kyoto Pharmaceutical University, Kyoto 607-8412, Japan;* S. Shimohama: *Department of Neurology, Graduate School of Medicine, Kyoto University, Kyoto 606-8507, Japan;* P.J. Gebicke-Haerter: *Department of Psychopharmacology, Central Institute of Mental Health, Mannheim D-68159, Germany.*

Mapping the Progress of Alzheimer's and Parkinson's Disease
Edited by Mizuno *et al.*, Kluwer Academic/Plenum Publishers, 2002

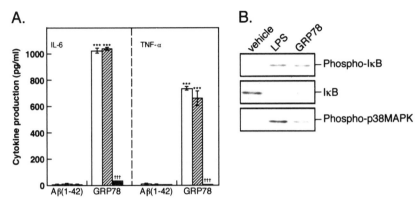

Fig. 1. BiP/GRP78-induced cytokine production from rat microglia and kinase activation. **A:** Isolated rat microglia were treated with 0.13 μM BiP/GRP78 or 10 μM Aβ(1-42) which were heat- (closed column) or vehicle-treated (open column), in the presence (hatched column) or absence of 10 μg/ml polymyxin B. After 24 h, the amount of IL-6 and TNF-α in the culture medium was determined by ELISA. **B:** After rat glial cells were treated with 0.06 μM BiP/GRP78 for 60 min, the phosphorylation of IκB and p38MAPK, and IκB degradation were determined.

2. BiP/GRP78-INDUCED CYTOKINE PRODUCTION

Expression level of BiP/GRP78 protein in the microglia was lower than that in the neurons. Therefore, we exogenously added recombinant BiP/GRP78 protein into a culture of isolated microglia from rat and mouse brains. Exogenous BiP/GRP78 induced production of interleukin (IL)-6 and tumor necrosis factor (TNF)-α. To exclude a possibility of endotoxin contamination, sample of recombinant BiP/GRP78 was heat-treated at 100°C for 20 min before incubation with the microglia. Heat-treatment completely abolished the ability of BiP/GRP78 to cytokine production. In addition, polymyxin B, an endotoxin inhibitor, did not inhibit the production (Fig. 1A). On the other hand, heat-treatment of lipopolysaccharide (LPS) did not inhibit the production, but polymyxin B completely inhibited LPS-induced production (data not shown). These results suggest that microglial activation by exogenous BiP/GRP78 may not due to endotoxin contamination.

3. BiP/GRP78-INDUCED Aβ CLEARANCE

Although Aβ(1-42) alone did not induce cytokine production (Fig. 1A), Aβ(1-42) at low concentration (less than 1 μM) was bound and/or phagocytosed to the rat microglia (Fig. 2). Oligomer (such as dimer, trimer, tetramer, etc.) of bound and/or phagocytosed Aβ(1-42) were detected by an immunoblotting of anti-Aβ antibody (Fig. 2A). By the laser confocal microscopy, Aβ immunoreactivity was detected in small vesicle of few microglia which expressed membrane protein CD11b. Subsequently, the amount of Aβ(1-42) in the microglia was significantly decreased after 3 days (Fig. 2B). Thus, extracellular Aβ(1-42) is phagocytosed and scavenged by the microglia even in physiological condition. At that time, exogenous BiP/GRP78 significantly induced the increase after 1 day and then the decrease after 3 days in the mount of Aβ(1-42). However, IL-6 and TNF-α did not change Aβ phagocytosis. Thus, BiP/GRP78 may facilitate both phagocytosis and clearance of Aβ(1-42) in the microglia, which are not mediated by produced cytokines such as IL-6 and TNF-α.

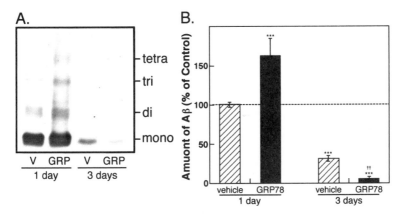

Fig. 2. BiP/GRP78-induced facilitation of Aβ(1-42) clearance. Rat microglia were incubated with 0.3 μM Aβ(1-42) in the presence or absence of 0.06 μM BiP/GRP78. After 1 and 3 days, the amount of Aβ(1-42) in the microglia were assessed. **A:** Oligomer such as mono-, di-, tri- and tetramer of Aβ(1-42) were detected after 1 day. **B:** Total amount of these oligomer of Aβ was measured by densitometric analysis. *** $p < 0.001$ vs. the vehicle value after 1 day. †† $p < 0.01$ vs. the vehicle value after 3 days.

4. MULTIFUNCTION OF BiP/GRP78 FOR NEUROPROTECTION

We have identified novel biological functions exerted by BiP/GRP78 that induced production of cytokines and enhanced phagocytosis of Aβ in the microglia. Interestingly, exogenous addition of recombinant BiP/GRP78 intensively increased its immunoreactivity in the perinuclear region and cytoplasm (Kakimura et al., 2001). Since BiP/GRP78 involves the carboxy-terminal sequence Lys-Asp-Glu-Leu (KDEL) that is the ER-retrieval signal, exogenous BiP/GRP78 may translocated to the ER. It is known that ER overload response (EOR) induces NF-κB activation and cytokine production (Kaufman, 1999). In our study, exogenous BiP/GRP78 induced the phosphorylation of IκB and p38 mitogen-activated protein kinase (MAPK) and then induced IκB degradation in rat glial cells, similar to LPS response (Fig. 1B). Thus, BiP/GRP78-induced microglial activation may be mediated by activation of NF-κB and p38MAPK. Recently, Katayama et al. reported that the level of BiP/GRP78 protein was reduced slightly in the brains of patients with sporadic AD and markedly in the brains with FAD linked to *PS1* mutations (Katayama et al., 1999). These observations suggest that: i) extracellular Aβ(1-42) is usually taken up by microglia, ii) Aβ(1-42) clearance is positively regulated by BiP/GRP78 which is expressed by ER stress in normal brain, and iii) the amount of Aβ(1-42) may be accumulated by downregulation of BiP/GRP78 in AD brains.

It is unknown whether BiP/GRP78 is extracellularly secreted *in vivo* brain. However, since BiP/GRP78 was constitutively and highly expressed in the normal neurons, we speculate that BiP/GRP78 may be extracellularly leaked from damaged, broken or dead neurons caused by neurodegeneration. IL-6 and TNF-α are known as proinflammatory cytokines in the immune system, furthermore, microglial activation and these cytokines are previously considered to participate the exacerbation of neurodegeneration (Dickson et al., 1993). In contrast, recent studies suggest that these cytokines may act as autocrine and paracrine mediators that induce glial proliferation and neuroprotective responses (Maeda et al., 1994; Barger et al., 1995; Sawada et al., 1995). Under ischemia-induced neurodegeneration, BiP/GRP78 functions as a chaperone for IL-6 (Hori et al., 1996). In addition, Aβ(1-42) could bind to BiP/GRP78. Thus, BiP/GRP78 has multifunction for neuroprotection such as chaperone, cytokine production and Aβ clearance.

It is known that Aβ clearance in the microglia is mediated by membrane receptors such as scavenger receptors (El Khoury et al., 1996; Paresce et al., 1996). In addition, anti-Aβ antibodies induced Fc receptor-mediated phagocytosis of Aβ (Bard et al., 2000). Thus, BiP/GRP78-induced clearance of Aβ in the microglia was a different mechanism from that of anti-Aβ antibodies. Therefore, BiP/GRP78 also may be other option of the target for AD therapy.

5. ACKNOWLEDGMENT

The present study was supported in part by the Frontier Research Program and Grants-in-Aid from the Ministry of Education, Science, Sports and Culture of Japan.

REFERENCES

Bard F., Cannon C., Barbour R., Burke R.-L., Games D., Grajeda H., Guido T., Hu K., Huang J., Johnson-Wood K., Khan K., Kholodenko D., Lee M., Lieberburg I., Motter R., Nguyen M., Soriano F., Vasquez N., Weiss K., Welch B., Seubert P., Schenk D., and Yednock T. (2000) Peripherally administered antibodies against amyloid β-peptide enter the central nervous system and reduce pathology in a mouse model of Alzheimer disease. *Nature Med.* **6,** 916-919.

Barger S.W., Hörster D., Furukawa K., Goodman Y., Krierglstein J. and Mattson M.P. (1995) Tumor necrosis factors α and β protect neurons against amyloid β-peptide toxicity: Evidence for involvement of a κB-binding factor and attenuation of peroxide and Ca^{2+} accumulation. *Proc. Natl. Acad, Sci. USA* **92,** 9328-9332.

Dickson D.W., Lee S.C., Mattiace L.A., Yen S.-H. and Bronsnan C. (1993) Microglia and cytokines in neurological disease, with special reference to AIDS and Alzheimer's disease. *Glia* **7,** 75-83.

Duff K., Eckman C., Zehr C., Yu X., Prada C.-M., Perez-tur J., Hutton M., Buee L., Harigaya Y., Yager D., Morgan D., Gordon M.N., Holcomb L., Refolo L., Zenk B., Hardy J., and Younkin S. (1996) Increased amyloid-β42(43) in brains of mice expressing mutant presenilin 1. *Nature* **383,** 710-713.

El Khoury J., Hickman S.E., Thomas C.A., Cao L., Silverstein S.C., and Loike J.D. (1996) Scavenger receptor-mediated adhesion of microglia to β-amyloid fibrils. *Nature* **382,** 716-719.

Hori O., Matsumoto M., Kuwabara K., Maeda Y., Ueda H., Ohtsuki T., Kinoshita T., Ogawa S., Stern D.M. and Kamada T. (1996) Exposure of astrocytes to hypoxia/reoxygenation enhances expression of glucose-regulated protein 78 facilitating astrocyte release of the neuroprotective cytokine interleukin 6. *J. Neurochem.* **66,** 973-979.

Kakimura J., Kitamura Y., Taniguchi T., S. Shimohama and Gebicke-Haerter P.J. (2001) BiP/GRP78-induced production of cytokines and uptake of amyloid-β(1-42) peptide in microglia. *Biochem. Biophys. Res. Commun.* **281,** 6-10.

Katayama T., Imaizumi K., Sato N., Miyoshi K., Kudo T., Hitomi J., Morihara T., Yoneda T., Gomi F., Mori Y., Nakano Y., Takeda J., Tsuda T., Itoyama Y., Murayama O., Takashima A., St. George-Hyslop P., Takeda M., and Tohyama M. (1999) Presenilin-1 mutations downregulate the signalling pathway of the unfolded-protein response. *Nature Cell Biol.* **1,** 479-485.

Kaufman R.J. (1999) Stress signaling from the lumen of the endoplasmic reticulum: Coordination of gene transcriptional and translational controls. *Genes Dev.* **13,** 1211-1233.

Maeda Y., Matsumoto M., Hori O., Kuwabara K., Ogawa S., Yan S.D., Ohtsuki T., Kinoshita T., Kamada T. and Stern D.M. (1994) Hypoxia/reoxygenation-mediated induction of astrocyte interleukin 6: A paracrine mechanism potentially enhancing neuron survival. *J. Exp. Med.* **180,** 2297-2308.

Mattson M.P., LaFerla F.M., Chan S.L., Leissring M.A., Shepel P.N., and Geiger J.D. (2000) Calcium signaling in the ER: Its role in neuronal plasticity and neurodegenerative disorders. *Trends Neurosci.* **23,** 222-229.

Nakagawa T., Zhu H., Morishima N., Li E., Xu J., Yankner B.A., and Yuan J. (2000) Caspase-12 mediates endoplasmic-reticulum-specific apoptosis and cytotoxicity by amyloid-β. *Nature* **403,** 98-103.

Paresce D.M., Ghosh R.N., and Maxfield F.R. (1996) Microglial cells internalize aggregates of the Alzheimer's disease amyloid β-protein via a scavenger receptor. *Neuron* **78,** 553-565.

Sawada M., Suzumura A. and Marunouchi T. (1995) Cytokine network in the central nervous system and its roles in growth and differentiation of glial and neuronal cells. *Int. J. Dev. Neurosci.* **13,** 253-264.

Selkoe D.J. (1999) Translating cell biology into therapeutic advances in Alzheimer's disease. *Nature* **399 Suppl.,** A23-A31.

Urano F., Wang X.Z., Bertolotti A., Zhang Y., Chung P., Harding H.P., and Ron D. (2000) Coupling of stress in the ER to activation of JNK protein kinases by transmembrane protein kinase IRE1. *Science* **287,** 664-666.

ETIOLOGY, PATHOGENESIS, AND GENETICS OF PARKINSON'S DISEASE

Yoshikuni Mizuno[1], Nobutaka Hattori[1], Hideki Shimura[1], Tohru Kitada[1], Shin-ichiro Kubo[1], Mei Wang[1], Ken-ichi Sato[1], Toshiaki Suzuki[2], Tomoki Chiba[2], Keiji Tanaka[2], Shuichi Asakawa[3], Shinsei Minoshima[3], and Nobuyoshi Shimizu[3]

1. INTRODUCTION

Parkinson's disease (PD) is the second most common neurodegenerative disease after Alzheimer's disease. Recent scientific interest has been focused on the molecular mechanism of nigral neuronal death and how to prevent it. In this communication, we review recent progress in familial PD and then we discuss what we can learn from familial PD to explore the pathogenesis of sporadic PD.

2. PARK1

PARK 1 is an autosomal dominant form of familial PD linked to the long arm of chromosome 4 (4q21-23), in which point mutations of the α-synuclein gene cause clinical phenotypes of PD.[1] Only two types of mutations are known, i.e., Ala53Thr and Ala30Pro.[1, 2] The Ala30Pro mutation was reported in a German family, but the Ala53Thr mutation was seen mainly in families of Greek origin. Both mutations are very rare. Clinical features of PARK1 are essentially similar to those of late onset sporadic PD, except for the earlier onset of age (mid-forties in average) and a higher incidence of dementia (more than 50%).[3] But in the German family with the Ala30Pro mutation, the age of onset is late fifties. They respond to levodopa. Pathologic features are marked neuronal degeneration of the substantia nigra and Lewy bodies not only in the substantia nigra but also in the cerebral cortex .[3]

α-Synuclein is a 16 kD protein expressed in presynaptic membranes and in the nucleus. Its function is not well understood. What is interesting is the observation that α-

[1] Juntendo University School of Medicine, Tokyo 113-8421, Japan, [2]Tokyo Metropolitan Institute of Medical Sciences, Tokyo, [3]Keio University School of Medicine, Tokyo. Correspondence to Y. M.

Mapping the Progress of Alzheimer's and Parkinson's Disease
Edited by Mizuno *et al.*, Kluwer Academic/Plenum Publishers, 2002

239

synuclein is a major component of Lewy bodies.[5] Therefore, α-synuclein appears to play an important role in sporadic PD as well. α-Synuclein is liable to self-aggregation, and mutated α-synucleins are more so.[6] Aggregated α-synuclein is insoluble and accumulates in cytoplasm and neuritic processes as well leading nigral neurons to death.[7] Such aggregates may interfere with transport of vital substances. α-Synuclein overexpressed transgenic mice and flies show Lewy body-like inclusions and dopaminergic neurodegeneration.[8, 9] But the precise molecular mechanism of α-synuclein accumulation and neuronal death is not yet known.

3. PARK2

PARK2 is an autosomal recessive young onset PD (AR-JP). Its clinical features were first delineated by Yamamura et al.[10] in 1973. The mean age of onset is late twenties and they show all the four cardinal features of PD. Initial symptom is more often gait disorder rather than rest tremor. Dystonic feature in feet while walking is often a presenting symptom in very young patients. Sleep benefit is another characteristic feature, but this becomes less conspicuous as the onset age becomes older. They respond to levodopa well, but show motor fluctuations easily. Dementia and autonomic failures are usually not seen. Progression is slow. Pathologic features are severe neuronal loss and gliosis in the substantia nigra and less so in the locus coeruleus.[11] Usually no Lewy bodies are found,

We found an AR-JP family, which showed co-segregation of the disease phenotype and a polymorphic mutation of the Mn SOD gene on chromosome 6[12] and we eventually mapped the disease gene within the 17 cM region at 6q25.2-27.[13] Then we found a patient who lacked one of the microsatellite markers (D6S305), which we used in the linkage analysis.[14] We thought that this microsatellite maker might be located within the disease gene and that turned out to be the case.

We started to screen KeioBAC library, a comprehensive human cDNA library, by the microsatellite marker, D6S305, and we were able to clone a cDNA consisting of 2960 bps, of which 1,395 bps comprised the coding region.[15] As this was a novel gene, we named it as *parkin*. The total size of *parkin* was more than 1.5 Mb, the second largest gene after the dystrophin gene. Deduced amino acid sequence of the parkin protein showed a unique feature, in that there was about 30% homology to ubiquitin in the amino-terminal region and a RING box (RING-IBR-RING; RING: rare interesting new gene, IBR: in-between RINGS) near the carboxy-terminal side.

Various kinds of exonic deletions and point mutations were found in AR-JP patients.[16, 17] AR-JP has been found to have a worldwide distribution[18] comprising the most common form of familial PD. No clear correlation has been found between the type of mutations and clinical features. But European patients, in whom point mutations are more common, have higher age of onset than Japanese patients, in whom exonic deletions are more common. Parkin mRNA is ubiquitously expressed in the brain as well as in systemic organs,[15] but at the protein level, it is more densely expressed in the nigrostriatal system; in AR-JP patients no parkin protein was detected at post-mortem.[19]

We showed that parkin protein is a ubiquitin ligase (E3) of the ubiquitin system.[20] The ubiquitin system is an energy-dependent proteolysis machinary working together with the 26S proteasome (see Suzuki et al., in this volume). The function of the ubiquitin ligase is to polyubituitinate a target protein, which is to be rapidly proteolysed in the 26S proteasome. Thus each ubiquitin ligase has its own specific substrates. In the absence of parkin protein in AR-JP, accumulation of such substrates may act toxic to the nigral neurons.

Three candidate substrates have been reported for Parkin, i.e., CDCrel-1,[21] 22-kDa glycosylated form of α-synuclein (αSp22),[22] and Pael receptor.[23] CDCrel-1 is a synaptic vesicle protein inhibiting exocytosis.[24] αSp22 accumulates in the brain of AR-JP patients. Normal α-synuclein has a molecular mass of 16 kDa. Pael (Parkin-associated endothelin receptor-like) receptor is a novel homologue of endothelin receptor type B expressed in the endoplasmic reticulum (ER). When Pael R is over expressed in cells, this receptor tends to become unfolded, insoluble and ubiquitinated in vivo. The insoluble Pael receptor leads to unfolded protein-induced cell death. Pael R also accumulates in AR-JP brains.[23] The pathogenesis of AR-JP may be related to the ER stress.

Parkin is localized in Golgi bodies and synaptic vesicle membranes.[25] Parkin colocalizes with synaptotagmin and synaptophysin. Another interesting observation was made by Schlossmacher et al.,[26] in that Parkin colocalizes α-synuclein in Lewy bodies of sporadic PD and dementia of Lewy body type. Parkin is localized in the center of Lewy bodies and α-synuclein in the more peripheral part, suggesting that Parkin may be necessary in the initial stage of Lewy body formation. In AR-JP, Lewy bodies may not be formed because of the absence of Parkin.

4. PARK3

PARK3 is an autosomal dominant familial PD linked to the short arm of chromosome 2 (2p13).[27] Clinical features are essentially similar to those of late onset sporadic PD. The penetrance is only 40 %. Dementia is documented in 2 out of 6 families reported. Pathologic features consist of nigral degeneration with Lewy bodies.[28] The disease gene has not been identified yet.

5. PARK4

PARK4 is an autosomal dominant familial FD linked to the short arm of chromosome 4.[29] Clinical features consist of levodopa-responsive parkinsonism and dementia. Pathologic characteristics are nigral degeneration and appearance of nigral as well as cortical Lewy bodies.[30, 31] What is interesting in this family is the presence of an essential tremor clinical phenotype within the same family; they have same haplotypes as those manifesting parkinsonism. The disease gene has not been identified yet.

6. PARK5

PARK5 is linked to the short arm of chromosome 1 (1p32-33), but no clinical or genetic information has been reported in the literature.

7. PARK6

PARK6 is an autosomal recessive familial PD linked to chromosome 1 (1p35-36). Recently, Valente et al.[32] found this new locus by analyzing autosomal recessive families not having *parkin* gene mutations. Clinical features are essentially similar to those of PARK2 with the age of onset usually under 45 years. PARK 6 may be the second most common autosomal recessive form of PD.

8. OTHER FORMS OF FAMILIAL PARKINSONISM

There are many other interesting familial forms of PD, which include a family linked to the UCH-L1 (ubiquitin caroboxyterminal hydrolase-L1),[33] parkinsonism, depression, and alveolar hypoventilation (Perry syndrome),[34] rapid onset dystonia-parkinsonism linked to chromosome 12,[35] and X-linked dystonia-parkinsonism (Lubag dystonia).[36] Finally frontotemporal dementia and parkinsonism linked to chromosome 17 is caused by mutations of the tau gene.[37]

9. SPORADIC PARKINSON'S DISEASE

Our ultimate goal is to elucidate the molecular mechanism of nigral degeneration and to find neuroprotective treatment of sporadic PD. What can we learn from the studies on familial PD? First of all, α-synuclein appears to be a key molecule. It accumulates both in autosomal dominant AD and sporadic PD. Furhtermore, α-synuclein-22 accumulates in autosomal recessive PD with *parkin* mutations. As nigral neurons are believed to die of apoptosis, [38] it appears to be important to explore α-synuclein's commitment to apoptosis cascades.

It has been believed that nigral degeneration in sporadic PD is induced by the interaction of genetic predisposition and environmental nigral neurotoxins inducing mitochondrial failure and oxidative stress within nigral neurons. Both mitochondrial failure and oxidative stress are important apoptosis-inducing signals. We recently showed marked iron accumulation in ARJP brains.[39] Thus there may be secondary defects in sporadic PD similar to those found in familial PD. Progress in familial PD will open new areas of research on sporadic PD.

10. ACKNOWLEDGEMENTS

This study was in part supported by Grant-in-Aid for Scientific Research on Priority Areas from Ministry of Education, Science, Sports, and Culture, Japan and by "Center of Excellence" Grant from National Parkinson Foundation, Miami.

REFERENCES

1. M. H. Polymeropoulos, C. Lavedan, E. Leroy, S. E. Ide, A. Dehejia, A. D. B. Pike, H. Root, J. Rubenstein, R. Boyer, E. S. Stenroos, S. Chandrasekharappa, A. Athanassiadou, T. Papapetropoulos, W. G. Johnson,

A. M. Lazzarini, R. C. Duvoisin, G. Di Iorio, L. I. Golbe, and R. L. Nussbaum, Mutation in the α-synuclein gene identified in families with Parkinson's disease. *Science* **276**,2045-2047(1997).

2. R. Krüger, W. Kuhn, T. Müller, D. Woitalla, M. Graeber, S. Kösel, H. Przuntek, J. T. Epplen, L. Schöls, and O. Riess, Ala30Pro mutation in the gene encoding α-synuclein in Parkinson's disease. *Nature Genet.* **18**,106-108(1998).

3. L. I. Golbe, G. Di Iorio, V. Bonavita, D. C. Miller, and R. C. Duvoisin, A large kindred with autosomal dominant Parkinson's disease. *Ann. Neurol.* **27**,276-282(1990).

4. L. Maroteaux, J. T. Campanelli, and R. H. Scheller, Synuclein: a neuron-specific protein localized to the nucleus and presynaptic nerve terminal. *J. Neurosci.* **8**, 2804-2815(1988).

5. M. G. Spillantini, M. L. Schmidt, A. M. Y. Le, J. Q. Trajanowski, R. Jakes, and M. Goedert, α-Synuclein in Lewy bodies. *Nature* **388**,839-840 (1997).

6. O. M. El Agnaf, R. Jakes, M. D. Curran, and A. Wallace, Effects of the mutations Ala30 to Pro and Ala 53 to Thr on the physical and morphological properties of α-synuclein protein implicated in Parkinson's disease. *FEBS Lett.* **440**,67-70(1998).

7. K. Wakabayashi, S. Hayashi, A. Kakita, M. Yamada, Y. Toyoshima, M. Yoshimoto, and H. Takahashi, Accumulation of α-synucleinMACP is a cytopathological feature common to Lewy body disease and multple system atrophy. *Acta Neuropathol. (Berl.)* **96**,445-52(1998).

8. E. Masliah, E. Rockenstein, I. Veinbergs, M. Mallory, M. Hashimoto, A. Takeda, Y. Sagara, A. Sisk, and L. Mücke. Dopaminergic loss and inclusion body formation in α-synuclein mice: implications for neurodegenegerative disorders. *Science* **287**,1265-1269(2000).

9. M.B. Feany and W. W. Bender, A Drosophial model of Parkinson's disease. *Nature* **404**,394-398(2000).

10. Y. Yamamura, I. Sobue, K. Ando, M. Iida, T. Yanagi, and C. Kono. Paralysis agitans of early onset with marked diurnal fluctuation of symptoms. *Neurology* **23**,239-244(1973).

11. H. Mori, T. Kondo, M. Yokochi, H. Matusmine, Y. Nakagawa-Hattori, T. Miyake, K. Suda, and Y. Mizuno, Pathologic and biochemical studies of juvenile parkinsonism linked to chromosome 6q. *Neurology* **51**,890-892(1998).

12. S. Shimoda-Matsubayashi, H. Matsumine, T. Kobayashi, Y. Nakagawa-Hattori, Y. Shimizu, and Y. Mizuno, Structural dimorphism in the mitochondrial targeting sequence in the human MnSOD gene. A predictive evidence for conformational change to influence mitochondrial transport and a study of allelic association in Parkinson's disease. *Biochem. Biophys. Res. Commun.* **226**, 561-565,(1996).

13. H. Matsumine, M. Saito, S. Shimoda-Matsubayashi, H. Tanaka, A. Ishikawa, Y. Nakagawa-Hattori, M. Yokochi, T. Kobayashi, S. Igarashi, H. Takano, K. Sanpei, R. Koike, H. Mori, T. Kondo, Y. Mizutani, A. A. Schaffer , Y. Yamamura, S. Nakamura, S. Kuzuhara, S. Tsuji, and Y. Mizuno, Localization of a gene for autosomal recessive form of juvenile parkinsonism (AR-JP) to chromosome 6q25.2-27. *Amer. J. Hum. Genet.* ;**60**,588-596(1997).

14. H. Matsumine, Y. Yammaura, N. Hattori, T. Kobayashi, and Y. Mizuno, A microdeletion spanning D6S305 co-segregates with autosomal recessive juvenile parkinsonism (ARJP). *Genomics* **49**,143-146(1998).

15. T. Kitada, S. Asakawa, N. Hattori, H. Matsumine, Y. Yamamura, S. Minoshima, M. Yokochi, Y. Mizuno, and N. Shimizu, Deletion mutation in a novel protein "Parkin" gene causes autosomal recessive juvenile parkinsonism (AR-JP). *Nature* **392**,605-608(1998).

16. N. Hattori, H. Matsumine, T. Kitada, S. Asakawa, Y. Yamamura, T. Kobayashi, M. Yokochi, H. Yoshino, M. Wang, T. Kondo, S. Kuzuhara, S. Nakamura, N. Shimizu, and Y. Mizuno, Molecular analysis of a novel ubiquitin-like protein (PARKIN) gene in Japanese families with AR-JP: evidence of homozygous deletions in the PARKIN gene in affected individuals. *Ann. Neurol.* **144**,935-941(1998).

17. N. Hattori, H. Matsumine, S. Asakawa, T. Kitada, H. Yoshino, B. Elibol, A. J. Brooks, Y. Yamamura, T. Kobayashi, M. Wang, A. Yoritaka, N. Shimizu, and Y. Mizuno, Point mutations (Thr240Arg and Gln311Stop) in the Parkin gene. *Biochem. Biohys. Res. Commun.* **249**,754-758(1998).

18. C. B. Lücking, A. Dürr, V. Bonifati, J. Vaughan, G. De Michele, T. Gasser, B. S. Harhangi, G. Meco, P. Deneêfle, N. W. Wood, Y. Agid, A. Brice, European Consortium of Genetic Susceptibility in PD, and French PD Genetics Study Group, Association between early-onset Parkinson's disease and mutations in the parkin gene. *New Engl. J. Med.* **342**,1560-1567(2000).

19. H. Shimura, N. Hattori, S. Kubo, M. Yoshikawa, T. Kitada, H. Matsumine, S. Asakawa, S. Minoshima, Y. Yamamura, N. Shimizu, Y. Mizuno. Immunohistochemical and subcellular localization of Parkin: absence of protein in AR-JP. *Ann. Neurol.* **45**,668-672(1999)

20. H. Shimura, N. Hattori, S. Kubo, Y. Mizuno, S. Asakawa, S. Minoshima, N. Shimizu, K. Iwai, T. Chiba, K. Tanaka, and T. Suzuki, Familial Parkinson's disease gene product, Parkin, is a ubiquitin-protein ligase. *Nature Genet.* **25**,302-305(2000)

21. Y. Zhang, J. Gao, K. K. Chung, H. Huang, V. L. Dawson, and T. M. Dawson, Parkin functions as an E2 dependent ubiquitin-protein ligase and promotes the degradation of the synaptic vesicle associated protein, CDCrel-1. *Proc. Natl. Acad. Sci. USA* **21**,13354-13359(2000)

22. H. Shimura, M. G. Schlossmacher, N. Hattori, M. P. Frosch, A. Trockenbacher, R. Schneider, Y. Mizuno, K. S. Kosik, and D. J. Selkoe, A novel form of α-synuclein is a substrate for Parkin and accumulate in autosomal recessive Parkinson disease *Science* (2001), in press.

23. Y. Imai, M. Soda, H. N. Inoue, N. Hattori, Y. Mizuno, and R. Takahashi, An unfolded putative transmembrane polypeptide, which can lead to endoplasmic reticulum stress, is a substrate of Parkin. *Cell* (2001), in press.

24. C. L. Beites, H. Xie, R. Browser, and W. Trimble, The septin CDCrel-a blinds syntaxin and inhibits exocytosis. *Nature Neurosci.* **2**,434-439(1999).

25. S. Kubo, T. Kitami, H. Shimura, S. Noda, Y. Uchiyama, S. Asakawa, S. Minoshima, N. Shimizu, Y. Mizuno, and N. Hattori, Parkin is transported to presynaptic terminals through the trans-Golgi network and associated with synaptic vesicles. *J. Neurochem.* (2001), in press.

26. M. G. Schlossmacher, M. P. Frosch, W. P. Gai , M. Medina, H. Shimura, T. Ochiishi, R. Sharon, N. Hattori, Y. Mizuno, D. J. Selkoe, and K. S. Kosik, Parkin and α-synuclein interact in normal brain and colocalize in Lewy bodies of Parkinson Disease. *Amer. J. Pathol.* (2001), in press.

27. T. Gasser, B. Müller-Myhsok, Z. K. Wszolek , R. Oehlmann, D. B. Calne , V. Bonifati, B. Bereznai, E. Fabrizio , P. Vieregge, and R. D. Horstmann, A susceptibility locus for Parkinson's disease maps to chromosome 2p13. *Nature Genet.* **18**,262-265(1998).

28. M. A. Denson, Z. K. Wszolek, R. F. Pfeiffer, E. K. Wszolek, P. M. Paschall, and R. D. McComb, Familial parkinsonism, dementia, and Lewy body disease: study of family G. *Ann. Neurol.* **42**,638-643(1997).

29. M. Farrer, K. Gwinn-Hardy, K. Muenter, F. W. De Vrièze, R. Crook, J. Prez-Tur, S. Lincoln, D. Maraganore, C. Adler, S. Newman, K. MacElwee, P. McCarthy, C. Miller, C. Walters, and J. A. Hardy, A chromosome 4p haplotype segregaating with Parkinson's disease and postural tremor. *Hum. Molecular Genet.* **8**,81-85(1998)

30. M. D. Muenter, L. S. Forno, O. Hornykiewicz, S. J. Kish, D. M. Maraganore,R. J. Caselli, H. Okazaki, F. M. Howard, Jr, B. J. Snow, and D. B. Calne, Hereditary form of parkinsoism-dementia. *Ann. Neurol.* **43**,768-781(1998).

31. C. H. Waters and C. A. Miller, Autosomal-dominant Lewy body parkinsonism in a four-generation family. Ann Neurol **35**,59-64(1994).

32. E. M. Valente, A. R. Bentivolglio, P. H. Dixon, A. Ferraris, T. Ialongo, M. Frontali, A. Albanese, N. W. Wood, Localization of a novel locus for autoromal recessive early-onset parkinsonims, PARK6, on human chromosome 1p35-36. *Amer. J. Hum. Genet.* **68**,895-900(2001).

33. E. Leroy, R. Boyer, G. Auburger, B. Leube, G. Ulm, E. Mezey, G. Harta, M. J. Brownstein, S. Jonnalagada, T. Chernova, A. Dehejia, C. Lavedan, T. Gasser, P. J. Steinbach, K. D. Wilkinson, M. H. Polymeropoulos, The ubiquitin pathway in Parkinson's disease. *Nature* **395**,451-452(1998).

34. T. L. Perry, P. J. A. Bratty, S. Hansen, J. Kennedy, N. Urquhart, C. L. Dolma, Hereditary mental depression and parkinsonism with taurine deficiency. *Arch. Neurol.* **32**,108-113(1975).

35. P. L. Kramer , M. Mineta, C. Klein, K. Schilling, D. de Leon, M. R. Farlow, X. O. Breakfield, S. B. Bressman, W. B. Dobyns, L. J. Ozelius, A. and Brashear, Rapid-onset dystonia-parkinsonism: linkage to chromosome 19q13. *Ann. Neurol.* **46**,176-182(1999).

36. K. C. Wilhelmsen, D. E. Weeks, T. G. Nygaard, C. B. Moskowitz, R. L. Rosales, D. C. dela Paz,, E. Sobrevega, S. Fahn, and T. C. Gilliam, Genetic mapping of "Lubag" (X-linked dystonia-parkinsonism) in a Filipino kindred to the pericentromeric region of the X chromosome. *Ann. Neuol.* **29**,124-131(1991).

37. M. Hutton, C. L. Lendon, P. Rizzu, M. Baker, S. Froelich, H. Houlden, S. Pickering-Brown, S. Chakraverty, A. Isaacs, A. Grover, J. Hackett, J. Adamson, S. Lincoln, D. Dickson, P. Davies, R. C. Petersen, M. Stevens, E. de Graaff, E. Wauters, J. van Baren, M. Hillebrand, M. Joosse, J. M. Kwon, P. Noworthy, L. K. Che, J. Norton , J. C. Morris, L. A. Reed, J. Trojanowski, H. Basun, L. Lannfelt, M. Neystat, S. Fahn, F. Dark, T. Tannenberg, P. R. Dodd, N. Hayward, J. B. J. Kwok, P. R. Schofield, A. Andreadis, J. Snowden, D. Craufurd, D. Neary, F. Owen, B. A. Oostra, J. Hardy, A. Goate, J. van Swieten, D. Mann, T. Lynch, and P. Heutink, Association of missense and 5'-splice-site mutations in *tau* with the inherited dementia FRDP-17. *Nature* **393**,702-705(1998).

38. H. Mochizuki, G. Goto, H. Mori, and Y. Mizuno, Histochemical detection of apoptosis in Parkinson's disease. *J. Neurol. Sci.* **137**,120-123(1996).

39. M. Takanashi, H. Mochizuki, K. Yokomizo, N. Hattori, H. Mori, Y. Yamamura, and Y. Mizuno, Iron accumulation in the substantia nigral of autosomal recessive juvenile parkinsonism (ARJP). *Parkinsonism Rel. Disord.* **7**,311-314(2001).

PARKINSON'S DISEASE, NEUROTOXINS AND THE DOPAMINE TRANSPORTER
Evidence from classical brain dopamine studies[§]

Oleh Hornykiewicz[*]

1. INTRODUCTION

Research on the etiology of Parkinson's disease (PD) has been greatly stimulated by mapping and molecular cloning of genes (α-synuclein; parkin) associated with familial (rare) forms of PD.[1] However, the cause of the most common, the sporadic – or idiopathic – form of PD is still unknown. Here some evidence points to genetic predisposition (genetic polymorphisms) and environmental or endogenous toxins, with the "final common mechanism" being mitochondrial dysfunction and/or oxidative stress.[2]

Whatever the PD initiating event may be, any biochemical hypothesis about nigral cell death should (a) account for the primary site of the dopamine (DA) neuron degeneration; and (b) offer an explanation for the specific pattern of striatal DA loss typical of idiopathic PD. In the following sections it will be argued that a neurotoxic agent with the properties of a DA transporter (DAT) substrate fulfils these criteria.

2. DOES THE DEGENERATION START IN THE STRIATAL TERMINALS?

2.1. Primary site of DA neuron death determines nigrostriatal DA patterns

When the neuronal perikaryon dies, the whole neuron, including its terminal tree, degenerates, and the neuron's specific neurotransmitter substance disappears from both the cell body region and the projection area. When, however, the degenerative process starts in the nerve terminals the cell bodies often survive, the result being a regional neurotransmitter "mismatch": the transmitter loss will be greater in the terminal area than in the cell body area.

---

[*] Brain Research Institute, University of Vienna, Spitalgasse 4, A-1090 Vienna, Austria
[§] In memory of Professor Hirotaro Narabayashi, M.D. (1922-2001)

Mapping the Progress of Alzheimer's and Parkinson's Disease
Edited by Mizuno *et al.*, Kluwer Academic/Plenum Publishers, 2002

Table 1. Comparison between the nigrostriatal DA pattern in a disorder with primary degeneration of DA perikarya versus a disorder with primary DA terminal loss [a]

| | Dopamine | |
	NIIB disorder [b]	Pick's disease
Caudate	3	7
Putamen	1	9
Nigra	> 1	85

[a] Data (in percent of control) calculated from ref. [3,4]
[b] Neuronal intranuclear inclusion body disorder

Table 1 shows data for the nigrostriatal DA system illustrating these two neuro-degenerative patterns. In neuronal intranuclear inclusion body disorder (NIIB) the neuronal (including DA) perikarya are primarily affected; here, the DA loss was equally severe in both the nigra and the caudate and putamen.[3] Conversely, in Pick's disease, that most likely starts as neuroaxonal degeneration, the DA loss in the striatum greatly exceeded the DA loss in the nigra.[4]

2.2. PD – Dying-back of striatal DA terminals?

Which of the two patterns applies to PD? Table 2 shows the nigrostriatal DA and tyrosine hydroxylase values in idiopathic PD. It is evident that in PD the striatum is more severely affected than the nigra. The mismatch "striatum - nigra" is confirmed by corresponding observations on tyrosine hydroxylase, the marker enzyme for intact DA neurons (Table 2).

The striking mismatch between the striatal vs nigral DA levels permits the conclusion that in PD the DA neuron degeneration may be a dying-back process, starting in the striatal terminals and not, as usually assumed, in the nigral cell bodies.

Table 2. Mismatch between the loss of striatal versus nigral DA and tyrosine hydroxylase (TH) in PD patients and the MPTP primate [a, b]

| | Parkinson's disease | | MPTP primate |
	DA	TH	DA
Caudate	3.7	17	5.4
Putamen	1.5	18	7.9
Nigra	15.3	35	31

[a] For references, see ref. [5]
[b] Data expressed in percent of control

2.3. Ultrastructural study of PD caudate favours dying-back of DA neurons

In an electron microscopic study of caudate biopsies from control and PD patients, the occurrence, specifically in the PD material, of numerous dystrophic neurites has been directly demonstrated.[6] Together with the concomitantly greatly reduced DA levels, the presence of the neuroaxonal dystrophic morphology was judged to be "consistent with a dying-back process" of the nigrostriatal DA neurons in PD.[6]

2.4. Evidence for dying-back from the MPTP primate model of PD

The PD-like condition produced by MPTP in the primate has greatly contributed to the understanding of pathophysiology and pathogenesis of PD proper. It is, therefore, significant that in the chronic MPTP parkinsonian rhesus monkey a "striatum - nigra" DA mismatch has been observed identical with that in patients with PD[7] (Table 2). This is highly relevant to PD because the MPTP-induced DA pattern can be directly related to the toxin's primary striatal site of action, as indicated by studies showing severe striatal axonal/and terminal pathology and/or DA loss, but only mild nigral changes (for references, see ref.[7,8]).

That MPTP acts primarily on the striatal terminals is not surprizing, because the number of DAT sites – which are the "gate" for MPTP's DA-toxic metabolite MPP^+ [9]– is considerably higher in the striatal terminals than nigral cell bodies.[10] Therefore, MPP^+ is accumulated preferentially in the striatal DA terminals where it triggers the degenerative process (spreading retrogradely to the rest of the DA neuron, incl. nigral cell bodies).

2.5. Significance for PD etiology

The above discussed evidence both from biochemistry and electron microscopy obtained directly in patients with PD, together with the respective observations in the primate with the PD-like neurotoxin (MPTP)-induced condition, supports the working hypothesis that the striatal DAT sites and a noxious agent, the putative "PD toxin", may indeed be instrumental in the etiology of PD.

3. HOW CAN WE EXPLAIN THE DA PATTERNS IN PD STRIATUM?

In PD the DA loss in putamen is consistently more marked than in the caudate, with the caudal and dorsal putamen being specially severely affected.[5, 11] A comparison of the striatal DA loss with the nigral patterns of cell loss shows that they are interrelated: Those parts of the nigra that project to the putamen, i.e. the ventrolateral cell groups, also have a more severe cell loss than the dorsal groups projecting to the caudate (Figure 1).

3.1. The striatal DA pattern and the "DAT-Neurotoxin" hypothesis

What may be the reason for the different vulnerability of DA neurons to the disease process? One of the most interesting differences between the respective DA neuron subgroups is the number of their DAT sites. In the primate nigra the ventrolateral DA cells are, relative to the dorsal cell groups, especially rich in DAT mRNA.[12] The same

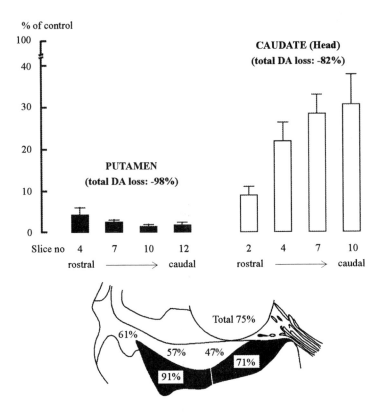

Figure 1. Comparison between the specific patterns of DA loss in the putamen and caudate (bar diagram) and the intranigral patterns of compacta cell loss (schematic drawing below the diagram). Note that the severely DA depleted putamen (black bars) receives its DAergic innervation from the highly depopulated ventral tier of the compacta (black area); whereas the moderately degenerated dorsal tier of the compacta (open area) innervates the less markedly DA depleted caudate (open bars). (Compiled from ref.[11,15])

intranigral DAT pattern may also be present in the human brain.[13] Most significantly, a study of DAT protein in the PD striatum has disclosed that "those areas expressing the highest levels of DAT are most susceptible [to degeneration] in PD".[14]

The DAT, in addition to its physiological role, allows DA toxic compounds (e.g. MPP[+]; see Section 1) to enter the DA neurons. The preferential loss, in PD, of DAT rich neurons is consistent with the involvement of a "PD toxin" with DAT substrate properties. By allowing a specific DA toxic agent (or agents) to enter the DA terminal, and eventually destroy the whole nigrostriatal neuron in the manner of a dying-back process (see Section 1), the DAT would by necessity imprint its own distribution pattern on the final pattern of striatal DA loss typic for idiopathic PD. It is noteworthy that the especially vulnerable DAT rich putamen terminals have, compared with the caudate, low levels of vesicular monoamine transporter (VMAT2), suggesting lower intracellular sequestration (i.e. detoxification) of noxious chemicals, as demonstrated for MPP[+].[16,17]

3.2. Do DA terminals with low number of DAT sites predominate in PD striatum?

The notion of a "PD toxin" affecting preferentially the DAT rich DA neurons implies that in the course of the disease those DA terminals with fewer DAT sites (as the more toxin "resistant") should become the predominant species. This appears to be indeed the case, as: (a) in the PD nigra, the surviving DA neurons have lower levels of DAT mRNA than nigral neurons of normal controls;[18] and (b) in the PD putamen the remaining DA neurons have abnormally low levels of DAT protein relative to their vesicular monoamine transporter content (marker of viable DA neurons)[19] (Table 3).

3.3. What are the chemical requirements for the postulated "PD toxin"?

Recently many potentially DA toxic compounds of exogenous and endogenous origin have been identified. In the context of the present discussion, the only chemical requirement for any of these compounds is that they should be substrates of the specific DAT. Here of special interest are endogenously formed compounds with neurotoxic potential, such as some isoquinoline and ß-carboline derivatives.[20-22] Given some affinity for the DAT, they may accumulate – especially in genetically predisposed individuals – to reach, over long periods of time, toxic levels and cause selective DA neuron death.

Table 3. Parkinson's disease – Remaining DA neurons in putamen (but not caudate) have, relative to their vesicular monoamine transporter (VMAT2), especially low DAT levels [a]

	DAT (protein)	VMAT2 (^{3}H-DTBZ-binding)
Caudate	31	49
Putamen	3.4	23

[a] Data, expressend in percent of control, calculated from ref.[19]

4. CLINICAL IMPLICATIONS OF THE "DAT-NEUROTOXIN" HYPOTHESIS

(1) The predominance, in PD, of DAT poor striatal terminals has two clinically important implications: (a) the L-dopa induced dyskinesias can be explained as being the result of progressive reduction of the striatal capacity for removal (via DAT) of L-dopa derived synaptic DA; and (b) the observation that in the beginning of PD the rate of clinical progression is faster than in the later stages of the disease can be explained by the progressive predominance of DAT poor, more "PD-toxin" resistant, DA terminals.

(2) A challenging therapeutic possibility can be deduced from the "striatum - nigra" mismatch on which the dying-back concept (as an important component of the "DAT-Neurotoxin" hypothesis) is based. The "striatum - nigra" DA mismatch shows that in PD, despite severe loss of striatal terminals, a significant proportion of nigral neurons remains viable. Consequently, enhancement of nigral neurotrophic mechanisms may prove successful in arresting, or even partly reversing, the parkinsonian disease process.

REFERENCES

1. Y. Mizuno, N. Hattori, T. Kitada, H. Matsumine, H. Mori, H. Shimura, S. Kubo, H. Kobayashi, S. Asakawa, S. Minoshima, and N. Shimizu, Familial Parkinson's disease. α-Synuclein and parkin, *Adv. Neurol.* **86**, 13-21 (2001).
2. A. H. V. Schapira, Causes of neuronal death in Parkinson's disease, *Adv. Neurol.* **86**, 155-162 (2001).
3. S. J. Kish, J. J. Gilbert, L. J. Chang, L. Mirchandani, K. Shannak, and O. Hornykiewicz, Brain neurotransmitter abnormalities in neuronal intranuclear inclusion body disorder, *Ann. Neurol.* **17**, 405-407 (1985).
4. I. Kanazawa, S. Kwak, H. Sasaki, O. Muramoto, T. Mizutani, A. Hori, and N. Nukina, Studies on neurotransmitter markers of the basal ganglia in Pick's disease, with special reference to dopamine reduction, *J. Neurol. Sci.* **83**, 63-74 (1988).
5. O. Hornykiewicz, Biochemical aspects of Parkinson's disease, *Neurology* **51** (Suppl. 2), S2-S9 (1998).
6. B. Lach, D. Grimes, B. Benoit, and A. Minkiewicz-Janda, Caudate nucleus pathology in Parkinson's disease: ultrastructural and biochemical findings in biopsy material, *Acta Neuropathol.* **83**, 352-360 (1992).
7. Ch. Pifl, G. Schingnitz, and O. Hornykiewicz, Effect of 1-methyl-4-phenyl-1,2,3,6-tetrahydropyridine on the regional distribution of brain monoamines in the rhesus monkey, *Neuroscience* **44**, 591-605 (1991).
8. M. Herkenham, M. D. Little, K. Bankiewicz, S.-C. Yang, S. P. Markey, and J. N. Johannessen, Selective retention of MPP$^+$ within the monoaminergic systems of the primate brain following MPTP administration: an in vivo autoradiographic study, *Neuroscience* **40**, 133-158 (1991).
9. J. A. Javitch, R. J. D'Amato, S. M. Strittmatter, and S. H. Snyder, Parkinsonism-inducing neurotoxin, N-methyl-4-phenyl-1,2,3,6-tetrahydropyridine: uptake of the metabolite N-methyl-4-phenylpyridine by dopamine neurons explains selective toxicity. *Proc. Natl. Acad. Sci. USA* **82**, 2173-2177 (1985).
10. M. J. Nirenberg, R. A. Vaughan, G. R. Uhl, M. J. Kuhar, and V. M. Pickel, The dopamine transporter is localized to dendritic and axonal plasma membranes of nigrostriatal dopaminergic neurons, *J. Neurosci.* **16**, 436-447 (1996).
11. S. J. Kish, K. Shannak, and O. Hornykiewicz, Uneven pattern of dopamine loss in the striatum of patients with idiopathic Parkinson's disease, *New Engl. J. Med.* **318**, 876-880 (1988).
12. S. N. Haber, H. Ryoo, C. Cox, and W. Lu, Subsets of midbrain dopaminergic neurons in monkeys are distinguished by different levels of mRNA for the dopamine transporter: comparison with the mRNA for the D$_2$ receptor, tyrosine hydroxylase and calbindin immunoreactivity, *J. Comp. Neurol.* **362**, 400-410 (1995).
13. A. M. Murray, F. B. Weihmueller, J. F. Marshall, H. I. Hurtig, G. L. Gottleib, and J. N. Joyce, Damage to dopamine systems differs between Parkinson's disease and Alzheimer's disease with parkinsonism, *Ann. Neurol.* **37**, 300-312 (1995).
14. G. W. Miller, J. K. Staley, C. J. Heilman, J. T. Perez, D. C. Mash, D. B. Rye, and A. I. Levey, Immunochemical analysis of dopamine transporter protein in Parkinson's disease, *Ann. Neurol.* **41**, 530-539 (1997).
15. J. M. Fearnley and A. J. Lees, Ageing and Parkinson's disease: substantia nigra regional selectivity, *Brain* **114**, 2283-2301 (1991).
16. G. R. Uhl, Hypothesis: the role of dopaminergic transporters in selective vulnerability of cells in Parkinson's disease, *Ann Neurol.* **43**, 555-560 (1998).
17. G. W. Miller, R. R. Gainetdinov, A. I. Levey, and M. G. Caron, Dopamine transporters and neuronal injury, *Trends Pharmacol. Sci.* **20**, 424-429 (1999).
18. G. R. Uhl, D. Walther, D. Mash, B. Faucheux, and F. Javoy-Agid, Dopamine transporter messenger RNA in Parkinson's disease and control substantia nigra neurons, *Ann. Neurol.* **35**, 494-498 (1994).
19. J. M. Wilson, A. I. Levey, A. Rajput, L. Ang, M. Guttman, K. Shannak, H. B. Niznik, O. Hornykiewicz, C. Pifl, and S. J. Kish, Differential changes in neurochemical markers of striatal dopamine nerve terminals in idiopathic Parkinson's disease, *Neurology* **47**, 718-726 (1996).
20. M. Naoi and W. Maruyama, N-methyl(R)salsolinol, a dopamine neurotoxin, in Parkinson's disease, *Adv. Neurol.* **80**, 259-264 (1999).
21. T. Nagatsu, Isoquinoline neurotoxins in the brain and Parkinson's disease, *Neurosci. Res.* **29**, 99-111 (1997).
22. K. Matsubara, T. Gonda, H. Sawada, T. Uezono, Y. Kobayashi, T. Kawamura, K. Ohtaki, K. Kimura, and A.Akaike, Ednogenously occurring β-carboline induces parkinsonism in nonprimate animals: a possible causative protoxin in idiopathic Parkinson's disease, *J. Neurochem.* **70**, 727-735 (1998).

GENETIC RISK FACTORS IN PARKINSON'S DISEASE

Kin-Lun Tsang, Zhe-Hui Feng, Hong Jiang, David B. Ramsden and Shu-Leong Ho*

1. INTRODUCTION

This chapter will take a broad approach to the role of genetic factors in Parkinson's disease (PD). It is important to distinguish between genetic and inherited factors. Inherited factors relate to those that are passed from generation to generation. Genetic factors relate to constitutive genes that can interact with each other and with the environment. There has been considerable debate as to the extent to which inherited factors contribute to the development of PD since the 1890s, when Gowers reported that up to 15% of secondary familial cases existed in PD.

2. TWIN STUDIES AND EPIDEMIOLOGICAL STUDIES

In the early 1980s, to resolve this debate, twin studies reported that concordance between monozygotic (MZ) and dizygotic (DZ) twins were similar, indicating that inheritance did not appear to play a significant role in PD.[1-3] At about that time, the discovery that a synthetic compound MPTP (1-methyl-4-phenyl-1,2,3,6-tetrahydropyridine) could cause parkinsonism only strengthened that view.[4]

However, the numbers of twin pairs were relatively small. Reports of families with autosomal dominant parkinsonism stimulated the reappraisal of these twin studies. To circumvent the problem of small sample size, Tanner and her colleagues identified 193 twins with PD out of 19842 white male twins enrolled in the National Academy of Sciences/National Research Council World War II Veteran Twins Registry.[5] In 71 MZ and 90 DZ pairs with complete diagnoses of PD, pairwise concordance was similar between MZ and DZ (0.155 MZ, 0.111 DZ; relative risk 1.39; 95% confidence interval, 0.63-3.1). In 16 twin pairs with onset of PD symptoms at or before age 50 years in at

---

* Kin-Lun Tsang, Zhe-Hui Feng, Hong Jiang and Shu-Leong Ho, Division of Neurology, University Department of Medicine, University of Hong Kong, Queen Mary Hospital, Hong Kong, PR China. David B. Ramsden, Department of Medicine, University of Birmingham, UK.

Mapping the Progress of Alzheimer's and Parkinson's Disease
Edited by Mizuno *et al.*, Kluwer Academic/Plenum Publishers, 2002

251

least 1 twin, MZ concordance was 1.0, and DZ was 0.167 (relative risk 6.0; 95% confidence interval, 1.69-21.26). A separate followup study of 9 MZ and 12 DZ twins showed that concordance was similar 8 years later.[6] None of the 5 out of 7 cotwins with abnormal PET but without motor features of parkinsonism developed overt motor features 6 years later. Hence, it appeared that inheritance did not have a major influence on PD from these twin studies, except in a small subgroup of early onset PD. However, twin studies do not exclude the influence of mitochondrial genetic abnormalities. Mitochondrial genes are inherited cytoplasmically and maternally, and known to have *de novo* or acquired mutations. They also cannot exclude the effects of environment on mitochondrial function, and interactions between various genes and the environment.

The EuroParkinson Study Group found that PD significantly aggregated in families.[7] A population study in Iceland based on comprehensive genealogic information on not just the nuclear family but also the extended family has shown evidence that inheritance can influence the development of PD.[8] Family members had increased risk of developing PD, if they had a relative affected with PD. This risk extended beyond the nuclear family. In siblings of the affected proband, the risk was 6.7x more than matched-controls. In offsprings, the risk was 3.2x. In 2nd degree relatives (nephews/nieces) the risk was 2.7x. In 3rd degree relatives, the risk becomes insignificant as were spouses. Hence, there was also a dilution effect on the influence of inheritance on the risk of developing PD, the more distant their relationship was from the affected index family member.

3. CANDIDATE GENE STUDIES

A number of candidate gene association studies have been performed to identify genes that may contribute to this inherited risk. It has been disappointing so far that many of these studies have not provided significant leads. Studies that had initially reported a link of one of these genes or a combination of these genes to sporadic PD, have not been reproduced in other populations. There were various reasons for these inconsistencies. They include different methodologies, different selection criteria for PD and control groups, and small sample sizes.

However, there may be another simple reason. Conventional molecular techniques used in carrying out many of these studies can only detect differences in gene structure but not gene expression or gene function. Gene expression can be affected by a variety of environmental factors and can change with time, unlike gene structure. Furthermore, although constitutive genes are present in all tissues, their expression may be different in different tissues even in the same individual. In this post-genomic era, we now know that there are about 30,000 human genes but there are at least 200,000 gene products.

4. LINKAGE ANALYSIS OF FAMILIAL PARKINSONISM

The breakthrough came from linkage studies of familial parkinsonism who have a "concentrated" form of inherited factors. At least five different loci and mutations have been identified in families with a spectrum of parkinsonism (Table 1). Three genes, α-synuclein, *Parkin* and ubiquitin C-terminal hydrolase (UCH-L1), will be briefly mentioned. Although α-synuclein, *Parkin* and UCH-L1 were initially linked to different forms of parkinsonism, their roles in their pathogenesis appear to be closely related. All three of them are expressed mainly in neurons. α-synuclein and *Parkin* are expressed in a restricted number of brain regions, but UCH-L1 appears to be expressed by neurons throughout the brain.[9]

Table 1. Genetic loci in Parkinson's disease

Locus	*Gene*	*Mutation*	*Phenotype*
4q21-23	α synuclein	Ala53T→Thr; Ala30→Pro	AD-PD
6q25.2-q27	Parkin	Multiple	AR-JP
4p14	Ubiquitin carboxy-terminal hydrolase	Ile93→Met	Early onset AD-PD
2p13	Unknown	Unknown	AD-PD
17q21-22	Tau	Splice site & mis-sense	FTDP-17
17q21-22	Tau	Haplotype association	PSP

The alanine to threonine mutations in the α-synuclein gene encoded in chromosome 4 was identified in a large Italian-American and several smaller Greek families,[10,11] and Ala30Pro mutation in a small German family with autosomal dominant parkinsonism with Lewy bodies.[12] However, these families with these mutations are rare.[13,14] They do not occur in sporadic PD. But the role of α-synuclein in the pathogenesis of PD is far greater than what gene association studies would suggest. α-synuclein appears to be important in synaptic function,[15] and is a major component of Lewy bodies.[16] Mutations in this gene result in a greater propensity for the protein to form insoluble aggregates, or disrupt binding of the protein to synaptic vesicles.[17] UCH-L1 is involved in the scavenging of ubiquitin after it has been tagged to damaged molecules that have been targeted for proteosomal degradation.[18] A mis-sense mutation was found in a German kindred with early onset autosomal dominant PD.[19]

The *Parkin* gene (encoded in chromosome 6) mutations are multiple, including exon deletion, multiplications and point mutations.[20-23] Apart from the neuronal loss, the distinctive histopathological feature was the lack of Lewy bodies. It was first identified in several Japanese families with autosomal recessive juvenile parkinsonism,[21] that had prolonged survival and levodopa responsiveness. This gene is now identified to be an ubiquitin-protein ligase, and mutations cause loss of its activity.[24] It then results in accumulation of proteins which causes cell death. *Parkin* gene mutations have been associated with sporadic juvenile-onset PD,[25] and adult-onset, clinically typical tremor dominant parkinsonism.[26] Similar distribution of mRNA expression with α-synuclein suggested a shared role in the pathogenesis of PD.[9]

5. GENDER AS A RISK FACTOR

Despite the fact that women live longer than men, a large proportion of epidemiological studies show a lesser female prevalence compared to males across different racial groups,[27] with male to female ratios ranging from 1.51 to 3.7. This gender difference may be due to a number of estrogenic effects on nigrostriatal neurons. Estrogen can stimulate dopamine synthesis by directly enhancing striatal tyrosine hydroxylase activity and has both presynaptic and postsynaptic effects. It also indirectly affects the nigrostriatal system via opioid, glutamatergic and aminobutyric acid systems. We have shown that estrogen can downregulate human COMT transcription, protein synthesis and activity via estrogen receptors,[28,29] as these effects can be blocked by a specific estrogen-receptor antagonist. In clinical settings, we also demonstrated that estrogen could help motor performance of post-menopausal PD patients.[30] Moreover, estrogen also confers neuroprotection.

6. INHERITED FACTORS VERSUS ENVIRONMENTAL FACTORS

It would be difficult to explain that inherited factors alone can cause most sporadic cases of idiopathic PD. Secondary and familial cases of parkinsonism are rare compared to the number of sporadic cases. Environmental factors, such as smoking appears to have a protective effect. Manganese and carbon monoxide poisoning, and environmental models of parkinsonism exist. Conversely, no geographical clusters of idiopathic parkinsonism exist. Conjugal cases are very rare, even in marriages lasting many years. Furthermore, a widely distributed environmental toxin has yet to be identified.

An increasingly held hypothesis is shown in Figure 1. At one extreme, inherited factors predominate to the extent that no matter whether the individuals are exposed to environmental toxins or not, they will develop parkinsonism. At the other end of the spectrum, MPTP-parkinsonism exists where environmental factor is so predominant that the individuals will develop parkinsonism whether or not they have an inherited susceptibility. In the vast majority of patients who have sporadic PD, both environment and inherited factors appear to play a role in determining the function of genes involved in its pathogenesis.

7. AN EXOGENOUS TOXIN CAN INDUCE MUTATIONS IN SUPEROXIDE DISMUTASE-1 (SOD1) GENE SIMILAR TO THAT FOUND IN FAMILIAL AMYOTROPHIC LATERAL SCLEROSIS (ALS)

We will mention an experiment to illustrate this hypothesis. This is based on the role of the SOD1 gene in a similar neurodegenerative disorder, in ALS. There are rare familial cases of ALS just like PD, and mutations in SOD1 have been linked to these familial cases of ALS. The hypothesis is that toxins can induce mutations in cultured neurons with deficient DNA mismatch repair. The extension of this hypothesis is that accumulated mutations in these neurons may result in neurodegeneration.

Our aim was to determine SOD1 mutations caused by methyl methane sulphonate (MMS), a synthetic methylating agent, on a neuroblastoma cell line (SH-SY5Y). We used a breast carcinoma cell line (MCF-7) as a comparison. The cells were exposed to

MMS to determine the concentration where 50% of death rates (LD_{50}) occurred in each cell line. We then exposed the cells to MMS at this concentration for a week. The RNA extracted was then reverse-transcribed and amplified to obtain the SOD1 cDNA. The

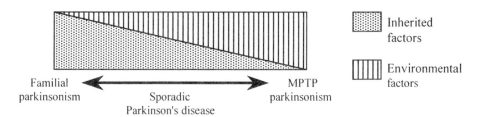

Figure 1. Inherited versus environmental factors in Parkinson's disease.

cDNA was then sequenced to identify any mutations. We also determined the DNA mismatch repair activities in an indirect manner after MMS exposure. PCR amplification of extracted DNA was carried out after MMS exposure using known markers.[31,32] The microsatellite instability pattern was analyzed with gel electrophoresis, which is indicative of microsatellite structure associated with genes encoding for DNA mismatch repair enzymes.

The LD_{50} in breast carcinoma cells was not surprisingly higher, almost twice, as they are more robust than the neuroblastoma cells. Seven mutations were found in the SOD1 gene, two of which were double mutations in the same mutant clone. These MMS-induced point mutations in SOD1 cDNA were G to A, or A to G substitutions, similar to those mutations found the SOD1 gene in familial ALS. One of them was identical. The mutations were also spread out within the SOD1 cDNA. No mutations were found in the breast carcinoma cells, suggesting that the DNA repair mechanism in these cells is more effective. The microsatellite instability pattern in both cell lines were not affected by MMS suggesting that the inherent DNA repair enzymes were not affected by MMS.

It appeared from our experiment that SOD1 gene in a neuronal cell line was more susceptible to MMS-induced mutation than non-neuronal cells. MMS can cause mutations in SOD1 gene similar to familial ALS. It is also conceivable that a lower DNA mismatch repair in neuronal cells may not be able to correct the mutations induced by MMS as effectively as in non-neuronal cells. Inheritance would determine how effective is the DNA repair mechanism to combat this recurrent environmental exposure.

8. CONCLUSIONS

Pathogenic mechanisms in familial PD is relevant to sporadic cases, and that they are more closely associated than we are led to believe from candidate gene association studies. Apart from those genes that have been mentioned above, there may well be more genes identified in future that could be linked to sporadic PD. The future lies with a shift to proteonomics from genomics. Proteins and their activities are indicators of gene function, and not just gene structure. *De novo* mutations may not just occur in constitutive genes but also in gene transcripts and proteins. And finally, interactions between environmental and inherited factors are important in the expression of genes involved in sporadic PD.

REFERENCES

1. Marttila, R.J., Kaprio, J., Koskenvuo, M. and Rinne, U.K. Parkinson's disease in a nationwide twin cohort. *Neurology* 38:1217-1219, 1988.
2. Ward, C.D., Duvoisin, R.C., Ince, S.E., Nutt, J.D., Eldridge, R. and Calne, D.B. Parkinson's disease in 65 pairs of twins and in a set of quadruplets. *Neurology* 33:815-824, 1983.
3. Duvoisin, R.C., Eldridge, R., Williams, A., Nutt, J. and Calne, D. Twin study of Parkinson disease. *Neurology* 31:77-80, 1981.
4. Burns, R.S., Chiueh, C.C., Markey, S.P., Ebert, M.H., Jacobowitz, D.M. and Kopin, I.J. A primate model of parkinsonism: selective destruction of dopaminergic neurons in the pars compacta of the substantia nigra by N-methyl-4-phenyl-1,2,3,6-tetrahydropyridine. *Proc Natl Acad Sci U S A* 80:4546-4550, 1983.
5. Tanner, C.M., Ottman, R., Goldman, S.M., et al. Parkinson disease in twins: an etiologic study. *JAMA* 281:341-346, 1999.
6. Vieregge, P., Hagenah, J., Heberlein, I., Klein, C. and Ludin, H.P. Parkinson's disease in twins: a follow-up study. *Neurology* 53:566-572, 1999.
7. Elbaz, A., Grigoletto, F., Baldereschi, M., et al. Familial aggregation of Parkinson's disease: a population-based case-control study in Europe. EUROPARKINSON Study Group. *Neurology* 52:1876-1882, 1999.
8. Sveinbjornsdottir, S., Hicks, A.A., Jonsson, T., et al. Familial aggregation of Parkinson's disease in Iceland. *N Engl J Med* 343:1765-1770, 2000.
9. Solano, S.M., Miller, D.W., Augood, S.J., Young, A.B. and Penney, J.B. Expression of alpha-synuclein, parkin, and ubiquitin carboxy-terminal hydrolase L1 mRNA in human brain: genes associated with familial Parkinson's disease. *Ann Neurol* 47:201-210, 2000.
10. Polymeropoulos, M.H., Lavedan, C., Leroy, E., et al. Mutation in the alpha-synuclein gene identified in families with Parkinson's disease. *Science* 276:2045-2047, 1997.
11. Athanassiadou, A., Voutsinas, G., Psiouri, L., et al. Genetic analysis of families with Parkinson disease that carry the Ala53Thr mutation in the gene encoding alpha-synuclein. *Am J Hum Genet* 65:555-558, 1999.
12. Kruger, R., Kuhn, W., Muller, T., et al. Ala30Pro mutation in the gene encoding alpha-synuclein in Parkinson's disease. *Nat Genet* 18:106-108, 1998.
13. Vaughan, J.R., Farrer, M.J., Wszolek, Z.K., et al. Sequencing of the alpha-synuclein gene in a large series of cases of familial Parkinson's disease fails to reveal any further mutations. The European Consortium on Genetic Susceptibility in Parkinson's Disease (GSPD). *Hum Mol Genet* 7:751-753, 1998.
14. Scott, W.K., Yamaoka, L.H., Stajich, J.M., et al. The alpha-synuclein gene is not a major risk factor in familial Parkinson disease. *Neurogenetics* 2:191-192, 1999.
15. Maroteaux, L., Campanelli, J.T. and Scheller, R.H. Synuclein: a neuron-specific protein localized to the nucleus and presynaptic nerve terminal. *J Neurosci* 8:2804-2815, 1988.
16. Spillantini, M.G., Schmidt, M.L., Lee, V.M., Trojanowski, J.Q., Jakes, R. and Goedert, M. Alpha-synuclein in Lewy bodies. *Nature* 388:839-840, 1997.
17. Perrin, R.J., Woods, W.S., Clayton, D.F. and George, J.M. Interaction of human alpha-Synuclein and Parkinson's disease variants with phospholipids. Structural analysis using site-directed mutagenesis. *J Biol Chem* 275:34393-34398, 2000.
18. Finley, D. and Chau, V. Ubiquitination. *Annu Rev Cell Biol* 7:25-69, 1991.
19. Leroy, E., Boyer, R., Auburger, G., et al. The ubiquitin pathway in Parkinson's disease. *Nature* 395:451-452, 1998.
20. Hattori, N., Matsumine, H., Asakawa, S., et al. Point mutations (Thr240Arg and Gln311Stop) [correction of Thr240Arg and Ala311Stop] in the Parkin gene. *Biochem Biophys Res Commun* 249:754-758, 1998.
21. Kitada, T., Asakawa, S., Hattori, N., et al. Mutations in the parkin gene cause autosomal recessive juvenile parkinsonism. *Nature* 392:605-608, 1998.
22. Leroy, E., Anastasopoulos, D., Konitsiotis, S., Lavedan, C. and Polymeropoulos, M.H. Deletions in the Parkin gene and genetic heterogeneity in a Greek family with early onset Parkinson's disease. *Hum Genet* 103:424-427, 1998.
23. Nisipeanu, P., Inzelberg, R., Blumen, S.C., et al. Autosomal-recessive juvenile parkinsonism in a Jewish Yemenite kindred: mutation of Parkin gene. *Neurology* 53:1602-1604, 1999.
24. Shimura, H., Hattori, N., Kubo, S., et al. Familial Parkinson disease gene product, parkin, is a ubiquitin-protein ligase. *Nat Genet* 25:302-305, 2000.

25. Lucking, C.B., Durr, A., Bonifati, V., et al. Association between early-onset Parkinson's disease and mutations in the parkin gene. French Parkinson's Disease Genetics Study Group. *N Engl J Med* 342:1560-1567, 2000.

26. Klein, C., Pramstaller, P.P., Kis, B., et al. Parkin deletions in a family with adult-onset, tremor-dominant parkinsonism: expanding the phenotype. *Ann Neurol* 48:65-71, 2000.

27. Tanner, C.M., Hubble, J.P. and Chan, P. Epidemiology and genetics of Parkinson's disease. In: *Movement disorders: neurologic principles and practice*, edited by Watts, R.L. and Koller, W.C. NY: McGraw-Hill, 1997, p. 137-152.

28. Xie, T., Ho, S.L. and Ramsden, D. Characterization and implications of estrogenic down-regulation of human catechol-O-methyltransferase gene transcription. *Mol Pharmacol* 56:31-38, 1999.

29. Tsang, K.L., Jiang, H., Ramsden, D.B. and Ho, S.L. The use of estrogen in the treatment of Parkinson's disease. *Parkinsonism & Related Disorders* (In Press) 2001.

30. Tsang, K.L., Ho, S.L. and Lo, S.K. Estrogen improves motor disability in parkinsonian postmenopausal women with motor fluctuations. *Neurology* 54:2292-2298, 2000.

31. Dietmaier, W., Wallinger, S., Bocker, T., Kullmann, F., Fishel, R. and Ruschoff, J. Diagnostic microsatellite instability: definition and correlation with mismatch repair protein expression. *Cancer Res* 57:4749-4756, 1997.

32. Boland, C.R., Thibodeau, S.N., Hamilton, S.R., et al. A National Cancer Institute Workshop on Microsatellite Instability for cancer detection and familial predisposition: development of international criteria for the determination of microsatellite instability in colorectal cancer. *Cancer Res* 58:5248-5257, 1998.

INFLAMMATORY CHANGES AND APOPTOSIS IN PARKINSON'S DISEASE

Etienne C. Hirsch*

1. INTRODUCTION

Parkinson's disease is characterized by a slow and progressive neuronal loss in the nigrostriatal pathway which is involved in the clinical manifestation of the disease. Yet, this loss of dopaminergic neurons is not linear over time. Indeed, PET scan studies show that dopaminergic neurons degenerate rapidly at early stages of the disease and that degeneration progresses more slowly as the disease evolves (Calne et al., 1997). This suggests that the mechanisms involved in the death of dopaminergic neurons may differ at the beginning and end stage of the disease. This concept is further supported by the analysis of human subjects intoxicated by 1-methyl-4-phenyl-1,2,3,6,-tetrahydropyridine (MPTP). PET scan studies have shown that after the initial insult with this toxic compound, the loss of dopaminergic neurons continued to progress despite the fact that the patients were no longer in contact with the drug (Calne et al., 1997). A recent autopsy study of the brain of 3 patients who had been in contact with MPTP several years before their death evidenced the presence of reactive microglial cells in their substantia nigra (Langston et al., 1999). These data suggest that MPTP, having induced a primary insult to dopaminergic neurons set in motion a self-perpetuating phenomenon leading to a progressive loss of dopaminergic neurons. Furthermore, the presence of reactive microglial cells in the substantia nigra of the patients suggests the possible involvement of these cells in the degenerative process. From a more general point of view, this indicates that the glial reaction could be involved in the slow and progressive degeneration of dopaminergic neurons occurring in Parkinson's disease after the initial phase of the disease.

*Etienne C. Hirsch, INSERM U 289, Experimental Neurology and Therapeutics, Hôpital de la Salpêtrière – 47, Boulevard de l'Hôpital – 75651 Paris Cedex 13 – France

Mapping the Progress of Alzheimer's and Parkinson's Disease
Edited by Mizuno *et al.*, Kluwer Academic/Plenum Publishers, 2002

259

2. GLIAL CELLS AND INFLAMMATORY PROCESSES IN PARKINSON'S DISEASE

The presence of activated microglial cells in the substantia nigra of patients with Parkinson's disease, reported by McGeer and coworkers in their pioneering work, suggests their involvement in the pathological process (McGeer et al., 1988). The mechanism by which these glial cells may participate in the death of the dopaminergic neurons is not known. Nevertheless, the glial cells are involved in the production of pro-inflammatory cytokines, which at high concentrations have been shown to play a deleterious role for neurons. Indeed, the presence of glial cells expressing tumor necrosis factor-α (TNF-α) interleukin-1β (Il-1β) and interferon-γ (IFN-γ) has been reported in the substantia nigra of aged normal human control subjects and patients with Parkinson's disease (Hunot et al., 1999). Yet, the density of these glial cells was dramatically and significantly increased in the substantia nigra of patients with Parkinson's disease. Furthermore, Mogi and co-workers have reported the presence of an increased concentration of pro-inflammatory cytokines in the CSF and the striatum of patients with Parkinson's disease, where the terminals of dopaminergic neurons degenerate (Mogi et al., 1994a; Mogi et al., 1994b; Mogi et al., 1996a; Mogi et al., 1996b; Mogi et al., 1999). These data suggest an involvement of the pro-inflammatory cytokines secreted by the glial cells in the degeneration of dopaminergic neurons in Parkinson's disease. This concept is supported by *in vitro* data showing that TNF-α can induce in a dose-dependent manner the loss of dopaminergic neurons in rat primary mesencephalic cultures (McGuire and Carvey, 2001; personal communication). The mechanism by which TNF-α could act on dopaminergic neurons is not fully understood. Yet, these cytokines may play a deleterious role for the dopaminergic neurons by at least two non exclusive mechanisms.

First, the pro-inflammatory cytokines may induce the production of a diffusible compound such as nitric oxide, which may penetrate the dopaminergic neurons, where were it could play a deleterious role (Hunot et al., 1996). Thus, *in vitro* studies on an astrocytoma cell line showed that pro-inflammatory cytokines can induce the expression of the low affinity IgE receptor CD23, and that when these receptors were activated by an appropriate ligand, it induced the expression of the inducible form of nitric oxide synthase (iNOS, responsible for the production of nitric oxide). In line with this, the density of glial cells expressing CD23 or iNOS was increased in the substantia nigra of patients with Parkinson's disease (Hunot et al., 1999; Hunot et al., 1996). This increased density of glial cells expressing inducible iNOS may account for the increased concentration of nitrotyrosine in the substantia nigra and cerebrospinal fluid of patients with Parkinson's disease (Qurechi et al., 1995). Thus, nitric oxide may alter normal neuronal functions by a nitration of their tyrosines. Nitric oxide may also play a deleterious role by interfering with iron metabolism. Thus, nitric oxide may interfere with the traduction of ferritin, which is the major iron-binding protein in biological systems. In line with this, nitric oxide has been shown to block the traduction of ferritin required in the event of an increase in iron concentrations (Klausner et al., 1993). Thus, the increased concentration of nitric oxide reported in the parkinsonian substantia nigra may also account for the lack of increase in ferritin concentrations despite the increased level of iron in the parkinsonian substantia nigra (Faucheux et al., 2001). So, by inducing the expression of a diffusible factor, the pro-inflammatory cytokines may play a deleterious role for dopaminergic neurons.

The second mechanism by which pro-inflammatory cytokines may play a deleterious role for dopaminergic neurons may be mediated by specific receptors located on the plasmic membrane of dopaminergic neurons. Type I and type II receptors for TNF-α have been shown to be expressed on dopaminergic neurons (Boka et al., 1994). Interestingly, TNF-α receptors are known to belong to the family of receptors with a death domain. When activated, these receptors induce a complex transduction pathway leading to the death of the target cells by apoptosis. When stimulated by its ligand, TNF receptors are trimerized and this induces a change in the conformation of its intracytoplasmic loop. This trimerization of TNF-α receptors in turn recruits adapter molecules such as FADD or MORT1, which under normal circumstances are free in the cytoplasm. When these adapter molecules are recruited, they in turn activate proteases of the caspase family. The first caspase to be activated is caspase-8, an upstream caspase in this pathway. This in turn activates other caspases including a major effector caspase, caspase-3. Interestingly, dopaminergic neurons in the human substantia nigra also express adapter molecules and procaspases (Hartmann et al., 2000; Hartmann et al., 2001). Using highly specific antibodies recognizing the activated form of caspase-8 or caspase-3, we evidenced caspase activation in Parkinson's disease (Hartmann et al., 2000; Hartmann et al., 2001). Thus, by acting on specific receptors, the pro-inflammatory cytokines may also play a deleterious role by activating a pro-apoptotic pathway.

3. DOPAMINERGIC NEURONS DEGENERATE BY APOPTOSIS IN PARKINSON'S DISEASE BUT CASPASE INACTIVATION DOES NOT REPRESENT A THERAPEUTIC STRATEGY

The involvement of apoptosis in the degeneration of dopaminergic neurons in Parkinson's disease is still debated (for review, see Hartmann et al., 2001). Indeed, several studies using the TUNNEL technique were unable to show a specific involvement of apoptosis in Parkinson disease (for example, see Kingsbury et al., 1998). Yet, using other types of methodology, such as electron microscopy and specific nuclear fluorescent dyes, other groups of investigators have shown that dopaminergic neurons degenerate by apoptosis in Parkinson's disease (Mochizuki et al., 1996; Anglade et al., 1997; Tatton et al., 1998). Thus, from these contradictory results, it is difficult to determine whether apoptosis is really involved in the death of dopaminergic neurons in Parkinson's disease. Nevertheless, recent molecular studies showing an activation of caspase-3 and 8 (see above) rather suggest that dopaminergic neurons do indeed degenerate by apoptosis in this disease (Hartmann et al., 2000; Hartmann et al., 2001). In this context, it was tempting to speculate that a blockade of caspase activation might represent an interesting therapeutic target for Parkinson's disease. In order, to test this hypothesis, we performed an *in vitro* study to test the neuroprotective capacities of caspase inhibitors in a primary mesencephalic culture of rat embryo (Hartmann et al., 2001). In this study, the apoptotic death of dopaminergic neurons was induced by the active metabolite of MPTP (MPP$^+$), this being evidenced by the presence of chromatin condensation, observed using a nuclear staining. In this paradigm, neither specific inhibitors of caspase-3 or broad-spectrum inhibitors were able to protect dopaminergic neurons against apoptosis.

Moreover, the caspase inhibitors induced a switch from an apoptotic type of cell death to a non apoptotic form, very likely corresponding to necrosis. These data indicate that not only do caspase inhibitors not protect dopaminergic neurons against apoptosis they actually induce necrotic death of the cells. As necrosis is associated with a cytoplasmic dysfunction and a strong inflammatory response, these data suggest that caspase inhibitors would probably not protect dopaminergic neurons in Parkinson's disease but even exacerbate the loss of such neurons. Nonetheless, recent studies have shown that caspase inhibitors can induce a switch from apoptosis to necrosis when cells display energy depletion. In line with this, caspase inhibitors were able to partially protect dopaminergic neurons against MPTP intoxication when energy metabolism was restored by a glucose treatment (Hartmann et al., 2001). Yet, the co-treatment by caspase inhibitors and glucose only induced cell survival and not a restoration of dopaminergic functions, as evidence by measurements of dopamine uptake. Taking all these results into account, it would seem more likely that neuroprotective strategies for Parkinson's disease will require a polytherapy acting on different cell-death pathways to block the degeneration of dopaminergic neurons. The concept of a polytherapy has already been validated for other diseases such as AIDS and could perhaps be applied to neurodegenerative disorders.

4. CONCLUSION

The data presented here suggest that glial cells and the inflammatory reaction may participate in the cascade of events leading to nerve cell death in Parkinson's disease. They may play a role in the apoptotic death of dopaminergic neurons but prevention of apoptosis may not represent the best strategy to block neuronal degeneration in this disease. More generally, it would appear that the neuroprotective strategy in Parkinson's disease should target altered cells rather than neurons that are already engaged in an apoptotic process.

REFERENCES

Anglade, P., Vyas, S., Javoy-Agid, F., Herrero, M. T., Michel, P. P., Marquez, J., Mouatt-Prigent, A., Ruberg, M., Hirsch, E. C., Agid, Y., 1997, Apoptosis and autophagy in nigral neurons of patients with Parkinson's disease, *Histol. and Histopathol.*, **12**: 25-31.

Boka, G., Anglade, P., Wallach, D., Javoy-Agid, F., Agid, Y., Hirsch, E. C., 1994, Immunocytochemical analysis of tumor necrosis factor and its receptors in Parkinson's disease, *Neurosci. Lett.*, **172**: 151-154.

Calne, D. B., De La Fuente-Fernandez, R., Kshore, A., 1997, Contributions of positron emission tomography to elucidating the pathogenesis of idiopathic parkinsonism and dopa responsive dystonia, *J. Neural Transm.*, **50**: 47-52.

Faucheux, B. A., Martin, M. E., Beaumont, C., Hunot, S., Hauw, J. J., Agid, Y., Hirsch, E. C., 2001, Iron accumulation and deregulation of ferritin expression in the substantia nigra of patients with Parkinson's disease, (Submitted).

Hartmann, A., Hunot, S., Michel, P. P., Muriel, M.-P., Vyas, S., Faucheux, B. A., Mouatt-Prigent, A., Turmel, E., Srinivasan, A., Ruberg, M., Evan, G. I., Agid, Y., Hirsch, E. C., 2000, Caspase-3: a vulnerability factor and final effector in apoptotic death of dopaminergic neurons in Parkinson's disease, *Proc. Natl. Acad. Sci. (U.S.A.)*, **97**: 2875-2880.

Hartmann, A., Troadec, J.-D., Hunot, S., Kikly, K., Faucheux, B. A., Mouatt-Prigent, A., Ruberg, M., Agid, Y., Hirsch, E.C., 2001, Caspase-8 is an effector in apoptotic death of dopaminergic neurons in Parkinson's disease, but pathway inhibition results in neuronal necrosis, *J. Neurosci.*, **21**: 2247-2255.

Hartmann, A., Hirsch, E. C., 2000, Parkinson's disease: the apoptosis hypothesis revisited, *Adv. Neurol.*, Calne D. ed., **86**: 143-153.

Hunot, S., Boissiere, F., Faucheux, B., Brugg, B., Mouatt-Prigent, A., Agid, Y., Hirsch, E. C., 1996, Nitric oxide synthase and neuronal vulnerability in Parkinson's disease, *Neuroscience*, **72**: 355-363.

Hunot, S., Dugas, N., Faucheux, B., Hartmann, A., Tardieu, M., Debre, P., Agid, Y., Dugas, B., Hirsch, E. C., 1999, Fc(epsilon)RII/CD23 is expressed in Parkinson's disease and induces, in vitro, production of nitric oxide and Tumor Necrosis Factor-α in glial cells, *J. Neurosci.*, **19**: 3440-3447.

Kingsbury, A. E., Marsden, C. D., Foster, O. J., 1998, DNA fragmentation in human substantia nigra: apotposis or perimortem effect? *Mov. Disord.*, **13**: 877-884.

Klausner, R. D., Rouault, T.A., Harford, J. B., 1993, Reguling the fate of mRNA: the control of cellular iron metabolism, *Cell*, **72**: 19-28.

Langston, J.W., Forno, L.S., Tetrud, J., Reeves, A.G., Kaplan, J.A., Karluk, D., 1999, Evidence of active nerve cell degeneration in the substantia nigra of humans years after 1-methyl-4-phenyl-1,2,3,6-tetrahydropyridine exposure, *Ann. Neurol.*, **46**: 598-605.

McGeer, P.L., Itagaki, S., Boyes, B. E., McGeer, E. G., 1988, Reactive microglia are positive for HLA-DR in the substantia nigra of Parkinson's and Alzheimer's disease brains. *Neurology*, **38**: 1285-1291.

Mochizuki, H., Goto, K., Mori, H., Mizuno, Y., 1996, Histochemical detection of apoptosis in Parkinson's disease, *J. Neurol. Sci.*, **137**: 120-123.

Mogi, M., Harada, M., Riederer, P., Narabayashi, H., Fujita, K., Nagatsu, T., 1994, Tumor necrosis factor-alpha (TNF-alpha) increases both in the brain and in the cerebrospinal fluid from parkinsoninan patients, *Neurosci. Lett.*, **165**: 208-210.

Mogi, M., Harada, M., Kondo, J., Riederer, P., Inagaki, H., Minami, M., Nagatsu, T., 1994, Interleukin-1 beta, interleukin-6, epidermal growth factor and transforming growth factor-alpha are elevated in the brain from parkinsoninan patients, *Neurosci. Lett.*, **180**: 147-150.

Mogi, M., Harada, M., Kondo, T., Riederer, P., Nagatsu, T., 1996, Interleukin-2 but not basic fibroblast growth factor is elevated in parkinsonian brain, *J. Neural. Transm.*, **103**: 1077-1081.

Mogi, M., Harada, M., Narabayashi, H., Inagaki, H., Minami, M., Nagatsu, T., 1996, Interleukin (IL)-1 beta, IL-2, IL-4, IL-6 and transforming growth factor-alpha levels are elevated in ventricular cerebrospinal fluid in juvenile parkinsonism and Parkinson's disease, *Neurosci. Lett.*, **211**: 13-16.

Mogi, M., Togari, A., Tanaka, K., Ogawa, N., Ichinose, H., Nagatsu, T., 1999, Increase in level of tumor necrosis factor (TNF)-alpha in 6-hydroxydopamine-lesioned striatum in rats without influence of systemic L-DOPA on the TNF-alpha iinduction, *Neurosci. Lett.*, **268**: 101-104.

Qureshi, G.A., Baig, S., Bednar, I., Sodersten, P., Forsberg, G., Siden, A., 1995, Increased cerebrospinal fluid concentration of nitrite in Parkinson's disease, *NeuroReport*, **6**: 1642-1644.

Tatton, N. A., Maclean-Fraser, A., Tatton, W.G., Perl, .D.P., Olanow, C. W., 1998, A fluorescent double-labeling method to detect and confirm apoptotic nuclei in Parkinson's disease, *Ann. Neurol.*, **44**: S142-S148.

CYTOKINES AND NEUROTROPHINS IN PARKINSON'S DISEASE: INVOLVEMENT IN APOPTOSIS

Toshiharu Nagatsu[*], Makio Mogi[**], Hiroshi Ichinose[*], and Akifumi Togari[**]

1. INTRODUCTION

Parkinson's disease (PD) is characterized by deficiency of neurotransmitter dopamine (DA) in the nerve terminals of the nigrostriatal DA neurons in the substantia nigra pars compacta. Many factors are speculated to operate in the mechanism of cell death of the nigrostriatal DA neurons in PD, e.g., oxidative stress and cytotoxicity of reactive oxygen species (ROS), disturbance of intracellular calcium homeostasis, endogenous or exogenous neurotoxins, and mitochondrial dysfunctions especially in complex I of the electron transport system (Foley and Riederer, 1999). An association of altered immune responsiveness with PD that may be related to the programmed cell death (apoptosis) has also recently been speculated (Nagatsu & Mogi, 1998; Nagatsu et al., 1999; Mogi & Nagatsu, 1999c; Hirsch et al., 1999; Hartmann et al., 2000, 2001). We have obtained, specifically in the nigrostriatal DA regions of the postmortem brains of patients with PD as well as in animal models of PD, the following evidences suggestively implicating the immune response and apoptosis in PD.: (1) an increased neopterin/biopterin ratio in the cerebrospinal fluid (CSF) from patients with PD, (2) increased levels of beta2-microglobulin, the light chain of the major histocompatibility complex class I (MHC-I) molecules, (3) increased levels of proinflammatory cytokines such as TNF-alpha and IL-1beta, (4) decreased levels of neurotrophins such as BDNF and NGF, and (5) increased levels of apoptosis-related proteins or increased activities of apoptosis-producing caspases. These results suggest that the nigrostriatal DA cell death

[*]Toshiharu Nagatsu, Hiroshi Ichinose, Institute for Comprehensive Medical Science, Fujita Health University, Toyoake, Aichi, Japan 470-1192. [**]Makio Mogi, Akifumi Togari, School of Dentistry, Aichi-Gakuin University, Nagoya, Aichi, Japan 464-8650.

Mapping the Progress of Alzheimer's and Parkinson's Disease
Edited by Mizuno *et al.*, Kluwer Academic/Plenum Publishers, 2002

in PD may be caused by a glial immune response and subsequent apoptosis (for reviews, see Nagatsu et al., 2000a, 2000b).   Although the concept of programmed cell death (apoptosis) in PD is still controversial in itself, these data from postmortem brains of PD and animal models of parkinsonism suggest the presence of a proapoptotic environment in the nigrostriatal DA region in PD.

2. CHANGES IN CYTOKINES, NEUROTROPHINS, AND APOPTOSIS-RELATED PROTEINS IN THE BRAIN AND CEREBROSPINAL FLUID IN PARKINSON'S DISEASE

The brain is generally considered to be a "privileged" site, i.e., one free from immune reactions. However, recent findings revealed that immune responses may occur in the brain, probably by microglia that produce proinflammatory cytokines (Benveniste, 1992; Sei et al., 1995).

We found a significant increase in the neopterin/biopterin ratio in the CSF (Fujishiro et al., 1990) and an increased level of beta2-microglobulin (the light chain of MHC-I molecules) in the striatum in PD (Mogi et al., 1995b), which suggest the increased production of neopterin and beta2-microglobulin from activated glial cells in the nigrostriatal region.   Therefore, we measured various cytokines such as TNF-alpha, IL-1beta, IL-2, IL-6, EGF, TGF- alpha, and TGF-beta1 in the brain (striatum) and in the ventricular and lumbar CSF (VCSF, LCSF; Mogi et al., 1994a, 1994b, 1995a, 1996a, 1996b, 1996c). The levels of all cytokines were elevated specifically in the striatum, but not in the cerebral cortex in PD.   Cytokine levels in the LCSF were generally low compared with those in the VCSF except for the TNF-alpha level.   Other cytokines were detected in the VCSF, and their levels were significantly higher in PD or juvenile parkinsonism. We also found that the level of theTNF receptor R1 (TNF-R1, p53) was elevated in the substantia nigra in PD in comparison with that of controls (Mogi et al., 2000a).   In agreement with our results obtained by the use of the enzyme-linked immunoadsorbent assay (ELISA), Boka et al. (1994) found TNF-alpha-immunoreactive glial cells and TNFR1-positive DA neurons in the substantia nigra in PD; and other workers also reported increased cytokine levels in PD, e.g., those of IL-1beta and IL-6 in LCSF (Blum-Degan et al., 1995) and of TGF-beta1 and TGF-beta2 in VCSF (Vawter et al., 1996).

In contrast to increased cytokine levels, the levels of neurotrophins, which support the differentiation and survival of DA neurons, were decreased specifically in the nigrostriatal region in PD; i.e., the concentrations of NGF (the order of pg/mg protein) and BDNF (the order of ng/mg protein) were markedly decreased (Mogi et al., 1999b). However, we did not find significant changes in the PD brain in the concentration of glial cell line-derived neurotrophic factor (GDNF), which has a potent neuroprotective effect on DA neurons (Mogi et al., 2001). These results suggest that the unchanged level of GDNF in PD could be due to a compensatory increase from glial cells, which occurs when neither BDNF nor NGF is produced by neurons.   Both the increases in proinflammatory cytokines and the decreases in neurotrophins could produce a proapoptotic environment in the nigrostriatal region in PD.

We also found increases in apoptosis-related factors specifically in the nigrostriatal regions in PD, i.e., in the contents of soluble Fas (Mogi et al., 1996d) and antiapoptotic protein Bcl-2 (Mogi et al., 1996b) and in the activities of caspase 1 and caspase 3 (Mogi et al., 2000a). Caspase 3 is thought to be a vulnerability factor and the final effector in the apoptotic death of DA neurons in PD (Hartmann et al., 2000). There were significantly positive correlations between the levels of soluble Fas and those of IL-1beta, IL-6, Bcl-2, and TNF-alpha. Caspase 8, which is an effector proximal to TNF-alpha action, was found to be activated in MPTP-treated parkinsonian mice early in the course of cell demise, suggesting that caspase 8 activation precedes and is not the consequence of cell death. However, caspase 8 inhibitors did not result in neuroprotection but in neuronal necrosis, probably due to ATP depletion (Hartmann et al., 2001).

3. CHANGES IN CYTOKINES AND NEUROTROPHINS IN THE BRAIN IN PARKINSONIAN ANIMAL MODELS

We examined the changes in cytokines and neurotrophins in the nigrostriatal region in parkinsonian animal models. In PD mice produced by repeated intraperitoneal injection of 1-methyl-4-phenyl-1,2,3,6-tetrahydropterin (MPTP), the IL-1beta concentration was increased 23-fold and NGF concentration was decreased to about 50%, but only in the striatum (Mogi et al., 1998). These results of the increased IL-1beta and decreased NGF levels specifically in the striatum of MPTP-treated PD mice agree with our data from postmortem human brains. In hemiparkinsonian rats produced by injecting 6-hydroxydopamine (6-OHDA) into one side of the ventrotegmental bundle without or with L-DOPA treatment, the levels of TNF-alpha were significantly increased only in the substantia nigra and striatum of the injected side. L-DOPA administration did not produce any significant changes in TNF-alpha levels in 6-OHDA-treated or control sides of any of the brains (Mogi et al., 1999a). These results agree with the changes in the TNF-alpha levels in the striatum and LCSF in PD patients, and also suggest that the increased cytokine levels in PD patients may not be due to the secondary effects of L-DOPA therapy. We found that the increase in the level of TNF-alpha in rats with a 6-OHDA-lesioned striatum was suppressed by immunophylin ligand FK506 (Mogi et al., 2000b). Immunosuppressant FK506 may prevent 6-OHDA-induced activation of microglial cells, resulting in a decrease in TNF-alpha induction in the nigrostriatal DA region of the rats.

MPTP produces cytotoxic cell death of the nigrostriatal DA neurons. There have been found various MPTP-like neurotoxins in the brain in PD, e.g., tetrahydroisoquinolines (TIQs) and beta-carbolines (for a review, see Nagatsu, 1997). R-N-Methyl-salsolinol (R-N-methyl-5,6-dihydroxy-TIQ) was reported to inhibit complex I in mitochondria, thus producing ROS, and to cause apoptotic cell death (Naoi et al., 1998). Chronic, systemic inhibition of mitochondrial complex I by the lipophilic pesticide rotenone used at a concentration (20-30 nM) high enough to partially inhibit complex I, but too low to significantly impair respiration of brain mitochondria and probably to produce a bioenergic defect with ATP depletion, was reported to cause highly selective nigrostriatal DA degeneration that was associated behaviorally with parkinsonism and was accompanied neuropathologically by Lewy body-like pale body, i.e., eosinophilic fibrillar cytoplasmic inclusions containing ubiquitin and alpha-synuclein (Betarbet et al., 2000). Evidence for complex I deficiency in PD has been well established (for a review, see Mizuno et al., 1998). Since an adequate intracellular ATP level determines cell death fate by apoptosis rather than by necrosis (Eguchi et al., 1997), partial inhibition of complex I may stimulate production of ROS even without ATP depletion, and may ultimately cause apoptotic cell death by inducing the release of cytochrome c from mitochondria to enhance a-synuclein aggregation into Lewy bodies (Hashimoto et al., 2000).

4. THE FUTURE PROSPECT OF APOPTOSIS THEORY IN PARKINSON'S DISEASE

We have shown the increased levels of proapoptotic cytokines, the decreased levels of antiapoptotic neurotrophins, and the increased levels of apoptosis-related factors, which may indicate the apoptotic cell death of the nigrostriatal DA neurons. However, there still remain several questions to be asked about the roles of cytokines and neurotrophins in PD. First, the origin of cytokines and neurotrophins should be clarified. Immune-activated glial cells, i.e., microglia and astrocytes, may produce cytokines and neurotrophins, but neurons may also do so. The increases in cytokines in the nigrostriatal DA region suggest the induction of apoptotic cell death of DA neurons, but glial cells may also die first by apoptosis, and then the DA neurons. Second, the increased levels of cytokines and the decreased levels of neurotrophins may cause apoptotic cell death, but could be secondary and compensatory responses.

The recent discovery of the causative gene of familial PD, i.e., alpha-synuclein for autosomal dominant PD (Polymeropoulos et al., 1997) and Parkin for autosomal recessive juvenile PD (Kitada et al., 1998), could give important clues for elucidating the mechanism of apoptotic neuronal cell death in sporadic PD. Parkin was identified as a ubiquitin ligase E3 (Shimura, 2000). The accumulation of abnormal proteins in the nigrostriatal DA neurons in familial PD may also be linked to the apoptotic pathways that may be activated by oxidative stress and mitochondrial dysfunction in sporadic PD. Drugs that increase the production of neurotrophins such as apomorphine (Ohta et al., 2000) may be neuroprotective by inhibiting apoptotic cell death. Immunophylin ligands may prevent a slow and progressive DA axonal degeneration and neuronal death in vivo and may have therapeutic value in PD (Ogawa, 1999; Costantini et al., 2001).

5. ACKNOWLEDGEMENTS

Our work cited in this review was supported by funding from the program Health Science Research Grants, Research on Brain Science of the Ministry of Health and Labor of Japan. We are grateful to our collaborators, Drs. Yoshikuni Mizuno, Norio Ogawa, Sadako Kuno, and their coworkers, and to Dr. Peter Riederer.

REFERENCES

Benveniste, E. N., 1992, Inflammatory cytokines within the central nervous system: source, function, and mechanisms of action, *Am. J. Physiol.* 263: C1-C16.

Betarbet, R., Sherer, T. B., Mackenzie, G., Garcia-Osuna, M., Panov, A. V., and Greenamyre, J. T., 2000, Chronic systemic pesticide exposure reproduces features of Parkinson's disease, *Nature Neurosci.* 3: 1301-1306.

Blum-Degan, D., Muller, T., Kuhn, W., Gerlach, M., Przuntek, H., and Riederer, P., 1995, Interleukin 1-beta and interleukin-6 are elevated in the cerebrospinal fluid of Alzheimer's and de novo Parkinson's disease patients, *Neurosci. Lett.* 202: 17-20.

Boka, G., Anglade, P., Wallach, D., Javoy-Agid, F., Agid, Y., and Hirsch, E. C., 1994, Immunocytochemical analysis of tumor necrosis factor and its receptor in Parkinson's disease, *Neurosci. Lett.* 172: 151-154.

Costantini, L. C., Cole, D., Chaturvedi, P., and Isacson, O., 2001, Immunophilin ligands can prevent progressive dopaminergic degeneration in animal models of Parkinson's disease, *Europ. J. Neurosci.* 13: 1085-1092.

Eguchi, Y., Shimizu, S., and Tsujimoto, Y., 1997, Intracellular ATP levels determine cell death fate by apoptosis or necrosis, *Cancer Res.* 57: 1835-1840.

Foley, P., and Riederer, P., 1999, Pathogenesis and preclinical course of Parkinson's disease, *J. Neural Transm. Suppl.* 56: 31-74.

Fujishiro, K., Hagihara, M., Takahashi, A., and Nagatsu, T., 1990, Concentrations of neopterin and biopterin in the cerebrospinal fluid of patients with Parkinson's disease, *Biochem. Med. Metab. Biol.* 44: 97-100.

Hashimoto, M., Takeda, A., Hsu, L. J., Takenouchi, T., and Masliah, E., 2000, Role of cytochrome c as a stimulator of alpha-synuclein aggregation in Lewy body disease, *J. Biol. Chem.* 274: 28849-28852.

Hartmann, A., Hunot, S., Michel, P. P., Muriel, M. P., Vyas, S., Faucheux, B. A., Mouatt-Prigent, A., Turmel, H., Srinivasan, A., Ruberg, M., Evan, G. I., Agid, Y., and Hirsch, E. C., 2000, Caspase-3: a vulnerability factor and a final effector in the apoptotic cell death of dopaminergic neurons in Parkinson's disease, *Proc. Natl. Acad. Sci. USA* 97: 2875-2880.

Hartmann, A., Troadec, J. -D., Hunot, S., Kikly, K., Faucheux, B. A., Mouatt-Prigent, A., Ruberg, M., Agid, Y., and Hirsch, E. C., 2001, Caspase-8 is an effector in apoptotic death of dopaminergic neurons in Parkinson's disease, but pathway inhibition results in neuronal necrosis, *J. Neurosci.* 21: 2247-2255.

Hirsch, E. C., Hunot, S., Faucheux, B., Agid, Y., Mizuno Y., Mochizuki, H., Tatton, W. G., Tatton, N., and Olanow, W. C., 1999, Dopaminergic neurons degenerate by apoptosis in Parkinson's disease, *Mov. Disord.* 14: 383-385.

Kitada, T., Asakawa, S., Hattori, N., Matsumine, H., Yamamura, Y., Minoshima, S., Yokochi, M., Mizuno, Y., and Shimizu, N., 1998, Mutations in the parkin gene cause autosomal recessive juvenile parkinsonism, *Nature* 392: 605-608.

Mizuno, Y., Yoshino, H., Ikebe, S., Hattori, N., Kobayashi T., Shimoda-Matsubayashi, S., Matsumine, H., and Kondo, T., 1998, Mitochondrial dysfunction in Parkinson's disease, *Ann Neurol.* 44 Suppl. 1: S99-S109.

Mogi, M., Harada, M., Riederer, P., Narabayashi, H., Fujita, K., and Nagatsu, T., 1994a, Tumor necrosis factor-alpha (TNF-alpha) increases both in the brain and in the cerebrospinal fluid from parkinsonian patients, *Neurosci. Lett.* 165: 208-210.

Mogi, M., Harada, M., Kondo, T., Riederer, P., Inagaki, H., Minami, M. and Nagatsu, T., 1994b, Interleukin-1beta, interleukin-6, epidermal growth factor and transforming growth factor-alpha are elevated in the brain from parkinsonian patients, *Neurosci. Lett.* 180: 147-150.

Mogi, M., Harada, M., Kondo, T., Narabayashi, H., Riederer, P., and Nagatsu, T., 1995a, Transforming growth factor-beta1 levels are elevated in the striatum and in ventricular cerebrospinal fluid in Parkinson's disease, *Neurosci. Lett.* 193: 129-132.

Mogi, M., Harada, M., Kondo, T., Riederer, P., and Nagatsu, T., 1995b, Brain beta2-microglobulin levels are elevated in the striatum in Parkinson's disease, *J. Neural Transm. [P-D Sect]* 9: 87-92.

Mogi, M., Harada, M., Narabayashi, H., Inagaki, H., Minami, M., and Nagatsu, T., 1996a, Interleukin (IL)-1beta, Il-2, IL-4, IL-6 and transforming growth factor-alpha levels are elevated in ventricular cerebrospinal fluid in juvenile parkinsonism and Parkinson's disease, *Neurosci. Lett.* 211: 13-16.

Mogi, M., Harada, M., Kondo, T., Mizuno, Y., Narabayashi, H., Riederer, P., and Nagatsu, T., 1996b, Bcl-2 protein is increased in the brain from parkinsonian patients, *Neurosci. Lett.* 215: 137-139.

Mogi, M., Harada, M., Kondo, T., Riederer, P., and Nagatsu, T., 1996c, Interleukin-2 but not basic fibroblast growth factor is elevated in parkinsonian brain, *J. Neural Transm.* 103: 1077-1081.

Mogi, M., Harada, M., Kondo, T., Mizuno, Y., Narabayashi, H., Riederer, P., and Nagatsu, T., 1996d, The soluble form of Fas molecule is elevated in parkinsonian brain tissues, *Neurosci. Lett.* 220: 195-198.

Mogi, M., Togari, A., Ogawa, M., Ikeguchi, K., Shizuma, N., Fan, D. -S., Nakano, I., and Nagatsu, T., 1998,

Effects of repeated systemic administration of 1-methyl-4-phenyl-1,2,3,6-tetrahydropyridine (MPTP) to mice on interleukin-1beta and nerve growth factor in the striatum, *Neurosci. Lett.* 250: 25-28.

Mogi, M., Togari, A., Tanaka, K., Ogawa, N., Ichinose, H., and Nagatsu, T., 1999a, Increase in level of tumor necrosis factor (TNF)-alpha in 6-hydroxydopamine-lesioned striatum in rats without influence of systemic L-DOPA on the TNF-alpha induction, *Neurosci. Lett.* 268: 101-104.

Mogi, M., Togari, A., Kondo, T., Mizuno, Y., Komure, O., Kuno, S., Ichinose, H., and Nagatsu, T., 1999b, Brain derived growth factor and nerve growth factor concentrations are decreased in the substantia nigra in Parkinson's disease, *Neurosci. Lett.* 270: 45-48.

Mogi, M., and Nagatsu, T., 1999c, Neurotrophins and cytokines in Parkinson's disease, *Adv. Neurol.* 80: 135-139.

Mogi, M., Togari, A., Kondo, T., Mizuno, Y., Komure, O., Kuno, S., Ichinose, H., and Nagatsu, T., 2000a, Caspase activities and tumor necrosis factor receptor R1 (p55) level are elevated in the substantia nigra from parkinsonian brain, *J. Neural Transm.* 107: 335-341.

Mogi, M., Togari, A., Tanaka, K., Ogawa, N., Ichinose, H., and Nagatsu, T., 2000b, Increase in level of tumor necrosis factor TNF-alpha in 6-hydroxydopamine-lesioned striatum in rats is suppressed by immunosuppressant FK506, *Neurosci. Lett.* 289: 165-168.

Mogi, M., Togari, A., Kondo, T., Mizuno, Y., Komure, O., Kuno, S., Ichinose, H., and Nagatsu, T., 2001, Glial cell line-derived neurotrophic factor in the substantia nigra from control and parkinsonian brains, *Neurosci. Lett.* 300: 179-181.

Nagatsu, T., 1997, Isoquinoline neurotoxins in the brain and Parkinson's disease, *Neurosci. Res.* 29: 99-111.

Nagatsu, T., and Mogi, M., 1998, Cytokines in Parkinson's disease, *Adv. Behav. Biol.* 49: 407-412.

Nagatsu, T., Mogi, M., Ichinose, H., Togari, A., and Riederer, P., 1999, Cytokines in Parkinson's disease, *Neurosci. News* 2: 88-90.

Nagatsu, T., Mogi, M., Ichinose, H., and Togari, A., 2000a, Cytokines in Parkinson's disease, *J. Neural Transm. Suppl.* 58: 143-151.

Nagatsu, T., Mogi, M., Ichinose, H., and Togari, A., 2000b, Changes in cytokines and neurotrophins in Parkinson's disease, *J. Neural Transm. Suppl.* 60: 277-290.

Naoi, M., Maruyama, W., Kasamatsu, T., and Dostert, P., 1998, Oxidation of N-methyl (R) salsolinol: involvement to neurotoxicity and neuroprotection by endogenous catechol isoquinolines, *J. Neural Transm. Suppl.* 52: 125-138.

Ogawa, N., 1999, Immunophilin ligands: possible neuroprotective agents, *NeuroSci. News* 2: 28-34.

Ohta, M., Mizuta, K., Ohta, K., Nishimura, M., Mizuta, E., Hayashi, K., and Kuno, S., 2000, Apomorphine up-regulates NGF and GDNF synthesis in cultured mouse astrocytes, *Biochem. Biophys. Res. Commun.* 272: 18-22.

Polymeropoulos, M. H., Lavedan, C., Leroy, E., Ide, S. E., Dehejia, A., Dutra, A., Pike, B., Root, H., Rubenstein, J., Boyer, R., Stenroos, E. S., Chandrasekharappa, S., Athanassiadou, A., Papapetropoulos, T., Johnson, W. G., Lazzarini, A. M., Duvoisin, R. C., Di Iorio, G., Golbe, L. I., and Nussbaum, R. L., 1997, Mutation in the alpha-synuclein gene identified in families with Parkinson's disease, *Science* 276: 2045-2047.

Shimura, H., Hattori, N., Kubo, S., Mizuno, Y., Asakawa, S., Minoshima, S., Shimizu, N., Iwai, K, Chiba, T., Tanaka, K., and Suzuki, T., 2000, Familial Parkinson disease gene product, parkin, is a ubiquitin-protein ligase, *Nature Genet.* 25: 302-305.

Sei, Y., Vitkovik, L., and Yokoyama, M. M., 1995, Cytokines in the central nervous system: regulatory roles in neuronal function, cell death and repair, *Neuroimmunomodulation* 2: 121-133.

Vawter, M. P., Dillon-Carter, O., Tourtellotte, W. W., Carvey, P., and Freed, W. J., 1996, TGF beta1 and TGF beta2 concentrations are elevated in Parkinson's disease in ventricular cerebrospinal fluid, *Exp. Neurol.* 142: 313-322.

CHRONIC COMPLEX I INHIBITION REPRODUCES FEATURES OF PARKINSON'S DISEASE
Pesticides and Parkinson's disease

Todd B. Sherer, Ranjita Betarbet, and J. Timothy Greenamyre[*]

1. INTRODUCTION

Parkinson's disease (PD) is a late-onset, progressive disease characterized by motor and postural defects. Pathologically, PD is marked by specific nigrostriatal dopaminergic degeneration and the formation of Lewy bodies, cytoplasmic inclusions containing ubiquitin and α-synuclein.[1] Current therapeutic approaches for PD focus on alleviating symptoms that occur in the late-stages. An understanding of PD pathogenesis and the development of accurate models of the disease are essential for the development of novel therapies targeted at preventing the progression of PD.

2. PD PATHOGENESIS AND THE MPTP MODEL

Genetic analyses of rare cases of familial PD have uncovered mutations in proteins such as parkin and α-synuclein that may play a role in PD pathogenesis.[2-4] However, the role of genetics in the more typically encountered sporadic PD is uncertain. Additionally, epidemiological studies suggest an association between exposure to pesticides and the risk of developing PD.[5,6] An example of an environmental toxin that leads to a parkinsonian-like syndrome in primates and mice is 1-methyl-4-phenyl-1,2,3,6-tetrahydropyridine (MPTP).[7]

MPTP exposure has been frequently used as a model system for studying PD pathogenesis. The selectivity of the active metabolite of MPTP, 1-methyl-4-pyridinium (MPP$^+$), for dopaminergic neurons is due to the fact that it is a substrate for the dopamine transporter.[8] MPP$^+$ is also a potent mitochondrial poison that inhibits mitochondrial respiration at complex I of the electron transport chain.[9] Following the demonstration of

[*] Todd B. Sherer, Ranjita Betarbet, J. Timothy Greenamyre, Department of Neurology, Emory University, Atlanta, GA 30022.

Mapping the Progress of Alzheimer's and Parkinson's Disease
Edited by Mizuno *et al.*, Kluwer Academic/Plenum Publishers, 2002

MPTP's toxicity, several laboratories reported a modest but reproducible reduction in complex I activity in PD patients. This defect appears to be systemic, affecting brain as well as other tissues including muscle and platelets.[10-12]

3. DEVELOPMENT OF A NOVEL MODEL OF PD: CHRONIC ROTENONE EXPOSURE

An accurate *in vivo* experimental model of PD should reproduce the neuropathological features seen in PD, examine the relevance of the systemic mitochondrial impairment, and address the potential involvement of pesticide exposure in PD pathogenesis. No such model exists incorporating all these characteristics of PD. MPTP exposure causes selective nigrostriatal degeneration through complex I inhibition. However, since MPP^+ is a substrate for the dopamine transporter, mitochondrial inhibition due to MPTP exposure is limited to dopaminergic neurons. Thus, the MPTP model does not test the involvement of *systemic* mitochondrial dysfunction in PD.

Due to the limitations of the MPTP model, we developed a novel *in vivo* model of PD based upon chronic, systemic exposure of rats to the common pesticide, rotenone.[13] Chronic drug exposure was chosen to mimic the progressive nature of PD. Rotenone is a naturally occurring compound used as an insecticide in gardens and to sample or kill fish populations in lake. Rotenone is generally considered a safe alternative to synthetic pesticides. Rotenone is also a well characterized, potent, specific inhibitor of mitochondrial complex I. Importantly, rotenone is highly lipophilic and crosses biological membranes easily independent of the expression of transporters. Therefore, rotenone, unlike MPTP, is suited for examining the relevance of systemic reductions in complex I activity in PD.

4. ROTENONE INFUSION SELECTIVELY AND UNIFORMLY INHIBITS COMPLEX I ACTIVITY

Male Lewis rats (250-275g) were infused continuously with 2-3mg/kg/day of rotenone by jugular vein cannula attached to a subcutaneous osmotic minipump. Rats received rotenone for 1-5 weeks. Rotenone-infused animals demonstrated about a 75% reduction in [^{3}H] dihydrorotenone binding[14] to complex I in the brain (Fig 1). Systemic rotenone infusion affected complex I throughout the brain uniformly. Rotenone infusion did not inhibit the enzymatic activities of complex II and complex IV, measured by succinate dehydrogenase and cytochrome oxidase histochemistry, respectively (Fig 1).

5. SELECTIVE DEGENERATION OF THE NIGROSTRIATAL DOPAMINERGIC PATHWAY

Systemic, partial inhibition of complex I caused progressive nigrostriatal dopaminergic degeneration. Rotenone-infused animals demonstrated varying degrees of striatal dopaminergic degeneration seen by immunocytochemistry against tyrosine hydroxylase (TH), a marker of dopaminergic cells (Fig 2). Lesions in the nigrostriatal dopaminergic pathway began in dopaminergic nerve terminals in the striatum (Fig 2a-d).

Only in severe cases were reductions in TH staining in cell bodies of the substantia nigra apparent (Fig 2 e-h). These results suggest that striatal nerve terminals were more sensitive to rotenone than cell bodies in the substantia nigra. Striatal nerve terminals have more profound losses of dopamine compared to substantia nigra in brains of PD patients as well.[15] Interestingly, dopaminergic cells of the ventral tegmental area were relatively spared, as in PD, following rotenone exposure.

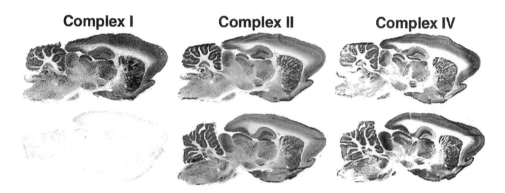

Figure 1. Rotenone inhibits complex I selectively and uniformly throught brain. Binding assay for complex I and histochemical analysis of complex II and IV. Top: Sections from a vehicle-infused rat. Bottom: Sections from a rotenone-infused animal. Complex I binding was reduced (73%) indicating that infused rotenone occupied these binding sites.

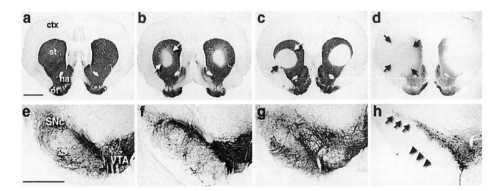

Figure 2. Systemic rotenone infusion caused nigrostriatal dopaminergic degeneration. Coronal sections from control (a,e) and rotenone-infused rats (b-d, f-h) were immunostained for TH. Rotenone-infused animals showed reduced TH staining initially in striatum (b-d) and then substantia nigra (f-h).

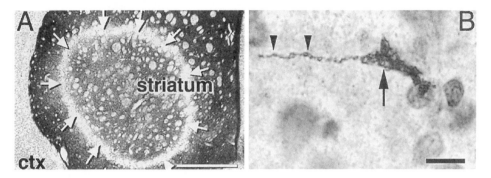

Figure 3. Rotenone infusion induced degeneration of the nigrostriatal dopaminergic pathway. Silver deposits were present in striatum (a) and in substantia nigra (b). Silver deposits were not seen in vehicle-infused rats.

To determine whether the loss of TH staining represented true neurodegeneration or simply the reduced expression of a phenotypic marker, we used silver staining, a marker of degenerating neurons. Silver deposits were present in striatal regions devoid of TH staining in rotenone-treated animals (Fig 3A). Additionally, cells with silver deposits were present in substantia nigra of rotenone-treated animals (Fig 3B).

Despite the loss of nigrostriatal dopaminergic nerve terminals, postsynaptic neuronal elements in the striatum were unaffected by chronic systemic rotenone infusion. Greater than 90% of striatal cells are GABAergic. In striatal regions devoid of TH staining, there was no loss of immunoreactivity against glutamic acid decarboxylase (GAD), a marker of GABAergic cells (Fig 4a-d). Additionally, rotenone-infusion did not alter acetylcholinesterase (AChE) staining. These results further demonstrate the selectivity of rotenone-induced neurodegeneration.

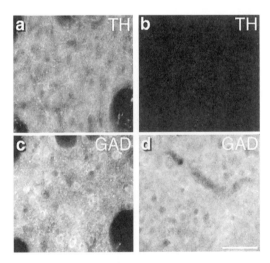

Figure 4. Rotenone-infusion resulted in selective dopaminergic degeneration. Striatal sections from a control animal (a,c) and rotenone-treated animal (b,d) were double labeled for TH and GAD. In regions of striatum that were completely devoid of TH immunoreactivity (b), GAD immunofluorescence remained intact (d).

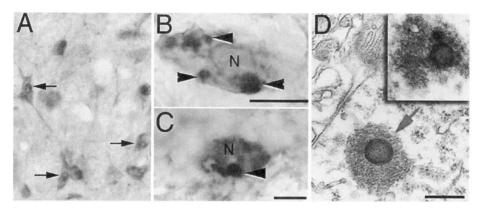

Figure 5. Cytoplasmic inclusions in nigral neurons of rotenone-treated animals. (a). Cytoplasmic inclusions in nigral neurons were positive for ubiquitin immunoreactivity (Arrow). (b,c). Nigral neurons with α-synuclein positive aggregates (Arrowheads). d). Electron micrograph showing inclusion with a dense core and fibrillar periphery. Inset, immuno-electron microscopy confirms that inclusions contain α-synuclein

6. FORMATION OF CYTOPLASMIC INCLUSIONS

The pathological hallmark of PD is the formation of cytoplasmic protein aggregates containing ubiquitin and α-synuclein known as Lewy bodies.[1] Rotenone- infused animals showed evidence of cytoplasmic inclusions in nigral cells that shared many characteristic features of Lewy bodies. These inclusions were ubiquitin and α-synuclein positive (Figure 5a-c). Ultrastructurally, these inclusions were fibrillar and were confirmed to contain α-synuclein by immuno-electron microscopy (Figure 5d).

7. BEHAVIOR

Rotenone-treated animals developed some behavioral characteristics reminescent of PD. All animals with a dopaminergic lesion became hypokinetic and showed unsteady movement and hunched posture. In some cases, animals developed rigidity.

8. SUMMARY

These results suggest that a systemic complex I defect induced by chronic rotenone infusion reproduced the behavioral, anatomical, neurochemical, and neuropathological features of PD. The fact that rotenone infusion uniformly inhibited complex I throughout brain but resulted in a selective nigrostriatal dopaminergic lesion suggests that this population of neurons had an intrinsic sensitivity to complex I defects. Although we have found that rotenone seems to have little toxicity when given orally, our results highlight the potential role of environmental toxins that influence mitochondrial activity in PD pathogenesis. Environmental exposures, combined with genetic variability in mitochondrial function and the ability to metabolize toxins, may contribute to the onset of many cases of typical idiopathic PD.

9. ACKNOWLEDGEMENTS

Figures 1-5 were taken from Betarbet, Sherer *et al.* (2000)[13] with the perimission of *Nature Neuroscience*.

REFERENCES

1. M. Irizarry, J. Growdon, T. Gomez-Isla, K. Newell, J. M. George, D. F. Clayton, *et al*, Nigral and cortical Lewy bodies and dystrophic nigral neurities in Parkinson's disease and cortical Lewy body disease contain α-synuclein immunoreactivity, *J. Neuropathol. Exp. Neurol.* **57(4)**, 334-337 (1998).
2. R. Kruger, W. Kuhn, T. Muller, D. Woitalla, M. Graeber, S. Kosel, *et al*, Ala³⁰Pro mutation in the gene encoding α-synuclein in Parkinson's disease, *Nature Genet.* **18(2)**, 106-108 (1998).
3. M.H. Polymeropoulos, C. Lavedan, E. Leory, S.E. Ide, A. Dehejia, A. Dutra, *et al*, Mutation in the α-synuclein gene identified in families with Parkinson's disease, *Science* **276(5321)**, 2045-2047 (1997).
4. T. Kitada, S. Asakawa, N. Hattori, H. Matsumine, Y. Yamamura, S. Minoshima, *et al*, Mutations in the parkin gene cause autosomal recessive juvenile parkinsonism, *Nature* **392(6676)**, 605-608 (1998).
5. A. Menegon, P. G. Board, A. C. Blackburn, G. D. Mellick, and D. G. LeCouteur, Parkinson's disease, pesticides, and glutathione transferase polymorphisms. *Lancet* **352(9137)**, 1344-1346 (1998).
6. P. G. Butterfield, B. G. Valanis, P. S. Spencer, C. A. Lindeman, and J. G. Nutt, Environmental antecedents of young-onset Parkinson's disease, *Neurol.* **43(6)**, 1150-1158 (1993).
7. J. W. Langston, P.A. Ballard, J. W. Tetrud, and I. Irwin, Chronic parkinsonism in humans due to a product of meridine-analog synthesis, *Science*, **219(4587)**, 979-980 (1983).
8. K. F. Tipton and T. P. Singer, Advances in our understanding of the mechanisms of the neurotoxicity of MPTP and related compounds, *J. Neurochem.*, **61(4)**, 1191-1206 (1993).
9. W. J. Nicklas, I. Vyas, and R. E. Heikkila, Inhibition of NADH-linked oxidation in brain mitochondria by 1-methyl-4-phenylpyridine, a metabolite of the neurotoxin 1-methyl-4-phenyl-1,2,3,6-tetrahydropyridine. *Life Sci.* **36(26)**, 2503-2508.
10. W. D. Parker, S. J. Boyson, and J. K. Parks, Electron transport chain abnormalities in idiopathic Parkinson's disease, *Ann. Neurol.* **26(6)**, 719-723 (1989).
11. A. H. Schapira, J. M. Cooper, D. Dexter, P. Jenner, J. B. Clark, and C. J. Marsden, Mitochondrial deficitis in Parkinson's disease, *Lancet* **1(8649)**, 1269 (1989).
12. Y. Mizuno, H. Yoshino, S, Ikebe, N, Hattori, T. Kobayashi, T. Shimoda-Mastubayashi, *et al.*, Mitochondrial dysfunction in Parkinson's disease, *Ann. Neurol.* **40(3 Suppl 1)**, S99-S109 (1998).
13. R. Betarbet, T. B. Sherer, G. MacKenzie, M. Garcia-Osuna, A. V. Panov, and J. T. Greenamyre, Chronic systemic pesticide exposure reproduces features of Parkinson's disease, *Nature Neurosci.* **3(12)**,1301-1306 (2000).
14. D. S. Higgins, Jr. and J. T. Greenamyre, [³H] dihydrorotenone binding to NADH: Ubiquinone reductase (complex I) of the electron transport chain: an autoradiographic study, *J. Neurochem.* **66(12)**, 3807-3816 (1996).
15. O. Hornykiewicz, Dopamine (3-hydroxytyramine) and brain function, *Pharmacol. Rev.* **18(2)**, 925-963 (1966).

NEUTRAL (*R*)SALSOLINOL *N*-METHYL-TRANSFERASE AS A PATHOGENIC FACTOR OF PARKINSON'S DISEASE

Wakako Maruyama[*], Takako Yamada, Yukihiko Washimi, Teruhiko Kachi, Nobuo Yanagisawa, Fujiko Ando, Hitoshi Shimokata, and Makoto Naoi

1. INTRODUCTION

The pathogenesis of idiopathic Parkinson's disease (PD) remains to be an enigma. PD at cellular level is characterized by selective degeneration of dopamine (DA) neurons in the substantia nigra. The discovery of MPTP suggested that endogenous or xenobiotic neurotoxins might account for the selective cell death. *N*-Methyl(*R*)salsolinol [*N*M(*R*)Sal] has a chemical structure similar to MPTP and the cytotoxicity selective to DA neurons was proved by *in vivo* experiments. The mechanism underlying the induction of apoptotic cell death was examined *in vitro*. The analyses of human materials suggest the increased level of *N*M(*R*)Sal in PD brains. Possible involvement of *N*M(*R*)Sal in the pathogenesis of idiopathic PD is discussed.

2. METABOLISM AND DISTRIBUTION OF DOPAMINE-DERIVED ISOQUINOLINES IN THE HUMAN BRAIN

*N*M(*R*)Sal is synthesized in DA neurons by 2 enzymatic steps, (*R*)salsolinol synthase and neutral (*R*)salsolinol *N*-methyltransferase (NMT) (Naoi et al., 1996b; Maruyama et al., 1992). In control human brain, DA, *N*M(*R*)Sal and the oxidized isoquinolinium ion (DMDHIQ$^+$) were detected in the nigro-striatum. Only the (*R*)-enantiomer of Sal and *N*-MSal were identified, suggesting their enzymatic synthesis. (*R*)Sal distributes ubiquitously, but *N*M(*R*)Sal accumulates in the nigro-striatum, and

[*] W. Maruyama, Department of Basic Gerontology, National Institute for Longevity Sciences, Obu, Aichi 474-8511, Japan. T. Yamada, Y. Washimi, T. Kachi, N. Yanagisawa, Department of Neurology, Chubu National Hospital, Obu. F. Ando, H. Shimokata, Department of Èpidemiology, National Institute for Longevity Sciences, Obu. M. Naoi, Institute of Applied Biochemistry, Mitake, Gifu, Japan.

DMDHIQ$^+$ only in the substantia nigra (Maruyama et al., 1997a). The activity of the enzymes involved in the metabolism of *N*M(*R*)Sal was examined in the human brain. It was found that the activity of NMT in the caudate-putamen well correlated with the amount of DMDHIQ$^+$ in the substantia nigra (Naoi et al., 1997). These results indicate that the activity of NMT may be the rate-limiting step in synthesis of *N*M(*R*)Sal in the brain.

3. TOXICITY OF DOPAMINE-DERIVED ISOQUINOLINES ON DOPAMINE NEURONS

The toxicity of dopamine-derived isoquinolines was examined by *in vivo* and *in vitro* experiments. *N*M(*R*)Sal and its related isoquinolines were continuously injected in the striatum of male Wistar rats for 3 days to 1 week (Naoi et al., 1996a). Only *N*M(*R*)Sal induced parkinsonism in rats. DA contents and the activity of tyrosine hydroxylase were decreased in the substantia nigra. Immunohistological observation confirmed selective depletion of DA neurons in the substantia nigra, indicating selective toxicity of *N*M(*R*)Sal.

The cytotoxicity of *N*M(*R*)Sal and related isoquinolines was examined using human dopaminergic neuroblastoma SH-SY5Y cells, as a model of DA neurons. *N*M(*R*)Sal was proved to induce apoptosis in a dose-dependent way, but neither its (*S*)-enantiomer, *N*M(*S*)Sal, nor other isoquinolines did (Maruyama et al., 1997; Maruyama et al., 2001). The mechanism of the cytotoxicity by *N*M(*R*)Sal was proved to be initiated by the decrease in mitochondrial membrane potential, followed by caspase-3 activation and nuclear translocation of glyceraldehyde-3-phospate dehydrogenase (GAPDH), and finally DNA fragmentation (Akao et al., 1999; Maruyama et al., 2001b). The results suggest that a molecule in mitochondria, which can distinguish the enantiometric configuration of *N*M(*R*)Sal, may initiate death program.

4. *N*-METHYL(*R*)SALSOLINOL AND ITS METABOLIZING ENZYMES IN PARKINSONIAN PATIENTS

The level of *N*M(*R*)Sal was determined using lumbar CSF obtained from patients with untreated PD and multiple system atrophy (MSA), and normal subjects. The level of *N*M(*R*)Sal was found to increase significantly in parkisonian CSF compared to that of MSA patients and control (Maruyama et al., 1996). To examine the mechanism of the increase in *N*M(*R*)Sal in PD, the activities of the enzymes related to its metabolism were analyzed using lymphocytes. The activity of NMT with the optical pH at 7 (neutral NMT) increased significantly in the lymphocytes prepared from PD patients compared to normal and disease control (Naoi et al., 1998). The significant positive correlation was confirmed between *N*M(*R*)Sal level in CSF and the activity of NMT in the lymphocyte prepared from the same PD patients (Fig. 1). These results suggest that the increased activity of the rate-limiting enzyme in the synthesis may account for the

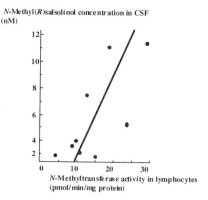

Fig. 1. Correlation between the level of NM(R)Sal in CSF and the activity of NMT in lymphocytes. The amount of NM(R)Sal and NMT activity was analyzed using HPLC with electrochemical detection.

increased level of NM(R)Sal in the brain. The activity of NMT was found to be slightly higher in PD patients with akinesia as the main symptom than that in those with tremor (Fig. 2). However, the sex, age, age of onset, duration of the illness, medication and Hohen and Yahr stage did not affect NMT activity.

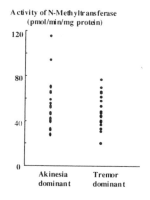

Fig. 2. The activity of NMT prepared from PD patients whose main symptoms are akinesia and tremor.

5. CONCLUSION

Our clinical and experimental data strongly suggest that a DA-derived endogenous isoquinoline, NM(R)Sal, and its synthesizing enzyme, neutral NMT, may be involved in the pathogenesis of idiopathic PD. Increased activity of NMT in the lymphocyte and probably also in the brain may be a causative factor of PD. The environmental and genetic factors, which regulate the activity of NMT, should be further investigated. In addition, the enantio-selective induction of apoptosis mediated by the disruption of mitochondrial membrane potential indicates that mitochondria may play a key role in the apoptotic cell death induced by NM(R)Sal. A molecule, which regulates the apoptotic signal transduction and is activated by NM(R)Sal or other neurotoxins, may be a target to control the cell death in PD and other neurodegenerative disorders. Further studied will bring us a clue to understand the role of neutral NMT in the pathogenesis of PD and to develop novel methods to rescue the declining of nigro-striatal DA neurons.

REFERENCES

Akao, Y., Nakagawa, Y., Maruyama, W., Takahashi, T., and Naoi, M., 1999, Apoptosis induced by endogenous neurotoxin, N-methyl(R)salsolinol, is mediated by activation of caspase 3. *Neurosci. Lett.* **267**: 153-156.

Maruyama, W., Nakahara, D., Ota, O., Takahashi, T., Takahashi, A., Nagatsu, T., and Naoi, M., 1992, N-Methylation of dopamine-derived 6,-dihydroxy1,2,3,4tetrahydroisoquinoline, (R)-salsolinol, in rat brains: In vivo microdialysis study. *J. Neurochem.*, **59**: 395-400.

Maruyama, M., Abe, T., Tohgi, H., Dostert, P., and Naoi, M., 1996, A dopaminergic neurotoxin, (R)-N-methylsalsolinol, increased in parkinsonian CSF. *Ann. Neurol.*, **40**: 119-122.

Maruyama, W., Sobue, G., Matsubara, M., Hashizume, Y., Dostert, P., and Naoi, M., 1997a, A dopaminergic neurotoxin, 1(R),2(N)-dimethyl-6,7dihydroxy1,2,3,4-tetrahydroisoquinoline, N-methyl(R)salsolinol, and its oxidation product, 1,2(N)-dimethyl-6,7-dihydroxyisoquinolinium ion, accumulate in the nigro-striatal system of the human brain. *Neurosci. Lett.* **223**: 61-64.

Maruyama, W., Naoi, M., Kasamatsu, T., Hashizume, Y., Takahashi, T., Kohda, K., and Dostert, P., 1997b, An endogenous dopaminergic neurotoxin, N-methyl(R)salsolinol, induces DNA damage in human dopaminergic neuroblastoma SH-SY5Y cells. *J. Neurochem.* **69**: 322-329.

Maruyama, M., Boulton, A. A., Davis, B. A., Dostert, P., and Naoi, M., 2001, Enantio-specific induction of apoptosis by an endogenous neurotoxin, N-methyl(R)salsolinol, in dopaminergic SH-SY5Y cells: Suppression of apoptosis by N-(2-heptyl)-N-methylpropargylamine. *J. Neural Transm.* **108**: 11-24.

Maruyama, W., Akao, Y., Youdim, M. B. H., Davis, B. A., and Naoi, M., 2001, Transduction-enforced Bcl-2 overexpression and an anti-Parkinson drug, rasagiline, prevent nuclear transduction of glyceraldehydes-3-phosphate dehydrogenase induced by an endogenous neurotoxin, N-methyl(R)salsolinol. *J. Neurochem.*, in press.

Naoi, M., Maruyama, W., Dostert, P., Hashizume, Y., Nakahara, D., Takahashi, T., and Ota, M., 1996a, Dopamine-derived 1(R), 2(N)-dimethyl-6,7-dihydroxyl-1,2,3,4-tetrahydroisoquinoline, N-methyl-(R)-salsolinol, induced parkinsonism in rat: Biochemical, pathological and behavioral studies. *Brain Res.*, **709**: 285-295.

Naoi, M., Maruyama, W., Dostert, P., Kohda, K., and Kaiya, T., 1996b, An novel enzyme enantio-selectively synthesizes (R)salsolinol, a precursor of a dopaminergic neurotoxin, N-methyl(R)salsolinol, *Neurosci. Lett.*, **212**: 183-186.

Naoi, M., Maruyama, W., Matsubara, K., and Hashizume, Y., 1997, A neutral N-methyltransferase activity in the striatum determines the level of an endogenous MPP$^+$-like neurotoxin, 1,2-dimethyl-6,7-dihydroxyisoquinolinium ion, in the substantia nigra of human brains. *Neurosci. Lett.* **235**: 81-84.

Naoi, M., Maruyama, W., Nakao, N., Ibi, T., Sahashi, K., and Strolin Benedetti, M., 1998, (R)Salsolinol N-methyltransferase activity increases in parkinsonian lymphocytes. *Ann. Neurol.* **43**:212-216.

FAMILIAL VARIANTS OF PARKINSON'S DISEASE

Katrina Gwinn-Hardy, Melissa Hanson, Zbigniew Wszolek, Amanda Adam, Andrew Singleton, Charles H. Adler, John N. Caviness, James Bower, Steven Hickey, Judy Ying Chen, Shamaila Waseem, Matt Farrer [*]

1. BACKGROUND

The familial occurrence of Parkinson's disease was noted as early as 1880, by Charcot's student, Leroux.[1] More recently, the discovery of several genetic causes and risk factors for parkinsonism has increased awareness of the importance of genetics in this and other complex neurological disorders. The phenotypic spectrum of pathogenic loci is wider than that of classic PD; it probably includes Lewy-body dementia, some forms of essential tremor, and other disorders. The discovery of causal mutations in the gene for _-synuclein[2], parkin[3], and the discovery of genetic linkages to chromosome (Ch) 2p[4], 4p[5], and 1q[6], have revolutionized the field of Parkinson's disease. Parkinsonism connotes a clinical syndrome including resting tremor, bradykinesia, rigidity, gait disturbance, and postural instability, along with other "secondary" features; the most common cause of parkinsonism is Parkinson's Disease (PD)[7]. Supranuclear gaze palsy, prominent autonomic symptoms, early dementia, or hallucinosis suggest other disorders, which may have important biological overlaps with PD (including Progressive Supranuclear Palsy, multiple system atrophy, and diffuse Lewy body disease, for example). Clinical diagnosis, despite validated criteria[8,9], remains accurate only ~80% percent of the time; pathological diagnosis remains the "gold-standard"[10]. Pathological hallmarks of PD include Lewy bodies, particularly in the substantia

[*] K Gwinn-Hardy, A Adam, A Singleton, NINDS, NIH, Bethesda MD, M Farrer, M Hanson, J Hardy, S Waseem, Z Wszolek, Mayo Clinic Jacksonville 32224, CH Adler, JN Caviness, Mayo Clinic Scottsdale AZ 85251, J Bower Mayo Clinic Rochester MN 55905, S Hickey, Jacksonville FL 32250, JY Chen UCLA, Los Angeles CA 90024

Mapping the Progress of Alzheimer's and Parkinson's Disease
Edited by Mizuno *et al.*, Kluwer Academic/Plenum Publishers, 2002

281

nigra pars compacta (SNpc)[10,11]. Of interest from a genetico-pathological standpoint, Lewy bodies stain with both α-synuclein and ubiquitin antibodies[11].

Epidemiological surveys reveal that early onset PD (< 50 years) is associated with a significant family history; late onset disease had a less evident genetic component[12,13]. Longitudinal evaluation of twins with late-onset PD via 18F-dopa positron emission tomography, conversely, suggests that cross-sectional studies are overlooking pre-symptomatic disease.[14] The finding of α-synuclein in pathological deposits, not only in PD, but also in other parkinsonian disorders[15,16,17], reveals the importance of α-synuclein in the mechanism of parkinsonism. The existence of clinical parkinsonism not related to a synuclein process, conversely, show that parkinsonism is not specific to a given underlying molecular abnormality, but rather, is a common endpoint of many pathological processes.[18] In order to better characterize the genetic contribution to Parkinson's disease, we have collected a series of cases with familial PD, and present the results of this clinical collection herein. This series reveals that age of onset is not a reliable cut-off for considering a genetic cause of PD. It also shows that familial PD is global, and that the phenotype is not reflective of age of onset or ethnic origin, as a general rule. Our collection also reveals that there is a wide phenotypic spectrum in familial PD, and that genetic characterization will be essential to allow further categorization of the causes of parkinsonism.

2. METHODS

Subjects were recruited under an IRB approved protocol. Inclusion criteria included a diagnosis of Parkinson's disease, and two or more known family members carrying a diagnosis of Parkinson's disease. All probands were examined by a Movement Disorders specialist and a clinical diagnosis of Parkinson's disease was made. Additional family members were evaluated by telephone interview, and review of medical records. Blood samples were collected and screened for known genetic mutations causal for parkinsonism including Parkin, alpha-synuclein, tau, and spinocerebellar ataxias (SCA) II and III. When possible, haplotype screening for 4p and 2p loci was done.

3. RESULTS

Results regarding ethnicity, apparent inheritance, age of onset, and other data is presented in Table 1. The families can be grouped into three general categories: 1) those with multiple affected members and apparent autosomal dominant inheritance: 2) those with a suggested Mendelian inheritance pattern with reduced penetrance[5,19]: 3) individual sib-pairs. The first group had an early average age of onset (under 50 years of age). A total of 25 collected families have only one living known affected individual. An alpha synuclein mutation has been found in one of these families, which is of Italo-American heritage. Eleven families are of Norwegian origin, with late onset disease, and a phenotype resembling that described for 2p[4]. Nine families had both Parkinson's disease and Essential tremor, but only one of these was found to have the 4p haplotype[5]. Four families had a SCA expansion in the borderline range causing their parkinsonism, and this phenotype appears to be more common in those of African heritage.

Table 1.

Phenotype	# Families	Ethnic Origin	# Affected	# Collected	Range of Onset (years)
Typical PD	38	Global	108	114	12-78
Typical and Atypical PD	13	Global	58	61	30-76
Typical PD, ET	17	Denmark, Germany, Ireland, England?	96	82	15-76
Typical PD, DLB	12	France, Germany, England, Scandinavia, Czech Republic, Pakistan, India, Italy, Ireland?	48	38	33-70
Typical PD, DLB, ET	4	Ireland, Sweden, Germany, Netherlands, Channel Islands, Norway, Finland	41	59	18-75

Table 1 Legend: PD= parkinsonism including Parkinson's disease; Typical=2 or > cardinal featuresplus asymmetry and dopamine responsiveness. DLB=Diffuse Lewy body disease; ET=essential tremor/isolated activation tremor. Atypical= supranuclear gaze palsy, dysautonomia, unresponsiveness to dopamine replacement, or other atypical features. Unaffected individuals are collected in families as well as affected; thus the number collected may be more than the number affected.

4. DISCUSSION

Parkinsonism is a heterogeneous disorder, clinically and pathologically, as has long been known. It is clear from our work, and that of others cited above, that PD is also genetically heterogeneous. The heterogeneity across samples has been the subject of much discussion. However, this work reveals that variation of the phenotype within families is also of importance, and in our experience, extremely common. For example, as demonstrated in Table 1, Typical Parkinson's disease can be seen in a single family, along with clinical Progressive Supranuclear Palsy, Diffuse Lewy body disease, and other disorders. One family has PD and PSP and DLB in same family, though in different members. While typical PD was the most common phenotype in the families studied, the combination of the other phenotypes totaled more than that of "Typical" PD in isolation (this is most likely due to ascertainment bias and not because the other phenotypes are truly more common than typical PD). Future studies will focus on what leads to genotype/phenotype variability within one family. Before that can be done, of course, most of these identified families need to have causal genes found. Pathological diagnosis remains to be done for most families however. Further study will include collections of family members through genealogical means, grouping of samples phenotypically, and genetically, and screening these and other familial databases for loci as they are found, as well as pathologica study. Multiple causes for parkinsonism exist and multiple genetic risk factors remain to be identified. The known genetic risk factors account for a small number of cases in our North American series. Listed here are ongoing familial ascertainment is being with a goal for genome wide searches for susceptibility loci.

REFERENCES

1. Leroux PD. Contribution á l'étude des causes de la paralysie agitante. 1880; *Thesis,* Parks.
2. Polymeropoulos MH, Lavedan C, Leroy E, et al. Mutation in the alpha-synuclein gene identified in families with Parkinson's disease *Science* 1997;276:2045-2047.
3. Kitada T, Asakawa S, Hattori N, et al. Mutations in the parkin gene cause autosomal recessive juvenile parkinsonism. *Nature* 1998;392, 605-608.
4. Gasser T, Muller-Myhsok B, Wszolek ZK, et al. A susceptibility locus for Parkinson's disease maps to chromosome 2p13. *Nat Genet* 1998;18:262-265.
5. Farrer M, Gwinn-Hardy K, Muenter M, et al. A chromosome 4p haplotype segregating with Parkinson's disease and postural tremor. *Hum Mol Genet* 1999;8:81-85.
6. Valente EM, Bentivoglio AR, Dixon PH, et al. Localization of a novel locus for autosomal recessive early-onset parkinsonism, PARK6, on human chromosome 1p35-p36. *Am J Hum Genet* 2001;68(4):895-900.
7. Rajput AH Epidemiology of Parkinson's disease. *Can J Neurol Sci* 1984;11(1 Suppl):156-159.
8. Gelb DJ, Oliver E, Gilman S. Diagnostic criteria for Parkinson disease. *Arch Neurol* 1999;56(1):33-39.
9. Calne DB, Snow JB, Lee C. Criteria for diagnosing Parkinson's disease. *Ann Neurol* 1992;32 Suppl:S125-S127.
10. Hughes AJ, Daniel SE, Kilford L, Lees AJ. Accuracy of clinical diagnosis of idiopathic Parkinson's disease: a clinico-pathological study of 100 cases. *J Neurol Neurosurg Psychiatry* 1999;55(3):181-184.
11. Spillantini MG, Schmidt ML, Lee VM, Trojanowski JQ, Jakes R, Goedert M Alpha-synuclein in Lewy bodies. *Nature* 1997;28:388(6645):839-840.
12. Tanner CM, Ottman R, Goldman SM, Ellenberg J, Chan P, Mayeux R, Langston JW. Parkinson disease in twins: an etiologic study. *JAMA* 1999;27:281(4):341-346.
13. Sveinbjornsdottir S, Hicks AA, Jonsson T, et al. Familial aggregation of Parkinson's disease in Iceland. *N Engl J Med* 2000;14:343(24)1765-1770.
14. Holthoff VA, Vieregge P, Kessler J, et al. Discordant twins with Parkinson's disease: positron emission tomography and early signs of impaired cognitive circuits. *Ann Neurol* 1994;36(2):176-182.
15. Dickson DW, Liu W, Hardy J, et al. Widespread alterations of alpha-synuclein in multiple system atrophy. *Am J Pathol* 1999;155(4):1241-1251.
16. Gwinn-Hardy K, Mehta ND, Farrer M, Maraganore D, Muenter M, Hardy J, Dickson D. Distinctive neuropathology revealed by α-synuclein antibodies in hereditary parkinsonism and dementia linked to chromosome 4p. *Acta Neuropathologica* 2000;99:663-672.
17. Tu, P.-H, Galvin, JE, Baba, M, et al. Glial cytoplasmic inclusions in white matter oligodendrocytes of multiple system atrophy brains contain insoluble α-synuclein. *Ann. Neurol* 1998;44:415-422.
18. Hardy J, Gwinn-Hardy K. Genetic classification of primary neurodegenerative disease. Review. *Science* 1998;282(5391)1075-1079 .
19. Gwinn-Hardy K, Crook R, Lincoln S, Adler CH, Caviness JN, Hardy J, Farrer M. A kindred with Parkinson's disease not showing linkage to established loci. *Neurology* 2000 Jan 25;54(2):504-507.

FAMILIAL PARKINSONISM WITH APATHY, DEPRESSION AND CENTRAL HYPOVENTILATION (PERRY'S SYNDROME)

Bülent Elibol[1], Tomonori Kobayashi[3], F.Belgin Ataç[2], Nobutaka Hattori[3], Gürdal Şahin[1], Günfer Gürer[1], Yoshikuni Mizuno[3]

INTRODUCTION

A rare and unique form of familial parkinsonism with autosomal dominant inheritance that is characterized by all cardinal features of parkinsonism, apathy, depression, central hypoventilation, weight loss, and rapid progression was first described by Perry et al.[1] Thereafter, four additional families from different ethnic background have been reported (Table 1).[2-6] In most of these affected individuals, disease presented between ages of 40-55, with symptoms of apathy, psychomotor slowness and/or depression, usually accompanied or followed by moderate degrees of parkinsonism and weight loss. Central type of hypoventilation developed eventually and caused death in most of the patients due to sudden apnea or complications of respiratory insufficiency. Pathology differs from idiopathic Parkinson's disease by severe neuronal loss and gliosis in substantia nigra with no or very few Lewy-body formation,[1,2,4,5] and in some patients, by extension of neuronal loss to locus ceruleus, caudate, pallidum and medulla.[2,4]

In this study, we present the clinical course of a new case with a history of affected mother and a possibly affected sister, and the results of molecular analysis of this family. After excluding two known mutations in alpha-synuclein gene, extensive analysis of the entire tau gene was performed for possible mutations, because of the signs of frontal, striato-pallidal and brainstem involvement.

---

[1] Hacettepe University School of Medicine, Department of Neurology, 06100, Ankara, Turkey.
[2] Baskent University School of Medicine, Department of Molecular Biology, 06530, Ankara, Turkey.
[3] Juntendo University School of Medicine, Dept. of Neurology, 2-1-1 Hongo, Bunkyo, Tokyo 113-8421, Japan

Mapping the Progress of Alzheimer's and Parkinson's Disease
Edited by Mizuno *et al.*, Kluwer Academic/Plenum Publishers, 2002

CASE REPORT

This 50 year-old man was first seen in Emergency Room (ER) because of severe respiratory failure and confusion in February 1999. History revealed progressive decrease in social interest and apathy for four years. For the last two years, staring, difficulty in speech, resting tremor and bradykinesia were added to his complaints. Meantime, respiration was noted to be superficial than usual and he had progressive weight loss of about 8 kgs. He was put on L-dopa/carbidopa 100/25 mg t.i.d. for almost a year, with a considerable good response in motor signs. When he was recently started on a higher dose of L-dopa with dopamine agonist, progressive confusion and dispnea developed and he was brought to ER. Arterial blood gases were as pH:7.1, pCO_2:106.6, and pO_2:48 mmHg. After respiratory support with mechanical ventilator, he became fully oriented. He was palilalic, apathetic and showed poor eye contact, although his cooperation was good. Mild to moderate bradykinesia (R>L) was present. There was no rigidity, tremor and postural instability. Two weeks after discharge from ER with recovery, he was re-hospitalized because of repeated attacks of respiratory insufficiency.

Mild cognitive impairment was detected by MMSE (score: 36/45, mainly memory and visuospatial subsets affected) and psychiatric interview revealed severe apathy without depression. Parkinsonian signs were well controlled by L-dopa/carbidopa 100/25 mg t.i.d. Complete blood count and chemistry, vitamin B12 and folic acid levels, ECG were all in normal range. MRI, SPECT, EEG, VEP, SEP, BAEP and CSF analysis were all normal. Alveolo-arterial gradient was compatible with hypoventilation of central origin. Thorax CT and fluoroscopy diaphragms were normal. Because of persistent hypoxia/hypercapnea, tracheostomy was performed and he was discharged with a home ventilator for intermittent respiratory support, especially during sleep.

Since his mother was said to have similar symptoms starting at age of 50 and was lost in three years with a sudden respiratory failure, this family was accepted to represent Perry's syndrome. During his 11 month-follow-up, he showed little or no progress in parkinsonism that remained responsive to L-dopa. However, his apathetic behavior did not change and he usually developed short periods of increased psychomotor activity after L-dopa doses. His final admission was due to aspiration pneumonia and he died in three weeks. Autopsy was not permitted. In the following 6 months, his younger sister was noticed to develop apathy and mental slowness. Because she refused examination, she was considered to have a probable diagnosis (Fig. 1).

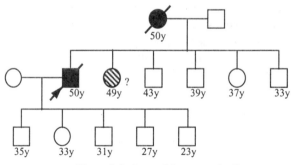

Figure 1. Pedigree of the present family.

Table 1. Clinical characteristics of previously reported families.

Reported Families	Patients (M/F)	Age of Onset	Characteristics of Parkinsonism	Additional Findings
Perry et al. (1975, 1990)	9 (7/2)	39-52	B, T, unresponsive to L-Dopa	Depression, dementia, weight loss, central hypoventilation,
Purdy et al. (1979)	5 (3/2)	43-46	B, R, unresponsive to L-Dopa	Depression, central hypo-ventilation, apathy
Roy et al. (1988)	6 (3/3)	44-56	B, R, T, responsive to L-Dopa	Depression, lethargia, Weight loss, central hypoventilation
Lechevalier et al. (1992)	5 (2/3)	45-56	B, R, T, responsive to L-Dopa	Depression, central hypoventilation, apathy, weight loss
Bhatia et al. (1993)	6 (3/3)	35-51	Rapidly progressive B, R, T, initially responsive to L-Dopa	Depression

B: Bradykinesia, R: Rigidity, T: Resting tremor

MATERIALS AND METHODS

We obtained an informed consent and collected peripheral blood samples of the patient and his family members (5 siblings and the 5 off springs). DNA was isolated from peripheral blood leukocytes by the standard methods under DNA-clone-free environment.

The known mutations of Ala53Thr and Ala30Pro were investigated in alpha-synuclein gene as previously described in detail.[7, 8] Genotyping of the tau gene was performed by direct sequencing according to Poorkaj et al.[9] PCR was carried out in 25 micro liter reaction mixtures, each of which contained 50 mM KCl, 10 mM Tris (pH 8.3), 1.5 mM $MgCl_2$, 0.01% gelatin with 5 pmol of each primer, 10 nmol of each dNTP, and 0.25 U of AmpliTaq Gold DNA polymerase (Applied Biosystems). Sequencing reactions of the PCR amplification products were carried out using Big Dye Terminator Cycle Sequencing Kit (PE Biosystems). Sequences were determined by ABI 310 Genetic Analyzer (PE Biosystems).

In the human brain, six tau isoform were identified. Exons 1, 4, 5, 6, 7, 8, 9, 11, 12, 13/14 are included in all the isoforms. Inclusion of exons 2, 3, and/or 10 are alternatives determining the six isoforms.[10] Exon 4A, 6, and 8 are known not to be expressed in the brain.[11] Exon 13 does not splice to exon 14, instead, a potential intron between exons 13 and 14 is retained in the brain.[11] This transcript has a stop codon in this potential intron shortly after exon 13. Thus, the post-transcriptional processing of the tau gene comprises many steps of alternative splicing. Nucleotide substitutions without causing amino acid changes or within intronic sequences may also affect these steps of alternative splicing. Therefore, we have sequenced all of the exons and the flanking intron segments of the tau gene.

RESULTS

The known mutations of Ala53Thr and Ala30Pro in alpha-synuclein gene were excluded in the proband and in the other family members.

Genotyping of the proband's tau gene was demonstrated in Table 2. His genotype was consistent with the heterozygous compound of H1 and H2 haplotypes, which were described by Baker et al.[12] As to the T to C change in exon 8, one of the siblings and one of the children of the proband had C allele homozygously (C/C). After we performed this genotyping, Higgins et al. [13] reported this polymorphism. According to them, this T to C transition in exon 8 was a common substitution.

As shown in Table 2, 8 of 11 (indicated by bold characters) polymorphisms found in this study were previously described.[9,12,13,14] Three of those nucleotide substitutions have not been reported yet. They were an A to G change at -139th before exon 3, a C to G change at 232nd in exon 4A, and 99th in exon 7.

Table 2. Genotypes of the patient's tau gene, extended haplotypes of H1 and H2, and other reported polymorphisms

	Genotypes	H1	H2	Nucleotide Substitution	Position	Amino acid Substitution
Exon 1	A/G	A	G	A to G	13th prior to initiation codon	NA[12,14]
Exon 2	C/T	C	T	C to T	18th 3' E2	NA[12,14]
Exon 3	A/G	A	G	A to G	-139th 5' E3	NA *
Exon 3	A/G	A	G	A to G	9th 3' E3	NA[12,14]
Exon 4A	C/G			C to G	232nd	Pro78Leu*
Exon 4A	G/G			G to A	480th	Asp161Asp[9]
Exon 4A	C/C			C to T	482nd	Asp161Asp[9]
Exon 4A	T/C			T to C	493rd	Val165Ala[9]
Exon 6	C/C			C to T	139th	His47Tyr[9]
Exon 6	T/T			T to C	157th	Ser53Pro[9]
Exon 7	G/A			G to A	99th	Pro176Pro*
Exon 7	G/G			G to A	103rd	Ala178Thr[14]
Exon 8	T/C			T to C	5th	Thr2Thr[13]
Exon 9	A/G	A	G	A to G	125th	Ala227Ala[12]
Exon 9	T/C	T	C	T to C	209th	Asn255Asn[9,12]
Exon 9	G/G	G, A	G	G to A	254th	Pro279Pro[12]
Exon 11	G/A	G	A	G to A	34th 3' Exon 11	NA[12,14]
Exon13/14	T/T	T	C	T to C	26th after stop codon	NA[9,12]

1) Positions of amino acids are based on the longest tau isoform expressed in central nervous system. In exons 4A, 6 and 8, positions are numbered from the first amino acid coded by each exon for they are not expressed in the central nervous system.
2) NA; not applicable.
3) * is not reported previously.

DISCUSSION

Clinical features of this family were consistent with the diagnosis of Perry's syndrome. As in most of the other reported families, apathy, psychomotor slowness/depression, central hypoventilation and progressive weight loss were the characteristic symptoms. The proband's parkinsonism was mild to moderate in severity and responded considerably well to L-dopa. Among the previously reported families, some patients responded to L-dopa in various degrees, while others were mostly unresponsive (Table1). The narrow range for age of onset, presentation and progression of the symptoms and death due to complications of central hypoventilation were found very similar in all these families, suggesting relatively homogeneous phenotype.

However, there were individuals who had only some of these clinical features, indicating phenotypic heterogeneity. For instance, all the reported members of the family from UK were presented mainly with rapidly progressive parkinsonism. Because of the associated depression and a similar pathology, this family was considered to have Perry's syndrome.[6] Indeed, some newly affected individuals have been found to show more characteristic signs in follow-up (Bathia, personal communicatation).

Pathological findings were variable, although severe neuronal loss and gliosis with no or few Lewy-bodies in substantia nigra (both in dorsal and ventral tiers) were in common. Interestingly, severity of parkinsonism was unproportionately milder in most of these cases.[2,4,5,6] Extension of degeneration to locus ceruleus in most families and to caudate head, pallidum and medulla in some cases, has been associated with mental depression and central hypoventilation.[2,4,5,6] Except the eosinophilic intranuclear inclusions seen in one study,[2] no specific finding was found suggesting any one of the groups in the current spectrum of neurodegenerative diseases.

This is the first reported molecular study on Perry's syndrome to our knowledge. After excluding the known mutations of alpha-synuclein gene, tau gene was analyzed for possible mutations, because of the clinical and pathological evidence of frontal, striatal and brainstem involvement. Mutations of the tau gene were identified to cause disinhibition-dementia-parkinsonism-amyotrophy complex, pallido-ponto-nigral degeneration, and frontotemporal dementia.[9,15,16] Since then, the tau gene mutations have been identified in familial neurological disorders with various phenotypes ranging from parkinsonism to disinhibition dementia, from depression to behavioral disorders.

We have found 8 previously identified and 3 novel polymorphisms in the proband's tau gene, all non-pathogenic in nature. Although the T to C change at 5th in exon 8 did not cause any amino acid change and exon 8 is known not to be expressed in brain, we thought it was noteworthy because of its close localisation to intron-exon boundary and that it might affect the alternative splicing at exon 8. Interestingly, in this position T/T homozygosity was reported to be an important susceptibility haplotype for sporadic progressive supranuclear palsy.[13] The fact that one of the siblings and one of the children of the proband had C allele homozygously in exon 8 indicated that both of their parents had C allele, thus one polymorphic allele came from healthy parents. Therefore, T to C change in exon 8 is not apparently associated with Perry's syndrome.

The three nucleotide substitutions, i.e. an A to G change at -139th before exon 3, a C to G change at 232nd in exon 4A, and 99th in exon 7 have not been reported yet. The first site is too far from the intron-exon boundary to consider its effect on the splicing. The second is within exon 4A that is not expressed in the brain. The third site is in the middle portion of exon 7 and the substitution does not bring any amino acid change. Still, segregation study whether or not other family members have those substitutions would be desirable.

The combination of the extended haplotypes of the proband was H1/H2. Baker et al.[12] reported that the prevalence of H1 haplotype is higher in PSP than in the control group. Together with those polymorphic genotypes shown in Table 2, these distinct haplotypes can be a useful marker for the haplotype analysis to examine whether families are compatible with linkage to the tau gene locus or not in the future analysis.

In conclusion, by demonstration of this new family from a different ethnic background we have confirmed that this rare but unique form of familial parkinsonism exhibits clinical homogeneity. By excluding known or potentially pathological mutations in α-synuclein and tau genes, we suggest further molecular studies, by preference, gathering reported families for an extensive linkage analysis.

REFERENCES

1. T.L. Perry, P.J.A. Bratty, S. Hansen, et al. Hereditary mental depression and parkinsonism with taurine deficiency. Arch Neurol, 32:108-113 (1975).
2. A. Purdy, A. Hahn, H.J.M. Barnett, et al. Familial fatal parkinsonism with alveolar hypoventilation and mental depression. Ann Neurol, 6:523-531 (1979).
3. E.P. Roy, J.E. Riggs, J.D. Martin, et al. Familial parkinsonism, apathy, weight loss and central hypoventilation: successful long-term management. Neurology, 38:637-639 (1988).
4. T.L. Perry, J.M. Wright, K. Berry, et al. Dominantly inherited apathy, central hypoventilation, and Parkinson's syndrome: clinical, biochemical and neuropathologic studies of 2 new cases. Neurology, 40:1882-7 (1990).
5. B. Lechevalier, C. Schupp, C. Fallet-Bianco, et al. Syndrome parkinsonien familial avec athymormie et hypoventilation. Rev Neurol (Paris), 148:39-46 (1992).
6. K.P. Bhatia, S.E. Daniel, C.D. Marsden. Familial parkinsonism with depression: a clinico-pathological study. Ann Neurol, 34:842-7 (1993).
7. M.H. Polymeropoulos, C. Lavedan, E. Leroy, et al. Mutation in the α-synuclein gene identified in families with Parkinson's disease. Science, 276:2045-2047 (1997).
8. R. Kruger, W. Kuhn, T. Muller, et al. Ala30Pro mutation in the gene encoding α-synuclein in Parkinson's disease. Nat Genet, 18:106-108 (1998).
9. P. Prookaj, T.D. Bird, E. Wijsman, et al. Tau is a candidate gene for chromosome 17 frontotemporal dementia. Ann Neurol, 43:815-825 (1998).
10. M. Goedert, M.G. Spillantini, R. Jakes, et al. Multiple isoforms of human microtubule-associated protein tau: sequences and localization in neurofibrillary tangles of Alzheimer's disease. Neuron, 3:519-526 (1989).
11. A. Andreadis, W.M. Brown, K.S. Kosik. Structure and novel exons of the human tau gene. Biochemistry, 31:10626-10633 (1992).
12. M. Baker, I. Litvan, H. Houlden, et al. Association of an extended haplotype in the tau gene with progressive supranuclear palsy. Hum Mol Genet, 8:711-715 (1999).
13. J.J. Higgings, L.I. Golbe, A. De Biase, et al. An extended 5'-tau susceptibility haplotye in progressive supranuclear palsy. Neurology, 14; 55:1364-1367 (2000).
14. H. Houlden, M. Baker, J. Adamson, et al. Frequency of tau mutations in three series of non-Alzheimer's degenerative dementia. Ann Neurol, 46:243-248 (1999).
15. M.G. Spillantini, J.R. Muriell, M. Goedert, et al. Mutation in the tau gene in familial multiple system tauopathy with presenile dementia. Proc Natl Acad Sci USA, 95: 7737-7741 (1998).
16. M. Hutton, C.L. Lendon, P. Rizzu, et al. Association of missense and 5'-splice-site mutations in tau with the inherited dementia FTDP-17. Nature, 393: 702-705 (1998).

UBIQUITIN-PROTEASOME PATHWAY IS A KEY TO UNDERSTANDING OF NIGRAL DEGENERATION IN AUTOSOMAL RECESSIVE JUVENILE PARKINSON'S DISEASE

Nobutaka Hattori,[1] Shuichi Asakawa,[2] Hideki Shimura,[1] Shin-ichiro Kubo,[1] Kenichi Sato,[1] Toshiaki Kitami,[1] Yoko Chikaoka-Kwamura,[1] Yuzuru Imai,[3] Ryosuke Takahashi,[3] Toshiaki Suzuki,[5] Keiji Tanaka,[5] Nobuyoshi Shimizu,[2] and Yoshikuni Mizuno[1]

1. INTRODUCTION

In most patients with Parkinson's disease (PD), the contribution of genetic and environmental factors remains to be elucidated. The importance of genes in PD was controversial for many years. It has become clear, however, genetic factors contribute to the pathogenesis of PD after identification of the distinct genetic loci for certain forms of familial PD (FPD) [1-6]. In addition, the role of genetic factors is supported by the high concordance in twins using positron emission tomography (PET)[1,2] and the increased risk among relatives of PD patients in case-control and family studies[3,4]. These findings indicate that genetic factors contribute to PD susceptibility. It is now clear that clinically defined PD represents a heterogeneous disorders that encompasses a small proportion of individuals with inherited disease and a large population with seemingly sporadic disease.

The discovery of genetic linkages for PD provided us promise for new insights into the pathogenesis of the disease. At present, two independent genes have been implicated in the clinical phenotypes of PD. α-Synuclein gene (PARK1) was first identified as the causative gene for autosomal form of FPD[5] and recently the parkin gene (PARK2) was done as the second causative gene for recessive form of FPD (autosomal recessive juvenile parkinsonism (AR-JP)[6].

1) Department of Neurology, Juntendo University School Medicine, Bunkyo, Tokyo 113-8421, Japan, 2)Department of Molecular Biology, Keio Univeresity, Shinjuku, Tokyo 160-8582, Japan, 3)Laboratory for Motor System Neurodegeneration, RIKEN Brain Science Institute, Wako, Saitama, Japan, 4)Tokyo Metropolitan Insitute of Medical Science and CREST, Bunkyo, Tokyo, 113-8613

Mapping the Progress of Alzheimer's and Parkinson's Disease
Edited by Mizuno *et al.*, Kluwer Academic/Plenum Publishers, 2002

Furthermore, a mutation in ubiquitin carboxy-terminal hydrolase L1 (UCH-L1) has been identified in one family with FPD[7], but it remains elusive whether or not mutation of UCH-L1 is tightly linked to FPD. UCH-L1 mutation seems to account for only one family. In addition, α-Synuclein mutations account for the several cases: i.e., about 10 familial cases of α-synuclein gene mutations. Among FPD, a number of newly-identified mutations of the parkin gene has been found in patients with young-onset PD (AR-JP).

Parkin contains 12 exons spanning 1.5 Mb and encodes a novel protein of 465 aminon acids with a molecular weight of approximately 52 kDa. Parkin consists of an N-terminal ubiquitin-like domain (Ubl) and a C-terminal RING box separated by a linker region (ref). The RING box encompasses three domains, termed RING1, IBR (in-between-RING) and RING2[8].

To date, various mutations such as exonic deletions, microdeletions, exon multiplications or point mutations resulting in missense and nonsense changes of parkins have been reported in AR-JP patients[9-12]. Furthermore, this form was distributed not only in Japan but also in the world wide. Thus, parkin mutations are now considered to be the major cause of FPD.

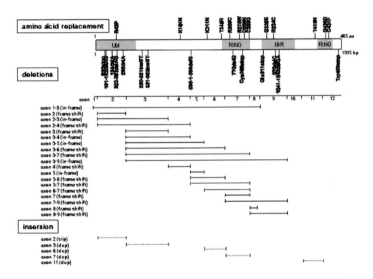

Figure 1. Mutations in the parkin gene that co-segregate with the disease phenotype reported in the literatures. Abbrevations show: dup, duplication; trip, triplication.

2. FUNCTION OF PARKIN: PARKIN IS DIRECT LINKED TO UBIQUITIN-PROTEASOME PATHWAY AS A UBIQUITIN LIGASE

Parkin is divided into three parts as above-mentioned. In addition, its sequence is conserved among three mammalian species such as mouse, rat, and human, particularly in the Ubl domain, the two RING-finger motifs, and IBR domain, suggesting that these domains are of functional importance. The presence of Ubl domain in parkin is an important clue to investigate the function of parkin.

The ubiquitin-proteasome pathway plays a central role in protein processing and degradation which contributes to protein quality control in cells[13, 14]. Ubiquitin (Ub) is attached covalently to target proteins. Protein ubiquitination is catalyzed by three enzymes, E1 (Ub-activating enzym), E2 (Ub-conjugating enzyme), and E3 (Ub-ligase), to form a poly-Ub chain that serves aas the degradation signal for proteolytic attack by the 26S proteasome [15, 16]. Because Ub appears to be involved in the pathogenic processes of some neurodegenerative disorders such as PD, AD, and polyglutamine diseases, and some prion diseases, parkin thus may be involved in the Ub-proteasome pathway. Indeed, immunohistochemical staining with anti-parkin and anti-Ub antibodies overlapped for Lewy bodies, indicating that parkin was localized in Lewy bodies[17]. Currently, a number of RING-finger proteins have been implicated in involvement of the Ub-proteasome pathway. In addition, a human homolog of *Drosophila ariadne (HHARI)*, which is mildly homologous to parkin and was a subset of the RING-finger/IBR proteins, interacted with the E2[18]. Therefore, we investigated whether or not parkin is linked to the Ub-proteasome pathway[8] and subsequently, parkin has been found to be linked to the Ub-proteasome pathway as a ubiquitin ligase. The high molecular-mass ubiquitinated proteins are co-immunoprecipitated with parkin when the human dopaminergic neuroblastoma SH-SY5Y cells are treated the proteasome inhibitor MG132. Mutant parkins from AR-JP patients with mutations in the RING-box are defective with regard to their binding ability to associate with UbcH7/8 and ubiquitinated cellular proteins, whereas mutations of the Ubl or linker region results in loss of interaction with ubiquitinated proteins without affecting UbcH7 recruitment.

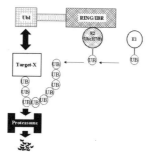

Figure 2. Model of parkin-directed ubiquitination pathway. Loss-of-function of parkin specifically causes AR-JP. Ub, ubiquitin; AR-JP, autosomal recessive juvenile parkinsonism. X is a substrate whose accumulation in non-ubiquitinated form could activate directly the cell death pathway.

Thus the target X protein(s) as a substrate for parkin may clarify the pathogenesis for young on set PD (AR-JP). The discovery of parkin function as E3 in the Ub-proteasome pathway has enhanced our standing of the molecular mechanisms underlying idiopathic PD as well as AR-JP, and suggests that failures of the Ub-proteasome pathway plays an essential role in the nigral degeneration of the disease.

3. THE MECHANISMS OF NIGRAL DEGENERATION:ONE OF SUBSTRATES FOR PARKIN IS A PAEL RECEPTOR

The loss of function of parkin in AR-JP should lead to the accumulation of non-ubiquitinated substrates that would otherwise be polyubiquitinated by the parkin and efficiently degraded by 26S proteasome. Thus, the identification of unknown substrate(s) must provide us good information for elucidating the pathogenesis of AR-JP. Very recently, a putative G protein-coupled transmembrane polypeptide, named Pael (parkin-associated endothelin receptor–like) receptor, was identified as an interacting protein with parkin by yeast two-hybrid screening for human adult brain cDNA libraries[19]. This protein, Pael receptor, is an authentic in vivo substrate for parkin. When overexpressed in culture cells, this receptor tends to become unfolded, insoluble, and ubiquitinated in vivo. The insoluble Pael receptor leads to unfolded protein-induced cell death. Parkin specifically ubiquitinated this protein in the presence of E2s (UBC6/7) resident in the endoplasmic reticulum and promoted the degradation of insoluble Pael receptor, resulting in suppression of the cell death induced by the overexpression of Pael receptor. Moreover, the insoluble form of Pael receptor accumulated in the AR-JP brains, suggesting that accumulation of this protein was direct linked to the nigral neurons in AR-JP (Fig. 3).

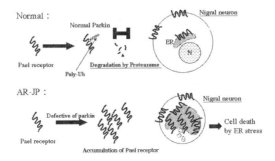

Figure 3. The schematic model of nigral neuronal death by accumulation of Pael receptor.

Parkin was upregulated by unfolded protein stress and it suppressed cell death induced by unfolded protein stress[20]. In response to stress in the ER, the unfolded protein response (UPR) that regulates the expression of a series of genes, including ER-associated protein degradation (ERAD)-related genes, is induced[21]. The ERAD system eliminates misfolding ER proteins including integral membrane and secretory proteins via degradation in the cytosol[22]. ERAD substrates are retrotranslocated across the ER membrane into cytosol, where they are ubiquitinated and degraded through the ub-proteasome pathway in collaborating with UBC6/7.which are localized in the ER. Indeed, parkin can interact with UBC6/7 as above-mentioned. Thus, ER stress could be a major event in the pathogenesis of AR-JP brains.

On the other hand, parkin was localized in the trans-Golgi network and the secretory vesicles by immunocytochemical analyses. In the subsequent subcellular fractionation studies of rat brain, parkin was co-purified with the synaptic vesicles (SVs)[23]. An immuno-electromicroscopic analysis indicated that parkin was present on the SV membrane. Based on the subcellular localization, the substrate(s) may be also localized in the synaptic terminals. CDCrel-1, which is localized in the synaptic terminals, is also one of substrates for parkin[24]. Furthermore, 22-kilodalton glycosylated form of α-synuclein (αSP22) identified as one of substrates for parkin[25]. α-Synuclein is also localized in the synaptic terminals. In addition, α-synuclein is a major component of Lewy bodies. Parkin could ubiquitinated the glycosylated α-synuclein (αSP22), suggesting that αSP22 accumulated as a non-ubiquitinated form in AR-JP brains. This finding may support the absence of Lewy bodies in AR-JP brains because the pathologic characteristic is usually the lack of Lewy bodies in AR-JP brains. αSP22 was observed not only in AR-JP brains but also in normal brains. These findings raise that the possibility that inherited form of PD such as FPD with α-synuclein gene mutations and AR-JP and common form of PD, that is, idiopathic PD involve etiologically distinct but biochemically related alterations of a shared metabolic pathway. As a next step, the relationship between Pael receptor, CDCrel-1, and modified form of α-synuclein should be clarified. Three molecules may also share the common cascades involving the nigral degeneration.

The recent explosion of genetic information has indicated that PD is not one entity, but is highly heterogeneous. There are several genetically, clinically, and pathologically distinct forms of PD that can be caused by mutations of α-synuclein, parkin, UCHL-1 and unknown causative genes. Although the mutations underlie a minority of the larger PD population, they nevertheless represent a cascade of events that culminates in the demise of nigral neurons. Indeed, α-synuclein, parkin, and UCHL-1 shared the ub-proteasome pathway as above-mentioned. Therefore, identification of the candidate genes will provide us with important clues to understand the single gene defect leading to selective nigral degeneration, as well as for developing methods to rescue diseased nigral neurons. Among them, three causative genes are involved in Ub-proteasome pathway. Therefore, ub-proteasome pathway is a key event for pathogenesis of neurodegeneration.

REFERENCES

1. N.Hattori, H. Shimura, S. Kubo, T. Kitada, M. Wang, S. Asakawa, S. Minoshima, Y. Shimizu, Y. Mizuno, Autosomal recessive juvenile parkinsonism: akey to understanding nigral degeneration in sporadic Parkinson's disease. *Neuropathology* [Suppl] 5, S85-S90 (2000).
2. N.Hattori, H. Shimura, S. Kubo, M. Wang, N. Shimizu, K. Tanaka, Y. Mizuno, Importance of familial Parkinson's disease and parkinsonism to the understanding of nigral degeneration in sporadic Parkinson's disease. *J Neural. Transm* [Suppl] 60, 85-100 (2000).
3. D.J. Burn, M.H. Mark, E.D. Playford, D.M. Maraganore, T.R. Zimmerman Jr, R. C. Duvoisin, A. E. Harding, C. D. Marsden, Parkinson's disease in twins studied with 18F-dopa and positron emission tomography. *Neurology* 42, 1894-1900 (1992).
4. V.A. Holthoff, P. Vieregge, J. Kessler, U. Pietrzyk, K. Herholz, J. Bonner, R. Wagner, K. Wienhard, G. Pawlik, W. D. Heiss, Discordant twins with Parkinson's disease: positron emission tomography and early signs of impaired cognitive circuits. *Ann. Neurol*, 36, 176-182 (1994).
5. M. H. Polymeropoulos, C. Lavedan, E. Leroy, S. E. Ide, A. Dehejia, A. Dutra, B. Pike, H. Root, J. Rubenstein, R. Boyer, E. S. Stenroos, S. Chandrasekharappa, A. Athanassiadou, T. Papapetropoulos, W. G. Johnson, A. M. Lazzarini, R. C. Duvoisin, G. Di Iorio, L. I. Golbe, R. L. Nussbaum, Mutation in the alpha-synuclein gene identified in families with Parkinson's disease. *Science* 276, 2045-2047 (1997).
6. T. Kitada, S. Asakawa, N. Hattori, H. Matsumine, Y. Yamamura, S. Minoshima, M. Yokochi, Y. Mizuno, N. Shimizu, Mutations in the parkin gene cause autosomal recessive juvenile parkinsonism, *Nature* 392, 605-608, (1998).
7. E. Leroy, R. Boyer, G. Auburger, B. Leube, G. Ulm, E. Mezey, G. Harta, M. J. Brownstein, S. Jonnalagada,

T. Chernova, A. Dehejia, C. Lavedan, T. Gasser, P. J. Steinbach, K. D. Wilkinson, M. H. Polymeropoulos, (1998b) The ubiquitin pathway in Parkinson's disease, *Nature* 395, 451-452 (1998).

8. H. Shimura, N. Hattori, S. Kubo, Y. Mizuno, S. Asakawa, S. Minoshima, N. Shimizu, K. Iwai, T. Chiba, K. Tanaka, T. Suzuki, Familial Parkinson disease gene product, parkin, is a ubiquitin-proteiin ligase. *Nat. Genet.* **25**, 302-305 (2000).

9. C. B. Lücking, A. Durr, V. Bonifati, J. Vaughan, G. DeMichele, T. Gasser, B. S. Harhangi, P. Pollak, A-M. Bonnet, D. Nichol, M. D. Mari, R. Marconi, E. Broussolle, O. Rascol, M. Rosier, I. Arnould, B. A. Oostra, M. M. B. Breteler, A. Filla, G. Meco, P. Denefle, N. W. Wood, Y. Agid, A. Brice, Association between early-onset Parkinson's disease and mutations in the parkin gene. *N. Engl. J. Med.* **342**, 1560-1567 (2000).

10. N. Abbas N, C. B. Lücking, S. Ricard, A. Dürr, V. Bonifati, G. De Michele, S. Bouley, J. R. Vaughan, T. Gasser, R. Marconi, E. Broussolle, C. Brefel-Courbon, B. S. Harhangi, B. A. Oostra, E. Fabrizio, G. A. Böhme, L. Pradler, N. W. Wood, A. Filla, G. Meco, P. Denefle, Y. Agid, A. Brice, A wide variety of mutations in the parkin gene are responsible for autosomal recessive parkinsonism in Europe. *Hum. Mol. Gen.* **8**, 567-574 (1999).

11. N. Hattori, H. Matsumine, T. Kitada, S. Asakawa, Y. Yamamura, T. Kobayashi, M. Yokochi, H. Yoshino, M. Wang, T. Kondo, S. Kuzuhara, S. Nakamura, N. Shimizu, Y. Mizuno, Molecular analysis of a novel ubiquitin-like protein (PARKIN) gene in Japanese families with AR-JP: evidence of homozygous deletions in the PARKIN gene in affected individuals. *Ann. Neurol.* **44**, 935-941 (1998).

12. N. Hattori, H. Matsumine, S. Asakawa, T. Kitada, H. Yoshino, B. Elibol, A. J. Brooks, Y. Yamamura, T. Kobayashi, M. Wang, A. Yoritaka, S. Minoshima, N. Shimizu, Y. Mizuno, Point mutations (Thr240Arg and Gln311Stop) in the Parkin gene. *Biochem. Biophys. Res. Commun.* **249**, 754-758, (1998).

13. A. Hershko, A. Ciechanover, The ubiquitin system, *Annu. Rev. Biochem.* **67**, 425-479 (1998).

14. M. Hochstrasser, Ubiquitin-dependent proetiin degradation, *Annu. Rev. Genet.* **30**, 405-439 (1996).

15. O. Coux, K. Tanaka, A. L. Goldberg, Structure and functions of 20S and 26S proteasomes, *Annu. Rev. Biochem.* **65**, 801-847 (1996).

16. W. Baumeister, J. Walz, F. Zühl, E. Seemüller, The proteasome: Paradigm of a self-compartmentalizing protease, *Cell* **92**, 367-380 (1998).

17. H. Shimura, N. Hattori, S. Kubo, M. Yoshikawa, T. Kitada, H. Matsumine, S. Asakawa, S. Minonshima, Y. Yamamura, N. Shimizu, Y. Mizuno Y, Immunohistochemical and subcellular localization of Parkin: Absence of protein in AR-JP Patients, *Ann. Neurol.* **45**: 655-658 (1999).

18. T. P. Moynihan, H. C. Ardley, U. Nuber, S. A. Rose, P. F. Jones, A. F. Markham, M. Scheffner, P. A. Robinson, The ubiquitin-conjugating enzymes UbcH7 and UbcH8 interact with RING finger/IBR motif-containing domains of HHARI and H7-AP1, *J. Biol. Chem.* **274**, 30963-30968 (1999)

19. Y. Imai, M. Soda, H. Inoue, N. Hattori, Y. Mizuno, R. Takahashi, An unfolded putative transmembrane polypeptide, which can lead to endoplasmic reticulum stress, is a substrate of parkin. *Cell* **105**, 891-902 (2001).

20. Y. Imai, M. Soda, R. Takahashi R, Parkin suppresses unfolded protein stress-induced cell death through its E3 ubiquitin-protein ligase activity. *J. Biol. Chem.* **275**, 35661-35664 (2000)

21. K. J. Travers, C. K. Patil, L. Wodicka, D. J. Lockhart, J. S. Weissman, P. Walter, Functional and genomic analyses reveal an essential coordianation between the unfolded protein response and ER-associated degradation, Cell **101**, 249-258 (2000).

22. S. Kubo, T. Kitami, S. Noda, H. Shimura, Y. Uchiyama, S. Asakawa, S. Minoshima, N. Shimizu, Y. Mizuno, N. Hattori, Parkin is associated with cellular vesicles. J Neurochem *J. Neurochem.* 78:42-54 (2001)

23. S. Wilhovsky, R. Gardner, R. Hampton, HRD gene dependence of endoplasmic reticulum-associated degradation, *Mol. Biol. Cell* 11, 1697-1708 (2000).

24. Y. Zhang, J. Gao, K. K. K. Chung, H. Huang, V. L. Dawson, T. M. Dawson, Parkin functions as an E2-dependent ubiquitin-protein ligase and promotes the degradation of the synaptic vesicle-associated protein, CDCrel-1, *Proc. Natl. Acad. Sci. USA.* **97**, 13354-13359 (2000).

25. H. Shimura, M. G. Schlossmacher, N. Hattori, M. P. Frosch, A. Trockenbacher, R. Schneider, Y. Mizuno, K. S. Kosik, D. J. Selkoe, *Science* in press (2001)

COLOCALIZATION OF PARKIN WITH α-SYNUCLEIN IN THE LEWY BODIES OF PARKINSON DISEASE

Michael G. Schlossmacher[1], Matthew P. Frosch[1,2], Wei Ping Gai[3], Nutan Sharma[4], Miguel Medina[1], Tomoyo Ochiishi[1], Hideki Shimura[1], Nobutaka Hattori[5], Yoshikuni Mizuno[5], Bradley T. Hyman[4], Dennis J. Selkoe[1] and Kenneth S. Kosik[1]

1. INTRODUCTION

Mutations in α-synuclein (αS) and parkin cause monogenic forms of inherited Parkinson disease (PD). We report here the unexpected finding that parkin is a constituent of the αS containing pathological neuronal inclusions (Lewy body) found in PD and in the closely related disorder, dementia with Lewy bodies (DLB).

Missense mutations in αS cause rare cases of autosomal dominant PD[1,2] and mutations in parkin are linked to autosomal recessive PD[3]. In the Lewy bodies (LB) of idiopathic PD and DLB, ubiquitin and αS are the principal known constituents[4]. Such inclusions are generally absent in parkin-linked PD[5]. Parkin was recently shown to function as an E3 ubiquitin ligase in cultured neuroblastoma cells[6] but its physiological substrates in normal brain are not yet known. The genotypes of parkin-linked autosomal recessive PD suggest a loss of this E3 ubiquitin ligase function. We hypothesized that parkin may play a role in the formation of LB through the ubiquitination of αS, and therefore examined whether the ubiquitin- and αS-rich LB from PD and DLB brains also contain parkin.

[1]Center for Neurologic Diseases, Departments of Neurology and [2]Pathology, Brigham & Women's Hospital, [4]Alzheimer's Disease Research Unit, Department of Neurology, Massachusetts General Hospital, Harvard Medical School, Boston, MA., USA; [3]Department of Physiology and Centre for Neuroscience, Flinders University, Bedford Park, Australia; and [5]Department of Neurology, Juntendo University Medical School, Tokyo, Japan.

Mapping the Progress of Alzheimer's and Parkinson's Disease
Edited by Mizuno *et al.*, Kluwer Academic/Plenum Publishers, 2002

297

2. PARKIN IN LEWY BODIES AT THREE SITES OF THE NERVOUS SYSTEM

We raised affinity-purified antisera to various peptides of human parkin and immunohi-stochemically probed formalin-fixed paraffin sections of midbrain from confirmed cases of PD, DLB and control human brain[7]. Anti-parkin antibody HP2A [to amino acids (aa) 342-353] faintly stained neuronal cytoplasmic parkin. In addition, it strongly and specifically labeled the core of classical brainstem LB in pigmented neurons of all midbrain sections from PD and DLB tested (Fig. 1). As expected, LB509, a well characterized αS antibody, stained the rim and core of LB in adjacent sections. Other neuroanatomical sites of LB pathology, i.e., anterior cingulate cortex in DLB and sympathetic gangliocytes in PD, also contained numerous HP2A-positive Lewy inclusions. Two additional parkin antibodies, HP7A (to aa 51-62) and HP1A (to aa 84-98) also specifically stained cytoplasmic parkin and occasionally LB (not shown). Generally, dystrophic Lewy neurites were not detected by anti-parkin antibodies. We conclude from these immunohistochemical data that parkin epitopes are present in Lewy inclusions at three distinct sites of the nervous system in two synucleinopathies, PD and DLB.

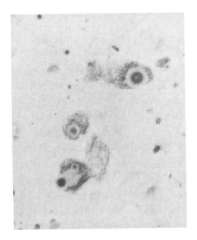

Figure 1: Immunhistochemical section from a brain with dementia with Lewy body disease stained with anti-parkin antibody HP2A (aa 342-353). Four intracellular round inclusions (Lewy bodies) of variable sizes are seen in neuromelanin-positive dopaminergic neurons of the substantia nigra. Note the preferential staining of Lewy body cores.

To estimate the number of anti-parkin antibody-positive LB in brainstem sections, we per-formed double labeling with anti-parkin HP2A and anti-αS H3C on paraformaldehyde-fixed, floating 40 um sections, using a previously employed protocol[8,9]. In midbrain sections from five DLB cases, which usually have more surviving neurons than are present in PD, 87-97% (~92%) of αS-positive compact LB were also detected by HP2A (n=120). No double labeling was observed with HP2A that was preabsorbed with its peptide antigen.

To further assess the proximity between parkin and αS epitopes within doublelabeled brainstem LB, we performed fluorescence resonance energy transfer (FRET) analysis, using a method developed for laser scanning confocal microscopy. In the absence of FRET, the inverse ratio of measured energy transfer before and after photobleaching (D/D_A) equals 1.0. In contrast, two distinct fluorophore-labeled epitopes that produce a positive FRET signal (with the ratio $D/D_A>1.0$ being proportional to proximity), are separated by <30 nm in situ[8,9]. We recently demonstrated such a positive FRET signal between antibodies directed against αS (H3C) and ubiquitin (anti-Ub) in LB ($D/D_A=1.33\pm0.16$; p<0.005), but not between antibodies to αS and other LB constituents, e.g. alpha-B crystallin[9]. In support of a close intermolecular association between parkin and αS epitopes in double-labeled LB identified in sections from our DLB cases, we obtained statistically significant FRET signals for the antibody pair HP2A and H3C ($D/D_A=1.38\pm0.27$; p<0.005; n=11). No double-labeling of LB was observed (and thus FRET analysis was not used) in sections stained with pre-absorbed HP2A and H3C. Furthermore, a similarly strong FRET signal was obtained for HP2A and anti-Ub ($D/D_A=1.29\pm0.14$; p<0.005; n=11)[8]. We conclude that parkin epitopes, particularly within the antigen for HP2A, are generally present in LB where they are in close proximity to both αS and ubiquitin.

To confirm biochemically the presence of intact parkin protein within LB, we prepared affinity-isolated LB samples from DLB cortex (LB+) and LB-negative samples from normal cortex (LB-), which were processed in parallel, as described in detail[10]. After urea/SDS-solubilization, the LB- and LB+ samples were immunoblotted, along with lysates of non-neural cells (e.g., HEK293, COS) that were transiently transfected with *myc-parkin* cDNA[6]. Anti-parkin antibodies, HP2A, HP6A (to aa 6-15) and HP1A all specifically detected the 54 kD myc-parkin fusion protein in extracts of transfected cells, as expected, and reacted with the 53 kD neural parkin in LB+ but not in LB-extracts. Also, in the LB+ samples, two anti-αS antibodies (i.e., syn-1 and LB509) identified monomeric αS (16 kD) and modified/ oligomeric αS isoforms (>34 kD), as expected.

3. SUMMARY

Our combined data from three distinct experimental methods indicate that parkin is an in-trinsic component of Lewy bodies identified in brainstem sections from PD and DLB and of cortical LB from DLB, as well as of Lewy inclusions found in peripheral sympathetic gangliocytes from PD. These findings in pathological tissue are paralleled by the specific co-localization of a portion of the 53 kD parkin protein and its reported E2 binding partner, UbcH7[6], with αS in highly purified (PSD95-negative) presynaptic as well as cytoplasmic fractions of normal adult rat brain (not shown). Such spatial proximity between the two principal gene products associated with heritable forms of PD, specifically within the pathological neuronal LB inclusions, reinforces our hypothesis that αS is a physiological substrate for parkin's ubiquitin ligase function in human brain[11].

4. ACKNOWLEDGMENT

M.G.S. was supported by the Grass Foundation (Robert S. Morison Fellowship), the Lefler Foundation, and the NIH (NS 02127); M.P.F. by a Beeson Scholar Award,; H.S. by a Pergolide Fellowship (Eli Lilly, Japan K.K.). Supported by the Morris R. Udall Center for Excellence in PD (NS 38375) at Brigham & Women's Hospital (M.P.F., D.J.S, and K.S.K.).

REFERENCES AND NOTES

1. M. H. Polymeropoulos, C. Lavedan, E. Leroy et al., Mutation in the alpha-synuclein gene identified in families with Parkinson's disese. *Science* **276**, 2045 (1997).
2. R. Krüger, W. Kuhn, T. Müller et al., Ala30Pro mutation in the gene encoding alpha-synuclein in Parkinson's disease. *Nat. Genet.* **18**, 106 (1998).
3. Y. Mizuno, N. Hattori, H. Mori, Genetics of Parkinson's disease. *Biomed. Pharmacother.* **53**, 109 (1999).
4. M. G. Spillantini, et al., *Nature* **388**, 839 (1997).
5. B. P. van de Warrenburg, M. Lammens, C. B. Lücking et al., Clinical and pathologic abnormalities in a family with parkinsonism and parkin gene mutations. *Neurology* **56**, 555 (2001).
6. H. Shimura, N. Hattori, S. Kubo et al., Familial Parkinson disease gene product, parkin, is a ubiquitin-protein ligase. *Nat. Genet.* **25**, 302 (2000).
7. Synthetic peptides to human parkin epitopes were HPLC purified, coupled to KLH and injected into rabbits. Affinity-purified antisera (Research Genetics, Inc.) were characterized by ELISA as well as dot- and immunoblotting on their respective immunogens, recombinant αS and ubiquitin. Clinically and pathologically confirmed PD (n=3), DLB (n=2) and control (n=2) specimens were collected at the Brigham & Women's Hospital Neuropathology Service in accordance with IRB-approved guidelines. Immunohistochemistry on 6 um thin sections was carried out as described [S. Stoltzner et al., *Am. J. Pathol.* **156**, 489 (2000)].
8. Pathologically confirmed DLB brains (n=5) were obtained from the Harvard Brain Tissue Resource Center. Substantia nigra sections were permeabilized with 0.5% Triton-X 100 (Sigma) and developed with HP2A (detected by Bodipy-conjugated goat α-rabbit IgG) and H3C (from D. Clayton) or anti-ubiquitin [(from Dako) detected by Cy3-conjugated donkey α-mouse IgG (Molecular Probes)]. Indirect immunofluorescent microscopy, FRET and statistical analysis (the ratio of D/D_A was compared to the null hypothesis value of 1.0 by one group t-tests) were carried out as described [R.B. Knowles, J. Chin, C.T. Ruff, B.T. Hyman, *J. Neuropathol. Exp. Neurol.* **58**, 1090 (1999)].
9. N. Sharma, P.J. et al., *Acta Neuropathol.*, in press (2001).
10. W. P. Gai, H.X. Yuan, X. Q. Li et al., In situ and in vitro study of colocalization and segregation of alpha-synuclein, ubiquitin, and lipids in Lewy bodies. *Exper. Neurol.* **166**, 324 (2000).
11. H. Shimura, M. G. Schlossmacher, N. Hattori et al., Ubiquitination of a new form of alpha-synuclein by parkin from human brain. *Science* **293**, 263-269 (2001).

UBIQUITINATION OF A NOVEL FORM OF
α-SYNUCLEIN BY PARKIN

Hideki Shimura,[1] Michael G. Schlossmacher,[1] Nobutaka Hattori,[2]
Matthew P. Frosch,[1,3] Alexander Trockenbacher,[4] Rainer Schneider,[4]
Yoshikuni Mizuno,[2] Kenneth S. Kosik,[1] and Dennis J. Selkoe[1]

The pathogenesis of PD remains unclear, but the existence of genetic susceptibility factors is well documented *(1)*. Three genes encoding parkin *(2)*, α-synuclein (αS) *(3,4)* and UCH-L1 *(5)*, respectively, were recently linked to familial forms of PD. Missense mutations in αS and UCH-L1 cause rare autosomal dominant forms of PD *(1)*. In contrast, mutations of *parkin* are a common cause of autosomal recessive PD (ARPD) *(6,7)*. The reported pathologic changes of *parkin*-linked ARPD are largely confined to the substantia nigra and locus coeruleus and include loss of monoaminergic neurons, gliosis and a lack of the αS-containing Lewy bodies (LB) that are a hallmark of idiopathic PD *(9-12)*.

Parkin consists of an N-terminal ubiquitin-like domain and a C-terminal RING box separated by a linker region (2, 13). Parkin was recently found in cell culture to act as an ubiquitin (Ub) ligase whose RING box recruits the E2 Ub conjugating enzymes, UbcH7 and UbcH8 (13-15). Ubiquitination is a vital cellular process by which a large variety of cellular proteins are conjugated with multimers of Ub, marking them for degradation by the proteasome. Conjugation requires a cascade of reactions that includes an E1 Ub activating enzyme, an E2 Ub conjugating enzyme and an E3 Ub ligase. The E3 specifies both E2 recruitment and the recognition and binding of the substrate. As an E3 ligase, parkin conjugates Ub onto its unknown substrate(s) for subsequent degradation by the proteasome (*13*).

The loss of functional parkin molecules in ARPD should lead to the gradual accumulation of non-ubiquitinated substrates that would otherwise be polyubiquitinated by this E3 ligase and efficiently degraded (*13*). In this regard, the absence of Lewy bodies in ARPD brains suggested to us that both parkin and its unknown target(s) may be

Center for Neurologic Diseases and Departments of [1]Neurology and [3]Pathology; Brigham and Women's Hospital, Harvard Medical School, Boston, USA; [2]Department of Neurology, Juntendo University School of Medicine, Tokyo, Japan; [4]Institute of Biochemistry, University of Innsbruck, Innsbruck, Austria

Mapping the Progress of Alzheimer's and Parkinson's Disease
Edited by Mizuno *et al.*, Kluwer Academic/Plenum Publishers, 2002

required for the formation of Lewy bodies. To test this hypothesis, we raised and purified high-affinity polyclonal antibodies to several regions of human parkin (*16*) and used these to probe Lewy bodies immunohistochemically. The results revealed that parkin co-localized with αS in subsets of classical brainstem Lewy bodies. This result led us to pursue the hypothesis that αS is a key substrate for the E3 ligase activity of parkin in human brain and that disease-causing parkin mutations prevent the ubiquitination of αS.

To determine whether parkin interacts with αS in human brain, we carried out co-immunoprecipitiation experiments on frozen frontal cortex homogenates of four normal subjects. In all cases, the 53 kD parkin protein was specifically precipitated by parkin antibodies. The reported E2 binding partner of parkin, UbcH7 (*13, 15*), co-precipitated with parkin, as expected. Parkin antibodies also co-precipitated a novel 22 kD isoform of αS (designated αSp22) but not the abundant 16 kD αS monomer (αSp16). These immunochemical results were confirmed by mass spectrometry of trypsin digests of the excised αSp22 band, which yielded multiple tryptic fragments of human αS(*16*).

In order to establish a functional role for parkin in human brain, we obtained immunoprecipitated parkin (IP-parkin) from frontal cortex homogenates and tested its E3 Ub ligase activity in a previously described in vitro assay (*13*). Normal brain-derived IP-parkin acted as E3 Ub ligase collaborating with UbcH7. More important, IP-parkin conjugated Ub onto αSp22, which co-immunoprecipitated with parkin, as expected, forming αSp22-Ubn.

These findings raised the possibility of abnormal substrate (αSp22) accumulation in parkin-linked ARPD brains. Immunoprecipitates obtained with the polyclonal anti-αS antibody, KC7071 from normal and ARPD brains were analyzed by WB with anti-αS antibody LB509. Under these experimental conditions, we detected αSp22 solely in homogenates of ARPD brains, while regular αSp16 monomer was seen in comparable amounts in both ARPD and normal tissue. These data suggest that αSp22 accumulates in ARPD brains (when compared with normal brain), while the amount of unmodified αSp16 remains similar. Therefore, the documented loss of parkin function in these ARPD brains apparently leads to the specific accumulation of its target substrate, αSp22.

Our study has shown a direct functional relationship in human brain between the two principal gene products associated to date with inherited forms of PD, parkin and αS. The data identify a multimeric ubiquitination complex that contains a novel, αSp22, as an Ub-accepting substrate for parkin. We confirm that parkin is an E3 Ub ligase and show that wild type but not mutant parkin binds αSp22 in vivo and generates high Mr species of αSp22-Ub$_n$ conjugates in vitro. The loss of normal parkin expression and thus function in ARPD leads to the accumulation of αSp22, as documented in all four of the parkin-genotyped ARPD brains we analyzed. Taken together, our findings raise the possibility that inherited and idiopathic forms of PD involve etiologically distinct but biochemically related alterations of a shared metabolic pathway. In this hypothetical model, the loss of E3 ligase function (that is, parkin) in ARPD leads to accumulation of the non-ubiquitinated αSp22, accelerated neuronal loss and a generally early clinical onset.

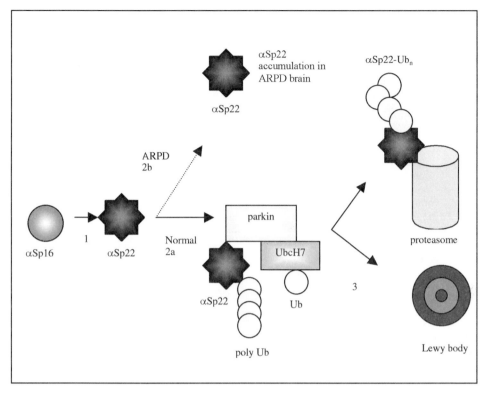

Figure 1. A model of parkin and αSp22 interaction in human brain. Step 1, αSp16 undergoes posttranslational modification to generate αSp22. Step 2a, human parkin recruits UbcH7 at its RING box domain and binds αSp22, and conjugates Ub onto αSp22 to generate αSp22-Ub$_n$. Step 3, αSp22-Ub$_n$ undergoes proteasomal degradation and/or sequestration in Lewy bodies. Interference with steps 2 will lead to accumulation of αSp22$_n$ in ARPD (Step 2b) and lack of αSp22-Ub$_n$ formation.

ACKNOWLEDGEMENT

We thank H. Mori, Y. Mizutani and K. Yamane for ARPD brain specimens, and J. Chan for providing normal human brain; K. Wynn and P. Howley for anti-E6-AP, (JH-16), T. Suzuki and K. Tanaka for a parkin cDNA-containing vector, and P. Lansbury for reagents.

REFERENCES

1. Y. Mizuno, N. Hattori, H. Matsumine, Neurochmical and neurogenetic correlates of Parkinson's disease, J. Neurochem 71, 893-902 (1998).
2. T. Kitada, et al.,Mutation in the parkin gene cause autosomal recessive juvenile parkinsonism, Nature 392, 605-608 (1998).

3.   M. H. Polymeropoulos, et al., Mutation in the alpha-synuclein gene identified in families with Parkinson's disease, Science 276, 2045-2047 (1997).
4.   R. Kruger, et al., Ala30Pro mutation in the gene encoding alpha-synuclein in Parkinson's disease, Nat. Genet. 18, 106-8. (1998).
5.   E. Leroy, et al., The ubiquitin pathway in Parkinson's disease, Nature 395, 451-2. (1998).
6.   C. B. Lucking, et al., Association between early-onset Parkinson's disease and mutations in the parkin gene. French Parkinson's Disease Genetics Study Group, N. Engl. J. Med. 342, 1560-7 (2000).
7.   N. Hattori, et al., Importance of familial Parkinson's disease and parkinsonism of the understanding of nigral degeneration in sporadic Parkinson's disease, J. Neural Transm. Suppl. , 101-16. (2000).
8.   H. Takahashi, et al., Familial juvenile parkinsonism: clinical and pathologic study in a family, Neurology 44, 437-41. (1994).
9.   H. Mori, et al., Pathologic and biochemical studies of juvenile parkinsonism linked to chromosome 6q, Neurology 51, 890-2. (1998).
10. S. Hayashi, et al., An autopsy case of autosomal-recessive juvenile parkinsonism with a homozygous exon 4 deletion in the parkin gene, Mov. Disord. 15, 884-8. (2000).
11. B. P. van de Warrenburg, et al., Clinical nad pathologic abnormalities in a family with parkinsonism and parkin gene mutations, Neurology 56, 555-7. (2001).
12. H. Shimura, et al., Familial Parkinson disease gene product, parkin, is a ubiquitin-protein ligase, Nat. Genet. 25, 302-5. (2000).
13. Y. Zhang, et al., Parkin functions as an E2-dependent ubiquitin-protein ligase and promotes the degradation of the synaptic vesicle-associated protein, CDCrel-1, Proc. Natl. Acad. Sci. U S A 97, 13354-9 (2000).
14. Y. Imai, M. Soda, R. Takahashi, Parkin suppresses unfolded protein stress-insdued cell death through its E3 ubiquitin-protein ligase activity, J. Biol. Chem. 275, 35661-4. (2000).
15. M. G. Schlossmacher, et al., submitted.
16. H. Shimura et al., Ubiquitination of a novel form of a-synuclein by parkin from human brain: implication for Parkinson disease, Science, in press

A SHORT HISTORY OF DOPAMINE AND LEVODOPA IN PARKINSON'S DISEASE

Reinhard Horowski[*]

1. INTRODUCTION: THE BEGINNING AND THE RACE

Arvid Carlsson, when learning about his Nobel Prize in 2000, told reporters: I have been thinking about this for 40 years[1] – a bold word but perfectly appropriate. Dopamine is the neurotransmitter which plays the major role in the central control of movements, posture, vigilance and cognition, emotions and reward, not to forget sexuality and fertility – this is now textbook knowledge but young Carlsson's proposal of a transmitter role for dopamine originally was met with mere aggression and utter disbelief. Such had been the influence of neurophysiologists that only acetylcholine, Otto Loewi's famous 'Vagus-stoff', with its fast electrophysiological action had been accepted as a chemical transmitter, and neurochemists had a very bad time for adding noradrenaline as a peripheral sympathetic neurohormone – however, both groups immediately joined forces to deny dopamine a right of its own (with very rare exceptions such as K.A. Montagu (M. Sandler, pers. comm.) and H. Blaschko (D.B. Calne, pers. comm.)). Had Carlsson yielded to this pressure, the development of effective anti-Parkinson, anti-psychotic and antidepressant therapies would have started much later and under less favourable conditions. But in this way his observation opened the route towards neuropharmacology, and especially to the fabulous career of dopamine – and to its precursor levodopa which was to become the gold standard for Parkinson therapy.

And thus the race was open with at least three major heavy competitors: the pioneers Isamu Sano, a lonely fighter from Japan, as well as Oleh Hornykiewicz and Walther Birkmayer from Vienna, backed by the solid tradition of Austrian pathology and strong curiosity, and independently the clinician André Barbeau from Montreal and especially his initial partner Theodore L. Sourkes, an experienced neurochemist, who had already discovered α-methyl-DOPA, a first rational antihypertensive drug. I. Sano deserves credit for fully describing his methodology of studying tissue dopamine, for describing brain distribution of dopamine and for establishing its great reduction in a first Parkinsonian

---

[*] Reinhard Horowski, SCHERING, Medical & Scientific Contacts, D-13342 Berlin, Email:
Reinhard.Horowski@Schering.de

Mapping the Progress of Alzheimer's and Parkinson's Disease
Edited by Mizuno *et al.*, Kluwer Academic/Plenum Publishers, 2002

patient.[2,3] Good evidence for the scientific climate in those days, however, gives his last remark: "I am quite embarassed to be a DA bug (freak?)."[4] Hornykiewicz with Ehringer proved dopamine depletion to be significant in the basal ganglia of Parkinsonian patients[5] whilst showing (with W. Birkmayer) that another interesting new candidate, serotonin, was much less affected.[6] W. Birkmayer, with O. Hornykiewicz, started levodopa therapy in Parkinsonian patients with spectacular, albeit short-lasting effects on the disease[7]; most fortunately by injecting levodopa i.v. he unknowingly overcame major bioavailability problems of the drug and his patients also had to suffer less from the emetic effect of the drug. This problem had already been described by the 'father' of levodopa, Max Guggenheim, from Basle who, based on T. Torquati's observation of dopa in beans[7] and on Casimir Funk's synthesis[9] had isolated enough levodopa to swallow 2.5 grams (prudently, he had given a single rabbit 1 gram before, with no odd consequences – rabbits (and rats) do not vomit, as we nowadays know!); he quite apologetically states in his paper: "... it became evident that the compound is not completely innocuous; some 10 mins after intake I experienced heavy nausea, I had to vomit twice and thus the substance could not be absorbed completely ...".[10] Walther Birkmayer openly admitted that his decision to inject levodopa (rather than to give it orally) was due to its extremely high price; in fact, he repeated an earlier experiment by Degkwitz and colleagues[11] who had been able to antagonize 'sedation' induced by reserpine (then an antipsychotic) in psychiatric patients quite successfully by injecting levodopa (i.e. they just repeated Carlsson's Nobel Prize-winning rabbit experiments in patients). The Montreal group, last but not least, showed low dopa excretion in Parkinsonian patients and, independently, observed favourable effects of levodopa in patients.[12]

Seen from the US, all these peoples worked in odd locations and, even worse, published their results (and ideas!) in strange places and in fringe languages such as Japanese, German and French ... thus, it had to be George C. Cotzias (and his co-workers) who had the guts, means and patience to use increasingly high oral doses of DOPA with increasing success in Chilean manganese miners[13] – manganese intoxication, indeed, has been known for many years to cause extrapyramidal symptoms comparable to Parkinson's disease, as had been described in a beautiful way, even before the times of F.H. Lewy and C. Tretiakoff, by Embden:[14]

Investigating four cases of manganese intoxication (of brown-stone millers who had been milling manganese dioxide for use in glass or ceramics), Heinrich Embden from Hamburg described a rapid appearance (after only months) of "enhanced muscle tone" ("which gives the face a mask-like appearance"), gait problems with pro- or retropulsion, tremor (not with intention), problems with writing, a low voice with dysarthria, but no nystagmus, no sensory or psychic alterations (but compulsive laughing), no Romberg sign, no reflex anomalies etc. The author rightly recommends to stop working in brown-stone mills (with hopes for recovery) but strangely enough does not discuss paralysis agitans resp. Parkinson's disease but puts his observation in a context of multiple sclerosis. In any case, however, he insists that Couper (1837), von Jaksch (1901) and he in the same year had independently contributed to establish the clinical syndrome of chronic manganese poisoning.

Incidentally, also Cotzias had to pay the very high price for levodopa – after he had started by using the much cheaper D,L-DOPA, where the D-form did nothing to contribute to anti-Parkinson activity but not unexpectedly caused additional side effects such as anemia. As Cotzias, although aware of earlier successful short-term use of levodopa,[15] intended to replenish nigral melanin levels, he might have felt that D-DOPA could do this job as well. Highly surprising, however, is it when an otherwise excellent historical review about levodopa and dopamine agonists by a former collaborator and friend of George Cotzias[16] opens with the simple words: "The levodopa era began in 1967, when Cotzias ....".

Obviously there are many more reviews about levodopa and Parkinson's disease as this is an outstanding success story of the last 40 years: with greatly improved quality of life for millions of patients it greatly contributed to changing neurology's sad image ('highly sophisticated art of diagnosis, with no therapeutic consequences whatsoever') and became a paradigm for many other therapeutic approaches. From the 'horse's mouth', there is a classic review paper by Arvid Carlsson already in 1972 which also forms the foundation for all subsequent therapeutic approaches in Parkinson's disease,[17] but also in psychiatric disorders; there is an account by O. Hornykiewicz ('starting point Vienna, a very simple story'),[18] a highly personalised report by A. Barbeau[19] and an excellent historical review by Th.O. Sourkes[20] for the Montreal group, just to name a few; but it is shocking that another pioneer is often neglected nowadays, and that reviews can be written without even quoting him,[21] and this is Walther Birkmayer from Vienna.

2. WALTHER BIRKMAYER AND HIS CONTRIBUTIONS TO THE TREATMENT OF PARKINSON'S DISEASE

Walther Birkmayer (1910-1996) was not only the first to demonstrate convincing effects of levodopa,[7] and to confirm this effect in an increasingly large patient population and against strong reservations; most convincingly, his colleagues showed in a hemi-Parkinsonian patient that dopamine was much more depleted on the contralateral side[22]. W. Birkmayer also was the first to introduce non-specific MAO-B inhibitors to increase and prolong levodopa effects;[7] he introduced together with P. Riederer and M.B.H. Youdim, and against all experts, again, the MAO-B inhibitor selegiline into Parkinson therapy[23] and proposed its neuroprotective properties, again with the same colleagues,[24] thus introducing a completely new concept into neurology; but still his most important contribution might be that he was the man who discovered the paradoxical levodopa-enhancing effect of peripheral decarboxylase inhibitors which made levodopa treatment really possible for the large majority of patients,[25] by insisting, against famous neurochemists, on his sharp clinical observation of the 'critical detail'.[26] He liked to recall[27] that he had been asked by A. Pletscher to investigate a decarboxylase inhibitor, benserazide, a potential antihypertensive and tranquilizer in 1965, but as he was convinced that dopamine was involved with movement, he decided to study this apparent inhibitor of dopamine synthesis in patients with increased movements (chorea). To his surprise, choreatic movements tended to become even worse, so with simple reasoning he switched to a hypomobile condition, i.e. Parkinson's disease, where he combined benserazide with his favourite drug levodopa. This combination clearly enhanced and prolonged the anti-kinetic effects of levodopa but at the same time also reduced the frequency and severity of side effects such as nausea, emesis and orthostatism (all this was confirmed by D.B. Calne et al. later on in a landmark prospective double-blind study[28] and became the basis of large-scale use of levodopa beyond clinical pharmacology – without Birkmayer's astute observation we would have had to wait until Paul Janssen came up with domperidone, another peripheral antagonist of levodopa- and dopamine-induced side effects, some 15 years later). W. Birkmayer loved to remember that his original observation on benserazide was met, by the neurochemists at Basle, "with quite more than the usual scepticism" against "soft" clinical observations. There, it became clear only after frenetic basic research that benserazide, by not crossing the blood-brain-barrier, did indeed prevent the peripheral formation of dopamine (and related side effects) but at the same time greatly enhanced central dopamine concentrations.[29] Subsequently again, every expert in the field could have told him that the addition of a MAO-B inhibitor would be of no good, dopamine being degraded in the neurons by MAO-A, but his curiosity induced him to test selegiline, as proposed by M.B.H. Youdim and P. Riederer, in his patients[23] – and, of course, as dopamine is also broken down by glial MAO-B and, furthermore, as selegiline increases not only brain dopamine, but also phenylethylamine levels (the brain's happiness transmitter, to quote Birkmayer again) and might act as a weak amphetamine prodrug, it became another spectacular success in Parkinsonian patients. And not enough: in the context of the selegiline success, Birkmayer et al.[24] were the first to suggest an additional neuroprotective effect of this drug and thus a completely new therapeutic concept. He had great pains to have his unorthodox concept published – in fact it took the multimillion Dollar DATATOP study to confirm his observations, including, unexpectedly, the symptomatic effects selegiline has on its own. This, of course, again started another big controversy in which we also took part.[30] Just to complement his success story: to my knowledge, the Vienna group was also the first to use parenteral dopaminergic therapies (parenteral lisuride) for the emergency treatment of life-threatening akinetic crisis in Parkinsonian patients,[31] and Birkmayer, by some controversial clinical claims of his, definitely has triggered highly valuable and important research into the role of iron[32] and mitochondrial function[33] in Parkinson's disease.

Unfortunately, however, this really great and active man had never considered it necessary to perform the correct controlled prospective randomized double-blind studies but did derive his new insights from closely observing just a few patients (who deeply trusted him, as was the case with George Cotzias). Thus, when I rather bluntly asked him whether he had ever considered winning a Nobel Prize, he correctly pointed to this problem (after giving the even more correct answer that there are so many brighter people in medical research). Furthermore, he added, at an encounter we had at Budapest in 1987, that he also would not deserve such a recognition, having been, at the age of 18, a member of a clandestine radical right-wing movement in the autocratic Austria of the Thirties – "but then", he added, "you perfectionist Germans took over Austria and you did immediately realise that I had a Jewish grandmother – such was the end of my political career, an error, for which I have to pay now ".

Walther Birkmayer, at least to me, appeared to be more than satisfied with other recognition his continuous endeavours had met, but it could be noticed that he was especially pleased with improving the situation of his many patients who were a selection of difficult cases. He reportedly even was called for consultation by a number of famous patients including Mao Ze Dong and Francisco Franco.

3. FAMOUS PEOPLE WITH PARKINSON'S DISEASE

It is, indeed, striking and A. Lieberman has elaborated on this, that we know of so many prominent people with this disease. It might be that Parkinson's disease is easy to diagnose from a distance (Walther Birkmayer liked to report that after twenty years of his work at Vienna, tramway drivers already told people, who were trembling and slow at entering their vehicles: "You must have got Parkinson's – go and see Professor Birkmayer!"). We have elaborated on Wilhelm von Humboldt to whom we owe a first and most comprehensive patient's description of Parkinson's disease[34] and, as others, have wondered whether early stages of dopamine deficiency could contribute to special personality traits, or even could lend some support to access prominence.[35] Amongst politicians, we also find Deng Xiaoping, Arafat, Franco and Hitler; religious leaders include Pope John Paul II and Billy Graham, but also Dalí, Muhammed Ali or Michael J. Fox are definitely prominent people. In an attempt to use 'matched' pairs, Garcia de Yebenes[36] compared Hitler with Stalin, Franco with Mussolini and Dalí with Picasso (the latter always being 'prominent controls'). He comes up with the conclusion that those who subsequently were to develop Parkinson's disease appeared to exhibit more rigid views, strategies and behaviours, whilst at the same time had poor or distorted personal relationships; in an essay asking 'Are prominent people more prone to develop Parkinson's disease?' we even went further,[37] by suggesting a positive relationship between successful acting, politics and Parkinson's disease, as far as to ask whether repressed emotions were instrumental in enhancing dopamine breakdown and thus favouring the development of Parkinson's disease. With this speculation, we are coming back to Walther Birkmayer[38] and his initial astonishment "that a single [and simple] chemically defined biological substance" could influence such complex traits and "symptoms of a neurological disease, a unique experience for all neurologists at that time".[27]

REFERENCES

1. M. Balter, Celebrating the Synapse, *Science* **290**, 424 (2000)
2. I. Sano, K. Taniguchi., T. Gamo, M. Takesada, Y. Kakimoto, Die Katechinamine im Zentralnervensystem, *Klin. Wochenschr.* **38**, 57-62 (1960)
3. P. Foley, Y. Mizuno, T. Nagatsu, A. Sano, M.B.H. Youdim, P. McGeer, E. McGeer, P. Riederer, The L-DOPA story – an early Japanese contribution, *Parkins. & Rel. Dis.* **6**, 1 (2000)
4. I. Sano, Biochemistry of the extrapyramidal system, *Parkins. & Rel. Dis.* **6**, 3-6 (2000)
5. E. Ehringer, O. Hornykiewicz, Verteilung von Noradrenalin und Dopamin (3-Hydroxytyramin) im Gehirn des Menschen und ihr Verhalten bei Erkrankungen des extrapyramidalen Systems, *Klin. Wochenschr.* **38**, 1236-1239 (1960)
6. H. Bernheimer, W. Birkmayer, O. Hornykiewicz, Verteilung des 5-Hydroxytryptamins (Serotonin) im Gehirn des Menschen und sein Verhalten bei Patienten mit Parkinson-Syndrom, *Klin. Wochenschr.* **39**, 1056-1059 (1961)
7. W. Birkmayer, O. Hornykiewicz, Der L-3,4-dioxyphenylalanin (=DOPA) Effekt bei der Parkinson Akinese. *Wien. Klin. Wochenschr.* **73**, 787-788 (1961)
8. T. Torquati, Über die Gegenwart einer stickstoffhaltigen Substanz in den grünen Hülsen von Vicia faba. *Arch. di farmacol. sperimentale* **15**, 308-312 (1913)
9. C. Funk, Synthese des d,1-3,4-Dioxyphenylalanins, *Journ. Chem. Soc.* **99**, 554-557 (1911)
10. M. Guggenheim, Dioxyphenylalanin, eine neue Aminosäure aus Vicia faba, *Hoppe-Seyler ZS Physiol. Chemie* **88**, 276-284 (1913)
11. R. Degkwitz, R. Frowein, C. Kulenkampff, U. Mohs, Über die Wirkungen des L-DOPA beim Menschen und deren Beeinflussung durch Reserpin, Chlorpromazin, Iproniazid und Vitamin B₆, *Klin. Wochenschr.* **38**, 120-123 (1960)
12. A. Barbeau, T.L. Sourkes, G.F. Murphy, Les catécholamines dans la maladie de Parkinson, In: *Monoamines et système nerveux central*, edited by J. DeAjuriaguerra (Masson, Paris, 1962), pp. 925-927
13. G.C. Cotzias, M.H. van Woert, L.M. Schiffer, Aromatic amino acids and modification of parkinsonism. *N. Engl. J. Med.* **276**, 374-378, 1967

14. H. Embden, Zur Kenntniss der metallischen Nervengifte – über die chronische Manganvergiftung der Braunsteinmüller, *Deutsche med. Wochenschr.* **73**, 795-796 (1901)

15. G.C. Cotzias et al., Phenothiazines: Curative or causative in regard to Parkinsonism? *Rev. Canad. Biol.* **20**, 193-198 (1961)

16. E. Tolosa, M.J. Marti, F. Valldeoriola, J.L. Molinuevo, History of levodopa and dopamine agonists in Parkinson's disease treatment, *Neurology* **50** (Suppl. 6), S2-S10 (1998)

17. A. Carlsson, Biochemical and Pharmacological Aspects of Parkinsonism, *Acta Neurol. Scand.* **48** (Suppl. 51), 11-42 (1972)

18. O. Hornykiewicz, Levodopa in the 1960s: starting point Vienna, In *20 years of Madopar® - New avenues*, edited by W. Poewe and A.J. Lees (Editiones Roche, Basel, 1993), pp. 11-27

19. A. Barbeau, The last ten years of progress in the clinical pharmacology of extrapyramidal symptoms, In *Psychopharmacology: A Generation of Progress*, edited by M.A. Lipton, A. DiMascio, K.F. Killam (Raven Press, New York, 1978), pp. 771-776

20. T.L. Sourkes, "Rational hope" in the early treatment of Parkinson's disease, *Can. J. Physiol. Pharmacol.* **77**, 375-382 (1999)

21. I.J. Kopin, Parkinson's Disease: Past, Present, and Future, *Neuropsychopharmacol.* **1**, 1-12 (1993)

22. G.S. Barolin, H. Bernheimer, O. Hornykiewicz, Seitenverschiedenes Verhalten des Dopamins (3-Hydroxytyramin) im Gehirn eines Falles von Hemiparkinsonismus, *Schweiz. Arch. Neurol. Neurochir. Psychiatr.* **94**, 241-248 (1964)

23. W. Birkmayer, P. Riederer, M.B.H. Youdim, W. Linauer, The potentiation of the anti-kinetic effect after L-DOPA treatment by an inhibitor of MAO-B, Deprenyl, *J. Neural. Transm.* **36**, 303 (1975)

24. W. Birkmayer, J. Knoll, P. Riederer et al., Increased life expectancy resulting from addition of L-deprenyl to Madopar treatment in Parkinson's disease: a long term study, *J. Neural. Transm.* **64**, 113-127 (1985)

25. W. Birkmayer, M. Mentasti, Weitere experimentelle Untersuchungen über den Catecholaminstoffwechsel bei extrapyramidalen Erkrankungen (Parkinson und Chorea-Syndrom), *Arch. Psychiat. Nervenkr.* **210**, 29 (1967)

26. W. Birkmayer, Das kritische Detail in der ärztlichen Diagnose, *Neurol. Psychiat. Schweiz* **2**, 83-84 (1989)

27. W. Birkmayer, J.G.D. Birkmayer, The L-Dopa Story, In *Parkinsonism and Aging. Aging Series Vol. 36*, edited by D.B. Calne, G. Comi, D. Crippa, R. Horowski, M. Trabucchi (Raven Press, New York, 1989), pp. 1-7

28. D.B. Calne, J.L. Reid, S.D. Vakil, S. Rao, A. Petrie, C.A Pallis, J. Gawler, P.K. Thomas, A. Hilson, Idiopathic Parkinsonism Treated with an Extracerebral Decarboxylase Inhibitor in Combination with Levodopa, *Brit. Med. J.* **25**, 729-732 (1971)

29. G. Bartholini, W.F. Burkhard, A. Pletscher, H.M. Bates, Increase in cerebral catecholamines caused by 3,4-dihydroxy-phenylalanine after inhibition of peripheral decarboxylase, *Nature* **215**, 852-853 (1967)

30. I. Runge, R. Horowski, Can we differentiate between symptomatic and neuroprotective effects in Parkinsonism?, *J. Neural. Transm. [P-D-Sect.]* **4**, 273-283 (1991)

31. W. Birkmayer, P. Riederer, Effects of Lisuride on Motor Function, Psychomotor Activity, and Psychic Behavior in Parkinson's Disease, In *Lisuride and Other Dopamine Agonists*, edited by D.B. Calne, R. Horowski, R.J. McDonald, W. Wuttke (Raven Press, New York, 1983), pp. 453-461

32. R. Sofic, P. Riederer, H. Heinsen, M.B.H. Youdim, Increased iron (III) and total iron content in post-mortem substantia nigra of parkinsonian brain, *J. Neural. Transm.* **74**, 199-205 (1988)

33. Y. Mizuno, S. Ohta, M. Tanaka, S. Takamiya, K. Suzuki, T. Sato, H. Oya, T. Ozawa, Y. Kagawa, Deficiencies in complex I subunits of the respiratory chain in Parkinson's disease, *Biochem. Biophys. Res. Commun.* **163**, 1450-1455 (1989)

34. R. Horowski, L. Horowski, S. Vogel, W. Poewe, F.-W. Kielhorn, An essay on Wilhelm von Humboldt and the shaking palsy: First comprehensive description of Parkinson's disease by a patient, *Neurology* **45**, 565-568 (1995)

35. W. Poewe, The preclinical phase of Parkinson's disease, In *Advances in Research on Neurodegeneration*, edited by D.B. Calne, R. Horowski, Y. Mizuno, W. Poewe, P. Riederer, M.B.H. Youdim (Birkhäuser, New York, 1993), pp. 43-54

36. G. de Yebenes, Notable Europeans with Parkinson's disease, in: *Parkinson's Disease: Advances in Neurology, Vol. 80*, edited by Gerald M. Stern (Lippincott Williams & Wilkins, Philadelphia, 1999), pp. 467-470

37. R. Horowski., L. Horowski, S.M. Calne, D.B. Calne, From Wilhelm von Humboldt to Hitler – are prominent people more prone to have Parkinson's disease? , *Park. Rel. Disorders* **6**, 205-214 (2000)

38. P. Foley, The L-DOPA story revisited. Further surprises to be expected? *J. Neural. Transm. [Suppl.]* **60**, 1-20 (2000)

FAMOUS PEOPLE WITH PARKINSON DISEASE

Abraham Lieberman[*]

1. INTRODUCTION

Muhammad Ali, Yasser Arafat, Jack Anderson (the political columnist), Jim Backus (the voice of Mr. Magoo), Habib Bourguiba (founder of modern Tunisia), Salvatore Dali, Michael J Fox, Francisco Franco, Adolf Hitler, Wilhelm Humboldt (the scientist), Arthur Koestler (the author), Sir Michael Redgrave, Pope John Paul II, Senator Claiborne Pell, Janet Reno, Terry Thomas (the Comedian) and Congressman Morris Udall are among the many famous people with Parkinson disease.

2. MEASURING FAME

Famous people get Parkinson disease. And infamous people get Parkinson disease. And ordinary people get Parkinson disease. But do a higher percent of famous people get Parkinson disease? Is there something about Parkinson disease—that leads to fame?

There are 1.2 million Americans with Parkinson disease, 0.4% of the population, 4 people with Parkinson disease for every 1,000 people. What's the percent among famous people? First we must ask: What is fame? How is it measured?

One way of measuring fame is to look at people who were chosen to be on the cover of *Time Magazine*. The man or woman so chosen is the person who in the opinion of *Time's* editors is, for a week, the most news-worthy, and by implication, the most famous. It's not a measure with each everyone may agrees, including the editors of *Reader's Digest, or Newsweek, or U.S. News & World Report*. And it's a fleeting measure, there are 52 issues of *Time* each year: 52 famous men and women each year, 796 men and women in the 73 years since 1927 when *Time*, the magazine, not the chronology, began.

A more enduring measure, is being on the cover of *Time Magazine* as *Time-Man (or Woman)-of-the Year*. Choosing the *Time Man (or Woman) -of-the Year* as the measure of fame involves relying on the professionalism, the objectivity, the knowledge of a particular group of editors. Ordinarily such choices, which aren't random, aren't compatible with the rules of statistical probability. There are other more objective

*Medical Director, National Parkinson Foundation, Miami, Florida
Mapping the Progress of Alzheimer's and Parkinson's Disease
Edited by Mizuno *et al.*, Kluwer Academic/Plenum Publishers, 2002

measures of fame such as Nobel Prize Laureates or Saints of the Catholic Church but *Time* it is—and probability will sort itself out.

Between 1927 and 1999, 59 single people appeared as *Time Man (or Woman) of-the Year*. If people who appeared more than once are counted as one, then 47 single people appeared as *Time Man (Or Woman) of-the Year*. Three of the 47 have Parkinson disease. The three are: Adolf Hitler, Deng XioPing, and Pope John Paul II. Three of 47 is 6.4% or 16 times the prevalence of Parkinson disease in the general population.

Among the 59 people who appear as *Time Man (or Woman) of-the Year*, are 11 who appear more than once. The 11 include, in alphabetical order: Winston Churchill (twice), Deng XioPing (twice), Dwight D Eisenhower (twice), Mikhail Gorbachev (twice), Lyndon B Johnson (twice), George C Marshall (twice), Richard M Nixon (twice), Ronald Reagan (twice), Franklin D Roosevelt (three times), Josef Stalin (twice), and Harry S Truman (twice). One of the 11, Deng XioPing, has Parkinson disease. One of 11 is 9.0% or 22 times the prevalence of Parkinson disease in the general population.

3. PARKINSON DISEASE AND FAME

Do famous people, like ordinary people, get Parkinson disease because it's a disease of maturity? A disease that allows its victims sufficient time to become famous--then suffer? It's not a disease of birth, or infancy, or childhood: a disease that doesn't allow its victims time to become famous. It's not Down's Syndrome, or childhood AIDS. Parkinson is not a disease that destroys the developing or maturing brain. The peak onset is age 60 years. And, by age 60 years, famous people, and ordinary people, have accomplished in life what they're going to accomplish. Or do famous people get Parkinson disease because the brain, unknowingly, makes a Faustian bargain? Is there something about Parkinson disease that leads to fame? Something that 'super-charges' or energizes the brain before slowing it down?

A note. While Parkinson disease may be your grandfather's disease (men more than women, 55% to 45% get Parkinson disease), 15% of people get Parkinson disease before they're 50. Muhammad Ali developed Parkinson at age 44 years. Michael J Fox who developed Parkinson at age 32 years. Adolf Hitler developed Parkinson at age 45 years.

4. SYMPTOMS OF PARKINSON DISEASE

Parkinson disease is defined by four classical or traditional symptoms:

(1) Stiffness.
(2) Tremor.
(3) Slowed movement.
(4) Impaired Balance.

The 4 classical symptoms result from a defect in brain regions that regulate movement. The major defect is a loss of dopamine in two brain regions: the substantia nigra and striatum. Dopamine is a chemical that increases the flow of impulses from nerve cells in the substantia nigra to the striatum.

5. THE BASIS OF PARKINSON DISEASE

The substantia nigra is in the brainstem, a region midway between the cortex, the thinking part of the brain and the spinal cord. There are 400,000 nerve cells in the substantia nigra, 0.04% of the 10,000,000,000 nerve cells in the brain. When because of Parkinson disease 240,000 of the cells, 60% of the total, die, the symptoms of Parkinson disease appear. It's estimated that *the process of Parkinson disease*, as distinct from it's *symptoms*, begins 5 to 15 years before the symptoms appear and the disease is diagnosed.

It's reasonable to ask how a disease that affects less than 0.04% of the brain's cells, a disease that, apparently, affects only mobility, can lead to greatness–to fame. The answer may be that Parkinson disease, as is apparent to patients, families, and friends, is more than a disease of movement. Disorders in mood, including depression, are common: 40% of all Parkinson patients are depressed.

6. DEPRESSION AND PARKINSON DISEASE

Depression, sadness, accompanied by feelings of guilt, remorse, hopelessness, pessimism, and gloom, is an integral part of Parkinson disease. The depression may be a *reactive depression*: a natural reaction to having Parkinson disease–a chronic, progressive, and debilitating disorder. A *reactive depression* is likely to occur in:

(1) A person who knows what is in store because a parent, a brother, a sister, or a close friend suffered from and was disabled by Parkinson disease.
(2) A person who, in addition to having Parkinson disease, is also undergoing a mid-life crisis related to retirement, job loss, or the illness of a spouse.
(3) A person who because of Parkinson disease suffers a loss of self-esteem related to tremor, or slowed movement, or lack of balance.

However, many if not most Parkinson patients suffer from an *inner depression,* a depression not in reaction to knowledge of the disease, a mid-life crisis, or disability. In fact in 20% of Parkinson patients depression may be the first symptom of Parkinson disease. Of course a depression appearing before the onset of the symptoms of tremor and slowed movement, an *inner depression,* can only, in retrospect, be appreciated as the start of Parkinson disease.

The inner depression of Parkinson disease is marked by apathy, indifference, passivity, and a loss of energy. These qualities are opposite to those which empowered, in a good way, Pope John Paul II, Salvatore Dali, Deng XioPing, and in an evil way, Adolf Hitler. What is the basis for the depression, the lack of energy in Parkinson disease? And, could the depression be preceded by a "mania", a surge of creative energy?

7. THE BASIS OF DEPRESSION

In Parkinson disease the symptoms of tremor and slowed movement result from a loss of *dopamine cells* in the brainstem. In addition to the loss of dopamine cells there is a loss of cells containing *nor-adrenalin* (a chemical involved in depression) and *serotonin* (another chemical involved in depression) cells. The nor-adrenalin and serotonin cells are also located in the brainstem, adjacent-to, or in the vicinity-of, the dopamine cells.

The depression associated with Parkinson disease, as well as depression NOT associated with Parkinson disease, the depression that affects as many as 27 million Americans, 10% of the population, results from a loss of nor-adrenalin and serotonin cells in the brainstem. These cells connect and interact with millions of cells in the cortex–perhaps 10% of all brain cells. A loss of nor-adrenalin or serotonin cells in the brainstem may depress one or more centers in the cortex resulting in depression.

The process of Parkinson disease begins 5 to 15 years before the symptoms of tremor and slowed movement appear. And it's possible, perhaps probable, that before depression appears, there's a stage of excitation, a surge of energy. And that's when the foundations of greatness, of fame, are laid. Credence for this view comes from recent observations utilizing deep brain stimulation (DBS).

Deep brain stimulation is a revolutionary treatment for Parkinson disease. During deep brain stimulation an electrode is inserted into the brainstem near the dopamine cells. Stimulating this region "revives" the dopamine cells. In a few patients this stimulation results in depression. And, in a few patients stimulating this region results in excitation–mania. Although these effects are temporary, they verify the presence of regions in the brain that can change the setting and mood of the way we think—all people, not just people with Parkinson disease.

HISTORICAL ASPECTS OF PARKINSON'S DISEASE IN JAPAN

Kunio Tashiro[*]

1. INTRODUCTION

It is well-known that James Parkinson published his book " An Essay on the Shaking Palsy " in 1817, and Charcot proposed to call this disease as Parkinson's disease in 1892.

In Table 1, the list of historical events from Parkinson 1817 to the AAN Treatment Guidelines published in 1998 is presented. The dotted horizontal line indicates before and after levodopa therapy .

Here, I discuss the history of Parkinson's disease or Parkisonism with regard to Japan, indicating some of major Japanese contributions on Parkinson' disease in italics (Table 1).

First of all, two Japanese neurologists, Professor Kinnosuke Miura (1864-1950)[1] and Professor Hiroshi Kawahara (1858-1918)[2] , who were introduced as Pioneers in Neurology in Journal of Neurology in 2000~2001 , are to be mentioned. Both Dr. Miura and Dr.Kawahara lived the same era as Charcot (1825-1893). The time Miura spent with Charcot was only 8 months, but he was deeply impressed by the clinical observation of Charcot, and also interested in anatomic-pathological research at the Charcot Laboratory. The detail of Dr. Miura's introduction appeared in the section of "Pionners in Neurology" of Journal of Neurology, published in September 2000, by Professor Makoto Iwata, Tokyo Women's Medical University. On the other hand, Dr. Kawahara had never been to Europe, but he had great knowledges on European neurological journals and writings, and was acquainted with the essence of modern European neurology. Dr. Akira Takahashi, Professor Emeritus, Nagoya University, introduced Dr. Kawahara in the section of " Pioneers in Neurology" , Journal of Neurology , in March issue of 2001. To our surprise, a typical and characteristic drawing of Parkinson's disease, which was almost equivalent to those of Gowers or Charcot, was clearly illustrated in Dr. Kawahara's Textbook of Neurology called Naika-ikou Shinkeikeitou-hen, published originally in 1897.

[*] Kunio Tashiro, Department of Neurology, Hokkaido University School of Medicine, Sapporo 060-8638, Japan

Mapping the Progress of Alzheimer's and Parkinson's Disease
Edited by Mizuno *et al.*, Kluwer Academic/Plenum Publishers, 2002

Table 1. Historical Aspects of Parkinson's Disease *with special reference to Japanese contributions*

1817	Parkinson J:	An Essay on the Shaking Palsy
		"Japanese encephalitis (epidemics since 1870s)"
1892	Charcot JM :	maladie de Parkinson
1897	*Kawahara H:*	*Naika Ikou-[shinkeikeitou-hen]*
1913	Lewy FH :	Lewy body
1917	von Economo C.	von Economo encephalitis & parkinsonism
1919	Trétiakoff C :	Parkinson's disease & substantia nigra
1922~1929		"post-encephalitic parkinsonism (von Economo)"
1928	*Miura K,ed :*	*Miura Shinkeibyo-gaku*
1949		——trihexyphenidyl (artane) ——
1953		——*pallidotomy (Narabayashi, H, Okuma, T)*—
1957	Carlsson A, et al.:	reserpine-induced hypokinesia & 3,4-dopa
1959	*Sano I, et al.:*	*distribution of catechol compounds in human brain*
		----------Beginning of Levodopa Therapy (1960~1961)--------------
1969~1970		——amantadine (Schwab,RS., et al) —
1973	*Yamamura Y, et al.:*	*paralysis agitans of early onset*
1974		—— bromocriptine (Calne,DB, et al) —
1981		——*l-threo-dops (droxidopa) (Narabayashi,H ,et al)* —
1983	Langston JW ,et al:	parkinsonism due to MPTP
1985		—— transplantation of adrenal medulla(Backlund E, et al)—
1987	*Mizuno Y, et al:*	*mitochondrial complex I subunits in PD*
1989		——MAO-B inhibitor (deprenyl) (DATATOP) ———
1990		——transplantation of fetal midbrain (Lindvall O, et al) ——
1992		——ventroposterolateral pallidotomy (Laitinen LV, et al) ——
1994	*Ichinose H, et al*	*GTP cyclohydrase I gene in hereditary progressive dystonia (Segawa Disease)*
1995		——DBS of subthalamic nucleus (Limousin P, et al) ———
1997	Polymeropoulos MH, et al:	alpha-synuclein gene in autosomal dominant parkinsonism
1998	*Kitada T, et al.:*	*parkin gene in autosomal recessive juvenile parkinsonism*
1998		——Treatment Guidelines (AAN, Olanow, CW, Koller, WC) —

2. PARKINSONISM DUE TO ENCEPHALITIS

It is famous that von Economo encephalitis caused post-encephalitic parkinsonism around the years of 1922 to 1929. Post-encephalitic parkinosnism of von Economo became famous again, when levodopa was just started to use for Parkinoson's disease in the latter half of 1960s.

Transient, but extraordinary clinical response to this parkinsonism was shown in the movie called "Awakening" by Oliver Sacks, which made a great hit and contributed to the propaganda of Neurology. Therefore, Dr. Sacks was awarded at the time of the 15th World Congress of Neurology held at Vancouver in 1993.On the other hand, not much have been mentioned about Japanese encephalitis causing parkinsonism, which brought about epidemics since 1870s, earlier than von Economo encephalitis. Japanese encephalitis is probably the only neurological disorder crowned with "Japanese", so I bring up this disease at this time.

Major epidemics was recorded in Japan since the 1870s, and the virus was isolated in 1934, which was named as Japanese B encephalitis virus to be differentiated from von Economo encephalitis virus A. But, at present, just Japanese encephalitis without B is well accepted. It is still listed as the most common encephalitis outside of North America, after the newest edition of *Principles of Neurology* by Victor and Ropper[3] in 2001. According to the paper in JNNP 2000[4], approximate global distributions of major neurotropic flaviruses are mapped, and Japanese encephalitis is still seen in the wide distribution among Asian countries.

Recently several papers are published on parkinsonism due to Japanese encephalitis, mostly from India[5] and Japan[6], showing MRI findings with localized lesions in the bilateral substantia nigra.

Ogata and his coworkers[7] recently succeeded in making a rat model of parkinsonism induced by Japanese encephalitis virus,publishing in Journal of NeuroVirology in 1997. Fischer rat infected with Japanese encephalitis virus at the 13th day after birth was crucial to develop parkinsonism at this experiments. Almost complete cell loss was induced at the substantia nigra , and its clinical presentation was bradykinesia , which responded to levodopa.

3. SUBSEQUENT CHRONOLOGICAL JAPANESE CONTRIBUTIONS

As shown in Table 1, Sano et al.[8] reported the distribution of catechol compounds in human brain in 1959. This was truly the pioneering work in this field, being probably the same values as works by Dr.Carlsson, who was a Novel Prize Winner in 2000.

The reports of "paralysis agitans of early onset" by Yamamura et al. in 1973, abnormalities of mitochondrial complex 1 subunits by Mizuno, et al. in 1987, GTP cyclohydrase 1 gene in Segawa disease by Ichinose, et al. in 1994, and the discovery of parkin gene in autosomal recessive juvenile parkinsonism by Kitada, et al, in 1998 are the some of the representaive Japanese contributions in Japanese research of Parkinson's disease and parkinsonism.

As for the contributions of therapeutic methods, we are proud to name Professor Narabayashi, who started pallidotomy in 1953, prior to levodopa treatment. He also introduced l-threo-dops (droxidopa) therapy in 1981.

At present, levodopa, dopamine receptor agonists, MAO-B inhibitors, anticholinergic agents, amantadine hydrochloride, or droxidopa are all available in Janan, except COMT inhibitors, and the Japanese Society of Neurology is in the process of making treatment guidelines for the Japanese.

4. PARKINSONISM IN ARTS

In 1988, the 9th International Symposium on Parkinson's Disease was held at Jerusalem, Israel. Professor Donald Calne[9] made a fantastic speech on Leonard da Vinci, which was later published in NEJM as Correspondence as "Did Leonardo describe Parkinson's Disease? " This report suggested that Parkinson's disease did exist before industrial revolution.

Then, one of my friends, Lance Fogan[10] at California wrote a paper entitled "The Neurology in Shakespeare", stating Shakespeare seemed to be describing paralysis agitans. In Part 2, Henry VI, Butcher queries,   "Why dost thou quiver, man? , then Lord Say responds, The palsy, and not fear, provokes me ".

According to Dr. Lakke's paper[11] " Images of Parkinson's disease, past and present " in Journal of the History and the Neurosciences, 1994, Rembrandt van Rijn figured an innkeeper suffering from Parkinson's disease.

So, prior to James Parkinson's *Shaking Palsy*, Parkinson's disease must have been existing.

Finally I mention about Japanese traditional arts, such as   Noh drama, and Kabuki, in their relations to "neurology ".

Kabuki was developed in Edo Era , showing very active performances.

Dr.Yasuo Toyokura, Professor Emeritus, University of Tokyo, published papers showing " Upgoing toe signs ", (Babinski sign), during actors' performances.

On the other hand, Noh drama , the other traditional art, started in the period of Muromachi Era (that is between 1336 and 1493). In contrast to Kabuki, Noh exhibits tranquility.   I abstracted and translated short phrases from the Book published in 1998, by Professor Shinpei Matsuoka[12].

According to his note, Zeami, Noh drama writer and actor, established Noh as the Japanese traditional art in the period of Muromachi Era. He died at age 81, having extreme longevity at that period. Around the age of 50, Zeami himself was definitely realizing his aging body, struggling to search for the new idea how Noh should be. Then he found the "the beauty of the aged ", by concentrating energy internally under the suppression of actual body movements. That was the moment Japanese art made discovery of the " beauty of the aged ".

One of the Noh drama called " Kasuga Ryujin " demonstrated a body of " Okina "(old man) metamorphosed to the body of " Ryujin " (god of dragon), suggesting the secret of the origin of Noh drama.. The actors in Noh show slight stooped posture with flexed arms and hands, similar to the aged persons, but quite resembling the posture of Parkinson's disease.

Just recently, an article entitled " The Noh mask effect: vertical view point dependence of facial expression perception " was published in the Proceedings of Royal Society of London[13]. This paper was summarized and introduced at the Nature News

Service on November 15, 2000, as *"Noh subtlety for the British"*. Many Japanese peoples might understand the performance of Noh as art, but if you observe its performance acting old people, you might remind of old people or Parkinson's disease. If so, Parkinson's disease was recognized in Japan, approximately 100 years prior to Leonardo da Vinci.

REFERENCES

1. M. Iwata, Kinnosuke Miura (1864-1950), J. Neurol. 247, 725-726 (2000).
2. A. Takahashi, Hiroshi Kawahara (1858-1918), J. Neurol. 248, 241-242 (2001).
3. M. Victor and A.H. Ropper, in: *Adams and Victor's Principles of Neurology*, 7th ed, (McGraw-Hill, New York 2001), p.792.
4. T. Solomon, N.M. Dung, R. Kneen, M. Gainsborough, D.W. Vaughn, and V.T. Khanh, Japanese encephalitis, J. Neurol. Neurosurg. Psychiatry 68, 405-415 (2000).
5. S. Pradhan, N. Pandey, S. Shashank, R.K.Gupta, and A. Mathur, Parkinsonism due to predominant involvement of substantia nigra in Japanese encephalitis, Neurology 53, 1781-1786 (1999).
6. H. Shoji, Japanese encephalitis and parkinsonism, Shinkei Naika (Neurol. Med.), 52, 369-373 (2000) (in Japanese).
7. A. Ogata, K. Tashiro, S. Nukuzuma, K. Nagashima, and W.W. Hall, A rat model of Parkinson's disease induced by Japanese encephalitis virus, J. NeuroVirology 3, 141-147 (1997).
8. I. Sano, T. Gamo, Y. Takimoto, K. Taniguchi, M. Takesada, and K. Nishimura, Distribution of catechol compounds in human brain, Biochem. Biophys. Acta, 32, 586-587 (1959).
9. D.B. Calne, A.Dubini, and G. Stern, Did Leonardo describe Parkinson's disease ?, N. Engl. J. Med.,320, 594 (1989).
10. L. Fogan, The neurology in Shakespeare, Arch. Neurol. 46, 922-924 (1898).
11. J.P.W.F. Lakke, Images of Parkinson's disease, past and present, J. Hist. Neurosci. 3, 132-138 (1994).
12. S. Matsuoka, Noh~Chu-sei karano Hibiki (Kadokawa Shoten, Tokyo, 1998) (in Japanese).
13. M.J. Lyons, R. Campbell, A. Plante, M. Coleman, M. Kamachi, and S. Akamatsu, The Noh mask effect: vertical viewpoint dependence of facial expression perception, Proc. R. Soc. Lond. 267, 2239-2245 (2000).

CLINICAL EVALUATION AND DIFFERENTIAL DIAGNOSIS OF PARKINSONISM

Philip D. Thompson[*]

1. INTRODUCTION

Parkinsonism or akinetic rigid syndromes are characterised by small, slow movements (bradykinesia), rigidity and tremor. Parkinson's disease is the commonest cause, accounting for at least 80% of akinetic rigid syndromes (1). Increasingly, attention has turned to delineating the clinical characteristics of Parkinson's disease and the conditions that commonly mimic Parkinson's disease. These conditions include progressive supranuclear palsy (Steele-Richardson-Olszewski syndrome), Multiple system atrophy and cerebrovascular disease. In the full expression of these conditions, the diagnosis is usually obvious, but in the early stages precise diagnosis may be difficult.

2. PARKINSON'S DISEASE

The clinical criteria for the diagnosis of Parkinson's disease are rest tremor, rigidity and bradykinesia with supporting features including a unilateral onset and gait disturbance. A unilateral onset and a dramatic response to levodopa are also important clues to diagnosis. The characteristic rest tremor is present in at least 80% of cases. Features such as postural instability (with frequent falls) and cognitive impairment (which develops in 30% of cases) occur late in the course of the disease and if present early should suggest alternative diagnoses. A progressive course with falls suggests progressive supranuclear palsy or multiple system atrophy. Significant early cognitive impairment should suggest diffuse Lewy body disease (especially with hallucinations) or Alzheimer's disease.

In a pathological study of 100 cases of akinetic rigid syndromes (1), Parkinson's disease was found to account for 76%. Of those, with a pathological diagnosis of Parkinson's disease, more than half had coexistent cerebral pathology, most commonly cerebrovascular disease and Alzheimer's pathology. The influence of these multiple pathologies on the rate of progression, extent of disability and response to therapy in the elderly parkinsonian has not been studied in a systematic manner.

[*]Department of Neurology and University Department of Medicine, Royal Adelaide Hospital and University of Adelaide, Adelaide 5000 South Australia

Mapping the Progress of Alzheimer's and Parkinson's Disease
Edited by Mizuno *et al.*, Kluwer Academic/Plenum Publishers, 2002

3. OTHER AKINETIC RIGID SYNDROMES

Progressive supranuclear palsy, multiple system atrophy and Alzheimer's disease or pathology accounted for the majority of the remaining 24% of the 100 cases studied by Hughes et al (1). Features that suggested diagnoses other than Parkinson's disease include a rapid or stepwise progression, symmetry of signs at disease onset, abnormal eye movements, early postural instability and falls, autonomic failure, early dementia, pseudobulbar palsy particularly dysphagia, pyramidal signs, ataxia and a poor response to levodopa

3.1. Progressive supranuclear palsy (Steele-Richardson-Olszewski Syndrome)

This condition begins in the 7th decade and usually progresses over a period of 6 to 10 years (2). Impaired balance and falls are common presenting features often leading to injury in the first year. Dysarthria is also an early feature. Visual complaints, often vague in the first instance are common. The typical supranuclear palsy of vertical gaze, affecting particularly down gaze, may not be present initially but becomes evident within three to four years in 90% of cases. The ocular motor disorder may comprise slow vertical saccades, hypometric saccades, square wave jerks, and disorders of eyelid opening with lid retraction, blepharospasm and apraxia of eyelid opening. Rigidity is predominantly axial, affecting the trunk and neck. The trunk is held in an upright posture and the neck is extended in contrast to be flexed posture of Parkinson's disease. These features are characteristic and distinguish this condition from Parkinson's disease. Speech is often soft and stuttering often punctuated by groaning sounds and perseveration. There may be marked slowness of thought, apathy and frontal lobe signs including inertia and grasp reflexes. The signs are usually symmetrical and the response to levodopa is poor. Tremor is uncommon but does occur in some cases. Pyramidal signs are present in two-thirds of cases.

3.2. Multiple System Atrophy

Multiple system atrophy accounts for approximately 5% of cases of akinetic rigid syndromes. The age of onset is in the sixth decade and the disease duration extends over seven or eight years. The condition may present as a pure akinetic rigid syndrome (striatonigral degeneration), with ataxia (olivopontocerebellar degeneration), striatonigral degeneration with autonomic failure (Shy-Drager syndrome) or postural hypotension due to pure autonomic failure. Accordingly, the clinical features include parkinsonism, autonomic failure, urinary dysfunction, cerebellar ataxia and corticospinal signs (3). A postural tremor is present in the majority of cases and the illness may begin unilaterally. With the passage of time an asymmetric akinetic rigid syndrome emerges with prominent falls. Autonomic failure manifests as postural hypotension, incontinence and impotence. Additional features include dysarthria, ataxia, myoclonus and laryngeal stridor. The response to levodopa is generally poor, but at least 30% of patients may notice a deterioration when the drug is withdrawn. Another characteristic feature is the appearance of cranial (oromandibular) dystonia during treatment with levodopa. MRI may show atrophy and hypointensity of the putamen with an hyperintense signal along the lateral border of the putamen, and brainstem and cerebellar atrophy.

3.3. Corticobasal degeneration

Corticobasal degeneration typically begins in the seventh decade with unilateral clumsiness, loss of dexterity, rigidity, apraxia and dystonia affecting an arm or leg. Postural instability, a parkinsonian gait, myoclonus develop as the disease progresses. Eventually, the ability to use the limb is lost. A supranuclear ophthalmoplegia, dysarthria, postural tremor, cortical sensory loss, alien limb behaviour and frontal lobe signs also may develop (4). Brain imaging may show asymmetric cortical atrophy affecting the frontoparietal cortex.

3.4. "Lower-half" parkinsonism in cerebrovascular disease and hydrocephalus

The term "lower half parkinsonism" refers to the clinical syndrome of a parkinsonian gait with small, shuffling steps ("marche á petit pas") and freezing caused by hemispheric white matter ischaemia due to small vessel cerebrovascular disease (5). Occasionally, hydrocephalus and other diseases affecting the frontal white matter produce a similar clinical picture. Clinical clues that distinguish this gait from that of Parkinson's disease include an upright body posture when walking, a wide stance base, reflecting a degree of ataxia, impaired postural reflexes, preservation of upper body movements with normal facial expression and normal upper limb movements often with exaggerated arm swing during walking (5). Tremor is minimal or absent. Additional signs include a pseudobulbar palsy, brisk tendon reflexes and extensor plantar responses. There is no appreciable response to levodopa.

3.5. Drug-induced parkinsonism

Dopamine receptor antagonists such as the major tranquillisers, antiemetics and dopamine storage depleting drugs produce and aggravate parkinsonism. The clinical picture of drug-induced parkinsonism may be identical to Parkinson's disease though tremor is not prominent. Improvement in drug induced parkinsonism follows withdrawal of the offending drug but may take up to 6 months. Persistence of parkinsonian symptoms more than 6 months after drug withdrawal raises the possibility that drug exposure has unmasked underlying idiopathic Parkinson's disease.

REFERENCES

1. Hughes AJ, Daniel SE, Kilford L, Lees AJ. Accuracy of clinical diagnosis of idiopathic Parkinson's disease: a clinico-pathological study of 100 cases. J Neurol Neurosurg Psychiatry 1992; 55:181-184.
2. Litvan I. Recent advances in atypical parkinsonian disorders. Curr Op Neurol 1999;12: 441-446.
3. Wenning GK, Benshlomo Y, Magalhaes M, Daniel SE, Quinn NP. Clinical features and natural history of multiple system atrophy- an analysis of 100 cases Brain 1994; 117: 835-845.
4. Rinne JO, Lee MS, Thompson PD, Marsden CD. Corticobasal degeneration: A clinical study of 36 cases. Brain 1994; 117: 1183-1196.
5. Thompson PD, Marsden CD The gait disorder of subcortical arteriosclerotic encephalopathy (Binswanger's disease). Movement Disorders 1987; 2: 1-8.

THE NATURAL HISTORY AND PROGNOSIS OF PARKINSON'S DISEASE

Werner Poewe[*]

1. INTRODUCTION

A significant proportion of basic research in the field of Parkinson's disease is related to identify underlying mechanisms responsible for nigral cell loss in this disorder. The ultimate goal of such research is to identify agents that will modify or hold the progression of the disease. At present, no candidate has yet emerged for which clinical studies would have demonstrated sustained and marked effects on the progressive course of Parkinson's disease. At the same time data concerning the natural history of Parkinson's disease are limited and there is considerable uncertainty in a number of important areas – including possible duration of a pre-clinical phase, linear versus exponential rates of progression, different progression rates in subtypes of the disease as well as risk factors and predictors for different rates of progression and mortality.

2. PROGRESSION OF MOTOR DISABILITY IN UNTREATED DISEASE

Two early studies have assessed the rates of progression of disability of patients with Parkinson's disease in the pre-levodopa area. Hoehn & Yahr, in their 1967 paper, studied a subgroup 183 patients using their newly developed Hoehn & Yahr scale (Hoehn and Yahr, 1967). Progression to stages IV and V had occurred in 16% of patients with disease duration of less than 5 years increasing to 37% and 42% after 10 and 15 years respectively. Median delays before reaching Hoehn & Yahr stages IV and V were 9 and 14 years respectively and even shorter in the series reported by Martilla and Rinne in 1977 (Martilla and Rinne, 1977). However, there was large inter-individual variability and about a third of patients with durations of disease for 10 years or longer were still in stages I and II. Similar percentages have also been reported in a more recent series by Hely and co-workers (1999).

---

[*]Werner Poewe, Department of Neurology, University Hospital Innsbruck, 6020 Innsbruck, Austria.

Mapping the Progress of Alzheimer's and Parkinson's Disease
Edited by Mizuno *et al.*, Kluwer Academic/Plenum Publishers, 2002

More recent studies have assessed the rate of progression of UPDRS motor scores in a prospective fashion. Data from the seminal DATATOP study comparing the impact of deprenyl and tocopherol versus placebo on the progression of Parkinson's disease have provided adequate data to estimate the annual rates of decline in untreated disease. Again these were not uniform and differed from 3.5% to 8% per year according to whether or not patients had reached the endpoint (Parkinson Study Group, 1998).

Besides such inter-individual variation there is also evidence that progression of motor scores when measured in treated patients during the off stage is dynamic with faster rates early versus later in the disease (Bonnet et al., 1987; Lee et al., 1994; Goetz et al., 1987). Nevertheless available data on annual rates of motor UPDRS scores in untreated PD would predict severe disability within 10 years of disease duration – rather in line with the early observations published by Hoehn and Yahr (1967).

3. IMPACT OF TREATMENT

When comparing disease progression in 282 patients receiving LD therapy to cohorts of patients from the pre LD-era Hoehn & Yahr found that LD-treated patients had latencies to reach successive Hoehn & Yahr stages that were prolonged by about 3-5 years per stage compared to untreated patients. In addition, the percentages of patients who had become severely disabled or had died in successive 5 year epochs of disease duration up to 15 years was reduced by 30-50% per epoch compared to patients from the pre-LD era. Several studies – but not all (Hely et al, 1999) – have found similar latencies to the different Hoehn & Yahr stages in PD patients receiving LD plus additional dopaminergic and/or non-dopaminergic drugs. The slowest rates of progression have been reported by Lücking et al (2000) in a cohort of young onset PD patients with and without the Parkin-mutation where latencies to Hoehn & Yahr IV or V were 40 years or longer. This is likely to reflect a different rate of progression in subtypes of PD rather than impact of treatment alone.

Recent studies have focussed on impact of drug treatment on rates of progression of UPDRS score after temporary drug washout or rates of decline of surrogate markers of nigrostriatal dysfunction using 18 FD PET or β-CIT-SPECT (see below). Deprenyl treatment was found to delay the need for L-Dopa by about 9 months and also appeared to reduce the annual rate of decline of UPDRS motor scores by up to 50% but these findings could also be explained by the drug's symptomatic effects and clearly do not seem to extend beyond 12 months of treatment. Quite contrary to the findings of reduced progression of disability by L-Dopa therapy there has been renewed debate regarding a neurotoxic potential of LD by virtue of its oxidative metabolism. A double-blind randomised prospective placebo-controlled trial designed to assess LD's effects on natural disease (ELLDOPA) is currently underway. Results from another randomised double-blind prospective 12 months trial have, however, failed to indicate any difference in the effects of Deprenyl versus L-Dopa on rates of progression of motor scores (Olanow et al., 1995).

Nevertheless, L-Dopa therapy has introduced an additional source of progressive disability into the natural evolution of PD through its potential to induce abnormal involuntary movements as well as motor response oscillations. A larger number of retrospective series in the 1970's and 1980's has found rates of these motor complications generally above 50% after more than 5 years of treatment (Poewe et al.,

1986). More recent prospective double-blind trials with durations of 2 to 5 years and L-Dopa as an active treatment arm suggest somewhat lower rates between less than 20 and up to 50%. A recent community-based study found incidences of about 30 to 40% (Schrag and Quinn, 2000).

4. PROGRESSION OF SURROGATE MARKERS

Fearnley and Lees (1991) have studied the subregional topography and progression rates of nigral cell loss in post mortem brain material at the UK PDS BRC and found a global rate of cell loss of 40% per decade. There was an exponential decline of the rate of attrition, which appeared much faster early in the disease compared to later stages – suggestive of a preclinical window of approximately 5 years. The latter was also suggested by sequential 18-FD-PET studies approximately 18 months apart which found an approximately decline of putaminal Ki values for tracer uptake of 9% per annum. Assuming linear decline extrapolation of these results would again imply a preclinical phase of approximately 5 to 6 years.

Subsequent studies of rate of progression of nigro-striatal dysfunction as assessed by either sequential 18-FD-PET or β-CIT-SPECT have all found annual rates of decline in the order of 6 - 10%. In addition, there is evidence in these studies, that progression rates may be faster early in the disease compared to later. All patients studied so far have been under treatment so that the impact of dopaminergic therapy on progression rates relative to untreated PD is unknown. Comparator studies between LD and dopamine agonist have so far failed to detect significant differences between these two approaches regarding functional imaging markers.

5. RISK FACTORS FOR RAPID PROGRESSION

A number of studies have assessed risk factors predicting faster progression of disability in idiopathic Parkinson's disease. They have consistently identified older age at on-set and cognitive decline and dementia as being associated with rapid progression of motor disability.

Age, dementia, and hallucinosis have also been found to be the most relevant predictors for institutional care in Parkinson's disease. Hely and co-workers (1999) in their 10 year follow-up of a prospectively studied cohort of 146 patients found 23% having been admitted to nursing homes by 10 years of diseases. 86% of these patients had died by the time of last follow-up and mean survival in nursing home care was 34 months.

6. MORTALITY

In their seminal paper Hoehn & Yahr found an excess mortality in PD patients over the age matched normal population of about threefold and concluded that " ..... the state of parkinsonism severely limits life expectancy" (Hoehn and Yahr, 1967). In a series of 200 cases with pathologically confirmed Parkinson's disease Jellinger and colleagues investigated the impact of clinical dementia and Alzheimer pathology on survival. Mean survival was around 10 years in patients whose brains were free of Alzheimer type pathology but only half that long (4.9 years) in patients with AD pathology of CERAD grade B or C (Jellinger et al., 2001). Most studies in the post-LD era have found reduced excess mortality but mortality ratios were still between 1.5 and 2.5. The most notable exception is the 9 year follow-up data of the DATATOP study cohort that found a death rate of 2.1% per year and a standard mortality ratio of 0.9 - most likely reflecting the initial selection criteria excluding all significant co-morbidity. Cognitive decline and dementia as well as older age at onset have been identified as predictors of decreased survival in several studies including our own and we and others have also found improved survival in patients presenting with tremor-dominant disease. Pneumonia is the leading cause of death in Parkinson's disease followed by cardiovascular events, stroke and cancer in most surveys. A recent population based study of causes of death in Parkinson's disease found that significantly more PD patients (20%) had died from pneumonia than controls (9%) while more controls had died from coronary heart disease (23%) compared to Parkinson's disease patients (13%).

The relevance of aspiration pneumonia as a contributor to the increased mortality in Parkinson's disease is also suggested by a post-mortem study showing a tight correlation between the onset of clinically relevant dysphagia and survival time across a group of 77 post-mortem confirmed cases of different degenerative parkinsonian disorders including IPD, MSA, PSP, CBD, and DLB (Müller et al., 2000).

REFERENCES

Bonnet, A.M., Loria, Y., Saint-Hilaire, M.H., Lhermitte, F., Agid, Y., 1987, Does long-term aggravation of Parkinson's disease result from nondopaminergic lesions? *Neurology* **37**:1539-1542.

Fearnley, J.M., Lees, A.J., 1991, Ageing and Parkinson's disease: substantia nigra regional selectivity. *Brain* **114**:2283-2301.

Goetz, C.G., Tanner, C.M., Shannon, K.M., 1987, Progression of Parkinson's disease without levodopa. *Neurology* **37**:695-698.

Hely, M.A., Morris, J.G., Traficante, R., Reid W.G., et al., 1999, The sydney multicentre study of Parkinson's disease: progression and mortality at 10 years. *J Neurol Neurosurg Psychiatry* **67**:300-307.

Hoehn, M.M., Yahr, M.D., 1967, Parkinsonism: onset, progression, and mortality. *Neurology* **15**:427-442.

Jellinger, K.A., Seppi, K., Wenning, G.K., Poewe, W., 2001, Impact of coexistent Alzheimer pathology on the natural history of Parkinson's disease. J Neural Transm (in press)

Lee, C.S., Schulzer, M., Mak, E.K., Snow B.J., et al., 1994, Clinical observations on the rate of progression of idiopathic parkinsonism. *Brain* **117**:501-507.

Lücking, C.B., Durr, A., Bonifati, V., Vaughan, J., et al., 2000, Association between early-onset Parkinson's disease and mutations in the parkin gene. French Parkinson's Disease Genetics Study Group. *N Engl J Med* **342**:1560-1567.

Martilla, P.J., Rinne, U.K., 1977, Disability and progression in Parkinson's disease. *Acta Neurol Scand* **56**: 159-169.

Müller, J., Wenning, G.K., Verny, M., McKee, A., Chauduri, K.R., Jellinger, K., Poewe, W., Litvan, I., 2001, Progression of dysarthria and dysphagia in post-mortem confirmed parkinsonian disorders. *Arch Neurol* **58**:259-264.

Olanow, C.W., Hauser, R.A., Gauger, L., et al., 1995, The effect of deprenyl and levodopa on the progression of Parkinson's disease. *Ann Neurol* **38**:771-777.

Parkinson Study Group, 1998, Mortality in DATATOP: a multicenter trial in early Parkinson's disease. *Ann Neurol* **43**:318-325.

Poewe, W.H., Lees, A.J., Stern, G.M., 1986, Low-dose l-dopa therapy in Parkinson's disease: a 6-year follow-up study. *Neurology* **36**:1528-1530.

Schrag, A., Quinn, N., 2000, Dyskinesias and motor fluctuations in Parkinson's disease: a community-based study. *Brain* **123**:2297-2305.

GAIT DISTURBANCES IN PARKINSON'S DISEASE

Nir Giladi[1], Jacov Balash [1], Jeffrey M. Hausdorff[1,2] [1]

1. INTRODUCTION

Gait disturbances have recently been added as the fifth cardinal motor symptom of parkinsonism. This addition was made due to the unique phenomenology of gait disturbances (e.g., in the form of freezing and festination) and their importance in the clinical spectrum of parkinsonism. In his famous "Essay on Shaking Palsy", James Parkinson first elegantly described how the parkinsonian patient is "irresistibly impelled to make much quicker and shorter steps, and thereby to adopt unwillingly a running pace...". Gait disturbances, like decreased arm swing, slower speed, and shorter steps appear early in the course of Parkinson's disease (PD), sometimes even as presenting symptoms. From the functional point of view, however, gait abnormalities are of clinical significance mainly in the advanced stages of the disease. With the current, modern treatment of PD, patients are able to maintain locomotion longer, but this is accompanied by motor response fluctuations and dyskinesias. "Off" state akinesia, severe lethargy,

Table 1. Types of Gait Disturbances in Parkinsonism

Continuous	Paroxysmal	
	Type I (freezing of gait)	**Type II (festination)**
- Shorten stride with increased cadence	- Start/turn hesitation	- Gait initiation festination
- Bradykinetic gait/turning	- Tight quarters hesitation	- Tight quarters festination
- Shuffling gait/turning	- Hesitation while reaching destination	- Festination while reaching destination
- Disequilibrium gait		
- Fear of falling gait/turning	- Hesitation during mental over-load or stressful situations	- Festination on "open-run-way"
- Dysrhythmic gait		- Festination during mental over-load or stressful situation
- Dyskinetic gait		
- Stiff gait/turning		

1) Movement Disorders Unit, Department of Neurology, Tel-Aviv Sourasky Medical Center, Tel-Aviv, Israel 64239, 2) Beth Israel Deaconess Medical Center, Boston, Mass., USA 02115

Mapping the Progress of Alzheimer's and Parkinson's Disease
Edited by Mizuno *et al.*, Kluwer Academic/Plenum Publishers, 2002

light headedness and dystonic postures of the legs or trunk and "on" state dyskinesias have created a new type of gait disturbance that was not seen prior to the levodopa era. Gait disturbances in PD exert profound effects on function and independence and haverecently been shown to be a major risk factor for institutionalization and death in patients with PD (Hely et al 1999).

In general, all patients will develop gait disturbances in the advanced stages of PD. Most neuro-physiological research on parkinsonian gait has focused on stride length, cadence, limb support and the gait cycle, i.e., gait changes that occur continuously during walking. Gait abnormalities in PD may also be transient (Table 1), however, little attention has been given to these paroxysmal disturbances.

2. NORMAL LOCOMOTION: OVERVIEW

The basic requirements for walking are: 1) **Equilibrium** – the capacity to assume an upright posture and maintain balance; 2) **Locomotion** – the ability to initiate and maintain rhythmic stepping; 3) **Non-neurological factors** – including, for example, good range of motion, strong bones, and efficacious muscles. Locomotion itself requires the **ability to initiate steps** by shifting the center of gravity laterally towards the support limb, thereby allowing the swing foot to leave the ground. This is accomplished via bilateral ankle dorsiflexion (moving the body forward), simultaneous abduction of the swing hip, and mild flexion of the support hip and knee. Once this classic strategy is performed, the swing leg lifts up and forward and the first step is taken. After stepping begins, a second motor program maintains rhythmic and effective locomotion while modifying the stepping in response to changes in internal plans and external/environmental circumstances. This motor plan regulates a complex temporal sequence of muscle activation, synchronized changes of joint positions, and linear and angular accelerations of the limbs and the body's center of gravity to produce locomotion while maintaining upright equilibrium.

Walking is controlled by cortical, basal ganglia, cerebellar and brain stem ascending and descending pathways. The final execution of the motor plan is carried out by the spinal motor neurons. At the level of the cortex, locomotion involves the primary motor cortex, pre-motor cortex, and the supplementary motor area (SMA) of the frontal lobes as well as the occipital and posterior parietal cortex. At a subconscious level, the basal ganglia (BG) influence locomotion through their main cortical output, the pre-motor cortex, but also through a descending pathway to the cholinergic pedunculopontine nucleus (PPN), an area that is considered the human brain stem locomotion center (Pahapill, 2000).

3. PATHOPHYSIOLOGY OF LOCOMOTION CONTROL IN PARKINSONISM

A major cause of gait changes in PD is the dopaminergic deficit in the basal ganglia. Decreased dopamine in the nigro-striatal pathway leads to excessive thalamic inhibition of the pre-motor cortex (Haslinger, 2001). Many of the gait changes in PD and especially the continuous ones (Table 1) can be explained by the hyopdopaminergic state. Nonetheless, the relationship between paroxysmal gait disturbances and the hypodopaminergic state is more complex and other factors such as the synchronization between different pathways might play a role.

Gait disturbances in PD are highly variable from one patient to the next, suggesting a heterogeneity in the pathophysiology. Dopaminergic treatment only partially improves gait disturbances; at times it may even produce a deleterious effect, suggesting the involvement of other brain structures or neurotransmitters. In fact, direct lesion of the frontal cortex by infarction in the region of both anterior cerebral arteries causes a peculiar disorder of locomotion, characterized by the inability to initiate stepping, like that seen in parkinsonism (Ueno, 1989). Similarly, selective lesion of the brain stem PPN in monkeys produces a contralateral akinesia similar to parkinsonism (Aziz, et al. 1998). Hemorrhage in the brain stem involving the PPN area in a patient caused severe parkinsonian locomotion disturbances including an inability to stand and generate stepping (Masdeu et al. 1993).

During a 20 meter long walk, regional **dopaminergic** activation was observed mainly in the putamen, reflected as excitation of the direct (putamino-internal pallidal-thalamic) pathway associated with cortically initiated movements (Fukuyama et al. 1997). In contrast, a 50 minute walk of PD patients without rest or assistance caused increased dopaminergic metabolism in the orbitofrontal cortex and the caudate nucleus but not in the putamen. Based on these observations, Ouchi et al (Ouchi et al. 2001) suggested that bipedal gait regulation in PD is influenced by the mesocortical-ventrostriatal dopaminergic system and not as much by the nigro-striatal motor pathway. The influence of this system might partially explain the relationship between mental loading and the paroxysmal gait disturbances in PD.

The caudate nucleus is one of the main BG target structures innervated by the nigro-striatal dopaminergic pathway, although it is also associated with the limbic system. Decreased dopaminergic activity at he level of the caudate nucleus is usually seen relatively late in the course of PD, like the disturbances to gait, suggesting a possible association. Indeed, patients with predominantly paroxysmal gait disturbances have an unusual decrease of levodopa uptake at the level of the caudate nucleus (Leenders et al, 2001 in press). In addition, after hemorrhage to both caudate nuclei, prolonged akinetic episodes, apathy and abulia were seen (Trillet et al.1990) while bilateral ablation of the caudate nuclei in cats caused the opposite behavior with obtrusive gait following a moving object (Villablance, Marcus, 1975).

Another important structure in the BG which might play an important role in gait disturbances is the internal globus pallidus (Gpi). The Gpi processes striatal inputs and transfers the information through the thalamus to the cortex and the brain stem. Pallidal necrosis induced by carbon monoxide causes axial bradykinesia, freezing of gait and postural disorders with retropulsion (Feve et al. 1993) while Gpi lesions or stimulation alleviates gait disorders.

4. GAIT INITIATION IN PARKINSON'S DISEASE

PD patients generally use the normal strategy for gait initiation (Rosin et al 1997). They tend, however, to initiate gait more slowly and less forcefully. In addition, due to impaired generation of postural shifts at the right degree and synchronization, frequent hesitation can be seen. Kinematics of joint motion have also been shown to be more variable as has the adaptation of the normal gait initiation strategy to the environment. The most striking and clinically significant gait initiation disturbance is start hesitation. After turning hesitation, start hesitation is the second most frequent type of transient gait disturbance, as scored out of 158 minutes of videotaped gait in 21 patients with parkinsonism in "on" and "off" states (18 with PD, 3 with primary freezing of gait) and 147 episodes of paroxysmal gait disorders (PGD) (Table 2). The basic mechanism of start hesitation is unknown, but disturbances in the shifting of the center of gravity as a result of postural reflex abnormalities (Elble et al 1996) and abnormal activation of leg muscles may be contributing factors (Yanagisawa et al 1994).

Table 2 Frequency of various kinds of PGD

Type of PGD	Number of episodes	Percent
Turning hesitation	85	57.8
Starting hesitation	27	18.4
Hesitation at tight quarters	19	12.9
Hesitation in open space	16	10.9
Total	147	100

5. GAIT DYNAMICS IN PARKINSON'S DISEASE

Many of the observed changes in gait are related to bradykinesia and hypokinesia (Table 1), however, another important aspect of gait disturbances in parkinsonism is related to rhythmicity. Study of the rhythmicity and dynamics of walking provide additional insight into the clinical features of PD, the patho-physiologic factors that contribute to gait disturbances in PD, as well as an alternative method for quantifying disease progression and therapeutic efficacy.

One property of rhythmicity in PD is well known. While stride length and gait speed are typically reduced, the cadence and the time between steps are typically and perhaps surprisingly *normal*, at least in the early stages. Alternatively put, the *average* rhythm or stride interval is normal in PD. Indeed, some have inferred from this that the "stepping mechanism", the neural process that produces sequential contraction and relaxation of muscle groups, is intact in PD. Actually, however, certain aspects of gait rhythm dynamics are altered in PD (Blin et al 1990; Hausdorff et al 1998, 2000).

When considering time ordered events or

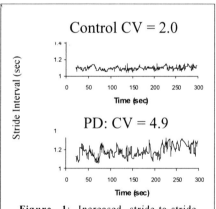

Figure 1: Increased stride-to-stride variability in the stride interval time series of a PD patient (below) compared to a healthy control (above).

time series, like the time between the stride interval, one can study not only the average value, but also how the stride interval fluctuates about its average value, the fluctuation magnitude. This parameter reflects how regular and consistent a process is. The fluctuation magnitude may be quantified in a number of different ways, for example, one can estimate the magnitude of the fluctuations in the gait rhythm by calculating the coefficient of variation, ($CV = 100*SD/mean,$) where SD is the standard deviation of the stride interval.

In PD, the problem with the rhythm is not with the average stride interval. There is, however, increased stride-to-stride variability such that the CV is significantly increased, perhaps by 100% or more, compared to control subjects (e.g., see Figure 1). Not only is the overall variability increased in PD, but the variability of the sub-phases of the gait cycle, e.g. swing and stance time, are also increased with respect to control values (Hausdorff et al, 1998). Moreover, some studies suggest that variability measures are a more sensitive marker of PD than bradykinesia and that measures of the fluctuation magnitude are significantly related to measures of disease severity, while certain average values are not. (See htt://www.physionet.org and Hausdorff et al, 2000 for more details).

The reasons for the inability of PD patients to accurately regulate gait timing from one stride to the next remain to be fully determined. The increased variability may be the result of impairment in anticipatory reflexes, a disruption in the normal internal cueing required to string together sub-movements, and/or a diminished capacity to perform automatic sequential movements. Conversely, the irregular rhythm may be related to a problem with muscle force generation rather than a deficit in timing per se. Alternatively, the impaired reflexes that are responsible for postural instability or changes in the basal ganglia's ability to integrate sensory stimuli may be involved.

Whatever the exact origins of the increased stride-to-stride variability, it is interesting to note that a number of factors suggest that the mechanisms responsible for the reduction in gait speed and the overall bradykinesia in PD are different from the basal ganglia pathology that gives rise to the increased variability. Rhythmic training apparently improves gait speed, but not variability. Similarly, L-dopa administration appears to improve gait speed without affecting the variability of the stride duration (Blin et al, 1991). This is another example that further supports the idea that different neural networks involved in PD apparently regulate different aspects of gait disturbances.

6. PAROXYSMAL GAIT DISTURBANCES

Paroxysmal gait disturbances (PGD's) are a relatively rare phenomenon in the early stages of PD, observed in about 7% of naive patients (Giladi at al., 1992, 2001). At this stage, PGD's are relatively brief lasting a second or two, typically in the form of turn or start hesitation and they respond to simple tricks. As the disease progresses, PGD's become more frequent and they are experienced by 26 % of the patients at the stage when significant symptomatic treatment is required (Giladi et al. 2001a). In the more advanced stages of the disease, PGD's become a frequent and very disabling symptom, affecting over 50% of the patients (Giladi et al. 2001b). At this stage, PGD's are characterized by frequent appearance, prolonged duration, poor and transient response to simple tricks. The combination of prolonged and complex PGD's with abnormal postural responses frequently leads to falls. Furthermore, as the disease progresses, patients may become more sensitive to mental overload while walking and the result is sudden hesitations in doorways, stressful situations, and while responding to the environment.

Initially, PGD's respond well to dopaminergic treatment but as the disease progresses, PGD's play a major role during the "off" state as well. It has been suggested that prolonged dopaminergic treatment can worsen PGD's (Giladi et al. 2001b). The pathophysiology of PGD's is not clear but similar transient disturbances in the performance of continuous motor acts have been reported in writing and speech. One can speculate that paroxysmal hesitation in introducing or maintaining motor acts originates from transient basal ganglia derangement in the control of internally driven complex acts. This speculation is based on the known benefit of the employment of externally driven motor acts (tricks) or external pacing like a metronome to overcome the paroxysmal blocks.

In summary, gait disturbances in parkinsonism are highly variable and can be viewed as a result of lesions or abnormalities at all levels of the CNS including the cortex, basal ganglia, thalamus, brain stem, cerebellum and their interconnecting pathways. Future work is needed to fully explain the variety variety of responses to dopaminergic therapy and the multiple factors that produce continuous as well as paroxysmal gait disturbances in PD.

REFERENCES

Aziz TZ., Davies L., Stein J., France S., 1998, The role of descending basal ganglia connections to the brainsetm in parkinsonian akinesia, *Br. J. Neurosurg.* 12,245-249.

Blin, O., Ferrandez A. M, and Serratrice G., 1990, Quantitative analysis of gait in Parkinson patients: increased variability of stride length, *J.Neurol.Sci.* 98,91-97.

Blin, O., Ferrandez A. M, Pailhous J., and Serratrice G., 1991, Dopa-sensitive and dopa-resistant gait parameters in Parkinson's disease, *J.Neurol.Sci.* 103,51-54.

Elble RJ, Cousins R, Leffler K, Hughes L., 1996, Gait initiation by patients with lower-half parkinsonism, *Brain* 119,1705-1716.

Feve AP., Fenelon G., Wallays, C., et al. 1993, Axial motor disturbances after hypoxic lesions of the globus pallidus. *Mov Disord.*, 8,321-326.

Fukuyama H., Ouchi Y., Matsuzaki S., et al., 1997, Brain functional activity during gait in normal subjects: A SPECT study. *Neurosci. lett.* 228,183-186.

Garcia-Rill E., 1986, The basal ganglia and the locomotor regions, *Brain Research Reviews*, 11,47-63.

Giladi N., D. McMahon, S. Przedborski, E. Flaster, S. Guillroy, V. Kostic, S. Fahn, 1992, Motor blocks ("freezing") in Parkinson's disease, *Neurology*, 42, 333-339.

Giladi N., Treves T.A., Simon E.S., et al., 2001, Freezing of gait in patients with advanced Parkinson's disease, *J. Neural Transm.,* 108, 53-61.

Giladi N., McDermott M., Fahn S., Przedborski S., Jankovic J., Stern M., Tanner C., The Parkinson Study Group, 2001, Freezing of gait in Parkinson's disease: Prospective assessment of the DATATOP cohort. *Neurology,* in press.

Grasso R., Peppe A., Stratta F., et al., 1999, Basal ganglia and gait control: apomorphine administration and internal pallidum stimulation in Parkinson's disease, *Exp Brain Res.* 126,139-148.

Hanakawa T, Katsumi Y, Fukuyama H, et al., 1999, Enhanced lateral premotor activity during paradoxical gait in Parkinson's disease, *Brain* 122(7),1271-1282.

Haslinger B., Erhard P., Kämpfe N. et al., 2001, Event-related functional magnetic resonance imaging in Parkinson's disease before and after levodopa, *Brain* 124(3),558-570.

Hausdorff JM, Lertratanakul A, Cudkowicz ME, Peterson AL, Kaliton D, Goldberger AL., 2000, Dynamic markers of altered gait rhythm in amyotrophic lateral sclerosis. *J Applied Physiology.* 88,2045-2053.

Hausdorff, J. M., M. E. Cudkowicz, R. Firtion, J. Y. Wei, and A. L. Goldberger, 1998, Gait variability and basal ganglia disorders: stride-to-stride variations of gait cycle timing in Parkinson's disease and Huntington's disease, *Mov Disord.* 13,428-437.

Hely MA, Morris JG, Traficante R, Reid WG, O'sullivan DJ, Williamson PM, 2000, The Sydney multicentre study of Parkinson's disease: progression and mortality at 10 years, *J Neurol Neurosurg Psychiatry.* 68(2), 254-255.

Masdeu JC., Alampur U., Cavaliere R., Tavoulareas G., 1994, Astasia and gait failure with damage of the pontomesencephalic locomotor region, *Ann. Neurol.* 35,619-621.

Nutt J.G. Peculiar gait and balance synergies. AAN 51st Annual Meeting, Toronto 1999, 2DS.003.

Ouchi Y., Kanno T., Okada H., et al., 2001, Changes in dopamine availability in the nigrostrial and mesocortical dopaminergic systems by gait in Parkinson's disease, *Brain*, **124**,784-792.

Pahapill PA, Lozano AM, 2000, The pedunculopontine nucleus and Parkinson's disease, *Brain*, **123**:1767-1783.

Rosin R., Topka H., 1997, Gait initiation in Parkinson's disease, *Mov Disord*, **12**(5),682-690.

Trillet M., Croisile B., Tourniaire D., Schott B., 1990, Disorders of voluntary motor activity and lesions of caudate nuclei, *Rev Neurol.* **146**(5),338-344.

Ueno E., 1989, Clinical and physiological study of apraxia of gait and frozen gait. *Rinsho Shinkeigaku* **29**(3),275-283.

Villablance J.R, Marcus RJ., 1975, Effects of caudate nuclei removal in cats. Comparison with frontal cortex ablation. UCLA Forum Med Sci. **18**,273-311.

Winn P., 1998, Frontal syndrome as a consequence of lesions in the pedunculopontine tegmental nucleus: A short theoretical review, *Brain Res Bull.* **476**,551-563 .

Yanagisawa N, Ueno E., Takami M., 1991. Frozen gait in Parkinson's disease and vascular parkinsonism – A study with floor research forces and EMG. In: *Neurobiological Basis of Human Locomotion*. M. Shimamura, S. Grillner, V.S. Edgerton Ed., Japan Scientific Societies Press, Tokyo. pp. 291-304.

Zijlstra W., Rutgers A.W.F., Van Weerden T.W., 1998, Voluntary and involuntary adaptation of gait in Parkinson's disease, *Gait and Posture*, 7,53-63.

LEVODOPA DOES NOT ENHANCE NIGRAL DEGENERATION

Ali H. Rajput[*]

1. INTRODUCTION

Levodopa is the most widely used symptomatic drug for Parkinson's disease in the industrialized countries. The metabolism of levodopa and of dopamine each release oxyradicals and quinones which are believed to be toxic to the substantia nigra neurons.[1] The reduced levels of striatal dopamine in parkinsonism leads to increased dopamine turnover,[2] resulting in larger quantities of the putative toxic metabolic products of levodopa. The substantia nigra neuronal loss in Parkinson's disease is estimated to occur at nearly twice the rate of the age matched controls.[3-5] Adding the exogenous levodopa to treat Parkinson's disease would further increases the levels of the levodopa metabolites (O. Hornykiewicz personal communication) and hence may enhance the nigral neuronal loss. To date, levodopa has improved more than 32 million person years quality of human life in the world.

If the exogenous levodopa were indeed toxic to the substantia nigra neurons, it may cancel the symptomatic benefit by enhancing the progression of the pathological process. In that event, LD therapy may shorten the life expectancy because the death in Parkinson's disease is related the degree of disability[6] which in turn indicates the severity of dopaminergic neuronal loss.[7]

In normal aging, there is a progressive decline in the substantia nigra neurons.[4, 5, 8] If levodopa were toxic to these neurons, the normal elderly individuals when exposed to levodopa may develop parkinsonian features. Fortunately, the use of levodopa in those with normal complement of substantia nigra neurons is extremely rare though patients with other movement disorder[9, 10] may be mistaken for Parkinson's disease and treated with levodopa. Individuals with restless leg syndrome who receive long-term levodopa therapy may also

---

[*]Ali H. Rajput, University of Saskatchewan, Saskatoon, Saskatchewan, S7N 0W8, ph: (306) 966-8009, fax: (306) 966-8030, email: rajputa@adh.sk.ca

Mapping the Progress of Alzheimer's and Parkinson's Disease
Edited by Mizuno *et al.*, Kluwer Academic/Plenum Publishers, 2002

337

have an increased risk of developing Parkinson's disease if the drug were toxic to the substantia nigra.

Considering all the above, it is important to identify if levodopa as used in Parkinson's disease is toxic to the human substantia nigra or not. The best method to address this issue would be a controlled study with one group of parkinsonian patients receiving levodopa while the drug is withheld for life from a comparable group of patients. Such study is ethically impossible as no one who needs the drug for symptomatic control can be denied that. Because the ideal study is not possible, we must rely on the best alternative evidence available in the human subjects.

We report our observations based on epidemiology and individual case studies.

2. MATERIALS AND METHODS

During a 22 year interval (1968-1990), we accumulated 3563 person years of follow-up data on 934 parkinsonian patients. These patients included a group that had onset of parkinsonism prior to the discovery of levodopa and came to our clinic in search of that treatment. The second group was made up of patients with parkinsonism onset when a Provincial Drug Plan allowed unrestricted access to all prescription drugs including levodopa at no cost (except the pharmacist fees). The Kaplan-Meier graphs[11] were made comparing survival in parkinsonian patients and the regional population of the same sex and the year of birth. Survival in 934 parkinsonian patients and the different subgroups based on the access to levodopa were compared with the respectively matched regional general population.

The second part of our study consisted of observations in individual patients. These included three elderly Essential tremor patients with one autopsy, two dopa-responsive dystonia patients with one autopsy, and one dopa-responsive nigral parkinsonism patient, all of whom received long-term levodopa.

3. RESULTS

The life expectancy in the entire 934 Parkinson patient population seen during 22 years was significantly shorter (P<0.001) compared to that expected in the regional general population of the same sex and the same year of birth. 215 patients, most of whom had onset prior to the discovery of levodopa had an extremely restricted levodopa access. The survival in this subgroup was considerably reduced compared to that expected (P<0.001) in the general population. The second subgroup included 565 patients who had onset of parkinsonism during an era when levodopa was readily accessible under the Provincial Drug Plan. When the survival in these cases was compared with the sex and year of birth matched regional general population, it was significantly reduced (P<0.029). When the earlier 215 cases were compared with the more recent 565 cases, the 10 year survival was significantly longer (P=0.0247) in the recent cases.

Most of the 215 patients that had highly restricted levodopa access came to the Movement Disorder clinic to obtain levodopa prescription. We evaluated the survival in these patients based on the degree of the disease severity at the time of levodopa initiation.[12] Those patients who received first levodopa prescription while at Stage 1 or Stage 2 Hoehn & Yahr[13] had no significant decline in survival. On the other hand, individuals who were at Stage 3, 4, or 5 of Hoehn & Yahr[13] at levodopa initiation had a significantly reduced survival compared to the expected (P<0.0001).

Three Essential tremor patients who received levodopa due to erroneous diagnosis of parkinsonism included a 92 year old female who received 24 kg (plain) levodopa over 26 years. Autopsy revealed normal number of substantia nigra neurons and there was no evidence of neuronal pathology.[14] The second Essential tremor patient received 22 kg levodopa over an interval of 22 years. The drug was discontinued 8 years ago without evidence of parkinsonian symptoms, although she continues to have mild resting tremor in the upper limbs. The third Essential tremor patient is now 102 years old. She received 8 kg of levodopa over a period of 13 years. She has been off levodopa for more than two years with no clinical features of parkinsonism.

We have followed two dopa-responsive dystonia patients. The first case is a woman aged 19 years who received 3 kg of levodopa over an interval of 11 years. After accidental death, the autopsy revealed that the substantia nigra neurons were hypo- pigmented compared to the controls of the same age. However, there was neither depletion in the number nor evidence of ongoing cell damage.[15] The second patient who has received levodopa in excess of 22 kg during a 30 year interval is functioning normally and has no evidence of parkinsonism.

One dopa-responsive nigral parkinsonian patient has received more than 22 kg levodopa over 30 years. His off levodopa disability measurement revealed a 0.5 UPDRS modified[16] Hoehn & Yahr scale[13] increase compared to the untreated status more than 30 years earlier.

4. DISCUSSION

Oxidative stress due to free radicals has been suspected in the pathogenesis of Parkinson's disease.[1] A recent study reported chronic exposure to rotenone produced parkinsonian features in rats[17] and the nigral neurons contained cytoplasmic fibrillar inclusions positive for _-synuclein and ubiquitin. That was attributed to the rotenone selectively reducing mitrochondrial complex I. These observations suggest that certain level of oxidative stress in animals may produce substantia nigra damage. However, it remains to be seen if a similar substance occurring in nature produces human Parkinson's disease.

The issue in relation to the levodopa effect on the substantia nigra neurons must consider two main factors: the dose of levodopa administered to Parkinson's disease patients and the concentrations of the metabolites produced.

In the human subjects, the levodopa treatment has a certain dose range to control the symptoms without producing major adverse effects. This concentration may or may not produce sufficient metabolites as may cause damage to the substantia nigra.

If levodopa were to enhance the nigral damage it would cause a more rapid clinical decline in Parkinson's disease. Such decline may, however, be masked by the symptomatic benefit of levodopa. The effect of levodopa on survival in Parkinson's disease may, therefore, be positive, negative, or neutral. However if levodopa were toxic to the substantia nigra, those patients who receive the drug at early stage of disease and for longer period of time (larger cumulative dose) would suffer more severe neuronal loss and hence are more likely to have a shortened survival. In our study, those initiated on the drug while at Stage 1 and 2 Hoehn & Yahr[13] had greater survival benefit than those treated at a later stage of parkinsonism. That indicates against the hypothesis that levodopa enhances substantia nigra damage.

Every patient with Parkinson's disease after reaching Stage 2 Hoehn & Yahr has his/her own rate of progression when untreated.[18] We have followed one nigral parkinsonism patient[19] and have repeatedly assessed his off levodopa disability to compare with the untreated status. Before treatment at age 31 years, he had Stage 3 Hoehn & Yahr[13] disability. Without treatment, our patient would have reached Stage 5 Hoehn & Yahr disability by age 50 or earlier. He has been well controlled on levodopa and off levodopa the disability at age 61 was Stage 3.5,[16] ie. 0.5 stage greater than 30 years earlier. Thus, levodopa did not accelerate his underlying disease process but, instead, seems to have retarded the disease progression.

The dopa-responsive dystonia patients one with autopsy study and the other with clinical follow-up showed no evidence of parkinsonism on prolonged use of levodopa.[15]

The three elderly Essential tremor individuals each of whom received levodopa for long periods of time did not develop clinical features of parkinsonism. One patient continued taking the drug until death. At age 92, she had normal substantia nigra.[20] Thus, levodopa did not enhance the age related decline in the number of substantia nigra neurons in these elderly individuals.

In summary, on the levodopa dose utilized in the treatment of parkinsonism, there is no evidence that the drug enhances the substantia nigra neuronal damage. While the dopaminergic adverse effects, notably dyskinesia, are less common with dopamine agonist,[21, 22] levodopa remains the best drug. Therefore, levodopa can be used without the fear of accelerating the disease process in Parkinson's disease.

REFERENCES

1.   S. Fahn, G. Cohen, The oxidant stress hypothesis in Parkinson's disease: Evidence supporting it. *Ann.Neurol.* 32:804-812 (1992).
2.   H. Bernheimer, O. Hornykiewicz, Herabgesetzte Konzentration der Homovanillinsaure im Gehirn von Parkinsonkranken Menschen als Ausdruck der Storung des zentralen Dopaminstoffwechsels. *Klin.Wochenschr.* 43:711-715 (1965).
3.   F. J. G. Vingerhoets, B. J. Snow, C.S. Lee, M. Schulzer, E. Mak, D. B. Calne, Longitudinal fluorodopa positron emission tomographic studies of the evolution of idiopathic parkinsonism. *Ann.Neurol.* 36:759-764 (1994).
4.   J. M. Fearnley, A. J. Lees, Ageing and Parkinson's disease: Substantia nigra regional selectivity. *Brain* 114:2283-2301 (1991).
5.   W. R. G. Gibb, Neuropathology of the substantia nigra. *Eur.Neurol.* 31(suppl 1):48-59 (1991).

6. R. K. Mosewich, A. H. Rajput, A. Shuaib, B. Rozdilsky, L. Ang, Pulmonary Embolism: An under-recognized yet frequent cause of death in Parkinsonism. *Mov.Disord.* 9(3):350-352 (1994).
7. E. C. Alvord, Jr, L.S. Forno, J. A. Kusske, R. J. Kauffman, J. S. Rhodes, C. R. Goetowski, The Pathology of Parkinsonism: A Comparison of Degenerations in Cerebral Cortex and Brainstem, in: *Advances in Neurology Volume 5*, edited by F. H. McDowell, A. Barbeau (Raven Press, New York, 1974), pp 175-193.
8. P. L. McGeer, E. G. McGeer, J. S. Suzuki, Aging and Extrapyramidal Function. *Arch.Neurol.* 34:33-35 (1977).
9. N. Quinn, D. Parkes, I. Janota, C. Marsden, Preservation of the substantia nigra and locus coeruleus in a patient receiving levodopa (2Kg) plus decarboxylase inhibitor over a four year period. *Mov.Disord.* 1(1):65-68 (1986).
10. D. Riley, Is levodopa toxic to human substantia nigra. *Mov.Disord.* 13(2):369-370 (1999).
11. E. L. Kaplan, P. Meier, Non-parametric estimation from incomplete observations. *J.Am.Stat.Assoc.* 53:457-481 (1958).
12. A. H. Rajput, R. J. Uitti, A. H. Rajput, K. P. Offord, Timely levodopa (LD) administration prolongs survival in Parkinson's disease. *Parkinsonism & Related Disorders* 3(3):159-165 (1997).
13. M. M. Hoehn, M. D. Yahr, Parkinsonism: Onset, Progression, and Mortality. *Neurology* 17:427-442 (1967).
14. A. H. Rajput, M. E. Fenton, S. Birdi, R. Macaulay, Is levodopa toxic to human substantia nigra? *Mov.Disord.* 12(5):634-638 (1997).
15. A. H. Rajput, W. R. G. Gibb, X. H. Zhong, K. S. Shannak, S. Kish, L. G. Chang, O. Hornykiewicz. Dopa-responsive dystonia: Pathological and biochemical observations in a case. *Ann.Neurol.* 35:396-402 (1994).
16. S. Fahn, R. L. Elton, UPDRS Development Committee. Unified Parkinson's disease rating scale. in: *Recent Developments in Parkinson's disease. Second Edition*, edited by S. Fahn, C. D. Marsden, D. Calne, M. Goldstein. (Macmillan Healthcare Information, Florham Park, N.J., 1987), pp 153-305.
17. R. Betarbet, T. B. Sherer, G. MacKenzie, M. Garcia-Osuna, A. V. Panov, J. T. Greenamyre, Chronic systemic pesticide exposure reproduces features of Parkinson's disease. *Nat.Neurosci.* 3(12):1301-1306 (2000).
18. R. J. Marttila, U. K. Rinne, Disability and progression of Parkinson's disease. *Acta Neurol Scand.* 56:159-169 (1977).
19. A. Rajput, A. Kishore, B. Snow, C. F. Bolton, A. H. Rajput, Dopa-responsive-non-progressive, Juvenile Parkinsonism: Report of a case. *Mov.Disord.* 12(3):453-455 (1997).
20. A. H. Rajput, The Protective Role of Levodopa in the Human Substantia Nigra, in:, *Parkinson's Disease: Advances in Neurology*, edited by D. Calne, S. Calne (Lippincott Williams & Wilkins, Philadelphia, 2001), pp 327-336.
21. O. Rascol, D. J. Brooks, A. D. Korczyn, P. P. De Deyn, C. E. Clarke, A. E. Lang, A five-year study of the incidence of dyskinesia in patients with early Parkinson's disease who were treated with ropinirole or levodopa. *N.Engl.J.Med.* 342:1484-1491 (2000).
22. Parkinson Study Group. Pramipexole vs Levodopa as Initial Treatment for Parkinson Disease. A Randomized Controlled Trial. *JAMA* 284:1931-1938. (2000).

DEMENTIA IN IDIOPATHIC PARKINSON'S SYNDROME

Heinz Reichmann and Ulrike Sommer

At present we estimate that about 250.000 patients have idiopathic Parkinson's syndrome (IPS) in Germany. Beside typical motor symptoms such as akinesia, rigidity and tremor these patients may present psychiatric and cognitive symptoms such as anxiety, depression and dementia. Dementia is characterised by a syndrome of progressive decline in multiple areas of cognitive function, eventually leading to an inability to maintain occupational and social performance. Typical causes of dementia are neurodegenerative disease, vascular disease, traumatic brain injury, viral disease and substance abuse.

Since most patients develop bradyphrenia during the disease progress it is sometimes difficult to detect early dementia, especially since there are no typical rating scales developed for dementia in IPS. Normal cognitive changes with aging include vision changes, hearing loss, benign memory impairment, slowed reaction time, and speed of processing declines. DSM-IV criteria for dementia include memory impairment and one or more of the following: aphasia, apraxia, agnosia, disturbance in executive functioning; cognitive deficits cause significant impairment in social and occupational function and represent a decline; gradual onset, continued decline; deficits not due to another condition; deficits do not occur only during delirium.

It is not yet solved what dementia in IPS really means and whether this is really a distinctive condition, since only some years ago senile dementia of Lewy body type was defined (McKeith et al. 1996) and there are patients which suffer from concomitant Alzheimer's and Parkinson's disease. It is thus rather controversial to state on first look that patients with IPS have their "own" kind of dementia. In this paper we want to present evidence that IPS patients may indeed suffer from a specific kind of dementia.

As mentioned above we have good criteria for senile dementia of Lewy body type (Table 1). With respect to this a recent paper by Perry and coworkers (1991) is of interest. The authors autopsied 89 patients from a psychogeriatric unit in Newcastle Upon Tyne

Klinik und Poliklinik für Neurologie, Universitätsklinikum Carl Gustav Carus, 01307, Dresden, Germany

Mapping the Progress of Alzheimer's and Parkinson's Disease
Edited by Mizuno *et al.*, Kluwer Academic/Plenum Publishers, 2002

and found out that based on neuropathologic examination 52% fulfilled the criteria of Alzheimer's dementia, 7% had typical vascular dementia, 3% Parkinson's disease and dementia and 20% fulfilled all criteria for dementia of Lewy body type. These and other authors state that nowadays dementia of Lewy body type is the second most common dementia after Alzheimer's ; this was not known to us even some years ago. We should therefore accept that dementia of Lewy body type is still underdiagnosed and may well be confused with so-called Parkinson-dementia in some of the papers that we will mention in this publication. We should suspect this condition in patients who present with visual hallucinations, episodic somnolence, cognitive fluctuations, great sensitivity to neuroleptics and dementia who develop Parkinonian features only later in their course of disease. Based upon the data from Perry et al. (1991) it is much less frequent to develop dementia after some years of Parkinsonism than the other way around.

James Parkinson himself had not detected any cognitive dysfunction in the six patients he described in his "Essay on the Shaking Palsy" which appeared in 1817. He stated in this booklet "the senses and intellect are uninjured". When we check more recent literature we find estimates for dementia between 8% (Taylor and Saint-Vyr, 1985) and 80% (Martin et al., 1973) while most authors consider a percentage between 20 and 40% to be more realistic. All of these studies have one drawback and that is that they are based on observations made in specialised Movement Disorder clinics which results in a selection bias. In addition, until lately Parkinsonian syndromes were not well defined, the samples used in the studies were too small and there were no real control groups. For this reason our personal view is that we may better rely on a study by Aarsland and colleagues in Rogaland County in Norway (Aarsland et al., 1996). These authors found out of 220.858 inhabitants in Rogaland County 400 who were suspect of IPS. In the end they were convinced that 245 patients really suffered from IPS and these patients were studied using the Mini-Mental State Examination, Gottfries, Brane and Steen scale and the intellectual subscale of the Unified Parkinson's Disease Rating Scale which tests for cognitive impairment. Using these scales they found 28% of all patients with IPS to be demented. These data also allow the conclusion that patients with IPS show a 2- 3fold higher risk to develop dementia than age-matched controls.

There are also data on whether risk factors for developing dementia after developed IPS exist (Aarsland et al., 1996). Typical risk factors seem to be old age at the start of the disease, familiar dementia, severe extrapyramidal symptoms, poor education and low intelligence, early development of hallucination and psychosis or initial bilateral Parkinsonian motor symptoms. In our view the latter two conditions may well indicate dementia of Lewy type.

As mentioned above we believe that in many patients dementia of Alzheimer's type with Parkinsonism can be differentiated from IPS with dementia. This can be done if we use the criteria of cortical and subcortical dementia established by Wilson et al. in 1912 and further developed by Albert and colleagues in 1974.

This is a neuropathologic or anatomic concept whereupon neuropathologic alterations are not found in the cortex as in Alzheimer' disease but subcortically. Typical diseases with dementia of subcortical type are Wilson' disease, Huntington's disease or progressive supranuclear palsy and especially IPS with dementia. While patients with Alzheimer's disease present with aphasia, agnosia and amnesia (cortical impairment) patients with subcortical dementia are characterised by forgetfulness, slow thinking, alterations in personality and mood and loss of ability to deal with already acquired knowledge. Patients with IPS and dementia show no neurofibrillary tangles while they have some plaques and typical Lewy bodies in the substantia nigra (SN). In contrast to normal IPS without dementia patients with IPS and dementia reveal not only Lewy bodies in the SN but also in the cortex such as hippocampus. The 60-80% percent cell loss of cholinergic neurones in the Nucleus basalis of Meynert seems to be more relevant for the development of dementia (Nakano and Hirano, 1984). This finding is important since it indicates that not only patients with Alzheimer's disease but also with IPS and consecutively developed dementia present with a cholinergic deficit.

The diagnosis of IPS with dementia is a clinical diagnosis, there are no specific tests available to confirm the diagnosis because diagnostic procedures are rather unspecific. Patients with IPS and dementia normally show a deceleration of basic rhythm in EEG and their late acoustic evoked potentials are delayed (P300). Cranial computer tomography reveals inner and outer brain atrophy and the width of the 3rd ventricle seems to correspond quite well with the degree of dementia (Lichter et al., 1988). Nonetheless, all these procedures are not at all specific for dementia and IPS. 18F Fluorodeoxyglucose-(FDG)-PET shows in patients with IPS and dementia a reduced metabolism in fronto-temporo-parietal areas which resembles the situation in Alzheimer's dementia (Turjanski and Brooks, 1997). In contrast patients with dementia with Lewy bodies show occipital hypoperfusion on SPECT (Lobotesis et al., 2001) which is not the case in AD. Patients with IPS and dementia show on SPECT a decrease in blood flow especially of the temporoparietal cortex (Jagust et al., 1992).

It is generally accepted that IPS is mostly due to a cell loss of dopaminergic neurones in the substantia nigra leading to a dopaminergic deficit in the striatum and causing motor dysfunction. Beside of this there are certainly more transmitters involved in the disease process such as a glutamatergic overstimulation at the striatum and also a cholinergic overstimulation. For this reason, many patients were formerly successfully treated with anticholinergics when no dopaminergics were available. Since they are extremely cheap anticholinergics are still world wide the most used antiparkinsonian drugs with a special benefit for patients with a tremor-dominant form. We and others use anticholinergics only for patients with tremor when we cannot improve the tremor by use of dopaminergic or antiglutamatergic treatment. Clozapine would be an alternative but impractical due to the obligation to perform regularly blood tests and the danger of agranulocytosis. A second condition in which anticholinergics are sometimes beneficial are biphasic dyskinesias. Our reluctance to use anticholinergics stems from the possibility that we may enhance the chance of development of dementia.

Unfortunately, not many studies adress the question whether anticholinergics may cause dementia or not. To our knowledge there are only two studies from Israel and France (Dubois et al., 1990). Dubois and colleagues studied 20 Parkinsonian patients with and 20 without anticholinergic treatment who were in the Hoehn and Yahr stadium III with a duration of the disease of 11 years. Patients were carefully selected for comparative motor complications, duration of disease and similar levo dopa dosage over the years. The average dosage of anticholinergics was 10 mg/day. The authors conducted detailed neuropsychological assessment to test for intellectual function, memory, concentration, visual reproduction, orientation, instrumental activities and frontal lobe-function. Patients who had received anticholinergics were significantly worse in frontal lobe performance which underlines that ascending cholinergic neurones may be relevant for these functions. As mentioned above anticholinergics have been longest of all antiparkinsonian drugs. Thus, it may seem problematic to treat the cholinergic deficit of the nucleus basalis Meynert with cholinergics. Nonetheless, since a new generation of cholinergic agents are available it seems appropriate to test them in IPS.

One of the first studies performed used the first new cholinesterase-inhibitor tacrine in a dosage of 4 x 10 to 3 x 20 mg daily in 7 patients aged between 66 and 82 years and with a duration of dementia between 3 months and 2 years (Hutchinson and Fazzini, 1996). All patients had developed Parkinsonian symptoms before the onset of dementia. By applying the Mini Mental State Score (Folstein) the authors showed a 3 to 13 point-increase. Improvement was mainly seen in orientation, attention, and visuospatial awareness. Surprisingly part II and III (cognitive function and activities of daily living and motor function) of the Unified Parkinson's Disease Rating Scale also improved (mostly gait) which could not only be based on cognitive improvement. We would however raise a point of caution since the improvement in UPDRS was so extremely high and only 7 patients were analysed. Similar results, however, were presented at the Congress of the European Neurological Society in Milan in 1999 when Foy and Sagar presented a poster reporting on 13 dyskinetic patients treated for 2 weeks with 5 mg donepezil which resulted in a significant reduction in dyskinesia and better reaction time. Motor abilities and time of off-periods remained unchanged (Foy and Sagar, 1999). In Germany, Henneberg established in a Parkinson Clinic the optimal tolerable dosage of donepezile to be 7,5 mg/d in 29 female and 55 male patients. Again, activities of daily living were improved after 4 weeks by 20-30%. Ten patients had developed side effects like delusions, uneasiness, diarrhea and paranoia.

We took advantage from a post-marketing surveillance study where donepezil was tested in Germany in 2092 patients with AD (Berger et al. 2000). Patients enrolled in thisstudy received 5 mg donepezil daily for more than 4 weeks and then based on clinical judgement 10 mg/day for approximately 3 months. Adverse events were recorded, and supportive data were obtained from global judgement of tolerability. Efficacy parameters were Mini Mental State Examination , and Nurses' Observation Scale for Geriatric patients (NOSGER). Sixty percent were women, MMSE was 17.8 on average and the patients had an average age of 73 years. Seventy-three of these patients additionally had a Parkinsonian syndrome. This subgroup too had a MMSE of 17.8 on average. The whole population developed an improvement of 1.4 points in MMSE while patients with IPS and AD showed a 1.7 point improvement. Similarly positive results were obtained for the NOSGER scale. More than half of the patients presented with a good efficacy and good tolerability for the treatment. Adverse events were reported in 11 patients being dizziness, gait disturbance, headache, stroke, weakness, swooning, urine incontinence and heart attack with resulting death in which a link with donepezil was doubtful. We can conclude from this study that in the general clinical practice setting donepezil was well tolerated in this population and improvement in cognition and function could be documented. From all these studies we received the impression that donepezil is safe in IPS and dementia.

Meanwhile McKeith and colleagues have shown improvement of cognitive function in patients with dementia of Lewy body type with rivastigmine (2000). In this open label study 11 patients with a mean age of 78.5 years and probable dementia of Lewy body type were treated with a mean of 9.6 mg of rivastigmine daily. After 12 weeks mean Neuropsychiatric Inventory scores fell by 73% for delusions, 63% for apathy, 45% for agitation and 27% for hallucinations. Parkinsonian symptoms tended to improve as well.

All these findings prompted us to initiate an open trial with 10 mg donepezil daily in patients with IPS who had thereafter developed dementia. We expect to find not only improvement in dementia but also good tolerability and maybe even improvement in Parkinsonian motor symptoms.

Table 1. Characteristics of Dementia of Lewy body type

Second most frequent cause of primary degenerative dementia following Alzheimer's disease.

1. Lewy bodies in the brain stem and cerebral cortex.
2. Progressive dementia, cognitive fluctuations, well-formed recurrent visual hallucinations, falls, great sensitivity to neuroleptic drugs, visuoconstructive and visuoperceptual disturbances, transitory alterations in consciousness, misnamings, spontaneous Parkinsonism.
3. Compared to Parkinson's disease there is less asymmetry of motor symptoms, less tremor, less favourable response to L-Dopa and more falls
4. Compared to Alzheimer's disease there are marked cognitive fluctuations and phases of reduced alertness, hallucinations and delirium.

REFERENCES

Aarsland, D., Tandberg, E. and Cummings, J.P., 1996, Frequency of dementia in Parkinson disease, *Arch.Neurol.* **53**:538-542.

Albert, M.L., Feldman, R.G. and Willis, A.L., 1974, The subcortical dementia of progressive supranuclear palsy, *J.Neuro.Neurosurg.Psych.* **37**:121-130.

Berger, F., Sramko, C., Baas, H., Fuchs, G. and Reichmann, H., 2000, Donepezil in the treatment of patients with concomitant Alzheimer's disease and Parkinson's disease: results from a post-marketing surveillance study, *Mov.Disord.* **Suppl.3**: 129.

Dubois, B., Pillon, B., Lhermitte, F. and Agid, Y., 1990, Cholinergic deficiency and frontal dysfunction in Parkinson's disease, *Ann.Neurol.***28**:117-121.

Foy C.M.L. and Sagar, H.J., 1999, Donepezil for the treatment of dyskinesias in patients with idiopathic Parkinson's disease (IPD), *J.Neurol.* **Suppl I:** 47.

Henneberg, A., 1999, Cognitive dysfunction in Parkinson's disease –(Parkinson-plus-dementia)- successful treatment by donepezile (Aricept). Presented at: Mental dysfunction in Parkinson's disease, Amsterdam 1999.

Hutchinson M. and Fazzini, E., 1996, Cholinesterase inhibition in Parkinson's disease. *J.Neurol.Neurosurg.Psych* **61**:324-325.

Jagust, W.J., Reed, B.R., Martin, E.M., Eberling, J.L. and Nelson-Abbott, R.A., 1992, Cognitive function and regional cerebral blood flow in Parkinson's disease, *Brain* **115**:521-537.

Lichter, D.G., Corbett, A.J., Fitzgibbon G.M., 1988, Motor, cognitive and CT correlates in Parkinson's disease, *Arch Neurol* **45**:854-960.

Lobotesis, K., Fenwick, J.D., Phipps, A., Ryman, A., Swann, A., Ballard, C., McKeith, I.G. and O'Brien, J.T., 2001, Occipital hypoperfusion on SPECT in dementia with Lewy bodies but not AD, *Neurology* **56**:643-649.

Martin, W.E., Loewenson, R.B., Resch, A. and Baker, A.B., 1973, Parkinson's disease: clinical analysis of 100 patients, *Neurology* **23**:783-790.

McKeith, I.G., Galasko, D., Kosaka, K., Perry, E.K., Dickson, D.W., Hansen, L.A., Salmon, D.P., Lowe, J., Mirra, S.S., Byrne, E.J., Lennox, G., Quinn, N.P., Edwardson, J.A., Ince, P.G., Bergeron, C., Burns, A., Miller, B.L., Lovestone, S., Collerton, D., Jansen, E.N.H., Ballard, C., de Vos, R.A.I., Wilcock, G.K., Jellinger, K.A. and Perry, R.H., 1996, Consensus guidelines for the clinical and pathologic diagnosis of dementia with Lewy bodies (DLB): report of the consortium on DLB international workshop, *Neurology* **47**:1113-1124.

McKeith, I.G., Grace, J.B., Walker, Z., Byrne, E.J., Wilkinson, D., Stevens, T. and Perry, E.P., 2000, Rivastigmine in the treatment of dementia with Lewy bodies: preliminary findings from an open trial, *Int. J. Geriat. Psychiatry* **15**:387-392.

Nakano I. and Hirano, A., 1984, Parkinson's disease: neuron loss in the nucleus basalis without concomitant Alzheimer's disease. *Ann. Neurol.* **15**:415-418.

Perry, E.K., McKeith, P., Thompson, P., Marshall, E., Kerwin, J., Jabeen, S., Edwardson, J.A. Ince, P., Blessed, G., Irving, D. and Perry, R.H., 1991, Topography, extent, and clinical relevance of neurochemical deficits in dementia of Lewy body type, Parkinson's disease, and Alzheimer's disease. *Ann. NY Acad. Sci.* **640**:197-202.

Taylor A. and Saint-Cyr, J.A., 1985, Dementia prevalence in Parkinson's disease. *Lancet* **1**: 1037.

Turjanski N. and Brooks, D.J., 1997, PET and the investigation of dementia in the parkinsonian patient, *J.Neurol.Transm.* **Suppl.51**:37-48.

Wilson, S.A.K., 1012, Progressive lenticular degeneration: a familial nervous disease associated with cirrhosis of the liver, *Brain* **34**:295-509.

AUTONOMIC DISORDERS AND MALIGNANT SYNDROME IN PARKINSON'S DISEASE

Sadako Kuno[*]

1. AUTONOMIC DISTURBANCES

Autonomic disturbances in Parkinson's disease (PD) are relatively modest compared with the disorders seen in multiple system atrophy. However, autonomic disorders in PD can lead to serious problems such as sudden death caused by cardiovascular disturbances or malignant syndrome developed following abrupt interruption of dopaminergic medications. The major autonomic disturbances in PD include constipation (Edwards et al., 1993), imparied thermoregulation (Goets et al., 1986), orthostatic hypotension (McTavish and Goa, 1989) and urinary problems (Stocchi et al., 1997). First, the management of these autonomic disorders will be described briefly below. For the detail of the treatment guidelines the reader may wish to consult an excellent review by Olanow and Koller (1998).

1.1. Constipation

Constipation may be seen in more than 90% of parkinsonian patients at the age older than 60 to 70. Patients should be encouraged to take a large amount of fluid and high-fiber foods. Also, physical exercise is helpful. If patients are treated with anticholinergic agents, these medications should be stopped. Instead, cholinomimetic agents should be given, if this treatment does not worsen parkinsonian symptoms. If the symptoms remain unimproved, patients should be treated with stool softeners, laxatives or enemas.

1.2. Abnormalities of thermoregulation

This problem may result from neuronal loss in the hypothalamus. Parkinsonian patients often manifest excessive sweating (hyperidrosis). An increase in the frequency of dopaminergic medications reduces wearing off or motor fluctuations, thereby improving excessive sweating associated with "off" periods. Excessive sweating durig "on" periods may be treated with ß-adrenergic blockers.

[*]Department of Neurology and Clinical Research Unit, Utano National, Hospital, Kyoto 616-8255, Japan

1.3. Orthostatic hypotension

Reduced vasoconstriction caused by a decrease in the sympathetic outflow may be responsible for orthostatic hypotension in PD. Patients should be advised not to stand suddenly and not restrict salt intake. If patients are treated with antihypertensive medications or diuretics, these medications should be withdrawn. Blood pressure may change considerably from one time to another, and some parkinsonian patients are very sensitive to small changes in blood pressure. Therefore, orthostatic hypotension must be treated with caution.

1.4. Urinary problems

Many parkinsonian patients manifest nocturia, pollakiuria and urinary urgency which are termed lower urinary tract symptoms. This may cause sleep disturbances and may impair the quality of life of patients. Nocturia can be improved simply by reducing fluid intake after the evening meal. If this is not effective, anticholinergic agents may be applied to reduce detrusor contractions. But these drugs may develop voiding problems.

1.4.1. Nocturia and D1/D2 Agonists

Recently, one of our female parkinsonian patients noted that substitution of pergolide, a D_1/D_2 agonist, for bromocriptine, a D_2 agonist, improved nocturia. Later, we found that nocturia was improved in 8 out of 10 female parkinsoinan patients following treatment with pergolide. We had an opportunity of measuring the bladder volume by cystometry in a female parkinsonian patient. The volume threshold at which bladder contractions were induced in this patient was 190 ml. The threshold bladder volume increased to 350 ml 90 min after administration of pergolide (500 µg, p.o.).

1.4.2. D1/D2 Agonists on the Bladder Volume in Parkinsonian Monkeys

We extended the above study further using MPTP-induced parkinsonian monkeys (*Macaca fasciularis*). The mean bladder volume was significantly smaller in parkinsonian monkeys (40 ml) than in normal monkeys (65 ml), reflecting lower urinary tract symptoms. Bromocriptine decreased the bladder volume, whereas pergolide increased the bladder volume (Yoshimura et al., 1998). These results were consistent with our clinical findings that a D1/D2 agonist, but not a D2 agonist, is beneficial in treating lower urinary tract symptoms in PD.

2. NEUROLEPTIC MALIGNANT SYNDROME

Neuroleptic malignant syndrome (NMS) is a serious complication of neuroleptic therapy, characterized by muscle rigidity, hyperpyrexia, disturbances in consciousness and autonomic instability. Parkinsonian patients may develop NMS following sudden withdrawal of antiparkinsonian medications (Henderson and Wooten, 1981; Toru et al., 1981). However, NMS can develop in parkinsonian patients even with continued dopaminergic treatment. This suggests the presence of some risk factors for NMS other than levodopa therapy withdrawal.

2.1. Seasonal Variability of the Incidence of NMS

A 57-year-old male parkinsonian patient developed NMS in summer without interruption of antiparkinsonian medications. Based on this experience, we examined whether the incidence of NMS shows seasonal variation. Our 7 parkinsonian patients who manifested NMS without interruption of levodopa therapy between 1980 and 1994 were found to develop NMS exclusively in the warm seasons (May to August). Following interruption of levodopa therapy, NMS could develop in any season, although the incidence was still higher in summer than in other seasons. These results suggest that the environmental high temperature or dehydration in hot weather can be a possible risk factor for NMS (Kuno et al., 1997).

2.2. NMS Coincided with the Premenstrual Period

We also found that two episodes of NMS in a 45-year-old female patient coincidedd with the premenstrual period. In this case, NMS again developed without interruption of antiparkinsonian medications (Mizuta et al., 1993). This finding must be interpreted with caution because many biological processes alter in association with the menstrual cycle. The severity of parkinsonism has been found to increase during the premenstrual period. Also, NMS tends to be developed during an "off" period. Thus, a reduction in the effectiveness of antiparkinsonian drugs or the resultant deterioration in parkinsonism may be the primary risk factor for NMS. This would account for the development of NMS following levodopa therapy withdrawal. Dehydration or hot whether may promote aggravation of parkinsonism, thereby possibly leading to the development of NMS. However, the incidence of NMS induced by neuroleptics has no seasonal variations (Caroff and Mann, 1988). At this stage, it would be still premature to elaborate the notion whether there are multiple, independent risk factors for NMS or whether the development of NMS can be accounted for by a single, unifying risk factor.

REFERENCES

Caroff, S. N., and Mann, S. C., 1988, Neuroleptic malignant syndrome, *Psychopharmacol Bull* **24**:25.

Edwards, L. L., Quigley, E. M. M., Hofman, R., and Pfeiffer, R. F., 1993, Gastrointestinal symptoms in Parkinson's disease: 18-month follow-up study, *Mov Disord* **8**:83.

Goets, C. G., Lutge, W., and Tanner, C. M., 1986, Autonomic dysfunction in Parkinson's disease, *Neurology* **36**:73.

Henderson, V. W., and Wooten, G. F., 1981, Neuroleptic malignant syndrome: A pathogenetic role for dopamine receptor blockade? *Neurology* **31**:132.

Kuno, S., Mizuta, S., and Yamasaki, S., 1997, Neuroleptic malignant syndrome in parkinsonian patients: Risk factors, *Eur Neurol* **38** (Suppl 2):56.

McTavish, D., and Goa, K. L.. 1989, Midodrine, a review of its pharmacological properties and therapeutic use in orthostatic hypotension and secondary hypotensive disorders, *Drugs* **38**: 757.

Mizuta, E., Yamasaki, S., Nakatake, M., and Kuno, S., 1993, Neuroleptic malignant syndrome in a parkinsonian woman during the premenstrual period, *Neurology* **43**:1048.

Olanow, C. W., and Koller, W. C., 1998, An algorithm (decision tree) for the management of Parkinson's disease: Treatment guidelines, *Neurology* **50** (Suppl 3):S1.

Stocchi, F., Carbone, A., Inghilleri, M., Monge, A., Ruggieri, S., Beradelli, A., and Manfredi, M., 1997, Urodynamic and neurophysiological evaluation in Parkinson's disease and mutiple system atrophy, *J Neurol Neurosurg Psychiatry* **62**:507.

Toru, M., Matsuda, O., Makiguchi, K., and Sugano, K., 1981, Neuroleptic malignant syndrome-like state following a withdrawal of antiparkinsonian drugs, *J Nerv Ment Dis* **169**:324.

Yoshimura, N., Mizuta, E., Yoshida, O., and Kuno, S., 1998, Therapeutic effects of dopamine D1/D2 receptor agonists on detrusor hyperreflexia in 1-methyl-4-phenyl-1,2,3,6-tetrahydropyridine-lesioned parkinsonian cynomolgus monkeys, *J Pharmacol exp Therap* **286**:228.

PROGRESS IN SEGAWA'S DISEASE

Masaya Segawa[*]

1. INTRODUCTION

Segawa Disease is an autosomal dominant inherited dystonia caused by mutation of the gene of GTP cyclohydrolase I (GCHI) located on 14q22.1 and q22.2. This was described originally as 'hereditary progressive dystonia with marked diurnal fluctuation (HPD)' by us[1] and later it is also called 'dopa responsive dystonia (DRD)'[2] which first proposed as a normenclature including all dystonia responded to levodopa[3]. Segawa syndrome proposed by Deonna[4] commonly used in Europian countries is a nomenclature including a recessive disorder which later revealed to be caused by mutation of the gene of tyrosin hydroxylase[5].

Segawa disease or HPD/DRD is characterized by early onset in childhood, age-dependent clinical course, diurnal fluctuation and marked sustained response to levodopa. The pathophysiology hypothesized by these clinical characteristics has been confirmed by PET studies and neuropathological-neurohistochemical investigation. Although a broad clinical variation is detected after the discovery of the causative gene, these can be explained by neuronal pathways involving in classical cases.

In this article I will show the progress of Segawa's disease by reviewing chronologically the clinical and laboratory investigation and I suggest the involvement of particular nigrostriatal (NS) - dopamine (DA) neuron and pathways of the basal ganglia which consequently cause this disease at the particular age.

2. DISCOVERY OF THE DISEASE, CLINICAL EVALUATION AND PATHOPHYSIOLOGICAL CONSIDERATION

I found this disorder in 1970 in two girls, who were cousins, and reported in 1971 in a Japanese journal as hereditary basal ganglia disease with marked diurnal fluctuation. However, it was after experienced a 51 year old woman with a clinical course of 43 years, the grandmother of one proband that I was certain that this disease is dystonia and not Parkinson's disease (PD), and published in 1976 under the nomenclature of hereditary progressive dystonia with marked diurnal fluctuation (HPD)[1].

[*]Segawa Pediatric Neurology Clinic, 2-8 Kannda-Surugadai, Chiyoda, Tokyo, Japan

Mapping the Progress of Alzheimer's and Parkinson's Disease
Edited by Mizuno *et al.*, Kluwer Academic/Plenum Publishers, 2002

Table 1 Main characteristics of Segawa's disease differentiationg
from other dopa ressponsive basal ganglia diseases

Early occurrence/Age related clinical course
Diurnal fluctuation
Postural dystonia throughout the course
Postural tremor develops later
No parkinsonian resting tremor
Interlimb coordination is preserved
No mental psychological abnormalities
No autonomic symptoms
Marked sustained ressponse to L-dopa without any side effects

This 51 year old patient also clarified age dependent clinical course: marked progression in the first one and one half decade, subsidence of it later with age and stationary without apparent progression from the 4th decade. The main feature is postural dystonia throughout the course of the illness but postural tremor appears later particularly from the 4th decade. The diurnal fluctuation is marked in childhood. However, it subsides in grade with age and becomes inapparent from the middle of the 3rd decade.

With the follow-up of early cases and accumulation of patients, I confirmed this disease to be a particular basal ganglia disorder different from those previously reported[6], including juvenile parkinsonism described by Yokochi[7] and by Yamamura et al[8], the latter was recently revealed to be caused by abnormalities of the parkin gene (Table 1). Marked and sustained effects without any unfavorable side effects was confirmed after evaluation of long term administration of levodopa for more than 15 years[9].

3. CONSIDERATION ON PATHOPHYSIOLOGY OF SEGAWA'S DISEASE (HPD/DRD)

Pathophysiology of this disease was speculated from clinical characteristics. The marked and sustained response to levodopa without side effects suggested this disease as a functional disorder of the NS-DA neuron. With correlating the development of the NS-DA neuron we found the similarity of the clinical course to the decremental age variation of the TH in the caudate nucleus shown by McGeer & McGeer[10]. This persuaded us to the hypothesis that TH is decreased at the terminal of the NS-DA neuron, and that it does not show progressive decrement but follows the normal decremental variation with a low level of around 20% of normal values. With this lesion, diurnal fluctuation of symptoms is also explained because the NS-DA neuron shows circadian oscillation at the terminal[10] but not in the perikarkyon[11].

Polysomnographies performed to detect the mechanism for sleep effects revealed normal preservation of all components modulated by the cholinergic, noradrenergic and serotonergic neurons of the brainstem, while they showed abnormalities in the pattern of occurrence and the number of phasic components; gross movements (GMs) and twitch movements (TMs)[1]. Selective sleep deprivation revealed that diurnal fluctuation was dependent on REM sleep (sREM)[1]. sREM is a stage for accumulating dopamine with expense of acetylcholine. Numbers of TMs per hour of sREM, which reflect the activities of the NS-DA neuron, decreased significantly in HPD/DRD[1,12] which followed decremental age variation and incremental nocturnal variation of normal children with low values[13].

Evaluation of the sleep stage dependent modulation of GMs revealed absence of DA receptor supersensitivity[12]. The preference of the direction of the horizontal REMs in HPD/DRD showed characteristics observed in cases with lesion in the NS-DA neuron not of those with lesion in the striatum[12]. These confirmed that in HPD the main lesion is in the NS-DA neuron with normal preservation of the striatal pathways but without DA receptor supersensitivity.

Examination of voluntary saccades revealed abnormalities though they were milder than those observed in Parkinson's disease (PD) destructive saccades. (See Hikosaka et al of this volume). These results showed involvement of the output pathways of the basal ganglia to the superior colliculus (SC). The increase in frequency of destructive saccades in HPD of childhood suggests hypofunctioning of the striatal indirect pathway but it aleviated with age as normal children.

From early 1990, neuroimaging studies with $[^{18}F]$ dopa and $[^{11}C]$ raclopride or $[^{11}C]$N-spiperone were performed extensively. $[^{18}F]$ dopa PET scan showed normal[14] or below normal[15] incorporation of DA at the terminal of the NS-DA neuron. $[^{11}C]$ raclopride PET scan revealed normal[16] or elevated D2 receptors[17]. $[^{11}C]$ N-spiperone PET showed mild upward regulation of the receptors[18]. However, as the second $[^{11}C]$ raclopride scan after 7 month of levodopa revealed no alteration of the results, Kishore et al[17] concluded that the increased D2 receptor binding in HPD/DRD was a result of a homeostatic response to the DA deficiancy state and a not the sole factor determining the clinical state.

These results implied that decrease of TH is the main pathology of HPD/DRD[14] and that the DA-D2 receptors are not involved or have no essential roles for pathophysiology of HPD/DRD. Furthermore Jeon et al[19] showed normal DA transporter density by [123I] β-CIT single-PET and confirm normal preservation of the structure of the NS-DA neuron.

Biochemical studies of cerebrospinal fluid (CSF) started in early 70s with estimation of DA metabolities and revealed marked decrease in homoranilic acid (see- ref No.7). However, discovery of decrease of neopterin in CSF by Fujita and Shintaku and Furukawa et al were truly epoch making[20]. They demonstrated clearly the decrease of GCH-1 as the cause of HPD/DRD and accelerated the studies of this disease.

Neuropathological[21] and neurohistochemical[22] studies on an 18-year old female died by accident, who later proved to be HPD/DRD by gene analysis[23], revealed no degenerative changes in the substantia nigra and other components of the basal ganglia and reduced TH in the striatum but not in the substantia nigra. Furthermore there were decrease in DA in both striatum and the substantia nigra. Although the grade of decrement was milder than that in idiopathic PD(iPD) their regional and subregional rostro-caudal gradient were identical to iPD suggesting the no primary lesion in the striatum[20]. However, subregional dorso-ventral gradient show marked decrease of it in the ventral area in difference from dorsal predominance in iPD. This suggested that the NS-DA neuron connecting to the D1-direct pathways are affected predominantly[20,22]. Recently Furukawa et al[24] revealed reduction of neopterin and biopterin in the striatum and normal preservation of the activities of dopamine transporter in other autopsied case.

4. STUDIES ON MOLECULAR BIOLOGY OPEN NEW ERA IN THE RESEARCH OF HPD/DRD

In 1993 Nygaard et al[25] detected linkage to the long arm of the chromosome 14 between 14q11.2-q22.2 and 14q11-q24.3 in Canadian families with DRD. At the same time Ichinose and his colleagues determined the chromosomal localization of the gene for GCH-1 and revealed that HPD is caused by mutations of GCH-1 gene located on 14q22.1 to q22.2[26]. Up to now more than 60 mutations were detected. Although distinct mutations were found among families, they were identical within an family[27]. Familial variation of symptoms seemed to be dependent on the loci of mutation[27]. Recently cases with compound heterozygotes were detected who showed severer symptoms than classic cases[28].

After detection of the causative gene, broad phenotypical variation was detected as shown in table 2. Besides those observed in cases with compound heterozygotes all of the clinical manifestations of dystonia are observed in HPD/DRD.

Table 2 Phenotypical Variation Observed after Detection of the Causative Gene

1)	Focal dystonia: Writer's cramp Guitalist's finger
2)	Paroxysmal Dystonia
3)	Spontaneous reduction and exacerbation of dystonia
4)	Dystonic spasm with or without pain
5)	Action dystonia
6)	Oculogyric crisis
7)	*Muscle hypotonia with delay in development in crawling
8)	*Delay in development of language

*observed in patients with compound heterozygote

5. DISCOVERY OF THE CAUSATIVE GENE PROVOKES QUESTION TO BE CLARIFIED

It is necessary to explain why a heterozygous mutation cause 1) decrease in GCH-1 activities to the levels of clinical manifestation, 2) inter and intrafamilial variation of clinical symptoms, 3) decrease in TH levels predominantly among hydroxylases, 4) preferential involvement of the D1-direct pathway, and 5) characteristic clinical symptoms including all variants.

There are no conclusive answers to these questions, but several ideas have been proposed. For the first question, one considered as classic dominant negative effect[29], but the others as destabilizing effect of the mutant subunit[30, 31]. For the second question, the rates of mutant/wild-type GCH-1 mRNA in lymphocytes vary in one mutation and also depending on the locus of the mutation[29]. These may cause intrafamilial variation including the rate of penetrance and inter-familial variation, respectively. For the third question, difference in distribution of GCH-1 in RNA in dopaminergic and serotonergic neurons[32] and destabilization of the molecule of TH or impairment of axonal transport[33] depending on the neurohistochemical findings of decrease in TH protein in the striatum[21,22,24]. We considered the diference of Km value for TH and phenylalanine hydroxylase[34]. That is, in HPD/DRD with heterozygotic mutant gene, tetrahydrobiopterin (BH4) is decreased partially, so TH with low affinity to BH4 may be selectively affected.

Actually muscle hypotonia and failure in locomotion observed in patients with compound heterozygote[28] are the symptoms caused by deficiency of serotonin.

Neopterin and bioperin in the striatum could have important roles for DA biosynthesis in the striatum, but there is no evidence that these pteridin metabolism relates to the D1-direct pathways. All of the inherited disorders of pteridine metabolism develop dystonia[35]. It is shown that the pteridine metabolism has critical period early in infancy to early childhood and the D1-direct pathways mature earlier while D2-indirect pathways later[36]. This suggests topographical specificity of the metabolism of the NS-DA neuron in the course of its development.

Possible pathophysiology of classic HPD/DRD based on early partial decrement of BH4 at the terminals of the NS-DA neuron has already shown[20]. Then, it should explain the symptoms observed after discovery of the causative gene.

Muscle cramp or dystonic cramp is not a rare symptom of HPD/DRD. With reference to a case treated with stereotactic operation I suggested that dystonic muscle cramp is severe state of dystonic hypertonus, which develops through the reticulo spinal pathways via the descending output of the basal ganglia[20].

Oculogyric crisis with transient aggravation after levodopa observed in a case with compound heterozygotes[28] suggests existence of D2 receptor supersensitivity and involvement of striatal indirect pathway due to marked decrease of BH4. However, this symptom as well as action retrocollis is observed in patients with heterozygotic mutation. In these reduction of GCH-1 activity of peripheral mononuclear cells were not necessarily low (Segawa, unpublished data). As in DYT-1 there are two phenotypes, one postural dystonia and the other action dystonia, and upper generation of the proband with action dystonia show focal dytonia. For the pathophysiologies of action dystonia, striatal indirect pathway and the ascending output pathways of the basal ganglia are suggested. Focal dystonia could be induced by the disinhibited thalamo-cortical pathway[37]. Similar neuronal pathways might be involved in action dystonia and focal dystonia of HPD/DRD.

These suggest there are particular NS-DA neuron and the related neuronal pathways of the basal ganglia for dystonia which are modulated by particular biochemical and morphological processes during development. Mutation of GCH-1 may cause particular disturbance of function on particular group of the NS-DA neuron and the particular neuronal pathways of the basal ganglia and develop HPD/DRD. In this process each locus of the GCH-1 gene might have particular roles and particular critical age for development and cause phenotypical variation of HPD/DRD. Collaborative studies between neurologists and molecular iologists are necessary not only, for clarification of the pahophysiology of HPD/DRD but also of the process of development of the NS-DA neuron and the neuronal pathways in the basal ganglia.

REFERENCES

1. M. Segawa, A. Hosaka, F. Miyagawa. Y. Nomura, H. Imai, Hereditary progressive dystonia with marked diurnal fluctuation, in : Advances in neurology, vol.14, edited by R. Eldridge, S. Fahn (Raven Press, New York, 1976), pp. 215-233.
2. D. B. Calne, Dopa-responsive dystonia, Ann Neurol. 35(4), 381-382 (1994)

3. T. G. Nygaard, C. D. Marsden, R. C. Duvoisin, Dopa responsive dystonia. In Advances in Neurology. Vol 50, edited by S. Fahn, C. D. Marsden, D. B. Calne (New York:Raven Press, 1988), pp. 377-384.

4. T. Deonna, Dopa-sensitive progressive dystonia of childhood with fluctuations of symptoms- Segawa's syndrome and possible variants, Neuroped. 17(2), 75-80 (1986).

5. B. Lüdecke, B. Dworniczak, K. Bartholome, A point mutation in the tyrosine hydroxylase gene associated with Segawa's syndrome, Human Genet. 95(1), 123-125 (1995).

6. M. Segawa, Y. Nomura, M. Kase, Diurnally fluctuating hereditary progressive dystonia, in: Handbook of Clinical Neurology : Extrapyramidal Disorders, 5(49), edited by P. J. Vinken and G. W. Bruyn, (Elsevier, Amsterdam, 1986), pp. 529-539.

7. M. Yokochi, Juvenile Parkinson's disease-Part I. Clinical aspects (in Japanese), Adv Neurol Sci. 23(5), 1060-1073 (1979).

8. Y. Yamamura, I. Sobue, K. Ando, M. Iida, T. Yanagi, C. Kono, Paralysis agitans of early onset with marked diurnal fluctuation of symptoms, Neurology. 23(3), 239-244 (1973).

9. M. Segawa, Y. Nomura, S. Yamashita, M. Kase, N. Nishiyama, S. Yukishita, et al, Long term effects of L-Dopa on hereditary progressive dystonia with marked diurnal fluctuation, in : Motor Disturbance II, edited by A. Berardelli, R. Benecke, M. Manfredi, C. D. Marsden (London:Academic Press, 1990), pp. 305-318.

10. E. G. McGeer, P. L. McGeer. Some characteristics of brain tyrosine hydroxylase, in : New Concepts in Neurotransmitter Regulation, edited by Mandel J (Plenum, New York, London, 1973), pp. 53-68.

11. G. F. Steinfels, J. Heym, R. E. Strecker, B. L. Jacobs, Behavioural correlates of dopaminergic unit activity in freely moving cats, Brain Res. 258(2), 217-228 (1983).

12. M. Segawa, Y. Nomura, S. Tanaka, S. Hakamada, E. Nagata, E. Soda, et al, Hereditary progressive dystonia with marked diurnal fluctuation. Consideration on its pathophysiology based on the characteristics of clinical and polysomnographical findings, in : Advances in Neurology.Vol 50, edited by S. Fahn, C. D. Marsden, D. B. Calne (New York:Raven Press, 1988), pp. 367-376.

13. M. Segawa, Y. Nomura, Hereditary progressive dystonia with marled diurnal fluctuation. Pathophysiological importance of the age of onset, in : Advances in Neurology Vol 60, edited by H. Narabayashi, T. Nagatsu, N. Yanagisawa, Y. Mizuno (New York: Raven Press, 1993), pp. 568-576.

14. B. J. Snow, A. Okada, W. R. W. Martin, R. C. Duvoisin, D. B. Calne, Positron-emission tomography scanning in dopa-responsive dystonia, parkinsonism-dystonia, and young-onset parkinsonism, in : Hereditary Progressive Dystonia with Marked Diurnal Fluctuation, edited by Segawa M (Parthenon, Carnforth UK, 1993), pp. 181-186.

15. G. V. Sawle, K. L. Lenders, D. J. Brooks, G. Harwood, A. J. Lee, R. S. J. Frackowiak, et al, Dopa-responsive dystonia: [F-18] Dopa positron emission tomography, Ann Neurol. 30(1), 24-30 (1991).

16. K. L. Leenders, A. Antonini, H-M. Meinck, A. Weind, Striatal dopamine D2 receptors in dopa-responsive dystonia and Parkinson's disease, in : Age-Related Dopamine-Dependent Disorders: Monogr Neural Sci.Vol 14, edited by M. Segawa, Y. Nomura (Basel:Karger, 1995), pp. 95-100.

17. A. Kishore, T. G. Nygaard, R. de la Fuente-Fernandez, A. B. Naini, M. Schulzer, E. Mak, et al, Striatal D2 receptors in symptomatic and asymptomatic carriers of dopa responsive dystonia measured with [11c]-raclopride and positron-emission tomography, Neurology, 50(4), 1028-32 (1998).

18. G. Kunig, K. L. Leenders, A. Antonini, P. Vontobel, A. Weindl, H. M. Meinck, D2 receptor binding in dopa-responsive dystonia. Ann Neurol. 44(5), 758-762 (1998).

19. B. S. Jeon, J-m. Jeong, S-S. Park, J-M. Kim, K-Y. Yoon, LeeN-C, et al, Dopamine transporter density measured by [123I] beta-CIT single –photon emission computed tomography is normal in dopa-responsive dystonia, Ann Neurol. 43(6), 792-800 (1998).

20. M. Segawa, Hereditary progressive dystonia with marked diurnal fluctuation, Brain & Development. 22 (supple 1), 65-80 (2000).

21. A. H. Rajput, W. R. G. Gibb, X. H. Zhong, K. S. Shannak, S. Kish, L. G. Chang, et al, Dopa- responsive dystonia: pathological and biochemical observations in one case, Ann Neurol. 35(4), 396-402 (1994).

22. O. Hornykiewicz, Striatal dopamine in dopa-responsive dystonia. Comparison, with idiopathic Parkinson's disease and other dopamine-dependent disorders, in : Age-Related Dopamine-Dependent Disorders:Monogr Neural Sci. Vol 14., edited by Segawa M, Nomura Y (Basel:Karger, 1995), pp. 101-108.

23. Y. Furukawa, M. Shimadzu, A. H. Rajput, FRCPC, Y. Shimizu, T. Tagawa, et al, GTP-cyclohydrolase I gene mutations in hereditary progressive and dopa responsive dystonia, Ann Neurol. 39(5), 609-617 (1996).

24. Y. Furukawa, T. G. Nygaard, M. Gutlich, A. H. Rajput, C. Pifl, L. DiStefano, L. J. Chang, K. Price, M. Shimadzu, O. Hornykiewicz, J. W. Haycock, S. J. Kish, Striatal biopterin and tyrosine hydroxylase protein reduction in dopa-responsive dystonia, Neurology. 53(5), 1032-41 (1999).

25. T. G. Nygaard, K. C. Wihelmsen, N. J. Risch, D. L. Brown, J. M. Trugman, T. C. Gilliam, et al, Linkage mapping of dopa-responsive dystonia (DRD) to chromosome 14q, Nat Genet. 5(4), 386-391 (1993).

26. H. Ichinose, T. Ohye, E. Takahashi, N. Seki, T. Hori, M. Segawa, et al, Hereditary progressive dystonia with marked diurnal fluctuation caused by mutations in the GTP cyclohydrolase I gene, Nature Genetics. 8(3), 236-242 (1994).

27. N. Nishiyama, S. Yukishita, H. Hagiwara, S. Kakimoto, Y. Nomura, M. Segawa, Gene mutation in hereditary progressive dystonia with marked diurnal fluctuation (HPD), strictly defined dopa-responsive dystonia, Brain & Development. 22(Suppl 1), 102-106 (2000).

28. Y. Furukawa, S. J. Kish, E. M. Bebinm R. D. Jacobson, J. S. Fryburg, W. G. Wilson, et al, Dystonia with motor delay in compound heterozygotes for GTP-cyclohydrolase I gene mutations, Ann Neurol. 44(1), 10-16 (1998).

29. M. Hirano, T. Yanagihara, S. Ueno, Dominant negative effect of GTP cyclohydrolase I mutations in Dopa-responsive hereditary progressive dystonia, Ann Neurol. 44(3), 365-371 (1998).

30. T. Suzuki, T. Ohye, H. Inagaki, T. Nagatsu, H. Ichinose, Characterization of wild-type and mutants of recombinant human GTP cyclohydrolase I: relationship to etiology of dopa-responsive dystonia. J. Neurochem. 73(6), 2510-2516 (1999).

31. W. L. Hwu, Y. W. Chiou, S. Y. Lai, Y. M. Lee, Dopa-responsive dystonia is induced by a dominant-negative mechanism, Ann Neurol. 48(4), 609-13 (2000).

32. M. Shimoji, K. Hirayama, K. Hyland, G. Kupatos, GTP cyclohydrolase I gene expression in the brains of male and female hph-1 mice. J. Neurochem. 72(2), 757-764 (1999).

33. Y. Furukawa, Dopa-responsive dystonia: recent advances and remaining issues to be addressed, Mov Disord. 14(5), 709-715 (1999).

34. M.D. Davis, P. Ribeiro, J. Tipper, S. K. Danis, "7-Tetrahydrobiopterin," A naturally occurring analogue of tetrahydrobiopterin, is a cofactor for and a potential inhibitor of the aromatic amino acid hydroxylases. Proc. Natl. Acad.Sci. U.S.A. 89(21), 10109-10113 (1992).

35. Y. Nomura, K. Uetake, S. Yukishita, H. Hagiwara, T. Tanaka, R. Tanaka, K. Hachimori, N.Nishiyama, M. Segawa, Dystonia responding to levodopa and failure in biopterin metabolism, in : Dystonia 3 Advances in Neurology Vo. 78, edited by S. Fahn, C. D. Marsden, Delong (Lippincott-Raven Publishers, Philadelphia, 1998), pp. 253-266.

36. M. Segawa, Development of the nigrostriatal dopamine neuron and the pathways in the basal ganglia, Brain & Development. 22(Suppl 1), 1-4 (2000).

37. Y. Nomura, T. Ikeuchi, S. Tsuji, M. Segawa, Two phenotypes and anticipation observed in Japanese cases with early onset torsion dystonia (DYT1) – pathophysiological consideration, Brain & Development. 22 (Suppl 1), 92-101.

PROGRESS IN THE TREATMENT OF PARKINSON'S DISEASE

Donald B. Calne[*]

Since other papers have dealt with such topics as algorithms for the use of current medications and preliminary results with new medications, I shall take three aspects of treatment that have not been discussed elsewhere in this Congress. Each of these topics has, or will be, reported in neurological journals over the course of the present year.

1. TRANSPLANTATION OF RETINAL PIGMENTED EPITHELIAL CELLS

Human retinal pigmented epithelial cells produce levodopa and dopamine. These cells adhere to surfaces so a preparation can be produced such that the cells attach to a gel microcarrier and the combination can be implanted in the brain. The microcarrier stops the retinal pigmented epithelial cells from wandering off into neural tissue, and the microcarrier also confers some protection from immune attack by the host. In monkeys with experimental lesions induced by MPTP, human retinal pigmented epithelial cells, inserted into the striatum, are capable of achieving a marked improvement in motor function (1).

With this encouraging background, six patients with Parkinson's disease have recently been treated with the same cell-carrier preparation in a pilot study (2). Three hundred and twenty five cells were inserted in each of 5 tracks in the post-commissural putamen contralateral to the more severe clinical deficits. The tracks were 5mm apart. All patients tolerated the procedure and follow up observations have now reached 9 months. The results are very encouraging. The total UPDRS scores, in CAPIT "on" and "off" states, improved in all patients by an average of some 20%. This change was significant at a level of p= 0.03.

[*] Neurodegenerative Disorders Centre, UBC Hospital, Purdy Pavilion, Room M36, 2221 Wesbrook Mall, Vancouver, BC, Canada, V6T 3B5, Telephone: 604-822-7048, Fax: 604-822-7866

Mapping the Progress of Alzheimer's and Parkinson's Disease
Edited by Mizuno *et al.*, Kluwer Academic/Plenum Publishers, 2002

361

Many patients were able to reduce their intake of dopaminomimetic therapy. Dyskinesia was not a problem, in contrast to the disturbing experience reported with transplantation of human fetal cells (3), though follow up may have been too short for reassurance on this point.

The retinal pigmented epithelial cells used in this study derived from a human fetal retina. Studies are now in progress to see if the same therapeutic results can be achieved with neonatal retinal pigmented epithelial cells. In theory, cell culture from one retina should yield enough tissue for treating some 10,000 patients.

2. THE MECHANISM OF "WEARING-OFF" REACTIONS

Conventional views of the mechanism underlying motor fluctuations attribute them to some form of toxicity deriving from the cumulative dose of levodopa therapy, or to loss of the buffering capacity for storing dopamine because of the loss of dopaminergic nerve endings in the Parkinsonian striatum. An alternative mechanism was proposed by de la Fuente-Fernandez et al. early last year (4) - that fluctuations might be related to increased dopamine turnover. Evidence in support of this hypothesis was presented by Rajput et al. (5), who reported that the postmortem ratio of HVA/dopamine was higher in the striatum of patients suffering from fluctuations, compared to patients who had stable responses to levodopa. Last month a prospective study, employing positron emission tomography, yielded evidence that a rapid rate of dopamine turnover preceded the appearance of motor fluctuations (6).

The evidence comprised measurement of the extent to which administration of levodopa displaced [11-C]raclopride bound to the D2 receptors in the striatum of Parkinsonian patients. Each of eight patients, with stable responses to levodopa, underwent the PET scans with raclopride. The first was undertaken before giving a standard dose of levodopa, the second one hour after the levodopa, and the third four hours after the levodopa. Four out of eight patients displayed a rapid turnover of dopamine. After following the patients, on chronic levodopa therapy, over a period of three years, four had developed "wearing-off" reactions - the same four who had initially had a rapid dopamine turnover. A further patient who was already experiencing "wearing-off" reactions underwent the same PET protocol, and was found to have the same rapid turnover of dopamine found in the patients who subsequently developed fluctuations. Taken together, these findings gave strong support to the rapid turnover hypothesis.

The observations carry further implications. Because the rapid turnover of dopamine was detected before clinical fluctuations appeared, after a mean duration of only 1.6 years of levodopa therapy, we can speculate that the increased turnover rate my have been an aberrant, excessive homeostatic attempt to compensate for the pathology underlying Parkinson's disease. This idea would be in keeping with the fact that fluctuations are more frequent in younger patients, who presumably have the capacity for a more vigorous effort to compensate for the loss of dopaminergic neurons.

3. SLEEP ATTACKS

In 1999 Frucht *et al.* (7) published a report suggesting that the new dopamine agonists, pramipexole and ropinirole, can cause sudden sleep during the day, which they termed "sleep attacks". Because of these episodes, they said that there was a significant risk that patients who were driving cars could cause road traffic accidents. This paper led to government regulatory agencies in several countries, such as Canada, insisting that patients who were taking either of the two new agonists should be warned not to drive. The report by Frucht *et al.* (7) gave no information on the frequency of the "sleep attacks" they described, and no information on the relative risk of daytime sleepiness with other antiparkinson drugs. Sanjiv *et al.* (8) have recently addressed these issues. They designed a study with a power of 80% to detect an 8% change in daytime somnolence at a significance level of p=0.05. Subjects fell into five groups - controls, patients taking levodopa alone, patients taking levodopa with bromocriptine, patients taking levodopa with ropinirole, and patients taking levodopa with pramipexole. There were forty subjects in each group. The patients were all taking similar doses of levodopa, and the doses of synthetic agonists were closely comparable to each other. Exclusion criteria included dementia, hallucinations, sedative drugs, hypnotics, antidepressant drugs, or antihistamines. Patients were also excluded if they had general medical diseases. A validated questionnaire was employed to measure the risk of dozing off to sleep during the day, the Epworth scale. A modified Epworth scale was also applied to measure the risk of sudden sleep without warning during the day.

The results were simple. When adjusted for age and disease severity, there was no significant difference for dozing or sudden sleep among the Parkinsonian patients in each category of treatment. In contrast the normal controls had less daytime dozing and sudden sleep. The situation seems to be comparable to the risk of daytime sleepiness among patients taking hypnotics, sedatives, antidepressants or antihistamines.

REFERENCES

1. Subramanian, T., Bakay, R.A.E., Cornfeldt, M.E., and Watts, R.L. Blinded Placebo-Controlled Trial to Assess the Effects of Striatal Transplantation of Human Retinal Pigmented Epithelial Cells Attached to Microcarriers (hRPE-M) in Parkinsonian Monkeys. *Parkinsonism and Related Disorders* 5:S1ll, 1999. (Abstract)
2. Watts, R.L., Raiser, C.D., Stover, N.P., Cornfeldt, M.L., Schweikert, A.W., Subramanian, T., and Bakay, R.A.E. Stereotaxic Intrastriatal Implantation of Retinal Pigment Epithelial Cells Attached to Microcarriers in Advanced Parkinson Disease (PD) Patients: A Pilot Study. *Neurology* 56:A283, 2001. (Abstract)
3. Freed, C.R., Greene, P.E., Breeze, R.E., Tsai, W.Y., DuMouchel, W., Kao, R., Dillon, S., Winfield, H., Culver, S., Trojanowski, J.Q., Eidelberg, D., and Fahn, S. Transplantation of embryonic dopamine neurons for severe Parkinson's disease. *New England Journal of Medicine* 344(10):710-719, 2001.
4. de la Fuente-Fernandez, R., Pal, P.K., Vingerhoets, F.J.G., Kishore, A., Schulzer, M., Mak, E.K., Ruth, T.J., Snow, B.J., Calne, D.B., and Stoessl, A.J. Evidence for impaired presynaptic dopamine function in parkinsonian patients with motor fluctuations. *Journal of Neural Transmission* 107:49-57, 2000.
5. Rajput, A.H., Fenton, M.E., Sitte, H., Pifl, C., George, D., Macaulay, R., and Hornykiewicz, O. Major Motor Complications of Levodopa Therapy Do Not Represent Spectrum of the Same Pathophysiology. *Mov Disord* 15:113-113, 2000. (Abstract)

6. de la Fuente-Fernandez, R., Lu, J., Sossi, V., Jivan, S., Schulzer, M., Holden, J.E., Lee, C.S., Ruth, T.J., Calne, D.B., and Stoessl, A.J. Biochemical variations in the synaptic level of dopamine precede motor fluctuations in Parkinson's disease: PET evidence of increased dopamine turnover. *Ann Neurol* 49:298-303, 2001.

7. Frucht, S., Rogers, J.D., Greene, P.E., Gordon, M.F., and Fahn, S. Falling asleep at the wheel: Motor vehicle mishaps in persons taking pramipexole and ropinirole. *Neurology* 52:1908-1910, 1999.

8. Sanjiv, C.C., Schulzer, M., Mak, E., Fleming, J., Martin, W.R.W., Brown, T., Calne, S.M., Tsui, J., Stoessl, A.J., Lee, C.S., and Calne, D.B. Daytime somnolence in patients with Parkinson's disease. *Parkinsonism and Related Disorders* 7(4):283-286, 2001.

THEORETICAL CONSIDERATION OF DRUG THERAPY IN PARKINSON'S DISEASE

Jin-Soo Kim and Young H. Sohn*

1. THE BASAL GANGLIA AND ITS CHANGE IN PARKINSON'S DISEASE (PD)

1.1 Basal Ganglia Anatomy

The basal ganglia comprise several nuclei in the forebrain, diencephalons, and midbrain thought to play a significant role in the control of posture and movement. They are composed of two primary input structures, two primary output structures, and two intrinsic nuclei. The input structures of the basal ganglia are the striatum (caudate and putamen) and the subthalamic nucleus (STN). The striatum receives excitatory inputs from virtually all areas of cerebral cortex and STN receives excitatory inputs from motor areas of the frontal lobe. The output structures are internal segment of the globus pallidus (GPi) and substantia nigra pars reticulata (SNr). The outputs from GPi and SNr are inhibitory and project to motor areas in the brainstem and thalamus. The intrinsic nuclei are the globus pallidus external segment (GPe) and substantia nigra pars compacta (SNc). SNc is the locus of dopaminergic neurons, which receives the bulk of its input from the striatum and sends the bulk of its output back to the striatum [1,2].

There are two major projections through the basal ganglia, i.e. from the the striatum to GPi and SNr. The direct pathway projects directly from a certain subpopulation of striatal neurons to GPi/SNr and involves GABAergic/substance P containing fibers. The indirect pathway, which has its origin in a different population of striatal neurons, proceeds through GPe and STN before reaching GPi/SNr. In the indirect pathway, striatal GABAergic/enkephalinergic fibers provide an inhibitory innervation to GPe. In turn, GABAergic projections from GPe provide an inhibitory innervation to STN. The indirect pathway ends with excitatory, glutamatergic fibers from STN to GPi/SNr. Since both direct and indirect pathways converge, with opposing effects, on GPi/SNr, a balance between the inhibitory influence of the direct pathway and the excitatory influence of the indirect pathway controls the output of GPi/SNr (figure 1A) [1-3].

*Department of Neurology and Brain Research Institute, Yonsei University College of Medicine, Seoul, Korea

Mapping the Progress of Alzheimer's and Parkinson's Disease
Edited by Mizuno *et al.*, Kluwer Academic/Plenum Publishers, 2002

1.2. Basal Ganglia Function: Focused Selection and Inhibition.

The current concept of basal ganglia function can be abstracted as focused selection and inhibition. There are multiple mechanisms intrinsic to the striatum that integrate inputs from wide areas of the cerebral cortex and focus the output. The striatum sends a focused inhibitory projection to the basal ganglia output nuclei, GPi and SNr. It also sends an inhibitory projection to GPe which, in turn, inhibits STN and GPi. This indirect pathway from striatum through GPe and STN could act in opposition to the direct pathway and result in further focusing of the information flow from striatum to GPi. This hypothesis states that tonically active inhibitory output of the basal ganglia act as a 'brake' on motor pattern generator s (MPGs) in the cerebral cortex and the brainstem (via thalamus). When a movement is initiated by a particular MPG, basal ganglia output neurons projecting to competing MPGs increase their firing rate, thereby increasing inhibition and applying a ' brake' on those generators. Other basal ganglia neurons

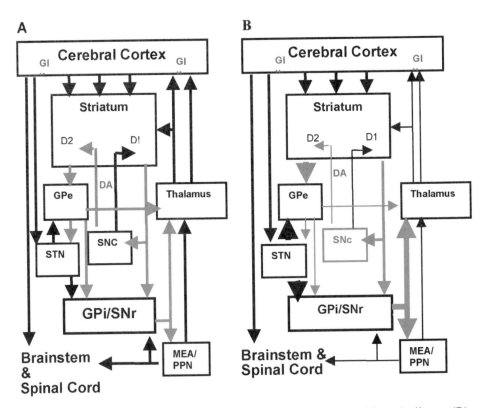

Figure 1. Diagrams of the basal ganglia circuit in normal (A) and Parkinson's disease (B).
(➤ Excitatory pathways, ➨ Inhibitory pathways)

projecting to the generators involved in desired movement decrease their discharge, thereby removing tonic inhibition and releasing the 'brake' from those desired generator.

Thus selected movements are enabled and competing posture and movements are prevented from interfering with the one selected [1].

When one makes voluntary movements, the motor cortex sends an excitatory corollary signal to STN. STN excitation results in widespread excitation of GPi, which in turn produces inhibition of thalamocortical and brainstem motor mechanisms. In parallel to the pathway through STN, signals are sent from all areas of cerebral cortex to striatum. The cortical inputs are transformed by the striatal integrative circuitry to a focused, context-dependent output that inhibits specific neurons in GPi. The inhibitory striatal input to GPi is slower, but more powerful, than the excitatory STN input. The resultant focally decreased activity in GPi selectively disinhibits the desired thalamocortical and brainstem MPGs. Indirect pathways from striatum to GPi result in further focusing of the output. The net result of basal ganglia activity during a voluntary movement is the braking of competing MPGs and focused release of the brake from the selected MPGs .

1.3. The Role of Dopamine in Basal Ganglia Function

Nigrostiriatal dopaminergic input activates D1 dopamine receptors in the striatum, which in turn enhances the activity of the direct pathway, but inhibits D2 dopamine receptor-mediated indirect pathway activity. This opposite action of dopaminergic input on direct and indirect pathways promotes focused selection of desired movement [1]. The dopaminergic neurons from the ventral tegmental area to the nucleus accumbens are considered to mediate natural and drug-induced reward and sensitization. The nigrostriatal dopamine system seems to have similar function that it may mediate reinforcement, which strengthens a given behavior when elicited subsequently. Dopamine neurons have slow conductance velocity that it may be released after the execution of movement. Thus they do not seem to transmit fast sensory and motor information, rather they appear to signal aspects of reinforcement and reward. Dopamine may also be related with strengthening of striatal long-term potentiation (LTP) or long-term depression (LTD), which are the underlying mechanisms for learning, reinforcement or extinction. After certain training, dopamine can be released in anticipation of a reward. Dopamine neurons fire in relation to attentional and incentive process. Based upon these observations, it can be assumed that a 'good' behavioral outcome enhances dopamine release and strengthens a given behavior when elicited subsequently, while a 'bad' behavioral outcome reduces dopaminergic activity and will weaken a given behavior [4]. In summary, nigrostriatal dopamine neurons appear to be related to reward and reinforcement of reproduction of well-fitted motor program. This concept can be supported by the fact that automated behaviors and movements are severely affected in PD.

1.4. Basal Ganglia Change in PD

Reduced nigrostriatal dopaminergic input results in a reduced activity of the direct pathway, but an enhanced activity of the indirect pathway. These changes cause increased inhibitory output from GPi and SNr, which in turn produce increased inhibition on the motor cortex and brainstem. Increased activity of the indirect pathway leads to overactivity of STN, which results in enhanced tonic inhibition. In contrast, decreased direct pathway activity results in reduced phasic disinhibition, i.e. focused selection of desired movement. Both direct and indirect pathway changes may cause disturbances in selecting desired motor program from unwanted movements, which eventually produce co-contraction of different muscles. This phenomenon may be responsible for the cardinal symptoms in PD, such as rigidity and bradykinesia (figure 1B).

2. DOPAMINERGIC THERAPY

Levodopa is still the most effective therapy in PD and is also the most ideal way of treatment. Usual daily doses of levodopa with dopa decarboxylase inhibitors are sufficient to replace physiologically required dopamine in the brains of PD patients. In most patients, however, chronic levodopa therapy is eventually complicated by motor fluctuation and levodopa-induced dyskinesia. With the progression of PD, the space (surviving dopaminergic neurons) for conversion of administered levodopa and storage of newly synthetized dopamine becomes severely limited that leads to nonphysiological way of dopaminergic stimulation in the striatum. Intermittent stimulation according to the dosing schedule instead of rather tonically active dopaminergic neurons in physiological state produces post-synaptic alterations, which in turn result in shortened, unpredictable responses as well as levodopa-induced involuntary movements such as dyskinesia and dystonia [5,6]. These complications definitely limit the effect and usability of levodopa, particularly in advanced patients.

Table1. Receptor affinity profiles of dopamine agonists.

Drugs	D1	D2
Bromocriptine	0/-	++
Pergolide	+	++
Lisuride	0/+	++
Apomorphine	+	++
Cabergoline	0/+	++
Ropinirol	0	++
Pramipexol	0	++

++, full agonist; +, moderate agonist; 0, little or no effect; -, antagonist

Since dopamine depletion causes imbalance between direct and indirect pathway activities, the balancing of these systems should be corrected by simultaneous stimulation of both D1 and D2 dopamine receptors. As shown in table 1, however, all currently available dopamine agonists are mainly targeting D2 dopamine receptor. Only pergolide and apomorphine show weak D1 affinity compared to D2. With this D1 affinity, these drugs may be more ideal than the others [7]. In fact, subcutaneous apomorphine is often effective in patients who failed to respond to levodopa. In MPTP-treated monkeys, D1 agonists itself also show strong anti-parkinson effect with less occurrence of dyskinesia compared to D2 agonists [8]. However, unfortunately, we do not have any clinically available D1 agonist now. Although Parkinson's disease is mainly the problem of dopaminergic transmission, these limitations of dopaminergic therapy lead many scientists to shift their attention to non-dopaminergic modulation.

3. CHOLINERGIC DRUG TREATMENT

There are two cholinergic systems in the basal ganglia. The first one is the intrastriatal interneurons, which is the main source of acetylcholine in the striatum. The other one is cholinergic inputs from the pedunculopontine tegmental nucleus. The cholinergic intrastriatal interneurons are mostly large (20-50 μm) aspiny neurons. They account only 1-2% of all striatal neurons, but have much more widespread dendritic trees than do the projecting neurons, suggesting their widespread connections. They receive inputs from cortical and thalamic glutamatergic projections and synapses with the striatal projection neurons. These connections suggest their role in processing of information forward from the cortex to projection cells [9]. The intrastiratal cholinergic neurons are thought to have opposing action to the nigrostriatal dopaminergic input. Thus it inhibits direct pathway and excites indirect pathway. Animals with unilateral ablation of striatal cholinergic neurons showed contralateral rotation in acute phase, but this behavior was transient and recovered in chronic phase. This eventual recovery appears to result from adaptive attenuation of dopamine transmission with down-regulation of both D1 and D2 dopamine receptors [10]. Currently available anticholinergic drugs are trihexyphenidyl and benztropin. They have only weak and mild antiparkinson effect. It should be cautious to use these drugs because persistent restraint of acetylcholine function may result in further reduction of dopamine transmission in PD patients.

Another possible cholinergic agents are nicotinic acetylcholine agonists. These drugs are able to stimulate release of dopamine in the striatum and frontal cortex. In MPTP-treated monkeys, this drug used with small dose of levodopa produced more improvement in both cognitive and motor function compared to levodopa alone [11]. Since these drugs can modulate noradrenaline and acetylcholine in addition to dopamine, they possibly improve affective and cognitive symptoms of Parkinson's disease as well.

4. GLUTAMATE ANTAGONISTS

Glutamate is major excitatory neurotransmitter in the central nervous system, including the basal ganglia. The striatum receives glutamatergic input from the whole cerebral cortex as well as the thalamus. And STN projects glutamatergic output to the pallidum and the substantia nigra. In Parkinson's disease, one of the most problematic changes is thought to be hyperglutamatergic activity in STN [4,12]. As we can see the antiparkinson effect of subthalamic deep brain stimulation, selective suppression of glutamatergic activity in this nucleus could be one of the most effective therapeutic approaches. This glutamatergic overactivity has been observed in MPTP-treated monkeys as well. In primates, however, the effects of glutamate antagonists were very limited and only showed some effect if they are administered with dopaminergic agents, which is probably due to glutamatergic hypoactivity of thalamic output. Administration of these agents with levodopa clearly shows synergistic effects. Increased density of striatal NMDA receptors was observed in the postmortem brains of Parkinson patients. This increase is correlated to the degree of dopamine neuronal loss and is compensatory response [12].

A non-competitive NMDA antagonist, amantadine has moderate antiparkinson effect, and both NMDA and AMPA antagonists can reverse levodopa-induced motor complications in experimental animals [13]. Also amantadine and another glutamate antagonist dextromethorphane can alleviate motor fluctuation as well as peak dose dyskinesia [14, 15]. However, we should be careful to use these agents in the clinic, because it may cause schizophrenic psychosis and learning and memory impairments. As mentioned earlier, glutamate is present in the whole CNS so that more selective blocker acting only on STN should be required. Low doses of glutamate agonists with dopaminergic agents are recommended [12].

Neurotoxicity of excitatory amino acid is well known and glutamate overactivity in STN appears to exacerbate dopaminergic neuronal degeneration in PD. In experimental animals, NMDA antagonists prevent MPTP-induced dopamine neuronal loss. These observations suggest the neuroprotective effect of glutamate antagonists in PD [12].

5. PROPOSED THERAPY ACTING ON OTHER TRANSMITTER SYSTEMS

There are also a number of afferent inputs to the medium spiny neurons, including opiate, adenosine and norepinephrine. Then, the integrated information in the medium spiny neuron is projected to the other basal ganglia structures using GABA. All of these transmitters can be possible candidates for therapeutic targets in Parkinson's disease.

5.1. Adenosine antagonists

Adenosine A2A receptors are highly localized to the striatum, nucleus accumbens and olfactory tubercle. Within the striatum, A2A receptors are selectively localized on D2 receptor bearing medium spiny neurons. A2A receptor stimulation exerts a negative effect on motor function. In dopamine denervated rodents, A2A agonists inhibit the motor stimulation induced by dopamine agonists, while antagonists induce rotational behavior and potentiate such behavior elicited by dopamine agonists [16]. A selective adenosine A2A antagonist, KW-6002, showed dose-dependent antiparkinson effect that reached the level of efficacy of levodopa, but induced little or no dyskinesia in MPTP-treated primates [17,18]. When co-administered, this agent potentiated the effect of levodopa on motor activity without affecting the severity of dyskinesia [18].

5.2. Noradrenergic antagonists

MPTP is also toxic to the noradrenergic neurons originating in the locus ceruleus, Disruption of the noradrenergic system reduced the basal release and turnover of dopamine in the rat striatum. A selective presynaptic α2-adrenoreceptor blocking agent, idazoxan, increased locomotor activity and improved disability score without dyskinesia in MPTP-treated monkeys. In combination with levodopa, it did not impair the antiparkinsonian response but significantly reduced dyskinesias in these animals [19]. Yohimbin, another α2-adrenoreceptor antagonist, also reduced levodopa-induced dyskinesia in MPTP-treated monkeys [20].

5.3. Opioid Agonists

Kappa-opioid receptors are located in the basal ganglia regions associated with increased glutamate transmission in parkinsonism and presynaptic kappa opioid receptor activation can reduce glutamate release in the basal ganglia. Kappa opioid agonists are proved to improve locomotor activity in dopamine depleted animals, and when co-administered with levodopa, the antiparkinson effect are synergistic [21].

5.4. GABA Agonists

GABA is also present in both direct and indirect pathway. So it is difficult to limit its effect selectively, but focal injection of GABA agonists into GPi showed symptom improvements in a patient as well as in MPTP monkeys [22, 23].

REFERENCES

1. Mink JW. The basal ganglia: focused selection and inhibition of competing motor programs. Prog Neurobiol 50:381-425, 1996.
2. Young AB, Penny JB. Biochemical and functional organization of the basal ganglia. In: Jankovic J, Tolosa E, eds. Parkinson's disease and movement disorders, 3rd edition. Baltimore: Williams and Wilkins, pp1-13, 1998.
3. Gerfen CR, Engber TM, Mahan LC, Sosel Z, Chase TN, Monsma FJ, Sibley DR. D1 and D2 dopamine receptor regulated gene expression of striatonigral and striatopallidal neurons. Science 250:1429-1432, 1990
4. Schmidt WJ. Dopamine-glutamate interactions in the basal ganglia. Amino Acids 14:5-10, 1998.
5. Chase TN, Oh JD, Blanchet PJ. Neostriatal mechanisms in Parkinson's disease. Neurology 51(supple 2):S30-S35, 1998.
6. Obeso JA, Rodriguez-Oroz MC, Rodrigeuz M, Lanciego JL, Artieda J, Gonzalo N, Olanow CW. Pathophysiology of the basal ganglia in Parkinson's disease. Trends Neurosci 23(suppl.):S8-S19, 2000.
7. Ogawa N. Early introduction of dopamine agonists in the long-term treatment of Parkinson's disease. Neurology 51(suppl 2):S13-S20, 1998.
8. Blanchet PJ, Grondin R, Bedard PJ. Dyskinesia and wearing-off following dopamine D1 agonist treatment in drug-naïve 1-methyl-4-phenyl-1,2,3,6-tetrahydropyridine-lesioned primates. Mov Disord 11:91-94, 1996..
9. Calabresi P, Centonze D, Gubellini P, Pisani A, Bernardi G. Acetylcholine-mediated modulation of striatal function. Trends Neurosci 23:120-126, 2000.
10. Kaneko S, Hikida T, Watanabe D, Ichinose H, Nagatsu T, Kreitman RJ, Pastan I, Nakanishi S. Synaptic integration mediated by striatal cholinergic interneurons in basal ganglia function. Science 289:633-637, 2000.
11. Schneider JS, Van Velson M, Menzaghi F, Lloyd GK. Effects of the nicotinic acetylcholine receptor agonist SIB-1508Y on object retrieval performance in MPTP-treated monkeys: comparison with levodopa treatment. Ann Neurol 43:311-317, 1998.
12. Lange KW, Kornhuber J, Riederer P. Dopamine/glutamate interactions in Parkinson's disease. Neurosci Biobehav Rev 21:393-400, 1997.
13. Chase TN, Oh JD. Striatal dopamine- and glutamate-mediated dysregulation in experimental parkinsonism. Trends Neurosci 23(suppl.):S86-S91, 2000.
14. Verhagen Metman L, Del Dotto P, Van den Munckhof P, Fang J, Mouradian MM, Chase TN. Amantadine as treatment for dyskinesia and motor fluctuations in Parkinson's disease. Neurology 50:1323-1326, 1998.
15. Verhagen Metman L, Del Dotto P, Natte R. Van den Munckhof P, Chase TN. Dextromethorphan improves levodopa-induced dyskinesias in Parkinson's disease. Neurology 51:203-206, 1998.
16. Moreau J-L, Huber G. Central adenosine A2A receptors: an overview. Brain Res Rev 31:65-82, 1999.

17. Kanda T, Jackson MK, Smith LA, Pearce RK. Nakamura J, Kase H, Kuwana Y, Jenner P. Adenosine A2A antagonist: a novel antiparkinson agent that does not provoke dyskinesia in Parkinson monkeys. Ann Neurol 43:507-513, 1998.
18. Grondin R, Bedard PJ, Tahar AH, Gregoire L, Mori A, Kase H. Antiparkinsonian effect of a new selective adenosine A2A receptor antagonist in MPTP-treated monkeys. Neurology 52:1673-1677, 1999.
19. Grondin R, Tahar AH, Doan VD, Ladure P, Bedard PJ. Noradrenergic antagonism with idazoxan improves L-dopa-induced dyskinesias in MPTP monkeys. Naunyn-Schmiedeberg's Arch Pharmacol 361:181-186, 2000.
20. Gomez-Mancilla B, Bedard PJ. Effects of nondopaminergic drugs on L-dopa-induced dyskinesias in MPTP-treated monkeys. Clin Neuropharmacol 16:418-427, 1993.
21. Hughes NR, McKnight AT, Woodruff GN, Hill MP, Crossman AR, Brotchie JM. Kappa-opioid receptor agonists increase locomotor activity in the monoamine-depleted rat model of parkinsonism. Mov Disord 13:228-233, 1998.
22. Baron MS, Wichmann T, DeLong MR. Inactivation of the sensorimotor territory in the internal pallidum reverses parkinsonian signs in MPTP-treated monkeys. Soc Neurosci Abstr 18:693, 1992.
23. Penn RD, Kroin JS, Reinkensmeyer A, Corcos DM. Injection of GABA-agonist into globus pallidus in patient with Parkinson's disease. Lancet 351:340-341, 1998.

NEUROPROTECTION IN PARKINSON'S DISEASE

L.V. Prasad Korlipara[*] and Anthony H.V. Schapira[*,**]

1. INTRODUCTION

The development of strategies to slow down or reverse the progression of Parkinson's disease (PD) presents a major therapeutic challenge to neuroscientists and clinicians. Pathologically PD is characterised by the degeneration of dopaminergic neurones of the substantia nigra pars compacta and Lewy bodies, eosinophilic inclusions in the surviving neurones. Dopaminergic neuronal degeneration causes a depletion of dopamine in the corpus striatum, upon which the dopaminergic neurones project. Treatments for Parkinson's disease can be considered as symptomatic, neuroprotective or restorative. The repletion of endogenous dopamine with levodopa yields dramatic symptomatic benefit and has remained the mainstay of treatment. Its long term use, however, is beset by reduced efficacy, disabling side effects and concerns that levodopa may itself be neurotoxic. Restorative therapies restore neurochemical or clinical deficits after neuronal death has occurred. An example of such an approach includes cell replacement with fetal nigral neurones. This chapter will focus on neuroprotective strategies, which protect vulnerable or sick neurones from the toxic or biochemical events that cause neuronal death and slow the progression of the disease. Rational approaches to neuroprotection require an understanding of aetiology, pathophysiology and the common pathways that result in neuronal death.

2. AETIOLOGY OF PARKINSON'S DISEASE

It is becoming increasingly apparent that PD has multiple environmental and/ or genetic aetiologies which result in common clinical, pathological, neurochemical and biochemical endpoints. The Parkinsonian syndrome induced in users of 1-methyl, 4-phenyl tetrahydropyridine (MPTP), as well as the association with pesticides, herbicides and rural living stimulated interest in environmental causes. The increased incidence of

[*] University Department of Clinical Neurosciences, Royal Free and University College Medical School, and [**]Institute of Neurology, University College London, London, NW3 2PF, UK

Mapping the Progress of Alzheimer's and Parkinson's Disease
Edited by Mizuno *et al.*, Kluwer Academic/Plenum Publishers, 2002

373

PD in the first degree relatives of sufferers, and twin studies measuring nigrostriatal dysfunction using 18 fluorodopa uptake positron emission tomography (PET) suggested a genetic component. More recently mutations in the α–synuclein[1] and parkin[2] genes have been identified in families exhibiting autosomal dominant and recessive parkinsonism respectively.

3. MECHANISM OF CELL DEATH IN PD

Biochemical abnormalities have been identified in PD brain, and in animal and cellular models, that provide clues as to how genetic and environmental factors may cause cell death in PD. Analysis of post mortem PD brain show higher levels of lipid, protein and DNA oxidation products than controls[3] indicating that increased oxidative stress plays a prominent role. The enzymatic and auto-oxidative metabolism of dopamine ensures that basal levels of oxidative stress in the substantia nigra are high, possibly explaining the selective vulnerability of dopaminergic neurones in PD. The oxidation of dopamine by monoamine oxidase (MAO) yields hydrogen peroxide, which normally reacts with reduced glutathione, but can also react with iron to generate highly reactive hydroxyl ions. A 30-60% decrease in reduced glutathione is present in PD substantia nigra, whilst increased iron levels have been detected. This reduction in anti-oxidant defences, coupled with increased levels of pro-oxidants moves the balance towards increased oxidative stress. Striatal dopamine depletion may causes a secondary increase in the rate of dopamine turnover, placing further demand on anti-oxidant systems, whilst levodopa therapy, through its conversion to dopamine, may also increase free radicals.

Mitochondrial dysfunction, resulting in incomplete electron transfer across the electron transport chain results in yet further generation of free radicals. Analysis of post-mortem PD substantia nigra pars compacta demonstrated a 30-40% deficiency in complex I activity,[4] whilst MPP^+, the toxic metabolite of MPTP is a complex I inhibitor. The presence of activated microglia in PD substantia nigra, together with increased levels of a number of cytokines, including TNF-α, interferon-γ and IL-1β, suggest an inflammatory aspect, which may then lead to further free radical formation.[3] The efficacy of N-methyl-D-aspartate (NMDA) antagonists and nitric oxide synthase (NOS) inhibitors in animal models of PD,[5] and the presence of NMDA receptors and glutaminergic inputs in the substantia nigra implicates excitotoxcity in PD. The persistent overstimulation of NMDA glutamate receptors leads to massive cellular calcium influx, activation of NOS, causing free radical generation and cell death. Mitochondrial dysfunction, causing a reduction in ATP and thus a reduced cell membrane potential is expected to render the neurones vulnerable to persistent overstimulation of NMDA receptors, which exhibit a voltage dependent block at resting membrane potential.

Neuronal death in neurodegenerative diseases was traditionally considered to occur by necrosis, which involves a rapid cellular respiratory failure, massive ionic shifts, cellular swelling and cell lysis. In vitro, animal models and post mortem examination of PD substantia nigra have provided evidence of apoptotic cell death in PD, characterised by the condensation of nuclear chromatin into electron dense masses, preservation of plasma and nuclear membranes and the appearance of apoptotic bodies.[6] Free radicals, reduced transmembrane potential and excitotoxic stimuli may induce opening of a large proteinaceous pore complex, the mitochondrial membrane permeability transition pore (MPT). Pore opening commences a sequence of events that includes mitochondrial swelling, translocation of cytochrome c into the cytosol and activation of apoptotic effector mechanisms by the sequential activation of caspases.

While it is not clear which of these findings are primary or secondary phenomena., it seems likely that all contribute to the toxic environment and potentiate each other. There is no clear explanation as to how known aetiological factors produce this biochemical environment. Mutations in α-synuclein are intriguing as aggregated α-synuclein is one of the dominant constituents of Lewy bodies in idiopathic PD. A soluble unfolded protein, it has the capacity to undergo self-aggregation at high concentrations and the mutant forms increase this tendency. The ubiquitin system which targets damaged proteins for proteosomal degradation may be important, given the description of a ubiquitin carboxyl hydrolase L1 mutation in familial PD,[7] the ubiquitin protein ligase function of parkin, and the strong immunoreactivity for ubiquitin exhibited by Lewy bodies. Therefore dysfunction of α-synuclein due to genetic or acquired factors or a mishandling of protein processing, resulting in protein aggregation, remains an attractive hypothesis. How protein aggregates could produce this cycle of toxicity, however needs to be further defined although their effects on dopamine handling, proteosomal function and free radical generation may be possible mechanisms.

4. NEUROPROTECTIVE THERAPIES: GENERAL CONCEPTS

One of the major problems facing researchers in PD neuroprotection is that once the clinical symptoms of PD develop, up to 70-80% of the nigral dopaminergic neurones may be dysfunctional or have already died. The early symptomatic phase of progressive nigral degeneration, however, may still provide a significant therapeutic window for the use of agents that retard the pathological process. Ideally, risk factors for asymptomatic individuals need to be defined as well as methods available that provide surrogate markers of nigrostriatal integrity. Such techniques include PET and single photon emission computed tomography (SPECT) to evaluate striatal dopamine function. Not only will these be important for the detection of presymptomatic degeneration, they will also be valuable in assessing the response of such processes to therapeutic intervention. The kinetics of neuronal loss in PD may vary according to the aetiology. Post mortem studies indicate that there is a 5% age related loss of dopaminergic neurones per decade. PD symptoms occur when a certain threshold of nigral degeneration or striatal dopamine depletion is exceeded. In environmental causes, an acute toxic insult may cause nigral cell death followed by the normal age related degeneration until the threshold is exceeded. In genetic causes, whilst the rate of cell death is unknown, patients tend to present at an earlier age and cell death may occur at an accelerated rate. However, in idiopathic PD, post mortem data assessing nigral cell death and PET and SPECT studies indicated a greatly increased rate of decline over controls, and predicted a pre-symptomatic phase of 5 years.[8] This suggested that the disease progression was not life long, was more rapid than previously thought and did not involve an insult followed by age-related decline.

4.1. Dopamine Agonists

Dopamine agonists have long been used in the treatment of PD and theoretical reasons suggesting a neuroprotective action make this class of drugs particularly attractive in PD treatment, given their beneficial effects on symptoms. The D2 dopamine agonists, bromocriptine, pergolide, apomorphine and lisuride, and the D3 preferring D2/D3 agonists, pramipexole and ropinirole, have demonstrated in vitro and in vivo neuroprotective properties to an array of neurotoxic insults. Dopamine agonists may be protective via a number of mechanisms. Firstly, it is accepted that dopamine agonist therapy both delays the need for levodopa therapy and, for those patients already on levodopa, causes a reduction in levodopa dosage requirements for equivalent symptom control. Therefore they may diminish any neurotoxic properties of levodopa, although this is an area of controversy. Both levodopa and dopamine are markedly toxic, causing necrosis and apoptosis, to a variety of cultured cells, including neuronal and non-neuronal cells. Signs of such effects, however, have not been reproduced in humans or most animal models. It is possible that the in vivo environment may involve different local concentrations of dopamine and levels of protection by anti-oxidant enzymes, anti-apoptotic molecules, neurotrophic factors and glial cells.

Secondly, the progressive depletion of such neurones in PD causes a secondary increase in dopamine synthesis, release and metabolism from surviving neurones, increasing the free radical load in the PD substantia nigra. The agonist stimulation of dopamine autoreceptors would be expected act upon this feedback mechanism reducing the free radical load. In addition, D2 and D2/D3 receptor agonists in clinical use are all able to scavenge free radicals at specific concentrations. The importance of direct anti-oxidant effects is not known as scavenging tends to occur at relatively high concentrations, which may not be achieved in the brain, while neuroprotective effects have been observed in some models at much lower concentrations. In other dopaminergic cell models, anti-oxidant effects were reported to be the dominant mechanism independent of receptor mediated effects.[9] The local drug concentration in individual neurones achieved in the substantia nigra is obviously a critical parameter.

Carvey found that pramipexole and ropinirole, but not D2 agonists, attenuated the typical loss of tyrosine hydroxylase positive cultured rat rostral mesencephalic tegmentum (RMT) neurones that occurs over time.[10] Additionally the attenuation of levodopa toxicity by pramipexole was augmented by co-incubation with other D3 agonists and attenuated by D3 antagonists, suggesting effects mediated specifically by D3 receptors. They went on to identify a 35 kDa protein that was produced by dopaminergic mesencephalic neurones, enhanced by D3 agonists and was neurotrophic to dopaminergic neurones.[10] Pramipexole also exerts anti-apoptotic effects in primary neuronal and dopaminergic SHSY-5Y cells, upregulating anti-apoptotic Bcl-2, whilst inhibiting cytochrome c release, caspase 3 activation and α–synuclein aggregation.[11] It is possible, however, that this is merely a consequence of its other described actions.

4.2 Selegiline (deprenyl)

The rationale behind the investigation of the MAO inhibitor, selegiline, as a neuroprotective agent was due to its predicted inhibition of the enzymatic dopamine oxidation, its well documented levodopa sparing effect and the possibility that an unknown neurotoxin, like MPTP may be metabolised by MAO. Selegiline and its metabolites exert anti-apoptotic effects in animal and *in vitro* models. It had been observed that patients treated with selegiline and levodopa seemed to survive longer and accumulate less disability. In the DATATOP study, selegiline and Vitamin E, as another potential neuroprotective anti-oxidant, were given alone and in combination to 800 PD patients.[12] In this and another similar study involving 54 patients,[13] selegiline signif-icantly delayed the onset of disability requiring levodopa. However, these interpretations were questioned when it became apparent that the effect may have been due to a symptomatic effect rather than retarding disease progression. To address this issue trials have since been conducted with a drug washout period before the end-point assessment and showed selegiline to slow progression.

4.3. Other Neuroprotective Options

As the pathways leading to nigral degeneration are increasingly dissected, the number of potential targets to protect neurones from such processes increases. The presence of activated microglia and the upregulation of cytokines in PD may suggest a role for anti-inflammatory or immunomodulatory agents. Coenzyme Q10, which may attenuate mitochondrial bioenergetic dysfunction, decreased the loss of striatal dopamine and neuronal loss in MPTP treated mice and caused a trend towards increased complex I in PD patients.[14] Molecules involved in excitotoxicity pathways are further key targets. Opening of the permeability transition pore can be regulated by a number of pro-apoptotic and anti-apoptotic proteins, in particular bcl-2 and bcl-X_L, and modulation of their expression may be useful. Also, the binding of the immunosuppressant cyclosporin (CsA) to cyclophilin D potently inhibits pore opening in a number of in vitro models, while the use of CsA in animal PD models has been encouraging.

CsA and FK-506 may also exert effects by complex formation with members of highly conserved families of proteins, called immunophilins, and the interaction of the drug-immunophilin complex with target proteins, such as the calmodulin dependent phosphatase, calcineurin.[15] Two examples of proteins whose phosphorylation levels are affected by CsA and FK-506 are nitric oxide synthase and GAP-43. Phosphorylated NOS is inactive and therefore calcineurin inhibition by CsA or FK-506 would potentially protect neurones from excitotoxic mechanisms, while GAP-43 has been linked to neuronal growth and neurite extension. Accordingly, FK506 and CsA potentiated neurite extension in PC-12 cells, while immunophilin ligands exert neuroprotective and restorative effects in dopaminergic neuronal and rodent models.[16] In animal PD models there are an array of other neurotrophic factors that can protect and restore dopaminergic neurones. Viral and somatic gene transfer of glial cell line derived neurotrophic factor (GDNF) and other such factors have successfully been used in rat PD models.[17]

5. CONCLUSION

It is possible that future neuroprotective strategies will be a cocktail of agents some of those tailored to specific pathophysiology, and others which target common processes. In the clinical sphere, ways of identifying pre-symptomatic PD patients and monitoring disease progression will be of paramount importance. Several longitudinal studies using PET and SPECT to assess the effects of dopamine agonists are underway, and hopefully will further clarify our understanding of the actions of these drugs in PD. Thus progress is being made and consequently there is much hope in PD therapeutics

REFERENCES

1. M.H. Polymeropoulos, C. Lavedan, E. Leroy, et al. Mutation in the α-synuclein gene identified in families with Parkinson's disease. *Science* **276,** 2045-2047 (1997).
2. T. Kitada, S. Asakawa, N. Hattori, et al. Mutations in the parkin gene cause autosomal recessive juvenile parkinsonism. *Nature* **392,** 605-608 (1998).
3. P. Jenner, and C.W. Olanow. Understanding cell death in Parkinson's disease. *Ann. Neurol* **44,** S72-84 (1998).
4. A.H.V. Schapira, J.M. Cooper, D. Dexter, P. Jenner, J.B. Clark, and C.D. Marsden. Mitochondrial complex I deficiency in Parkinson's disease. *Lancet* **1,** 1269 (1989).
5. M.F. Beal. Excitotoxicity and nitric oxide in Parkinson's disease pathogenesis, *Ann. Neurol.* **44,** S110-S114(1998).
6. W.G. Tatton, and C.W. Olanow. Apoptosis in neurodegenerative diseases: the role of mitochondria. *Biochim. Biophys. Acta* **1410,** 195-213 (1999).
7. E. Leroy, R. Boyer, G. Auburger, et al. The ubiquitin pathway in Parkinson's disease. *Nature* **395,** 415-45 (1998).
8. D.J. Brooks. The early diagnosis of Parkinson's Disease. *Ann. Neurol.* **44,** S10-S18 (1998).
9. W.D. Le, J. Jankovic, W. Xie, and S.H. Appel. Antioxidant property of pramipexole independent of dopamine receptor activation in neuroprotection. *J. Neural. Transm.* **107,** 1165-1173 (2000).
10. P.M. Carvey, S.O. McGuire, and Z.D. Ling. Neuroprotective effects of D3 dopamine receptor agonists. *Parkinsonism Relat. Disord* **7,** 213-223 (2001).
11. J. Kakimura, Y. Kitamura, K. Takata et al. Release and aggregation of cytochrome c and alpha-synuclein are inhibited by the antiparkinsonian drugs, talipexole and pramipexole. *Eur. J. Pharmacol.* **417,** 59-6(2001).
12. I. Shoulson. DATATOP: a decade of neuroprotective inquiry. Parkinson Study Group. Deprenyl and tocopherol antioxidative therapy of parkinsonism. *Ann. Neurol.* **44,** S160-S166 (1998).
13. J.W. Tetrud, and J.W. Langton. The effect of deprenyl in the natural history of Parkinson's disease, *Science* **245,** 519-522 (1989).
14. C.W. Shults, R.H. Haas, and M.F. Beal MF. A possible role of coenzyme Q10 in the etiology and treatment of Parkinson's disease. *Biofactors* **9,** 267-272 (1999).
15. G.S. Hamilton, and J.P. Steiner. Immunophilins: Beyond immunosuppression. *J. Medicinal Chem.* **41,** 5119-43 (1998).
16. J.P. Steiner, G.S. Hamilton, D.T. Ross, et al. Neurotrophic immunophilin ligands stimulate structural and functional recovery in neurodegenerative animal models. *Proc. Natl. Acad. Sci. U. S. A.* **94,** 2019-2024(1997).
17. D.L. Choi-Lundberg, Q. Lin, T. Schallert, et al. Dopaminergic neurons protected from degeneration by GDNF gene therapy. *Science* **275,** 838-841 (1997).

SUSTAINED DOPAMINE AGONISM WITH CABERGOLINE IN PARKINSON'S DISEASE

Evolution of therapy from animal models to man

Linda S. Appiah-Kubi and K. Ray Chaudhuri[*]

1. INTRODUCTION

Cabergoline (1- [(6-allylergolin-8beta-yl)carbonyl] -1- [3-(dimethyloamino)propyl] -3- ethylurea) is an ergoline dopamine agonist with a high affinity for the dopamine D2 receptor. Cabergoline is unique among other available dopamine agonists as it has the longest mean elimination half life of 65-110 hours calculated on the basis of urinary excretion rates in healthy volunteers and parkinsonian patients.[1] Cabergoline has high selectivity for dopamine receptors, in vitro and in vivo and binds mainly to D2 receptors in vitro and to D1 receptors with affinity similar to pergolide and 3.6 times greater than bromocriptine.[2] Owing to its long half life, in clinical practice, cabergoline need only be given once a day, in comparison to other agonists which usually require three times-a-day dosing. The long duration of action of cabergoline and the resultant sustained rather than pulsatile dopamine receptor stimulation in PD, is of interest given the possible role of pulsatile dopaminergic therapy in the development of dyskinesias. This review focuses on work with cabergoline in animal models and man related to its beneficial therapeutic effects in PD.

2. PRECLINICAL STUDIES WITH CABERGOLINE

In 6-hydroxydopamine striatally lesioned rats, acute treatment with cabergoline did not produce psychotomimetic effects while lisuride (a short acting agonist) produced

[*] Linda S. Appiah-Kubi, Guy's King's & St Thomas' Medical School, London, UK. K Ray Chaudhuri, Regional Movement Disorders Unit, University Department of Neurology, King's College Hospital, University Hospital Lewisham, and Guy's King's & St Thomas' Medical School, London, UK.

Mapping the Progress of Alzheimer's and Parkinson's Disease
Edited by Mizuno *et al.*, Kluwer Academic/Plenum Publishers, 2002

379

significant stereotyped behaviour. Chronic treatment with cabergoline but not lisuride, upregulated striatal dopamine receptors and did not affect apomorphine induced sniffing.[3] In levodopa primed cynomolgus monkeys with severe dyskinesias, chronic cabergoline therapy improved parkinsonism and dyskinesias,[4] while in drug naïve monkeys, cabergoline therapy showed a strong motor response and reversal of increased expression of delta-fosB proteins induced by dopaminergic denervation.[5] In another study, comparing bromocriptine, pergolide and cabergoline in levodopa naïve monkeys, excitability, irritability and aggressiveness (measures of hyperkinesia) were not seen after cabergoline at therapeutic doses (0.05 – 0.2 mg/kg, sc).[6] These studies, therefore, support the notion that the sustained dopaminergic agonism induced by cabergoline may be beneficial for levodopa-induced dykinesias, while providing optimal anti-parkinsonian effects.

3. NEUROPROTECTION AND CABERGOLINE

Recent evidence in animal models suggests that dopamine agonists may have neuroprotective or neuro-rescuing properties although there are, as yet, no convincing studies to prove neuroprotection by dopamine agonists in humans. Increased oxidative stress, secondary to generation of toxic free radicals, and apoptosis are implicated in the process of neurodegeneration. In PC12 cells, levodopa administration leads to apoptosis in up to 75% cells and these changes can be reversed by free radical scavengers.[7] Cabergoline, in addition to bromocriptine and pergolide, can scavenge free radicals and nitiric oxide and thus protect cells from levodopa induced apoptosis. Repeated administration of cabergoline also raises striatal glutathione levels in mice. Sander et al have reported increased (33%)-tyrosine hydroxylase positive cell survival after cabergoline pre-treatment in 6-hydroxydopamine treated rats.[8] These results suggest that cabergoline therapy has the potential for neuroprotection in vivo.

4. CLINICAL STUDIES: ADJUNCTIVE AND MONO THERAPY

Several placebo-controlled, open trials with adjunctive cabergoline in advanced PD patients with levodopa induced fluctuations showed that cabergoline (a) significantly redued percentage of "off" time and improved motor function and (b) reduced levodopa requirement.[9, 10] Cabergoline, is the only dopamine agonist apart from ropinirole to have been examined in a five year levodopa-controlled, multi-centre double-blind randomised monotherapy trial in 419 drug naïve PD patients. In spite of restriction of cabergoline dose to ≤ 4 mg/day (in clinical practice doses up to 6 or 8 mg are often used), results are similar to the ropinirole study and indicate significant improvement in motor scores and a reduction (5.3% vs 31.9% with levodopa) in all aspects of fluctuations and dyskinesias in those maintained on cabergoline alone at five years (76 out of 211).[11, 12] In particular, the percentage of early morning akinesia (1.3%) was significantly reduced in those receiving cabgeroline monotherapy in comparison to those on levodopa monotherapy (5.5%).[13] This study indicated that cabergoline shares a similar side effect profile to levodopa and in particular, rates of hallucination were not different between cabergoline and levodopa.

4.1 Comparison With Other Dopamine Agonists

Trials comparing bromocriptine and cabergoline suggested that cabergoline (up to 6 mg/day) improved motor scores similar to bromocriptine (5-40 mg/day) given thrice a day.[14, 15] Proportional duration of "off" periods was reduced to a greater extent in the cabergoline treated group, and improvements in Clinical Global Impression scores persisted throughout a three month period of stable-dose treatment.[14, 15] Work from our group, auditing the effect of (a) cabergoline and pergolide and (b) cabergoline and controlled release levodopa on nocturnal symptoms in PD in an unselected group of patients suggested that evening dosing with cabergoline led to better subjective reports of sleep, improved early morning dystonia, nocturnal akinesia and restless legs type symptoms in comparison to pergolide or controlled release levodopa. This observation has been supported by Ulm *et al* (1999) in a German multi-centre study (the MODAC study) comparing daytime effects of cabergoline and pergolide.[16] This cross-over study involving 43 patients investigated the two agonists on the basis that their similar binding affinities for D2 and D1 receptors could reveal differences in pharmacokinetic properties, such as the difference in half-life (65 hours vs 7-16 hrs respectively). Cabergoline was better tolerated with respect to all adverse events, and in particular produced a statistically significant improvement in daytime dyskinesias ($p < 0.05$). Furthermore, the once daily dosing was preferred by most patients and domperidone requirement (to prevent nausea) was infrequent.[16]

4.2 Usefulness in PD-related Sleep Problems and Restless Legs Syndrome

Using a newly developed visual analogue scale for detecting nocturnal disabilities in PD (Parkinson's Disease Sleep Scale, PDSS), we have reported a range of motor and urinary problems causing sleep disruption in PD.[17] Studies indicate that the long half life of cabergoline allows the drug to be efficacious throughout the night, particularly when given at night-time. The resultant improved sleep also has a beneficial effect on daytime activity and in some patients, successful treatment of night-time problems improve excessive daytime sleepiness. We have also compared the excessive daytime sleepiness scores (using Epworth Sleepiness Scale) in three groups of patients on cabergoline, pramipexole and levodopa mono and combination therapy.[17] Cabergoline therapy does not appear to be associated with excessive daytime sleepiness which may predispose susceptible patients to sudden daytime dozing. We and others have also reported the beneficial effect of cabergoline on idiopathic and PD related RLS and in particular its efficacy in RLS patients exhibiting levodopa related augmentation.[18, 19, 20] Actigraphic studies, monitoring motor activity via a wrist-worn acceleration sensor, also demonstrate improved motor symptoms, associated with better sleep after cabergoline treatment in PD.[21]

4.3 Dual Agonist Therapy

In some patients, severe levodopa induced diphasic and inter-dose dyskinesias may need surgical treatment. In those, not suitable for surgery there is limited therapeutic option and we have developed the use of daytime apomorphine infusion and evening dosing with cabergoline with up to 80% reduction in oral levodopa therapy in these patients. The strategy is tolerated by most patients and leads to up to 60% reduction in troublesome dyskinesias. In some, levodopa can be stopped completely. There appears to be little problem with tolerance at 18 months follow up although the procedure is labour intensive and costly.

4.4 Tolerability in the Young and Elderly

Dopamine agonists are often avoided in the elderly because of side effects. We have recently reported our experience of treating over 200 PD subjects with an age range from 39-88 years with cabergoline over a mean follow up period of two years (range 0.5 –12 years). Cabergoline has been used as monotherapy in the young (24 of 92 patients under 65 years) and the elderly (17 of 110 patients 65 years and over) and also as adjunctive therapy with daily doses of up to 8 mg. Using the Shulman et al[23] criteria for tolerability (sustained use of the agonist for a minimum of 6 months) we have found cabergoline to have excellent (>80%) and comparable tolerability in the young and the elderly, even in those who were intolerant to other agonists.[22] A rapid switchover from other agonists is possible. The tolerability rates in the elderly (>75 yrs old), including monotherapy, are markedly better than bromocriptine, pergolide, pramipexole or ropinirole as based on the study reported by Shulman et al.[23]

5. SIDE EFFECTS

The side effect profile of cabergoline appears to be similar to other ergoline agonists. In fact, monotherapy trials suggest that rates of hallucination and nausea with cabergoline are comparable to levodopa.[11] Our studies also indicate that sleepiness is not a major problem with the use of cabergoline and as such it appears to be excellently tolerated by the elderly. Prolonged use may lead to ankle swelling and daytime fatigue which may be dose dependent. As cabergoline is an ergoline product, therefore, high dose long term therapy may require monitoring for fibrosis although we have never come across this problem.

6. SUMMARY

Cabergoline is an unique ergoline dopamine agonist providing oral sustained dopaminergic stimulation in PD. In theory, this is the most physiological method of replacing intra-synaptic striatal dopamine which in turn leads to lesser chance of developing dyskinesias. Pre-clinical and clinical studies with cabergoline monotherapy in PD support this observation. In addition, our work suggests that cabergoline is beneficial for nighttime motor problems in PD and tolerated well by PD patients of all age groups.

ACKNOWLEDGEMENTS

We thank S Pal and V Singh-Curry for assisting with studies and use of the PD sleep scale, Alison Forbes, PD nurse specialist, for collecting patient information, Kevin Bridgman from Pharmacia for supporting database and publication of sleep scale. KRC has received honorarium from Pharmacia, Glaxo-SmithKline, Britannia and Athena pharmaceuticals for acting as advisor and guest lectures.

REFERENCES

1. H. M. Brecht, A comparison of dopamine agonists. *Aktuel Neurol*, **25**, S310-S316 (1998).
2. R. G. Fariello, Pharmacodynamic and pharmacokinetic features of cabergoline. Rationale for use in Parkinson's Disease, *Drugs* **55** (suppl 1), 10-16 (1998).
3. V. Pedersen, K. Double, W. J. Schmidt, P. Riederer, and M. Gerlach, Effect of cabergoline and lisuride on motor behaviour and dopamine receptor binding in 6-OHDA-lesioned rats, *J Neural Transm* **106**(1), (1999) (abstract).
4. N. Arai, M. Isaji, M. Kojima, E. Mizuta, and S. Kuno. Combined effects of cabergoline and L-dopa on parkinsonism in MPTP-treated cynomolgus monkeys. *J Neural Transm* **103**(11), 1307-16 (1996).
5. J. P. Doucet, Y. Nakabeppu, P. J. Bedard, B. T. Hope, E. J. Nestler, B. J. Jasmin, J. S. Chen, M. J. Iadarola M. St-Jean, N. Wigle, P. Blanchet, R. Grondin, and G. S. Robertson. Chronic alterations in dopaminergic neurotransmission produce a persistent elevation of deltaFosB-like protein(s) in both the rodent and primate striatum, *Eur J Neurosc.* **8**(2), 365-81 (1996).
6. N. Arai, M Isaji, H. Miyata et al. Differential effects of three dopamine receptor agonists in MPTP-treated monkeys. *J Neural Transm.* **10**, 55-62 (1995).
7. K. Koshimura, J. Tanaka, Y. Murakami and Kato Y, Effects of dopamine and L-DOPA on survival of PC12 cells, *J Neurosc Res.* **62**(1), 112-9 (2000).
8. J. Sander, R. Dengler, and P. Odin. Neuroprotective effects of the dopamine agonist cabergoline in the rat 6-OHDA Parkinson model, Preclinical data on dopamine agonists, in: *Focus on Medicine No. 14,* edited by T. Chase and P. Bedard, (Blackwell Science, Oxford, 1999), pp. 7-13.
9. M.J. Steiger, T. El Debas, T. Anderson , L. J. Findley, and C. D. Marsden, Double blind study of the activity and tolerability of cabergoline versus placebo in parkinsonians with motor fluctuations, *J Neurol* **243**(1), 68-72 (1996).
10. J. T. Hutton, W. C. Koller,J. E. Ahlskog, R. Pahwa, H. I. Hurtig, M. B. Stern, B. C. Hiner, A. Lieberman, R. F. Pfeiffer, R. L. Rodnitzky, C. H. Waters, M. D.Muenter, C. H. Adler, and J. L. Morris, Multicentre placebo-controlled trial of cabergoline taken once daily in the treatment of Parkinson's Disease. *Neurology* **46**, 1062-65 (1996).
11. U.K. Rinne, F. Bracco, C. Chouza et al. Early treatment of Parkinson's Disease with cabergoline delays the onset of motor complications . Results of a double-blind levodopa controlled trial. *Drugs.* **55** (suppl 1), 23-30 (1998).
12. O Rascol, DJ Brooks, AD Korczyn, PP De Deyn, CE Clarke and AE Lang. A five-year study of the incidence of dyskinesia in patients with early Parkinson's disease who were treated with ropinirole or levodopa. 056 Study Group, *N Engl J Med* 18:**342**(20), 1484-1491 (2000).
13. B. Musch. The long-acting dopamine agonist cabergoline in the treatment of early Parkinson's Disease. Presented at the 5th International Conference on Progress in Alzheimers, Kyoto, Japan (2001) p117 (abstract).
14. C. D. Marsden. Clinical experience with cabergoline in patients with advanced Parkinson's Disease treated with levodopa. *Drugs* **55** (suppl 1), 17-22 (1998).
15. O Gershanik, R Dom, J Tichy et al. Multicentre multinational double-blind study of the activity and tolerability of cabergoline vs bromocriptine in parkinsonian patients suffering from L-dopa associated motor complications, not on treatment with dopamine agonist agents. Pharmacia & Upjohn, Peapack, NJ, FCE report 21336/739i, (1994).
16. G. Ulm, and P. Schuler, on behalf of the MODAC Study Group. Cabergoline versus Pergolide a video-blinded, randomised, multicentre cross-over study. *Akt Neurologie.* **25**, 360-365(1999).
17. S. Pal, KF Bhattacharya, C. Agapito, K. Ray Chaudhuri. A study of excessive daytime sleepiness and its clinical significance in three groups of Parkinson's disease patients taking pramipexole, cabergoline and levodopa mono and combination therapy, *J Neural Transm.***108**(1), 71-77 (2001).
18. K. Ray Chaudhuri, K. Bhattacharya, C. Agapito, M. C. Porter, J. Mills and C. Clough. The use of cabergoline in nocturnal parkinsonian disabilities causing sleep disruption: a parallel study with controlled-release levodopa. *Eur J Neurol.* **6** (suppl), S11-S15 (1999).
19. Ghatani T, Agapito C, Bhattacharya K, Clough C, Chaudhuri K Ray. Comparative audit of pergolide and cabergoline therapy in the treatment of nocturnal "off" periods causing sleep disruption in Parkinson's disease. *Eur J Neurol.* **8**(suppl 1):8-11(2001)

20. K.Stiasny, J. Robbecke, P. Schuler and W. H. Oertel. Treatment of idiopathic restless legs syndrome (RLS) with the D2-agonist cabergoline--an open clinical trial. *Sleep.* **23**(3), 349-54 (2000).
21. S. Katayama, T. Miyamoto, M. Tatsumoto and K. Watanabe. Averaging actigraphic analysis of diurnal motor fluctuations during dopamine agonist therapy. (abstract) Presented at the International Congress of European Federation of Neurological Societies Copenhagen, Denmark. *Eur J Neurol* 7 (suppl 3) S75. (2000).
22. K. Ray Chaudhuri, L. S. Appiah-Kubi, S. Pal, A. Forbes and D. Gunarwardena. The efficacy and tolerability of sustained dopamine agonism using cabergoline in young and elderly Parkinson's Disease patients with nocturnal disability, (abstract P-11-2) Presented at the 5th International Conference on progress in Alzheimer's and Parkinson's Disease, Kyoto, Japan (2000) p28.
23. Shulman LM, Minagar A, Rabinstein A, Weiner WJ The use of dopamine agonists in very elderly patients with Parkinson's disease, *Mov Disord* **15**(4), 664-668 (2000).

SLEEPINESS AND SLEEP ATTACKS IN PARKINSON'S DISEASE

Anette V. Nieves and Anthony E. Lang[*]

1. INTRODUCTION

The publication from Frucht et al[1] describing "sleep attacks" brought to the neurology community's attention the potential risks of abnormal sleepiness in Parkinson's disease. Following this publication drastic measures were taken in many countries throughout the world to curtail driving privileges in patients treated with the new D2/D3 receptor agonists, pramipexole and ropinirole. Physicians were often asked to report patients taking these medications to their respective driving and vehicle licensing authorities and in some countries patients taking these drugs were not permitted to drive regardless of whether they had experienced "sleep attacks" or not. However, recent reports have shown that these medications used early in the disease are associated with a lower incidence of dyskinesias than levodopa.[2, 3] These are the very individuals who are most likely to require driving privileges to maintain a normal or highly functional lifestyle. But how common are sleepiness and sleep attacks in Parkinson's disease? What causes these disturbances in this population and how can we treat them?

2. NORMAL SLEEP IN THE ELDERLY

Five percent of the general population may experience excessive daytime sleepiness (EDS).[4] Sleepiness may be missed by a clinician since the most common

[*] Anette V. Nieves and Anthony E. Lang, Morton and Gloria Shulman Movement Disorders Centre, Toronto Western Hospital, MP-11, 399 Bathurst Street, Toronto, Ontario, Canada M5T 2S8.

Mapping the Progress of Alzheimer's and Parkinson's Disease
Edited by Mizuno *et al.*, Kluwer Academic/Plenum Publishers, 2002

385

complaints patients have are loss of energy, fatigue, lethargy, weariness, lack of initiative, memory lapses, or difficulty concentrating rather than sleepiness per se.[4]

The elderly as a group have been found to be sleepier than other age groups. Aging is associated with a decrease in slow wave sleep (SWS), i.e., stages 3 and 4.[5, 6] Although the latency of the first rapid eye movement (REM) sleep decreases, total REM sleep is maintained.[5, 6] Transient arousals, which may consist of brief episodes of alpha rhythm in sleep lasting between 2 and 15 seconds, are common. These may be partly due to sleep-related respiratory disturbances and periodic leg movements.[5] There is a known biphasic pattern of objective sleep in healthy young adults and elderly, which occurs from 2am to 6am and from 2pm to 6pm.[4] This may result in naps, given the opportunity, during the afternoon. Sedation is also a common side effect of some of the medications that the elderly are commonly exposed to, therefore contributing to sleepiness.

3. SLEEP IN PARKINSON'S DISEASE

Sleep disturbances are extremely common in PD.[7] In a controlled series of 78 PD patients, 66.6% (vs 53.5% in the control group) were found to have problems initiating sleep while 88.5% (vs 74.4% in the control group) had trouble maintaining sleep.[8] In another study,[9] although there was no difference for initiating sleep, female patients experienced more difficulty with sleep maintenance (87.5%) and excessive dreaming (68.4%) than men (64% and 31.6% respectively).

Sleep fragmentation is common with up to 40% of the night spent awake.[10, 11] Sleep latency and the number of awakenings increase with increasing severity of parkinsonian symptoms while awake.[11, 12] The amount of stage 1 sleep can be increased and amounts of SWS and REM sleep decreased.[11] Patients with hypertonia and rigidity tend to have less REM sleep[11] and those with nocturnal hallucinations have markedly decreased REM sleep and total sleep time.[13]

Tremor usually disappears with the onset of stage 1 sleep, but may reappear in stages 1 and 2, during sleep stage changes for 15 seconds or less, during bursts of rapid eye movements and shortly before or after REM sleep.[11] Although the amplitude of the tremor tends to be half of the waking tremor,[11] it is sometimes enough to wake the patient. The table details the common causes of sleep disturbances in patients with PD.

4. EXCESSIVE DAYTIME SLEEPINESS AND SLEEP ATTACKS

EDS is inappropriate and undesirable sleepiness during waking hours. It can be measured objectively by correlation with the Multiple Sleep Latency Test (MSLT) or subjectively with the Epworth Sleepiness Scale (ESS), which has been validated with the MSLT. EDS must be differentiated from feeling sleepy all day without episodes of inappropriate sleep due to other causes (e.g., depression). Fatigue, a common and often disabling feature of PD, should also be excluded.

A questionnaire study of 78 PD patients and 43 healthy elderly control subjects[8] showed that spontaneous dozing was more common in PD patients but daytime napping was equal in both groups. Tandberg et al[14] found that PD patients (15.5%) experienced more EDS,

Table. Causes of Sleep Disturbances in Parkinson's Disease (PD)

PD-related	Non-PD-related
▪ Tremor	▪ Age-related sleep disturbances
▪ Difficulty turning / Stiffness	▪ Nocturia (may also be PD related)
▪ Dystonia / Pain	▪ Coincidental sleep disorders (e.g.,
▪ Akathisia	sleep apnea)
▪ Depression / Anxiety	▪ Co-existing medical and psychiatric
▪ RLS[a], PLMS[b]	conditions (including dementia)
▪ REMBD[c]	▪ Other medications (e.g.,
	antidepressants, anxiolytics, drugs
Treatment-related	which might reduce metabolism of
▪ Nocturnal myoclonus	other sedating drugs)
▪ Dyskinesias	
▪ Alerting effects of drugs	
▪ Vivid dreams / Nightmares	

RLS[a] = Restless Leg Syndrome, PLMS[b] = Periodic Leg Movements of Sleep, REMBD[c] = Rapid Eye Movement Behavior Disorder

defined as falling asleep 3 times or more during the day or a total daytime sleep of more than 2 hours, than healthy controls (1%) and patients with diabetes (4%). Those with EDS were found to have a higher parkinsonian staging, more disability as reflected on the motor subscale of the Unified Parkinson's Disease Rating Scale (UPDRS), longer duration of levodopa treatment, more depressive symptoms and presence of hallucinations, and a lower score on the mini-mental status examination. There was no significant difference between patients with EDS and controls regarding nocturnal sleep problems and the use of sleeping medications. This correlated with a recent multicenter Canadian study restricted to patients who were nondemented and highly functional (off-period Schwab and England ADLS scores ≥ 70%),[15] which still showed that a higher ESS score correlated with the severity of the disease as evaluated by the Hoehn and Yahr (H&Y) stage.

Another study using MSLT showed no significant difference between PD patients and controls, although there was a considerable variation between subjects in the PD group.[16] Six out of the 27 PD patients had sleep onset REM (SOREM) episodes. These patients were significantly sleepier, despite exhibiting longer total sleep time, higher sleep efficiency, and shorter nocturnal sleep latencies than the rest of the PD patients. They also had less SWS and longer disease duration. None reported cataplexy. The MSLT result did not correlate with average daily levodopa intake.

A sleep attack, as defined by Frucht et al[17] is "an event of overwhelming sleepiness that occurs without warning, or that occurs with a prodrome that is sufficiently short or overpowering to prevent the patient from taking appropriate protective measures." They reported 8 patients who experienced "sleep attacks" while taking pramipexole, 1 who also experienced them on ropinirole. This report has created considerable controversy. It is not clear whether these episodes represent true abrupt onset sleep or are really due to EDS in patients ignoring the premonitory warning signs. It is unclear whether these are truly a new phenomenon, experienced largely or exclusively with the new dopamine agonists (possibly due to the D3 agonist properties) or whether they are a nonspecific consequence of the broad group of dopaminergic drugs used in PD.

Dopamine D1 receptor agonists produce behavioral arousal and suppression of REM sleep, while antagonists produce sedation and increase the amount and duration of REM sleep. Low doses of dopamine D2 receptor agonists induce sleep and increase SWS. On the other hand, high doses may reduce total sleep time and suppress REM sleep.[11] Recent studies have revealed that EDS is not exclusive to the non-ergot dopamine agonists, but is also commonly seen with other dopaminergic drugs including levodopa, apomorphine, and ergot-derived dopamine agonists.[15, 18-22]

The term "sleep attacks" was originally used in patients with narcolepsy but later abandoned when it was found that patients were not falling asleep without an initial latency or prodrome period.[23, 24] It is known that people may fall asleep without recalling the prodrome of feeling sleepy.[25, 26] Studies with electroencephalogram have detected evidence of initial stages of sleep even before the patient has the perception of falling asleep.[26] Most investigators believe that the majority of patients experiencing "sleep attacks" and falling asleep in inappropriate circumstances, including driving a car, are really experiencing EDS, ignoring the warning sleepiness and then not remembering it upon awakening. However, there is probably a small subgroup of PD patients with extremely short sleep latencies[27] or even SOREM, as shown by Rye et al.[16] Future studies will have to separate these and other possible subgroups in order to determine the role of specific medications or classes of medications. Other predisposing factors (e.g., pre-existing sleep disorders such as sleep apnea and HLA-DR2/DQw1) also have to be ruled out.

The multicenter Canadian study[15] showed that inappropriate falling asleep (as opposed to EDS) was not associated with a higher H&Y stage or score for part 1 of the UPDRS. The ESS score had adequate sensitivity for predicting prior episodes of falling asleep while driving and its specificity was increased by the use of an "Inappropriate Sleep Composite Score." The presence of features of other sleep disorders (e.g., sleep apnea, REMBD, PLMS) did not predict the risk of falling asleep at the wheel either. There was no association between a high ESS score or falling asleep at the wheel and the total equivalent dopaminergic dose. In a different study,[22] a high score on a modified ESS, which inquired about falling asleep without warning, correlated with the standard ESS in patients with PD and there was no difference between different dopaminergic medications.

Of 420 drivers in the multicenter Canadian survey,[15] only 16 had experienced sudden onset sleep (SOS) while driving since the onset of PD. Only 3 denied preceding symptoms of drowsiness or other warnings when specifically asked about them. The 3 patients attributed the event to alteration in their medications, although it did not correlate to any specific dopaminergic drug.

5. RECOMMENDATIONS

Initial assessment must include a careful history and physical exam. The physician should inquire about symptoms of EDS and SOS routinely especially when starting, titrating or tapering a dopaminergic drug. Special attention should be given to the use of other medications that may cause drowsiness. It is mandatory to educate patients and their family members about the potential sedating effects of antiparkinsonian medications and warn them not to drive or participate in similar dangerous activities if they are experiencing EDS.

Driving should be formally restricted in patients who are experiencing clear warning signs such as dozing or SOS in inappropriate circumstances. Antiparkinsonian medication should be adjusted and an attempt should be made to improve the quality of nighttime sleep. Light or fragmented sleep sometimes caused by parkinsonian symptoms may improve with dopaminergic therapy[10, 28] including a longer acting dopamine agonist or controlled-release levodopa preparation given in the evening or before bed. On the other hand, high doses of dopaminergic drugs may cause insomnia.

Psychiatric disturbances that may be contributing to the sleep disturbance should be ruled out. Management of depression, anxiety, or hallucinations may improve the patient's sleep complaints. The potential contribution of autonomic dysfunction, particularly orthostatic hypotension (post-prandial and other), to EDS should be evaluated and treated appropriately.

If the problem persists the patient should be evaluated in greater detail with polysomnographic studies. This may disclose evidence for a primary sleep disorder that could be contributing to EDS. Lacking other treatable conditions or if reduction of causative dopaminergic drugs results in a disabling increase in parkinsonism then consideration should be given to the concomitant use of alerting drugs. Modafinil has been recently released for the treatment of narcolepsy. The exact mechanism of action of this drug is uncertain, however it does seem to be better tolerated than older stimulants and has a low potential for abuse.[29] This drug has been reported to be safe and effective in patients with PD up to doses of 200mg/day.[30, 31] Our experience indicates that modafinil is well tolerated in doses up to 600 mg/day (mean 172mg/day) (unpublished data) improving somnolence measured on the ESS. Some patients who had been unable to increase the dose of dopamine agonists due to predictable somnolence were able to do so while taking modafinil. Further larger double-blind, placebo controlled studies are needed to confirm these observations. Studies are also needed to assess modafinil in the treatment of fatigue associated with PD given reports of efficacy for fatigue associated with multiple sclerosis.[32, 33]

REFERENCES

1. S Frucht, JD Rogers, PE Greene, MF Gordon, and S Fahn, Falling asleep at the wheel: motor vehicle mishaps in persons taking pramipexole and ropinirole. *Neurology* 52(9), 1908-1910 (1999).
2. O Rascol, DJ Brooks, AD Korczyn, PP De-Deyn, CE Clarke, AE Lang, A five-year study of the incidence of dyskinesia in patients with early Parkinson's disease who were treated with ropinirole or levodopa. 056 Study Group. *NJEM* 342(20), 1484-1491 (2000).
3. Parkinson Study Group, Pramipexole versus levodopa in early Parkinson's disease: A randomized controlled trial. *JAMA* 284(15), 1931-1938 (2000).
4. T Roth, TA Roehrs, MA Carskadon, and WC Dement, Daytime Sleepiness and Alertness. In: Principles and practice of sleep medicine, edited by MH Kryger, T Roth, and WC Dement (WB Saunders Co, Philadelphia, 1994), pp. 40-49.
5. D Bliwise, Normal Aging. In: Principles and practice of sleep medicine, edited by MH Kryger, T Roth, and WC Dement (WB Saunders Co, Philadelphia, 1994), pp. 26-39.
6. PK Pal, S Calne, A Samii, and JAE Fleming, A review of normal sleep and its disturbances in Parkinson's disease. *Parkinsonism Relat. Disord.* 5, 1-17 (1999).
7. AJ Lees, NA Blackburn, and VL Campbell, The nighttime problems of Parkinson's disease. *Clin Neuropharmacol.* 11, 512-519 (1988).
8. SA Factor, T McAlarney, JR Sanchez-Ramos, and WJ Weiner, Sleep disorders and sleep effect in Parkinson's disease. *Mov. Disord.* 5(4), 280-285 (1990).
9. JJ van Hilten, M Wegge,am, EA van der Velde, GA Kerkhof, JG van Dijk, RAC Roos, Sleep, excessive daytime sleepiness and fatigue in Parkinson's disease. *J. Neural. Transm.* 5, 235-244 (1993).
10. P Bergonzi, C Chiurulla, D Gabi, G Mennini, and F Pinto, L-dopa plus dopa decarboxylase inhibitor: sleep organization in Parkinson's syndrome before and after treatment. *Acta Neurol. Belgium* 75, 5-10 (1975).

11. M Aldrich, Parkinsonism. In: Principles and practice of sleep medicine, edited by MH Kryger, T Roth, and WC Dement (WB Saunders Co, Philadelphia, 1994), pp. 783-789.

12. A Friedman, Sleep pattern in Parkinson's disease. *Acta Med. Pol.* 21, 193-199 (1980).

13. CL Comella, CM Tanner, and RK Ristanovik, Sleep disturbances and hallucinations in Parkinson's disease: results of quantitative sleep studies. *Ann. Neurol.* 30, 295 (1991).

14. E Tandberg, JP Larsen, and K Karlsen, Excessive Daytime Sleepiness and Sleep Benefit in Parkinson's Disease: A Community-Based Study. *Mov. Disord.* 14(6), 922-927 (1999).

15. AE Lang, DE Hobson, W Martin, and J Rivest, Excessive Daytime Sleepiness and Sudden Onset Sleep in Parkinson's Disease: A Survey from 18 Canadian Movement Disorders Clinics. *Neurology* 56 (8, Suppl 3), A307 (2001).

16. DB Rye, DL Bliwise, B Dihenia, and P Gurecki, Daytime sleepiness in Parkinson's disease. *J. Sleep Res.* 9, 63-69 (2000).

17. SJ Frucht, PE Greene, and S Fahn, Sleep Episodes in Parkinson's Disease: A Wake-Up Call. *Mov. Disord.* 15(4), 601-603 (2000).

18. S Pal, KF Bhattacharya, C Agapito, and KR Chaudhuri, A study of excessive daytime sleepiness and its clinical significance in three groups of Parkinson's disease patients taking pramipexole, cabergoline and levodopa mono and combination therapy. *J. Neural. Transm.* 108, 71-77 (2001).

19. JJ Ferreira, M Galitzky, JL Montastruc, and O Rascol, Sleep attacks and Parkinson's disease treatment. *Lancet* 355, 1333-1334 (2000).

20. AHV Schapira, Sleep attacks (sleep episodes) with pergolide. *Lancet* 355, 1332-1333 (2000).

21. JJ Ferreira, C Thalamas, JL Montastruc, A Castro-Caldas, and O Rascol, Levodopa monotherapy can induce "sleep attacks" in Parkinson's disease patients. *J. Neurol.* 248, 426-427 (2001).

22. CC Sanjiv, M Schulzer, E Mak, J Fleming, WR Martin, T Brown, SM Calne, J Tsui, AJ Stoessl, CS Lee, and DB Calne, Daytime somnolence in patients with Parkinson's disease. *Parkinsonism Relat. Disord.* 7(4), 283-286 (2001).

23. P Clarenbach, Parkinson's disease and sleep. *J. Neurol.* 247(Suppl 4), IV/20-23 (2000).

24. CW Olanow, AHV Schapira, and T Roth, Waking Up to Sleep Episodes in Parkinson's Disease. *Mov. Disord.* 15(2), 212-215 (2000).

25. LA Reyner and JA Horne, Falling asleep whilst driving: are drivers aware of prior sleepiness? *Int. J. Legal Med.* 111(3), 120-123 (1998).

26. A Itoi, R Cilveti, M Voth, Can Drivers Avoid Falling Asleep at the Wheel? Relationship Between Awareness of Sleepiness and Ability to Predict Sleep Onset. In: AAA Foundation for Traffic Safety (Washington, DC, 1993).

27. F Tracik and G Ebersbach, Sudden Daytime Sleep Onset in Parkinson's Disease: Polysomnographic Recordings. *Mov. Disord.* 16(3), 500-506 (2001).

28. JJM Askenasy and MD Yahr, Reversal of sleep disturbance in Parkinson's disease by antiparkinsonian therapy: a preliminary study. *Neurology* 35, 527-532 (1985).

29. DR Jasinski and R Kovacevic-Ristanovic, Evaluation of the abuse liability of modafinil and other drugs for excessive daytime sleepiness associated with narcolepsy. *Clin. Neuropharmacol* 23(3),149-156 (2000).

30. RA Hauser, MN Wahba, TL Zesiewicz, and WM Anderson, Modafinil treatment of pramipexole-induced sleep somnolence. *Mov. Disord.* 15(S3), 133 (2000).

31. CH Adler, JN Caviness, JG Hentz, M Lind, and J Tiede, Modafinil for the Treatment of Excessive Daytime Sleepiness in Patients with Parkinson's Disease. *Neurology* 56(Suppl 3), A308 (2001).

32. KW Rammohan, JH Rosenberg, CP Pollack, DJ Lynn, A Blumenfeld, and HN Nagaraja, Modafinil – Efficacy and Safety for the Treatment of Fatigue in Patients with Multiple Sclerosis. *Neurology* 54(7, Suppl 3), A24 (2000).

33. M Terzoudi, P Gavarielidou, G Heilakos, K Visuiki, and CE Karageorgiou, Fatigue in Multiple Sclerosis: Evaluation and a New Pharmacological Approach. *Neurology* 54(7, Suppl 3), A61-62 (2000

NEUROPROTECTION BY DOPAMINE AGONISTS AND APPLICATION OF IMMUNOPHILIN LIGANDS

Tomoyoshi Kondo[*]

1. INTRODUCTION

A current therapeutic interest in Parkinson's disease (PD) is neuroprotection of dopaminergic neurons in the substantia nigra. Several hypotheses on the etiology of PD, such as the free radical theory, glutamate excitation toxic theory, and mitochondrial energy crisis theory, have been proposed. Based on current basic research, dopamine (DA) agonists act in many ways that lead to neuroprotection [1-9]. Free radical scavenging action, activation of an antioxidation mechanism and induction of antiapoptotic proteins could account for their actions. The actions of a nonimmunosuppressive immunophilin ligand, GPI 1046, with respect to neuroprotection or neuroaxonal regrowth in earlier reports, are interesting[10]. Although previous reports on the actions of GPI 1046 are conflicting, [10-14] a previous report using cell culture and our present results of an animal experiment suggest that GPI 1046 lessens neurotoxic insults on DA neurons.

Based on the above viewpoints, I describe in this article the potential application of the drugs for neuroprotective therapy of PD.

2. ANIMAL EXPERIMENT ON NUROPROTECTION BY DA AGONISTS

Numerous in vitro and in vivo experiments have demonstrated neuroprotective action of DA agonists against neurotoxins such as MPTP and 6-hydroxydopamine indicating their potential for use in the neuroprotection therapy of patients with PD. All DA agonists that have already been clinically used have some kind of neuroprotective action. The neuroprotective actions of each DA agonist differ in each report. However, this may be due in part to the time when the experiments were performed. A previous interest in the mechanism

---

[*]Tomoyoshi Kondo, Wakayama Medical University, 641-8510 Kimiidera, Wakayama, Wakayama 641-8510, Japan

Mapping the Progress of Alzheimer's and Parkinson's Disease
Edited by Mizuno *et al.*, Kluwer Academic/Plenum Publishers, 2002

391

of neuroprotection appears to have shifted to the study of antiapoptotic actions of DA agonists4.

The earliest report which indicates a potential for neuroprotection was the free radical scavenging action of bromocriptine against 6-OHDA toxicity in the nigra of rats1. Bromocriptine also protects DA neurons against methamphetamine6 or MPTP toxicity[2]. A DA agonist, talipexole,[7] which is commercially available only in Japan, has also been reported to have neuroprotection against methamphetamine toxicity in the nigrostriatal system of mice.

3. ARE THE CLINICAL BENEFITS OF DA AGONIST THERAPY ATTRIBUTED TO NEUROPROTECTION?

A clinical benefit of DA agonist therapy may be a reduction in the incidence and degree of motor complications in long-term of therapy. Several reports proposed the possibility of lessening incidence and degree of the wearing-off phenomenon or dyskinesia by DA agonist monotherapy or its combination therapy with levodopa from the initiation of antiparkinsonian therapy [15, 16]. Is the reduction in incidence or degree of motor complications caused by the neuroprotective action of DA agonists? Although the above-mentioned basic research indicates that the possible mechanism of reduction in motor complications is the neuroprotective action of DA agonists,[1-9] other mechanisms are possibly derived from specific characteristics of the drugs.

According to a hypothesis on the mechanism of the wearing-off phenomenon, severe reduction in the number of DA nerve terminals resulting in the loss of buffer action against fluctuations of the levodopa level in the striatum is an important cause [17]. Under such condition, an intermittent stimulation of DA receptors using a short-half-life agent primes the occurrence of dyskinesia. Compared to levodopa, in general, DA agonists have longer plasma half-lives and have modest antiparkinsonian effects. These appear to be important features that lead to the reduction in' motor complications such as the wearing-off phenomenon and dyskinesia in patients who were treated by DA agonist monotherapy or combination therapy with levodopa and DA agonist from the beginning of the antiparkinsonian therapy.

4. HOW DOES THE NEUROPROTECTION BY DA AGONISTS ASSUME?

In the case of DA agonist therapy, as mentioned above, it is difficult to separate neuroprotective effects from symptomatic effects of a drug.

Two interim reports of long-term trials of DA agonist therapy, aiming to quantify the function of dopaminergic nerve terminals in the striatum during the course of the therapy were published. One report is on the application of SPECT using beta-CIT[18], and the other is on the application of PET using fluorodopa[19] for evaluation of the degree of degeneration of DA nerve terminals in the striatum. Both methods have failed to show the gradual reduction in the function of DA nerve terminals at a statistically significant level. Although these quantitative methods, which measure the degree of degeneration of DA nerve terminals, failed in this respect, these appear to be essential for demonstrating the neuroprotective action of DA agonists.

5. IIMMUNOPHILIN LIGAND AND NEUROPROTECTION

Immunophilin is highly expressed in the CNS. This protein is particularly expressed in injured neurons and is known to have some neurotrophic actions on damaged neurons. Administration of immunophilin ligands induces recovery of neurons or nerves, such as damaged siatic nerve, serotonergic neurons damaged by p-chloroamphetamine, and dopaminergic neurons damaged by 6-OHDA[10]. However, previous data on neuroprotection or neuroregeneration by a nonimmunosuppressive immunophilin ligand, GPI 1046, are conflicting [10-14].

What is the expression pattern of immunophilin in the human brain, particularly in the case of chronic degenerative diseases. Regarding the therapeutic application of immunophilin ligands to parkinsonian patients, expression of immunophilins in the nigra is essential. According to our data, the expression of the anticyclophilin antibody-reactive protein in the human brain of both the normal control and PD patients can be determined by immunoblotting analysis (in the striatum) or immunohistochemical method (in the susbtantia nigra). Table 1 shows the number of patients whose striatal and/or nigral tissues positively reacted against the anti-human cyclophilin antibody. The expression of immunophilin in the nigrostriatal system of a parkinsonian brain indicates a possibility of neuroprotective or neuroreganerative actions of immunophilin ligands.

Table 1. Number of patients whose striatal and/or nigral tissues positively reacted against an anti-human cyclophilin antibody

	nigra	striatum
	(positive samples) / cases examined	
PD patients	(1) / 4	(2) / 2
Control	(2) / 4	(3) / 4

Immunoblotting analysis of the striatal and nigral tissues of a patient with Parkinson's disease and control subjects was carried out using an anti-human cyclophilin antibody. The number of patients whose nigral tissues positively reacted against anti-cyclophilin antibody is shown in the table.

6. NEUROPROTECTION BY A NON-IMMUNOSUPPRESSIVE IMMUNOPHILIN LIGAND, GPI 1046, AGAINST MPTP TOXICITY

The author examined a neuroprotective potential of the nonimmunosuppressive immunophilin ligand, GPI 1046, against MPTP toxicity in the nigrostriatal system of C57BL6 mice. In the experiment, the extent of decrease in the DA content in the striatum following MPTP administration was reduced by 30 % by GPI 1046 administration (Figure 1).

A reduction in the TH-positive cell population in the substantia nigra due to MPTP toxicity was also suppressed by GPI 1046 administration (Table 2).

Although the mechanism of neuroprotection by GPI 1046 has not yet been clarified, an antioxidative action caused by an increase in the intracellular glutathione level after GPI 1046 administration has been proposed[11].

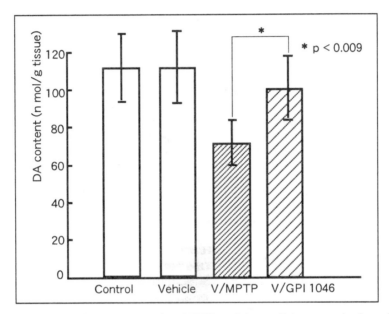

Figure 1. Effect of GPI 1046 against MPTP toxicity on DA content in the striatum

The figure shows the DA content in the striatum 4 days after the end of MPTP administration. GPI 10 46 (10 mg/kg) was administered subcutaneously 3 days prior to the start of MPTP administration for 9 days. Mice were sacrificed on the tenth day of the experiment.　The extent of reduction in the DA content in the striatum of the GPI 1046-treated group was significantly smaller than that of the vehicle-treated group (p < 0.009, Mann-Whitney U-test).

Table 2. Protection of nigral cells against MPTP toxicity by GPI 1046

Vehicle control (n=5)	66.6 +/- 30.6
MPTP + vehicle (n=3)	40.7 +/- 14.6
GPI 1046 (n=5)	63.5 +/- 25.2

GPI 1046 was administered 3 days prior to MPTP administration (bolus, 50 mg/kg, once) for 7 days. Mice were sacrificed 5 days after MPTP administration.　The sections were 6 micrometers thick. TH-positive cells in ten sections, at intervals of 30 micrometers, of the midbrain were counted. The average cell number in the ten sections is shown in the figure.　The number of TH-positive cell in the GPI 1046-treated group was significantly larger than that in the vehicle-treated group (p < 0.0001, Mann-Whiteny U-test).

At present, the mechanism of neuroprotection by immunophilin ligands, particularly nonimmunosuppressive immunophilin ligands, is still under debate. However, the neuroprotective therapy using these ligands for PD is very promissing. Therefore, further study are needed.

7. CONCLUSION

The potential neuroprotective actions of by DA agonists or immunophilin ligands were described.

As mentioned above, the demonstration of the neuroprotective action of the drugs is limited in animal experiments. At present, compared to the neuroprotective actions of immunophilin ligands, those of DA agonists appear to be more promising based on the experimental results.

For the demonstration of the effectiveness of neuroprotective therapy for patients with PD, the development of methods, such as SPECT or PET, that can quantify the progression of the disease other than instead of basing on symptoms is essential. Further advances in the development of new analogs of the above-mentioned drugs are also expected.

REFERENCES

1. N. Gogawa, K. Tanaka, M. Asanuma, M. Kawai, T. Masumizu, M. Kohno, and A Mori, Bromocriptine protects mice against 6-hydroxydopamine and scavenges hydroxyl free radicals in vitro, Brain Res 657, 207--213 (1994).
2. T. Kondo, T Ito, and Y.Sugita, Bromocriptine scavenges methamphetamine-induced hydroxyl radicals and attenuates dopamine depletion in mouse striatum, Ann NY Acad Sci 738, 222--229 (1994).
3. D. Muralikrishnan and K.P.Mohanakumar, Related Articles Neuroprotection by bromocriptine against 1-methyl-4-phenyl-1,2,3,6-tetrahydropyridine-induced neurotoxicity in mice, FASEB J 12, 905--912 (1998).
4. H. Sawada, M. Ibi, T. Kihara, M. Urushitani, A. Akaike, J. Kimura, and S. Shimohama, Dopamine D2-type agonists protect mesencephalic neurons from glutamate neurotoxicity: mechanisms of neuroprotective treatment against oxidative stress, Ann Neurol 44, 110--119 (1998).
5. L. Zou, J. Jankovic, D. B. Rowe, W. Xie, S. H. Appel, and W. Le, Neuroprotection by pramipexole against dopamine- and levodopa-induced cytotoxicity, Life Sci 64, 1275--1285 (1999).
6. M. Iida, I. Miyazaki, K. Tanaka, H. Kabuto, E. Iwata-Ichikawa, and N. Ogawa, Dopamine D2 receptor-mediated antioxidant and neuroprotective effects of ropinirole, a dopamine agonist, Brain Res 838, 51--59 (1999).
7. T. Kondo, H. Shimada, K. Hatori, H. Sugita, and Y. Mizuno, Talipexole protects DA neurons from methamphetamine toxicity in C57BL/6N mouse, Neurosci Lett 247, 143--146 (1998).
8. M. Gomez-Vargas, S. Nishibayashi-Asanuma, M. Asanuma, Y. Kondo, E. Iwata, and N. Ogawa, Pergolide scavenges both hydroxyl and nitric oxide free radicals in vitro and inhibits lipid peroxidation in different regions of the rat brain, Brain Res 790, 202--208 (1998).
9. J. Sander, R. Dengler, and P. Odin, Neuroprotective effects of the dopamine agonist cabergoline in the rat 6-OHDA Parkinson model, in Focus on Medicine 14, Preclinical data on dopamine agonists, edeted by Chase T (Blackwell Science, 1999) pp7--14
10. J.P. Steiner, G.S. Hamilton, D.T. Ross, H.L. Valentine, H. Guo, M.A. Connolly, S. Liang, C. Ramsey, J.H. Li, W. Huang, P. Howorth, R. Soni, M. Fuller, H. Sauer, A.C. Nowotnik, and P.D. Suzdak, Neurotrophic immunophilin ligands stimulate structural and functional recovery in neurodegenerative animal models, Proc Natl Acad Sci U S A 94, 2019—2024 (1997).

11. K. Tanaka, N. Fujita, M. Yoshioka, and N. Ogawa, Immunosuppressive and non-immunosuppressive immunophilin ligands improve H(2)O(2)-induced cell damage by increasing glutathione levels in NG108-15 cells, Brain Res 889, 225—228 (2001).

12. H. Sauer, J.M. Francis, H. Jiang, G.S. Hamilton, and J.P. Steiner, Systemic treatment with GPI 1046 improves spatial memory and reverses cholinergic neuron atrophy in the medial septal nucleus of aged mice, Brain Res 842, 109--118, 1999

13. E.M. Parker, A. Monopoli, E. Ongini, G. Lozza, C.M. Babij, Rapamycin, but not FK506 and GPI-1046, increases neurite outgrowth in PC12 cells by inhibiting cell cycle progression, Neuropharmacology 39, 1913--1919 (2000).

14. C. Winter, J. Schenkel, E. Burger, C. Eickmeier, M. Zimmermann, and T. Herdegen, The immunophilin ligand FK506, but not GPI-1046, protects against neuronal death and inhibits c-Jun expression in the substantia nigra pars compacta following transection of the rat medial forebrain bundle, Neuroscience 95, 753--762 (2000).

15. U.K. Rinne, F. Bracco, C. Chouza, E. Dupont, O. Gershanik, J.F. Marti Masso, J. L. Montastruc, and C.D. Marsden, Early treatment of Parkinson's disease with cabergoline delays the onset of motor complications. Results of a double-blind levodopa controlled trial, The PKDS009 Study Group, Drugs 55 Suppl 1, 23—30 (1998).

16. O. Rascol, D.J. Brooks, A.D.Korczyn, P.P. De Deyn, C.E. Clarke, A.E. Lang, A five-year study of the incidence of dyskinesia in patients with early Parkinson's disease who were treated with ropinirole or levodopa, 056 Study Group, N Engl J Med 342, 1484—1491 (2000).

17. J.A. Obeso, M.C. Rodriguez-Oroz, M.Rodriguez, M.R. DeLong, and C.W. Olanow, Pathophysiology of levodopa-induced dyskinesias in Parkinson's disease: problems with the current model, Ann Neurol 47 (Suppl 1), S22--32 (2000).

18. Parkinson Study Group, A randomized controlled trial comparing pramipexole with levodopa in early Parkinson's disease: design and methods of the CALM-PD Study, Clin Neuropharmacol 23, 34—44 (2000).

19. K.L. Leenders, Neuroimaging with F-Dopa PET in monotherapy, Pergolide Global Medical Conference, Barcelona, Spain, June 10, 2000

PROBLEMS OF LONG-TERM
LEVODOPA TREATMENT

Fabrizio Stocchi[*]

1. INTRODUCTION

There is no question about the fact that levodopa remain the most effective treatment for PD. It greatly improved the quality of life and the life expectancy of patients and this is close to that of general population now[1, 2]. Nonetheless, it has been clearly demonstrated that various problems may arise in the long-term use of levodopa[3-5]. The first sign of the appearance of motor complication is the progressive shortening of levodopa duration of action. Furthermore, the benefit of levodopa can be associated with involuntary movements or dyskinesias. Often these first become apparent at the peak effect of each dose, but then more complicated patterns emerge with ballistic, stereotypic and dystonic, often painful, dyskinesias appearing at the beginning and the end of each dose effect (diphasic dyskinesias), or in OFF periods. As time passes, the fluctuation from mobility (ON) to immobility (OFF) occur more frequently and become more abrupt and more disabling. The pattern of response to treatments becomes more unpredictable, with many doses of levodopa having a delayed effect or even no effect at all[6, 7].

After approximately five years of treatment, more than 50% of patients experience motor fluctuations and dyskinesias. The prevalence of motor fluctuations and dyskinesias increases with the duration and severity of the disease as well as with the duration of treatment. Thus, after ten years of levodopa therapy, motor complications have developed in 70-80% of patients and in almost 100% of those with disease onset below the age of 45-50 years[8].

2. PHARMACODYNAMIC OF LEVODOPA AND MOTOR FLUCTUATIONS

During long-term levodopa treatment, there is no change in the peripheral pharmacokinetic handling of the drug[9]. However, the central paharmacodynamic response to levodopa does alter[10-12]. The rate of motor deteroration after withdrawal of a steady-state optimal-dose levodopa infusion becomes significantly faster in patients with

[*] Dip. Neurosciences, University of Rome "La Sapienza" and "Neuromed" Viale dell'Università 30, 00185 Rome. Italy.

Mapping the Progress of Alzheimer's and Parkinson's Disease
Edited by Mizuno *et al.*, Kluwer Academic/Plenum Publishers, 2002

advanced disease exhibiting on-off fluctuations, compared with those in the early stages of the illness[13]. The change in duration of levodopa action with disease progression could be due either to loss of nigral changes, with consequent further loss of presynaptic storage, or to changes in postsynaptic events, or to both.

The long duration of action of levodopa responsible for the sustained motor effect observed in early parkinsonian patients was supposed to be due to presynaptic storage of dopamine[14, 15]. Indeed recent studies showed that dopamine agonists can induce long duration of action similar to that induced by levodopa[16, 17]. These data indicate that changes in postsynaptic mechanisms are associated with alterations in levodopa dose-response and the appearance of dyskinesias.

When levodopa is first introduced, the magnitude of benefit tends to reflect the dose employed, with a roughly linear pharmacologic dose-response relation with plasma levodopa concentration, up to the limit of maximal benefit. However, as fluctuation and dyskinesias emerge, this pharmacodynamic response changes[11]. The linear dose-response relation switches to a sigmoid curve, with the emergence of a critical threshold. Doses below this threshold do not turn the patients ON. Doses at or above the threshold turn the patients fully ON. Increasing the dose above the threshold now does not increase the magnitude of the benefit in ON, but increases its duration. However, the threshold for the production of dyskinesias also changes.

During the long-term levodopa treatment, the dose required to produce dyskinesias progressively gets less and less until it becomes almost the same as that required to turn the patient ON[18]. Dyskinesias then become an almost inevitable consequence of improvement in mobility[19]. Furthermore, although increasing the dose of levodopa above the threshold will prolong the duration the duration of the ON period, often this is at the expense of increasing dyskinesias. The development of the two different threshold lead to the appearance of a "therapeutic window" which is the range of levodopa plasma level sufficient to keep the patient ON without dyskinesias. Therapeutic window correlate most closely with the duration of levodopa treatment and is narrower in patients with ON-OFF phenomenon than in those with wearing-off type fluctuations.

3. THE PATHOGENESIS OF MOTOR COMPLICATIONS

In PD, intermittent oral levodopa administration and its short half life leads to intermittent absorption with resulting fluctuations in plasma levels[20]. It has been shown that the striatal dopamine levels that result from the peripheral administration of the drug go through similar fluxes and consequently repeated cycles of relative dopaminergic activation and withdrawal occur[21, 22]. This generate a non physiologic pulsatile stimulation. Studies in MPTP monkeys have shown that long-acting dopamine agonists are associated with a reduced frequency and severity of motor complications in comparison to regular formulations of levodopa[23, 24].

These observations suggest that long-acting dopaminergic agents that provide more continuous stimulation of dopamine receptors might be associated with a reduced risk of inducing motor complications[25-29]. There is considerable evidence indicating that, in comparison to levodopa, long-acting dopaminergic agents are associated with a reduced frequency and severity of motor complications in MPTP monkeys[23, 24]. Indeed, the same short-acting dopamine agonist induces gene changes in striatal neurons and dyskinesia when it is administered intermittently, but not when it is infused in a continuous manner[30, 31]. Prospective double blind clinical trials in PD patients similarly demonstrate that the risk of inducing motor complications is markedly reduced if therapy is initiated with a long-acting dopamine agonist compared to a short-acting formulation of levodopa[32, 33].

It is less clear if motor complications can be reversed with long-acting or continuous dopaminergic therapies. In MPTP monkeys, cabergoline has been shown to reverse established motor complications[34]. In PD patients with motor complications, several open label non-controlled studies have examined therapies designed to provide more continuous dopaminergic stimulation. Continuous infusions of levodopa, apomorphine, or lisuride have been shown to consistently reduce OFF time and the severity of motor fluctuations[35-38]. In some, this improvement in OFF time was associated with a marked reduction in the severity and duration of dyskinesia[36].

Mouradian et al.[39] demonstrated that giving complicated fluctuating parkinsonian patients a continuous intravenous infusion of levodopa (7 to 12 days round-the-clock), fluctuations in motor performances gradually diminished. Moreover the authors reported a prolongation of the duration of levodopa antiparkinsonian action and a widening of the therapeutic window.

ON-OFF fluctuations and dyskinesias can also be improved by continuous dopaminergic stimulation provided by dopamine agonist. Baronti et al.[40] gave 7 patients with advanced Parkinson's disease lisuride by continuous subcutaneous infusion for 3 months using a small wearable pump. The duration of antiparkinsonian action of levodopa nearly doubled after 3 months of continuous lisuride infusion and the efficacy half time for levodopa (i.e., the time required for motor scores to decline to 50% of their maximum severity) increased by approximately 90% (p 0.005). The difference between the levodopa threshold dose for toxicity (dyskinesias) and for efficacy (antiparkinsonian response) increased by approximately threefold, indicating a widening of the therapeutic window for levodopa .

Similar results have been obtained by our group in a long-term study. 46 parkinsonian patients have been treated for a various lengths of time (6-72 months), with a continuous subcutaneous infusion of Lisuride in our department. This therapeutic strategy induced a dramatic improvement of fluctuations and a better control of dyskinesias[41, 42]. In a randomised long term study we demonstrated that patients randomised to receive treatment with continuous subcutaneous daytime infusions of the dopamine agonist lisuride experienced significant improvement in both OFF time and dyskinesia in comparison to patients randomised to receive oral levodopa treatment with or without other medical therapies. Patients randomised to receive lisuride infusions experienced significant improvement in both OFF time and dyskinesia in comparison to baseline that persisted throughout the 4-year duration of the study. In contrast, patients randomised to receive oral levodopa experienced significant worsening of both dyskinesia and OFF time in comparison to baseline and UPDRS scores tended to worsen.

These results lend further suggest that alterations at both presynaptic and postsynaptic levels contributing to motor complications tend to normalise with the more physiological stimulation afforded by continuous replacement strategies, especially when given chronically. Thus the denervation of the dopaminergic system and the pulsatile stimulation of the receptor site eventually lead to a narrowing of the therapeutic window.

4. PERIPHERAL CONTRIBUTION TO FLUCTUATIONS

Beyond intermittent absorption, several pharmacokinetic proprieties of levodopa contribute to fluctuations in plasma. The first is levodopa short distribution and elimination half-lives. The short half-lives of levodopa is a reflection of peripheral metabolism, which depends mainly on O-methylation when levodopa is administered with peripheral dopa decarboxylase inhibitor.

Levodopa is most effective when absorbed quickly in the proximal small intestine, the segment of maximum absorption. Factors that delay gastric emptying also delay and blunt peak plasma levodopa levels and may cause a delay or a complete failure of the clinical response to the dose[43]. Excessive gastric acidity also delay gastric emptying but excessive neutralisation of stomach contents may lead to incomplete dissolution of the levodopa tablets and thus incomplete absorption[44-46]. The importance of gastric emptying was underscored by studies of intraduodenal infusion of levodopa that bypass gastric emptying and result in significant smoothing of plasma levodopa variability and a concomitant stabilisation of response fluctuations[47, 48].

In different studies we showed that a solution of levodopa methyl-ester, a highly soluble prodrug produced by esterification of the carboxylic moiety of the levodopa molecule, is absorbed very quickly and consequently produces an earlier plasma peak inducing a more predictable clinical effect than standard levodopa preparation[49]. Moreover different author have shown that giving a continuous intravenous infusion of levodopa is possible to ensure a constant sufficient plasma levodopa plasma level giving the patients prolonged periods free from fluctuation in motor performances[50, 51].

Levodopa is transported through the intestinal wall using the intestinal carrier transport system for LNAAs. Under physiological conditions this carrier system cannot be saturated; thus it is unlikely to limit levodopa absorption or to contribute to plasma levodopa variability[52, 53]. Unlike the intestinal LNAA transport system, LNAA transport at the blood-brain barrier can be saturated at physiologic postprandial concentrations of plasma amino acids. Blockage of levodopa entry into brain was demonstrated indirectly by showing that intravenous boluses of LNAAs can abruptly terminate the antiparkinsonian response, and more directly by PET scanning studies in which pretreatment with LNAAs leads to exclusion of levodopa from the brain[54, 55].

However the role played by the LNAAs is the physiopatology of fluctuations is still unclear. The competitions in the transport may be important but more important seems to be the role played by the LNAAs into the brain. Woodward et al.[56] showed that levodopa entry into the brain despite the large amount of the LNAA present in plasma but the patients turned OFF. Thus phenylalanine may play a role in the brain (receptor site?) inducing OFF phenomena.

REFERENCES

1. Martilla RJ, Rinnie UK, Sirtola T, Sonninen V. Mortality of patients with Parkinson's disease treated with levodopa. *J Neurol* 216: 147-153 (1977)
2. Scigliano G, Musicco M, Soliveri P, Piccolo I, Girotti P, Caraceni T. Mortality associated with early and late levodopa therapy initiation in Parkinson's disease. *Neurology* 40: 265-269 (1990)
3. Sweet RD, McDowell FH. Five years' treatment of Parkinson's disease with levodopa. *Ann Inter Med* 83: 456-463 (1975)
4. Lesser RP, Fahn S, Snider SR, et al. Analysis of the clinical problems in parkinsonism and the complications of long- term levodopa therapy. *Neurology* 29: 1253-1260 (1979)
5. Marsden CD, Parkes JD. "On-off" effects in patients with Parkinson's disease on chronic levodopa therapy. *Lancet* i: 292-296 (1976)
6. Marsden CD, Parkes JD, Quinn N. Fluctuations of disability in Parkinson's disease - clinical aspects. In: Marsden CD. and Fahn S. eds. *Movement disorders*. Butterworth Scientific, London (1982) 123-145
7. Luquin MR, Scipioni O, Vaamonde J, Gershanik O, Obeso JA. Levodopa-induced dyskinesias in Parkinson's disease: clinical and pharmacological classification. *Mov Disord* 7: 117-124 (1992)
8. Schrag A, Quinn NP. Dyskinesias and motor fluctuations in Parkinson's disease. A community-based study. *Brain* 123; 2297-305 (2000)

9. Fabbrini G, Juncos JL, Mouradian MM, Serrati C, Chase TN. Levodopa pharmacokinetic mechanism and motor fluctuations in Parkinson's disease. *Ann Neurol* 21: 370-376 (1987)

10. Bravi D, Mouradian MM, Roberts JW, Davis TL, Sohn YH, Chase TN. Wearing-off fluctuations in Parkinson's disease: contribution of postsynaptic mechanisms. *Ann Neurol* 36: 27-31 (1994)

11. Mouradian MM, Juncos JL, Fabbrini G, Schlegel J, Bartko JJ, Chase TN. Motor fluctuations in Parkinson's disease: central pathophysiological mechanisms, Part II. *Ann Neurol* 24: 372-378 (1988)

12. Mouradian MM, Heuser IJE, Baronti F, Fabbrini G, Juncos JL, Chase TN. Pathogenesis of dyskinesias in Parkinson's disease. *Ann Neurol* 25: 523-526 (1989)

13. Fabbrini G, Mouradian MM, Juncos JL, et al. Motor fluctuations in Parkinson's disease: Central pathophysiologic mechanism, Part I. *Ann Neurol* 24: 366-371 (1988)

14. Nutt JG, Carter JH, Woodward WR. Long-duration response to L-Dopa. *Neurology* 45: 1613-1616 (1995)

15. Nutt JG, Holford NHG. The response to L-Dopa in Parkinson's disease: imposing pharmacological law and order. *Ann Neurol* 39: 561-573 (1996)

16. Barbato L, Stocchi F, Monge A, Vacca L, Ruggieri S, Nordera G, Marsden CD. The long-duration of action of levodopa may be due to a postsynaptic effect. *Clin Neuropharmacol* 20:394-401 (1997)

17. Stocchi F, Vacca L, Berardelli A, DePandis F, Ruggieri S. Long-Duration effect of postsynaptic compartment: study using a dopamine agonist with a short half-life. *Mov Disord* 16: 301-305 (2001)

18. Mouradian MM, Juncos JL, Fabbrini G, Chase TN. Motor fluctuations in Parkinson's disease. *Ann Neurol* 25: 633-634 (1989)

19. Nutt JC. Levodopa-induced dyskinesia: review, observations and speculations. *Neurology* 40: 340-345 (1990)

20. Procelinko R, Thomas GB, Solomon HM. The effect of an antiacid on the absorption and metabolism of levodopa. *Clin Pharmacol Ther* 13: 149 (1972)

21. Garnett ES, Fimau G, Chan PKH, et al. (18F)fluorodopa, an analogue of dopa, and its use in direct external measurements of strage degradation and turnover of intracerebral dopamine. *Proc Natl Sci* 75: 464-467 (1978)

22. Spencer SE, Wooten GF. Altered pharmacokinetics of L-dopa metabolism in rat striatum deprived of dopaminergic innervation. *Neurology* 34: 1105-1108 (1984)

23. Bédard PJ, Di Paolo T, Falardeau P, Boucher R. Chronic treatment with L-dopa, but not bromocriptine induces dyskinesia in MPTP-parkinsonian monkeys. Correlation with [³H]spiperone binding. *Brain Res* 379: 294-299 (1986)

24. Pearce, R.K. Banerji T, Jenner P, Marsden CD. De novo administration of ropinirole and bromocriptine induces less dyskinesia than L-dopa in the MPTP-treated marmoset. *Mov Disord* 13: 234-241 (1998)

25. Olanow CW, Schapira AHV, Rascol O. Continuous dopaminergic stimulation in the early treatment of PD. *Trends Neurosci* 23: 117-126 (2000)

26. Abercrombie ED, Bonatz AE, Zigmond MJ. Effects of L-DOPA on extracellular dopamine in striatum of normal and 6-hydroxydopamine-treated rats. *Brain Res*525: 36-44 (1990)

27. Tedroff J, Pedersen M, Aquilonius S, Hartvig P, Jacobsson G, Langstrom B. Levodopa-induced changes in synaptic dopamine in patients with Parkinson's disease as measured by [11C]raclopride displacement and PET. *Neurology* 46: 1430-1436 (1996)

28. Calon F, Grondin R, Morissette M et al. Molecular basis of levodopa-induced dyskinesias. *Ann Neurol* 47: 70-78 (1998)

29. Filion M, Tremblay L, Bedard PJ. Effects of dopamine agonists on the spontaneous activity of globus pallidus neurons in monkeys with MPTP-induced parkinsonism. *Brain Res.* 547: 152-161 (1991)

30. Blanchet PJ, Calon F, Martel JC, et al. Continuous administration decreases and pulsatile administration increases behavioral sensitivity to a novel dopamine D2 agonist (U-91356A) in MPTP-exposed monkeys. *J Pharmacol Exp Ther* 272: 854-9 (1995)

31. Morissette M, Goulet M, Soghomonian JJ, et al. Preproenkephalin mRNA expression in the caudate-putamen of MPTP monkeys after chronic treatment with the D2 agonist U91356A in continuous or intermittent mode of administration: comparison with L-DOPA therapy. *Brain Res Mol* 49: 55-62 (1997)

32. Rascol O, Brooks DJ, Korczyn AD, et al. A five year study of the incidence of dyskinesia in patients with early parkinson's disease who were treated with ropinirole or levodopa. *N Engl J Med* 342: 1484-91 (2000)

33. Parkinson Study Group. Pramipexole vs levodopa as initial treatment for Parkinson's disease. *JAMA* 284: 231-238 (2000)

34. Tahar AH, Grégoire L, Bangassoro E, Bedard PJ. Sustained cabergoline treatment reverse dyskinesias in parkinsonian monkeys. *Clin Neuropharmacol* 23:195-202 (2000)

35. Vaamonde J, Luquin MR, Obeso JA. Subcutaneous lisuride infusion in Parkinson's Disease: Response to chronic administration in 34 patients. *Brain* 114: 601-614 (1991)

36. Colzi A, Turner K, Lees AJ. Continuous subcutaneous waking day apomorphine in the long term treatment of levodopa induced interdose dyskinesia in Parkinson's disease. *J Neurol Neurosurg Psychiatry* 64: 573-576 (1998)
37. Sage JI, Trooskin S, Sonsalla PK, Heikkila RE, Duvoisin RC. Long-term duodenal infusion of levodopa for motor fluctuations in parkinsonism. *Ann Neurol* 24: 87-89 (1988)
38. Nutt JG, Obeso JA, Stocchi F. Continuous dopamine receptor stimulation in advanced Parkinson's disease. *Trends Neurosci* 23(suppl): S109-115 (2000)
39. Mouradian MM, Heuser IJE, Baronti F, Chase TN. Modifications of central dopaminergic mechanisms by continuous levodopa therapy for advanced Parkinson's disease. *Ann Neurol* 27: 18-23 (1990)
40. Baronti F, Mouradian MM, Davis T, et al. Continuous lisuride effects on central dopaminergic mechanisms in Parkinson's disease. *Ann Neurol* 32: 776-781 (1992)
41. Stocchi F, Ruggieri S, Antonini A, et al. Subcutaneous lisuride infusion in Parkinson's disease: clinical results using different modes of administration. *J Neural Transm* (suppl): 27-33 (1988)
42. Stocchi F, Ruggieri S, Viselli F, et al. Subcutaneous lisuride infusion. In: Lakke PWFJ, Delhaas EM, Rufgers AWF (eds): Parenteral Drug Therapy in Spasticity and Parkinson's Disease. *New Trens Neurol* 205-211 (1992)
43. Kurlan R, Rothfield K, Woodward WR, et al. Erratic gastric emptying of levodopa may cause "random" fluctuations of parkinsonian mobility. *Neurology* 38: 419-421 (1988)
44. Rivera-Calimlim L, Dujovne CA, Morga JP, et al. L-Dopa treatment failure: Explanation and correction. *Br Med J* 4: 93-94 (1970)
45. Leon AS, Speigel H. The effect of antiacid administration on the absorption and metabolism of levodopa. *J Clin Pharmacol* 12: 263-267 (1972)
46. Melamed E, Bitton V, Zelig O. Episodic unresponsiveness to single doses of L-dopa in parkinsonian fluctuators. *Neurology* 36: 100-103 (1986)
47. Kurlan R, Rubin AJ, Miller C, et al. Duodenal delivery of levodopa for on-off fluctuations in parkinsonism: preliminary observations. *Ann Neurol* 20: 262-265 (1986)
48. Ruggieri S, Stocchi F, Carta A, et al. Jejunal delivery of levodopa methyl ester. *Lancet* ii: 45-46 (1989)
49. Stocchi F, Barbato L, Bramante L, Nordera G, Vacca L, Ruggieri S. Fluctuating parkinsonism: a pilot study of a single afternoon dose of levodopa methyl ester. *J Neurol* 243: 377-380 (1996)
50. Marion MH, Stocchi F, Quinn NP, Jenner P, Marsden CD. Repeated L-Dopa infusions in fluctuatins Parkinson's Disease: clinical and pharmacokinetic data. *Clin Neuropharmacol* 9: 165-181 (1986)
51. Stocchi F, Ruggieri S, Carta A, Ryatt J, Quinn NP, Jenner P, Marsden CD, Agnoli A. Intravenous boluses and continuous infusions of L-Dopa Methyl Ester in fluctuating patients with Parkinson's disease. *Mov Disord* 7: 249-256 (1992)
52. Wade DN, Mearrick PT, Morris JL. Active transport of L-dopa in the intestine. *Nature* 242: 463-465 (1973)
53. Nutt JG, Woodward WR, Carter JH, et al. Influence of fluctuations of plasma large neutral amino acids with normal diets on the clinical response to levodopa. *J Neurol Neurosurg Psychiatry* 52: 481-487 (1989)
54. Leenders KL, Poewe WH, Palmer AJ, et al. Inhibition of L-1(19F)fluorodopa uptake into human brain by amino acids demonstrated by positron emission tomography. *Ann Neurol* 20: 258-262 (1986)
55. Nutt JG, Woodward WR, Hammerstad JP, et al. "On-off" phenomenon in Parkinson's disease: relationship to L-dopa and transport. *N Engl J Med* 310: 483-488 (1984)
56. Woodward WR, Olanow CW, Beckner BS, et al. The effect of L-dopa infusion with and without phenylalanine challenges in parkinsonian patients: plasma and ventricular CSF L-dopa levels and clinical responses. *Neurology* 43: 1704-1708 (1993)

EVALUATION OF ABNORMAL INVOLUNTARY MOVEMENTS IN PARKINSON'S DISEASE

Håkan Widner [*]

1. INTRODUCTION

Motor fluctuations and drug-induced dyskinesias are common complications of the disease process in Parkinson's disease (PD) and the treatment. As reviewed elsewhere in these proceedings, approximately 10% patients develop these complications per treatment year, and the younger the onset of disease, the more pronounced are the developments of abnormal involuntary movements (AIMs). Much of the current pharmacological treatment schedules and novel neurosurgical techniques are aimed at preventing or directly treat these AIMs. Measurements of PD related AIMs have proved to be difficult.

There are several factors that contribute to these difficulties. There are different temporal patterns in relation to drug intake: peak-dose dyskinesias (on-phase dyskinesias); bi-phasic dyskinesias (beginning and end-of-dose dyskinesias); and off-phase dyskinesias, usually off-dystonia. There are also different maximal levels of AIMs in the various forms, with off-dystonia usually at its maximum when the longest time has elapsed since the last drug intake, eg. in the early morning, wheras bi-phasic dyskinesias often have a very severe period of end-of-dose dyskinesias in the evening, whereas peak-of dose dyskinesias can vary its maximum mornings and afternoons.

The phenotypic variation is also great, with both hyperkinesias and reduced movements with dystonia, either in on or off conditions. Choreiform AIMs are the most comment type, but athetosis, stereotypies and myoclonus are other phenotypes, sometimes in complex combinations. Dystonia may also take different patterns and expressions, with a cramp-like, fixed often painful and prolonged AIM, manifested as a odd postures. Mobile dystonic movements, alone or in combination with twisting, choreathetotic movements appear more commonly as on-phase dystonias. The different phenotypes are usually associated with the different patterns or on- or off-phases, but there are several transitional phases when symptoms may be occurring simultaneously

---

[*] Håkan Widner, MD, PhD, Associate Professor of Neurology, Dept. Clinical Neurosciences, Lund University Hospital, S-221 85 Lund, Sweden

but in different sections of the body. In order to be able to describe, quantify and assess treatment effects, tools are needed for evaluation of these symptoms (Luquin, et al. 1992; Kidron, and Melamed., 1987; Poewe et al. 1988; Marconi et al. 1994).

2. PREVIOUS RATING SCALES

There are several proposed rating systems for AIMs in PD and other disorders, and the plethora of scales may indicate that none of these have been optimal and gained ubiquitous use. A rating scale should be able to assess intensity (ie. varying degrees of dyskinesias), duration, possibly phenotype and distribution, and degree of disability ie. the degree of interference with activities of daily living, and the degree of handicap, imposed by the AIMs on the patient's ability to function in the society. In addition, an ideal scale should be reliable, ie its precision should be high and the possibility for a random error should be small. The validity, ie. the accuracy of the intended measurement should be high, and the scale should be responsive, in it should be able to detect clinical important changes. Finally a scale should be relevant in relation to PD.

Most scales are modifications of the Abnormal Involuntary Movement Scale, AIMS, developed for assessment of tardive dyskinesias (Guy 1976). This scale is simple to use, is lateralised and responsive to changes with a sufficiently high score. It does not assess impairment or degree of handicap. However, in its original form, it is highly biased towards oro-facial dyskinesias. There are many modifications of the scale, with various omissions and some additions (Marsden and Schaster 1981).

There are only two scales that have been subjected to reliability testing. One of them is a scale developed at the Salpetriere hospital (Marconi et al. 1994), largely based on the AIMs scale. Another scale, originally developed by Obeso was recommended in the CAPIT, Core Assessment Programme for Intracerebral Transplantations, protocol (Langston et al. 1991). The scale has been modified subsequently by Goetz and co-workers (Goetz et al. 1994), and has a high reliability and validity. However, its responsiveness has proved to be low, as demonstrated by a study by Utti et al. 1997, where one rated AIMs prior to and after pallidotomy. When AIMs were quantified with a AIMS-like scale, significant reduction of AIMs were detected, but when the same patients were rated with the Goetz scale, no significant differences were detected. This demonstrates that the design of a scale is of importance to detect differences.

3. A NEW SCALE

Within the framework of the NIPD, Neurosurgical Interventions in Parkinson's disease, an European Union Biomed II sponsored programme, a revision of the CAPIT protocol was performed and among the issues needed review were dyskinesia ratings (Defer and Widner 1999) 1. A scale to assess AIMs should: (1) easy to use and apply to any situation, e.g., while performing standardized motor tests for parkinsonism; (2) separate ratings of different body parts, including lateralization; (3) separate ratings of dystonia and hyperkinesias. There were several pilot versions tested and after evaluations the following ratings scale was developed, called CDRS, Clinical Dyskinesia Rating Scale (Hagell and Widner 1999).

There are separate ratings of dystonia and hyperkinesias, and the total score is separated for both these forms of AIMs. The highest severities of dystonia and hyperkinesias observed in the face (including tongue), neck, trunk, and left and right upper and lower extremities are scored according to a 5-grade scale (0 = none; 1 = mild; 2 = moderate; 3 = severe; 4 = extreme, with the possibility to use 0.5-intervals), giving a maximal total score of 28 for hyperkinesias and dystonia, respectively. Ratings are to be based on observations of the patient during activation and at rest. The scores are assigned to the body part where the dyskinesias are observed, e.g., hemiballism in an arm is assigned to the arm, and face dyskinesias (e.g., tongue movements, blepharospasm or grimaces) are assigned to the face. The exception is dyskinesias causing head movements, which are assigned to the neck. An examiner may add more detailed information on, e.g., the phenotypic expressions of the dyskinesias or their relation to activation in a comments-section of the scale.

The scale was validated by 10 rates (neurologists, neurosurgeons, nurses in movement disorders units) by assessment one or two times of videotapes of two patients during a single-dose L-dopa test. There were 23 sequences, consisting of (1) the patient at rest, silent as well as speaking; (2) rigidity testing of the patients' wrists, elbows and knees, performed by an examiner with the patient sitting in a chair; (3) bilateral performance of 20 pronations/supinations of the hands against the ipsilateral knee with the patient sitting in a chair; and (4) the patient standing up, walking 7 meters, turning around and walking back again. The initial sequence of each patient did also feature bilateral performance of 20 fist clenches, 10 finger dexterity movement cycles and 20 foot taps with the patient sitting in a chair. The tests were performed in the defined off-phase, i.e., in the morning after 12 h fasting and withdrawal of all anti-parkinsonian medications. The patients standard morning doses of L-dopa (50 and 200 mg, respectively) was given and the patients developed AIMs.

There was high and statistically significant intra-rater consistency when the ratings were repeated, in particular for hyperkinesias, and significant by less so for dystonia, as determined by Kendall rank order correlation coefficient, with a Chi-square value as a significance level (Siegel and Castellan, 1988).

There was a high inter-rater concordance in particular for hyperkinesias ($W = 0.903$; $p<0.001$) Kendall coefficient of concordance and the Chi-Square value for statistical significance), and less so for dystonia ($W = 0.420$; $p<0.001$).

4. CAPSIT-PD RECOMMENDATIONS

The CAPIT protocol aimed at unifying an assessment protocol for effects of neural tissue transplantation and became a powerful and widely used set of recommendations. However, for other types of interventional treatments, and because of experience that the full CAPIT protocol was very laborious with many patients were assessed, a revision of the protocol was instigated in the NIPD programme.

Several simplifications and clarifications were made, among the definition of practically defined off and on periods and a reduced number of timed tests and clearer instructions performing movements. The recommendations were published under the acronym CAPSIT-PD, Core Assessment Programme for Sugical Interventions and Transplantations in Parkinson's Disease (Defer et al. 1999).

In order to assess parkinsonian symptoms and AIMs the following recommendations were made.

1/. For global assessment of symptoms, the Unified Parkinson's disease rating scale, UPDRS version 3.0 should be used. In the IV part there are patient estimations of the percentage of the day that the patient experiences AIMs.

2/ A three or four graded diary, off, partial on, on and on with dyskinesias should be filled out by the patients, for every half hour for one week prior to a visit. If the patients experiences any time during the 30 min. period painful dystonia, falls or freezing, this is recorded.

3/ The Clinical Dystonia Rating Scale should be used when the UPDRS part III exam is made in practically defined off (after 12 hours medication interruption) and in practically defined on (the best condition obtained with a test dose), and on any other occasion that adequately reflects the patients symptoms. Ratings of hyperkinesias and dystonias should be performed in parallel. The ratings should be performed when the patient is at rest, talking and performing movements such as those in the UPRDS part III or timed tests.

4/ If the patient has an abnormal or rare pattern of AIMs such as bi-phasic types, this should be characterised further, possibly by generating an AIMs pattern after a single-dose-L-dopa challenge test, with regular assessments and generation of an Area-under-the-Curve of AIMs, or to perform additional ratings when the patient is experiencing the maximal amount of AIMs.

The objective of the CAPSIT-PD is to allow for comparison of various types of interventions and the protocol for AIMs assessment can readily be use in drug studies as well as in comparative trials using different surgical techniques.

The combined approach with diary, UPDRS and CRDS will allow characterisation of a patient's AIMs with a minimum of efforts, but with techniques that will allow for a sensitive and reliable assessment.

REFERENCES

Luquin, M.R., Scipioni, O., Vaamonde,J., Gershanik, O., and Obeso.,J.A., 1992, Levodopa-induced dyskinesias in Parkinson's disease: clinical and pharmacological classification. *Mov Disor..*, 7:117-124.

Kidron, D., and Melamed., E., 1987, Forms of dystonia in patients with Parkinson's disease. *Neurology*, 37:1009-1011.

Poewe, W.H., Lees, A.J., and Stern, G.M., 1988, Dystonia in Parkinson's disease: clinical and pharmacological features. *Ann Neurol.*, 23:73-78.

Marconi, R., Lefebvre-Caparros, D., Bonnet, A-M., Vidailhet, M., Dubois, B., and Agid, Y. 1994, Levodopa-induced dyskinesias in Parkinson's disease phenomenology and pathophysiology. *Mov Disord.*, 9:2-12.

Guy. W., AIMS. In: Guy W, ed. 1976, ECDEU assessment manual. Rockville: Department of Health, Education and Welfare, 1976:534-537.

Marsden, C.D., and Schachter, M., 1981, Assessment of extrapyramidal disorders. *Br J Clin Pharmac.*, 11:129-151.

Langston, J.W., Widner, H., Goetz, C.G., Brooks, D., Fahn, S., Freeman, T.B., Watts, R., 1992, Core assessment program for intracerebral transplantations (CAPIT). *Mov Disord.*, 7:2-13.

Goetz, C.G., Stebbins, G.T., Shale, H.S., et al., 1994, Utility of an objective dyskinesia rating scale for Parkinson's disease: inter- and intrarater reliability assessment. *Mov Disord.*, 9:390-394.

Uitti, R.J., Wharen Jr, R.E., Turk M.F., et al. 1997, Unilateral pallidotomy for Parkinson's disease: comparison of outcome in younger versus elderly patients. *Neurology* 49:1072-1077.

Goetz, C.G., Stebbins, G.T., Blasucci, L.M., Grobman, M.S., 1997, Efficacy of a patient training videotape on motor fluctuations for On-Off diaries in Parkinson's disease. *Mov Disord.*, 12:1039-1041.

Widner H., and Defer, G-L., 1999, Dyskinesias assessment: from CAPIT to CAPSIT. *Mov Disord.*, **14**: suppl. 1, 60-66.

Hagell P. and Widner, H., 1999, Clinical rating of dyskinesias in Parkinson's disease: use and reliability of a new rating scale. *Mov Disord.*, **14**: 448-455.

Siegel, S., Castellan, N.J., 1988, Non parametric statistics for the behavioral sciences. New York: McGraw-Hill.

Defer, G-L., Widner, H., Marie, R-M., Remy, P., Levivier M., and Conference Participants, 1999, Core Assessment Program for Surgical Interventional Therapies in Parkinson's Disease (CAPSIT-PD), *Mov Disord.*, **14**:572-584.

MEDICAL TREATMENT OF L-DOPA-INDUCED DYSKINESIAS

Olivier Rascol[1], Christine Brefel-Courbon[1], Pierre Payoux[2], Joaquim F. Ferreira[3]

1. INTRODUCTION

L-dopa-induced dyskinesias are frequent. We shall limit our topic to "peak-dose dyskinesias " and exclude from our scope other abnormal movements (Luquin et al, 1992 a ; Marconi et al, 1994). Dyskinesia generally occur in patients who are also suffering from the "ON-OFF" phenomenon, which make therapeutic management a difficult challenge, due to the narrowing of the therapeutic window.

The L-dopa-treated MPTP-monkey model, the development of new antiparkinsonian drugs and surgical treatments of Parkinson's disease allowed substantial progresses in our understanding of dyskinesia. Both L-dopa and the severity of the underlying dopamine denervation are major contributing factors. This combination leads to a number of functional and biological changes in the basal ganglia which provoke dyskinesia. Such changes are not limited to the dopamine systems, and also involve non dopaminergic transmitters (Brotchie et al, 1998). Once primed, the message for dyskinesias is stored in the brain for long periods of time (at least several months). This phenomenon is know as the "priming", possibly related to L-dopa short elimination half-life : the intermittent administration of the drug (several times a day) stimulates dopamine receptors in a non-physiological pulsatile way which is supposed to induce molecular and behavioral supersensitivity changes (Olanow et al, 2000). According to our present understanding of dyskinesias and in respect with this "priming" phenomenon, we can propose 3 main strategies to manage L-dopa-induced dyskinesias:

- the first is preventive, to reduce or delay the risk of occurrence of the priming phenomenon,

- the second is curative, to reverse, once primed, the changes which produce dyskinesias,

- the third is symptomatic, to avoid the expression of dyskinesias, regardless of the underlying priming processes.

[1] Department of Clinical Pharmacology, Clinical Investigation Center, INSERM U455, and [2] Department of Nuclear Medicine, Toulouse University Hospital, France and [3] Department of Neurology, Lisbon University Hospital, Portugal, Correspondance to Pr Olivier Rascol, Laboratoire de Pharmacologie, Faculté de Médecine, 37 Allées J Guesde, 31073 TOULOUSE Cedex, FRANCE, Tel.: 33 561 14 59 62, Fax : 33 561 25 51 16 , E-mail : rascol@cict.fr

Mapping the Progress of Alzheimer's and Parkinson's Disease
Edited by Mizuno *et al.*, Kluwer Academic/Plenum Publishers, 2002

2. THE PREVENTION OF DYSKINESIAS

Several strategies have been proposed : to delay time to L-dopa, to reduce L-dopa dose, to reduce the pulsatile stimulation of dopamine receptors...

The early use of selegiline in the " DATATOP " study (Parkinson Study Group, 1993) showed that it can significantly delay the need for L-dopa and spare L-dopa dose. However, on long-term follow-up the risk for dyskinesias was not reduced (Parkinson Study Group, 1996).

Early treatment with CR L-dopa, supposed to reduce the " pulsatility " of L-dopa stimulation, also failed to reduce the risk for dyskinesias (Block et al, 1997). No data are available with the early combination of a COMT inhibitor.

The only drugs which demonstrated positive results are the dopamine D2 agonists. The first pilot reports with bromocriptine were published in the late seventies (Rascol et al, 1979). This was confirmed in the MPTP-treated monkey model (Bedard et al, 1986). To maintain antiparkinsonian efficacy over time, low doses of L-dopa are adjuncted soon, (Rinne, 1985) or late (Montastruc et al, 1994) with still significantly less dyskinesias. Similar findings were reported with cabergoline (Rinne et al, 1998), ropinirole (Rascol et al, 2000) and pramipexole (Parkinson Study group, 2000).

3. THE REVERSAL OF THE PRIMING OF DYSKINESIAS

In humans, like in monkeys, the low propensity of some dopamine agonists, like bromocriptine or ropinirole, to induce few, if any, dyskinesias is only observed if L-dopa has not previously primed the phenomenon. Once the priming has occured, all agonists, including these two reproduce typical dyskinesias. If such a priming is not fully reversible, this should encourage us to use as much as possible the preventive strategy. Priming reversibility remains a matter of debate. Available animal data suggest that certain biochemical and behavioral abnormalities last for long periods of time, even when L-dopa has been washed-out for several months. Clinical data suggest that the phenomenon is probably partly reversible, but this is far from rapid, and may require months of interruption of dopatherapy. This delay is poorly compatible with the usual management of patients in most instances, because it is impossible to withhold L-dopa for long periods of time in severely affected patients.

After short-term " drugs holidays " L-dopa-induced dyskinesias were somewhat less severe when resuming L-dopa therapy, but soon reappeared, within few weeks of dopatherapy (Goetz et al, 1982). It is not possible to withdraw L-dopa therapy, without the concomitant help of other interventions. The adjunction of a MAO-B or a COMT inhibitor, or that of a dopamine agonist, remains largely disappointing. In some cases, however, large doses of dopamine agonists allow to substantially reduce L-dopa dose, and partly improves dyskinesias. L-dopa slow-release formulations also proved to be of little, if no help.

Several investigators have tested the effects of the continuous intravenous infusion of L-dopa (Schuh et al, 1993). Within few days, such an infusion shifts the L-dopa-induced dyskinesias threshold to the right, reducing maximum dyskinesias activity, without significantly altering dose-response for relief of parkinsonism. Dopamine agonists, like apomorphine, have been used on long-term subcutaneous infusions in severely fluctuating dyskinetic patients. After several months of infusion, the "therapeutic window" between antiparkinsonian and dyskinetic thresholds was enlarged, suggesting that the reduction of L-dopa dose, or the continuous delivery of dopamine stimulation by the agonist induced a partial reversal of the priming phenomenon (Colzi et al, 1998).

The renewed interest in the surgical treatments of patients with Parkinson's disease provided further interesting observations. The high-frequency deep brain stimulation of the subthalamic nuclei markedly ameliorates parkinsonian symptoms (Limousin et al, 1995). This effects allows to substantially reduce L-dopa doses, and after some months of a such regimen, it has been reported that a standard dose of L-dopa, which induced formally marked dyskinesias, induces much less abnormal movements (Agid, 1998).

4. THE HINDRANCE TO THE EXPRESSION OF DYSKINESIA IN ALREADY PRIMED PATIENTS

The priming of dyskinesias involves changes at the level of the dopaminergic synapse, and affects down-flow non-dopaminergic relays. Interventions at both levels may block the expression of dyskinesias in primed patients with agents acting on dopaminergic or non-dopaminergic receptors.

4.1. Dopaminergic medications

Reducing the dose of L-dopa may reduce dyskinesias, but at the inevitable cost of an unacceptable increase in the duration and severity of OFF periods. To compensate for this problem, one can give more frequent and smaller doses of L-dopa. The result is usually disappointing because more frequent but lower doses of L-dopa aggravate ON-OFF fluctuations.

Blocking D2 receptors neuroleptics, suppresses dyskinesias, but again at the cost of an unacceptable aggravation of OFF periods (Klawans and Weiner, 1974).

Partial agonists act as agonists or antagonists in different pharmacological conditions : they exert an agonist activity at denervated supersensitive sites, and competitively inhibit full agonists at fully innervated normosensitive receptors. Partial agonists might possess a better therapeutic-to-toxic ratio than full agonists in dyskinetic patients with "supersensitive" dopamine receptors. Terguride, a partial D2 agonist, exhibits both antiparkinsonian and antihyperactivity effects in MPTP-treated primates (Tetsuo et al, 1993). Inconsistent results have however been reported in patients (Barontin et al, 1992b ; Critchley and Parkes, 1987).

It has been proposed that L-dopa-induced dyskinesias were the consequence of D1 rather than D2 stimulation. Bromocriptine has a lesser propensity than L-dopa to induce dyskinesias, and has mild D1 antagonistic effects (it is a partial D1 agonist). Clozapine, an "atypical" neuroleptic, has some D1 antagonistic effects, among other properties. Small open trials have shown that clozapine reduces L-dopa-induced dyskinesias without significant deterioration of the antiparkinsonian potency of L-dopa (Durif et al, 1997) but controlled trials are missing. The same is true with other "atypical" neuroleptics, and a recent study with olanzapine in psychotic parkinsonian patients showed an unacceptable worsening of parkinsonism (Goetz et al, 2000). In fact, some short-acting selective D2 agonists, like PHNO, induce dyskinesias in L-dopa naïve non-primed MPTP-monkeys, challenging the D1 hypothesis (Luquin et al, 1992b). In MPTP-treated monkeys, D1 agonists relieved parkinsonian symptoms to the same extent than L-dopa or D2 agonists, but with less dyskinesias (Grondin et al, 1997). In patients, ABT-431, a D1 agonist, induced a full antiparkinsonian response, but at the cost of the same extent of abnormal movements (Rascol et al, 2001). It is therefore impossible to draw clear conclusions.

4.2. Non-dopaminergic agents

The concept of using non-dopaminergic adjuncts to the usual dopamine treatments of Parkinson's disease to reduce the problem of dyskinesias is relatively recent. It should be possible to manipulate non-dopaminergic systems and to reduce dyskinesias without compromising L-dopa antiparkinsonian efficacy (Brotchie, 1998).

Glutamate plays a crucial role in different relays of the motor cortico-subcortical loops involved in Parkinson's disease. Increased synaptic efficacy of NMDA receptors expressed on basal ganglia neurons have been proposed to play a role in L-dopa-induced dyskinesias (Chase and Oh, 2000). Blocking NMDA receptors, diminished choreic dyskinesias in the monkey model (Papa and Chase 1996). Amantadine has mild antiglutamate properties. It decreased L-dopa-induced dyskinesias in patients with Parkinson's disease, without concomitant worsening of the parkinsonian symptoms (Verhagen Metman et al, 1998a). Less conclusive results have been reported with other potentially antiglutamate compounds, like dextrometorphan (Verhagen Metman et al, 1998b). Other drugs which act on the glutamate synapses, like riluzole, might also be interesting (Merims et al, 1999).

Experimental and clinical studies suggest that dopaminergic transmission in the brain is influenced by serotonin. In patients, ritanserin, a selective 5HT2 antagonist, improved dyskinesias in a small single-blind placebo-controlled cross-over study (Meco et al, 1988). Fluoxetine, a selective serotonin reuptake inhibitor, also decreased dyskinesias triggered by an acute apomorphine challenge in an open pilot study (Durif et al, 1995). It is difficult to extend the results of such small uncontrolled short-term studies. There is also a number of case reports on drug-induced parkinsonism and tardive dyskinesias syndromes with selective serotonine reuptake inhibitors. These drugs, might thus exert neuroleptic-like effects. Buspirone, acts as a 5HT1A agonist. It has been inconsistently reported to decrease L-dopa-induced dyskinesias (Ludwig et al, 1986; Kleedorfer et al, 1991; Bonifati et al, 1994; Hammerstad et al, 1986). Its anxiolytic property could account for a minor non-specific anti-dyskinetic effect.

Encouraging findings have been reported in the monkey model with alpha-2 antagonists like yohimbine (Gomez-Mancilla and Bedard, 1993) and idazoxan (Henry et al, 1999). Positive preliminary data were obtained in a pilot study in dyskinetic patients (Rascol et al, 2001), but such data have not been reproduced (Manson et al, 2000).

There are changes in enkephaline and endorphins levels in the basal ganglia of dyskinetic monkeys (Henry and Brotchie, 1996). Inconsistent effects have been reported in patients receiving one of the currently available opiate antagonists in clinical practice : naloxone and naltrexone (Trabucchi et al, 1982 ; Sandyk and Snides, 1986; Nutt et al, 1978 ; Rascol et al, 1994, Manson et al, 2001).

Striatal adenosine A_{2A} receptors are also candidates : since an antagonist induced a significant antiakinetic effect with little or no dyskinesias in animals previously primed to exhibit dyskinesias (Kanda et al, 1998). Many clinical case reports have incidently described the beneficial effects of a variety of drugs on dyskinesias in MPTP-treated monkeys or in patients with Parkinson's disease. These reports must replicated in controlled trials before warranting any clinical application.

5. CONCLUSION: THE MANAGEMENT OF L-DOPA-INDUCED DYSKINESIAS IN EVERYDAY CLINICAL PRACTICE

It is presently recommended to try to prevent as much as possible the priming of dyskinesias, specially in young patients, because they are at greater risk for motor complications. This is possible with the early use of dopamine agonists, but this strategy is however only partly successful.

Once primed, dyskinesias remain difficult to treat. Reducing L-dopa dose and using more frequent daily drug intakes is only temporarily and incompletely useful, often leading to an unacceptable worsening of the ON-OFF phenomenon. Replacing the largest possible dose of L-dopa by the adjunction of a dopamine D2 agonist is also only partly and transitorily useful, and requires further validation. One can also try to add a non dopaminergic drug, for symptomatic antidyskinetic purposes, and amantadine is then the best therapeutic option. The adjunction of clozapine cannot be recommended in the absence of well-controlled studies with regard to its haematological toxicity, while the usefulness of other " atypical " neuroleptics, remains questionable. If none of these options improves the patient, it is then reasonable to offer a treatment with a subcutaneous pump of apomorphine, in the countries where the drug is available, and to discuss as a final step, the indication of pallidotomy or subthalamic stimulation.

REFERENCES

Agid Y. Levodopa. Is toxicity a myth ? Neurology 1998 ; 50 : 858-863.

Baldwin D, Fineberg N, Montgomery S. Fuoxetine, fluvoxamine and extrapyramidal tract disorders. Int Clin Psychopharmacol 1991 ; 6 : 51-58.

Baronti F, Mouradian MM, Conant KE, Giuffra M, Brughitta G, Chase TN. Partial dopamine agonist therapy of levodopa-induced dyskinesias. Neurology 1992b ; 42 : 1241-1243.

Bedard PJ, DiPaolo T, Falardeau P, Boucher R. Chronic treatment with levodopa but not bromocriptine induces dyskinesias in MPTP parkinsonian monkeys : correlations with [³H] spiperone binding. Brain Res 1986 ; 379 : 294-299.

Block G, Liss C, Reines S, Irr J, Nibbelink D. Comparison of immediate-release and controlled release carbidopa/levodopa in Parkinson's disease. A multicentre 5-year study. The CR First Study Group. Eur Neurol 1997 ; 37 : 23-27.

Bonifati V, Fabrizio E, Cipriani R, Vanacore N, Meco G. Buspirone in levodopa-induced dyskinesias. Clin Neuropharmacol 1994 ; 17 : 73-82.

Chase TN, Oh JD. Striatal dopamine- and glutamate-mediated dysregulation In experimental parkinsonism. Trends Neurosci 2000 ; 23 : S86-91.

Colzi A, Turner K, Lees AJ. Continuous subcutaneous waking day apomorphine in the long term treatment of levodopa induced interdose dyskinesias in Parkinson's disease. J Neurol Neurosurg Psychiatry 1998 ; 64 : 573-576.

Critchley P, Parkes D. Transdihydrolisuride in Parkinsonism. Clin Neuropharmacol 1987 ; 10 : 57-64.

Durif F, Vidailhet M, Assal F, Roche C, Bonnet AM, Agid Y. Low-dose clozapine improves dyskinesias in Parkinson's disease. Neurology 1997; 48 : 658-662.

Durif F, Vidailhet M, Bonnet AM, Blin J, Agid Y. Levodopa-induced dyskinesias are improved by fluoxetine. Neurology 1995 ; 45 : 1855-1858.

Goetz CG, Blasucci LM, Leurgans S, Pappert EJ. Olanzapine and clozapine : comparative effects on motor function In hallucinating PD patients. Neurology 2000 ; 55 : 789-794.

Goetz CG, Tanner CM, Klawans HL. Drug holiday in the management of Parkinson's disease. Clin Neuropharmacol 1982 ; 5 : 351-364.

Gomez-Mancilla B, Bédard PJ. Effect of Nondopaminergic drugs on L-DOPA-induced dyskinesias in MPTP-treated monkeys. Clin Neuropharmacol 1993 ; 16 : 418-427.

Grondin R, Bedard PJ, Britton DR, Shiosaki K. Potential therapeutic use of the selective dopamine D1 receptor agonist, A-86929 : an acute study in parkinsonian levodopa-primed monkeys. Neuroloy 1997 ; 49 : 421-426.

Hammerstad JP, Carter J, Nutt JG, Casten GC, Shrotriya RC, Alms DR, Temple D. Buspirone in Parkinson's disease. Clin Neuropharmacol 1986 ; 9 : 556-560.

Henry B, Brotchie JM. Potential of opioid antagonists in the treatment of levodopa-induced dyskinesias in Parkinson's disease. Drug Aging 1996; 9 : 149-158.

Henry B, Fox SH, Peggs D, Crossman AR, Brotchie JM. The alpha2-adrenergic receptor antagonist idazoxan reduces dyskinesia and enhances anti-parkinsonian actions of L-dopa in the MPTP-lesioned primate model of Parkinson's disease. Mov Disord 1999 ; 14 : 744-753.

Kanda T, Jackson MJ, Smith LA, Pearce RK, Nakamura J, Kase H, Kuwana Y, Jenner P. Adenosine A_{2A} antagonist : a novel antiparkinsonian agent that does not provoke dyskinesias in parkinsonian monkeys. Ann Neurol 1998 ; 43 : 507-513.

Klawans HL, Weiner WJ. Attempted use of halaperidol in the treatment of L-dopa-induced dyskinesias. J Neurol Neurosurg Psychiatry 1974 ; 37 : 427-430.

Kleedorfer B, Lees AJ, Stern GM. Buspirone in the treatment of levodopa induced dyskinesias. J Neurol Neurosur Psychiatry 1991 ; 54: 376-377.

Limousin P, Pollak P, Benazzouz A, Hoffmann D, Le Bas JF, Broussolle E, Perret JE, Benabid AL. Effect of parkinsonian signs and symptoms of bilateral subthalamic nucleus stimulation. Lancet 1995 ; 345 (8942) : 91-5.

Ludwig CL, Weinberger DR, Bruno G, Gillespie M, Bakker K, LeWitt PA, Chase TN. Buspirone, Parkinson's disease, and the locus coeruleus. Clin Neuropharmacol 1986 : 9 : 373-378.

Luquin MR, Laguna J, Obeso JA. Selective D-2 receptor stimulation induces dyskinesias in parkinsonian monkeys. Ann Neurol 1992b ; 31 : 551-554.

Luquin MR, Scipioni O, Vaamonde J, Gershanik O, Obseso JA. Levodopa-induced dyskinesias in Parkinson's disease : clinical and pharmacological classification. Mov Disorders 1992a ; 7 : 117-124.

Manson AJ, Iakovidou E, Lees AJ. Idazoxan is ineffective for levodopa-induced dyskinesias In Parkinson's disease. Mov Disord 2000 ; 15 : 336-337.

Manson AJ, Katzenschlager R, Hobart J, Lees AJ. High dose naltrexone for dyskinesias induced by levodopa. J Neurol Neurosurg Psychiatry 2001 ; 70 : 554-556.

Marconi R, Lefebvre-Caparros D, Bonnet AM, Vidailhet M, Dubois B, Agid Y. Levodopa-induced dyskinesias in Parkinson's disease Phenomenology and pathophysiology. Mov Disorders 1994 ; 9 : 2-12.

Meco G, Marini S, Lestingi L, Linfante I, Modarelli FT, Agnoli A. Controlled single-blind cross-over study of ritanserin and placebo in L-dopa-induced dyskinesias in Parkinson's disease. Cur Ther Res 1988 ; 43 : 262-270.

Merims D, Ziv I, Djaldetti R, Melamed E. Riluzole for levodopa-Induced dyskinesia In advanced Parkinson's disease. Lancet 1999 ; 353 ; 1764-1765.

Montastruc JL, Rascol O, Senard JM, Rascol A. A randomised controlled study comparing bromocriptine to which levodopa was later added, with levodopa alone in previously untreated patients with Parkinson's disease : a five year follow up. J Neurol Neurosurg Psychiatry 1994 ; 57 : 1034-1038.

Nutt JG, Rosin AJ, Eisler T, Calne DB, Chase TN. Effect of an opiate antagonist on movement disorders. Arch Neurol 1978 ; 35 : 810-811.

Olanow W, Schapira AH, Rascol O. Continuous dopamine-receptor stimulation In early Parkinson's disease. Trends Neuroscie 2000 ; 23 : S117-126.

Papa SM, Chase TN. Levodopa-induced dyskinesias improved by a glutamate antagonist in parkinsonian monkeys. Ann Neurol 1996 ; 39 : 574-578.

Parkinson Study Group. Effects of tocopherol and deprenyl on the progression of disability in early Parkinson's disease. Ann Neurol 1993; 328 : 176-183.

Parkinson Study Group. Impact of deprenyl and tocopherol treatment in Parkinson's disease in DATATOP patients requiring levodopa. Ann Neurol 1996 ; 39 : 37-45.

Parkinson Study Group. Pramipexole vs levodopa as initial treatment for Parkinson disease : a randomized controlled trial. Parkinson Study Group. JAMA 2000 ; 284 : 1931-1938.

Rascol A, Guiraud B, Montastruc JL, David J, Clanet M. Long-term treatment of Parkinson's disease with bromocriptine. J Neurol Neurosurg Psychiatry, 1979 ; 42 : 143-150.

Rascol O, Arnulf I, Peyro Saint Paul H, Brefel-Courbon C, Vidailhet M, Thalamas C, Bonnet AM, Descombes S, Bejjani B, Fabre N, Montastruc JL, Agid Y. Noradrenergic antagonism with idazoxan improves L-dopa-induced dyskinesias in parkinsonian patients. Mov Disord 2001 ; 16 : 708-713.

Rascol O, Brooks DJ, Korczyn AD, De Deyn PP, Clarke CE, Lang AE. A five-year study of the incidence of dyskinesias in patients with early Parkinson's disease who were treated with ropinirole or levodopa. 056 Study Group. N Engl J Med 2000 ; 342 : 1484-1491.

Rascol O, Fabre N, Blin O, Poulik J, Sabatini U, Senard JM, Ané M, Montastruc JL, Rascol A. Naltrexone, an opiate antagonist, fails to modify motor symptoms in patients with Parkinson's disease. Mov Disorders 1994 ; 9 : 437-440.

Rascol O, Nutt JC, Blin O, Goetz CG, Trugman JM, Soubrouillard C, Carter JH, Currier LJ, Fabre N, Thalamas C, Giardina WW, Wright S. Induction by dopamine D1 receptor agonist ABT-431 of dyskinesia similar to levodopa in patients with Parkinson disease. Arch Neurol 2001 ; 58 : 249-254.

Rinne UK, Brocco F, Chouza C, Dupont E, Gershanik O, Marti Mosso JF, Montastruc JL, Marsden CD, and the Parkinson's disease Study Group. Early treatment of Parkinson's disease with cabergoline delays the onset of motor complications. Results of a double-blind levodopa controlled-trial. Drugs 1998 ; 55 : S23-S30.

Rinne UK. Combined bromocriptine-levodopa therapy early in Parkinson's disease. Neurology 1985 ; 35 : 196-198.

Sandyk R, Snider SN. Naloxone treatment of L-dopa-induced dyskinesias in Pakinson's disease. Am J Psychiatry 1986 ; 143 : 118.

Schuh LA, Bennett JP. Suppression of dyskinesias in advanced Parkinson's disease : continuous intravenous levodopa shifts dose response for production of dyskinesias but not for relief of parkinsonism in patients with advanced Parkinson's disease. Neurology 1993 ; 43 : 1545-1550.

Tetsuo A, Yamaguchi M, Mizuta E, Kunos S. Effects of terguride, a partial D2 agonist, on MPTP-lesioned parkinsonian cynomolgus monkeys. Ann Neurol 1993 ; 33 : 507-511.

Trabucchi M, Bassi S, Frattola L. Effects of naloxone on the " on-off " syndrome in patients receiving long-term levodopa therapy. Arch Neurol 1982 ; 39 : 120-121.

Verhagen Metman L, Del Dotto P, Natte R, van den Munckhof P, Chase TN. Dextromethorphan improves levodopa-induced dyskinesias in Parkinson's disease. Neurology 1998b ; 51 : 203-206.

Verhagen Metman L, Del Dotto P, van den Munckhof P, Fang J, Mouradian MM, Chase TN. Amantadine as treatment for dyskinesias and motor fluctuations in Parkinson's disease. Neurology 1998a ; 50 : 1323-1326.

MANAGEMENT OF ANXIETY AND DEPRESSION IN PARKINSON'S DISEASE

Bhim S. Singhal, Jamshed A. Lalkaka and Ashit Sheth[*]

1. INTRODUCTION

Depression and anxiety are common in Parkinson's disease (PD). Nearly 40% of PD patients suffer from depression. Of these about half have major depression and the other half have minor depression and dysthymia.[1] Anxiety frequently coexists with depression and manifests either as generalized anxiety disorder (GAD) or panic disorder with or without agoraphobia.[2] Depression is often under diagnosed and under treated as features of depression and PD overlap. Anxiety and depression may also be subsyndromic. Recognition and treatment of anxiety and depression in PD are essential to improve the quality of life (QoL) in PD patients.

2. MANAGEMENT

2.1. Non-Pharmacologic Strategies

Parkinson's disease affects not just the patient, but also the spouse and caregivers. Anxiety, stress, diminished sleep and concerns about the future are often experienced by caregivers. These get aggravated if PD patients are also depressed and anxious.

Counseling and Behavioral Therapy are effective in improving symptoms of depression and anxiety in patients and caregivers. Patients should be encouraged to participate in support groups and community self help programs. PD nurse specialists have an important role to play. They assess problems, counsel patients and act as a link between the patient and the physician.[3]

[*] Bhim S. Singhal, Jamshed A. Lalkaka and Ashit Sheth, Bombay Hospital Institute of Medical Sciences, Mumbai, India

2.2. Drug Therapy for Anxiety in PD

The drugs commonly used for anxiety in PD include benzodiazepines, azapirones (buspirone) and antidepressants. *Benzodiazepines* are full agonists at GABA-benzodiazepine receptor complex. Commonly used benzodiazepines include alprazolam, clonazepam, diazepam, oxazepam, chlordiazepoxide, midazolam and clobazam. Benzodiazepines (especially alprazolam and clonazepam) have a rapid onset of action and are effective for short-term use in anxiety and panic disorders. In chronic anxiety states like GAD, the long-term use of these agents may cause dependence and withdrawal effects. Hence, other drugs like buspirone or tricyclic antidepressants should be preferred.

Azapirones: Buspirone is a 5HT 1A partial agonist. It is mainly indicated in GAD. It is also useful in major depressive disorder and mixed anxiety depression.[4] The main advantages of buspirone over the benzodiazepines are that it does not produce dependence or withdrawal effects and there are no drug interactions with alcohol, benzodiazepines and sedative-hypnotics. Its disadvantage is that it has a slow onset of action, (3-4 weeks). Newer related compounds like tandospirone and controlled release formulations are under trial.

2.3. Treatment of Panic Attacks

During an acute panic attack, benzodiazepines like alprazolam and clonazepam are drugs of choice. They are short-acting and cause rapid improvement of symptoms. If panic attacks are related to the off period of PD, antiPD treatment should be directed to reducing the off periods by antiparkinsonian medications.[5] For long term treatment, antidepressants like selective serotonin reuptake inhibitors (SSRIs) or tricyclic antidepressants (TCAs) are indicated. Though their onset of action is slow (3-8 weeks), they do not cause tolerance or withdrawal effects. To avoid the initial worsening of anxiety on initiation of these agents, benzodiazepines can be used concomitantly for the initial few weeks. The newer antidepressant drugs like venlafaxine and nefazodone have also been found useful in the long term treatment of panic attacks.[6]

2.4. Treatment of Depression in PD

Can AntiPD Drugs and Surgical Procedures Help in Depression?

Levodopa, the mainstay of antiPD treatment, has not proved effective as an antidepressant. The only situation in which levodopa may alleviate depression, is when depression is clearly linked to the off state of PD. Anticholinergic drugs like *trihexyphenidyl* and *biperiden* have shown antidepressant efficacy in some PD patients.[4] *Amantidine* is also known to have antidepressant properties.[4] *Bromocriptine* (dopamine receptor agonist) was shown to have antidepressant activity at high doses.[7] *Pergolide* has no effect on depression in PD. The role of newer dopamine receptor agonists like pramipexole as antidepressant in PD patients still needs to be evaluated.

Selegiline is a selective MAO-B inhibitor. It has useful antiparkinsonian and antidepressant activity especially when given in early PD. A French, multicentre, double blind, placebo controlled trial showed a significant improvement in Hamilton Depression Rating Scale scores after 3 months of treatment with selegiline 10mg/day.[8] *Deep brain stimulation* (DBS) is gaining popularity as a surgical procedure for advanced PD. De novo depression following subthalamic nucleus DBS has been reported recently.[9] *Human foetal mesencephalic transplant* caused long term depression and behaviour changes in patients without a prior psychiatric history.[10] There is a need for well defined studies of mood changes and cognitive function in surgical procedures for PD.

2.5. Antidepressant Drugs

The two most commonly used compounds for depression are the TCAs and the SSRIs. The *TCAs* exert their therapeutic effects through blockade of the reuptake pumps for both serotonin and norepinephrine. However, they also block 3 other receptors viz. muscarinic cholinergic, H_1 histamine and alpha1 adrenergic receptors resulting in the side effects of TCAs. Anticholinergic effects include dry mouth, blurred vision, urinary retention, constipation and confusion, especially in the elderly. H_1 receptor blockade causes sedation and weight gain. Effect on adrenergic receptor results in orthostatic hypotension.

Amitriptyline, nortriptyline, imipramine, desipramine and doxepin are the commonly used TCAs. Nortriptyline was found to improve mood with no change in parkinsonian signs.[11] Imipramine and desipramine, besides improving depression, were also found to have beneficial effect on the motor signs of PD.[12,13] In a study comparing amitriptyline and fluvoxamine in depressed PD patients, both drugs were found equally effective with comparable side effect profiles.[14]

SSRIs selectively inhibit the serotonin reuptake without blocking the cholinergic, histaminic or adrenergic receptors. Therefore they do not cause sedation, anticholinergic effects or orthostatic hypotension.[6] The other advantages over TCAs are that SSRIs do not affect cardiac conduction and do not decrease seizure threshold. The SSRIs commonly used for depression are fluoxetine, sertraline, paroxetine, fluvoxamine and citalopram. Unlike TCAs there are no large scale double blind trials of SSRIs for depression in PD. Open label studies, however, have demonstrated good efficacy of SSRIs in depressed PD patients.[15,16]

Adverse reactions have been described with SSRIs. Fluoxetine has been reported to cause extrapyramidal symptoms in depressed patients and also worsen the parkinsonian motor features in PD patients.[17,18] However, a retrospective chart review of depressed patients with PD treated with fluoxetine identified no worsening of parkinsonism.[19] SSRIs may sometimes cause akathisia and agitation.[20] Sexual dysfunction and sleep disturbance are other side effects. The "serotonin syndrome" though rare, can be caused by SSRIs, either alone or in combination with selegiline.[21] Selegiline should be discontinued prior to starting antidepressant therapy with SSRI or TCA.

2.6. Newer Antidepressant Drugs in PD

Newer antidepressants have appealing pharmacokinetic profiles and may be effective in patients with PD. *Venlafaxine* is a serotonin-norepinephrine reuptake inhibitor (SNRI). It shares the norepinephrine and serotonin reuptake inhibitory properties of classical TCAs, but without alpha 1, cholinergic or histamine receptor blocking properties.[20] The onset of therapeutic efficacy may be quicker (about 2 weeks) than conventional antidepressants. Its side effects are similar to those of SSRIs and it also has a tendency to diastolic hypertension.[22]

Trazodone and Nefazodone are serotonin antagonist-reuptake inhibitors (SARI). They have a strong action as 5 HT-2 receptor antagonists and weak action as 5 HT reuptake inhibitor. Being strong HT-2 receptor antagonists, they have an advantage over SSRIs in that they lack the undesirable side effects like agitation, anxiety, akathisia and sexual dysfunction. *Trazodone,* the first member of this group, is an effective antidepressant. Due to its histamine receptor blocking property, it is extremely sedating. It may be useful for depressed patients who have significant sleep disturbance. It may occasionally cause priapism. *Nefazodone* has a weak histamine receptor blocking property and hence is far less sedating than trazodone.[20] Nefazodone also has weak alpha antagonist properties compared to trazodone and hence, no significant orthostatic hypotension or priapism is known to occur with this drug. In a clinical trial, nefazodone was found to be as efficacious as imipramine in antidepressant efficacy.[23]

Bupropion is a norepinephrine and dopamine reuptake blocker (NDRI). Besides its modest antidepressant activity, it improves the motor symptoms of PD.[24] Since it lacks significant serotonergic activity, it does not have the bothersome side effects of SSRIs. The side effects of bupropion include nausea, vomiting, elevated blood pressure, hallucinations, confusion and also an increased tendency to seizures.[20]

Mirtazapine (alpha-2 antagonist) increases noradrenergic neurotransmission by direct antagonism of alpha-2 adrenergic autoreceptors. Its efficacy as an antidepressant is similar to that of amitriptyline and better tolerated, without significant anticholinergic, serotonergic or adrenergic side effects.[25] Mirtazapine causes sedation and weight gain.

Moclobemide is a reversible MAO-A inhibitor. It does not produce the hypertensive crises (cheese effect) which occurs with other MAO inhibitors.[26] It is thus a safe drug and its efficacy is similar to other MAO-inhibitors.

2.7. Principles of Antidepressant Drug Therapy

The commonly used antidepressants, (TCAs and SSRIs) take 3-6 weeks for onset of clinical benefit. Hence, each drug should be given for at least 6 weeks in adequate dose to assess therapeutic efficacy. Once clinical benefit occurs, the therapy should be continued for at least 6 to 12 months to minimize chances of a relapse. Antidepressant drugs, especially TCAs, must be used with caution in the elderly in view of the anticholinergic side effects on cognition and sphincters.

2.8. Electroconvulsive Therapy (ECT)

The main indications for ECT in depressed PD patients are major depression refractory to optimal antidepressant drug therapy and depression with suicidal ideation where quick action is required. ECT also significantly improves the motor symptoms of PD. Adverse effects of ECT include memory loss, confusion, slurred speech, and tremors. Despite these side effects, ECTs have shown better efficacy and tolerability than TCA therapy.[27] Improvement has to be sustained with the use of antidepressant drugs to prevent relapse.

2.9. Transcranial Magnetic Stimulation (TMS)

This procedure is still under trial. Rapid rate stimulation provides temporary relief from depression under experimental conditions.[28] TMS exposes focal regions of the brain to magnetic and electrical energy that does not cause structural damage, but induces functional changes. Drawbacks of TMS include a temporary therapeutic effect, muscle tension headache at the site of stimulation and a risk of seizures.

3. CONCLUSION

Depression and anxiety are common and frequently coexist in PD. They are often under diagnosed and poorly treated. As they significantly affect QoL in PD, they must be treated. Besides counseling, the common antidepressants used are TCAs and SSRIs. Prior to use of these drugs, selegiline should be withdrawn to avoid the risk of serotonin syndrome. For patients refractory to standard antidepressants, newer drugs like venlafaxine, nefazodone and mirtazapine may be tried. ECT should be used for patients refractory to drug therapy and those having major depression with suicidal thoughts.

REFERENCES

1. J. L. Cummings. Depression and Parkinson's disease: A review. *Am. J. Psychiatry.* **149**, 443-454, (1992).
2. M. B. Stein, I. J. Heuser, J. L. Juncos, T. W. Uhde. Anxiety disorders in patients with Parkinson's disease. *Am. J. Psychiatry.* **147**, 217-220 (1990).
3. D. G. MacMahon. Parkinson's disease nurse specialists. An important role in disease management. *Neurology.* **52 (Suppl.3)**, S21-S25 (1999)
4. J. M. Silver, S. C. Yudofsky. Drug Treatment of Depression in Parkinson's disease. In: Parkinson's Disease: Neurobehevioral Aspects, edited by S. J. Huber and J. L. Cummings (Oxford University Press, New York,1992), pp 240-254.

5.   A. Lieberman. Managing the neuropsychiatric symptoms of Parkinson's disease. *Neurology.* **50 (Suppl.6)**, S33-S38 (1998)

6.   T. Tom, J. L. Cummings. Depression in Parkinson's disease. Pharmacological characteristics and treatment. *Drugs and Aging.* **12(1)**, 55-74 (1998)

7.   R. Jouvent, P. Abensur, A. M. Bonnet et al. Antiparkinsonian and antidepressant effects of high doses of bromocriptine. *J. Affect. Disord.* **5**, 141-145 (1983).

8.   H. Allain, P. Pollak, H. C. Neukirch. Symptomatic effect of selegiline in de novo parkinsonian patients: the French Selegiline Multicentre Trial. *Mov. Disord.* **8 (Suppl.1)**, S36-S40 (1993).

9.   B. P. Bejjani, P. Damier, I. Amulf et al. Transient acute depression induced by high frequency deep-brain stimulation. *N. Engl. J. Med.* **340**, 1476-1480 (1999).

10. L. H. Price, D. D. Spencer, K. L. Marek et al. Psychiatric status after human foetal mesencephalic tissue transplantation in Parkinson's disease. *Biol. Psychiatry.* **38**, 498-505 (1995).

11. J. Andersen, E. Aabro, N. Gulmann, A. Hjelmsted, H. E. Pedersen. Anti-depressant treatment in Parkinson's disease : a controlled trial of the effect of nortriptyline in patients with Parkinson's disease treated with L-dopa. *Acta. Neurol. Scand.* **62**, 210-219 (1980).

12. R. R. Strang. Imipramine in treatment of parkinsonism : a double-blind placebo study. *Brit. Med. J.* **2**: 33-34 (1965).

13. L. Laitenen. Desipramine in treatment of Parkinson's disease. *Acta. Neurol. Scand.* **45**, 109-113 (1969).

14. J. M. Rabey, E. Orlov, A. D. Korczyn. Comparison of fluvoxamine and amitriptyline for the treatment of depression in PD. *Neurology.* **96**, 374 (1996).

15. R. A. Hauser, T. A. Zesiewicz. Sertraline for the treatment of depression in Parkinson's disease. *Mov. Disord.* **12**, 756-759 (1997).

16. W. Wittgens, O. Donath, U. Trenckmann. Treatment of depressive syndromes in Parkinson's disease with paroxetine. Mov. *Disord.* **12**, 128 (1997).

17. E. N. Steur. Increase of Parkinson disability after fluoxetine medication. *Neurology.* **43**, 211-213 (1993).

18. R. H. Bouchard, E. Pourcher, P. Vincent. Fluoxetine and extrapyramidal side effects. *Am. J. Psychiatry.* **146**, 1352-1353 (1989).

19. C. F. Caley, J. H. Friedman. Does fluoxetine exacerbate Parkinson's disease? *J. Clin. Psychiatry.* **53**, 278-282 (1992).

20. A. Stoudermire. New antidepressant drugs and the treatment of depression in the medically ill patient. *Psychiar. Clin. North Am.* **19**, 495-514 (1996).

21. I. Richard, R. Kurlan, C. Tanner et al. Serotonin syndrome and the combined use of deprenyl and an antidepressant in Parkinson's disease. Parkinson Study Group. *Neurology.* **48(4)**, 1070-1077 (1997).

22. L. A. Cunningham. Depression in the medically ill: choosing an antidepressant. *J. Clin. Psychiatry.* **55(Suppl.A)**, 90-97 (1994).

23. R. Fontaine, A. Ontiveros, R. Elie et al. A double blind comparison of nefazodone, imipramine and placebo in major depression. *J. Clin. Psychiatry.* **55**, 234-241 (1994).

24. C. G. Goetz, C. M. Tanner, H. L. Klawans. Bupropion in Parkinson's disease. *Neurology.* **34**, 1092-1094 (1984).

25. S. Kasper. Clinical efficacy of mirtazapine: A review of metaanalyses of pooled data. *Int. Clin. Psychopharmacol.* **10 (Suppl. 4)**, 25-35 (1995).

26. W. Haefely, W. P. Burkard, A. Cesura et al. Pharmacology of moclobemide. *Clin. Neuropharmacol.* **16 (Suppl. 2)**, S8-S18 (1993)

27. R. Faber, M. R. Trimble. Electroconvulsive therapy in Parkinson's disease and other movement disorders. *Mov. Disord.* **6**, 293-303 (1991).

28. M. S. George, E. M. Wassermann, R. M. Post. Transcranial magnetic stimulation: a neuropsychiatric tool for the 21st century. *J. Neuropsychiatry Clin. Neurosci.* **8**, 373-382 (1996).

TANDOSPIRONE CITRATE, A NEW SEROTONERGIC ANXIOLYTIC AGENT: A POTENTIAL USE IN PARKINSON'S DISEASE

Yukihiro Ohno *

1. INTRODUCTION

Anxiety is one of the symptoms frequently observed in patients with Parkinson's disease. The incidence rate of anxiety in Parkinson's disease has been reported to be up to 40%, which is higher than in normal or other disease population[1]. In addition, anxiety often coexists with depression. It is also suggest that anxiety and depression may not psychological reaction to the illness, but rather they may be linked to certain neurobiological basis associated with Parkinson's disease, such as functional alterations of dopaminergic and serotonergic systems[1]. However, mechanisms or optimal pharmacological treatment for anxiety and depression in the Parkinson's disease have not been established.

Recently, 5-HT_{1A} receptor agonists were developed as the novel anxiolytic agents that lack the benzodiazepine-like side effects. Buspirone is the prototype of this class of agents whereas it possesses potent dopamine D_2 blocking actions. Tandospirone citrate (Sediel®) is the newer 5-HT_{1A} receptor-related anxiolytic agent that is widely used in Japan for the treatment of anxiety and depression in patients with psychosomatic disorders and neurosis. In this paper, the pharmacological profile of tandospirone will be reviewed in comparison with diazepam and buspirone, and its potential benefits for the treatment of patients with Parkinson's disease will be discussed.

Yukihiro Ohno, Discovery Research Laboratories I, Research Center, Sumitomo Pharmaceuticals Co., Ltd., 3-1-98 Kasugade-naka, Konohana-ku, Osaka, Japan 554-0022

Mapping the Progress of Alzheimer's and Parkinson's Disease
Edited by Mizuno *et al.*, Kluwer Academic/Plenum Publishers, 2002

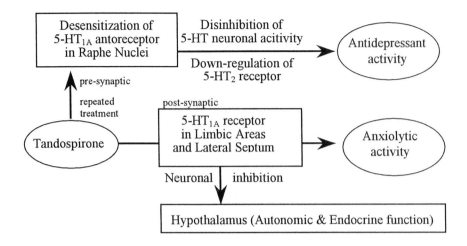

Fig. 1 Mechanisms for anxiolytic and antidepressant actions of tandospirone

2. SELECTIVE INTERACTION WITH 5-HT$_{1A}$ RECEPTORS

Tandospirone shows a high affinity for 5-HT$_{1A}$ receptors with a Ki value of 25 nM. It negligibly interact with other neurotransmitter receptors including 5-HT$_{1B}$, 5-HT$_2$, dopamine D$_1$, D$_2$, noradrenergic α_1, α_2, β, GABA and benzodiazepine receptors[2]. Autoradiography studies[3] revealed that the specific binding sites of tandospirone mainly localized in the three brain regions, the limbic areas (e.g., hippocampus, entorhrinal cortex and amygdala), lateral septum and raphe nuclei (Fig.1). Distribution pattern of the binding sites of tandospirone corresponded well to that of 8-OH-DPAT, a selective 5-HT$_{1A}$ receptor agonist, but not to benzodiazepine receptors which distributed throughout the brain. Tandospirone acts as an agonist at 5-HT$_{1A}$ receptors with an intrinsic activity of about 80-90%, and it inhibits the neuronal activity in the hippocampus, the lateral septum and the dorsal raphe nucleus (Fig. 1)[4-6].

3. ANXIOLYTIC AND ANTIDEPRESSANT EFFECTS

Tandospirone shows significant ameliorative effects in various models of anxiety and stress. It significantly increased the number of punished responding in the Vogel or Geller-Seifter conflict test in a dose-dependent manner (Table 1)[7, 8]. These actions mimicked those of diazepam and buspirone, suggesting that tandospirone is a clinically effective anxiolytic agent. It is likely that stimulation of the post-synaptic 5-HT$_{1A}$ receptors in the hippocampus play an important role in its anxiolytic activity (Fig. 1), since the anticonflict action of tandospirone was observed in animals treated with

5,7-dihydroxytryptamine[9], which destroyed the pre-synaptic $5-HT_{1A}$ receptors, and a microinjection of tandospirone into the dorsal hippocampus exerted a significant anticonflict effect[10].

Unlike benzodiazepines, tandospirone possesses significant antidepressant actions. It reduces the immobility time of rats in the forced swim test and produces a down-regulation of the cortical $5-HT_2$ receptors following its repeated administration (Table 1)[11, 12]. Godbout et al.[5] showed that chronic treatment with tandospirone caused desensitization of the pre-synaptic $5-HT_{1A}$ autoreceptors in the dorsal raphe without affecting the post-synaptic $5-HT_{1A}$ receptors in the hippocampus, which consequently disinhibits the activity of 5-HT neurons (Fig. 1). These actions were similar to those of the antidepressant agents, suggesting that tandospirone would be effective in treating depression.

Table 1. Pharmacological actions of tandospirone

Pharmacology tests	ED50 or MED (mg/kg)	
	Tandospirone	Diazepam
Anxiolytic action		
Vogel conflict test	20 (p.o.) 5 (i.p.)	20 (p.o.) 2.5 (i.p.)
Geller-Seifter conflict test	20 (p.o.) 1.25 (i.p.)	- 2.5 (i.p.)
Antidepressant action		
Muricide test	5 (i.p.)	-
Forced swim test	10 (s.c.)	NE
$5-HT_2$ receptor down regulation	20 (s.c.)	-
Antiparkinsonian action		
Haloperidol-induced catalepsy	10 (i.p.)	-
Rotation behavior in 6-OH-DA treated rats	10 (i.p.)	-
Reserpine-induced hypolocomotion	10 (i.p.)	-

NE: No effects

4. ANTI-PARKINSONIAN EFFECTS

In our recent studies[13], tandospirone has been shown to improve the motor symptoms in several animal models of Parkinson's disease (Table 1). Tandospirone significantly inhibited the extrapyramidal sign (i.e., catalepsy) induced by the D_2 antagonist haloperidol. In the hemi-lesion model of dopamine neurons by 6-hyrdoxydopamine, tandospirone elicited the rotation behavior contralaterally to the lesion site. It also ameliorated hypolocomotion induced by monoamine depletion with reserpine. These actions mimicked those of bromocriptine or L-DOPA, suggesting that tandospirone would be effective in treating parkinsonian symptoms. In addition, the anti-parkinsonian action of tandospirone was blocked by the 5-HT_{1A} antagonist (i.e., WAY-100635) but not by the D_2 antagonist (i.e., haloperidol), indicating that its anti-parkinsonian effects are mediated by non-dopaminergic mechanisms.

Table 2.   Safety profile of tandospirone

Pharmacology tests	ED50 or MED (mg/kg, p.o.)		
	Tandospirone	Buspirone	Diazepam
Muscle relaxation Traction test	>300	>300	3.9

Impaired motor coordination
    Rota-rod test (mice)
    Rota-rod test (rats)
	207
	>300
		172
		125
			2.1
			9.1

Potentiation of anesthesia
    Hexobarbital test
    Ethanol test
	212
	>300
		263
		>300
			1.8
			3.0

Cognitive impairment Passive avoidance test	>80	-	5

Anti-dopamine action
    Apomorphine climing test
    Apomorphine stereotypy test
	188
	305
		48
		44
			>100
			-

5. SAFETY PROFILE

Tandospirone is different from benzodiazepines in the safety profile. Diazepam has potent CNS depressive actions and induces hypnotic/sedative effects, impairment of motor coordination, muscle relaxation and cognitive dysfunction at doses which exert anticonflict effects (Table 2)[5]. Benzodiazepines are also known to possess liability for drug dependence and abuse. By contrast, 5-HT$_{1A}$ receptor-related anxiolytic agents, tandospirone and buspirone, do not produce the benzodiazepine-like side effects even at high doses[5]. Furthermore, tandospirone was much weaker than buspirone in inducing dopamine D$_2$ blocking action (Table 2)[5].

6. POTENTIAL BENEFITS FOR PARKINSON'S DISEASE

Potential benefits of tandospirone for the treatment of Parkinson's disease are supported by the following characteristics; 1) it produces both anxiolytic and antidepressant activity, 2) it lacks the benzodiazepines-like side effects including sedation, muscle relaxation, impairment of motor coordination and cognitive function, drug dependence, 3) it is much weaker than buspirone, a prototype of 5-HT$_{1A}$ receptor-related anxiolytic agent, in inhibiting dopaminergic system, 4) it ameliorates motor symptoms in animal model of Parkinson' disease. Recently, Nomoto et al.[14] demonstrated that tandospirone improved motor symptoms (e.g., walking stability and activity of daily life) in about 40% of the patients with Parkinson's disease. These patients were suffered from Parkinson's disease longer than the non-responders and were in more advanced stage on their daily life activity. Although psychotic symptoms (e.g., anxiety and depression) were not specifically monitored in that study, these findings are consistent with our results and suggest that tandospirone exerts anti-parkinsonian effects and is useful for the treatment of Parkinson's disease. Further studies are required to confirm the clinical benefits of tandospirone and to elucidate the mechanisms of its action in Parkinson's disease.

References

1. I. H. Richard, R. B. Schiffe, and R. Kurlan, Anxiety and Parkinson's disease, J. Neuropsychiat., 8, 383-392 (1996).
2. H. Shimizu, N. Karai, A. Hirose, T. Tatsuno, H. Tanaka, Y. Kumasaka, and M. Nakamura, Interaction of SM-3997 with serotonin receptors in rat brain, Jpn. J. Pharmacol., 46, 311-314 (1988)
3. H. Tanaka, H. Shimizu, Y. Kumasaka, A. Hirose, T. Tatsuno, and M. Nakamura, Autoradiographic localization and pharmacological characterization of [^{3}H]-tandospirone binding sites un the rat brain, Brain Res., 546, 181-189 (1991)
4. H. Tanaka, T. Tatsuno, H. Shimizu, A. Hirose, Y. Kumasaka, and M. Nakamura,

Effects of tandospirone on second messenger systems and neurotransmitter release in the rat brain, Gen. Pharmacol., 26, 1765-1772 (1995)

5. R. Godbout, Y. Chaput, P. Blier, and C. De Montigny, Tandospirone and its metabolite, 1-(2-pyrimidinyl)-piperazine. I. Effects of acute and long-term administration of tandospirone on serotonin neurotransmission, Neuropharmacology, 30, 679-690 (1991)

6. Y. Ohno, K. Ishida, T. Ishibashi, H. Tanaka, H. Shimizu, and M. Nakamura, Effects of tandospirone, a selective 5-HT$_{1A}$ receptor agonist, on activities of the lateral septal nucleus neurons in cats, in: *Serotonin in the Central Nervous System and Periphery*, edited by A. Takada and G. Cuzon (Elsevier Science B.V., New York), pp. 159-165 (1995)

7. H. Shimizu, A. Hirose, T. Tatsuno, Y. Kumasaka, M. Nakamura, and J. Katsube, Pharmacological properties of SM-3997; A new anxioselective anxiolytic candidate, Jpn. J. Pharmacol., 45, 493-500 (1987)

8. H. Shimizu, Y. Kumasaka, T. Tatsuno, A. Hirose, and M. Nakamura, Anticonflict action of tandospirone in a modified Geller-Seifter conflict test in rats, Jpn. J. Pharmacol., 58, 283-289 (1992)

9. H. Shimizu, T. Tatsuno, H. Tanaka, A. Hirose, Y. Araki, and M. Nakamura, Serotonergic mechanisms in anxiolytic effects of tandospirone in the Vogel conflict test, Jpn. J. Pharmacol., 59, 105-112 (1992)

10. Y. Kataoka, K. Shibata, A. Miyazaki, Y. Inoue, K. Tominaga, S. Koizumi, S. Ueki, and M. Niwa, Involvement of the dorsal hippocampus in mediation of the antianxiety action of tandospirone, a 5-hydroxytryptamine$_{1A}$ agonistic anxiolytic. Neuropharmacology, 30, 475-480 (1991)

11. S. Wieland, and I. Lucki, Antidepressant-like activity of 5-HT$_{1A}$ agonists measured with the forced swim test, Psychopharmacology, 101, 497-504 (1990)

12. S. Wieland, C. T. Fischette, and I. Lucki, Effect of chronic treatments with tandospirone and imipramine on serotonin-mediated behavioral responses and monoamine receptors, Neuropharmacology, 32, 561-573 (1993)

13. T. Ishibashi, Y. Ohno, and H. Noguchi, Ameliorative effects of tandospirone citrate on motor disturbances in Parkinson's disease models, Jpn. J. Psychopharmacol. (in Japanese), 18, 348 (1998)

14. M. Nomoto, S. Iwata, S. Kaseda, M. Mitsuda, T. Shimizu, and T. Fukuda, A 5-HT$_{1A}$ receptor agonist, tandospsirone improves gait disturbance of patients with Parkinson's disease, JMDD, 7, 65-70 (1997)

STEREOTACTIC SURGERY OF PARKINSON'S DISEASE OVER 30 YEARS

Chihiro Ohye

1. INTRODUCTION

Stereotactic surgery of Parkinson's disease has already some 50 years' history since the idea of stereotaxy was first applied to the human brain by Spiegel, Wycis, et al., (1947). Thereafter, the stereotactic surgery developed together with the advent of Neuroscience in one hand, the Technology in the field of computer system in the other.

In this paper, a short survey of the development of stereotactic surgery over (beyond) these 30 years, with its up and down, will be given. Interested readers may refer to the more comprehensive historical survey on stereotactic surgery for Parkinson's disease by Redfern (1989).

2. NAISSANE OF STEREOTACTIC PALLIDOTOMY

The first stereotactic surgery using modified Horsley–Clarke apparatus was performed for a case with psychiatric problems targeting the thalamic dorsomedial nucleus with the idea to minimize the undue damage of overlapping brain tissue. Not only the idea of minimal invasion into the human brain, but also already from the very beginning, we can recognize another important principle of stereotaxy: an application of the knowledge of basic science in those days, for this case, that the thalamic dorsomedial nucleus has rich connection with the frontal lobe, the target for lobotomy or lobectomy. Then, they tried pallidotomy also for cases with choreoathetotic movement (Spiegel and Wycis, 1950a, 1950b).

At almost the same period after the World War II, there were several similar tendencies to use own stereotactic apparatus to the human brain in the world wide scale, in America, Sweden, Germany, France and Japan. In this regard, we cannot go into this field without mentioning a great contribution of Narabayashi, who invented almost independently his own stereotactic apparatus and performed the first pallidotomy for Parkinson's disease as

Chihiro Ohye, Functional and Gamma Knife Surgery Center, Hidaka Hospital Takasaki, Gunma, Japan

Mapping the Progress of Alzheimer's and Parkinson's Disease
Edited by Mizuno *et al.*, Kluwer Academic/Plenum Publishers, 2002

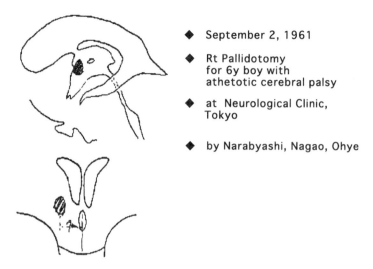

◆ September 2, 1961

◆ Rt Pallidotomy
  for 6y boy with
  athetotic cerebral palsy

◆ at Neurological Clinic,
  Tokyo

◆ by Narabyashi, Nagao, Ohye

Figure 1. An example of a lesion (hatched area) by stereotactic pallidotomy of old style, performed in 1961 by Narabayashi et al. Note that the lesion is more anterior and medial than the present style. Upper: Lateral view with ventricles. Lower: Frontal view.

early as in 1952 (Narabayashi and Okuma, 1953). After graduating Tokyo University, Faculty of Medicine, he entered in the Department of Psychiatry, where Professor Uchimura was the Chairman. Among relatively free and liberal atmosphere of the Department, Narabayashi was interested in problems of movement disorders. According to him, he was enchanted to the field of ivoluntary movement after reading monograph on Extrapyramidal Disease by A. Jacob, a distinguish German Neurologist. But, in fact, the first lesion to the relatively young Parkinsonian patient was not in the globus pallidus but caudate nucleus without noticeable change in parkinsonian symptoms of tremor nor rigidity. So, Narabayashi asked for the second operation about one month later and the patient kindly accepted to try again. Then, Narabayashi made a procaine oil-wax mechanical lesion in the globus pallidus and to their surprise, tremor and rigidity disappeared immediately without motor paresis. It was really an exciting moment, on June 4th, 1952. Professor Uchimura had keen insight into this remarkable result and encouraged Narabayashi to continue the Stereotactic Operation as it contained something new and true scientific importance. In this sense, Narabayashi was lucky to start his work under such generous guidance.

Until 1956, he modified twice his stereotactic apparatus, and with his third model, he tried 25 cases and reported the results in Archives of Neurology (Narabayashi, Okuma, Shikiba, 1956). They soon noticed that pallidal lesion (pallidal target at that time was anterior part of the internal segment of globus pallidus as shown in Figure 1), had excellent continuous effect on rigidity but on tremor, it was only transient. This conclusion hold true even today (Narabayashi, 1997, Narabayashi et al., 1997), in the modern era of newly revived pallidotomy (Laitinen et al, 1992). Since this pioneer work, Narabayashi devoted whole his life for the development of stereotactic surgery and for the understanding of movement disorders by stereotactic surgery. (To our regret, he died on March 18, 2001, just before this Meeting in Kyoto.)

It is interesting to tell an anecdote here. Almost at the same time, Spiegel and Wycis performed pallidotomy also for Parkinson's disease, and between Narabayashi they have debated about the priority of the pallidotomy. After several conflicts, Spiegel and Wycis recognized that Narabayashi did a couple of month ealier, in reality. Thereafter, they kept long lasting very good friendship each other.

3. THALAMOTOMY AND SELECTIVE THALAMOTOMY

Because of the transient effect of pallidal surgery on parkinsonian tremor, Hassler and Riechert (1954) claimed after anatomo-pathological study to change the target from pallidum to thalamic nucleus ventralis oralis (Vo). According to Hassler (1949), the Vo nucleus was divided in two parts, ventrooralis anterior (Voa) and ventrooralis posterior (Vop), the former received pallidal afferent and projected to cortical area 6, and the latter received cerebellar afferent to cortical area 4. Therefore, a coagulation of Voa is good for rigidity and a coagulation of Vop is effective for tremor. And in fact, clinically, relatively large lesion in Vo area ameliorated rigidity as well as tremor. Many attempts were made to make a thalamic lesion in and around Vo nucleus covering really a wide area of ventro-lateral part of the thalamus.

In order to get more constant and effective clinical results, we started to make selective lesion after physiological identification of each thalamic nuclei by depth recording using microelectrode, the technique of which was introduced by French group (Albe-Fessard, Arfel, Guiot, 1963). Systematic recording of the background activity and spike discharges from thalamic target area, mainly Vo, Ventralis intermedius (Vim) and ventralis caudalis (Vc), led us to integrate into a conclusion about the functional organization of the human thalamus (Ohye, 1982, 1997). It is essentially the same as Hassler's parcellation (1977) and modified by physiological findings obtained by depth recording in the human thalamus during the course of stereotactic operation (Ohye, 1998a). For parkinsonian tremor and other kinds of tremor, only a small area identified by rhythmic discharge and kinesthetic neurons is essential to get almost permanent arrest of the tremor of whatever kind (Ohye et al, 1989, Ohye, 1998b). This particular zone is supposed to be the most lateral part of the Vim nucleus, thus we call it as the selective Vim thalamotomy (Ohye, 1990).

For the treatment of parkinsonian rigidity, we do Vo thalamotomy after physiological identification of Vim nucleus first and then turned the coagulation needle toward anterior to cover the Vop nucleus. By this method of Vo thalamtomy, we can treat not only parkinsonian rigidity but also so-called Dopa-induced dyskinesia (Narabayashi et al., 1984) and several types of dystonia.

4. NEW HORIZON OF STEREOTAXY AFTER L-DOPA ERA

Since the end of 1960 or the beginning of 1970, the drug therapy by L-Dopa was introduced and it rapidly prevailed all over the world because of its miracle but theoretical effects on most of the parkinonian symptoms. Consequently, the stereotactic surgery was once shaded behind the medical treatment. Impact of the initial effect of L-Dopa was so strong that many centers of stereotactic surgery almost stopped their activity, except some centers. Narabayashi's group was one of them which continued constantly the stereotactic operations as before with the idea of selective thalamotomy with microrecording as mentioned above.

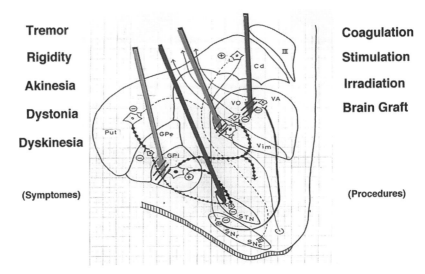

Figure 2. Schematic illustration of alternatives of modern stereotactic surgery for PD. Center: Simplified neural circuits of the basal ganglia   are shown with arrows toward different targets. Arrow with short oblique bars means stimulation or destruction, only arrow meaans stimulation. Choice of different   operative procedures (right column) and target depends   on symptomes (left column) of each patient.

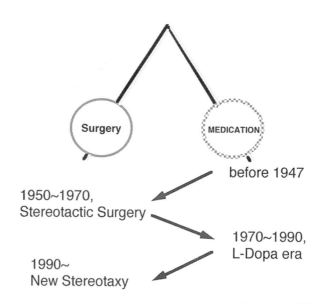

Figure 3.   To and fro, pendulum movement between medical and surgical treatment of PD since the beginning of the stereotactic surgery.   Recent tendency seems to be in favor of surical side.

Meanwhile, a long-term use of L-Dopa revealed unexpected clinical hazards such as L-Dopa-induced dyskinesia, on-off phenomena, wearing off etc. On the other hand, there emerged new techniques of computerized imaging. Together with the rapid progress of Neuroscience including understanding of Parkinson's disease itself, a new horizon of stereotactic surgery opens since around 1990. In fact, nowadays, we have several different alternative ways to treat Parkinson's disease surgically. For example, as the target of surgery, we have not only thalamic nuclei but also internal segment of posteroventral pallidum and the subthalamic nucleus. Also selective Vim thalamotomy is now extended to selective Vo thalamotomy as mentioned above and combined Vim-Vo thalamotomy for particular types of involuntary movement. Method of surgical treatment is increasing also, from usual stereotactic radiofrequency coagulation to chronic stimulation by implanted electrode, stereotactic radiosurgery, and brain graft for replacing the damaged nervous tissue (Figure 2).

There remain many problems to be solved in each modern stereotactic surgery. For example, we do not know yet which target is better to get best clinical result, which operative procedure should be chosen: destruction or chronic stimulation etc. There are certainly advantage and disadvantage in each method. For the future, among them, a new trend of brain graft using stem cell seems to be the most promising in the near future in the sense that it is the only method to give new energy to damaged nervous system (Isacson et al., 2001).

Looking back the development of the stereotactic surgery of Parkinson's disease, we see to and fro pendulum like movement between medical (drug) and surgical treatment with the cycle of about 20 years as shown in Figure 3. At present, we are optimistic about the present state, which is in favor of surgical treatment with several new ideas and techniques.

REFERENCES

Albe-Fessard, D., Arfel, G., et Guiot, G., 1963, Activites electriques caracteristiques de quelques structures ceebrales chez l'homme. Ann Chir.,17:1185

Hassler, R., 1949, Uber die afferenten Bahnen und Thalamuskerne des motorischen Systems des Gehirns I und II, Arch Psychiat., 182:759 und 786

Hassler, R., 1977, Architectonic organization of the thalamic nuclei. in: *Atlas for Stereotaxy of The Human Brain.* Ed 2, G. Schaltenbrand. and W. Wahren, eds., G. Thieme, Stuttgart.

Hassler, R., Riechert, T., 1954, Indikationen und Lokalizationsmethode der gezielten Hirnoperationen. *Nervenarzt,* 25:441-447.

Isacson, O., Costantini, L., Schumacher, J. M., Cicchetti, F., Chung, S., Kim K., 2001, Cell implantation therapies for Parkinson's disease using neural stem, transgenic or xenogeneic donor cells. *Parkinsonism Relat. Disord.,* 3:205.

Laitinen, L.V., Bergenheim, A.T., and Mariz, M.I., 1992, Leksell's posteroventral pallidotomy in the treatment of parkinson's disease. *J. Neurosurg .* 76: 53.

Narabayashi H., 1997, Pallidotomy revisited. Analysis of posteoventral pallidotomy *Stereotac func Neurosurg.* 69:54.

Narabayashi, H., and Okuma T., 1953, Procaine oil blocking of the globus pallidus for the treatment of rigidity and tremor of Parkinsonism. *Proc. Jap. Acad.* 29(3): 33 Narabayashi, H., Okuma, T., and Shikiba S., Procaine oil blocking of the globus pallidus. *Arch. Neurol.* 75:36.

Narabayashi, H., Miyashita, N., Hattori, Y., Saito, K., and Endo, K., 1997 Posteroventral pallidotomy :its effect on motor symptoms and scores of MMPI test in patients with Parkinson's disease. *Parkinsonism & Related Disord.* 3:7-20

Narabayashi, H., Yokochi, F., and Nakajima, Y., 1984, Levodopa-induced dyskinesia and thalamotomy. *J. Neurol. Neurosurg. Psychiat.* 47:831.

Ohye, C., 1982, Depth microelectrode studies. in *Stereotaxy of the Human Brain.* G. Scaltenbrand and A.E.Walker eds., G. Thieme, Stuttgart, New York, pp372-389.

Ohye, C., 1997, Functional organization of the human thalamus: Stereotactic intervention. In *Thalamus. Vol. II,Experimental and Clinical Aspects*. M. Steride, E.G. Jones and D.A. McCormick, eds., Elsevier Science, Amsterdam, pp517-542

Ohye C: 1998a, Thalamotomy for Parkinson's disease and other types of tremor. Part 1, Historical background and technique. in *Textbook of Stereotactic and Functional Neurosurgery*. P.L. Gildenberg, R.R. Tasker, eds., MacGraw-Hills, New York. pp 1167-1178.

Ohye, C, 1998b, Neural Noise Recording in Functional Neurosurgery. in *Textbook of Stereotactic and Functional Neurosurgery*. P.L. Gildenberg, R.R. Tasker, eds., MacGraw-Hill, New York. pp 941-947.

Ohye, C., Shibazaki, T., Hirai, T., Wada, H., Hirato, M.,Kawashima,Y., 1989, Further physiological observations on the ventralis intermedius neurons in the human thalamus. *J. Neurophysiol.*, 61:488.

Ohye, C., Shibazaki, T., Hirato, M., Kawashima, Y., Matsumura, M., 1990, Strategy of selective VIM-thalamotomy guided by microrecording. *Stereotac. Funct. Neurosurg.*, 54/55:186.

Redfern R.M., 1989, History of stereotactic surgery for Parkinson's disease. *British J Neurosurg.*, 3:271.

Spiegel, E.A., Wycis, H.T., Marks, M., and Lee,A.J., 1947, Stereotaxic apparatus for operation on the human brain. *Science*, 106:349.

Spiegel,E.A., Wycis, H.T., 1950a, Pallidothalamotomy in chorea. *Arch. Neurol.Psychiat* 64:295.

Spiegel E.A., Wycis, H.T., 1950b, Thalamotomy and pallidotomy for reatment of choreic movements. *Acta neurochir.*, 2:417.

SURGICAL TREATMENT FOR DYSKINESIAS

Andres M. Lozano and Hazem A. Eltahawy[1]

1. INTRODUCTION

L-dopa therapy reverses parkinsonism early in the course of the disease but commonly induces involuntary choreic and dystonic movements (dyskinesias) with disease progression and long-term therapy (Lozano et al., 2000). A number of pharmacologic interventions have been developed to reduce the severity and duration of dyskinesias including changing the dosage schedule, administration of dopamine agonists, use of controlled release or liquid preparations of l-dopa and usage of amantidine. While these strategies may be helpful, dyskinesias can remain medically intractable. When these manipulations fail, some patients may be helped by surgery. Indeed, Parkinson's disease (PD) patients with ongoing motor disability with drug induced dyskinesias can be among the best surgical candidates.

2. SURGICAL PATHOPHYSIOLOGY

Experimental studies in laboratory animals have shown that dopaminergic agents have striking effects on neural activity in basal ganglia circuits (Filion et al., 1991; Papa et al., 1999). These effects are particularly pronounced in animals with dopamine depletion secondary to lesions in the nigrostriatal system. In animals treated with the neurotoxin MPTP, there is heightened neuronal firing in the subthalamic nucleus (STN) and internal segment of the globus pallidus(GPi), and a decreased rate of neuronal discharge in the external segment of the globus pallidus(GPe) (Bergman et al., 1994; Filion et al., 1985; Filion and Tremblay, 1991). These disruptions in neural activity result in pathological outflow from the basal ganglia output nuclei GPi and substantia nigra reticulata (SNr). This pathological activity is projected to the thalamus and brain stem where it is thought to disrupt function in thalamocortical and brain stem motor circuits which is considered to be responsible for the major clinical manifestations of parkinsonism (Delong, 1990; Lang and Lozano, 1998).

How the administration of l-dopa or dopamine agonists reverses parkinsonism but leads to the generation of involuntary movements is poorly understood. Studies in

[1] Department of neurosurgery, Toronto Western Hospital

Mapping the Progress of Alzheimer's and Parkinson's Disease
Edited by Mizuno et al., Kluwer Academic/Plenum Publishers, 2002

parkinsonian non-human primates have shown that the administration of dopamine agonists reduce spontaneous firing rates in GPi (Filion *et al.*, 1991). Recently, a series of experiments involving administration of the dopamine agonist apomorphine in PD patients undergoing functional neurosurgery have been reported (Hutchinson *et al.*, 1997; Lozano *et al.*, 2000). The striking findings are a reduction in the spontaneous firing rate of GPi neurons with apomorphine, a change in the pattern of neuronal firing and a focusing of GPi neurons receptive fields for movement related activity(Levy R, 2001).

In these studies the reversal of parkinsonism (patients noting "on" sensation) after apomorphine coincided with approximately 50% reduction of GPi neuronal firing rate from the high spontaneous firing rate in the off state (fig. 1). There was also a rise in GPe firing rates. The onset of dyskinesias coincided with the maximal reduction in GPi firing rate. As the patient returns to the parkinsonian state over time as the effect of the drug wears off, the GPi firing rates increased back into the parkinsonian range, producing a return of symptoms. Thus the dyskinesia state is accompanied by a pathological output from GPi which displays an altered pattern of activity and is characterized by a profound reduction in firing rate.

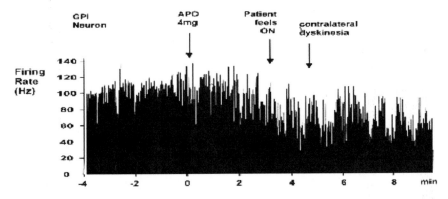

Figure 1. Response of a single internal globus pallidus (GPi) neuron in a patient with Parkinson's disease before and after the administration of 4 mg subcutaneous apomorphine (APO). Bin width is 1 sec. (From Hutchison and colleagues with permission.)

This model suggests that interventions in GPi act by either abolishing (lesions) or temporarily disrupting (deep brain stimulation) the neural substrate responsible for l-dopa induced dyskinesias. With most of its neural element destroyed or physiologically disrupted, GPi does not alter its firing rate or pattern from the elevated and synchronized parkinsonian rate to the much lower rate associated with the generation of dyskinesias. The disrupted GPi no longer provides the necessary component of the network responsible for the generation of dyskinesias (Lozano *et al.*, 2000). What is remarkable is the paradox that having no output is better than having pathologically output from a dysfunctional GPi.

3. SURGICAL PROCEDURES

The past decade has not only seen the revival of stereotactic surgical management for PD after the emergence of intractable drug induced dyskinesia became apparent but also witnessed a considerable widening of the armamentarium of procedures available. Each has its own advantages, drawbacks and indications. The rationale for surgical treatment of dyskinesia the disconnection of the abnormally firing basal ganglia from the normal executive portion of the motor system. This disconnection may occur, at least in theory, at any point along the cortico-striato-pallido-brainstem/thalamic-cortical pathway including its amplifier GPe-subthalamic-GPi sub-circuit. It can be effected by direct lesions using for example radio-frequency thermo-coagulation or high frequency deep brain stimulation (DBS). Targets along the pathway include GPi, STN, and thalamus and the axons interconnecting these structures.

3.1 Thalamic Procedures

That thalamic interventions can influence l-dopa dyskinesias has been described. What is less clear is the relationship between the thalamic target and the effect on dyskinesias. Targets that are said to influence l-dopa dyskinesias include the pallidal receiving thalamus and the centromedian parafascicular (CM-PF) nucleus complex. There is experimental evidence that pallidal receiving areas have a role in choreiform dyskinesias (Narabayashi et al., 1984; Page et al., 1993). The CM-pf complex of the medial thalamus may also be important in controlling choreiform dyskinesias (Caparros-Lefebvre et al., 1999). The relative importance of the changes in thalamocortical function that occur with interventions at thalamocortical relay nuclei versus the CM-Pf complex that has strong striatal projections in not well characterized.

3.2 Pallidotomy and Pallidal Stimulation

Pallidotomy is extremely effective in controlling l-dopa dyskinesias. A meta-analysis of the effects of pallidotomy has been recently reported (Alkhani and Lozano, 2001). These investigators reviewed 85 articles providing details on 1959 patients with PD who were treated with radiofrequency pallidotomy. For the 263 patients in whom pre- and postoperative on period contralateral dyskinesia scores were documented, the mean score improved by 67.7 % after pallidotomy. In the majority of cases, there was little or no reduction daily doses of l-dopa after pallidotomy. This indicates that pallidotomy has a direct effect against dyskinesias.

The improvements in dyskinesia with pallidotomy are immediate, predominantly contralateral and are long lasting, at least up to 5 years (Fine J, 2000). The ipsilateral effect on dyskinesias is lost in the first year after surgery. The magnitude of improvement was dependent on lesion location in the pallidum with more anterior-medial lesions in GPi being most effective against dyskinesias (Gross et al., 1999). This may reflect the segregation of different motor outflow loops directed to primary motor cortex, supplementary motor cortex or premotor cortex within GPi (Hoover and Strick, 1993).

The effects of GPi DBS on dyskinesias can also be striking but are more complex. With DBS as with pallidotomy, the effects of stimulation both on dyskinesias and on parkinsonism are a function of the site of stimulation within the pallidal complex. It is possible to obtain seemingly opposite effects on akinesia and dyskinesias in the same patient depending on which of 4 DBS contacts is being stimulated and whether this contact lies in ventral GPi or within the GPe ((Bejjani et al., 1998; Krack et al., 1998)

3.3 Subthalamic Nucleus Lesions and DBS

The classical teaching is that STN lesions produce hemiballism. The concept that reduced STN activity leads to involuntary movements may also underlie the pathogenesis of l-dopa induced dyskinesias. Several groups however, have now suggested that so-called STN lesions, to the contrary, reduce L-dopa dyskinesias, in some cases, by as much as 70% (Alvarez *et al.*, 2001; Gill and Heywood, 1997; Gill SS, 2000). Only one patient among the combined 13 patients in the 2 groups had persistent hemiballismus 7 days after surgery associated with an infarction of the subthalamic area and part of the thalamus.

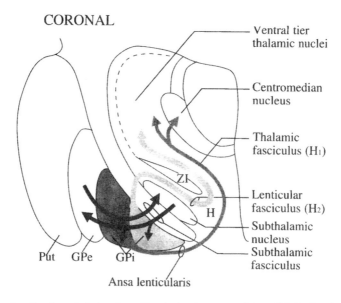

Figure 2. Coronal section through the basal ganglia sohwing the course of the pallidofugal projection. Lesions in the area of the subthalamic nucleus (STN) that extend dorsally or medially beyond the nucleus will interrupt the pallidal outflow coursing through the lenticular fasiculus and ansa lenticularis. (With permission, Parent et al., Chapter 2: Movement Disorder Surgery In: Lozano AM, ed. Prog Neurol Surg, Basel, Krager,2000; 15:21-30).

Carpenter and Mettler's group (Carpenter *et al.*, 1950; Whittier JR, 1949) examined the relationship between lesions in the area of the STN and the production of contralateral chorea in otherwise normal monkeys. When the lesions were restricted to the STN, animals developed contralateral involuntary movements. If the lesions were misplaced or larger than anticipated, extending to involve either the internal segment of the pallidum or the lenticular fasiculus, then no choreic movements were seen. This suggests that lesions in the STN proper are dyskinesiogenic but concomitant interruption of the pallidofugal system as often occurs with STN surgery because of the close apposition and intimate relationship of this pathway to the STN is sufficient to abolish dyskinesias (Lozano, 2001).

As shown in Figure 2, the pallido-fugal fibres are apposed to the dorsal surface of STN. In clinical practice, so-called "STN lesions" diminish STN outflow to the reticulata and the GPi (a procedure that would be expected to produce hemiballism) but at the same time almost invariably also disrupt the pallidofugal pathways and suppress the otherwise anticipated chorea.

Why does STN stimulation improve l-dopa dyskinesia? Two mechanisms seem likely. First the inevitable concomitant effects on the adjacent pallidofugal projection. This mechanism may be particularly important given the short chronaxie and the enhanced and preferential susceptibility of axons versus cell bodies to electrical stimulation (Ranck, 1975). Another mechanism through which STN deep brain stimulation reduces l-dopa dyskinesias is indirect. The improvement in akinesia and rigidity with STN DBS allows reduction of l-dopa intake with as expected, a concomitant reduction in dyskinesias. It is interesting that after long term STN DBS and decreased l-dopa, the threshold for the induction of dyskinesias in response to l-dopa challenges is elevated, suggesting the possibility of post-synaptic desensitization (Benabid AL, 2000) or neuralplasticity.

4. CONCLUSIONS

The cellular mechanism underlying l-dopa dyskinesia are poorly understood. It is clear that dopaminergic therapy produces striking changes in the physiology of the parkinsonian basal ganglia. The GPi and its connections appear to be crucial in acting as the voice for the manifestation of l-dopa dyskinesias. Disrupting the flow of information to or from the dyskinetic GPi is highly effective in suppressing l-dopa induced dyskinesias.

REFRENCES

Alkhani, A. and Lozano, A. M., 2001. Pallidotomy for parkinson disease: a review of contemporary literature. *J Neurosurg* **94**, 43-9.
Alvarez, L., Macias, R., Guridi, J., Lopez, G., Alvarez, E., Maragoto, C., Teijeiro, J., Torres, A., Pavon, N., Rodriguez-Oroz, M. C., Ochoa, L., Hetherington, H., Juncos, J., DeLong, M. R. and Obeso, J. A., 2001. Dorsal subthalamotomy for Parkinson's disease. *Mov Disord* **16**, 72-8.
Bejjani, B. P., Damier, P., Arnulf, I., Papadopoulos, S., Bonnet, A. M., Vidailhet, M., Agid, Y., Pidoux, B., Cornu, P., Dormont, D. and Marsault, C., 1998. Deep brain stimulation in Parkinson's disease: opposite effects of stimulation in the pallidum. *Mov Disord* **13**, 969-70.
Benabid AL, K. A., Benazzouz A, Pialat B, Van Blerkom N, Fraix V, Pollak P, 2000. *Subthalamic nucleus deep brain stimulation.* Karger: Basel.
Bergman, H., Wichmann, T., Karmon, B. and DeLong, M. R., 1994. The primate subthalamic nucleus. II. Neuronal activity in the MPTP model of parkinsonism. *J Neurophysiol* **72**, 507-20.
Caparros-Lefebvre, D., Pollak, P., Feltin, M. P., Blond, S. and Benabid, A. L., 1999. [The effect of thalamic stimulation on levodopa induced dyskinesias-- evaluation of a new target: the center parafascicular median]. *Rev Neurol (Paris)* **155**, 543-50.
Carpenter, M. B., Whittier, J. R. and Mettler, F. A.,1950. Analysis of choreoid hyperkinesia in the rhesus monkey. Surgical and pharmacology analysis of hyperkinesia resulting from lesions of the subthalamic nucleus of Luys. *Journal of Comparative Neurology* **92**, 293-331.
Delong, M. R.,1990. Primate models of movement disorders of basal ganglia origin. *TINS* **13**, 281285.
Filion, M., Boucher, R. and Bedard, P. (1985) Globus pallidus unit activity in the monkey during the induction of parkinsonism by 1-methyl-4-phenyl-1,2,3,6-tetrahydropyridine (MPTP). **11**, 1160.
Filion, M. and Tremblay, L., 1991. Abnormal spontaneous activity of globus pallidus neurons in monkeys with MPTP-induced parkinsonism. *Brain Research* **547**, 142-151.
Filion, M., Tremblay, L. and Bedard, P. J.,1991. Effects of dopamine agonists on the spontaneous activity of globus pallidus neurons in monkeys with MPTP-induced parkinsonism. *Brain Research* **547**, 152-161.

Fine J, D. J., Chen R, Hutchison WD, Lozano AM, Lang AE.2000. Long term follow up of unilateral pallidotomy in advanced Pparkinson's disease. *submitted.*

Gill, S. S. and Heywood, P.,1997. Bilateral dorsolateral subthalamotomy for advanced Parkinson's disease. **350,** 1224.

Gill SS, H. P.,2000. Subthalamic nucleus lesions. In: *Movement disorder surgery Vol. 15,* pp. 196-226. Ed. L. AM. Karger: Basel.

Gross, R. E., Lombardi, W. J., Lang, A. E., Duff, J., Hutchison, W. D., Saint-Cyr, J. A., Tasker, R. R. and Lozano, A. M.,1999. Relationship of lesion location to clinical outcome following microelectrode-guided pallidotomy for Parkinson's disease. *Brain* **122,** 405-16.

Hoover, J. E. and Strick, P. L.,1993. Multiple output channels in the basal ganglia. **259,** 819-821.

Hutchinson, W. D., Levy, R., Dostrovsky, J. O., Lozano, A. M. and Lang, A. E.,1997. Effects of apomorphine on globus pallidus neurons in parkinsonian patients. *Ann Neurol* **42,** 767-75.

Krack, P., Pollak, P., Limousin, P., Hoffmann, D., Benazzouz, A., Le Bas, J. F., Koudsie, A. and Benabid, A. L.,1998. Opposite motor effects of pallidal stimulation in Parkinson's disease. *Annals of Neurology* **43(2),** 180-192.

Lang, A. E. and Lozano, A. M.,1998. Parkinson's disease. First of two parts. *New England Journal of Medicine* **339,** 1044-1053.

Levy R, D., Lang AE, Sime E, Lozano AM,2001. The effects of apomorphine on subthalamic nucleus and globus pallidus internus neurons in patients with Parkinson's disease. *Brain* **in press.**

Lozano, A. M.,2001. The subthalamic nucleus: Myth and opportunities. *Mov Disord* **16,** 183-4.

Lozano, A. M., Lang, A. E., Levy, R., Hutchison, W. and Dostrovsky, J.,2000. Neuronal recordings in Parkinson's disease patients with dyskinesias induced by apomorphine. *Ann Neurol* **47,** S141-6.

Narabayashi, H., Yokochi, F. and Nakajima, Y.,1984. Levodopa-induced dyskinesia and thalamotomy. *J Neurol Neurosurg Psychiatry* **47,** 831-9.

Page, R. D., Sambrook, M. A. and Crossman, A. R.,1993. Thalamotomy for the Alleviation of Levodopa-Induced Dyskinesia: Experimental Studies in the 1-Methyl-4-Phenyl-1,2,3,6-Tetrahydropyridine-Treated Parkinsonian Money. **55,** 147-165.

Papa, S. M., Desimone, R., Fiorani, M. and Oldfield, E. H.,1999. Internal globus pallidus discharge is nearly suppressed during levodopa- induced dyskinesias [In Process Citation]. *Ann Neurol* **46,** 732-8.

Parent A, C. M., Levesque M (2000) Anatomical considerations in basal ganglia surgery. In: *Movement Disorder Surgery Vol. 15,* pp. 21-30. Ed. L. AM. Karger: Basel.

Ranck, J. B., Jr.,1975. Which elements are excited in electrical stimulation of mammalian central nervous system: a review. *Brain Res* **98,** 417-40.

Whittier JR, M. F.,1949. Studies on the subthalamus of the Rhesus monkey. *Journal of Comparative Neurology* **90,** 319-372.

DEEP BRAIN STIMULATION AND NEUROPROTECTION

Alim-Louis Benabid, Brigitte Piallat, Abdelhamid Benazzouz, Adnan Koudsié, Stéphan Chabardes, Pierre Pollak, Doris Lenartz, and Christian Andressen[*]

1. INTRODUCTION

Parkinson's Disease (PD) is characterized by a nigral degenerescence of dopaminergic neurons, which are responsible for the deregulation of parallel processing systems of loop from the cortex to the striatum and the basal ganglia. Palliative treatment by administration of dopamine agonists or dopamine precursors is the current medical treatment. Inactivation by different means (lesion or high-frequency stimulation which is supposed to be inhibitory) are the current surgical methods to palliate the symptoms. Replacement of the missing neurotransmitter dopamine by grafts of competent cells, mainly into the striatum, close to the dopamine receptors of the striatal cells, is another approach which is still at the experimental level although a large number of preliminary clinical trials have been done. The question of how to find ways to slow down, stop, or reverse this degenerescence, instead of treating Parkinson's disease symtoms, at the level of the consequences of dopaminergic nigral degenerescence, is another totally different approach, which will be either a preventive treatment or a restorative treatment instead of a palliative treatment. This can be achieved either for pharmacological approaches, or by designing a surgical procedure, which would be able to interfere with this degenerative process.

[*] Alim-Louis Benabid, Brigitte Piallat, Abdelhamid Benazzouz, Adnan Koudsié, Stéphan Chabardes, Pierre Pollak Neuroscience Department, Joseph Fourier University, Grenoble, France. Doris Lenartz, Christian Andressen Neuroscience Department, Köln University, Germany.

2. RATIONALE FOR HIGH-FREQUENCY STIMULATION (HFS) AS A NEUROPROTECTION METHOD

HFS of STN is being used as a treatment of advanced stage of Parkinson's Disease (PD) (Limousin et al,1999). It mimics symptomatic effects of levodopa. One hypothetical mechanism is the inhibition of subthalamic nucleus (STN) cellular activity, which reduces a glutamatergic output of STN or to target structures such as substantia nigra (SN). Could this palliative antiglutamate action of HFS of STN be able to slow down the degenerescence of dopaminergic neurons in SN of PD patients is the question which is addressed by the experimental approach described in this paper, and by the preliminary clinical data, which could be examined to try to substantiate this hypothesis.

3. PUTATIVE MECHANISM OF ACTION OF HFS

From previous works in our laboratory, (Benazzouz et al, 1996) it has been suggested that HFS in STN could be directly inhibitory, as microrecording following STN stimulation shows that in STN itself, cells present post-stimulus inhibition of several hundred milliseconds in duration. Simultaneous recording in the substantia nigra reticulata (SNr) and into the entopedoncular nucleus (EP, equivalent of the human internal pallidum GPi), shows that the putative arrest of the glutamatergic output of STN, which is inhibited by stimulation, induces the suppression of activation into these two target structures, which exhibit also a significant and long post-stimulus inhibition. As these two structures are gabaergic and project onto different structures, one may expect that the suppression of this gabaergic output, which is inhibitory, might induce a release of the inhibition, or an equivalent excitation, which is actually what is observed in the ventro-lateral (VL) cells of the thalamus (equivalent of the motor-thalamus in humans) as well as in substantia nigra compacta (SNc) which receives projection from those different structures including STN. The recording into the GP cells (equivalent in human to the external pallidum GPe structure) shows on the contrary a post-stimulus strong excitation, which would be by itself delivering a strong gabaergic input to GP and to EP and SNr and account for their observed inhibition. This hyperactivity of GPe could be due to an antidromic activation through its projection onto SNR, and in return could also, using this projection, be responsible for the observed excitation in STN. Lesioning GPe does not suppress the inhibitory effect observed in SNR and EP after STN stimulation, although it seems that this inhibitory period is somewhat shortened.

In addition, it is known that the glutamatergic projection from STN to SNr are also involving SNc. However, the strong anatomical intrication of SNc and SNr and the extracellular release of glutamate make it possible that the hyperactivity in STN could be responsible for an increase in local concentration of glutamate into these structures. Glutamate is known to exerce his neurotoxicity through NMDA receptors. NMDA enhances neurotoxicity of metamphetamine. Neurotoxicity on the other side is attenuated by NMDA antagonists such as MK801, CPP, etc. We have developed three programs to try to demonstrate this hypothesis in rats, in monkeys and in human patients.

4. NEUROPROTECTION RAT PROGRAM

4.1. Paradigm of thePROTECT and RESCUE protocols

It is based on the 6-hydroxydopamine (6-OHDA) model of Sauer and Oertel (1994) where 6-OHDA is injected into the caudate. Its retrograde transport, or the lesioning axotomy effect on the terminals, is responsible for a progressive and partial loss of dopaminergic cells into SNc. Our program in the rat is made of two different paradigms, the PROTECT and the RESCUE protocols. In the paradigm of PROTECT protocol, prior to injection of 6-OHDA (20mg) into the right striatum, kainic acid (KA) is injected into the ipsilateral STN to induce a lesion of the nucleus, 6 days before injection of 6-OHDA, (Piallat et al, 1996). Following this, the animal is sacrificed, and the immuno-histochemistry (IHC) is performed to show the enzyme tyrosine-hydroxilase (TH immuno-reactivity or TH-IR). The RESCUE protocol is meant to represent more closely the clinical situation where the treatment follows the development of the disease. Therefore, the KA lesion of STN is performed after the intrastriatal injection of 6-OHDA within a delay ranging from 0 to 10 days. The animals are kept alive for the same duration and then sacrificed and IHC is performed.

4.2. Results of the PROTECT protocol.

The IHC study of the striatum shows an almost complete disappearance of the 6-OHDA terminals, and at the level of SNc, the loss of TH as compared to the contro-lateral side taken as a reference ranges between 40 to 50 %. In the situation when kainic acid has been injected 6 days before injection of 6-OHDA, this pattern disappears and the differences in TH-IR in the caudate and in the SNc, are not significantly different from the control side. The time-course of this cell loss performed at 15 days and 90 days shows at 90 days that the difference between the treated and non treated side is smaller suggesting that even if cells have been acutely protected, on the long term, some of them die from the injury they have received at the time of injection. Using HPLC, observation of the chemical time-course into the ipsi and contro-lateral sides at 10 and 90 days shows that the amount of striatal DA on the normal side is about 8 to 10 ng/mg while on the side injured by 6-OHDA, this reaches the value of 5 ng, meaning a decrease of about 50 %. In the animals who have received STN lesion prior to 6-OHDA injection, there is no significant difference in the DA loss, which even is slightly higher at 15 days, but at 90 days the amount of striatal DA on the STN protected side is non significantly different from the contro-lateral normal side. This pattern is reproduced also for striatal DOPAC which in the non-STN treated situation is strongly decreased at 15 days by a ratio of 4 as compared to the normal side. This returns to a higher ratio of 5 to 8 at 90 days. In the KA treated rats, the loss of striatal DOPAC is also visible on the 6-OHDA injected side almost equally as in the non-STN treated animals, but at 90 days, the amount of striatal DOPAC is not significantly different on the normal side and on the 6-OHDA STN-lesioned side. Similarly, changes in the striatal HVA parallel strictly the changes in DOPAC.

4.3. Results of the RESCUE protocol

In order to know what was the best time to perform the STN lesion after the 6-OHDA injection, a time-course study has been done of the evolution of TH immuno-reactivity after 6-OHDA injection. Surprisingly, 3 hours after 6-OHDA injection, this loss of TH-IR, which is already 40 %, will progressively increase until 60 % at 30 days. In the meantime, the cells treated by cresyl-violet show that although they do not express TH, the neurons are still present and that there is a progressive decrease in cells due to their apoptotic process until they match the value of the loss of TH-IR at 30 days, showing that after 6-OHDA injection, the neurons are still alive, but have lost very quickly their functional properties and their ability to produce DA, its metabolites, and also to express TH-IR. On this basis, the time for lesion has been chosen to be at day 3, lesions performed after day 5 being much less or non efficient in the protective process. The protection of STN after STN lesion made 3 days after a 6-OHDA injection is less than when the STN is done 6 days before, but still extremely highly significantly different from the control group, showing that the process of death can be stopped or cells could even be rescued. This leads to the neuroprotection concept in 6-OHDA injection where we could consider that following the administration of 6-OHDA if an STN lesion has been made prior, there are no change or no permanent changes. When the lesion of STN is done after 6-OHDA injection, then the sooner it happens the more protective it is. After 6-OHDA, the cells will immediately loose at the time the dopamine production while the TH expression is still present, and later the TH expression will disappear, and if nothing comes to further protect the cells, these neurons will die. Reversely, if an STN lesion is made soon after 6-OHDA injection, the protection is better the sooner and the cells will go less and less further in their process of degeneration.

5. MPTP MONKEY PROGRAM

In this program, the MPTP are treated systemically by injection of 0.2 mg/kg/day IP or IM of MPTP under clinical control until the time they start to express clinical symptoms of akinesia and/or rigidity and less frequently tremor. The administration of MPTP is then discontinued. All animals are examined by a SPECT DAT-scan testing for striatal DA transporter by a FPCIT SPECT before and after MPTP administration and just prior the sacrifice at 30 days, which is followed by examination of the brains using immuno-histochemistry. In the MPTP program as well as on the rat 6-OHDA program, two branches are considered one called PROTECT (KA injection in STN prior to the administration of MPTP) and a second called RESCUE (KA injection after MPTP). Two other protocols of MPTP monkey programs, not yet started, including also two PROTECT (STN-HFS for one week prior to MPTP and continued all over down to the time of sacrifice) and RESCUE (STN-HFS is started after discontinuation of MPTP) branches, use STN chronic HFS instead of KA lesion, by unilateral STN electrode. All other steps are similar.

5.1. Preliminary results of the PROTECT branch of the KA MPTP monkey program

Five monkeys have been done, two monkeys have been already processed, showing that there is a significant reduction of the TH-IR cell loss on the STN lesioned side. The ratio of TH-IR cells of the protected to the non-protected side being equal to 1.9 in monkey n° 1 and to 3.1 in monkey n° 2.

6. HUMAN NEUROPROTECTION PROJECT

This project has two branches. One is retrospective analysis of patient and the other one a prospective study. The retrospective analysis is based on the OFF/OFF UPDRS III observation of PD patients implanted in STN during the regular follow-up. The value of the OFF/OFF UPDRS III rating at that time can be considered, although it might be biased by the washout phenomenon, can be considered as a measure of the background level of severity of the disease. The preliminary results of the follow-up of 80 patients followed up for five years showed that 25 % of them (20) impair although the pre-operative background level (41.85 ± 17.94) remains non significantly different from the UPDRS score at 5 years (58.25 ± 21.57). 36 % of the patients (29) are stable. 39% (31) of the patients are on the contrary improved at 1 year. (pre-operative UPDRS 57.48 ± 12.86 and at one year 37.94 ± 14.09). However, this group splits in two sub-groups: one made of 15 patients who, after this improvement at 1 year, return to the pre-op value without any significant difference. This re-impairment being due to the appearance of midline symptoms. 16 patients persist to improve very significantly. These data might be due to a bias induced by the persisting effect after the offset of stimulation as it may happen also after the arrest of a levodopa. However, if this does not allow to consider these data as a demonstration of neuroprotection, they are still compatible with the neuroprotection hypothesis. In order to investigate the possible neuroprotective effect, a clinical prospective trial project is being set including young patients having a short evolution of their disease, who would be evaluated clinically and by SPECT DAT-scan and fluorodopa PET, pre and post STN bilateral implantation, followed up every 2 years until 6 years. The population will be split in two groups, the first one being operated in STN at the beginning of the study, the second group on year 3.

7. CONCLUSION

In conclusion, HFS of STN alleviates motor symptoms of advanced PD patients. The mechanism is allegedly related to the inhibition of STN neurons. The putative anti-glutamate effect of STN-HFS may slow down the dopamine degenerescence in the substantia nigra. STN lesion in 6-OHDA rats prevents nigral degenerescence or restore altered nigro-striatal functions. These results are being confirmed in MPTP-treated monkeys. Long-term follow-up of bilaterally implanted PD patients is coherent with the hypothesis of STN-HFS induced neuroprotection. Clinical trial with functional imaging or nigrostriatal transmission is being proposed in patients at early stage of PD. It is of primary importance to understand what is the reality of this putative neuroprotection. If it were demonstrated that STN-HFS not only alleviates reversibly the symptoms, but also is able to influence the evolution rate of the disease, possibly slowing down the process of neurodegenerescence or even stopping it, this would be a profound effect on the management of parkinson patients, and would raise the question of when to operate these patients.

REFERENCES

Benazzouz, A., Piallat, B., Pollak, P., Benabid, A.L., 1995, Responses of substantial nigra reticulata and globus pallidus complex to high frequency stimulation of the subthalamic nucleus in rats : electrophysiological data. *Neuroscience Lett.* **189**:77-80.

Limousin, P., Krack, P., Pollak, P., Benazzouz, A., Ardouin, D., Hoffmann, D., Benabid, A.L. , 1998, Electrical stimulation of the subthalamic nucleus in advanced Parkinson's disease. *N Engl J Med.* **339**:1105-1111.

Piallat, B., Benazzouz, A., Benabid, A.L., 1996, Subthalamic nucleus lesion in rat prevents dopaminergic nigral neuron degeneration after striatal 6-OHDA injection : behavioral and immunohistochemical studies. *Eur J Neurosci.* **8**:1408-1414.

Sauer, H. and Oertel, W.H., 1994, Progressive degeneration of nigrostriatal dopamine neurons following intrastriatal terminal lesions with 6-OHDA: a combined retrograde tracing and immunocytochemical study in the rat. *Neuroscience.* **59**:401-415

CYCLIC GMP-MEDIATED PRECONDITIONING GENE INDUCTION AS A TREATMENT OF ALZHEIMER'S DEMENTIA AND PARKINSON'S DISEASE

Chuang Chin Chiueh* and Tsugunobu Andoh

1. SUMMARY

With advancements in molecular biology and genetics it is now feasible to employ gene technology as a treatment of diseases including some of the neurodegenerative brain disorders. Currently, gene transfer and multipurpose stem cells have been targeted for their future use in the management of Alzheimer's dementia and Parkinson's disease. We are investigating whether gene induction evoked by either sublethal stress or subtoxic chemicals, can induce adaptation or compensatory protection through a polygenic induction of cytoprotective genes and proteins. We have developed a human brain cell model for investigating hormesis or preconditioning neuroprotective mechanism mediated by multiple gene induction and new protein synthesis (Andoh, Lee and Chiueh, 2000; http://fasebj.org/cgi/doi/10.1096/fj.00-0151fje). Preconditioning induction of neuronal nitric oxide synthase and associated nitric oxide-cGMP-PKG pathway increases antioxidative (thioredoxin and MnSOD), anti-apoptotic (Bcl-2 and Ref-1), and cell-repairing proteins (BDNF) for protecting human brain cells against oxidative stress caused by serum deprivation and 1-methyl-4-phenylpyridinium (MPP^+). This nitric oxide-cGMP mediated gene induction and cytoprotection is likely to becoming the mechanism in common in some of the atypical

* C. C. Chiueh, Unit on Neurodegeneration and Neuroprotection, Laboratory of Clinical Science, NIMH, NIH Bldg. 10, Rm. 3D-41, Bethesda, MD 20892-1264, USA and T. Andoh Department of Applied Pharmacology, Toyama Medical and Pharmaceutical University, Toyama 930-0194, Japan. *E-mail address: "C. C. Chiueh, Ph.D." <chiueh@helix.nih.gov>

Mapping the Progress of Alzheimer's and Parkinson's Disease
Edited by Mizuno *et al.*, Kluwer Academic/Plenum Publishers, 2002

447

neuroprotective agents such as 17ß-estradiol and statins since both of them induce NOS1 or NOS3 and nitric oxide. Without complications of brain surgery and intracerebral injection of viral vectors, gene induction may be more practically than gene therapy as a non-invasive treatment in Parkinson's disease and Alzheimer's dementia, especially for slowing progression in neurodegeneration and clinical deterioration. (Key words: *gene therapy; tolerance; thioredoxin or Trx; 17ß-estradiol; statin; neuronal nitric oxide synthase or NOS1;* 1-methyl-4-phenylpyridinium or *MPP⁺; nitric oxide; stroke; neuroprotection*)

2. INTRODUCTION

In addition to antioxidants, gene therapy and stem cell transplantation have emerged as possible strategies for the treatment of neurodegenerative diseases such as Alzheimer's dementia and Parkinson's disease. By virtue of unique biological properties and their presence in the adult brain, neural stem cells represent good candidates for applying these cells to not only structural brain repair but also gene therapy. However, one of the major hindering problems in gene therapy is the requirement of invasive brain surgery of cell transplantation including the injection of foreign cells and viral vectors into patient's brain that can lead to unforeseen complications in the brain. In this study, we are investigating an alternative gene technology--gene induction--to see whether preconditioning induction of multiple genes and their proteins could enhance neuroprotection through the induction of cyto-protective proteins/factors in the brain (Andoh et al., 2000; 2001; Chiueh et al., 2001).

2.1. Gene Induction as an Alternative Procedure in Gene Therapy

Preconditioning cyto-protection is achieved by using either sublethal stressors or subtoxic chemicals to induce multiple genes and their proteins for developing the adaptive process in cells for protecting against subsequent severe oxidative stress and/or toxic insults. This preconditioning process can be induced *in vivo* by applying a brief period of sublethal physiological and/or environmental stressors. Preconditioning can also be initiated in brain cells by the administration of sublethal dose of drugs that interrupt DNA repair, mitochondrial respiratory process, and/or energy production (i.e., 2-deoxyglucose, 3-nitropropionic acid, and LPS/diphosphoryl lipase A). This process of adaptation is acquired through out the evolution era and existed in most living cells including myocardial, hepatic, and neuronal cells. Without developing natural adaptation or preconditioning protection cells and/or neurons can not survive from repeated neurotoxic insults and/or environmental stress.

In the past decade, several preconditioning procedures have been identified and tested *in vitro* and *in vivo* including the application of hypoxia/ischemia, hyperbaric oxygen, and/or hypoglycemic shock for a brief sublethal period (Liu et al., 1992; Kato et al., 1994; Andoh et al., 2000). Recently, it has been shown that drug tolerance or hormesis can be induced by the administration of sublethal dose of chemicals (Riepe et al., 1997; Guo and Mattson, 2000; Toyoda et al., 2000; Chen et al., 2001). Based on the preconditioning results obtained from the ischemia/reperfusion animal model it is known that preconditioning increases hypoxia tolerance likely through the induction of heat shock proteins (HSP-27, -70, and/or -72) in the brain. However, other cytoprotective proteins (i.e., BDNF, adenosine, glycogen, ceramide, NADPH diaphorase, nitric oxide synthase or NOS, MnSOD, and K(ATP) channels are also induced in the brain of preconditioned animals. Therefore, recent results suggest that preconditioning protection may be mediated by the synthesis of new proteins that is maintained by multiple genes rather than a single gene.

2.2. The Role of Nitric Oxide-Cyclic GMP-PKG Pathway in Preconditioning Responses

The present working hypothesis is derived from early studies that nitric oxide ($\cdot$NO) may play a dual role in late preconditioning against myocardial stunning, acting initially as the trigger and subsequently as the mediator of cardioprotection (Bolli et al., 1997). Because it has shown NOS inhibitors, given 24 hours after the preconditioning ischemia, consistently abrogate preconditioning cardioprotection. These results suggest a possible role for cGMP as a trigger in preconditioning protection. Significantly, $\cdot$NO produced by the endothelial NOS (NOS3) and/or neuronal NOS (NOS1) may also support preconditioning neuroprotection in brain (Kato et al., 1994; Giddy et al., 1999; Centeno et al., 1999; Andoh et al., 2000). We have recently developed a preconditioning cell model for investigating the putative role of $\cdot$NO and cGMP in precondition protection of neurons and other brain cells against oxidative stress evoked by serum deprivation and MPP$^+$ (Andoh et al., 2000; Andoh et al., 2001; Chiueh et al., 2001). The results of our preclinical studies have supported the hypothesis that preconditioning neuroprotection in human brain cells is triggered by the production of sublethal concentration of oxidants, maintained by the induction of NOS1 and related $\cdot$NO-cGMP-PKG signaling pathway.

3. NITRIC OXIDE-MEDIATED PRECONDITIONING NEUROPROTECTION AGAINST MPP$^+$-INDUCED OXIDATIVE STRESS

It has been shown that oxidative stress may cause neurodegenerative disorders including parkinsonism evoked by 1-methyl-4-phenyl-1,2,3,6-tetrahydropyridine (MPTP). We employed the preconditioning cell model to investigate whether $\cdot$OH or $\cdot$NO mediates MPTP-induced neurotoxicity. A brief two hour sublethal serum deprivation preconditioning stress induces human NOS1 in neurotrophic SH-SY5Y cells, activates cGMP, and inhibits oxidative stress evoked by 1-methyl-4-phenylpyridinium ion (MPP$^+$)--the toxic metabolite of MPTP (Chiueh et al., 2001). Consistently, S-nitrosoglutathione (GSNO) and 8-Br-cGMP inhibit lipid peroxidation and apoptotic cell death caused by MPP$^+$. The protective effects of preconditioning, GSNO, and 8-Br-cGMP are blocked by inhibitors of NOS1 and PKG. Moreover, gene transfer of human NOS1 cDNA into CHO-K1 cells renders these cells more resistant to MPP$^+$-induced cytotoxicity by approximately 30-fold. $\cdot$NO and cGMP seem to mediate the preconditioning protection against severe oxidative stress caused by MPP$^+$ (Andoh, Chock and Chiueh, in preparation). It is $\cdot$OH that mediates toxicity caused by MPTP/MPP$^+$ whereas the induction of NOS1 and $\cdot$NO is protective. We therefore propose that the •NO/cGMP-mediated signaling pathway may be part of cellular adaptive system for enhancing survival and tolerance against oxidative stress and apoptosis.

4. PRECONDITIONING MECHANISM: A CYCLIC GMP MEDIATED NEW PROTEIN SYNTHESIS

Based on the results obtained from the above preconditioning animal and cellular models, it is likely that •NO and cGMP trigger and maintain the preconditioning cardioprotection and also neuroprotection. However, It has been extremely difficult to use *in vivo* animal model for investigating underlying molecular mechanism of preconditioning neuroprotection. Using the present human neurotrophic cell model, we have investigated the cGMP-mediated gene induction and new protein synthesis in preconditioning responses (Andoh et al., 2000; 2001). A brief non-lethal serum deprivation for 2 hours increases the maximal expression of redox factor-1 (Ref-1) and the mRNA of NOS1 at the end of preconditioning stress. The expression of newly synthesized NOS1 protein is delayed for at least 6 hours and peaked 24 hours thereafter. Sequential increases of NOS1 mRNA, NOS1 protein, •NO, cGMP, and PKG can be demonstrated in these preconditioned human neurotrophic cells. Moreover, pharmacological elevation in tissue cGMP levels by •NO donors before sustained ischemia enhances tolerance against MPP^+ that is comparable to response produce by preconditioning procedures (Andoh, Chock and Chiueh, submitted). Western blotting procedure reveals a delayed up-regulation of multiple proteins including a redox protein thioredoxin (Trx), an anti-apoptotic protein Bcl-2, and an antioxidative protein MnSOD in the preconditioned SH-SY5Ycells for maintaining adaptive responses for neuroprotection (**Scheme 1**). Prior studies have shown that preconditioning also increases the

A. Preconditioning Stress (Sublethal Stress)
↓

(2)
Ref-1
(1) NOS1 → •NO → cGMP → PKG → Trx   → ? → (3) Neuroprotection
Bcl-2
BDNF
MnSOD

↑
B. Chemical Preconditioning (Hormesis)

Scheme 1. Preconditioning Gene Induction: the Role of Nitric Oxide-cGMP-PKG Signaling Pathway in Induction of Cytoprotective Proteins and Factors. (1) Sublethal preconditioning procedures activate the •NO-cGMP-PKG signaling pathway. (2) cGMP mediates new protein synthesis through induction of multiple genes. (3) Neuroprotection is mediated by up regulation of antioxidative (Trx; MnSOD), anti-apoptotic (Ref-1; Bcl-2) and repair (BDNF) proteins/factors (Andoh et al., 2000; 2001).

expression of BDNF (Yanamoto et al., 2000). Preconditioning induced new protein synthesis is the results of multiple gene induction and the •NO-cGMP-PKG pathway may regulate preconditioning-mediated gene induction since inhibition of PKG abrogates preconditioning neuroprotection and also prevents new protein synthesis, especially a cytosolic redox protein thioredoxin (Trx). Trx is a potent antioxidant, anti-apoptotic agent, and a modulator of transcription factors (i.e., AP-1) that can activate new protein synthesis for enhancing cell survival and adaptation following toxic insults and oxidative stress (Andoh, Chock and Chiueh, submitted).

5. PHARMACOGENOMICS: GENE INDUCTION THROUGH HORMESIS

It has been documented that some drugs evoke unexpected neuroprotective action, which is not related to their original specific pharmacological actions. For example, *l*-deprenyl (MAO-B inhibitor), estrogen (female sex hormone), and statins (cholesterol lowering agent) have been clinically proven that they may process unexpected neuroprotective action for the prevention/treatment of neurodegenerative disorders such as Alzheimer's dementia and Parkinson's disease. Through the investigation of the neuroprotective mechanisms of these drugs it is now clear that these atypical neuroprotectants have a common mechanism of action--increase of isoforms of NOS at low nanomolar concentrations. It is feasible that these drugs can increase •NO that activates the c-GMP-mediated gene induction and new protein synthesis, a mechanism is similar to that of preconditioning neuroprotection. It is not clear whether sublethal doses of chemical preconditioning agents (i.e., 2-deoxyglucose, 3-nitropropionic acid, diethyldithiocarbamate, and LPS/diphosphoryl lipase A) can activate the •NO-cGMP-PKG pathway and induce hormesis in cell/neuron cultures and animal models.

5.1. 17ß-Estradiol

The beneficial effects of estrogen replacement therapy have been associated with a low incidence of age-related diseases such as Alzheimer's disease and heart attack. Recent experimental results using murine microglias suggest that the activation of MAP kinase by estrogen ß receptors may be involved in estrogen-mediated anti-inflammatory pathways (Bruce-Keller et al., 2000). Moreover, treatment with 17ß-estradiol increases calmodulin and activates NOS1 and NOS3, •NO, and cellular cGMP in a dose-related manner (EC_{50}, approximately 1 nM), and these effects are inhibited by tamoxifen and ICI-182780 (Hayashi et al., 1994; Ceccatelli et al., 1996). Our preliminary results indicate that at nanomolar concentrations 17ß-estradiol can induce the expression of NOS1 mRNA and protein not just the enzyme activity. However, at high concentrations (>μM) estrogen acts like antioxidants. Furthermore, our recent findings indicate that priming of human brain cells with 17ß-estradiol (<10 nM) for a few days increases the tolerance of brain cells against subsequent oxidative stress acting through the induction of NOS1 (Lee, S. Y., Andoh, T., and Chiueh, C.C., in preparation). Therefore, 17ß-estradiol may induce the following signaling steps: (1) the activation of estrogen ß receptors, (2) the increase in the transcription of NOS1 and 3 proteins and up-regulation of NOS activity. (3) •NO increases the production of cGMP after binding to the heme moiety of guanylate cyclase. (4) cGMP activates PKG. Finally, (5) 17ß-Estradiol induces new protein synthesis as discuss above in preconditioning mechanism.

5.2. Statins (HMG-Co A Inhibitors)

Experiments with statins (inhibitors of ß-hydroxy-ß-methylglutaryl coenzyme A reductase or HMG-CoA) derived from fungus have yielded several clinical useful drugs (i.e., lovastatin, atorvastatin, simvostatin, and pravastatin) for treating hypercholesterolemia by lowering low-density lipoprotein (LDL) or the synthesis of cholesterol and thus the incidence of heart attacks and strokes. Statins also produce unexpected neuroprotective action that can drastically and significantly reduce to one-third the risk of developing dementia in statin-treated patients with or without hyperchlesterolemia since it can decrease the ß-amyloid levels (Fassender et al., 2001). Inhibition of HMG-CoA or mevalonate synthesis by statins blocks not only the formation of cholesterol but also of isoprenoids. Treatment of human endothelial cells with statins increases the expression of NOS3 mRNA and protein by blocking the geranylgeranylation of the GTPase Rho (Laufs et al., 2000). Treatment with statins in normocholesterolemic mice augments cerebral blood flow and reduces cerebral infarcts (Vaughan and Delanty, 1999). These effects of statins are completely absent in NOS3-deficient mice indicating that enhanced NOS3 activity by statins is the predominant mechanism by which these agents protect against cerebral injury. Furthermore, statins possess some antioxidant properties that may help to ameliorate oxidative stress in the brain. Finally, the production of •NO and the activation of cGMP-mediated preconditioning protection of the cerebrovascular system including slowing the formation of clots and increasing blood flow that can prevent strokes and minimize the stroke's neuronal loss. The proposed induction of NOS1 and the activation of cGMP-PKG signaling pathway by statins inside the brain could lead to inducing cytoprotective proteins/factors and thus decreasing incidence of dementia.

6. CONCLUDING REMARKS

Based on the oxidative stress hypothesis in neurodegenerative disorders, several clinical trials have been attempted for retarding the progression in Parkinson's disease and Alzheimer's dementia using a high dosage of single antioxidant. Despite of early encouraging reports most of these clinical trials can't substantiate the expected beneficial effects of antioxidants identified through pre-clinical studies using animal models. It is not known what causes this discrepancy since these antioxidants are effective in cellular and/or animal models but not human patients. Multiple antioxidants may be more effective than a single drug to activate the redox pathway of the endogenous antioxidative defense system to remove harmful oxidants from the brain. Unexpectedly, during the past decade, clinical trials of estrogen and statins for heart attacks and strokes have revealed a significant neuroprotective action such as lowering the risk of developing dementia. We propose that these atypical neuroprotective drugs may induce multiple genes for the synthesis of redox cellular defense proteins such as antioxidative enzymes, anti-apoptotic proteins, and neurotrophic factors in a way similar to •NO-mediated hormesis or preconditioning neuroprotection. For the treatment of neurodegenerative disorders including Parkinson's disease and Alzheimer's dementia, the induction of cyto-protective genes and beneficial proteins/factors via either hormesis or preconditioning procedures may become a practical alternation to gene transfer and stem cell transplantation.

REFERENCES

Andoh, T., Lee, S. Y., and Chiueh, C. C., 2000, Preconditioning regulation of bcl-2 and p66[shc] by human NOS1 enhances tolerance to oxidative stress. *FASEB J.* **14:** 2144 (on line 10.1096/fj.00-0151fje).

Andoh, T. , Chock, P. B., and Chiueh, C. C., 2001, The role of neuronal nitric oxide synthase (NOS1) and related cyclic GMP-PKG signaling pathway on preconditioning-mediated neuroprotection. *Soc. Neurosci. Meeting Abstract* **26:** in press.

Bolli, R., Manchikalapudi, S., Tang, X. L., Takano, H., Qiu, Y., Guo, Y., Zhang, Q. and Jadoon, A. K., 1997, The protective effect of late preconditioning against myocardial stunning in conscious rabbits is mediated by nitric oxide synthase. Evidence that nitric oxide acts both as a trigger and as a mediator of the late phase of ischemic preconditioning. *Circ. Res.* **81,** 1094.

Bruce-Keller, A. J., Keeling, J. L., Keller, J. N., Huang, F. F., Camondola, S., and Mattson, M. P. , 2000, Antiinflammatory effects of estrogen on microglial activation. *Endocrinology* **141:** 3646.

Ceccatelli, S., Grandison, L., Scott, R. E., Pfaff, D. W., and Kow, L. M., 1996, Estradiol regulation of nitric oxide synthase mRNAs in rat hypothalamus. *Neuroendocrinology* **64:** 357.

Centeno, J. M., Orti, M., Salom, J. B., Sick, T. J., and Perez-Pinzon, M. A., 1999, Nitric oxide is involved in anoxic preconditioning neuroprotection in rat hippocampal slices. *Brain Res.* **836:** 62.

Chen, Y., Ginis, I., and Hallenbeck, J. M., 2001, The protective effect of ceramide in immature rat brain hypoxia-ischemia involves up-regulation of bcl-2 and reduction of TUNEL-positive cells. *J. Cereb. Blood Flow Metab.* **21:** 34.

Chiueh, C. C., Andoh, T. and Chock, P. B., 2001, Protection against MPP[+]-induced toxicity via preconditioning induction of cytoprotective genes and proteins:Ref-1, NOS1, and Trx. *Soc. Neurosci. Meeting Abstract* **26:** in press.

Fassbender, K., Simons, M., Bergmann, C., Stroick, M., Lutjohann, D., Keller, P., Runz, H., Kuhl, S., Bertsch, T., von Bergmann, K., Hennerici, M., Beyreuther, K., and Hartmann, T., 2001, Simvastatin strongly reduces levels of Alzheimer's disease ß-amyloid peptides Aß 42 and Aß 40 *in vitro* and *in vivo*. *Proc. Natl. Acad. Sci. U.S.A.* **98:** 5856.

Gidday, J. M., Shah, A. R., Maceren, R. G., Wang, Q., Pelligrino, D. A., Holtzman, D. M., and Park, T. S., 1999, Nitric oxide mediates cerebral ischemic tolerance in a neonatal rat model of hypoxic preconditioning. *J. Cereb. Blood Flow Metab.* **19:** 331.

Guo, Z. H. and Mattson, M. P., 2000, *In vivo* 2-deoxyglucose administration preserves glucose and glutamate transport and mitochondrial function in cortical synaptic terminals after exposure to amyloid ß-peptide and iron: evidence for a stress response. *Exp. Neurol.* **166:** 173.

Hayashi, T., Ishikawa, T., Yamada, K., Kuzuya, M., Naito, M., Hidaka, H., and Iguchi, A., 1994, Biphasic effect of estrogen on neuronal constitutive nitric oxide synthase via Ca^{2+}-calmodulin dependent mechanism. *Biochem. Biophys. Res. Commun.* **203:** 1013.

Kato, H., Kogure, K., Liu, Y., Araki, T., and Itoyama, Y., 1994, Induction of NADPH-diaphorase activity in the hippocampus in a rat model of cerebral ischemia and ischemic tolerance. *Brain Res.* **652:** 71.

Laufs, U., Endres, M., Stagliano, N., Amin-Hanjani, S., Chui, D. S., Yang, S. X., Simoncini, T., Yamada, M., Rabkin, E., Allen, P. G., Huang, P. L., Bohm, M., Schoen, F. J., Moskowitz, M. A., and Liao, J. K., 2000, Neuroprotection mediated by changes in the endothelial actin cytoskeleton. *J. Clin. Invest.* **106:** 15.

Liu, Y., Kato, H., Nakata, N., and Kogure, K., 1992, Protection of rat hippocampus against ischemic neuronal damage by pretreatment with sublethal ischemia. *Brain Res.* **586:** 121.

Riepe, M. W., Esclaire, F., Kasischke, K., Schreiber, S., Nakase, H., Kempski, O., Ludolph, A. C., Dirnagl, U., and Hugon, J., 1997, Increased hypoxic tolerance by chemical inhibition of oxidative phosphorylation: "chemical preconditioning". *J. Cereb. Blood Flow Metab.* **17:** 257.

Toyoda, T., Kassell, N. F., and Lee, K. S., 2000, Induction of tolerance against ischemia/reperfusion injury in the rat brain by preconditioning with the endotoxin analog diphosphoryl lipid A. *J. Neurosurg.* **92:** 435.

Vaughan, C. J. and Delanty, N., 1999, Neuroprotective properties of statins in cerebral ischemia and stroke. *Stroke* **30:** 1969.

Yanamoto, H., Mizuta, I., Nagata, I., Xue, J., Zhang, Z., and Kikuchi, H., 2000, Infarct tolerance accompanied enhanced BDNF-like immunoreactivity in neuronal nuclei. *Brain Res.* **877:** 331.

NEUROPROTECTIVE PROPERTIES OF GPI1046, A NON-IMMUNOSUPPRESSIVE IMMUNOPHILIN LIGAND

Ken-ichi Tanaka and Norio Ogawa

1. INTRODUCTION

Immunophilin ligands that bind to immunophilins are known to exhibit neuroprotective properties [1-6], in addition to their immunosuppressive effects. However, immunosuppressive immunophilin ligands induce several undesirable side effects, including immune deficiency, when used in long-term continuous treatment. Resent studies have reported that non-immunosuppressive immunophilin ligands are attracting attention as new candidate drugs for neuroprotection and neurorestoration particularly since they do not have the adverse effects of immunosuppressants [3,5-12]. Furthermore, the neurotrophic actions of immunophilin ligands are restricted to damage neurons, in contrast to neurotrophic factors, which also elicit neurite outgrowth in naive neurons [6].

In the case of Parkinson's disease (PD), several researchers have reported that FK506 and related non-immunosuppressive immunophilin ligands such as GPI1046 and V10367 promote the regeneration of striatal DAergic innervation in 1-methyl-4-phenyl-1,2,3,6-tetrahydropyridine (MPTP)-intoxicated and 6-hydroxydopamine (6-OHDA)-lesioned PD animal model [6,7,13]. However, GPI1046 was not shown to be neuroprotective against apoptotic cell death in the substantia nigra following medial forebrain bundle lesion [14], although FK506 and V10367 were shown to be effective in this model [14,15]. On the other hand, Ross et al. reported recently that GPI1046 had partial neuroprotective effects against intranigral 6-OHDA lesion in rats [16]. Therefore, it is not yet enough to conclude that non-immunosuppressive immunophilin ligands are useful drugs for treating PD. Furthermore, the molecular mechanism of non-immunosuppressive immunophilin ligands and target cells in the brain remains to clarify. We examined to assess the antioxidant activities and neuroprotective properties of immunophilin ligands in cell cultures and *in vivo*.

Ken-ichi Tanaka, Norio Ogawa, Dept. of Brain Science, Okayama Univ. Grad. Sch. of Med. & Dent.
2-5-1 Shikatacho, Okayama 700-8558, Japan

Mapping the Progress of Alzheimer's and Parkinson's Disease
Edited by Mizuno *et al.*, Kluwer Academic/Plenum Publishers, 2002

2. CELL CULTURES
2.1. Cell Death
2.1.1. Cell Viability

In the assessment of cell viability, cells were treated with immunophilin ligands, FK506 or GPI1046, at a final concentration of 1-300 nM, to evaluate their protective effects against H_2O_2-induced cell damage in NG108-15 (H_2O_2 conc.; 500 μM), Neuro 2A (300 μM) and C6 (1 mM) cells. Pretreatment with FK506 and GPI1046 prevented the reduction of cell viability in a dose-dependent manner, although the significant dosage was different among each used cell lines. Therefore, our results indicate that the neuroprotective properties of GPI1046 against H_2O_2-induced reduction of cell viability may be almost equipotent to those of FK506.

2.1.2. Apoptosis

Morphological changes in the cells were observed by fluorescence microscopy after staining with Hoechst 33342 to assess whether the cells had undergone apoptosis. Both Neuro 2A and C6 cells positively stained with Hoechst 33342 showed condensed and fragmented nuclei typical of H_2O_2-induced apoptotic cell death in the 0.1% DMSO-treated group, although H_2O_2 did not induce necrotic cell death since propidium iodine staining was not investigated. Therefore, H_2O_2-induced cell death in Neuro 2A and C6 cells was clarified as having undergone apoptosis cell death. However, both FK506 and GPI1046 (100 nM) prevented H_2O_2-induced apoptotic cell death.

2.2. Antioxidant Parameter
2.2.1. Intracellular Glutathione Content

For preparation of the glutathione (GSH) sample, each naive cells of three types, NG108-15 (300 nM), Neuro 2A (100 nM) and C6 (100 nM), were grown for 24 hr in the presence or absence of FK506 or GPI1046, to clarify the mechanism of protective effects of immunophilin ligands. Both FK506 and GPI1046 significantly increased GSH contents compared with 0.1% DMSO treatment in all cell lines. In addition, the GSH activating effect of immunophilin ligands in the C6 glioma was somewhat stronger than that in the Neuro 2A neuroblastoma, although the efficacy of FK506 and GPI1046 on the H_2O_2-induced reduction of cell viability was not recognized any differences between two cells. Therefore, our results suggest that the target brain cells of immunophilin ligands may be primarily astrocytes.

2.2.2. Other Antioxidants

In C6 cell line, we also examined the effects of FK506 and GPI1046 on catalase activity and thiobarbituric acid-reactive substances (TBA-RS) level. Catalase activity was not significantly changed among the three groups, although GSH contents were increased by pretreatment with FK506 and GPI1046 compared with 0.1% DMSO treatment. However, FK506 (100 nM) or GPI1046 (100 nM) significantly reduced intracellular TBA-RS level. Therefore, both FK506 and GPI1046 prevent against H_2O_2-induced lipid peroxidation mainly due to increasing GSH content, not catalase activity.

3. IN VIVO EXPERIMENTS

3.1. Striatum Levels of Dopamine and Its Metabolites

6-OHDA markedly reduced the striatal DA, DOPAC and HVA concentrations compared with the vehicle-treated sham-operated mice. In addition, in vehicle + 6-OHDA mice, the striatal (DOPAC + HVA)/DA ratio, a parameter of DA turnover, was significantly higher than in vehicle + sham-operated mice. However, FK506 or GPI1046 prevented the reduction of striatal DA, DOPAC and HVA concentrations and normalized completely the acceleration of striatal DA turnover induced by 6-OHDA lesions. Therefore, the 6-OHDA-induced DAergic dysfunction was significantly normalized by both FK506 (0.5 mg/kg/day) and GPI1046 (10 mg/kg/day) in a bell-shaped manner. Our present results and previous reports [5,7] suggest that the neuroprotective effect of GPI1046 may be related to the degree of DAergic dysfunction. In fact, in the present study, 6-OHDA injection reduced the DA level to about 40-50% (from mild to moderate degree) of that in sham-operated mice. While GPI1046 was not shown to be effective in PD models when the DA level in PD models was about less than 20% (severe degree) of that in normal controls [14,15].

3.2. Glutathione System

Striatal GSH content was significantly increased by administration of FK506 (0.5 mg/kg/day) and GPI1046 (10 mg/kg/day). Furthermore, the results showed that the mRNA expression level of γ-glutamylcysteine synthetase (γ-GCS) increased following 7-day injection both of FK506 and GPI1046, although the levels of other four mRNAs did not change by FK506 or GPI1046 treatment. Therefore, both FK506 and GPI1046 increased GSH contents due to activation of γ-GCS mRNA expressions, which is limiting enzyme on the GSH synthesis pathway. Of the various antioxidant systems in the brain, the GSH system is particularly important in controlling cellular redox states and is the primary defense mechanism for removal of peroxide from the brain [17-19]. Moreover, apoptosis has been prevented by increasing intracellular GSH levels in several models of apoptosis [17,20-23]. Therefore, the GSH-activated effect of immunophilin ligands could be regarded as a beneficial property for these neuroprotection and/or neurorestration against oxidative stress-induced apoptotic cell death.

3.3. Other Antioxidant and Lipid Peroxidation

Striatal catalase and superoxide dismutase (SOD) activities were not changed by administration of immunophilin ligands at the same dosage, although striatal GSH content was significantly increased by administration both of FK506 (0.5 mg/kg/day) and GPI1046 (10 mg/kg/day). On the other hand, FK506 and GPI1046 reduced lipid peroxidation in mouse striatal homogenate as assessed by increased TBA-RS level, although striatal catalase and SOD activities did not change significantly. Therefore, immunophilin ligands have antioxidant properties mainly due to activation of glutathione system.

4. CONCLUSION

Our results suggest that GPI1046 may have neuroprotective effects both in cell cultures and *in vivo* [11,12,24,25]. Non-immunosuppressive immunophilin ligands, such as GPI1046, might have a beneficial effect on neurodegenerative diseases, particularly since they do not cause immune deficiency or other serious side effects. In addition, the GSH-activating effects could be regarded as a beneficial property of neuroprotective drugs.

REFERENCES

1. N. Ogawa, K. Tanaka, Y. Kondo, M. Asanuma, K. Mizukawa and A. Mori: The preventive effect of cyclosporin A, an immunosuppressant, on the late onset reduction of muscarinic acetylcholine receptors in gerbil hippocampus after transient forebrain ischemia. Neurosci. Lett. 152, 173-176 (1993).
2. J. Sharkey and S. P. Butcher: Immunophilins mediate the neuroprotective effects of FK506 in focal cerebral ischaemia. Nature 371, 336-339 (1994).
3. B. G. Gold: FK506 and the role of immunophilins in nerve regeneration. Mol. Neurobiol. 15, 285-306 (1997).
4. K. Matsuura, H. Makino and N. Ogawa: Cyclosporin A attenuates the decrease in tyrosine hydroxylase immunoreactivity in nigrostriatal dopaminergic neurons and in striatal dopamine content in rats with intrastriatal injection of 6-hydroxydopamine. Exp. Neurol. 146, 526-535 (1997).
5. J. P. Steiner, G. S. Hamilton, D. T. Rose, H. L. Valentine, H. Guo, M. A. Connolly, S. Liang, C. Ramsey, J. H. Li, W. Huang, P. Howorth, R. Soni, M. Fuller, H. Sauer, A. C. Nowotnik and P. D. Suzdak: Neurotrophic immunophilin ligands stimulate structural and functional recovery in neurodegenerative animal models. Proc. Natl. Acad. Sci. USA. 94, 2019-2024 (1997).
6. S. H. Snyder, D. M. Sabatini, M. M. Lai, J. P. Steiner, G. S. Hamilton and P. D. Suzdak: Neural actions of immunophilin ligands. TiPS. 19, 21-26 (1998).
7. L. C. Costantini, P. Chaturvedi, D. M. Armistead, P. G. McCaffrey, T. W. Deacon and O. Isacson: A novel immunophilin ligands: distinct branching effects on dopaminergic neurons in culture and neurotrophic actions after oral administration in an animal model of Parkinson's disease. Neurobiol. Dis. 5, 97-108 (1998).
8. H. Sauer, J. M. Francis, H. Jiang, G. S. Hamilton and J. P. Steiner: Systemic treatment with GPI 1046 improves spatial memory and reverses cholinergic neuron atrophy in the medial septal nucleus of aged mice. Brain Res. 842, 109-118 (1999).
9. B. G. Gold: Neuroimmunophilin ligands: evaluation of their therapeutic potential for the treatment of neurological disorders. Exp. Opin. Invest. Drugs 9, 2331-2342 (2000).
10. T. Herdegen, G. Fischer and B. G. Gold: Immunophilin ligands as a novel treatment for neurological disorders. TiPS. 21, 3-5 (2000).
11. K. Tanaka, N. Fujita, M. Asanuma and N. Ogawa: Immunophilin ligands prevent H_2O_2-induced apoptotic cell death by increasing glutathione levels in Neuro 2A neuroblastoma cells. Acta Med. Okayama 54, 275-280 (2000).
12. K. Tanaka, N. Fujita, M. Yoshioka and N. Ogawa: Immunosuppressive and non-immunosuppressive immunophilin ligands improve H_2O_2-induced cell damage by increasing glutathione levels in NG108-15 cells. Brain Res. 889, 236-239 (2001).
13. Y. Kitamura, Y. Itano, T. Kubo and Y. Nomura: Suppressive effect of FK-506, a novel immunosuppressant against MPTP-induced dopamine depletion in the striatum of young C57BL/6 mice. J. Neuroimmunol. 50, 221-224 (1994).
14. S. Happer, J. Bilsland, L. Young, L. Bristow, S. Boyce, G. Mason, M. Rigby, L. Hewson, D. Smith, R. O'Donell, D. O'Connor, R. G. Hill, D. Evans, C. Swain, B. Williams and F. Hefti: Analysis of the neurotrophic effects of GPI-1046 on neuron survival and regeneration in culture and in vivo. Neuroscience 88, 257-267 (1999).
15. C. Winter, J. Schenkel, E. Bürger, C. Eickmeier, M. Zimmermann and T. Herdegen: The immunophilin ligand FK506, but not GPI-1046, protects against neuronal death and inhibits c-Jun expression in the substantia nigra pars compacta following transection of the rat medial forebrain bundle. Neuroscience 95, 753-762 (2000).
16. D. T. Ross, H. Guo, P. Howorth, Y. Chen, G. S. Hamilton and J. P. Steiner: The small molecule FKBP ligand GPI 1046 induces partial striatal re-innervation after intranigral 6-hydroxydopamine lesion in rats. Neurosci. Lett. 297, 113-116 (2001).
17. A. J. L. Cooper and B. S. Kristal: Multiple roles of glutathione in the central nervous system. Biol. Chem. 378, 793-802 (1997).
18. R. Dringen, L. Kussmaul, J. M. Gutterer, J. Hirrlinger and B. Hamprecht: The glutathione system of peroxide detoxification is less efficient in neurons than in astroglial cells. J. Neurochem. 72, 2523-2530 (1999).
19. A. Hall: The role of glutathione in the regulation of apoptosis. Eur. J. Clin. Invest. 29, 238-245 (1999).
20. J. P. Beaver and P. Waring: A decrease in intracellular glutathione concentration procedes the onset of apoptosis in murine thymocytes. Eur. J. Cell Biol. 68, 47-54 (1995).
21. P. A. Sandstrom, M. D. Mannie and T. M. Buttke: Inhibition of activation-induced death in T cell hybridomas by thiol antioxidants: oxidative stress as a mediator of apoptosis. J. Leukoc. Biol. 55, 221-226 (1994).
22. R. W. G. Watson, O. D. Rotstein, A. B. Nathens, P. B. Dackiw and J. C. Marshall: Thiol-mediated redox regulation of neutrophil apoptosis. Surgery 120, 150-158 (1996).
23. X. F. Wang and M. S. Cynader: Astrocytes provide cysteine to neurons by releasing glutathione. J. Neurochem. 74, 1434-1442 (2000).
24. K. Tanaka, N. Fujita, Y. Higashi and N. Ogawa: Effects of immunophilin ligands on hydrogen peroxide-induced apoptosis in C6 glioma cells. Synapse (submitted).
25. K. Tanaka, I. Miyazaki, N. Fujita, M. Yoshioka and N. Ogawa: Immunophilin-mediated neuroprotection is based on activation of glutathione system in mice. FEBS Lett. (submitted).

GENE THERAPY FOR PARKINSON'S DISEASE USING AAV VECTORS

Keiya Ozawa,[1] Shin-ichi Muramatsu,[2] Ken-ichi Fujimoto,[2] Kunihiko Ikeguchi,[2] Yang Shen,[1, 2] Lijun Wang,[1, 2] Takashi Okada,[1] Hiroaki Mizukami,[1] Yutaka Hanazono,[1] Akihiro Kume,[1] Hiroshi Ichinose,[3] Toshiharu Nagatsu,[3] Keiji Terao,[4] and Imaharu Nakano[2*]

1. INTRODUCTION

Parkinson's disease (PD) is a common disorder characterized by progressive disturbance in the control of movement, and the symptoms result from a progressive loss of dopaminergic neurons in the substantia nigra. These neurons send dopamine down to their nerve endings that are located in the striatum. The severity of PD is proportional to the decline in the amount of dopamine in the striatum. Dopamine is produced from tyrosine, which is first hydroxylated and converted to L-dopa by tyrosine hydroxylase (TH) in the presence of tetrahydrobiopterin (BH_4) as a cofactor. L-dopa is subsequently decarboxylated into dopamine by a second enzyme called aromatic L-amino acid decarboxylase (AADC). The BH_4 is synthesized from guanosine triphosphate (GTP) in the TH-containing dopamine neurons where GTP cyclohydrolase I (GCH) catalyzes the first and rate limiting step of BH_4 biosynthesis. The GCH is considered to regulate the TH activity via regulation of BH_4 biosynthesis, and thus controls indirectly dopamine production.

Patients with PD can be treated with oral administration of L-dopa, but unfortunately, this therapy usually becomes less effective with progression of the disease and must often be discontinued due to the numerous serious side effects. Local production of dopamine in the striatum can be obtained by grafting of fetal mesencephalic neurons into the patient's brain, but serious practical and ethical issues are associated with this approach. Gene therapy offers a novel potential alternative to the treatment of PD. Initial gene therapy strategies for PD have focused on modifying cells ex vivo by delivering the gene for TH to these cells. However, a potentially simpler and more practical gene therapy approach involves delivering the gene directly into the brain. A promising approach for achieving this is the use of a gene delivery vehicle based on adeno-associated virus (AAV), a common virus that has never been associated with human disease.

* [1]Division of Genetic Therapeutics, Center for Molecular Medicine, and [2]Department of Neurology, Jichi Medical School, Tochigi 329-0498, Japan; [3]Institute for Comprehensive Medical Science, Fujita Health University, Aichi 470-1192, Japan; [4]Tsukuba Primate Center, National Institute of Infectious Diseases, Ibaraki 305-0843, Japan

Mapping the Progress of Alzheimer's and Parkinson's Disease
Edited by Mizuno *et al.*, Kluwer Academic/Plenum Publishers, 2002

2. GENE TRANSFER INTO NEURONS USING AAV VECTORS

AAV vectors and lentivirus vectors can transduce non-dividing neurons and long-term gene expression can be obtained. As for lentivirus vectors, human immunodeficiency virus (HIV) vectors are currently well investigated. However, the safety issue of HIV vectors is serious problem when chronic diseases are target. On the other hand, AAV vectors, derived from non-pathogenic virus, possess several unique properties and are potentially useful gene transfer vehicles for gene therapy. Therefore, AAV vectors are currently most appropriate for gene therapy of neurological diseases.

The AAV genome consists of 4681 nucleotides (nt) of single-stranded DNA, including two inverted terminal repeats (ITRs) of 145 nt long, which serve as an origin of replication and are also essential for packaging and rescue. The advantages of AAV vectors include lack of any associated disease with a wild-type virus, broad host cell range, integration of the gene into the host genome, and the ability to transduce non-dividing cells. AAV vectors are suitable for in situ transduction, and therefore, neurons, muscles, and hepatocytes are considered to be appropriate target cells. Most part of the transferred genes with AAV vectors are initially present as an episomal form, and the transgenes are expressed after the conversion to double-stranded DNA. Transgenes are believed to integrate gradually into the host cell genome, leading to their stable expression. Since the AAV vectors do not contain viral genes, immune response should not occur against the transduced cells. It is also important that cotransduction with multiple AAV vectors is efficient, because the capacity of gene insertion is limited for small-sized AAV vectors (approximately, up to 4.5 kb of a fragment including promoter and polyadenylation signal sequences can be inserted between the ITRs).

When primary cultured rat striatal cells were transduced with AAV vectors containing LacZ gene (AAV-LacZ), roughly 30% of striatal cells were positive for ß-galactosidase activity, and about one week was required for maximum expression[1]. The AAV-LacZ vector was also injected into the striatum of normal adult rats. When the rats were sacrificed at 1 week postinjection, LacZ-positive cells were clearly identified at the injection site within the striatum without any background staining using X-gal histochemistry.

3. DOPAMINE-SUPPLEMENT GENE THERAPY USING AAV VECTORS IN PARKINSONIAN RATS

To prepare a rodent model of PD, neurotoxic 6-hydroxydopamine (6-OHDA) was stereotaxically injected into the left medial forebrain bundle (MFB) to destroy the nigro-striatal pathway of dopaminergic neurons unilaterally. This MFB lesion causes the decrease of dopamine in the striatum like PD. The dopamine receptors of striatal cells in the lesioned side become supersensitive to dopamine. After the intraperitoneal administration of dopamine agonist, apomorphine, unbalance of motor function is induced, and the rats show rotational behavior. The dopamine content in the striatum can be estimated by counting the rotation after apomorphine administration. As a model experiment of gene therapy, AAV vectors were stereotaxically injected into the denervated striatum of 6-OHDA-lesioned rats. We showed that the coexpression of TH and AADC using two AAV vectors resulted in more effective dopamine production and more remarkable behavioral recovery in 6-OHDA-lesioned parkinsonian rats, than the expression of TH alone[1]. Not only TH and AADC but also BH_4 and GCH levels are reduced in parkinsonian striatum. Therefore, we next investigated whether transduction with separate AAV vectors expressing TH, AADC, and GCH was far more effective for gene therapy of PD than the dual transduction. First, in vitro experiments showed that the triple transduction with AAV-TH, AAV-AADC, and AAV-GCH resulted in greater dopamine production than the double transduction with AAV-TH and AAV-AADC in 293 cells. In parkinsonian rats treated with the triple transduction, dual immunofluorescence study showed that the striatal TH-positive cells were also positive for MAP2 (a neuronal marker), but not for GFAP (a glial marker), indicating that the transduced cells were neurons. In addition, efficient co-expression was confirmed by a double-labeling with both anti-TH and anti-AADC antibodies, or with both anti-TH and anti-GCH antibodies, which showed that more than 90% TH-IR cells were also positive for AADC and GCH. Furthermore, the triple transduction enhanced the BH_4 and dopamine production in denervated striatum of parkinsonian rats and improved the rotational behavior of the rats more efficiently than did the double transduction[2]. Behavioral recovery persisted for at least 18 months after stereotaxic intrastriatal injection. These results suggest that GCH as well as TH and AADC is important for effective gene therapy of PD. This finding would be valuable for developing more efficacious gene therapy strategies for PD. It is worthy of note that multiple genes can be efficiently delivered and expressed in neurons in vivo using separate AAV vectors as expected.

4. BEHAVIORAL RECOVERY IN A PRIMATE MODEL OF PARKINSON'S DISEASE BY GENE THERAPY WITH AAV VECTORS EXPRESSING DOPAMINE-SYNTHESIZING ENZYMES

To examine whether the above findings observed in parkinsonian rats can be extrapolated to humans, primate model experiments are very important as a preclinical study. The most critical difference between rats and monkeys is the size of striatum. Cynomolgus macaques (Macaca fascicularis) were chronically treated with 1-methyl-4-phenyl-1, 2, 3, 6-tetrahydropyridine (MPTP) to make bilateral striatal lesions. After animals became behaviorally stable for two months, mixture of three separate AAV vectors expressing TH, AADC, and GCH, respectively, were injected into the unilateral putamen stereotaxically. Coexpression of TH, AADC, and GCH in the unilateral putamen resulted in remarkable improvement in manual dexterity on the contralateral to the AAV-TH/-AADC/-GCH-injected side (S. Muramatsu, et al.: submitted). Fine motor task consisting of capturing four raisins improved on the contralateral hand. Systemic administration of apomorphine induced turning of body toward the ipsilateral side, suggesting that supersensitivity in the AAV-vector-injected putamen was reduced. TH-immunoreactive (TH-IR), AADC-IR, and GCH-IR cells were present in a large region of the putamen. Microdialysis demonstrated that concentrations of DA and its metabolites, 3,4-dihydroxyphenylacetic acid (DOPAC) and 3 methoxytyramine, in the AAV-TH/-AADC/-GCH-injected putamen were remarkably higher compared with the control side. No adverse behavioral or histopathological effects were observed. Our results show that AAV vectors efficiently introduce DA-synthesizing enzyme genes into the striatum of primates with restoration of motor functions. This triple transduction strategy is a feasible and novel treatment of PD.

5. FUTURE PERSPECTIVES

There are two major approaches to the treatment of PD; i.e. gene therapy and cell therapy. Cell-mediated gene therapy is a combination of both approaches, and is based on the amplification and genetic modification of cells in vitro prior to transplantation to the brain. Therapeutic genes can be introduced into the brain either by direct in vivo injection of vectors, or by ex vivo transduction.

Although gene therapy may be effective to improve the clinical manifestations of PD, the treatments aimed at augmenting striatal dopamine production do not halt the progressive degeneration of nigral dopaminergic neurons. Therefore, another approach seems to be important to obtain the long-term therapeutic effects of gene therapy for PD. The glial cell line-derived neurotrophic factor (GDNF), a potent neurotrophic factor for dopaminergic neurons, is a promising therapeutic molecule for limiting the neuronal damage caused by PD[3]. The combination of dopamine-supplement gene therapy and GDNF gene therapy would be a logical approach for the treatment of PD.

As for cell therapy, besides embryonic mesencephalic cells, other sources of catecholamine producing cells, such as adrenal medulla, sympathetic ganglia, and carotid body, have also been considered potentially useful for cell replacement therapy of PD. Neural stem cells derived from the cerebellum of the mouse can be induced to develop into dopaminergic neurons by the overexpression of the nuclear receptor Nurr1 and factors derived from local type 1 astrocytes. Neural progenitors may be applied to cell-mediated therapy, since they are brain-derived and can be expanded and genetically modified in vitro. When transplanted to the brain, neural progenitors undergo glial and neuronal differentiation depending on local environment. In the future, dopaminergic neurons will be generated from human embryonic (pluripotent) stem cells for clinical use in cell replacement therapy of PD. Genetic manipulation will promote the conversion of human stem cells to dopaminergic neurons, and further augment the ability of dopamine production with an appropriate regulatory system. Several approaches should be explored toward the realization of novel and efficient therapy of PD.

REFERENCES

1. D. Fan, M. Ogawa, K. Fujimoto, K. Ikeguchi, Y. Ogasawara, M. Urabe, M. Nishizawa, I. Nakano, M. Yoshida, I. Nagatsu, H. Ichinose, T. Nagatsu, G.J. Kurtzman, and Ozawa K, Behavioral recovery in 6-hydroxydopamine-lesioned rats by cotransduction of striatum with tyrosine hydroxylase and aromatic L-amino acid decarboxylase genes using two separate adeno associated virus vectors, *Hum. Gene Ther.* **9**, 2527-2535 (1998).
2. Y. Shen, S. Muramatsu, K. Ikeguchi, K. Fujimoto, D. Fan, M. Ogawa, H. Mizukami, M. Urabe, A. Kume, I. Nagatsu, F. Urano, T. Suzuki, H. Ichinose, T. Nagatsu, J. Monahan, I. Nakano, and K. Ozawa, Triple transduction with adeno-associated virus vectors expressing tyrosine hydroxylase, aromatic L-amino acid decarboxylase, and GTP cyclohydrolase I for gene therapy of Parkinson's disease, *Hum. Gene Ther.* **11**, 1509-1519 (2000).
3. D. Fan, M. Ogawa, K. Ikeguchi, K. Fujimoto, M. Urabe, A. Kume, M. Nishizawa, N. Matsushita, K. Kiuchi, H. Ichinose, T. Nagatsu, G.J. Kurtzman, I. Nakano, and K. Ozawa, Prevention of dopaminergic neuron death by adeno-associated virus vector-mediated GDNF gene transfer in rat mesencephalic cells in vitro, *Neurosci. Lett.* **248**, 61-64 (1998).

FK506 PROTECTS DOPAMINERGIC DEGENERATION THROUGH INDUCTION OF GDNF IN RODENT BRAINS

New Treatments on the Horizon in Parkinson's Disease

Atsumi Nitta, Rina Murai, Keiko Maruyama, and Shoei Furukawa[*]

1. INTRODUCTION

Parkinson'disease (PD) results from the progressive degeneration of dopaminergic (DA) neurons that innervate the striatum (Bernheimer et al., 1973; Hornykiewicz and Kish, 1986). With the progression of the disease, the available pharmacotherapy, involving use of the dopamine precursor L-dopa, becomes less effective and also leads to significant side effects (Hornykiewicz and Kish, 1986; Leenders et al., 1990). Therefore, recent advances in this field have concentrated on neuroprotective therapy to rescue the dopamine neurons.

Many kinds of neurotrophic factors may rescue adult and developing neurons from degeneration. A factor from a glial cell line (rat B49) (Schubert et al., 1974) was found to affect dopamine neurons in cultured neurons and was cloned and named glial cell line-derived neurotrophic factor (GDNF) (Lin et al, 1993). GDNF can promote survival and function of dopamine neurons *in vivo*, both the intact rat brain (Hudson et al., 1995) and in adult DA neurons after nigrostriatal lesions (Hoffer et al., 1994; Bowenkamp et al., 1995; Hundson et al., 1995; Johansson et al., 1995; Lindner et al., 1995; Tomac et al., 1995; Gash et al., 1996; Granholm et al., 1997 a, b). It has also been shown that GDNF is secreted in the target (striatum) and transported retrogradely to the DA cell bodies in the mesencephalon (Tomac et al., 1996). These results suggest that GDNF may be effective for dopaminergic degeneration. Therefore, GDNF is expected to be useful as a therapeutic tool for dopaminergic neurological disorders such as PD. However, there is an important obstacle against the therapeutic application of GDNF to PD. GDNF is a macromolecule that cannot pass through the blood-brain barrier, making it difficult to deliver GDNF from the periphery to brain. This drawback may force consideration of

[*] Atsumi Nitta, Rina Murai, Keiko Maruyama, Shoei Furukawa, Laboratory of Molecular Bioilogy, Gifu Pharmaceutical University, Gifu, Japan 502-8585

Mapping the Progress of Alzheimer's and Parkinson's Disease
Edited by Mizuno *et al.*, Kluwer Academic/Plenum Publishers, 2002

intraventricular infusion of GDNF as therapy, although this approach involves serious technical and/or ethical problems. Transfection of cells in vivo with the GDNF gene delivered by viral vectors and the transplantation of cells engineered to contain the normal GDNF gene may be promising approaches because a few reports demonstrate their effective protection against dopaminergic neurotoxins (Kirik et al., 2000). However, the clinical safety of these applications has not yet been fully established. Another promising approach to use neurotrophic actions for the therapeutic purposes is the stimulation of synthesis of GDNF.

FK 506 is one of immunosuppressant drugs. Immunosuprresion is used therapeutically for a variety of purpose. One of the most important is the treatment of patients undergoing organ transplantation (Morris, 1991). Further additional action in brains has been reported recently (Snyder, et. al., 1998). Fk506 can reduce ischemic brain damage in rats, the drug cannot protect animals against quinolinate-induced excitotoxicity (Snyder, et. al., 1998). These suggest the neuroprotective effects of FK506 may involve mechanisms distinct from NMDA-mediated signaling pathways. FK506 administration diminishes neural tissue damage following middle cerebral artery occlusion in rats (Snyder, et. al., 1998). FK506 derivatives provide pronounced protection against neurotoxicity elicited by the beta-amyloid peptide and serum deprivation of cortical cultures. The ability of FK506 to block neurotoxicity in numerous models of important neurological diseases may have clinical relevance. FK506 penetrate the blood-brain barrier reasonably well.

In this study, we demonstrate that FK506 increase GDNF in cultured brain cells and in the mice brains, further, protects dopaminergic denervation induced by neurotoxisity.

2. MATERIALS AND METHODS

Materials. FK506 and GDNF were donated by Fjisawa Pharmaceutical Co., Ltd. (Tokyo. Japan) and Amgen (CA, USA), respectively. 6-hydrexidopamine was purchased from Sigma Chemical Co. Ltd. (St. Louis, MO). All other materials used were reagent grade. Rats and mice were purchased from Nippon SLC (Shizuoka, Japan), and they were treated according to the Guideline of Experimental Animal Care issue from the Office of the Prime Minister of Japan.

Cell culture. Neurons were cultured from the hippocampi of 18-day-old rat embryos as described previously (Nitta et al., 1999a, b).

Enzyme immunoassay. Antibodies against GDNF antiserum was produced by immunizing rabbits with purified human recombinant GDNF. Protein (0.5 mg) in phosphate-buffered saline (5 ml) was emulsified with an equal volume of Freund's adjuvant and injected intradermally into rabbits four times at two-week intervals. All blood was collected one week after the final injection. Each antiserum (1 ml) was loaded on to GDNF-linked column (1ml bed volume; affi-Gel 10, Bio-Rad, CA, U.S.A.). After extensive sequential washing with three types of loading buffer, i.e. (I) 0.1 M Tris-HCl (pH 7.4) containing 0.9 % NaCl and (II) 0.05 M borate 0.05 M borate buffer (pH 8.0) (III) 0.05 mM sodium acetate buffer (pH5.0), bound antibodies were eluted with 0.1 M glycine-HCl buffer (pH 2.0). A part of the anti-GDNF antibody was eluted thus purified was biotinylated and used as the secondary antibody.

Measurement of GDNF Contents. GDNF contents in the conditioned medium (CM) and brain tissue was determined two-site enzyme immunoassay (EIA). The EIA system for GDNF was based on the method originally developed for NGF, BDNF and NT-3 (Nitta et al., 1999c). In short, multi-well plates (Falcon 3910, New Jersey, U.S.A.) were incubated with 5 μl of anti-GDNF antibody in 0.1 M Tris-HCl buffer (pH 9.0) (10 μg/ml) per well overnight, washed with washing buffer (0.1 M Tris-HCl buffer, pH 7.4, containing 0.4 M NaCl, 0.02% Na_3N 0.1 % bovine serum albumin, and 1mM $MgCl_2$), and then blocked with washing buffer containing 1 % (w/v) skim milk.

Sample or standard in washing buffer was then added to antibody-coated well; and the plate was incubated for 12-18 h at 4 ℃. The biotinylated secondary antibodies were reacted in avidin-conjugated β-galactosidase (Boehringer Mannheim, Germany) for 1 hr. Then, following thorough washing with washing buffer, enzyme activity retained in each well was measured by incubation with fluorogenic substrate; 4-methylumbelliferyl-β-D-galactoside (100 μM) in the washing buffer. The intensity of fluorescence was monitored with 360 nm excitation and 448 nm emissions. Standard curves of human recombinant GDNF at concentration between 5.08 pg and 3.3ng/ml was used for determination of concentration was determined by the EIA system described previously.

Unilateral 6-hydroxydopamine lesions and amphetamine-induced circling behavior; Young adult male mice, weighing 15-20 g at the start of the experiment were used. The mice were maintained under a 12 hr dark/light Twenty mice received 6-OHDA (20 μg/2 μl/mouse, calculated as free base; Sigma, St. Louis, MO) dissolved in ascorbate-saline (0.05 %) in the right striatum. The injection rate was 0.4 μl/min, and the tip of microsyringe (Hamilton 3020) was left in place for an additional 3 min before it was slowly retracted.

Behavioral analysis: All rotational testing was performed in the glass cylinder (diameter was 20 cm). The mice were allowed to rest for 15 min to adapt to the testing environment and then were injected intraperioneally (i. p.) with 10 mg /kg amphetamine sulfate (Dainippon Co, Pharmaceutical Ltd.) dissolved in PBS. Measurement of rotational activity began 10 min after injection. The animals were tested for 10 min. The rotometer recorded the number of turns the animals made during the test period. Clockwise turns (ipsilateral to the lesion) were counted as cycling turns.

2. RESULTS

Effects of FK506 on neuronal survival
Most of the hippocampal neurons could survive for 2 days after serum removal but then gradually died by 5 days when the cultures were started at a cell density of 10^4 cells/cm^2. FK506 (10pM) protected neuronal death as early as 1 day in neuronal cultures sustained in serum-free medium for 1 day before FK506 administration.

Effects of FK506 on GDNF content
In neuronal cultures, the GDNF content was 5.2 $\pm$0.4 pg/ml in the medium

conditioned without FK506 (10pM), whereas it increased to 22.7 ± 0.90 pg/ml with addition of 1 ng/ml of FK506. The content rose in a dose-dependent manner from 1.0 to 1,000 pg/ml FK506.

Effects of intraperioneally administration of FK506 in GDNF contents in the striatum

FK506 (1.5mg/kg) was intraperioneally injected for 10 days in straitum-lesioned by 6-OHDA rats. Twenty-four hours after the last administration, their brains were removed and used for EIA samples. The GDNF content in the control striatum was 278.5 pg/g. GDNF contents in the striatum were dramatically reduced to 8.5 pg/g in the striatum-lesioned rats. Repeated administration of FK-506 recovered GDNF contents to 245.3 pg/g in the lesioned-striatum.

Effects of intraperioneally administration of FK506 on the cycling behavior

Mice were allowed to rest for 15 min to adapt to the testing environment and then were injected intraperitoneally with 10mg/kg amphetamine sulfate dissolved in PBS. Measurement of rotation activity began 10 min after injection. FK 506 reduced the amphetamine-induced number of turns experienced by animals 10 days after the siriatum lesion (25 ± 3), compared with control animals that were received vehicle only (97 ± 4).

4. CONCLUSIONS

FK506 protects dopaminergic degeneration through induction of GDNF *in vivo* and *in vitro*. Immunosapression drugs such as FK506 may be candidates for a novel therapeutic agent for PD.

5. ACKNOWLEDGEMENTS

This work was supported in part by a grant from the Smoking Research Foundation, and Ministry of Grant-in Aid for Scientific Research of the Ministry of Education, Science, Culture, Sports and Technology of Japan.

REFERENCES

Bernheimer, H., Birkmayer, W., Hornykiewicz, O., Jellinger, K., Seitelberg, F., 1973, Brain dopamine and the syndromes of Parkinson and Huntington, J. Neurol. Sci. 20:415
Bowenkamp, K. E., Hoffman, A.F., Gerhardt, G.A., Henry, M.A., Biddle, P.T., Hoffer, B.J., Granholm, A.C., 1995, Glial cell-line derived neurotrophic factor supports survival of injured midbrain DA neurons. J. Comp. Neurol. 355:479
Gash, D. M., Zhang, Z., Ovadia, A., Cass, W. A., Simmerman, AY-L., Russell, D., Martin, D., Lapchak, P.A., Collins, F., Hoffer, B.J., Gerhardt, G.A., 1996, Functional recovery in parkinsonian monkeys treated with GDNF. Nature 380: 252
Granholm, A-C. Mott, J. L., Bowenkamp, K., Eken, S., Henry, S., Hoffer, B. J., Lapchak, P.A., Palmer, M.R., van Horne, C., Gerhardt, G.A., 1997a, Glial cell line-derived neurotrophic factor improves survival of ventral mesencephalic grafts to the 6-hydroxydopamine lesioned striatum. Exp Brain Res 116: 29
Granholm, A-C, Srivastava, N., Mott, J.L. Henry, S., Henry, M., Westphal, H., Pichel, J. G., Shen, L., Hoffer, B. J., 1997b, Morphological alterations in the peripheral and central nervous systems of mice lacking glial cell line-derived neurotrophic factor (GDNF): immunohistchemical studies. J. Neurosci. 17:1168-1178
Hoffer, B.J., Hoffman, A., Bowenkamp, K., Huettl, P., Hudson, J., Martin, D., Lin, L-FH, Gerhardt, G, 1994,

Glial cell line-derived neurotrophic factor reverses toxin-induced injury to midbrain DA neurons in vivo. Neurosci. Lett 182: 107-111

Hornykiewicz, O., and Kish, S., 1986, Biochemical pathology of Parkinson's disease. Adv. Neurol.45: 19

Johansson, M., Friedmann, M., Hoffer, B., Strombrg, I, 1995, Effects of GDNF on developing and mature mesencephalic grafts in oculo. Exp. Neurol.134: 25

Kirik, D., Rosenblad, C., Bjorkund, A., and Mandel, R. J., 2000, Long-term rAAV-mediated gene Transfer of GDNF in the rat Parkinson's model: intrastriatal but not intranigral transudation promotes functional regeneration in the lesioned nigrostrialtal system, 20: 4686

Leenders, K., Salmon, E., Tyrrell, P., Perani, D., Brooks, D., Sager, H., Jones, T., Marsden, C., and Frackowiac, R., 1990, The nigrostriatal DA system assessed in vivo by positron with Parkinson's disease. Arch. Neurol.47: 1290

Lin, L. F., Doherty, D., Lile, J., Bektesh, S., and Collins, F., 1993, GDNF: a glial cell line-derived neurotrophic factor for midbrain DA neurons. Science260: 1130

Lindner, M.D., Winn, S.R., Baetge, E.E., Hammang, J.P., Gentile, F.T., Doherty, E., McDermott, P.E., Frydel, B., Ullman, M. D., Schallert, T., 1995, Impalmtation of encapsulated catecholamine and GDNF-producing cells in rats with unilateral dopamine depletion and parkinsonian symptoms. Exp. Neurol. 132, 62

Morris, P.J., 1991, Cyclosporine, FK-506 and other drugs in organ transplantation. Curr. Opin. Immunol. 3:748

Nitta, A., Ito, M., Fukumitsu, H., Ohmiya, M., Ito, H., Sometani, A., Nomoto, H., Furukawa, Y., Furukawa, S., 1999a, 4-Methylcatechol increases brain-derived neurotrophic factor content and mRNA expression in cultured brain cells and in rat brain in vivo, 291, 1276

Nitta, A., Ohmiya, M., Sometani, A., Itoh, M., Nomoto, H., Furukawa, Y., and Furukawa, S., 1999b, Brain-derived neurotrophic factor preventsnneuronal cell death induced by corticosterone, 57:227

Nitta, A., Ohmiya, M., Jin-nouchi, T., Sometani, A., Asami, T., Kinukawa, H., Fukumitsu, H., Nomoto, H., and Furukawa, S., 1999c, Endogenous neurotrophin-3 is retrogradely transported in the rat sciatic nerve. Neuroscience, 88:678

Schubert, D., Heinemann, S., Carlisle, W., Tarikas, H., Kimes, B., Patrick, J., Steinbach, J.H., Culp, W., Brandt, B.L., 1974, Clonal cell lines from the rat centaral nervous system. Nature 249:224

Snyder, S. H., Lai, M. M., and Burnett, P. E., 1998, Immunophilins in the nervous syatem, Neuron, 21:283

Tomac, A., Lindqvist, E., Lin, L. FH, Ogren, S.O., Young, D., Hoffer, B.J., Olson, L., 1995, Protection and repair of the nigrostriatal DA system by GDNF in vivo, Nature 373, 335

Tomac, A., Widenfalk, J., Lin, L. H., Kohno, T., Ebendal, T., Hoffer, B. J., Olson, L., 1996, Retrograde axonal transport of glial cell line-derived neurotrophic factor in the adult nigrostriatal system suggests a trophic role in adult. Proc. Natl.Acad. Sci. USA 92:8274-8278

ANTI-APOPTOTIC THERAPY FOR PARKINSON'S DISEASE: OVEREXPRESSION OF AN APAF-1-DOMINANT-NEGATIVE INHIBITOR CAN BLOCK MPTP TOXICITY

Hideki Mochizuki [1]*, Hideki Hayakawa [1], Makoto Migita [2], Takashi Shimada [2], Masayuki Miura [3], and Yoshikuni Mizuno [1]

1. Introduction

The neurotoxin 1-methyl-4-phenyl-1,2,3,6-tetrahydropyridine (MPTP) causes a parkinsonian syndrome in humans and primates after selective uptake of its metabolite, MPP^+, into dopaminergic neurons. We sought to determine the major pathway of MPTP toxicity from among several apoptotic pathways. In this study, we established in vivo models of the inhibition of caspase cascade using adeno-asociated virus (AAV) vectors. We showed persistent high levels of focal Apaf-1 CARD or caspase-1 C285G mutant expression were available for further *in vivo* studies and for possible anti-apoptotic gene therapy in patients with Parkinson's disease.

2. Methods

Recombinant AAV Vector

To construct a plasmid bearing the dominant-negative truncated Apaf-1 transgene (Apaf-1 DN), the DNA fragment corresponding to the caspase-recruitment domain (CARD) of mouse Apaf-1 was obtained by RT-PCR. For the generation of recombinant AAV, subconfluent HeLa cells were cotransfected with the vector plasmid and the AAV helper plasmid pAAV/Ad using the calcium phosphate method. After concentration, the final titer of the AAV vectors was 5×10^{10} viral particles /ml.

[1] Department of Neurology, Juntendo University School Medicine 2-1-1, Hongo, Bunkyo-ku, Tokyo 113-8421, Japan

[2] Department of Biochemistry and Molecular Biology, Nippon Medical School, Tokyo 113-8602, Japan.

[3] Laboratory for Cell Recovery Mechanisms, RIKEN Brain Science Institute, 2-1 Hirosawa, Wako, Saitama, Japan

* Hideki Mochizuki, Department of Neurology, Juntendo University School of Medicine, 2-1-1, Hongo, Bunkyo-ku, Tokyo, Tel: 81-3-3813-3111, FAX: 81-3-5684-0476, e-mail: moc@sepia.ocn.ne.jp

2.1. Intrastriatal injection

We injected, into the striatum of C57 black mouse, the wild-type caspase-recruitment domain (CARD) of Apaf-1 as a dominant negative inhibitor of Apaf-1 (rAAV-Apaf-1-DN-FLAG-EGFP) using AAV virus vector, and then treated with 1-methyl-4-phenyl-1,2,3,6-tetrahydropyridine (MPTP).   Two weeks after the virus injection, the mice received four intraperitoneal injections of MPTP-HCl (30 mg/kg) in saline at 24-hr intervals.

3. Results

For in vitro experiments, we first made a rAAV-CAG-CARD-B19-EGFP or a rAAV-CAG-caspase-1-DN-B19-EGFP.   Mice received an injection of each AAV vector into the striatum.   As a control model, we injected rAAV-EGFP in the striatum (n=3)   and after 2 weeks we then treated the animals with MPTP (n=3) as in the first experiment. On the side contralateral to the vector injection, all mice displayed a dramatic loss of TH-immunoreactive neurons within the substantia nigra when compared with the Mock infection (Fig.A).

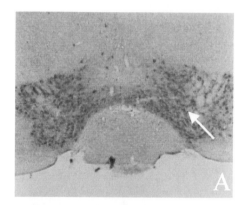
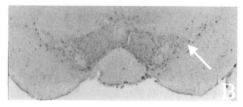
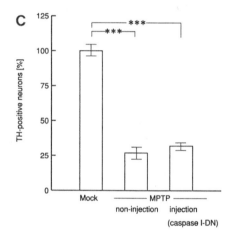

Fig.1-A) Photomicrographs of TH immunostaining at the substantia nigra of an AAV-EGFP-injected mouse as a
Mock infection (arrow).

B) Photomicrographs of TH immunostaining at the substantia nigra of MPTP treated AAV-caspase-1-
DN-injected mouse (arrow).

C) Ratio of TH-positive cells between the non-injected and the AAV-caspase-1-DN injected sides.

The expression level of caspase-1-DN was similar with that of Apaf-1-DN, but the numbers of TH-positive cells did not differ between the injected and non-injected sides (Fig. 1-B, C). In the same mouse, the number of dopaminergic neurons on the rAAV-Apaf-1-DN-EGFP–injected side (Fig.1-D, arrow) was significantly greater than the number of neurons on the non-injected side (Fig. 1-D, E).

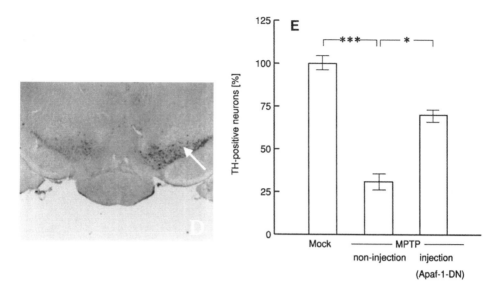

Fig.1-D) Photomicrograph of AAV-Apaf-1-DN-injected mouse (arrow).

E) Ratio of TH-positive cells between the non-injected and the AAV-Apaf-1-DN injected sides.

4. Discussion

The neurotoxin 1-methyl-4-phenyl-1,2,3,6-tetrahydropyridine (MPTP) causes a parkinsonian syndrome in humans and primates after the selective uptake of its metabolite, MPP^+, into dopaminergic neurons. MPP^+ concentrates with mitochondria according to the electrochemical gradient, where it selectively inhibits complex I of the electron transport chain. Complex I inhibition results in inhibition of ATP production and loss of the mitochondrial membrane potential, thus leading to nigral neuron death (1). MPP^+ also induces the opening of the mitochondrial transitional pore (MTP) and the release of cytochrome c via a complex I-dependent, free radical-mediated process, and this too may lead to cell death (2). Various signals mediating cell death may be initiated through the release of cytochrome c followed by mitochondrial damage (3). This pathway requires Apaf-1, which is responsible for the recruitment of procaspase-9. In the presence of dATP and cytochrome c, a 1:1 complex of Apaf-1 and procaspase-9 leads to oligomerization-induced activation of caspase-9 (4) and the subsequent activation of downstream caspases (5).

In this study, we generated overexpression of Apaf-1 CARD as an Apaf1 –dominant negative inhibitor in MPTP parkinsonian mice, and clearly showed that inhibition of this major mitochondrial apoptotic cascade, rather than the caspase-1 cascade, could prevent MPTP toxicity in vivo. TH-positive neurons in substantia nigra were dramatically increased at the rAAV-Apaf-1-DN-FLAG-EGFP injection site compared with those at the non-injected site. We also examined the effect of caspase 1 C285G mutant (6) as a dominant negative inhibitor of caspase 1 (rAAV-caspase-1-DN) in same model. However, there was no difference in the numbers of TH-positive neurons between the rAAV-caspase-1-DN-FLAG-EGFP injection site and the non-injected site. These data indicate that Apaf-1-DN delivery using an AAV vector system can prevent nigrostriatal degeneration in MPTP mice, suggesting that it might be a viable therapeutic strategy for PD patients. The major cascade of the dopaminergic cell death of MPTP is the mitochondrial apoptotic signaling pathway, and our present results suggest that Apaf-1-DN is potentially useful as an anti-mitochondrial apoptotic gene therapy for PD (7, 8).

5. Acknowledgements

ICE dominant negative construct (pS33; mouse ICE C285G Flag construct) was gifted by Prof. Yuan of Harvard Medical School. This study was supported by a High Technology Research Center Grant from the Ministry of Education, Culture, and Sports, Japan.

References

1. Y.Mizuno, S. Ikebe, S. Hattori, Y. Nakagawa-Hattori, H. Mochizuki *et al.* Role of mitochondria in the etiology and pathogenesis of Parkinson's disease. *Biochim. Biophys. Acta* **24**, 265-274 (1995).
2. D.S. Cassarino, J.K. Parks, W.D. ParkerJr, J.P. BennettJr. The parkinsonian neurotoxin MPP+ opens the mitochondrial permeability transition pore and releases cytochrome c in isolated mitochondria via an oxidative mechanism. *Biochim Biophys Acta* **6**,49-62 (1999).
3. G. Kroemer & J.C. Reed Mitochondrial control of cell death. *Nature Med* **6**, 513-519 (2000)
4. H. Zou, W.J. Henzel, X. Liu, A. Lutschg & X. Wang. Apaf-1, a human protein homologous to C. elegans CED-4, participates in cytochrome c-dependent activation of caspase-3. *Cell* **90**, 405–413 (1997).
5. S.M. Srinivasula, M. Ahmad, T. Fernandes-Alnemri & E.S. Al-nemri. Autoactivation of procaspase-9 by Apaf-1-mediated oligomerization. *Mol. Cell* **1**, 949–957. (1998)
6. P. Klevenyi, O.Andreassen, R.j. Ferrante, J.R. SchleicherJr, R.M. Friedlander *et al.* Transgenic mice expressing a dominant negative mutant interleukin-1beta converting enzyme show resistance to MPTP neurotoxicity. *Neuroreport* **25**,635-638 (1999)
7. D.W. Nicholson. From bench to clinic with apoptosis-based therapeutic agents *Nature* **407** 810 - 816 (2000).
8. H. Mochizuki, H. Hayakawa, M. Migita, M. Shibata, R. Tanaka et al. An AAV-derived Apaf-1 dominant negative inhibitor prevents MPTP toxicity as anti- apoptotic gene therapy for Parkinson's disease. *Proc. Natl. Acad. Sci. USA* (2001) in press.

CLINICAL DIAGNOSTIC CLASSIFICATION OF DEMENTIA WITH LEWY BODIES

Daniel I. Kaufer[*]

1, INTRODUCTION

Over the last two decades there has been emerging recognition of a degenerative brain disorder with Lewy bodies in cortical, as well as subcortical brain areas. Various terms including "diffuse Lewy body disease",[1] "senile dementia of the Lewy body type",[2] and "Lewy body variant of Alzheimer's disease"[3] have been used to describe the clinicopathological syndrome of dementia associated with cortical and subcortical Lewy bodies. Although the clinicopathological classification of this disorder, now referred to as Dementia with Lewy bodies (DLB),[4] recognizes overlapping features of Alzheimer's disease (AD) and Parkinson's disease (PD), a specific constellation of clinical signs and symptoms have been suggested to distinguish DLB.

In 1995, a panel of international experts (Consortium on DLB) proposed consensus guidelines for clinical and pathological diagnosis. The clinical diagnostic criteria for DLB represent a hiearchy of *central*, *core*, and *supportive* features, from which are derived categories for "Probable" and "Possible" DLB. The number of core features present is the determining factor in this framework; the presence of one core feature warrants the designation "Possible", and two or more constitute "Probable" DLB. This classification scheme represents an important milestone in research on DLB, and has been widely adopted worldwide. It bears emphasizing that the diagnoses of AD and DLB are not mutually exclusive, as many patients who meet diagnostic criteria for "Probable" or "Possible" DLB will also warrant the diagnosis of AD. From a research perspective, this strategy allows for overlapping clinical diagnoses to be sorted out on pathological grounds.

[*]Daniel I. Kaufer, M.D., University of Pittsburgh, Pittsburgh, PA, USA.

Mapping the Progress of Alzheimer's and Parkinson's Disease
Edited by Mizuno *et al.*, Kluwer Academic/Plenum Publishers, 2002

473

2. CONSENSUS CLINICAL DIAGNOSTIC CRITERIA FOR DLB

Central Feature. An essential requirement for diagnosing DLB is the presence of a dementia--an acquired and persistent decline in memory and other cognitive domains that interferes with social or occupational function. Compared to AD, DLB patients tend to be more impaired on attentional, executive, and visuospatial tasks,[3] and may exhibit mixed features of a "cortical" and "subcortical' dementia syndrome. Short-term recall is often relatively spared in DLB relative to AD. However, a specific cognitive profile distinguisghing AD and DLB is not explicitly stated in the consortium diagnostic criteria.

Core Features

Fluctuating attention. Fluctuations in attention or level of arousal is probably the most distinctive characteristic of DLB. However, there is no standardized way of assessing cognitive and behavioral fluctuations, making it somewhat difficult to reliably ascertain.[5] Explicit questioning of informants and direct observation of patients is crucial. Clinical fluctuations in DLB may take many forms, ranging from periods of confusion interspersed with periods of lucidity, to marked decreases in level of arousal and extended periods of daytime somnolence. The time course of variability may range from minute-to-minute changes, or last as long as weeks at a time. Diurnal variations in behavior referred to as "sundowning" are a common feature of dementia, particularly in more advanced stages, and are not specific for DLB. The fluctuations characteristic of DLB bear a strong resemblance to delirium, and the history and evaluation must carefully account for possible toxic and medical factors known to produce acute confusional states. This is particularly important, as demented patients are more sensitive to delirium-producing insults.

Visual Hallucinations. Psychotic symptoms are often a prominent feature of DLB, with florid visual hallucinations being most characteristic. Well-formed, recurrent visual hallucinations in DLB most commonly involve animals, children, or small people, but may also include inanimate objects or abstract perceptions. Although psychotic symptoms occur in about one-half of all patients with AD at some time during the course of the disease,[6] delusional beliefs, often paranoid in nature, are more common in AD than hallucinations. Phenomenologically, hallucinations in DLB are similar to those seen in delirium states, particularly delirium related to anticholinergic toxicity. Dopaminergic agents used to treat PD may also provoke similar hallucinatory experiences; marked sensitivity to this side effect may help discriminate DLB from the latter.

Parkinsonian Motor Signs. Extrapyramidal dysfunction is the *sine qua non* of PD and typically includes limb rigidity, bradykinesia (slowing of and difficulty initiating movements), a rest tremor, and postural instability. In DLB, parkinsonian motor features are present in about 80% of affected individuals, but tend to be less severe than in PD. Rigidity and bradykinesia are the most common parkinsonian features of DLB; decreased facial expression (masked facies), action or intention tremors, stooped posture, and a shuffling gait may also be present, as in PD. Extrapyramidal signs in PD patients typically respond well to levodopa and other antiparkinsonian agents, whereas the effect of such therapies on motor dysfunction in DLB patients may be more variable. It is estimated that 20 to 50% of AD patients will develop one or more extrapyramidal motor sign during the course of the disease, most often in the later stages.

Supportive Features. Several other clinical manifestations observed in pathologically-documented cases of DLB may raise suspicion for the diagnosis. Unexplained losses of consciousness may occur, with or without complete loss of muscle tone. Motor paralysis or general loss of muscle tone accompanying syncope is likely to reflect pathological involvement of brainstem autonomic nuclei, as opposed to a transient ischemic attack, which typically involves focal or lateralized neurological deficits. Syncope may also contribute to the high frequency of otherwise unexplained falls observed in patients with DLB. Neuroleptic sensitivity and the presence of delusions and nonvisual hallucinations are other features supporting the diagnosis of DLB. As in PD, extrapyramidal motor dysfunction accompanying DLB may render these individuals more susceptible to parkinsonian side effects of standard antipsychotic agents. Confusion, agitation, or stupor are other possible manifestations of neuroleptic sensitivity in DLB.[7] Delusions in DLB, as with visual hallucinations, tend to have a complex and bizarre flavor, and may be accompanied by auditory, tactile, or olfactory hallucinations.

3. RESEARCH EVALUATION OF DLB CLINICAL DIAGNOSTIC CRITERIA

Over the last five years a number of retrospective, and more recently, prospective Autospy-based validation studies of the consensus diagnostic criteria for DLB have been published (Table 1).[5,8-12] The retrospective studies conducted have typically observed high specificity for the clinical diagnosis of DLB, but have reported more modest diagnostic sensitivity, with a mean of 0.46. That is, as applied to retrospective, autopsy-confirmed DLB cases, less than half of the subjects with DLB, on average, meet the clinical diagnostic criteria for "probable DLB". One common limitation of the retrospective studies is that many, if not most of the subjects comprising the study samples were clinically evaluated prior to the publication of the consensus diagnostic criteria. The assessment of fluctuating attention and level of consciousness, in particular, was noted in several studies to be difficult to ascertain retrospectively. Two more recent validation studies utilizing prospective clinical assessment based on the consensus criteria for DLB show mixed results. In one study,[11] diagnostic sensitivity was observed to be much higher (0.83), whereas another study reported a lower sensitivity (0.31) akin to the majority of retrospective studies.[12]

4. CRITIQUE AND FUTURE DIRECTIONS

Studies to date suggest that the consensus clinical diagnostic criteria for DLB are robust in predicting a subset of subjects with cortical Lewy body pathology. Difficulties in assessing fluctuations in attention and level of consciousness are one of the major limitations of the consensus criteria, although standardized assessments under development may improve diagnostic sensitivity. Another area where diagnostic sensitivity may be improved is to define threshold and scaling criteria for the core features of parkinsonian motor signs and visual hallucinations. A third limitation of the current criteria is that supportive diagnostic features are not explicitly factored into the diagnostic algorithm for "possible" and "probable" DLB. Further research into the differential diagnostic utility of supportive features and the possible contribution of sleep disturbances, particularly REM Sleep Disorder, may help improve diagnostic accuracy for DLB.[13] Occipital hypoperfusion on SPECT functional neuroimaging has also been observed in DLB patients and may provide an ancillary means of helping to clinically distinguish DLB from AD.[14]

Table 1. Autopsy studies assessing the consensus diagnostic criteria for DLB.

Study	N	Sensitivity	Specificity
Retrospective:			
Mega et al [5]	18	0.40	1.00
Litvan et al [8]	105	0.57	0.87
Holmes et al [9]	80	0.22	1.00
Luis et al [10]	56	0.65	0.90
Mean		0.46	0.94
Prospective:			
McKeith et al [11]	50	0.83	0.95
Lopez et al [12]	26	0.31	1.00
Mean		0.57	0.98

Despite their limitations, the consensus clinical criteria for DLB have provided a standardized nomenclature and clinicopathological benchmarks for evaluating degenerative dementia syndromes with Lewy body pathology. Future work will improve diagnostic sensitivity and help clarify the nosological relationship of DLB to AD and PD.

REFERENCES

1. K. Kosaka, Diffuse Lewy body disease in Japan, *J. Neurol.* **237**, 197-204 (1990).
2. R. Perry , D. Irving, G.Blessed , et al, Senile dementia of the Lewy body type: A clinically and neuropathologically distinct form of dementia in the elderly, *J. Neuro. Sci.* **95**, 119-139 (1990).
3. L. Hansen, D. Salmon, D. Galasko, et al,. The Lewy body variant of Alzheimer's disease: A clinical and pathological entity, *Neurology* **40**,1-8 (1990).
4. I.G. McKeith, D. Galasko, K. Kosaka, et al, Consensus guidelines for the clinical and pathological diagnosis of Dementia with Lewy bodies (DLB): Report on the consortium on DLB international workshop, *Neurology* **47**, 1113-1124 (1996).
5. M.S. Mega, D.L. Masterman, D.F. Benson, et al, Dementia with Lewy bodies: Reliability and validity of clinical and pathologic criteria, *Neurology* **47**,1403-1409 (1996).
6. J.S. Paulsen, D.P. Salmon, L.J. Thal, et al, Incidence and risk factors for hallucinations and delusions in patients with probable AD, *Neurology* **54**,1965-1971 (2000).
7. I.G McKeith, A. Fairbairn, R. Perry, et al, Neuroleptic sensitivity in patients with senile dementia of the Lewy body type, *Brit. Med J.*,**305**;673-678 (1992).
8. I. Litvan, A. Macintyre, C.G. Goetz, et al, Accuracy of the clinical diagnoses of Lewy body disease, Parkinson's disease, and dementia with Lewy bodies, *Arch. Neurol.* **55**,969-978 (1998).
9. C. Holmes, N. cairns, P. Lantos, et al, Validity of current clinical criteria for Alzheimer's disease, vascular dementia, and dementia with Lewy bodies, *Br. J. Psychiatry* **174**, 45-51 (1999).
10. C.A. Luis, W.W. barker, K. Gajara, et al, Sensitivity and specificity of three clinical criteria for dementia with Lewy bodies in an autopsy-verified sample, *Int. J. Geriatr. Psychiatry* **14**, 526-533 (1999).
11. I.G. Mckeith, C.G. Ballard, R.H. Perry, et al, Prospective validation of consensus criteria for the diagnosis of dementia with Lewy bodies, *Neurology* **54**,1050-1058 (2000).
12. O.L. lopez, J.T. Becker, D.I.Kaufer, et al, Research evaluation and prospective diagnosis of dementia with Lewy bodies, *Arch Neurol.*, in press.
13. I.G. McKeith, E.K. Perry, R.H. Perry, et al, Report of the second dementia with Lewy bodies international workshop, *Neurology* **53**, 902-905 (1999).
14. K. Lobotesis, J.D. Fenwick, A. Phipps, et al, Occipital hypoperfusion on SPECT in dementia with Lewy bodies but not AD, *Neurology* **56**, 643-649 (2001).

NEUROPATHOLOGY OF DEMENTIA
WITH LEWY BODIES

Kenji Kosaka[*] and Eizo Iseki

1. INTRODUCTION

The term "dementia with Lewy bodies(DLB)" was proposed at the First International Workshop on Lewy body dementia, and the results of the Workshop were reported(McKeith et al, 1996). DLB is a generic term, covering diffuse Lewy body disease(DLBD) which we proposed in1984.

2. NEUROPATHOLOGICAL FEATURES OF DLB

The essential pathological feature for the diagnosis of DLB is the presence of numerous Lewy bodies in the central nervous system, including the cerebral cortex and amygdala. The detailed characteristics and distribution pattern of cortical Lewy bodies in DLBD were reported by us (Kosaka,1978) for the first time. Lewy bodies are widely distributed not only in the brain stem and diencephalon nuclei, but also in the cerebral cortex and amygdala. The brain stem and diencephalon findings are consistent with those of Parkinson's disease. Therefore, all DLBD cases have Parkinson's pathology. In preferential Lewy body sites, neuronal loss with astrocytosis is usually present.

Other neuropathological features include Lewy neurites in the CA2-3 of the hippocampus, ubiquitin-positive spheroids in the stratum lacunosum-moleculare of the hippocampus and central nucleus of the amygdala, and spongy state in the entorhinal cortex. In addition, senile plaques and neurofibrillary tangles(NFT) of varying severity are also frequently found.

3. DLBD WITHIN THE SPECTRUM OF LEWY BODY DISEASE

Our group (Kosaka et al, 1980) proposed the term Lewy body disease, and classified

---

[*] Kenji Kosaka, Department of Psychiatry, Yokohama City University School of Medicine, 65 Fukuura, Kanazawa-ku, Yokohama 256-0002, Japan

Mapping the Progress of Alzheimer's and Parkinson's Disease
Edited by Mizuno *et al.*, Kluwer Academic/Plenum Publishers, 2002

the disease into three types; a brain stem type, a transitional type and a diffuse type(Kosaka et al,1984). Furthermore, we(Kosaka et al, 1996) also added a cerebral type. Therefore, we now classify Lewy body disease into four types. The diffuse type of Lewy body disease is called DLBD(Kosaka et al, 1984, Kosaka, 1990).

DLB is a dementing illness closely associated with Lewy bodies, and DLBD is the representative form of DLB. The cerebral type of Lewy body disease(Kosaka et al,1996) is also a form of DLB. In the cerebral type. numerous Lewy bodies are found in the cerebral cortex and amygdala as is also the case in the diffuse type, but Lewy bodies are quite rare in the brain stem. The occurrence of the cerebral type suggests that Lewy bodies can first appear in the cerebral cortex, and not in the brain stem nuclei.

4. ALFA-SYNUCLEINOPATHY IN DLB

The subject of α-synuclein is now the most topical in DLB research. Monoclonal antibodies against α-synuclein can immunostain Lewy bodies in DLB and Parkinson's disease. This α-synuclein immunoreactivity has been associated with Lewy body filaments at the outer zone of the brain stem type of Lewy body. In the cortical Lewy body, granulo-filamentous components have also been found to be α-synuclein-immunoreactive(Iseki et al,2000). The finding of Lewy neurites in CA2-3 of the hippocampus is of great importance in DLB. Lewy neurites are also known to be α-synuclein-immunopositive. As seen with immunoelectron microscopy, Lewy neurites have α-synuclein-positive filamentous components in presynaptic terminals (Iseki et a, 2000).In addition, many ubiquitin- and α-synuclein-positive spheroids are also found in the stratum lacunosum-moleculare of the hippocampus and in the central nucleus of the amygdala in all DLB cases.

By means of electron microscopy, these spheroids were found to form in the distal axons(Iseki et al,1997,1998). From these studies, these Lewy neurites and spheroids are thought to arise from selective degeneration of the terminal or distal axons of the non-perforant and perforant pathway from the entorhinal cortex, respectively.

5. GROWTH AND DEGENERATION PROCESSES OF LEWY BODIES

Our group(Kosaka,1978) reported on the process of growth and degeneration of neurons containing cortical Lewy bodies as revealed using H.E. staining. Six stages were discernible. In the initial stage, a very small part of the neuronal cytoplasm stains homogeneously with eosinophilic staining and is encircled by Nissl substacnes and lipofuscin granules(stage 1). As the body increases in size, it gradually compresses both the cytoplasmic components and the nucleus towards the periphery of the cell (stage 2). In the third stage, the neuron containing a Lewy body appears swollen and shows nuclear eccentricity. The eccentric nucleus then undergoes necrosis(stage 4), and only nuclear debris can be found adjacent to the body(stage 5). The Lewy body itself also breaks down, becoming pale and containing vacuoles of varying size(stage 6). Finally, it vanishes, leaving only a small vacant space.

Recently, our group(Iseki et al,2000b) investigated the origin of α-synuclein-immunoreactive components in Lewy bodies. Cortical Lewy bodies were immunostained with an anti-phosphorylated neurofilament antibody. Double immunostaining revealed that the center of some cortical Lewy bodies is predominantly neurofilament-immunoreactive, and the periphery is predominantly α-synuclein-immunoreactive. Immuno-electron microscopy using the anti-phosphorylated neurofilament antibody showed that the center was intensely studded with nanogold particles. while fuzzy filamentous components of the periphery were less intensely represented. In addition, α-synuclein-immunoelectron microscopy revealed fuzzy filamentous components in the periphery of the Lewy body that were intensely studded with nanogold particles. On the other hand, some cortical Lewy bodies were tubulin-immunopositive. Those cortical Lewy bodies were diffusely double-immunostained with both tubulin and α-synuclein antibodies. α-Synuclein-immunoelectron microscopy revealed that the Lewy body was diffusely and intensely studded with nanogold particles. At higher magnification, loosely intertwined tubular components with fuzzy coating materials measuring 25-30 nm in diameter could be seen. These findings suggest that these tubular components originate from microtubules with an overall reduction in immunoreactivity and α-synuclein accumulation. The proportion of phosphorylated neurofilament-positive Lewy bodies and tubulin-positive Lewy bodies differs according to the individual and the affected region. The results of this study suggest that α-synuclein accumulates in different areas of the cytoskeleton of the cortical Lewy body, presumably due to a blockage in axonal transport.

Recently, our group(Iseki et al,2000a) also examined the degeneration of Lewy bodies using a double immunostaining method. We were able to find some degenerated neurons with weakly α-synuclein-immunopositive poorly-defined Lewy bodies(ghost Lewy bodies) engulfed by HLA-DR-positive microglias. Furthermore, we also found occasional GFAP-positive astroglial processes attached to the ghost Lewy bodies, which were weakly positive for α-synuclein antibody. Thus, Lewy body-containing neurons degenerate, leaving behind extracellular ghost Lewy bodies, followed by microglias and astrocytes. Then, they too disappear. Obtained with recent immunohisto-chemical methods, these findings illustrate the relationship between the Lewy body and neuronal cell death.

In addition, our recent study(Takahashi et al,1999) demonstrates that some intracellular cortical Lewy bodies are immunoreactive to an anti-C4d antibody. We can find occasional cortical Lewy bodies double-immuno-stained with both anti-C4d antibody and anti-α-synuclein antibody.

Occasional α-synuclein-positive Lewy bodies are also immuno-positive to chromogranin A antibody(Togo et al,2000). About half of chromogranin A-positive cortical Lewy bodies are intensely and evenly immunostained with α-synuclein antibody. Most chromogranin-A-positive Lewy bodies are surrounded by the processes of HLA-DR-positive microglias. These findings suggest that chromgranin A accumulates during the mature stage of the Lewy body, as the interruption of axonal flow proceeds.

6. COEXISTENCE OF CORTICAL LEWY BODIES AND TANGLES IN THE SAME NEURONS

Our group(Kosaka,1990) divided DLBD into two forms: a pure form and a common form. The common form is characterized by more or less concomitant presence of senile plaques and neurofibrillary tangles. Lewy bodies are α-synuclein-positive, but tau-negative, while tangles are tau-positive, but α-synuclein-negative. It has usually been pointed out that the coexistence of Lewy bodies and tangles within the same neurons is quite rare. Recently, our group(Iseki et al 1999) examined the frequency of neurons with coexistent Lewy bodies and tangles in brains of patients with the common form of DLBD by means of a double-immunostaining method using an anti-α-synuclein antibody and an anti-tau antibody. Double-positive neurons are frequently observed in the limbic areas. These neurons have the distinguishing feature of intermingled α-synuclein- and human tau-positive substances. Thus, the coexistence of Lewy bodies and tangles in the same neurons is not as rare as previously reported. Immunoelectron microscopy revealed α-synuclein-positive Lewy bodies surrounded by small paired helical filament bundles. In some neurons, randomly oriented paired helical filaments are scattered within Lewy bodies composed of α-synuclein-positive components which are either granular or amorphous. Thus, in addition to the coexistence of Lewy bodies and tangles in the same neurons, paired helical filaments are occasionally found inside or outside Lewy bodies. In addition, α-synuclein-positive components amorphous or granular in nature are also found among paired helical filament bundles. Our group(Marui et al,1999, 2000) also examined the coexistence of α-synuclein-positive components and tau-positive paired helical filaments in the same neurons of Alzheimer and Down syndrome brains. We then found the occasional coexistence of both components in the same neurons. In the Alzheimer brain, all neurons with α-synuclein-positive components have tau-positive tangles. On the other hand, all neurons with tau-positive tangles are not always α-synuclein-positive. Our hypothesis is as follows. In DLB, α-synuclein-positive components form the primary aggregates which are sometimes followed by tau-positive paired helical filaments; while in Alzheimer type dementia , α-synuclein-positive components aggregate only secondarily, following the formation of tau-positive paired helical filaments.

REFERENCES

Iseki, E., Li, F., Odawara, T., Kosaka, K.,1997, Hippocampal pathology in diffuse Lewy body disease using ubiquitin immunohisto-chemistry. J Neurol Sci 149:165-169

Iseki E, Marui W, Kosaka K, et al, 1998, Degenerative terminals of the perforant pathway human α-synuclein - immunoreactive in the hippocampus of patients with diffuse Lewy body disease. Neurosci Lett 258:81-84.

Iseki, E., Marui, W., Kosaka, K., et al, 1999, Frequent coexistence of Lewy bodies and neurofibrillary tangles in the same neurons of patients   with diffuse Lewy body disease. Neurosci Lett 265:9-12.

Iseki, E., Marui, W., Akiyama, H., et al., 2000a, Degeneration process of Lewy bodies in the brains of patients with dementia with Lewy bodies using α-synuclein-immunohistochemistry. Neurosci Lett 286:69-73

Iseki, E., Marui, W., Sawada, H., et al., 2000b, Accumulation of human α-synuclein in brains of dementia with Lewy bodies. Neurosci Lett290:41-44

Kosaka, K.,1978, Lewy bodies in cerebral cortex.Report of three cases. Acta Neuropathol 42:127-134

Kosaka, K.,1990, Diffuse Lewy body disease in Japan. J Neurol 237:197-204

Kosaka, K., 2000, Diffuse Lewy body disease. Neuropathology 20 Suppl:73-78

Kosaka, K., Iseki, E.,1997, Diffuse Lewy body disease within the spectrum of Lewy body disease. in Dementia with Lewy Bodies. ed

Perry, R. H., McKeith,I., Perry, E., , Cambridge University Press, Cambridge, pp. 238-247

Kosaka, K., Iseki, E. 1998, Recent advances in dementia research in Japan: Non-Alzheimer-type degenerarive dementias. Psychiatry Clin Neurosci 52: 367-378.

Kosaka, K., Iseki, E., 2000, Clinicopatholgical studies on diffuse Lewy body disease. Neuropathology 20:1-7

Kosaka, K., Iseki, E., Odawara, T., Yamamoto, T.,1996, Cerebral type of Lewy body disease. Neuropathology 16:72-75.,

Kosaka, K, Yoshimura, M., Ikeda, K., et al, 1984 Diffuse type of Lewy body disease: progressive dementia with abundant cortical Lewy bodies and senile changes of varying degree. A new disease? Clin Neuropathol 3:185-192

Marui W, Iseki E, Kosaka K, et al, 1999 An autopsied case of Down syndrome with Alzheimer pathology and α-synuclein immunoreactivity. Neuropathology 19:410-416

Marui W, Iseki E, Ueda K, Kosaka K, 2000 Occurrence of human α-synuclein immunoreactive neurons with neurofibrillary tangle formation in the limbic areas of patients with Alzheimer's disease. J Neurol Sci 174: 81-84

McKeith, I.G., Galasko,D., Kosaka, K., et al 1996 Consensus guidelines for the clinical and pathologic diagnosis of dementia with Lewy bodies(DLB):report of the consortium on DLB international workshop. Neurology 47:1113-1124

Takahashi, M., Iseki, E., Kosaka, K., 2000 Cyclin-dependent linase 5(Cdk 5) associated with Lewy bodies in diffuse Lewy body disease. Brain Res 862:253-256

Togo, T., Iseki, E., Marui, W., Akiyama, H., et al, 2001 Glial involvement in the degeneration process of Lewy body-bearing neurons and the degradation process of Lewy bodies of dementia with Lewy bodies. J Neurol Sci 184: 71 -75

MOLECULAR BIOLOGY OF LEWY BODY FORMATION

α-Synucleinopathies

Maria Grazia Spillantini and George K Tofaris[*]

1. INTRODUCTION

Parkinson's disease (PD) is the most common movement disorder, clinically characterized by tremor, rigidity and bradykinesia[1]. Neuropathologically it is defined by nerve cell loss in the substantia nigra and the presence of Lewy bodies (LB) and Lewy neurites (LN)[2-4]. LB and LN are also the characteristic neuropathological features of dementia with Lewy body (DLB), a common late-life dementia that exists in a pure form or overlaps with the neuropathological characteristics of Alzheimer's disease[5-7]. Unlike PD, DLB is characterised by a large number of LB in cortical brain areas. Ultrastructurally, LB and LN consist of abnormal filamentous material[3,4].

Although LB were first described in 1912, their composition became known only in 1997 when the discovery of a point mutation in the α-synuclein gene, in a small group of families with early-onset PD, led us to identify α-synuclein as the major component of LB and LN in idiopathic PD and DLB[8-10]. Moreover, α-synuclein deposits can be found in other diseases such as Alzheimer's disease, Down's syndrome, neurodegeneration with brain iron accumulation type 1 and in Gerstmann-Sträussler-Scheinker disease[11-16]. Filamentous glial and neuronal inclusions in multiple system atrophy (MSA) as well as neuronal inclusions in pure autonomic failure and in some cases of Parkinsonism-Dementia complex of Guam also contain α-synuclein[17-23]. The identification of α-synuclein in filamentous inclusions has provided a previously unsuspected link between PD, DLB and MSA, which we now collectively refer to as the α-synucleinopathies.

---

[*] Maria Grazia Spillantini and George K Tofaris, Brain Repair Centre and Department of Neurology, University of Cambridge, Cambridge CB2 2PY, UK.. mgs11@cam.ac.uk

Mapping the Progress of Alzheimer's and Parkinson's Disease
Edited by Mizuno *et al.*, Kluwer Academic/Plenum Publishers, 2002

2. α-SYNUCLEIN

In humans, there are at least 3 different proteins, α-, ß-, and γ-synuclein, which are expressed by different genes[24-29]. These are natively unfolded proteins that range from 127-140 amino acids in length and are 55-62% identical in their amino acid sequence. The amino-terminal half contains imperfect amino acid repeats with the consensus sequence KTKEGV. By immunohistochemistry synucleins are concentrated in nerve terminals with little staining in cell bodies and dendrites[26]. Ultrastructurally they are found in close proximity to synaptic vesicles[30]. Although the function of α-synuclein is not clear, experimental studies have shown that it binds to lipid membranes through its amino-terminal repeats, suggesting that it may be a lipid-binding protein [30-35]. α- and ß-Synuclein can selectively inhibit phospholipase D2[36]. This isoform of phospholipase D is present in plasma membranes where it may play a role in signal-induced cytoskeletal regulation and endocytosis. It is therefore possible that α- and ß-synuclein regulate vesicular transport processes. Accordingly, α-synuclein knock-out mice exibit increased release of dopamine with paired stimuli, suggesting that α-synuclein could be an activity-dependent negative regulator of neurotransmission[37]. Although there is a high degree of homology in the α-synuclein sequence across vertebrates, no synuclein homologues have been identified in *Saccharomyces cerevisiae*, *Caenorhabditis elegans* or *Drosophila melanogaster*[38].

3. α-SYNUCLEINOPATHIES

In PD brainstem LB appear as round, intracytoplasmic inclusions with a dense eosinophilic core and a clearer surrounding corona[2]. Ultrastructurally, the core is composed of filamentous and granular material that is surrounded by radially oriented filaments 10-20 nm in diameter[3,4]. LN have the same immunohistochemical and ultrastructural characteristics of LB. The biochemical nature of these filaments was unknown until genetic studies in early-onset familial PD cases identified two missense mutations in the α-synuclein gene[8,39]. The first mutation A53T was identified in a large Italian-American kindred and three smaller Greek families[8]. The second mutation A30P, was found in a German pedigree[42]. The identification of the A53T mutation was quickly followed by the discovery that α-synuclein is the major component of LB and LN in idiopathic PD as well as in families with the A53T mutation[9,40]. Lewy pathology is stained by anti-α-synuclein but not by anti ß- or γ-synuclein antibodies[9,10,41] (Fig.1a)

Ultrastructurally, α-synuclein filaments isolated from substantia nigra of PD brain are straight and unbranched, with a length of 200-600 nm and a width of 5-10 nm[41]. Antibodies against the carboxy-terminus of α-synuclein label the filaments along their entire lengths while antibodies against the amino terminus label only one end of the filament (Fig 1 b,c), indicating that the amino terminal region of the protein is buried in the body of the filament [41].

LB and LN are also the defining neuropathological characteristics of DLB. Unlike PD, DLB is characterised by the presence of numerous LB and LN in the cerebral cortex as well as in the substantia nigra and other subcortical areas [7]. As in PD, LB and LN in DLB are strongly immunoreactive for α-synuclein[9,10] as well as isolated filaments extracted from DLB brains[10,42]. These α-synuclein stained filaments show various morphologies, including a 5 nm wide filament, a less regular, 10 nm wide filament with dark-stain-penetrated central line, a 10 nm filament with 5 nm extensions at one or both ends and a twisted 5-10 nm filament with a crossover spacing of about 90 nm. The majority of filaments are 10 nm wide. Antibodies directed against the carboxy or amino-terminus of α-synuclein label the filaments with a specific pattern similar to filaments from PD[10].

MSA comprises cases of olivopontocerebellar atrophy, Shy-Drager syndrome and striatonigral degeneration. Neuropathologically, glial cytoplasmic inclusions present in oligodendrocytes, are the defining features of MSA[43]. Filamentous inclusions are also present in some nerve cells. Both glial and neuronal inclusions are strongly immunoreactive for α-synuclein[17-20] and filaments isolated from the brains of patients with MSA are strongly labeled with α-synuclein antibodies with a pattern similar to those from PD and DLB[18]. Two types of filaments are observed, some twisted, alternating in width between 5 and 18 nm, with a period of 70-90 nm and others with a uniform width of about 10 nm, as in PD and DLB[18].

4. SYNTHETIC α-SYNUCLEIN FILAMENTS

Although anti-α-synuclein antibodies stain LB and LN in human tissue and isolated filaments, the final proof that α-synuclein is the major component of these structures came from the reconstruction of filaments *in vitro* using recombinant α-synuclein. Initial studies were performed using truncated α-synuclein, which was reported to be present in LB[42,44] but later full length wild-type and mutant α-synuclein were also found to assemble into filaments with morphological and staining characteristics similar to those extracted from disease brains[44-49]. Furthermore, the A53T mutation was found to potentiate the rate of filament formation[45-49] whereas the effect of the A30P mutation on filament assembly is less clear[46,48]. The assembly of α-synuclein into filaments is a nucleation dependent process[50] that is accompanied by a transition from random coiled to ß-pleated sheet conformation[45,51]. Under similar conditions, ß- and γ-synuclein assemble into filaments much more slowly.

5. *In vivo* MODELS OF α-SYNUCLEIN AGGREGATION.

The identification of α-synuclein in LB[9,10] supports the relevance of neuropathological observations in our understanding of neurodegenerative processes. However, the drawback of neuropathological studies is that on post-mortem tissue, firm conclusions can only be drawn regarding end-stage disease, with very little insight into the underlying patho-physiological mechanisms. The obvious alternative is to reproduce the neuropathological process in animal or cellular models. However, the study of α-synuclein aggregation into LB like inclusions has achieved partial success in cellular models or transgenic animals.

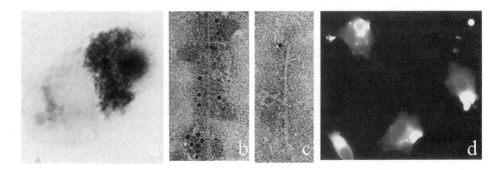

Fig. 1. α-Synuclein positive LB (A). Filaments isolated from PD brain decorated by antibodies against the carboxy- (B) and amino-terminus (C) of α-synuclein. α-Synuclein positive cytoplasmic inclusions in transfected SH-SY5Y cells (D).

The ultrastructural signature of LB is the presence of α-synuclein filaments which can also be reproduced *in vitro* using recombinant wild-type or mutant protein[44,46,48,49]. However, formation of LB in cells or transgenic mice simply by over-expression of wild-type or mutant protein is inherently resistant. This is surprising for such a fibrillogenic protein, especially since filamentous inclusions are readily reproduced when other aggregation-prone proteins are over-expressed in cell lines or transgenic mice[51,52]. Over-expression of α-synuclein in cell models leads to formation of a small number of α-synuclein positive inclusions with varying morphologies and immunohistochemical profiles with respect to ubiquitin, a hallmark of classical LB[55,56]. Similarly, we found that stable over-expression of wild-type α-synuclein in SH-SY5Y cells is sufficient to lead to accumulation of the protein in ubiquitin-negative cytoplasmic aggregates[57] (Fig 1d) made of amorphous rather than filamentous material .

The same is also true for transgenic mice, which show granular α-synuclein deposits, and subtle biochemical abnormalities but no fibrillar structures or overt cell death[58-60]. The only exception is the presence of filamentous α-synuclein inclusions in transgenic *Drosophila*, which are associated with an age-dependent loss of dopamine cells and locomotor defects[61]. Interestingly, *Drosophila* lack endogenous α-synuclein[38] and therefore, they may also be deficient in the machinery that normally limits the tendency of this protein to aggregate *in vivo*. In contrast, eukaryotic cells may be well equipped to prevent unsuitable interactions of natively unfolded proteins and filament formation. Bypassing, and/or inactivating this machinery, may provide the missing link between eukaryotic cell models and neuropathology. In this context, oxidative stress appears to be an attractive candidate. For example, under certain conditions, free radical generators, can stimulate the production of α-synuclein and ubiquitin positive intra-cytoplasmic inclusions in cells over-expressing α-synuclein, which contain mixtures of fibrillar and amorphous material[62]. Similarly, inhibition of mitochodrial complex I in rats by chronic intravenous infusion of the pesticide rotenone, induces specific neurodegeneration in the substantia nigra and formation of α-synuclein inclusions, which closely resemble LB[63]. Accordingly, complex I inhibition, iron accumulation and nitrated α-synuclein species[64], all taken as evidence of oxidative damage have been reported in LB.

However, an understanding of the mechanisms that lead to α-synuclein filament formation *in vivo*, is still lacking. The experimental work carried out suggests a 'two-step model' in LB formation. Under certain pathological conditions, unbound α-synuclein may accumulate into amorphous globs similar to those found in transgenic mice and cell culture (step one) with increased tendency to form filaments in the presence of a further, all-or-none, triggering event (step two). What causes α-synuclein to accumulate in the first place is not well understood, but impaired degradation[65] and/or aberrant protein modification are primary suspects.

Biochemically, this model is supported by the finding that α-synuclein can be transformed into a partially folded intermediate with increased propensity to form filaments[66], and by the kinetics of α-synuclein fibrillation, where nucleation (all-or-none) is the rate limiting step[52]. Neuropathologically, this model is supported by the evolution of α-synuclein deposits in PD which start as ubiquitin-negative cloud-like deposits found in morphologically normal neurons, progressing to more compact classical LB immunoreactive for both α-synuclein and ubiquitin[9,67]. The above hypothesis is also consistent with recent statistical analysis which predicts that in neurodegenerative disorders, affected neurons are in an abnormal state in which there is increased probability that a single catastrophic event will lead to cell death[68]. In this respect, the abnormal state can be paralleled to accumulation of α-synuclein intermediates and the single-catastrophic event to filament formation. Whether filament formation is the cause or marker of this single catastrophic event, or even a protective reaction to it remains to be seen. Future work on LB formation will be instrumental in our understanding of the neurodegenerative process in α-synucleinopathies, and in providing targets for therapeutic intervention.

REFERENCES

1. Parkinson J. (1817) An essay on the shaking palsy. London: Sherwood, Neely and Jones.
2. Lewy F. H. (1912) Paralysis agitans. I. Pathologische Anatomie. In: Lewandowsky M., Abelsdorff G., editors. Handbuch der Neurologie, vol. 3. Berlin: Springer Verlag, 920-933.
3. Forno L. S. (1996) Neuropathology of Parkinson's disease. J Neuropathol Exp Neurol 55: 259-272.
4. Duffy P. E. and Tennyson V. M. (1965) Phase and electron microscopic observations on Lewy body and melanin granules in the substantia nigra and locus coeruleus in Parkinson's disease. J Neuropathol Exp Neurol 24: 398-414.
5. Okazaki H., Lipkin L. E. and Aronson S. M. (1961) Diffuse intracytoplasmic ganglionic inclusions (Lewy type) associated with progressive dementia and quadriparesis in flexion. J Neuropathol Exp Neurol 21: 442-449.
6. Hansen L. A., Salmon D., Galasko D., Masliah E., Katzman R., De Teresa L., et al. (1990) The Lewy body variant of Alzheimer's disease: a clinical and pathological entity. Neurology 40: 1-8.
7. Perry R. H., Irving D., Blessed G., Fairbairn A. and Perry E. K. (1990) Senile dementia of the Lewy body type. A clinically and neuropathologically distinct form of Lewy body dementia in the elderly. J Neurol Sci 85: 119-139.
8. Polymeropoulos M. H., Lavedan C., Leroy E., Ide S. E., Dehejia A., Dutra A., Pike B., Root H., Rubenstein J., Boyer R. et al. (1997) Mutation in α-synuclein gene identified in families with Parkinson's disease. Science 276: 2045-2047.
9. Spillantini M. G., Schmidt M. L., Lee V. M-Y., Trojanowski J. Q., Jakes R. and Goedert M. (1997) α-Synuclein in Lewy Bodies. Nature 338: 839-840.
10. Spillantini M. G., Crowther R. A., Jakes R., Hasegawa M. and Goedert M. (1998) α-Synuclein in filamentous inclusions of Lewy bodies from Parkinson's disease and dementia with Lewy bodies. Proc Natl Acad Sci USA 95: 6469-6473.
11. Lippa C. F., Fujiwara H., Mann D. M. A., Giasson B. et al. (1998) Lewy bodies contain altered α-synuclein in brains of many families in Alzheimer's disease patients with mutations in presenilin and amyloid precursor protein genes. Am J Pathol 153: 1365-1370.
12. Spillantini M. G., Tolnay M., Love S. and Goedert M. (1999) Microtubule-associated protein tau, heparan sulphate and α-synuclein in several neurodegenerative diseases with dementia. Acta Neuropathol 97: 585-594.
13. Lippa C. F., Schmidt M. L., Lee V. M-Y. and Trojanowski J. Q. (1999) Antibodies to α-synuclein detect Lewy bodies in many Down's syndrome brains with Alzheimer's disease. Ann Neurol 45: 353-357.
14. Arawaka S., Saito Y., Murayama S. and Mori H. (1998) Lewy body in neurodegeneration with brain iron accumulation type 1 is immunoreactive for α-synuclein. Neurology 51: 887-889.
15. Wakabayashi K., Yoshimoto M., Fukushima T., Koide R., Horikawa Y., Morita T. et al. (1999) Widespread occurrence of α-synuclein/NACP-immunoreactive neuronal inclusions in juvenile and adult-onset Hallervorden-Spatz disease with Lewy bodies. Neuropathol Appl Neurobiol 25: 363-368.
16. Piccardo P., Mirra S. S., Young K., Gearing M., Dlouhy S. R. and Ghetti B. (1998) α-Synuclein accumulation in Gerstmann-Sträussler-Scheinker disease(GSS) with prion protein gene (PRNP) mutation F198S. Neurobiol of Aging, 19: S172
17. Wakabayashi K., Yoshimoto M., Tsuji S. and Takahashi H. (1998) α-Synuclein immunoreactivity in glial cytoplasmic inclusions in multiple system atrophy. Neurosci Lett 249: 180-182.
18. Spillantini M. G., Crowther R. A., Jakes R., Cairns N. J., Lantos P. L. and Goedert M., (1998) Filamentous α-synuclein inclusions link multiple system atrophy with Parkinson's disease and dementia with Lewy bodies. Neurosci Lett 251: 205-208.
19. Gai W. P., Power J. H. T., Blumbergs P. C. and Blessing W. W. (1998) Multiple system atrophy: a new α-synuclein disease? Lancet 352: 546-548.
20. Tu P-H., Galvin J. E., Baba M., Giasson B., Tomita T., Leight T. et al. (1998) Glial cytoplasmic inclusions in white matter oligodentrocytes of multiple system atrophy brains contain insoluble α-synuclein. Ann Neurol 44: 415-422.
21. Arai K., Kato N., Kashiwado K. and Hattori T. (2000) Pure autonomic failure in association with human α-synucleinopathy. Neurosci Lett 296: 171-173.
22. Kaufmann H., Hague K. and Perl D. (2000) Accumulation of α-synuclein in autonomic nerves in pure autonomic failure. Neurology 56: 980-981.
23. Yamazaki M., Arai Y., Baba M., Iwatsubo T., Mori O., Katayama Y. and Oyanagi K. (2000) α-Synuclein inclusions in amygdala in the brains of patients with the Parkinsonism-Dementia complex of Guam. J Neuropathol Exp Neurol 59: 585-591.
24. Maroteaux L., Campanelli J.T. and Scheller R.H. (1998) Synuclein: a neuron specific protein localised in the nucleus and presynaptic nerve terminal. J Neurosci 8: 2804-2815.
25. Ueda K., Fukushima H., Masliah E., Xia Y., Iwai A., Yoshimoto M, Otero D.A.C, Kondo J., Ihara Y. and Saitoh T. (1993) Molecular cloning of cDNA encoding an unrecognised component of amyloid in Alzheimer disease. Proc Natl Acad Sci USA 90: 11282-11286.

26. Jakes R., Spillantini M.G. and Goedert M. (1994) Identification of two distinct synucleins from the human brain FEBS Lett 345: 27-32.

27. Ji H., Liu Y. E., Jia T., Wang M., Liu J., Xiao G. et al. (1997) Identification of a breast cancer specific gene, BCSG1, by direct differential cDNA sequencing. Cancer Res 57: 759-764.

28. Buchman V. L., Hunter H. J. A., Pinon L. G. P., Thompson J., Privalova E. M., Ninkina N. N. et al. (1998) Persyn, a member of the synuclein family, has distinct pattern of expression in the developing nervous system. J Neurosci 18: 9335-9341.

29. Clayton D. F. and George J. M. (1999) Synucleins in synaptic plasticity and neurodegenerative disorders. J Neurosci Res 58: 120-129.

30. Davidson W.S., Jonas A., Clayton D.F. and George J.M. (1998) Stabilization of α-synuclein secondary structure upon binding to synthetic membranes. J. Biol. Chem. 273: 9443-9449.

31. Jensen P.H., Nielsen M.H., Jakes R., Dotti C.G. and Goedert M (1998) Binding of α-synuclein to rat brain vesicles is abolished by familial Parkinson's disease mutation. J. Biol. Chem. 273: 26292-26294.

32. McLean P. J., Kawamata H., Ribich S. and Hyman B. T. (2000) Membrane association and protein conformation of α-synuclein in intact neurons. J Biol Chem 275: 8812-8816.

33. Jo E., McLaurin J., Yip C. M., St George-Hyslop P. and Fraser P. E. (2000) α-Synuclein membrane interactions and lipid specificity. J Biol Chem 275: 34328-34334.

34. Perrin R. J., Woods W. S., Clayton D. F. and George J. M. (2000) Interaction of human α-synuclein and Parkinson's disease variants with phospholipids. J Biol Chem 275: 34393-34398.

35. Sharon R., Goldberg M. S., Bar-Josef I., Betensky R. A., Shen J. and Selkoe D. J. (2001) α-Synuclein occurs in lipid-rich high molecular weight complexes, binds fatty acids, and shows homology to the fatty acid-binding proteins. Proc Natl Acad Sci USA 98: 9110-9115.

36. Jenco J. M., Rawlingson A., Daniels B. and Morris A. J. (1998) Regulation of phospholipase D2: selective inhibition of mammalian phospholipase D isoenzymes by α- and β-synucleins. Biochemistry 37: 4901-4909.

37. Abeliovich A., Schmitz Y., Farinas I., Choi-Lundberg D., Ho W.H., Castillo P.E., Shinsky N., Verdugo J.M., Armanini M. et al. (2000) Mice lacking α-synuclein display functional deficits in the nigrostriatal dopamine system. Neuron 25: 239-252.

38. Goedert M. (2001) Alpha-synuclein and neurodegenerative diseases. Nature Rev Neurosci 2: 492-501.

39. Kruger R., Kuhn W., Muller T., Woitalla D., Graeber M., Kosel S., Przuntek H., Epplen J. T., Schos L. and Riess O. (1998) Ala30Pro mutation in the gene encoding α-synuclein in Parkinson's disease. Nature Genet 18: 106-108.

40. Spira P. S., Sharpe D. M., Halliday G., Cavanagh J. and Nicholson G. A. (2001) Clinical and pathological features of a parkinsonian syndrome in a family with an Ala 53 Thr α-synuclein mutation. Ann Neurol 49: 313-319.

41. Crowther R. A., Daniel S. E. and Goedert M. (2000) Characterisation of isolated α-synuclein filaments from substantia nigra of Parkinson's disease brain. Neurosci Lett 292: 128-130.

42. Baba M., Nakajo S., Tu P-H., Tomita T., Nakaya K., Lee V. M-Y., Trojanowski J. Q. and Iwatsubo T. (1998) Aggregation of α-synuclein in Lewy bodies of sporadic Parkinson's disease and dementia with Lewy bodies. Am J Pathol 152: 879-884.

43. Papp M. I., Kahn J. E. and Lantos P. L. (1989) Glial cytoplasmic inclusions in the CNS of patients with multiple system atrophy. J Neurol Sci 94: 79-100.

44. Crowther R.A., Jakes R., Spillantini M.G. and Goedert M. (1998) Synthetic filaments assembled from C-terminally truncated α-synuclein. FEBS Lett. 436: 309-312.

45. Serpell L.C., Berriman J., Jakes R., Goedert M. and Crowther R.A. (2000) Fiber diffraction of synthetic α-synuclein filaments shows amyloid-like cross beta conformation. Proc Natl Acad Sci USA 97: 4897-4902.

46. Conway K A., Harper J.D. and Lansbury P.T. (1998) Accelerated in vitro fibril formation by a mutant -synuclein linked to early-onset Parkinson's disease. Nat. Med. 4: 1318-1320.

47. El-Agnaf O. M. A., Jakes R., Curran M. D. and Wallace A. (1998) Effects of the mutations Ala30 to Pro and Ala53 to Thr on the physical and morphological properties of α-synuclein protein implicated in Parkinson's disease. FEBS Lett 440: 67-70.

48. Narhi L., Wood S.J., Steavenson S., et al. (1999) Both familial Parkinson's disease mutations accelerate α-synuclein aggregation. J. Biol. Chem. 274: 9843-9846.

49. Giasson B.I., Uryu K., Trojanowski J.Q. and Lee V. M-Y. (1999) Mutant and wild type human α-synucleins assemble into elongated filaments with distinct morphologies in vitro. J. Biol. Chem. 274: 7619-7622.

50. Wood S. J., Wypych J., Steavenson S., Louis J-C., Citron M. and Biere A-J. (1999) α-Synuclein fibrillogenesis is nucleation-dependent. J Biol Chem 274: 19509-19512.

51. Conway K. A., Harper J. D. and Lansbury P. T. (2000) Fibrils formed from α-synuclein and two mutant forms linked to Parkinson's disease are typical amyloid. Biochemistry 39: 2552-2563.

52. Davies S. W., Turmaine M., Cozens B. A., DiFiglia M., Sharp A. H., Ross C. A., Scherzinger E., Wanker E. E., Mangiarini L. and Bates G. (1998) Formation of neuronal intranuclear inclusions underlies the neurological dysfunction in mice transgenic for the HD mutation. Cell 90: 537-48.

53. Waelter S., Boeddrich A., Lurz R., Scherzinger E., Lueder G., Lehrach H. and Wanker E. E. (2001) Accumulation of mutant huntingtin fragments in aggresome-like inclusion bodies as a result of insufficient protein degradation. Mol Biol Cell. 12:1393-407.

54. Hsu L.J., Sagara Y., Arroyo A., Rockenstein E., Sisk A., Mallory M., Wong J., Takenouchi T., Hashimoto M. and Masliah E. (2000) α-Synuclein promotes mitochondrial deficit and oxidative stress. Am J Pathol 157: 401-410.

55. Tabrizi S. J., Orth M., Wilkinson J. M., Taanman J-W., Warner, T. T., Cooper J. M. and Schapira A. H. V. (2000) Expression of mutant α-synuclein causes increased susceptibility to dopamine toxicity. Hum Mol Gen 9: 2683-2689.

56. McLean P. J., Kawamata H. and Hyman B. T. (2001) Alpha-Synuclein-enhanced green fluorescent protein fusion proteins form proteasome sensitive inclusions in primary neurons. Neuroscience. 104: 901-12.

57. Tofaris G. K., Layfield R. and Spillantini M. G. (2001) α-Synuclein Metabolism and Aggregation is Linked to Ubiquitin-Independent Degradation by the Proteasome. Submitted.

58. Masliah E., Rockenstein E., Veinbergs I., Mallory M., Hashimoto M., Takeda A., Sagara Y., Sisk A. and Muche L. (2000) Dopaminergic loss and inclusion body formation in α-synuclein mice: Implications for neurodegenerative disorders. Science 287: 1265-1269.

59. Kahle P. J., Neumann M., Ozmen L., Muller V., Jacobson H., Schindzielorz A. et al. (2000) Subcellular localization of wild-type and Parkinson's disease-associated mutant α-synuclein in human and transgenic mouse brain. J Neurosci 20: 6365-6373.

60. van der Putten H., Wiederhold K. H., Probst A., Barbieri S., Mistl C. et al. (2000) Neuropathology in mice expressing human α-synuclein. J Neurosci 20: 6021-6029.

61. Feany M. B. and Bender W. W. (2000) A Drosophila model of Parkinson's disease. Nature 404: 394-398.

62. Ostrerova-Golts N., Petrucelli L., Hardy J., Lee J. M., Farer M. and Wozolin B. (2000) The A53T α-synuclein mutation increases iron-dependent aggregation and toxicity. J Neurosci 20: 6048-6054.

63. Betarbet R., Sherer T. B., McKenzie G., Garcia-Osuna M., Panov A. V. and Greenamyre J. T. (2001) Chronic systemic pesticide exposure reproduces features of Parkinson's disease. Nature Neurosci 3: 1301-1306.

64. Giasson B. I., Duda J. E., Murray I. V. J., Chen Q., Souza J. M., Hurtig H. I., Ischiropoulos H., Trojanowski J. Q. and Lee V. M-Y. (2000) Oxidative damage linked to neurodegeneration by selective α-synuclein nitration in synucleinopathy lesions. Science 290: 985-989.

65. McNaught K.. St.P and Jenner P. (2001) Proteasomal function is impaired in substantia nigra in Parkinson's disease. Neurosci. Lett. 297: 191-194.

66. Uvesky V., Li J. and Fink A. L. (2001) Evidence for a partially folded intermediate in α-synuclein fibril formation. J Biol Chem 276: 10737-10744.

67. Gomez-Tortosa E., Newell K., Irizarry M. C., Sanders J. L. and Hyman B. T. (2000) α-Synuclein immunoreactivity in dementia with Lewy bodies: morphological staging and comparison with ubiquitin immunostaining. Acta Neuropathol 99: 352-357.

68. Clarke G, Collins R. A., Leavitt B. R., Andrews D. F., Hayden M. R., Lumsden C. J. and McInnes R. (2000) A one-hit model of cell death in inherited neuronal degenerations. Nature 406: 195-199.

CHOLINERGIC AND MONOAMINERGIC CORRELATES OF CLINICAL SYMPTOMS IN DEMENTIA WITH LEWY BODIES

Elaine K Perry,[1] Margaret A Piggott,[1] Mary Johnson,[1] Clive G Ballard,[1] Ian G McKeith,[2] Evelyn Jaros,[3] Robert H Perry[3]

1, INTRODUCTION

Dementia with Lewy Bodies is the second most common cause of dementia in the elderly after Alzheimer's disease. The clinical diagnostic criteria which appear to work well in prospectively evaluated cohorts of patients, (McKeith et al, 2000) (1) include in particular the presence of visual hallucinations and fluctuations in cognition and attention together with extrapyramidal features. Since symptoms fluctuate and no relation has been found between symptoms with Lewy body pathology (Gomez-Tortosa et al, 2000 (2); Stern et al, 2001 (3)), functional abnormalities are likely to be important. This chapter contains a summary of neurotransmitter correlates of particular symptoms in prospectively assessed DLB patients.

2. NEUROTRANSMITTER CORRELATES

2.1 Cholinergic Correlates

The cholinergic system is affected in DLB to a greater extent than in Alzheimer's disease (AD) (Lippa et al, 1999 (4); Tiraboschi et al, 2000 (5)). Neocortical presynaptic activities are reduced to a greater extent. There are also losses of cholinergic activity in the basal ganglia including the striatum and substantia nigra, and also in the pedunculopontine pathway that projects to such areas as the thalamus. These no doubt

[1] Elaine K Perry, Margaret A Piggott, Mary Johnson, Clive G Ballard, MRC/Univ of Newcastle Development in Clinical Brain Ageing: MRC Building, [2]Ian G McKeith, Dept of Old Age Psychiatry, Wolfson Research Centre, [3]Evelyn Jaros, Robert H Perry, Dept of Neuropathology, Newcastle General Hospital, Westgate Road, Newcastle upon Tyne, NE4 6BE

relate to α-synuclein pathology in all of these areas, which include not only Lewy bodies and Lewy neurites but also neuronal inclusion bodies.

Originally in a retrospective cohort of DLB patients it was determined that choline acetyltransferase (ChAT) is lost to a greater extent in patients with, as opposed to without, hallucinations. In a prospective series of patients, the same significant trend was observed, and it was also noted that the majority of patients experiencing hallucinations were receiving L-dopa therapy, whereas patients without hallucinations were not. This suggests that hallucinations in DLB are associated with low cholinergic activity and L-dopa medication. In this more recent cohort, it was also noted that the nicotinic receptor, measured using α-bungarotoxin binding (which detects the alpha-7 subunit) was lower in hallucinating compared with non-hallucinating patients. The nicotinic receptor has not previously been implicated in psychotic features in DLB, and it remains to be determined whether nicotinic antagonist agents blocking the alpha -7 subtype can induce such psychotic features. The same receptor subtype was also associated with the symptom of delusional misidentification. This involves misinterpreting images such as television images or mirror images. In patients with this symptom the alpha-7 was again lower than in those without the symptom.

Not only are there nicotinic receptor abnormalities in DLB, but also abnormalities in the muscarinic receptor subtypes. It has been reported elsewhere using immunoprecipitation methodology, that there is a higher proportion of M1 receptor type and lower proportion of M2 receptor type in DLB (Shiozaki et al, 1999) (6). In the present prospectively assessed cases, an elevation in the muscarinic M1 subtype was detected using the radioligand pirenzepine. In a comparison between two subgroups of DLB with and without delusions there was no difference in medication using agents with anticholinergic effects or in L-dopa treatment. The two subgroups were also matched for age and severity of cognitive impairment. The muscarinic M1 receptor was however more elevated in individuals with delusions than those without (Ballard et al, 2000) (7). This suggests that delusions may be associated with a loss of presynaptic cholinergic activity and consequent compensatory elevation in the M1 receptor subtype.

The associations between these neuropsychiatric symptoms and disturbances in cholinergic neurotransmission are likely to be relevant to pharmacotherapy. In a recent multi-centre trial of Exelon (rivastigmine) there were significant improvements in delusions and hallucinations over a 20 week period in the group receiving rivastigmine, compared with those on placebo (McKeith et al, 2000) (8). The other symptoms which improved markedly and significantly with rivastigmine were apathy and agitation, although these particular clinical features have not been assessed in the prospective series.

One symptom that was not assessed in the rivastigmine clinical trial was fluctuations in attention or cognition. These occur in conjunction with various disturbances in consciousness which can include the patient appearing to be unresponsive to external stimuli while still awake, for periods of varying times. Disturbances in consciousness in DLB include episodes of stupor and unresponsiveness, clouding of consciousness, variability in cognitive function of over 50% from day to day, and variability in attentional performance correlated to fluctuations in quantitative EEG. In addition, using

EEG measurements, it has been determined that fluctuations in attention can occur across a very short time span within for example 90 seconds (Walker et al, 2000) (9). Such objective measurements using EEG relate to clinical evaluatioris of disturbances in consciousness. These features have been assessed in the prospective DLB cohort, and found to relate not to presynaptic cholinergic activities nor to the muscarinic receptor, but to the nicotinic receptor subtype binding agonists with high affinity. Epibatidine binding is reduced in many areas of the cortex in DLB, and in temporal association cortex was reduced to a lesser extent in patients experiencing disturbances in consciousness than in those with no disturbances in consciousness. This suggests that there may be a subgroup of neurones expressing the high affinity nicotinic receptor subtype (which includes the alpha-3 and alpha-4 combined with beta-2 subunits), which may be lost in those patients experiencing this symptom. It has been proposed that acetylcholine is an important neural correlate of consciousness (Perry et al, 1999) (10). The nicotinic receptor has been implicated in conscious awareness on account of its involvement in one of the mechanisms of general anaesthesia. Volatile inhalational anaesthetic agents interact with the alpha-4 beta-2 receptor with high affinity at clinically relevant doses (Perry et al, 1999) (10).

2.2 Dopaminergic Correlates

In the basal ganglia, the dopaminergic system is affected in DLB, though to a lesser extent in general than in PD. The presynaptic dopaminergic marker - the transporter or reuptake site, is reduced in the striatum, but in contrast to PD there is also reduction in the dopaminergic D2 receptor subtype in the striatum. These abnormalities no doubt relate to extrapyramidal dysfunction which in some respects may differ between PD and DLB. A reduction in dopamine uptake has been detected in DLB both pre – and post-mortem as a distinguishing feature from AD (Hu et al, 2000 (11); Piggott et al, 1999 (12)). In a recent survey of dopaminergic markers in prospectively assessed DLB patients, the dopamine transporter was assessed using mazindol binding, the D2 receptor using epidepride binding, and also the D1 and D3 receptors were evaluated. The only significant changes in the cerebral cortex amongst these markers were a moderate increase in the dopamine transporter and a significant reduction in D2 receptor binding. Neither of these changes was associated with hallucinations, delusional misidentification or fluctuations in consciousness. However there was a relationship with auditory hallucinations. In Brodmann areas 20 and 22 in the temporal lobe, the reduction in D2 receptor binding was not as great in patients experiencing auditory hallucinations, compared with those who were not experiencing auditory hallucinations. It may be of interest that the only symptom which related to dopaminergic markers in the cortex in DLB was auditory hallucinations, since this is one of the symptoms of psychosis in DLB which most closely resembles the psychosis in schizophrenia, in which dopaminergic overactivity has long been implicated.

2.3 Serotonergic Correlates

As in the basal ganglia and substantia nigra, there is also in DLB α-synuclein pathology and neuronal loss in the serotonergic raphé nuclei including dorsal raphé. There is, in both the striatum and the cortex an extensive loss of the 5-HT transporter or reuptake site. Cyanoimipramine binding to the transporter was examined in prospectively assessed patients, and found to relate to depression, but not to any of the other neuropsychiatric symptoms discussed above. In patients with depression the transporter was paradoxically less reduced than in patients without depression. There is some evidence that depression may be associated with compensatory regenerative activity in 5-HT neurones and this finding of higher transporter binding in depressed compared to non depressed patients maybe consistent with such a concept.

3. SLEEP ABNORMALITIES

Sleep abnormalities are increasingly being considered as an important symptom in diseases like Parkinson's disease and DLB. In DLB, REM behaviour disorder has been reported. In addition increased daytime sleepiness, sleep fragmentation and nightmares have been reported (Grace et al, 2000 (13)). In DLB, compared with AD patients, significantly more limb movements during sleep, unpleasant dreams and confusion on waking were apparent. In six DLB patients treated with rivastigmine compared to six untreated there was a trend towards normalisation of sleep patterns. This preliminary finding suggests a cholinergic component of sleep disturbances in DLB.

Mechanisms involved in controlling sleep onset, duration, REM sleep and dreaming directly involve alterations in neurotransmission. There is a progressive reduction in noradrenergic and 5-HT activity from sleep onset to REM onset where these neurons are silent. In contrast dopaminergic transmission in the ventral area is constant throughout the sleep/wake cycle. Cholinergic transmission in the nucleus basalis of Meynert is most active during waking and REM sleep, and less active during non REM sleep, whereas pedunculopontine cholinergic activity is absent during non REM but maximal during REM sleep. Many other transmitter systems are involved in or affected during the sleep-wake cycle, most notably adenosinergic transmission. Adenosine levels increase during waking and inhibitory effects on the basal forebrain neurons are thought to trigger sleep.

In attempting to understand mechanisms responsible for sleep abnormalities in DLB, adenosinergic transmission has not been evaluated, but there are clearly abnormalities in all of the other transmitter systems implicated above. How these relate to sleep disturbances can only be a matter of speculation at this stage. Daytime sleepiness could relate to impaired cholinergic transmission the forebrain and/or reductions in serotonergic and noradrenergic transmission. REM behaviour disorder, during which the normal inhibition of movement is disrupted, could relate to brainstem cholinergic pathology. In addition some of the neuropsychiatric symptoms such as hallucinations or delusions could involve disruption of the mechanisms controlling dreaming with intrusion of dream mentation into the waking state.

4, CONCLUSION

It seems likely on the basis of evidence summarised above, and also the fact that many symptoms in DLB fluctuate and so are not so likely related to irreversible morphological pathology, that disturbances in neurotransmission are responsible for key symptoms. In prospectively assessed cases, visual hallucinations are associated with lower ChAT and the nicotinic α7 receptor, whereas auditory hallucinations are associated with higher D2 receptor binding. Delusions relate to higher muscarinic M1 receptor, disturbances in consciousness to relatively preserved high affinity nicotinic receptor, and depression to relatively preserved 5-HT transporter. Therapeutically, cholinesterase inhibitors are effective in relieving hallucinations and delusions, and other important symptoms such as apathy. Neuroleptic agents are not the most appropriate antipsychotic agents to the elderly dementia population including DLB patients, on account of the risk of increasing Parkinsonism and it is likely that cholinergic drugs may emerge as a safer antipsychotic option. The possibility that such drugs may exacerbate syncope needs however to be evaluated.

Cholinergic therapy may not only be of symptomatic benefit in DLB but may also be disease stabilising. Accumulating evidence in animal models suggest nicotinic receptor stimulation protects dopaminergic neurons from age-related changes and in the human brain, substantia nigra neuron numbers are higher in individuals chronically exposed to nicotine (tobacco users) (Perry et al, 2000 (14)).

5. ACKNOWLEDGEMENTS

We are extremely grateful to Lorraine Hood for the preparation of this manuscript.

REFERENCES

1. McKeith, I.G., Ballard, C.G., Perry, R.H., Ince, P.G., O'Brien, J.T., Neill, D., Lowery, K., Jaros, E., Barber, R., Thompson, P., Swann, A., Fairbairn, A.F., and Perry, E.K., 2000, Prospective validation of consensus criteria for the diagnosis of dementia with Lewy bodies, Neurol. 54:1050-8.
2. Gomez-Tortosa, E., Irizarry, M.C., Gomez-Isla, T., and Hyman, B.T., 2000, Clinical and neuropathological correlates of dementia with Lewy bodies, Ann NY Acad Sci. 920:9-15.
3. Stern, Y., Jacobs, D., Goldman, J., Gomez-Tortosa, E., Hyman, B.T., Liu, Y., Troncoso, J., Marder, K., Tang, M.X., Brandt, J., and Albert, M., 2001, An investigation of clinical correlates of Lewy bodies in autopsy-proven Alzheimer disease, Neurol. 58:460-5.
4. Lippa, C.F., Smith, T.W., and Perry, E., 1999, Dementia with Lewy bodies: choline acetyltransferase parallels nucleus basalis pathology, J Neural Transm. 106:525-35.
5. Tiraboschi, P., Hansen, L.A., Alford, M., Sabbagh, M.N., Schoos, B., Masliah, E., Thal, L.J., and Corey-Bloom, J., 2000, Cholinergic dysfunction in diseases with Lewy bodies, Neurol. 54:407-11.
6. Shiozaki, K., Iseki, E., Uchiyama, H., Watanabe, Y., Haga, T., Kameyama, K., Ikeda, T., Yamamoto, T., and Kosaka, K., 1999, Alterations of muscarinic acetylcholine receptor subtypes in diffuse lewy body disease: relation to Alzheimer's disease, J Neurol Neurosurg Psychiat. 67:209-13.
7. Ballard, C., Piggott, M., Johnson, M., Cairns, N., Perry, R., Mckeith, I., Jaros, E., O'Brien, J., Holmes, C., and Perry, E., 2000, Delusions associated with elevated muscarinic binding in dementia with Lewy bodies, Ann Neurol. 48:868-76.
8. McKeith, I., Del Ser, T., Spano, P., Emre, M., Wesnes, K., Anand, R., Cicin-Sain, A., Ferrara, R., and Spiegel, R., Efficacy of rivastigmine in dementia with Lewy bodies: a randomised double-blind, placebo-controlled international study, Lancet. 356:2031-6.
9. Walker, M.P., Ayre, G.A., Cummings, J.L., Wesnes, K., McKeith, I.G., O'Brien, J.T., and Ballard, C.G., 2000, Quantifying fluctuation in dementia with Lewy bodies, Alzheimer's disease and vascular dementia, Neurol. 54:1616-25.
10. Perry, E., Walker, M., Grace, J., and Perry, R., 1999, Acetylcholine in mind: a neurotransmitter correlate of consciousness?, Trends Neurosci. 22:273-80.

11. Hu, X.S., Okamura, N., Arai, H., Higuchi, M., Matsui, T., Tashiro, M., Shinkawa, M., Itoh, M., Ido, T., and Sasaki, H., 2000, 18F-fluorodopa PET study of striatal dopamine uptake in the diagnosis of dementia with Lewy bodies, Neurol. **55**:1575-7.

12. Piggott, M.A., Marshall, E.F., Thomas, N., Lloyd, S., Court, J.A., Jaros, E., Burn, D., Johnson, M., Perry, R.H., McKeith, I.G., Ballard, C., and Perry, E.K., 1999, Striatal dopaminergic markers in dementia with Lewy bodies, Alzheimer's and Parkinson's diseases: rostrocaudal distribution, Brain. **122**:1449-68.

13. Grace, J.B., Walker, M.P., and McKeith, I.G., 2000, A comparison of sleep profiles in patients with Dementia with Lewy bodies and Alzheimer's disease, Int J Geriatr Psychiat. **15**:1028-33.

14. Perry, E., Martin-Ruiz, C., Lee, M., Griffiths, M., Johnson, M., Piggott, M., Haroutunian, V., Buxbaum, J.D., Näsland, J., Davis, K., Gotti, C., Clementi, F., Tzartos, S., Cohen, O., Soreq, H., Jaros, E., Perry, R., Ballard, B., McKeith, I., and Court, J., 2000, Nicotinic receptor subtypes in human brain ageing, Alzheimer and Lewy body diseases, Europ J Pharmacol. **393**:215-222.

MOLECULAR BIOLOGY OF α-SYNUCLEIN
Dementia with Lewy Bodies

Olaf Riess[*], Rejko Krüger, Hirokazu Kobayashi, Carsten Holzmann, Nobutaka Hattori, and Yoshikuni Mizuno

1. INTRODUCTION

The discovery of point mutations in the α-synuclein gene in autosomal dominant forms of Parkinson's disease (ADPD) initiated numerous immunohistochemical and biochemical studies to gain insights into the pathogenesis of this neurodegenerative disorder. Subsequently, α-synuclein has been identified as a major component of Lewy bodies (LB), a pathological hallmark of PD, in familial and sporadic cases. However, α-synuclein has also been identified in intracellular aggregates of other neurodegenerative disorders such as dementia with LB, multiple system atrophy (MSA), and amyotrophic lateral sclerosis (ALS), implicating a more general role of α-synuclein in neurodegeneration. This hypothesis is further supported by neuropathological and phenotypical features of transgenic animal models overexpressing α-synuclein. Despite this progress the function of α-synuclein, the mode of dysfunction of the mutated protein, and the pathogenesis of ADPD are still unknown to a great extent. Here, we discuss the present knowledge on the pathogenesis of α-synuclein in PD.

2. ROLE OF α-SYNUCLEIN IN PD

2.1. Phenotypical Manifestation Of Patients With α-Synuclein Mutations

Until today, only two mutations in the α-synuclein gene have been described in ADPD. The group of M.H. Polymeropoulos identified an alanine to threonine substitution

---

[*]Olaf Riess and Carsten Holzmann, Department of Medical Genetics, University Rostock, Rembrandt Street 16/17, D-18055 Rostock, Germany. Rejko Krüger, Department of Neurology, University Tübingen, Hoppe-Seyler-Str. 3, 72076 Tübingen, Germany. Hirokazu Kobayashi, Nobutaka Hattori, and Yoshikuni Mizuno, Department of Neurology, Juntendo University School of Medicine, 2-1-1 Hongo, Bunkyo-ku, Tokyo 113-0033, Japan.

Mapping the Progress of Alzheimer's and Parkinson's Disease
Edited by Mizuno *et al.*, Kluwer Academic/Plenum Publishers, 2002

at position 53 of α-synuclein (Ala53Thr) in a large Italian kindred which is known as the Contursi family (Golbe et al. 1990) and in three families of Greek origin (Polymeropoulos et al. 1997). Since then thirteen families with the Ala53Thr have been described all of them are of Greek origin (Polymeropoulos et al. 1997, Papadimitriou et al. 1999, Athanassiadou et al. 1999, Veletza et al. 1999, Spira et al. 2001). In addition, we described an Ala30Pro mutation in an ADPD family of German ancestry (Krüger et al. 1998). The age at onset of gene carriers has a high variability and varies between 33 and 59 years for the Ala53Thr mutations although unaffected mutation carriers older than have been described (Papadimitriou et al. 1999) and between 54 to 76 years for the Ala30Pro mutation (Krüger et al. 2001). All of the Ala53Thr patients suffer from bradykinesia and muscular rigidity but not from tremor whereas tremor is present in some Ala30Pro individuals. In all patients treatment with levodopa was excellent. A high proportion of patients developed dementia (Spira et al. 2001). Although both mutations are extremely rare their relevance to have initiated numerous biochemical and pathological studies cannot be overstated. As a consequence abnormal deposition of α-synuclein in form of Lewy bodies has been demonstrated in patients with the Ala53Thr mutation (Spira et al. 2001) but also in sporadic PD (Spillantini et al. 1997).

2.2. The Synuclein Gene Family

Alpha-synuclein belongs to a gene family consisting of four proteins (Table 1) which are expressed predominantly in nervous tissue and are strikingly conserved in vertebrates. Besides α-synuclein we currently distinguish β- (phosphoneuroprotein-14-kDa, PNP-14; Nakajo et al. 1993) and γ-synuclein (also known as breast cancer-specific gene 1 and persyn; Akopian and Wood 1995; Ji et al. 1997), and synoretin which is closely related to γ-synuclein (Surguchov et al. 1999). The amino-terminal half of each synuclein contains six incomplete repeats of 11 amino acids with the consensus sequence KTKEGV. Towards the carboxy-terminus each protein contains a number of negatively charged residues. Each synuclein is encoded by a different gene on chromosome 4q21, 5q35, and 10q23, respectively. The gene locus for synoretin has not been ascribed yet. Synuclein has been named due to initial reports on its subcellular localization in synapsis and in the nucleus (Maroteaux et al. 1988). However, nuclear localization has not been confirmed in subsequent studies (Iwai et al. 1995).

The functions of synucleins are poorly understood. Due to their enrichment in presynaptic terminals synucleins have been suggested to play a role in synaptic transmission. The hydrophobic domain of α-synuclein might be responsible for its putative association with synaptic vesicles. Recent observation suggest an involvement of α-synuclein in axonal transport (Jensen et al. 1998), in neuronal plasticity (George et al. 1995), and for α-and β-synuclein in signal transduction as they regulate phospholipase D (Jenco et al. 1998). All synucleins were shown to be a substrate for G protein-coupled receptor kinases (GRK) with GRK2 preferentially phosphorylating α- and β-synuclein (Pronin et al. 2000). GRK-mediated phosphorylation inhibits synuclein's interaction with both phospholipids and phospholipase 2 *in vitro* (Pronin et al. 2000).

γ-Synuclein might be a useful marker in breast and ovar cancer as high expression was shown in infiltrating malignant breast cancer carcinomas (Ji et al. 1997) and ovarian tumors (Lavedan et al. 1998) but not in normal or benign breast tissue. However, the significance of γ-synuclein in cancer is still unknown but it clearly raises the question of a more universal function of synucleins than currently assumed. This is supported by the recent cloning of synoretin (Surgichov et al. 1999) which is most significantly expressed in the retina and might be involved in phototransduction.

Table 1: The synuclein gene family

Name	Chromosomal localization	Protein length	Sequence identity
α-synuclein (SNCA)	4q21	140aa	Identical to NACP
β-synuclein (PNP14)	5q35	134aa	62% to NACP
γ-synuclein (BCSG1, Persyn)	10q23	127aa	55% to NACP
Synoretin	Unknown	127aa	84% to BCSG1

2.3. Biological Function Of α-Synuclein

In brain homogenates, 0.5-1% of the cytosolic protein is synuclein. It has been implicated in neuronal plasticity as it is upregulated during the critical period for song learning in birds (George et al. 1995). A role of α-synuclein in neuronal development and in synaptogenesis has been suggested due to its expression in murine and human brain development and its redistribution from the cytosol to the nerve terminals (Thal et al.1998, Bayer et al. 1999). Mice lacking α-synuclein show abnormal dopamine release under certain conditions suggesting that α-synuclein is a presynaptic regulator for dopamine (Abeliovich et al. 2000).

In the effort to decipher the biological function of α-synuclein synphilin-1 has been identified as interacting protein by the yeast two hybrid method (Engelender et al. 1999). Synphilin-1 harbors ankyrin domains suggesting a function as a cytoskeletal protein. Overexpression of both proteins promotes the formation of cytoplasmic eosinophilic inclusions in cell culture. Synphilin-1 has been also been identified as a component of Lewy bodies (Wakabayashi et al. 2000). Microtubule-associated protein 1B has been identified as another interacting protein of α-synuclein (Jensen et al. 2000) further supporting its interaction with the cytoskeleton. Its homology to 14-3-3 proteins (Ostrerova et al. 1999) further suggested a function as a chaperone which has been further supported by biochemical analysis (Souza et al. 2000) demonstrating that all three synucleins are capable to suppress aggregation of thermally denatured alcohol dehydrogenase and chemically denatured insulin. As 14-3-3 proteins, α-synuclein interacts also with protein kinase C, BAD, extracellular regulated kinase (Ostrerova et al. 1999), and tau (Jenssen et al. 1999).

Structural analysis indicated the absence of significant amounts of secondary structure in aqueous solution indicating that α-synuclein belongs to the class of "natively unfolded" protein (Weinreb et al.1996). Upon binding to acidic phospholipid vesicles α-synuclein forms α-helical structures and forms insoluble fibrils with high β-sheet content (Davidson et al. 1998).

2.4. Pathobiological Significance Of Mutated α-Synuclein

The identification of mutations, the Ala53Thr and the Ala30Pro mutation, in the α-synuclein gene (Polymeropoulos et al. 1997, Krüger et al. 1998) support the role in the pathogenesis of PD. Two potential ways of synuclein dysfunction might be predispose the cells to neurodegeneration: (i) loss of function e.g. through disruption of interaction with other proteins, through reduced expression, or (ii) through gain of function caused by structure abnormalities. A variety of observations currently provide evidence for all three mechanisms.

2.4.1. Disturbed Protein Interaction

As described before α-synuclein binds to vesicles in the brain through its amino-terminal repeat region. This vesicle-binding activity is significantly reduced by the Ala30Pro (Jensen et al. 1998) and the Ala53Thr mutation (Jo et al. 2000) implicating a disturbed vesicle transport in neuronal axons by the mutation. Mutated (Ala53Thr) protein forms in combination with ERK-2 a complex with transcription factor Elk-1 which mediates gene activity in response to growth factors (Iwata et al. 2001). The nuclear translocation of Elk-1 might be inhibited and subsequently the MAP kinase pathway might be insufficiently activated. Other functions of α-synuclein are disturbed by the mutations as its chaperone activity (Ostrerova et al. 1999, Kim et al. 2000) and its antiapoptotic effects (da Costa et al. 2000). All these pathways, however, do not explain the selectivity of the cell loss of dopaminergic neurons. In this respect it might be important that α-synuclein binds to the carboxyl-terminal tail of the human dopamine transporter (hDAT) accelerating cellular dopamine uptake and dopamine-induced cellular apoptosis (Lee et al. 2001). Earlier association studies revealed conflicting data on the significance of a variable tandem repeat (VNTR) in the 3'-untranslated region with PD (Kim et al. 2000; Leighton et al. 1997; LeCouteur et al. 1997) or of silent single nucleotide polymorphisms (Morino et al. 2000; Kimura et al. 2001). However, a detailed sequence analysis for amino acid polymorphisms and their relation to PD needs to be performed as studies whether the synuclein mutations affect binding to hDAT. A common link to the oxidative stress hypothesis in PD (Coyle and Puttfarcken 1993) could then be established. Clearly, mutations in α-synuclein in particular the Ala30Pro mutation enhance vulnerability of neuronal cells to oxidative stress (Ko et al. 2000)and induce the formation of aggregates (Hashimoto et al. 1999).

2.4.2. Altered α-Synuclein Expression

Reduced α-synuclein expression of the mutated allele in lymphoblastoid cell lines of Ala53Thr patients has been observed (Markopoulou et al. 1999). For carriers of the Ala30Pro mutation we could confirm this finding (Kobayashi et al. unpublished) suggesting self-regulating mechanisms of the cells on the expression of the altered allele. Upon administration of the parkinsonian toxin MPTP in mice an upregulation of α-synuclein has been shown (Vila et al. 2000) may be prevent induction of apoptotic pathways. It is in particular intriguing that dopaminergic neurons which are most prone to neuronal cell death in PD are also most susceptible to the neurotoxic effect of α-synuclein (Forloni et al. 2000), or, alternatively, that susceptibility to cell death of HEK293 cells is enhanced through dopamine (Tabrizi et al. 2000).

2.4.3. Altered Protein Structure And/Or Degradation

A subfragment of α-synuclein termed the non-β-amyloid component (NACP) was first identified as a component of Alzheimer's disease plaques (Ueda et al. 1993) although recent studies do not replicate these early findings (Culvenor et al. 1999). Subsequent studies, however, clearly show α-synuclein aggregation in PD, MSA, PSP, and Pick's disease (Tu et al. 1998, and reviewed in Dickson 1999). Mutated α-synuclein as identified in PD develops more β–sheet structure, is more prone to self-aggregation and to the formation of amyloid-like filaments (El-Agnaf et al. 1998a). The fibril formation is significantly accelerated by the mutations, most rapidly by Ala53Thr (Conway et al. 1998). This may lead to the formation of abnormal deposits in form of the Lewy bodies altering axonal transport of vesicles or proteins (reviewed in Riess et al. 1998) and inducing apoptotic cell death (El-Agnaf et al. 1998b). Most interestingly, the age at onset of Ala53Thr carriers (Papadimitriou et al. 1999) is significantly reduced compared to Ala30Pro patients and to idiopathic PD.

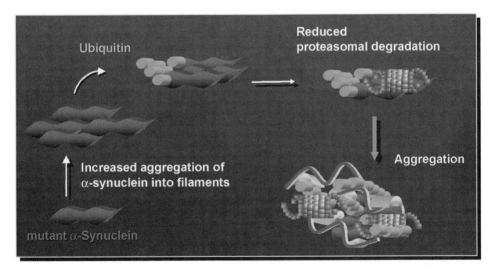

Figure 1

Missfolded proteins are commonly degarded by the ubiquitin-proteasome degradation pathway. In PD proteasomal function is impaired in the substantia nigra (McNaught and Jenner 2001). Mutant Ala53Thr protein was shown to have a 50% longer half-life time than wildtype α-synuclein providing a kinetic basis for its intracellular accumulation (Bennett et al. 1999). By administration of iron, the aggregation is most dramatically enhanced in Ala53Thr mutated α-synuclein (Ostrerova-Golts et al. 2000). Enhanced vulnerability of neuronal cells is also found with mutated α-synuclein after sensitization to oxidative stress (Ko et al. 2000). It has also been shown by immunohistochemical studies that nitrated α-synuclein is present in Lewy bodies (Giasson et al. 2000) and that nitrated α-synuclein forms oligomers (Souza et al. 2000) linking oxidative stress and nitrative damage to PD.

2.5. Role Of α-Synuclein In Sporadic PD

We and others (Krüger et al. 1999, Tan et al. 2000) identified an association of an imperfect dinucleotide repeat (NACP-Rep1) in the promoter region of the α-synuclein gene with sporadic PD. This element is highly conserved between human and mouse and is necessary for normal expression of α-synuclein implicating biological importance of the region (Touchman et al. 2001). Interestingly, overexpression of wildtype α-synuclein leads to its aggregation in transgenic animals (Masliah et al. 2000, Feany and Bender 2000, Kahle et al. 2000), promotes the aggregation of other proteins in cell culture (Engelender et al. 1999, Furlong et al. 2000) and causes dopamine neuron death (Zhou et al. 2000). The neuronal cell death appears to be correlated with the Rab5A-specific endocytosis of α-synuclein (Sung et al. 2001). However, it has still to be shown that polymorphisms in the promoter region of α-synuclein influence the level of expression and might therefore be related to neurodegeneration in sporadic PD. Undoubtedly, α-synuclein is a major component of the Lewy bodies (Spillantini et al. 1997) and might if not directly involved be part of a common pathway leading to neuronal cell ceath. Interestingly, MPTP application induces α-synuclein aggregation in baboons as seen in humans with idiopathic PD (Kowall et al. 1999). Also, α-synuclein liberates hydroxyl radicals upon incubation *in vitro* followed by the addition of Fe(ii) (Turnbull et al. 2001). Other metal ions as Copper(II) or iron also promote self-oligomerization of α-synuclein (Paik et al. 1999, Ostrerova-Golts et al. 2000). These results stress the link between α-synuclein and the generation of free radicals and oxidative stress in the pathogenesis of PD.

2.6. Role Of β- And γ-Synuclein In PD

Due to its high homology to α-synuclein β- and γ-synuclein were also considered as candidates for familial and sporadic PD. β-Synuclein was shown to bind to amyloid Aβ as α-synuclein (Jensen et al. 1997). Both synucleins stimulate Aβ-aggregation *in vitro* to the same extent. Also, α-and β-synuclein are expressed predominantly in the same areas of the brain, with both being present in nerve terminals (Jakes et al. 1994). For γ-synuclein a biochemical association with neurofilaments, a major component of Lewy bodies, has been reported (Buchman et al. 1998). Recently, involvement of β- and γ-synuclein in neurodegeneration in addition to α-synuclein has been implicated by immunohistochemical studies (Galvin et al. 1999). Limited mutation analysis of the β- and γ-synuclein genes, however, did fail to find mutations in ADPD (Lavedan et al. 1998, Lincoln et al. 1999a,b). This is in agreement with biochemical studies which have shown that β- and γ-synuclein is less fibrillogenic than α-synuclein (Biere et al. 2000) and that a

12-amino acid stretch (71-VTHVTAVAQKTV-82) of α-synuclein is essential for filament assembly (Giasson et al. 2001).

3. CONCLUSION

Since identification of point mutations in the α-synuclein gene at least one other protein named parkin has been shown to be involved in the pathogenesis of PD (Table 2). Parkin mutations are the major cause of juvenile parkinsonism (Kitada et al. 1998, Lücking et al. 2000). The N-terminal region of parkin is highly homologous to ubiquitin (Kitada et al. 1998). Subsequently some of us (N.H. and Y.M.) have shown that parkin mediates E3 ubiquitin-ligase activity interacting with the ubiquitin-conjugating enzyme UbcH7 (Shimura et al. 2000). As an E3 ligase, parkin conjugates Ub onto its substrates for subsequent degradation by the proteasome. Most interestingly, α-synuclein has been identified as on of the substrates for parkin (Shimura et al. in press). The synaptic vesicle-associated protein CDCrel-1 was identified as another substrate for parkin (Zhang et al. 2000). The identification of an Ile93Met mutation in the ubiquitin carboxy-terminal hydrolase L1 (UCH-L1) gene reducing the catalytic activity of UCH-L1 (Leroy et al. 1998) fits into our current pathogenic model of a disturbed protein degradation pathway in PD (Figure 2). UCH-L1 is thought to cleave polymeric ubiquitin to monomers and to hydrolyse bonds between ubiquitin and other small molecules. One could speculate that a reduction of the catalytic activity of UCH-L1 could subsequently lead to aggregation of proteins. However, the involvement of UCH-L1 in the pathogenesis of PD is still under debate as a M124L substitution did not coseggregate with the disease in a French family (Farrer et al. 2000) and as mutations in UCH-L1 have not been identified in another ADPD family (PARK4, Table 2) mapping to the same chromosomal region in 4p15 (Farrer et al. 1999). It will be interesting to see whether other gene loci responsible for PD, as PARK3 and PARK6 (Table 2), will be involved in the ubiquitin-mediated proteasome degradation pathway (Gasser et al. 1998; Valente et al. 2001).

Table 2: Gene loci in familial PD

Locus name	MIM Nr.	Mode of Inheritance	Chromosomal localization	Gene	Type of Mutation
PARK1	601508	AD	4q21-23	α-Synuclein	Point mutation
PARK2	600116	AR	6q25.2-27	Parkin	Deletion, Duplication, Point mutation
PARK3	602404	AD	2p13	unknown	unknown
PARK4	605543	AD	4p14-16.3	unknown	unknown
PARK5	191342	AD	4p14-15	Ubiquitin C-terminal Hydrolase L1	Point mutation
PARK6	605909	AR	1p35-36	unknown	unknown

MIM: *Mendelian Inheritance in Man* (http://www.ncbi.nlm.nih.gov/omim/); AD: autosomal dominant; AR: autosomal recessive

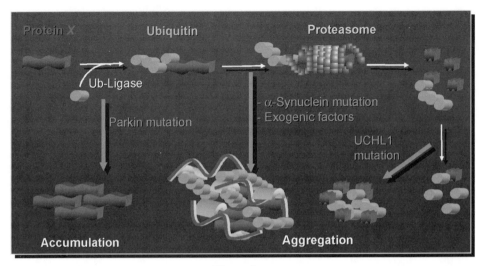

Figure 2.

REFERENCES

Abeliovich, A., Schmitz, Y., Farinas, I., Choi-Lundberg, D., Ho, W. H., Castillo, P. E., Shinsky, N., Verdugo, J. M., Armanini, M., Ryan, A., Hynes, M., Phillips, H., Sulzer, D., and Rosenthal, A., 2000, Mice lacking alpha-synuclein display functional deficits in the nigrostriatal dopamine system, *Neuron* **25** (1):239-252.

Akopian, A. N., and Wood, J. N., 1995, Peripheral Nervous System-specific Genes Identified by Subtractive cDNA Cloning, *J Biol Chem* **270**, No. 36:21264-21270.

Athanassiadou, A., Voutsinas, G., Psiouri, L., Leroy, E., Polymeropoulos, M. H., Ilias, A., Maniatis, G. M., and Papapetropoulos, T., 1999, Genetic Analysis of Families with Parkinson Disease that Carry the Ala53Thr Mutation in the Gene Encoding α-Synuclein, *Am J Hum Genet* **65**, No 2:555-558.

Bayer, T. A., Jäkälä, P., Hartmann, T., Egensperger, R., Buslei, R., Falkai, P., and Beyreuther, K., 1999, Neural expression profile of α-synuclein in developing human cortex, *NeuroReport* **10**:2799-2803.

Bennett, M. C., Bishop, J. F., Leng, Y., Chock, B., Chase, T. N., and Mouradian, M. M., 1999, Degradation of α-Synuclein by Proteasome, *J Biol Chem* **274**, No. 48:33855-33858.

Biere, A. L., Wood, St. J., Wypych, J., Steavenson, S., Jiang, Y., Anafi, D., Jacobsen, F. W., Jarosinski, M. A., Wu, G.-M., Louis, J.-C., Martin, F., Narhi, L. O., and Citron, M., 2000, Parkinson's Disease Associated α-Synuclein is More Fibrillogenic Than β- and γ-Synuclein and Cannot-Seed its Homologs, *J Biol Chem* **275**:34574-34579.

Conway, K. A., Harper J. D., and Lansbury, P. T., 1998, Accelerated in vitro fibril formation by a mutant α-synuclein linked to early-onset Parkinson disease, *Nat Med* **4**, No. 11:1318-1320.

da Costa, C. A., Ancolio, K., and Checler, F., 2000, Wild-type but not Parkinson's disease-related ala-53→Thr mutant alpha-synuclein protects neuronal cells from apoptotic stimuli, *J Biol Chem* **275** (31):24065-24069.

Coyle, J. T., and Puttfarcken, P., 1993, Oxidative Stress, Glutamate, and Neurodegenerative Disorders, *Science* **262**:689-695.

Culvenor, J. G., McLean, C. A., Cutt, S., Campbell, B. C. V., Maher, F., Jäkälä, P., Hartmann, T., Beyreuther, K., Masters, C. L., and Li, Q.-X., 1999, Non-Aβ Component of Alzheimer's Disease Amyloid (NAC) Revisited, *Am J Pathol* **155**, No. 4:1173-1181.

Davidson, W. S., Jonas, A., Clayton, D. F., and George, J. M., 1998, Stabilization of α-Synuclein Secondary Structure upon Binding to Synthetic Membranes, *J Biol Chem* **273**, No. 16:9443-9449.

Dickson, D. W., 1999, Tau and Synuclein and Their Role in Neuropathology, *Brain Pathol* **9**:657-661.

El-Agnaf, O. M. A., Jakes, R., Curran, M. D., and Wallace, A., 1998, Effects of the mutations Ala30 to Pro and Ala53 to Thr on the physical and morphological properties of α-synuclein protein implicated in Parkinson's disease, *FEBS Lett* **440**; 67-70.

El-Agnaf, O. M. A., Jakes, R., Curran, M. D., Middleton, D., Ingenito, R., Bianchi, E., Pessi, A., Neill, D., and Wallace, A., 1998, Aggregates from mutant and wild-type α-synuclein proteins and NAC peptide induce apoptotic cell death in human neuroblastoma cells by formation of β-sheet and amyloid-like filaments, *FEBS Lett* **440**: 71-75.

Engelender, S., Kaminsky, Z., Guo, X., Sharp, A. H., Amaravi, R. K., Kleiderlein, J. J., Margolis, R. L., Troncoso, J. C., Lanahan, A. A., Worley, P. F., Dawson, V. L., Dawson, T. M., and Ross, C. A., 1999, Synphilin-1 associates with α-synuclein and promotes the formation of cytosolic inclusions, *Nat Genet* **22**:110-114.

Farrer, M., Gwinn-Hardy, K., Muenter, M., DeVrieze, F. W., Crook, R., Perez-Tur, J., Lincoln, S., Maraganore, D., Adler, C., Newman, S., MacElwee, K., McCarthy, P., Miller, C., Waters, C., and Hardy, J., 1999, A chromosome 4p haplotype segregating with Parkinson's disease and postural tremor, *Hum Mol Genet* **8**, No. 1:81-85.

Farrer, M., Destée, A., Becquet, E., Wavrant-De Vrièze, F., Mouroux, V., Richard, F., Defebvre, L., Lincoln, S., Hardy, J., Amouyel, P., and Chartier-Harlin, M.-C., 2000, Linkage Exclusion in French Families With Probable Parkinson's Disease, *Mov Disord* **15**; No. 6:1075-1083.

Feany, M. B., and Bender, W. W., 2000, A Drosophila model of Parkinson's disease, *NATURE* **404**, No. 23:394-398.

Forloni, G., Bertani, I., Calella, A. M., Thaler, F., and Invernizzi, R., 2000, α-Synuclein and Parkinson's Disease: Selective Neurodegenerative Effect of α-Synuclein Fragment on Dopaminergic Neurons in Vitro and In Vivo, *Ann Neurol* **47**, No. 5:632-640.

Furlong, R. A., Narain, Y., Rankin, J., Wyttenbach, A., and Rubinsztein, D. C., 2000, α-Synuclein overexpression promotes aggregation of mutant huntingtin, *Biochem. J* **346**:577-581.

Galvin, J. E., Uryu, K., Lee, V. M.-Y., and Trojanowski, J. Q., 1999, Axon pathology in Parkinson's disease and Lewy body dementia hippocampus contains α-, β- and γ-synuclein, *PNAS* **96**, No. 23: 13450-13455.

Gasser, T., Müller-Myhsok, B., Wszolek, Z. K., Oehlmann, R., Calne, D. B., Bonifati, V., Bereznai, B., Fabrizio, E., Vieregge, P., and Horstmann, R. D., 1998, A susceptibility locus for Parkinson's disease maps to chromosome 2p13, *Nat Genet* **18**:262-265.

George, J. M., Jin, H., Woods, W. S., and Glayton, D. F., 1995, Characterization of a Novel Protein Regulated during the Critical Period for Song Learning in the Zebra Finch, *Neuron* **15**:361-372.

Giasson, B. I., Duda, J. E., Murray, I. V., Chen, Q., Souza, J. M., Hurtig, H. I., Ischiropoulos, H., Trojanowski, J. Q., and Lee, V. M., 2000, Oxidative damage linked to neurodegeneration by selective alpha-synuclein nitration in synucleinopthy lesions, *Science* **290** (5493):985-989.

Giasson, B. I., Murray, I. V., Trojanowski, J. Q., and Lee, V. M., 2001, A Hydrophobic Stretch of 12 Amino Acid Residues in the Middle of alpha-Synuclein Is Essential for Filament Assembly, *J Biol Chem* **276** (4):2380-2386.

Golbe, L. I., Iorio, G. D., Bonavita, V., Miller, D. C., and Duvoisin, R. C., 1990, A Large Kindred with Autosomal Dominant Parkinson's Disease, *Ann Neurol* **27**, No. 3: 276-282.

Hashimoto, M., Hsu, L. J., Xia, Y., Takeda, A., Sisk, A., Sundsmo, M., and Masliah, E., 1999, Oxidative stress induces amyloid-like aggregate formation of NACP/α-synuclein in vitro, *NeuroReport* **10**:717-721.

Iwai, A., Masliah, E., Yoshimoto, M., Ge, N., Flanagan, L., Rohan de Silva, H. A., Kittel, A., and Saitho, T., 1995, The Precursor Protein of Non-Aβ Component of Alzheimer's Disease Amyloid Is a Presynaptic Protein of the Central Nervous System, *Neuron* **14**:467-475.

Iwata, A., Miura, S., Kanazawa, I., Sawada, M., and Nukina, N., 2001, Alpha-Synuclein forms a complex with transcription factor Elk-1, *Journal Neurochem* **77**, No. 1: 239-252.

Jakes, R., Spillantini, M. G., and Goedert, M., 1994, Identification of two distinct synucleins from human brain, *FEBS Lett* **345**: 27-32.

Jenco, J. M., Rawlingson, A., Daniels, B., and Morris, A. J., 1998, Regulation of Phospholipase D2: Selective Inhibition of Mammalian Phospholipase D Isoenzymes by α- and β-Synucleins, *Biochemistry* **37**:4901-4909.

Jensen, P. H., Nielsen, M. S., Jakes, R., Dotti, C. G., and Goedert, M., 1998, Binding of α-Synuclein to Brain Vesicles Is Abolished by Familial Parkinson's Disease Mutation, *J Biol Chem* **273**, No. 41: 26292-26294.

Jensen, P. H., Højrup, P., Hager, H., Nielson, M. S., Jacobsen, L., Olesen, O. F., Gliemann, J., and Jakes, R., 1997, Binding of Aβ to α- and β-synucleins: identification of segments in α-synuclein/NAC precursor that bind Aβ and NAC, *Biochem. J* **323**:539-546.

Jensen, P. H., Islam, K., Kenney, J., Nielsen, M. S., Power, J., and Gai, W. P., 2000, Microtubule-associated protein 1B is a component of cortical Lewy bodies and binds alpha-synuclein filaments, *J Biol Chem* **275** (28):21500-21507.

Jo, E., McLaurin, J., Yip, C. M., St George-Hyslop, P., and Fraser, P. E., 2000, Alpha-Synuclein membrane interactions and lipid specificity, *J Biol Chem* **275** (44):34328-34334.

Kahle, P. J., Neumann, M., Ozmen, L., Muller, V., Jacobsen, H., Schindzielorz, A., Okochi, M., Leimer, U., van der Putten, H., Probst, A., Kremmer, E., Kretzschmar, H. A., and Haass, C., 2000, Subcellular localization of wild-type and Parkinson's disease-associated mutant alpha-synuclein in human and transgenic mouse brain, *J Neurosci* **20**(17): 6365-6373.

Kim, T. D., Paik, S. R., Yang, C. H., and Kim, J., 2000, Structural changes in alpha-synuclein affect its chaperone-like activity in vitro, *Protein Sci* **9**, (12):2489-2496.

Kim, J. W., Kim, D.-H., Kim, S.-H., and Cha, J.-K., 2000, Association of the Dopamine Transporter Gene with Parkinson's Disease in Korean Patients, *J Korean Med Sci* **15**:449-451.

Kimura, M., Matsushita, S., Arai, H., Takeda, A., and Higuchi, S., 2001, No evidence of associtation between a dopamine transporter gene polymorphism (1215A/G) and Parkinson's disease, *Ann Neurol* **49** (2):276-277.

Kitada, T., Asakawa, S., Hattori, N., Matsumine, H., Yamamura, Y., Minoshima, S., Yokochi, M., Mizuno, Y., and Shimizu, N., 1998, Mutations in the parkin gene cause autosomal recessive juvenile parkinsonism, *Nature* **392**, 9, 605-608.

Ko, L.-W., Mehta, N. D., Farrer, M., Easson, C., Hussey, Yen, S., Hardy, J., and Y., S.-H. C., 2000, Sensitization of Neuronal Cells to Oxidative Stress with Mutated Human α-Synuclein, *J Neurochem* **75**, No. 6:2546-2554.

Kowall, N. W., Hantraye, P., Brouillet, E., Beal, M. F., McKee, A. C., and Ferrante, R. J., 2000, MPTP induces alpha-synuclein aggregation in the substantia nigra of baboons, *Neuroreport* **11**, No. 1:211-213.

Krüger, R., Vieira-Saecker, A. M., Kuhn, W., Berg, D., Muller, T., Kuhnl, N., Fuchs, G. A., Storch, A., Hungs, M., Woitalla, D. et al, 1999, Increased susceptibility to sporadic Parkinson's disease by a certain combined alpha-synuclein/apolipoprotein E genotype, *Ann Neurol* **45**: 611-617.

Krüger, R., Kuhn, W., Leenders, K. L., Sprengelmeyer, R., Müller, T., Woitalla, D., Portmann, A. T., Maguire, R. P., Veenma, L., Schröder, U., Schöls, L., Epplen, J. T., Riess, O., and Przuntek, H., 2001, Familial parkinsonism with synuclein pathology, *Neurology* **56**:1355-1362.

Lavedan, C., Buchholtz, S., Auburger, G., Albin, R. L., Athanassiadou, A., Blancato, J., Burguera, J. A., Ferrel, R. E., Kostic, V., Leroy, E., Leube, B., Mota-Vieira, L., Papapetropoulos, , T. Pericak-Vance, M. A., Pinkus, J., Scott, W. K., Ulm, G., Vasconcelos, J., Vilchez, J. J., Nussbaum, R. L., and Polymeropoulos, M. H., 1998, Absence of mutation in the beta- and gamma-synuclein genes in familial autosomal dominant Parkinson's disease, *DNA Res* **5** (6):401-402.

Le Couteur, D. G., Leighton, P. W., McCann, S. J., and Pond, S. M., 1997, Association of a polymorphism in the dopamine-transporter gene with Parkinson's disease, *Mov Disord* **12** (5):760-763.

Lee, F. J., Liu, F., Pristupa, Z. B., and Niznik, H. B., 2001, Direct binding and functional coupling of alpha-synuclein to the dopamine transporters accelerate dopamine-induced apoptosis, *FASEB J* **15** (6): 916-926.

Leighton, P. W., Le Couteur, D. G., Pang, C. C. P., McCann, S. J., Chan, D., Law, L. K., Kay, R., Pond, S. M., and Woo, J., 1997, The dopamine transporter gene and Parkinson's disease in a Chinese population, *Neurology* **49**:1577-1579.

Leroy, E., Boyer, R., Auburger, G., Leube, B., Ulm, G., Mezey, E., Harta, G., Brownstein, M. J., Jonnalagada, S., Chernova, T., Dehejia, A., Lavedan, C., Gasser, T., Steinbach P. J., Wilkinson, K. D., and Polymeropoulos, M. H., 1998, The ubiquitin pathway in Parkinson's disease, *Nature* **395**, 1: 451-452.

Lincoln, S., Gwinn-Hardy, K., Goudreau, J., Chartier-Harlin, M. C., Baker, M., Mouroux, V., Richard, F., Destée, A., Becquet, E., Amouyel, P., Lynch, T., Hardy, J., and Matt Farrer, 1999, No pathogenic mutations in the persyn gene in Parkinson's disease, *Neurosci Lett* **259**:65-66.

Lincoln, S., Crook, R., Chartier-Harlin, M. C., Gwinn-Hardy, K., Baker, M., Mouroux, V., Richard, F., Becquet, E., Amouyel, P., Destée, A., Hardy J., and Farrer, M., 1999, No pathogenic mutations in the beta-synuclein gene in Parkinson's disease, *Neurosci Lett* **269** (2):107-109.

Lücking, C. B., Dürr, A., Bonifati, V., Vaughan, J., De Michele, G., Gasser, T., Harhangi, B. S., Meco, G., Denéfle, P., Wood, N. W., Agid, Y., and Brice, A., 2000, Association Between Early-Onset Parkinson's Disease and Mutations in the Parkin Gene, *N Engl J Med* **342**: 1560-1567.

Markopoulou, K., Wszolek, Z. K., Pfeiffer, R. F., and Chase, B. A., 1999, Reduced Expression of the G209A α-Synuclein Allele in Familial Parkinsonism, *Ann Neurol* **46**:374-381 .

Masliah, E., Rockenstein, E., Veinbergs, I., Mallory, M., Hashimoto, M., Takeda, A., Sagara, Y., Sisk, A., and Mucke, L., 2000, Dopaminergic Loss and Inclusion Body Formation in α-Synuclein Mice: Implications for Neurodegenerative Disorders, *Science* **287**:1265-1269.

McNaught, K. S., and Jenner, P., 2001, Proteasomal function is impaired in substantia nigra in Parkinson's disease, *Neurosci Lett* **297**(3):191-194.

Morino, H., Kawarai, T., Izumi, Y., Kazuta, T., Oda, M., Komure, O., Udaka, F., Kameyama, M., Nakamura, S., and Kawakami, H., 2000, A single nucleotide polymorphism of dopamine transporter gene is associated with Parkinson's disease, *Ann Neurol* **47** (4):528-531.

Nakajo, S., Tsukada, K., Omata, K., Nakamura, Y., and Nakaya, K., 1993, A new brain-specific 14-kDa protein is a phosphoprotein, *Eur. J. Biochem.* **217**:1057-1063.

Ostrerova, N., Petrucelli, L., Farrer, M., Mehta, N., Choi, P., Hardy, J., and Wolozin, B., 1999, α-Synuclein Shares Physical and Functional Homology with 14-3-3 Proteins, *J Neurosci* **19** (14):5782-5791.

Ostrerova-Golts, N., Petrucelli, L., Hardy, J., Lee, J. M., Farer, M., and Wolozin, B., 2000, The A53T α-Synuclein Mutation Increases Iron-Dependent Aggregation and Toxicity, *J Neurosci* **20** (16):6048-6054.

Paik, S. R., Shin, H.-J., Lee, J.-H., Chang, C.-S., and Kom, J., 1999, Copper(II)-induced self-oligomerization of α-synuclein, *Biochem. J.* **340**: 821-828.

Papadimitriou, A., Veletza, V., Hadjigeorgiou, G. M., Patrikiou, A., Hirano, M., and Anastasopoulos, I., 1999, Mutated α-synuclein gene in two Greek kindreds with familial PD: Incomplete penetrance? *Neurology* **52**:651-654.

Polymeropoulos, M. H., Lavedan, C., Leroy, E., Die, S. E., Dehejia, A., Dutra, A., Pike, B., Root, H., Rubenstein, J., Boyer, R., Stenroos, E. S., Chandrasekharappa, S., Athanassiadou, A., Papapetropoulos, T., Johnson, W. G., Lazzarini, A. M., Duvoisin, R. C., Di Iorio, G., Golbe, L. I., and Nussbaum, R. L., 1997, Mutation in the α-Synuclein Gene Identified in Families with Parkinson's Disease, *Science* **276**: 2045-2047.

Pronin, A. N., Morris, A. J., Surguchov, A., and Benovic, J. L., 2000, Synucleins Are A Novel Class Of Substrates for G Protein-Coupled Receptor Kinases, *J Biol Chem* **275**:26515-26522.

van der Putten, H., Wiederhold, K.-H., Probst, A., Barbieri, S., Mistl, C., Danner, S., Kauffmann, S., Hofele, K., Spooren, W. P. J. M., Ruegg, M. A., Lin, S., Caroni, P., Sommer, B., Tolnay M., and Bilbe, G., 2000, Neuropathology in Mice Expressing Human α-Synuclein, *J Neurosci* **20** (16):6021-6029.

Riess, O., Jakes, R., and Krüger, R., 1998, Genetic dissection of familial Parkinson's disease, *Mol Med* **4**, No 10: 438-444.

Shimura, H., Hattori, N., Kubo, S.-I., Mizuno, Y., Asakawa, S., Minoshima, S., Shimizu, N., Iwai, K., Chiba, T., Tanaka, K., and Suzuki, T., 2000, Familial Parkinson disease gene product, parkin, is a ubiquitin-protein ligase, *Nat Genet* **25**: 302-305.

Shimura, H., Schlossmacher, M. G., Hattori, N., Frosch, M. P., Trockenbacher, A., Schneider, R., Mizuno, Y., Kosik, K. S., and Selkoe, D. J., 2001, Ubiquitination of a Novel Form of α-Synuclein of Parkin from Human Brain: Implications for Parkinson Disease, *Science*, in press.

Souza, J. M., Giasson, B. I., Lee, V. M., and Ischiropoulos, H., 2000, Chaperone-like activity of synucleins, *FEBS Lett* **474** (1):116-119.

Souza, J. M., Giasson, B. I. Chen, Q., Lee, V. M., Ischiropoulos, H., 2000, Dityrosine Cross-linking Promotes Formation of Stable alpha-Synuclein Polymers. Implication Of Nitrative And Oxidative Stress In The Pathogenesis Of Neurodegenerative Synucleinopathies, *J Biol Chem* **275** (24):18344-18349.

Spillantini, M. G., Schmidt, M. L., Lee, V. M. Y., Trojanowski, J. Q., Jakes, R., and Goedert, M., 1997, α-Synuclein in Lewy bodies, *Nature* **388**:839-840.

Sung, J. Y., Kim, J., Paik, S. R., Park, J. H., Ahn, Y. S., and Chung, K. C., 2001, Induction of Neuronal Cell Death by Rab5A-dependent Endocytosis of α-Synuclein, *J Biol Chem*, in press.

Surguchov, A., Surgucheva, I., Solessio, E., and Baehr, W., 1999, Synoretin – A New Protein Belonging to the Synuclein Family, *Mol Cell Neurosci* **13**:95-103.

Tabrizi, S. J., Orth, M., Wilkinson, J. M., Taanman, J.-W., Warner, T. T., Cooper, J. M., and Schapira, A. H. V., 2000, Expression of mutant α-synuclein causes increased susceptibility to dopamine toxicity, *Hum Mol Genet* **9** (18):2683-2689.

Tan, E. K., Matsuura, T., Nagamitsu, S., Khajavi, M., Jankovic, J., and Ashizawa, T., 2000, Polymorphism of NACP-Rep1 in Parkinson's disease: an etiologic link with essential tremor?, *Neurology* **54**: 1195-1198.

Touchman, J. W., Dehejia, A., Chiba-Falek, O., Cabin, D. E., Schwartz, J. R., Orrison, B. M., Polymeropoulos, M. H., and Nussbaum, R. L., 2001, Human and Mouse α-Synuclein Genes: Comparative Genomic Sequence Analysis and Identification of a Novel Gene Regulatory Element, *Genome Res* **11**:78-86.

Tu, P.-H., Galvin, J. E., Baba, M., Giasson, B., Tomita, T., Leight, S., Nakajo, S., Iwatsubo, T., Trojanowski, J. Q., and Lee, V. M. Y., 1998, Glial Cytoplasmic Inclusions in White Matter Oligodendrocytes of Multiple System Atrophy Brains Contain Insoluble α-Synuclein, *Ann Neurol* **44**, No 3:415-423.

Turnbull, S., Tabner, B. J., El-Agnaf, O. M., Moore, S., Davies, Y., and Allsop, D., 2001, Alpha-synuclein implicated in Parkinson's disease catalyses the formation of hydrogen peroxide in vitro, *Free Radic Biol Med* **30** (10):1163-1170.

Uéda, K., Fukushima, H., Masliah, E., Xia, Y., Iwai, A., Yoshimoto, M., Otero, D. A. C., Kondo, J., Ihara, Y., and Saitoh, T., 1993, Molecular cloning of cDNA encoding an unrecognized component of amyloid in Alzheimer disease, *Neurobiology* **90**:11282-11286.

Valente, E. M., Bentivoglio, A. R., Dixon, P. H., Ferraris, A., Ialongo, T., Frontali, M., Albanese, A., and Wood, N. W., 2001, Localization of a Novel Locus for Autosomal Recessive Early-Onset Parkinsonism, PARK6, on Human Chromosome 1p35-p36, *Am. J. Hum. Genet.* **68**:895-900.

Veletza, S., Bostatzopoulou, S., Hantzigeorgiou, G. et al., 1999, Alpha-synuclein mutation associated with familial Parkinson's disease in two new Greek kindred, *J Neurol* **246** (Suppl. 1)I/43.

Vila, M., Vukosavic, S., Jackson-Lewis, V., Neystat, M., Jakowec, M., and Przedborski, S., 2000, Alpha-synuclein up-regulation in substantia nigra dopaminergic neurons following administration of the parkinsonian toxin MPTP, *J Neurochem* **74** (2):721-729.

Wakabayashi, K., Engelender, S., Yoshimoto, M., Tsuji, S., Ross, C. A., and Takahashi, H., 2000, Synphilin-1 Is Present in Lewy Bodies in Parkinson's Disease, *Ann Neurol* **47**:521-523.

Weinreb, P. H., Zhen, W., Poon, A. W., Conway, K. A., and Lansbury jr., P. T., 1996, NACP, A Protein Implicated in Alzheimer's Disease and Learning, Is Natively Unfolded, *Biochemistry* **35**, No 43: 13709-13715.

Zhang, Y., Gao, J., Chung, K. K. K., Huang, H., Dawson, V. L., and Dawson, T. M., 2000, Parkin functions as an E2-dependent ubiquitin-protein ligase and promotes the degradation of the synaptic vesicle-associated protein, CDCrel-1, *PNAS* **97**, No. 24:13354-13359.
Zhou, W., Hurlbert, M. S., Schaack, J., Prasad, K. N., and Freed, C. R., 2000, Overexpression of human alpha-synuclein causes dopamine neuron death in rat primary culture and immortalized mesencephalon-derived cells, *Brain Res* **866** (1-2):33-43.

ACCUMULATION OF INSOLUBLE α-SYNUCLEIN IN HUMAN LEWY BODY DISEASES IS RECAPITULATED IN TRANSGENIC MICE

Philipp J. Kahle, Manuela Neumann, Laurence Ozmen, Takeshi Iwatsubo, Hans A. Kretzschmar, and Christian Haass[*]

1. INTRODUCTION

The synaptic phosphoprotein α-synuclein (αSYN) is genetically and pathologcially linked to Parkinson's disease and related disorders. Two missense mutations in the αSYN gene cause autosomal-dominant hereditary PD.[1, 2] αSYN can be detected immunohistochemically in Lewy bodies (LBs) and Lewy neurites that characterize Parkinson's disease (PD), LB dementia (DLB), LB variant Alzheimer's disease, and neurodegeneration with brain iron accumulation type 1 (NBIA1).[3] Moreover, αSYN is the major fibrillar component of glial cytoplasmic inclusions in multiple system atrophy.[3] The formation of LB-like fibrils was found to be an intrinsic property of αSYN. Purified recombinant αSYN, but not βSYN aggregated *in vitro* to amyloid fibrils resembling those extracted from LBs.[4] PD mutations accelerated αSYN aggregation.[5] The causal relationship between αSYN fibrillization dopaminergic (DA) neurotoxicity and PD are therefore subject to intense research.

Transgenic animals expressing human wild-type (wt) and PD-associated mutant αSYN were recently presented. Wt and mutant αSYN assembled into LB-like fibrils in transgenic *Drosophila*, and a locomotor deficit became apparent with increasing age.[6] Somal and neuritic accumulations of wt and mutant αSYN were observed in pan-neuronally expressing transgenic mouse brain.[7-9] Ubiquitination was occasionally detected, but the αSYN accumulations did not meet ultrastructural criteria of LBs.[7, 8] Masliah *et al.*[7] reported a modest reduction of DA markers and locomotor performance. Van der Putten *et al.*[8] found that age-dependent degeneration of neuromuscular junctions

[*] Philipp J. Kahle, Manuela Neumann, Hans A. Kretzschmar, Christian Haass, Ludwig Maximilians University, Munich, Germany D-80336. Takeshi Iwatsubo, University of Tokyo, Tokyo, Japan 113-0033. Laurence Ozmen, F. Hoffmann – La Roche Ltd, Basel, Switzerland CH-4002.

caused a severe locomotor deficit and premature death in their mice. More recently, we have generated transgenic mice expressing human [A30P]αSYN under the control of tyrosine hydroxylase promoter elements. Again, somal and neuritic accumulation of transgenic αSYN in nigral DA neurons was detected. However, no effect on DA neuronal viability and no sensitization to the DA neurotoxin 1-methyl-4-phenyl-1,2,3,6-tetrahydropyridine was found in these mice.[10]

Sequential detergent extraction methods have been successfully employed to detect specifically αSYN in brains of patients with α-synucleinopathies. We have adapted the method of Culvenor *et al.*[11] to investigate SYN solubility in brains of human LB disease patients and transgenic mice. The animals expressed human wt and PD-associated mutant [A30P]αSYN under control of the brain neuron-specific promoter, Thy1. Most of the αSYN was highly soluble in aqueous buffer and the remainder easily extractable with sodium dodecylsulfate. However, detergent-insoluble αSYN monomers and aggregates were detected in urea extracts from LB disease patient brains, but not in controls. Likewise, some of the human αSYN was detergent-insoluble in transgenic mouse brains, in sharp contrast to the endogenous mouse αSYN and βSYN. Thus, the specific accumulation of detergent-insoluble αSYN in transgenic mice recapitulates a pivotal feature of human LB diseases.

2. RESULTS

Buffer- and detergent-soluble, monomeric αSYN was detected in brains from human controls as well as from LB disease patients. In contrast to control brain urea extracts that were virtually devoid of αSYN, strong immunoreactivity was found in urea extracts from PD, DLB, and NBIA1 patients. The detergent-insoluble αSYN was characterized using 4 different antibodies raised against distinct epitopes. Monomeric αSYN migrated as 16-19kDa band. Anti-NAC, but not carboxy-terminal antibodies detected a previously unrecognized αSYN species with slightly retarded electrophoretic motility. This band was unlikely to be crossreactive βSYN, because specific anti-βSYN did not reveal any signal in urea extracts. In addition, all 4 anti-αSYN antibodies recognized approximately 40kDa double bands (consistent with αSYN dimers), multiple bands in the 60-80kDa range (putative αSYN oligomers), and higher molecular weight aggregates. Interestingly, an approximately 25kDa band was consistently observed in urea extracts from LB disease patients. Limited amino-terminal degradation to a 14kDa band was occasionally noted, but the bulk of insoluble αSYN was full-length protein.

As the presence of SDS-insoluble αSYN appeared to be a diagnostic criterion for LB diseases in human brain, we applied the above method to transgenic mouse brains expressing human [wt]- and [A30P]αSYN in brain neurons. A portion of human αSYN was specifically detected in urea extracts of detergent-insoluble fractions from transgenic mouse brains. The detergent-insoluble αSYN species were monomers and higher molecular weight oligomers. Both transgenic [A30P]αSYN and [wt]αSYN were found in detergent-insoluble fractions. Insoluble transgenic αSYN became detectable between 4-6wk and persisted for at least 1yr. The onset of expression of insoluble transgenic αSYN was concomitant with the appearance of cytosolic accumulations. In sharp contrast, endogenous mouse αSYN as well as βSYN were entirely soluble in buffer and detergent.

3. DISCUSSION

PD, DLB, and NBIA1 are characterized immunohistochemically by αSYN-immunoreactive intraneuronal inclusions (LBs) and dystrophic neurites. Biochemically, detergent-insoluble αSYN was found to be diagnostic for these diseases. We have performed differential detergent extractions to evaluate the potential development of α-synucleinopathy in transgenic mice expressing human αSYN in brain neurons. Like in human LB diseases, detergent-insoluble human αSYN was detected in transgenic mouse brain. In striking contrast, endogenous mouse αSYN and βSYN were not found in the urea extracts. These results demonstrate that a transgenic mouse model recapitulates some specific features of α-synucleinopathies.

Both PD-associated [A30P]αSYN and human [wt]αSYN were detected in detergent-insoluble fractions. This is of note because the overwhelming majority of PD patients have no mutation in the αSYN gene.[12] The faster *in vitro* aggregation rate of concentrated solutions of mutant αSYN[5] was apparently not reflected by greater pathology of human mutant αSYN compared to [wt]αSYN in transgenic mice.[8, 9] A very crude estimate of transgenic αSYN concentrations in neuronal cytosol (0.5µM) indicated that the expression levels *in vivo* are approximately 50 times below the critical concentration of recombinant αSYN aggregation *in vitro*.[13] Perhaps the differences in aggregation kinetics between wt and mutant αSYN are not evident at concentrations reached in transgenic mouse neurons. Since αSYN expression is not elevated enough in PD patients to allow spontaneous aggregation, additional risk factors are likely to exist that favor the aggregation at subcritical αSYN concentrations. Potentially aggregation-promoting post-translational modifications of αSYN include phosphorylation,[14] nitration,[15] and glycation.[16] Moreover, ubiquitination and perturbation of proteosomal degradation should be considered. It is possible that post-translational modifications characteristic for human LB diseases have to occur in transgenic mice to allow true LB formation in an animal model.

References

1. M. H. Polymeropoulos, C. Lavedan, E. Leroy, S. E. Ide, A. Dehejia, A. Dutra, B. Pike, H. Root, J. Rubenstein, R. Boyer, E. S. Stenroos, S. Chandrasekharappa, A. Athanassiadou, T. Papapetropoulos, W. G. Johnson, A. M. Lazzarini, R. C. Duvoisin, G. Di Iorio, L. I. Golbe, R. L. Nussbaum, Mutation in the α-synuclein gene identified in families with Parkinson's disease, *Science* **276**(5321), 2045-2047 (1997).
2. R. Krüger, W. Kuhn, T. Müller, D. Woitalla, M. Graeber, S. Kösel, H. Przuntek, J. T. Epplen, L. Schöls, O. Riess, Ala30Pro mutation in the gene encoding α-synuclein in Parkinson's disease, *Nat. Genet.* **18**(2), 106-108 (1998).
3. J. E. Duda, V. M.-Y. Lee, J. Q. Trojanowski, Neuropathology of synuclein aggregates: New insights into mechanisms of neurodegenerative diseases, *J. Neurosci. Res.* **61**(2), 121-127 (2000).
4. A. L. Biere, S. J. Wood, J. Wypych, S. Steavenson, Y. Jiang, D. Anafi, F. W. Jacobsen, M. A. Jarosinski, G.-M. Wu, J.-C. Louis, F. Martin, L. O. Narhi, M. Citron, Parkinson's disease-associated α-synuclein is more fibrillogenic than β- and γ-synuclein and cannot cross-seed its homologs, *J. Biol. Chem.* **275**(44), 34574-34579 (2000).

5. L. Narhi, S. J. Wood, S. Steavenson, Y. Jiang, G. M. Wu, D. Anafi, S. A. Kaufman, F. Martin, K. Sitney, P. Denis, J.-C. Louis, J. Wypych, A. L. Biere, M. Citron, Both familial Parkinson's disease mutations accelerate α-synuclein aggregation, *J. Biol. Chem.* **274**(14), 9843-9846 (1999).
6. M. B. Feany, W. W. Bender, A *Drosophila* model of Parkinson's disease, *Nature* **404**(6776), 394-398 (2000).
7. E. Masliah, E. Rockenstein, I. Veinbergs, M. Mallory, M. Hashimoto, A. Takeda, Y. Sagara, A. Sisk, L. Mucke, Dopaminergic loss and inclusion body formation in α-synuclein mice: Implications for neurodegenerative disorders, *Science* **287**(5456), 1265-1269 (2000).
8. H. van der Putten, K.-H. Wiederhold, A. Probst, S. Barbieri, C. Mistl, S. Danner, S. Kauffmann, K. Hofele, W. P. J. M. Spooren, M. A. Ruegg, S. Lin, P. Caroni, B. Sommer, M. Tolnay, G. Bilbe, Neuropathology in mice expressing human α-synuclein, *J. Neurosci.* **20**(16), 6021-6029 (2000).
9. P. J. Kahle, M. Neumann, L. Ozmen, V. Müller, H. Jacobsen, A. Schindzielorz, M. Okochi, U. Leimer, H. van der Putten, A. Probst, E. Kremmer, H. A. Kretzschmar, C. Haass, Subcellular localization of wild-type and Parkinson's disease-associated mutant α-synuclein in human and transgenic mouse brain, *J. Neurosci.* **20**(17), 6365-6373 (2000).
10. S. Rathke-Hartlieb, P. J. Kahle, M. Neumann, L. Ozmen, S. Haid, M. Okochi, C. Haass, J. B. Schulz, Sensitivity to MPTP is not increased in Parkinson's disease-associated mutant α-synuclein transgenic mice, *J. Neurochem.* **77**(4), 1181-1184 (2001).
11. J. G. Culvenor, C. A. McLean, S. Cutt, B. C. V. Campbell, F. Maher, P. Jäkälä, T. Hartmann, K. Beyreuther, C. L. Masters, Q.-X. Li, Non-Aβ component of Alzheimer's disease amyloid (NAC) revisited: NAC and α-synuclein are not associated with Aβ amyloid, *Am. J. Pathol.* **155**(4), 1173-1181 (1999).
12. J. R. Vaughan, M. J. Farrer, Z. K. Wszolek, T. Gasser, A. Durr, Y. Agid, V. Bonifati, G. DeMichele, G. Volpe, S. Lincoln, M. Breteler, G. Meco, A. Brice, C. D. Marsden, J. Hardy, N. W. Wood, Sequencing of the α-synuclein gene in a large series of cases of familial Parkinson's disease fails to reveal any further mutations, *Hum. Mol. Genet.* **7**(4), 751-753 (1998).
13. S. J. Wood, J. Wypych, S. Steavenson, J.-C. Louis, M. Citron, A. L. Biere, α-Synuclein fibrillogenesis is nucleation-dependent. Implications for the pathogenesis of Parkinson's disease, *J. Biol. Chem.* **274**(28), 19509-19512 (1999).
14. M. Okochi, J. Walter, A. Koyama, S. Nakajo, M. Baba, T. Iwatsubo, L. Meijer, P. J. Kahle, C. Haass, Constitutive phosphorylation of the Parkinson's disease associated α-synuclein, *J. Biol. Chem.* **275**(1), 390-397 (2000).
15. B. I. Giasson, J. E. Duda, I. V. J. Murray, Q. Chen, J. M. Souza, H. I. Hurtig, H. Ischiropoulos, J. Q. Trojanowski, V. M.-Y. Lee, Oxidative damage linked to neurodegeneration by selective α-synuclein nitration in synucleinopathy lesions, *Science* **290**(5493), 985-989 (2000).
16. G. Münch, H. J. Lüth, A. Wong, T. Arendt, E. Hirsch, R. Ravid, P. Riederer, Crosslinking of α-synuclein by advanced glycation endproducts - an early pathophysiological step in Lewy body formation?, *J. Chem. Neuroanat.* **20**(3-4), 253-257 (2000).

ALTERNATIVE TREATMENTS FOR LEWY BODY DISEASE IN TRANSGENIC MICE

Dementia with Lewy bodies

Manfred Windisch*, Edward Rockenstein , Makoto Hashimoto , Margaret Mallory, Eliezer Masliah

1. INTRODUCTION

The 140 amino acid synaptic protein α-synuclein [1] has first been described in the human brain as the precursor of the non amyloid component [2-5] of Alzheimer's disease (AD). belongs to a family of proteins including ß-synuclein[6], γ-synuclein[7] and synoretinin[8]. The α-synuclein molecule is capable of self-aggregating to form both oligomers and potentially neurotoxic fibrillar polymers with amyloid-like characteristics[9]. In Lewy body disease (LBD), a common cause of dementia and parkinsonism in the elderly, neuronal accumulation of α-synuclein has been proposed to be involved in the pathogenesis of this disorder.[10,11]Supporting this possibility, recent studies have shown that: i) this synaptic-associated molecule is the most abundant component of Lewy bodies (LBs)[12,13] [14] ii) mutations in the α-synuclein gene (A30P and A53T) are associated with rare familial forms of parkinsonism,[15] and iii) expression in transgenic (tg) mice[16] and *Drososphila* [17]mimics several aspects of this disorder.

Studies have shown that different stress conditions, like free radicals or increased concentration [18,19] are promoting the formation of amyloid fibrils consisting of α-synuclein, while ß-synuclein does not aggregate[20]. This is most likely due to the fact that ß-synuclein is missing the highly hydrophobic NAC region of α-synuclein. ß-synuclein is inhibiting the formation of α-synuclein aggregates in vitro in a dose dependent manner, what in accordance

*Manfred Windisch, JSW-Research Ltd, 8020 Graz, Austria, Edward Rockenstein , Makoto Hashimoto , Margaret Mallory, Eliezer Masliah , University of California at San Diego, La Jolla, CA 92093-0624, USA

with previous findings[21]. In order to determine if the anti-aggregation effects of ß- synuclein in vitro are also relevant in vivo, hß-synuclein transgenic (tg) mice were generated and crossed with hα-synuclein tg mice[16]. It was previously already shown that the hα-synuclein tg mice show dopaminergic loss, formation of α-synuclein aggregates and develop motor deficits, qualifying them as an appropriate model to test anti-parkinsonian and anti-aggregation effects of ß-synuclein.

2. METHODS

The highest expressor line of wildtype hα-synuclein tg mice, producing the protein under control of platelet derived growth factor ß promoter were chosen for the experiment. The hß-synuclein cDNA fragment (405 nt, Genbank # S69965) was generated by RT/PCR from human brain mRNA and inserted into the Thy-1 expression cassette (provided by Dr. H. van der Putten, Novartis, Basel) between exon 2 and 4, purified and microinjected into one cell embryos (C57BLXDBA/2 F1) according to standard procedures, to produce the hß-synuclein tg mice. For the experiments single tg hα-synuclein, single tg hß-synuclein, bigenic and non tg mice were used. Motor performance was investigated at the age of 12 months in the rotarod instrument (5 trials on day 1; 7 trial at day two; 40 rpm). Length of time mice were able to balance at the rod was recorded as a measure of motor function. Thereafter the right hemibrain was snap frozen for RNA and protein analysis, the left hemibrain was immersion fixed in 4% paraformaldehyde for later neuropathological investigation. Levels of α- and ß-synuclein were analysed by Western blot in the brain homogenates. The immunohistochemical analysis was performed in blinded way using Quantimet 570C to determine numbers of hα-synuclein positive inclusions. Additional mmunohistochemical determination of hß-synuclein, synaptophysin and tyrosine-hydroxylase positive neurons was performed.

3. RESULTS

The evaluation of motor function in the rotarod showed that the bigenic animals did not display any significant disturbance, while the hα-synuclein tg mice are significantly impaired compared to their non tg littermates (Fig.1).

Interestingly the hß-synuclein tg mice display no functional alterations in spite that they are over-expressing ß-synuclein in similar extend as the hα-synuclein tg mice do α-synuclein. It is obvious that ß-synuclein expression in the bigenic mice is counteracting the functional disturbances caused by α-synuclein overexpression. This effect is associated with decreased α-synuclein immunoreactive neuronal inclusions in the bigenic mice. They have significantly lower numbers of α-synuclein inclusion bodies in the neocortex (Fig.2).

In some occasions ß-synuclein was co-localized in these depositions. Also the reduced density of neocortical anti-synaptophysin immunoreactive terminals in the hα-synuclein tg

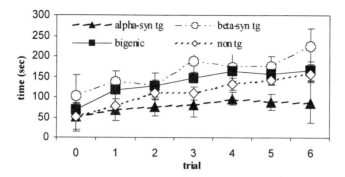

Figure 1. Characterization of motor functions in h α-synuclein tg mice crossed with hß-synuclein mice on the rotarod.

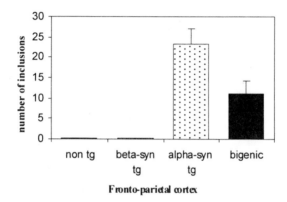

Figure 2. Reduction of neocortical hα-synuclein immunoreactive neuronal inclusion bodies in bigenic mice expressing hß-synuclein

animals was not found in the bigenic mice, displaying comparable results to the non-tg controls. In the bigenic mice also no decrease in tyrosin hydroxylase, a marker of the dopaminergic system, in the caudoputamen could be detected.

4. CONCLUSIONS

The results in the bigenic mice suggest, that the anti-aggregation effect of ß-synuclein is able to ameliorate the neurodegenerative events caused by α-synuclein accumulation. There is a clear correlation between decrease in α-synuclein inclusion bodies and an improvement in motor function. This suggest a potential of ß- synuclein or peptide derivatives of this protein as treatment for neurodegnerative disorders, which are characterized by Lewy bodies, like Lewy body dementia or Parkinson's disease. Because of the increasing body of evidence about the role of α-synuclein in the synaptic degeneration in AD, further directions of therapeutic investigations seem logical. The effects of ß-synuclein underline also the importance of protein-protein interaction for maintenance of normal function. These finding may help better understand the onset of other "protein-pathies" underlying onset and progression of neurodegenrative disorders.

REFERENCES

1. A. Iwai, Properties of NACP/a-synuclein and its role in Alzheimer's disease. *Biochimica et Biophysica Acta* 1502, 95-109 (2000).
2. A. Iwai, M. Yoshimoto, E. Masliah and T.Saitoh, Non-A beta component of Alzheimer's disease amyloid (NAC) is amyloidogenic. *Biochemistry* 34(32), 10139-10145 (1995).
3. A. Iwai, E. Masliah, M. Yoshimoto, L. Flanagan and H.A. de Silva, The precursor protein of non-A beta component of Alzheimer's disease amyloid is a presynaptic protein of the central nervous system. *Neuron* 14(2),467-475 (1995).
4. E. Masliah, A. Iwai, M. Mallory, K. Ueda and T. Saitoh, Altered presynaptic protein NACP is associated with plaque formation and neurodegeneration in Alzheimer's disease. *American Journal of Pathology* 148[1], 201-210 (1996).
5. K. Ueda, H. Fukushima, E. Masliah, Y. Xia, A. Iwai and M. Yoshimoto, Molecular cloning of cDNA encoding an unrecognized component of amyloid in Alzheimer disease. *Proc.Natl.Acad.Sci.USA* 90, 11282-11286 (1993).
6. R. Jakes, M. G. Spillantini and M. Goedert, Identification of two distinct synucleins from human brain. *FEBS Letters* 345, 27-32 (1994).
7. H. Ji, Y.E. Liu, T. Jia, M. Wang, J. Liu and G. Xiao, Identification of a breast cancer-specific gene, BCSG1, by direct differential cDNA sequencing. Cancer Research 57, 759-764 (1997).
8. A. Surguchov,I. Surgucheva, E. Solessio and W. Baehr , Synoretin-A new protein belonging to the synuclein family. *Mol Cell Neurosci.* 13(2), 95-103 (1999).
9. P. T. Lansbury Jr., Evolution of amyloid: what normal protein folding may tell us about fibrillogenesis and disease. *Proc Natl Acad Sci U S A* 96(7), 3342-3344 (1999).
10. J. Q. Trojanowski, M. Goedert, T. Iwatsubo and V.M. Lee, Fatal attractions: abnormal protein aggregation and neuron death in Parkinson's disease and Lewy body dementia. *Cell Death Differ.* 5(10), 832-837 (1998).
11. M. Hashimoto and E. Masliah, Alpha-synuclein in Lewy body disease and Alzheimer's disease. *Brain Pathol* 9(4), 707-720 (1999).
12. M.G. Spillantini, M. L. Schmidt, V. M. Lee, J.Q. Trojanowski, R. Jakes and M. Goedert, Alpha-synuclein in Lewy bodies. *Nature*; 388(6645), 839-840 (1997).
13. A. Takeda, M. Mallory, M. Sundsmo, W. Honer, L. Hansen and E. Masliah, Abnormal accumulation of NACP/alpha-synuclein in neurodegenerative disorders. *Am J Pathol* 152(2), 367-372 (1998).
14. K. Wakabayashi, K. Matsumoto, K. Takayama, M. Yoshimoto and H. Takahashi, NACP, a presynaptic protein, immunoreactivity in Lewy bodies in Parkinson's disease. *Neurosci Lett.* 239(1), 45-48 (1997).
15. M. H. Polymeropoulos, C. Lavedan, E. Leroy, S. E. Ide, A. Dehejia and A. Dutra, Mutation in the alpha-synuclein gene identified in families with Parkinson's disease. *Science*; 276(5321), 2045-2047 (1997).
16. E. Masliah, E. Rockenstein, I. Veinbergs, M. Mallory, M. Hashimoto and A. Takeda, Dopaminergic loss and inclusion body formation in alpha-synuclein mice: implications for neurodegenerative disorders. *Science* 287, 1265-1269 (2000).
17. M. B. Feany and W. W. Bender, A drosophila model of Parkinson's disease. *Nature* 404, 394-398. (2000).
18. M. Hashimoto, L. J. Hsu, A. Sisk, Y. Xia, A. Takeda and M. Sundsmo, Human recombinant NACP/alpha-synuclein is aggregated and fibrillated in vitro: relevance for Lewy body disease. *Brain Res.* 799(2), 301-306 (1998).
19. M. Hashimoto, L.J. Hsu, Y Xia, A. Takeda, A. Sisk and M. Sundsmo M, Oxidative stress induces amyloid-like aggregate formation of NACP/alpha- synuclein in vitro. *Neuroreport* 10(4), 717-721 (1999).
20. A. L. Biere, S. J. Wood, J. Wypych, S. Steavenson, Y.J. Jiang and D. Anafi, Parkinson's disease-associated a-sylnuclein is more fibrillogenic than b- and gamma-synuclein and cannot cross-seed its homologs. *The Journal of Biological Chemistry* 275[44], 34574-34579 (2000).
21. P.H. Jensen, E.S. Sorensen, T.E. Petersen, J. Gliemann and L.K. Rasmussen, Residues in the synuclein consensus motif of the alpha-synuclein fragment, NAC, participate in transglutaminase-catalysed cross-linking to Alzheimer-disease amyloid beta A4 peptide. *Biochem J.* 310(Pt 1),91-94 (1995).

FAMILIAL FRONTOTEMPORAL DEMENTIA AND PARKINSONISM (FTDP-17)

A consensus on clinical diagnostic criteria

Zbigniew K. Wszolek and Yoshio Tsuboi[*]

1. HISTORICAL OVERVIEW

The term "frontotemporal dementia and parkinsonism linked to chromosome 17" (FTDP-17) was introduced during the International Consensus Conference in Ann Arbor, Michigan in 1996. At that time, 13 kindreds were linked definitively or probably to the *wld* locus on chromosome 17.[1] Clinical diagnostic criteria for FTDP-17 were established on the basis of available information from these families, but only some kindreds had been described in full-length articles. The Conference venue created an opportunity to exchange clinical, molecular genetic, and pathological data among the 47 attending scientists. The clinical symptoms of FTDP-17 were grouped into three major categories: 1) behavioral disturbance, 2) cognitive disturbance, and 3) motor disturbance.

Behavioral disturbances in FTDP-17 include impaired social conduct ranging from aggressiveness to apathy, hyperorality, hyperphagia, obsessive stereotyped behavior, and psychosis. The most frequently seen *cognitive disturbances* are disturbed executive functions; visuospatial orientation and memory functions were preserved relatively well until late in the illness. The essential *motor sign* is parkinsonism, which appears either early or late in the course of the disease, is characterized by bradykinesia, axial and limb rigidity, and postural instability, and is usually unresponsive to levodopa therapy. The term "FTDP-17" was coined to emphasize these three major clinical signs.

*Zbigniew K. Wszolek, MD, Yoshio Tsuboi, MD, Mayo Clinic, Department of Neurology, 4500 San Pablo Rd, Jacksonville, Fl 32224, USA.

Mapping the Progress of Alzheimer's and Parkinson's Disease
Edited by Mizuno *et al.*, Kluwer Academic/Plenum Publishers, 2002

517

2. THE PROGRESS MADE SINCE 1996 IN UNDERSTANDING FTDP-17

Since the Consensus Conference, major progress has been made in the area of molecular genetics. The missense and deletion mutations have been identified in the *tau* gene on chromosome 17 (reviewed in [2]). To date, 25 different mutations located on exons 1, 9, 10, 12, and 13 and on the intron after exon 10 in the *tau* gene have been reported (Table 1). These mutations were described in 57 separately ascertained families.

Table 1. FTDP-17 mutations

Non–exon 10 mutations: *Exon 1* (R5W, R5H), *Exon 9* (K257T, I260V, G272V), *Exon 12* (V337M, E342V), *Exon 13* (G389R, R406W).
Exon 10 mutations: *Exonic* (N279K, _280K, L284L, delN296, N296H, N296N, P301L, P301S), *5' splice site* (S305N[-2], S305S[-1], +3, +11, +12, +13, +14, +16).

Progress has also been made in understanding clinical aspects of FTDP-17. Previously unpublished clinical information on some of the original 13 families has now become available, additional clinical and pathological details on already published families have been presented, and new families have been identified. It is now clear that FTDP-17 occurs worldwide, with 14 families identified in the USA, 10 in Japan, 8 in Great Britain, 7 in France, 7 in the Netherlands, 4 in Canada, 2 in Australia, 2 in Italy, and 1 each in Germany, Spain, and Sweden. FTDP-17 is still an extremely rare condition and it is impossible to estimate prevalence and incidence indices. Based on published reports, the estimated total number of affected patients is 456, including deceased individuals from all 57 known families. The most commonly occurring mutation is the exon 10 P301L missense mutation identified in 21 separately ascertained kindreds containing 148 affected individuals. The second common mutation is the exon 10 5' splice-site +16 intronic mutation described in 8 separately ascertained families including 78 affected individuals. The third common mutation is the exon 10 N279K missense mutation present in 4 separately ascertained kindreds with 48 affected family members.

3. PHENOTYPIC AND GENOTYPIC CORRELATION

The average age at symptomatic disease onset of FTDP-17 is 50 years, with a range from 25 to 65 years (data available from 40 families). The average duration of the disease is 9 years, with a range from 2 to 26 years (data available from 27 families).[2]

The phenotypic presentation of FTDP-17 kindreds varies not only with different mutations but also between kindreds with the same mutation. Affected individuals from the same family may also present with different clinical phenotypes. The personality and behavioral changes are probably the most consistent features, whereas dementia and parkinsonism are more variable. Depending on clinical presentation, FTDP-17 families can be divided into two major groups: 1) families with a dementia-predominant phenotype and 2) families with a parkinsonism-predominant phenotype.

In general, non–exon 10 missense mutations (Table 1) lead to a dementia-predominant phenotype. The motor symptoms such as parkinsonism, eye-movement abnormalities, dystonia, and upper or lower motor neuron dysfunction are rarely seen in patients with these mutations. In contrast, many affected individuals with the exon 10 missense or intronic mutations develop the parkinsonism-predominant phenotype, quite frequently in the early stages of their illness. However, some of these families, particularly those with the P301L mutation, show considerable phenotypic variation.

It is plausible to speculate that phenotypic variations are due to different molecular mechanisms. Available autopsy data in families with non–exon 10 missense mutations usually demonstrate widespread cortical neuronal pathology with the presence of straight filaments composed of six tau isoforms. The autopsy data from families with the exon 10 missense or 5' splice-site intronic mutations reveal cortical and subcortical neuronal and glial pathology with the presence of filamentous formations containing four repeat tau isoforms. However, pathological data is not available from all families.

4. CLINICAL PRESENTATION OF SELECTED MUTATIONS

The exon 10 P301L missense mutation has been identified in 21 separately ascertained kindreds distributed worldwide.[3-10] A founder effect may be involved in some Dutch and French-Canadian families (M. Hutton, personal communication, March 2001). The presenting symptoms usually include personality and behavioral changes, especially disinhibition and language difficulties. In some cases, parkinsonism is present from the onset of the illness (Seattle Family D), but parkinsonian signs usually develop late in the course of the disease. At autopsy, neuronal loss is almost always evident in the substantia nigra, even in cases with minimal or no parkinsonian symptoms.[4-8]

The exon 10 P301S missense mutation affects three kindreds: 1 Italian, 1 German, and 1 Japanese.[11-13] The phenotype of the German kindred is dominated by the occurrence of severe seizures unresponsive to conventional anticonvulsive therapy.[12] Some affected individuals from this kindred died of status epilepticus. Thus, the clinical presentation of this mutation extends our knowledge of FTDP-17 phenotypes. Interestingly, seizures were not observed in the Italian and Japanese kindreds. Myoclonus was seen in the German and Italian families but not in the Japanese family. Rapidly progressive dementia and parkinsonism were present in all three families.

The exon 10 5'splice-site +14 intronic mutation has been described in a family with dementia-disinhibition-parkinsonism-amyotrophy complex (DDPAC). This was the first kindred to be linked to chromosome 17[14,15] and is still only one that harbors this specific mutation. The DDPAC family exhibits signs of personality change, frontotemporal dementia, parkinsonism, and amyotrophy. There is a long (up to about 20 years) prodromal phase characterized by behavioral aberrations such as excessive religiosity, inappropriate sexual advances, overeating, shoplifting, and others. It is also the only kindred (mutation) with amyotrophy.

The exon 10 N279K missense mutation was originally described in the family with pallido-ponto-nigral degeneration (PPND),[16-18] the largest of all the FTDP-17 kindreds. The affected individuals usually present with parkinsonian features, including bradykinesia, rigidity, and postural instability. About one third of affected individuals present with either personality changes or dementia, alone or in combination with parkinsonism. The clinical features also include dystonia unrelated to medications, eye movement abnormalities, pyramidal tract dysfunction, frontal lobe release signs, perseverative vocalizations, and bladder incontinence. Response to levodopa therapy is seen only in the initial stage of the disease. This family has 4 branches, established by the children of the first known affected individual, who was born in 1854 and died at age 32 years. Thus, the 4 branches of the family are separated by more than 130 years, and geographical residences as well as profession also vary among the branches. Thus, environmental factors may have influenced the significant differences in phenotypic presentation observed among the 4 branches of this family. Another explanation of the phenotypic differences may be a currently unknown susceptibility or the presence of modifying genes. Recently, the exon 10 N279K missense mutation was found in 3 additional families, 1 in France and 2 in Japan.[19-21] The haplotype analysis revealed that the PPND, French and Japanese families are not related to each other, but both Japanese families share a common founder.[22]

5. WHEN SHOULD THE CLINICIAN SUSPECT THE PRESENCE OF FTDP-17?

FTDP-17 remains an extremely rare disorder. However, it should be considered in the presence of the following:

- age of disease onset between the third and the fifth decades
- rapid disease progression
- frontotemporal dementia, behavioral and personality changes with relatively preserved memory retention; similar to sporadic frontotemporal dementia
- parkinsonism-plus disorder (bradykinesia, rigidity, postural instability, paucity of resting tremor, and poor or no response to levodopa), frequently with early falls and supranuclear gaze palsy; somewhat similar to sporadic progressive supranuclear palsy; less commonly apraxia, dystonia and lateralization; similar to sporadic corticobasal ganglionic degeneration
- sometimes, progressive speech difficulties
- at times, severe seizure disorder superimposed on dementia and parkinsonism
- occasionally, amyotrophy in combination with dementia and parkinsonism
- positive family history of FTDP-17
- variability in clinical presentation among family members from the same kindred and among different families, even with the same mutation.

6. FUTURE CLINICAL RESEARCH DIRECTIONS

Despite significant progress in understanding the FTDP-17 phenotype, many clinical areas require further exploration. Because urinary incontinence, excessive salivation, and sweating abnormalities have been described in many affected individuals, future clinical research should investigate the involvement of the *autonomic nervous system*. Sleep studies are needed to examine the *sleep disturbances,* particularly insomnia, reported by some patients (eg, in the PPND family) and to investigate whether the rapid eye movement (REM) behavior disorder is associated with FTDP-17. The *personality and cognitive deficits* await more detailed neuropsychological examination, and the *eye movement and pupillary abnormalities* also await detailed description. S*tructural and functional imaging* has already been performed on some but not all FTDP-17 kindreds. However, the correlation studies comparing the findings on different families (mutations) have not been fully explored. Functional magnetic resonance imaging could provide inside into the cortical abnormalities observed in this disorder. Positron emission tomography (PET) examinations using 2-deoxy-2-fluoro-$[^{18}F]$-D-glucose (FDG), 6-$[^{18}F]$-fluoro-L-dopa (FD) and $[^{11}C]$-raclopride (RAC) have been performed on the PPND kindred, demonstrating decreased FDG metabolism and severe reduction of FD uptake with normal or increased RAC binding. The FDG findings are consistent with the presence of dementia. The FD and RAC findings are consistent with parkinsonism due to presynaptic dopaminergic dysfunction.[23] FDG PET studies performed on the DDPAC family support the clinical findings of dementia in this kindred as well.[24] Further PET studies with FDG, FD, and RAC tracers could be performed on other families. PET studies with other tracers will undoubtedly be performed.

Presymptomatic testing has been performed in the PPND family. However, the majority of at risk family members choose not to be tested, mainly because of the lack of symptomatic or curative therapies. This lack of interest in presymptomatic testing is quite similar to that found among individuals at risk for Huntington disease.[25] It remains to be seen how at-risk individuals from other kindreds react to the possibility of such testing.

Effort should be directed at the study of *phenotypic, genotypic,* and *pathological correlations* to provide better understanding of the underlying mechanisms of familial and sporadic tauopathies. Frank discussions with affected patients and their families regarding the prospect of autopsy should be encouraged.

There are no known therapies for FTDP-17. It is hoped that the development of transgenic mice [26] will provide an opportunity to test therapeutic agents in the near future.

We believe there is an urgent need for the *Second International Conference on FTDP-17.* Such a forum can provide the opportunity to discuss clinical, molecular genetic, and pathological progress on this disorder and delineate further research directions.

REFERENCES

1. N.L. Foster, K. Wilhelmsen, A.A. Sima, M.Z. Jones, C.J. D'Amato and S. Gilman, Frontotemporal dementia and parkinsonism linked to chromosome 17: a consensus conference. Conference Participants. *Ann Neurol* **41**, 706-715 (1997).
2. L.A. Reed, Z.K. Wszolek, M. Hutton, Phenotypic correlations in FTDP-17. *Neurobiol Aging* **22**, 89-107 (2001).
3. M. Hutton, C.L. Lendon, P. Rizzu, M. Baker, S. Froelich, H. Houlden, S. Pickering-Brown, S. Chakraverty, A. Isaacs, A. Grover, J. Hackett, J. Adamson, S. Lincoln, D. Dickson, P. Davies, R.C. Petersen, M. Stevens, E. de Graaff, E. Wauters, J. van Baren, M. Hillebrand, M. Joosse, J.M. Kwon, P. Nowotny, P. Heutink *et al.*, Association of missense and 5'-splice-site mutations in tau with the inherited dementia FTDP-17. *Nature* **393**, 702-705 (1998).
4. T.D. Bird, D. Nochlin, P. Poorkaj, M. Cherrier, J. Kaye, H. Payami, E. Peskind, T.H. Lampe, E. Nemens, P.J. Boyer and G.D. Schellenberg, A clinical pathological comparison of three families with frontotemporal dementia and identical mutations in the tau gene (P301L). *Brain* **122**, 741-756 (1999).
5. Z.S. Nasreddine, M. Loginov, L.N. Clark, J. Lamarche, B.L. Miller, A. Lamontagne, V. Zhukareva, V.M. Lee, K.C. Wilhelmsen and D.H. Geschwind, From genotype to phenotype: a clinical pathological, and biochemical investigation of frontotemporal dementia and parkinsonism (FTDP-17) caused by the P301L tau mutation. *Ann Neurol* **45**, 704-715 (1999).
6. P. Heutink, M. Stevens, P. Rizzu *et al.*, Hereditary frontotemporal dementia is linked to chromosome 17q21-q22; a genetic and clinicopathologic study of three Dutch families. *Ann Neurol* **41**, 150-159 (1997).
7. M.G. Spillantini, R.A. Crowther, W. Kamphorst, P. Heutink and J.C. van Swieten, Tau pathology in two Dutch families with mutations in the microtubule- binding region of tau. *Am J Pathol* **153**, 1359-1363 (1998).
8. S.S. Mirra, J.R. Murrell, M. Gearing, M.G. Spillantini, M. Goedert, R.A. Crowther, A.I. Levey, R. Jones, J. Green, J.M. Shoffner, B.H. Wainer, M.L. Schmidt, J.Q. Trojanowski and B. Ghetti, Tau pathology in a family with dementia and a P301L mutation in tau. *J Neuropathol Exp Neurol* **58**, 335-345 (1999).
9. C. Dumanchin, A. Camuzat, D. Campion, P. Verpillat, D. Hannequin, B. Dubois, P. Saugier-Veber, C. Martin, C. Penet, F. Charbonnier, Y. Agid, T. Frebourg and A. Brice, Segregation of a missense mutation in the microtubule-associated protein tau gene with familial frontotemporal dementia and parkinsonism. *Hum Mol Genet* **7**, 1825-1829 (1998).
10. K. Kodama, S. Okada, E. Iseki, A. Kowalska, T. Tabira, N. Hosoi, N. Yamanouchi, S. Noda, N. Komatsu, M. Nakazato, C. Kumakiri, M. Yazaki and T. Sato, Familial frontotemporal dementia with a P301L tau mutation in Japan. *J Neurol Sci* **176**, 57-64 (2000).
11. O. Bugiani, J.R. Murrell, G. Giaccone, M. Hasegawa, G. Ghigo, M. Tabaton, M. Morbin, A. Primavera, F. Carella, C. Solaro, M. Grisoli, M. Savoiardo, M.G. Spillantini, F. Tagliavini, M. Goedert and B. Ghetti, Frontotemporal dementia and corticobasal degeneration in a family with a P301S mutation in tau. *J Neuropathol Exp Neurol* **58**, 667-677 (1999).

12. A.D. Sperfeld, M.B. Collatz, H. Baier, M. Palmbach, A. Storch, J. Schwarz, K. Tatsch, S. Reske, M. Joosse, P. Heutink and A.C. Ludolph, FTDP-17: an early-onset phenotype with parkinsonism and epileptic seizures caused by a novel mutation. *Ann Neurol* **46**, 708-715 (1999).

13. M. Yasuda, K. Yokoyama, T. Nakayasu, Y. Nishimura, M. Matsui, T. Yokoyama, K. Miyoshi and C. Tanaka, A Japanese patient with frontotemporal dementia and parkinsonism by a tau P301S mutation. *Neurology* **55**, 1224-1227 (2000).

14. T. Lynch, M. Sano, K.S. Marder, K.L. Bell, N.L. Foster, R.F. Defendini, A.A. Sima, C. Keohane, T.G. Nygaard, S. Fahn, R. Mayeux, L.P. Rowland and K.C. Wilhelmsen, Clinical characteristics of a family with chromosome 17-linked disinhibition-dementia-parkinsonism-amyotrophy complex. *Neurology* **44**, 1878-1884 (1994).

15. K.C. Wilhelmsen, T. Lynch, E. Pavlou, M. Higgins and T.G. Nygaard, Localization of disinhibition-dementia-parkinsonism-amyotrophy complex to 17q21-22. *Am J Hum Genet* **55**, 1159-1165 (1994).

16. L.N. Clark, P. Poorkaj, Z.K. Wszolek, D.H. Geschwind, Z.S. Nasreddine, B. Miller, D. Li, H. Payami, F. Awert, K. Markopoulou, A. Andreadis, I. D'Souza, V.M. Lee, L. Reed, J.Q. Trojanowski, V. Zhukareva, T. Bird, G. Schellenberg, K.C. Wilhelmsen. Pathogenic implications of mutations in the tau gene in pallido-ponto- nigral degeneration and related neurodegenerative disorders linked to chromosome 17. *Proc Natl Acad Sci USA* **95**,13103-7 (1998).

17. L.A. Reed, M.L. Schmidt, Z.K. Wszolek, B.J. Balin, V. Soontornniyomkij, V.M. Lee, J.Q. Trojanowski and R.L. Schelper, The neuropathology of a chromosome 17-linked autosomal dominant parkinsonism and dementia ("pallido-ponto-nigral degeneration"). *J Neuropathol Exp Neurol* **57**, 588-601 (1998).

18. Z.K. Wszolek, R.F. Pfeiffer, M.H. Bhatt, R.L. Schelper, M. Cordes, B.J. Snow, R.L. Rodnitzky, E.C. Wolters, F. Arwert and D.B. Calne, Rapidly progressive autosomal dominant parkinsonism and dementia with pallido-ponto-nigral degeneration. *Ann Neurol* **32**, 312-320 (1992).

19. M.B. Delisle, J.R. Murrell, R. Richardson, J.A. Trofatter, O. Rascol, X. Soulages, M. Mohr, P. Calvas, B. Ghetti. A mutation at codon 279 (N279K) in exon 10 of the Tau gene causes a tauopathy with dementia and supranuclear palsy. *Acta Neuropathol (Berl)* **98**, 62-77 (1999).

20. M. Yasuda, T. Kawamata, O. Komure, S. Kuno, I. D'Souza, P. Poorkaj, J. Kawai, S. Tanimukai, Y. Yamamoto, H. Hasegawa, M. Sasahara, F. Hazama, G.D. Schellenberg, C. Tanaka. A mutation in the microtubule-associated protein tau in pallido-nigro-luysian degeneration. *Neurology* **53**, 864-868 (1999).

21. K. Arima, A. Kowalska, M. Hasegawa, M. Mukoyama, R. Watanabe, M. Kawai, K. Takahashi, T. Iwatsubo, T. Tabira, N. Sunohara. Two brothers with frontotemporal dementia and parkinsonism with an N279K mutation of the tau gene. *Neurology* **54**, 1787-1795 (2000).

22. Y. Tsuboi, M. Baker, M. Hutton, R.J. Uitti, O. Rascol, M-B. Delisle, L. Reed, C. Brefel-Courbon, J.R. Murrell, B. Ghetti, M. Yasuda, O. Komure, S. Kuno, K. Arima, N. Sunohara, Z.K. Wszolek. Families with N279K Mutation on the Tau Gene; Clinical, Molecular Genetic and Pathological Studies. *Neurology* **56**(Supple 3): A126 (2001).

23. P.K. Pal, Z.K. Wszolek, A. Kishore, R. de la Fuente-Fernandez, V. Sossi, R.J. Uitti, T. Dobko, and A.J. Stoessel, Positron emission tomography in pallido-ponto-nigral degeneration (PPND) family (frontotemporal dementia with parkinsonism linked to chromosome 17 and point mutation in *tau* gene), *Parknsonism Related Disorders* **7**, 81-88 (2001).

24. Z.K. Wszolek, T. Lynch and K. Wilhelmsen, Rapidly progressive autosomal dominant parkinsonism and dementia with pallido-ponto-nigral degeneration (PPND) and Disinhibition-dementia-parkinsonism-amyotrophy complex (DDPAC) are clinically distinct conditions that are both linked to 17q21-22. *Parkinsonism Related Disorders* **3**, 66-76 (1997).

25. C.A. McRae, G. Diem, T.G. Yamazaki, A. Mitek, Z.K. Wszolek. Interest in genetic testing in pallido-ponto-nigral degeneration (PPND): a family with frontotemporal dementia with Parkinsonism linked to chromosome 17. *Eur J Neurol.* **8**, 179-83 (2001).

26. J. Lewis, E. McGowan, J. Rockwood, H. Melrose, P. Nacharaju, M. Van Slegtenhorst, K. Gwinn-Hardy, M. Paul Murphy, M. Baker, X. Yu, K. Duff, J. Hardy, A. Corral, W.L. Lin, S.H. Yen, D.W. Dickson, P. Davies, M. Hutton. Neurofibrillary tangles, amyotrophy and progressive motor disturbance in mice expressing mutant (P301L) tau protein. *Nat Genet.* **25**, 402-405 (2000).

THE NEUROPATHOLOGY OF FRONTOTEMPORAL LOBAR DEGENERATION

David M.A. Mann and Stuart M. Pickering-Brown[*]

1. INTRODUCTION

Frontotemporal Lobar Degeneration (FTLD)[1] refers to a clinically and pathologically group of non-Alzheimer forms of dementia, generally with onset well before 65 years of age, which arise from a degeneration of the frontal and temporal lobes. The most common clinical syndrome, frontotemporal dementia (FTD), is characterised by personality and behavioural changes progressing into apathy and mutism.[1-3] Related disorders, like semantic dementia (SD), progressive aphasia (PA) and progressive apraxia (PAX),[1] are characterised by disturbances of language function or skilled movement, respectively.[1] Other clinical variants include FTD with Motor Neurone Disease.[1] Two histopathological profiles are associated with these disorders,[1] though neither *per se* predicts the precise clinical syndrome to emerge from the FTLD umbrella. In one form (microvacuolar-type), the pathology is that of large neocortical nerve cell loss with microvacuolation of the superficial cortical laminae and mild subpial reactive astrocytosis. In the other form (Pick-type), the severe nerve cell loss is accompanied by a florid astrocytosis with intraneuronal inclusions (Pick bodies) and swollen cells (Pick cells). The varying clinical presentations stem from differing topographical distributions of a similar underlying pathology, be this of the microvacuolar-, or Pick-, type.[1] In this report, using tau immunohistochemistry and western blotting, we review the pathology in a consecutive series of 52 patients with FTLD, coming to autopsy between years of 1987 and 2000.

[*] David MA Mann and Stuart Pickering-Brown, University of Manchester, Stopford Building, Oxford Rd, Manchester, M13 9PT, UK.

Mapping the Progress of Alzheimer's and Parkinson's Disease
Edited by Mizuno *et al.*, Kluwer Academic/Plenum Publishers, 2002

523

2. MATERIALS AND METHODS

Brains were obtained at autopsy from a consecutive series of 52 patients dying with sporadic and familial FTLD, diagnosed according to Lund-Manchester criteria,[2] comprising clinically of 31 patients with FTD, 7 with PA, 3 with SD, 8 with MNID and 3 with PAX.

Wax sections (5μm) of frontal cortex (Brodmann area 8/9) were immunostained by a conventional avidin-biotin peroxidase method using primary antibodies AT8 (Ser 202/Thr 205) (1:200), AT180 (Ser 231) (1:200), AT270 (Ser 181) (1:200) (all from Innogenetics, Belgium), CP13 (Ser 202) and PHF-1 (Ser396/404) (Gift of Dr P Davies, 1:2000 and 1:1000) and 12E8 (ser 262) (Gift of Dr P Seubert, 1:200) to detect phosphoepitopes of tau. Additional immunoreactions used ubiquitin (Dako, 1:750) and α-synuclein (gift of Dr D Hanger, 1:200) antibodies. All primary incubations were performed overnight at 4°C. Western blotting was performed using AT8 and CP27 (Gift of Dr. P Davies) to detected phosphorylated insoluble tau deposits and soluble non-phosphorylated species of tau, respectively.

3. RESULTS

Of the 52 cases of FTLD, only 22 patients (43%) had demonstrable tau pathology, at least as evidenced by immunohistochemistry. These could be sub-divided into 4 histopathological categories.

6 cases with microvacuolar histology displayed intraneuronal deposits of pathological tau, mostly in the form of amorphous granular deposits, or occasionally as discrete neurofibrillary structures (Figure 1a). Affected nerve cells were widely present throughout the frontal cortex, but mostly within the pyramidal neurones of layers 3 and 5. As well as being AT8 and CP13 positive, the neuronal tau deposits were strongly immunoreactive with AT180 and AT270 antibodies, but were less strongly so with 12E8 and PHF-1 antibodies where the immunoreaction had a definite filamentous appearance. These latter cells, in which the tau was present as a tangle-like structure, were also occasionally, but only weakly, ubiquitin-immunoreactive. Ballooned neurones were variably present, chiefly within layers 5 and 6. These were (usually weakly) immunoreactive to most tau antibodies (including AT8 and CP13) and ubiquitin, but were strongly reactive with AT270 antibody (Figure 1b). A widespread glial cell tau pathology was seen (Figure 1c), being most prominent with CP13, AT8, AT180 and AT270 antibodies but only occasional cells, in some cases, displayed 12E8 immunoreactivity. The tau-positive glial cells appeared to be mostly oligodendrocytes, since immunostaining with GFAP detected only a few of these tau positive cells. These glial cells were also immunoreactive to ubiquitin antibody. This neuronal immunohistochemical profile is identical to that of AD, except that few cells are immunopositive with 12E8 and PHF-1 antibodies, or ubiquitin reactive. Western blotting showed that, on dephosphorylation, tau was present as 4-repeat isoforms (Figure 2a). The tangles are in the form of broad, flat twisted ribbons (see also reference 3). The relative lack of PHF-1 and 12E8 immunoreactivity is consistent with their being less phosphorylated at these sites, compared to the PHF of AD. These 6 cases were associated with splice site mutations in tau gene, 5 being due to +16 splice site mutation and 1 due

to +13 splice site mutation.[4] Three other patients showed a similar, though much less severe, tau pathology, though in none were splice site, or missense, mutations in tau gene detected. A family history of a similar disorder was present in two of these latter cases but no information about past history was available in the other case. There might be other so far unrecognised mutational events within the tau gene that can lead to similar pathological changes as those seen in cases bearing mutations in and around exon 10. Hence, FTDP-17 is not a common cause of FTLD; frequency ranges from 0-18% depending upon the source of the clinical material and diagnostic inclusion criteria employed.[5]

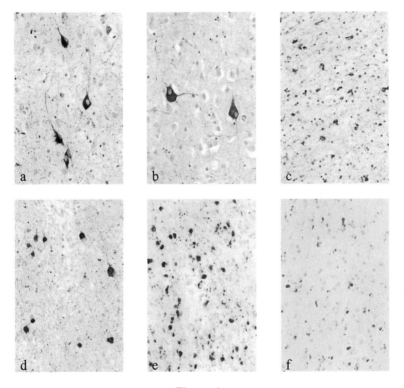

Figure 1.

Tau pathology in nerve cells of the frontal cortex in cases of frontotemporal dementia due to +16 splice site mutation in tau gene (FTDP-17) (a-c) or associated with Pick-type histology (d). In FTDP-17 there are numerous pyramidal cells (a) and glial cells (c) containing amorphous tau deposits, and others where the tau protein is aggregated into neurofibrillary structures. Nerve and glial cells containing amorphous and fibrillary tau deposits are immunoreactive with AT8 (a,c), AT180, AT270 and CP13 antibodies, whereas only aggregated into fibrillary structures is tau immunoreactive with 12E8 and PHF-1 antibodies. Ballooned neurones are strongly immunoreactive with AT270 antibody (b). Pick bodies are strongly reactive with AT8 (d), AT180, AT270 and CP13 antibodies, but not reactive, or only weakly so with 12E8 antibody. In a 61 year-old patient with FTD, (e,f) there are numerous, apparently extracellular, deposits of tau protein within white matter and deeper cortical layers that are strongly immunoreactive to all tau antibodies (e) and ubiquitinated in many instances (f). All immunoperoxidase-haematoxylin. AT8 antibody (a,c,d,e); AT270 antibody (b); Ubiquitin antibody (f). Magnification all x400, except (b) x100.

12 cases displayed Pick-type histology. In these, tau immunoreactive neurones were present in the smaller neurones of layer 2 and outer layer 3 and also within the pyramidal neurones of deeper layers 5 and 6. Most neurones contained a single discrete, rounded inclusion body, strongly immunoreactive with AT8 (Figure 1d), AT180, AT270 and CP13 antibodies. In most, but not all, instances they were also strongly immunoreactive with PHF-1 antibody, but were generally not immunoreactive (or were only weakly so) with 12E8 antibody and ubiquitin. In some of these cells, and in many other neurones not containing an inclusion body, amorphous granular deposits of tau were seen. Ballooned neurones immunoreactive (as above) to tau and ubiquitin were widely present throughout all cortical layers, but chiefly in layers 5 and 6. There were no tau immunoreactive glial cells present within the white matter, except in one PA case where there were abundant immunopositive cells, probably oligodendrocytes. The Pick bodies (and ballooned neurones) partially share an immunohistochemical profile with the NFT of AD. Both structures are strongly immunoreactive with CP13, AT8, AT180, AT270 and (usually)

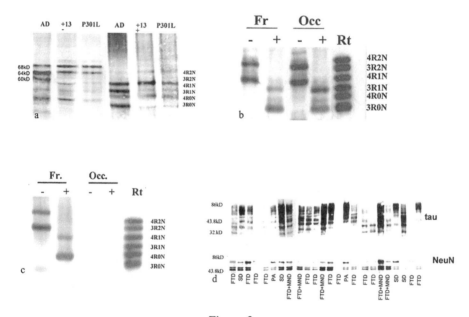

Figure 2.

Western blots of insoluble tau (a-c) and soluble tau (d) proteins in case of FTDP-17 due to +13 splice site mutation (a), case of FTD due to Pick-type histology (b), case of unusual FTD (c) and "tau-negative" FTLD cases (d). Fr and Occ indicate frontal and occipital cortex; + and − indicate with and without dephosphorylation.

PHF-1 antibodies. In most instances Pick bodies are unstained or only weakly stained with 12E8 antibody, and are not detected with ubiquitin. 12E8 antibody recognises phosphorylated Ser 262. Present data are consistent with previous reports[6] indicating that the tau fibrils in Pick bodies may be poorly phosphorylated at this site. However, it may be that the tertiary structure of the Pick body renders the epitope less accessible to 12E8, than it does in the case of tangles in AD or (some of) the tau inclusions of FTDP-17. Likewise, the inconsistent staining of Pick bodies with PHF-1 antibody, which recognises phosphorylated Ser 396/404, suggests variable epitope phosphorylation or antibody accessibility. As reported previously, [6] western blotting showed that, on dephosphorylation, tau was mostly present as 3-repeat isoforms though minor, and in some instances major, bands of 4-repeat tau were seen (Figure 2b). Clinically, 8 cases were of typical FTD, whereas one had suffered from PA, and 3 others from PAX. In none of these cases was a mutation in the tau gene present.[4,7]

Unusual tau pathology was seen in one FTD case, comprising of abundant, apparently extracellular, deposits of tau protein chiefly within white matter (Figure 1e), but some within deeper layers of the grey matter. These tau deposits were strongly immunoreactive with all tau antibodies (including PHF-1 and 12E8), and also to ubiquitin antibody (Figure 1f). No tau or ubiquitin immunopositive (inclusion bearing) neurones or glial cells were present, either in grey or white matter, nor were any ballooned neurones seen. There was severe loss of nerve cells from grey matter and a very heavy reactive astrocytosis throughout all cortical layers. Western blotting showed that tau was present as 4-repeat isoforms (Figure 2c). The fact that the tau deposits in this case were strongly detected by 12E8 and ubiquitin antibodies contrasts with the tau deposits of Picks and FTDP-17 cases and suggests a separate nosological status. There was no previous family history of dementia in this instance. We analysed the tau gene and found no missense or intronic mutations[4,7].

Despite the growing tendency to regard FTLD as a tauopathy, the majority (32/52) of patients (57%) did not have tau pathology in their brains. These all had a microvacuolar type of histology. Clinically, 13 patients had suffered from FTD, 3 from SD and 6 from PA. The other 8 patients had MNID, and in these cases typical ubiquitin-positive, synuclein negative inclusions were seen within layer 2 neurones of the cerebral cortex, granule cells of the dentate gyrus of the hippocampus and motor neurones of the brainstem (not shown). A previous family history of dementia was present in 10 of the 30 tau-negative patients (6 with FTD, 3 with PA and 1 with MNID), though no mutations in the tau gene were found in these, or in any of the other 20 tau-negative cases.[4] Western blotting of whole brain homogenates (Figure 2d) showed variable loss of soluble tau, sometimes virtually complete. These was good agreement between the degree of loss of tau and that of a neuronal nuclear marker protein, NeuN suggesting that loss of soluble tau was unselective and reflected the extent of neurodegeneration.

It was possible to genotype 45 of the 52 pathologically confirmed cases of FTD and in these the APOE ε4 allelic frequency was 19%. No associations were noted between the possession of APOE ε4 allele and the presence of tau pathology, either overall or when subdivided into FTDP-17 or Pick-type profiles. However, possessors of ε4 allele were more likely to accumulate deposits of Aβ, mostly as diffuse $A\beta_{42(43)}$ containing plaques, in their brains than those who did not bear this allelic variant, especially if disease onset was late. Indeed, 14/52 patients showed Aβ deposits in their brains, 9 of whom were possessors of APO E ε4 allele **and** had onset of disease later than 55 years of age.

4. DISCUSSION

There may be several reasons for the lack of insoluble tau deposits in most cases of FTLD.

Firstly, there may be alternate genetic loci for FTLD, sited on chromosome 17 or elsewhere, that dictate a pathological outcome via a route not involving the formation of insoluble tau deposits. This would be consistent with the observations that 33% of the tau-negative cases studied here demonstrated a family history of the disorder in one or more first-degree relatives, but did not show mutations in tau gene.[4] In this context, it has to be remembered that a single family with a symptomatology reminiscent of FTD but an absence of tau pathology has been reported and in this pedigree linkage to chromosome 3 has been determined.[8] Moreover, other recent genetic linkage studies[9] have pointed to a locus on chromosome 9 associated with the MNID phenotype. In this study, members of several families showed either clinical FTD or MND, raising the possibility that the phenotypic range associated with this locus may be wide and not necessarily limited to individuals showing a strict MNID symptomatology.

Secondly, the tissue pathology observed in the tau-negative FTD cases (which apart from the absence of tau positive neurones and glial cells is otherwise identical to that of FTDP-17) might result from a dysfunction of tau, involving a misregulation of translation or a decreased stability of message. This could compromise levels of functional tau without leading to a (visible) accumulation of (insoluble) tau within the brain. It has recently been reported[10] that the brains of tau-negative cases of FTD contain only low levels of soluble (normal) tau. This latter finding we confirm here. However, it appears this loss of soluble tau may (partially or wholly) reflect the loss of neurones from severely affected regions in end-stage autopsy cases, and is not necessarily lost preferentially relative to other neuronal perikaryal components.

Thirdly, it is possible that the microvacuolar pathology seen in these tau-negative cases stems from the disturbance of a cellular mechanism unrelated to tau or the microtubule network, but produces an otherwise 'identical' disease. Given that it is not possible, on the basis of clinical presentation, demography or gross brain morphometrics[1] to distinguish cases with FTDP-17 and others with Pick histology from the tau negative FTD cases, this latter possibility would seem unlikely.

It is indisputable that molecular genetic advances have greatly added to our understanding of the pathological mechanisms of these non-Alzheimer forms of dementia, and have led the recent refocusing of attention upon the pathological tau molecules and structures that characterise so many cases of neurodegenerative dementia. Yet the very absence of this kind of pathology in the majority of cases of FTLD means there is still much to be discovered concerning how the structure and function of tau protein is regulated and its impact upon the aetiology of FTLD. The full range of disorders that stem from a disordering of these processes remains to be determined. Moreover, caution must exercised when applying the term "tauopathy" to the frontotemporal dementias, either as a diagnostic label or as a means of classification. It is clear that a substantial proportion (perhaps even the majority) of cases do not accumulate pathological tau within their brains, at least as is detectable by immunohistochemical techniques. The increasing use of tau immunostaining as a standard investigative tool for the neuropathological diagnosis of the dementias could lead to the erroneous exclusion of many cases of FTLD were this molecular classification based on the presence of pathological tau to be rigorously adopted. In most instances of FTLD the (anticipated) finding of tau-positive nerve (and glial) cells will not be fulfilled.

REFERENCES

1. J. S. Snowden, D. Neary and D. M. A. Mann, *Frontotemporal lobar degeneration: frontotemporal dementia, progressive aphasia, semantic dementia*, (Churchill-Livingstone, New York, 1996).
2. Brun, E. Englund, L. Gustafson, et al, Clinical, neuropsychological and neuropathological criteria for fronto-temporal dementia, J Neurol Neurosurg Psychiatr 57,416-418 (1994).
3. M. G. Spillantini, R. A. Crowther, W. Kamphorst, et al, Tau pathology in two Dutch families with mutations in the microtubule-binding region of tau, Am J Pathol 153,1359-1363 (1998).
4. M. Hutton, C. L. Lendon, P. Rizzu, et al, Coding and 5' splice mutations in tau associated with inherited dementia (FTDP-17), Nature 393,702-705 (1998).
5. H. Houlden, M. Baker, J. Adamson, et al, Frequency of tau mutations in three series of non-Alzheimer's degenerative dementia, Ann Neurol 46,243-248 (1999).
6. Delacourte, N. Sergeant, J. Wattez, et al, Vulnerable neuronal subsets in Alzheimer's and Pick's disease are distinguished by their tau isoform distribution and phosphorylation, Ann Neurol 43,193-204 (1998).
7. S. M. Pickering-Brown, M. Baker, S.- H.Yen, et al, Pick's disease is associated with mutations in the tau gene, Ann Neurol 48,859-867 (2000).
8. J. Brown, A. Ashworth, S. Gydesen, et al, Familial non-specific dementia maps to chromosome 3, Hum Molec Genet 4,1625-1628 (1995).
9. Hosler, T. Siddique, P. C. Sapp, et al, Linkage of familial Amyotrophic Lateral Sclerosis with Frontotemporal Dementia to chromosome 9q21-22, JAMA 284,1664-1669 (2000).
10. V. Zhuckareva, V. Vogelsberg-Ragaglia, V. M. D. Van Deerlin, et al, Loss of brain tau defines novel sporadic and familial tauopathies with frontotemporal dementia. Ann Neurol 49,165-175 (2001).

PROGRESS IN PSP AND CBD

D.W. Dickson, K. Ishizawa, T. Togo, M. Baker, J. Adamson, M. Hutton, W.-K. Liu, and S.-H. Yen[*]

1. ABSTRACT

While progressive supranuclear palsy (PSP) and corticobasal degeneration (CBD) are neurodegenerative disorders with shared clinical and pathologic features, there are notable differences that warrant their continued separation as clinicopathologic entities. They are both hypokinetic extrapyramidal disorders, but unlike idiopathic Parkinson's disease, neither PSP nor CBD are responsive to levodopa. Extrapyramidal signs are usually symmetrical in PSP and asymmetrical in CBD, but there are exceptions. Focal cortical signs, such as apraxia and aphasia, are common in CBD, but not in PSP. A frontal type dementia is more common in CBD than PSP. PSP has downward gaze palsy early in the disease, and this is a means of differentiating PSP and CBD. Pathologically, both PSP and CBD are associated with neuronal and glial filamentous inclusions composed of tau protein, but the morphology of neuronal and glial lesions and their distribution differ in PSP and CBD. Pathology of the cortical gray and white matter are prominent in CBD, but usually mild in PSP, while deep gray matter lesions are more marked in PSP than CBD. PSP and CBD are both associated with accumulation of insoluble tau protein in the brain, with more in supratentorial areas in CBD than PSP. They are usually nonfamilial or sporadic "tauopathies," but genetic studies suggest that polymorphisms in the tau gene are similar in PSP and CBD. The present review describes some recent progress in the neuropathology, genetics and biochemistry of PSP and CBD.

2. NEUROPATHOLOGY OF PSP AND CBD

PSP is a multisystem degeneration involving the extrapyramidal motor system. While original descriptions of PSP drew attention to neurofibrillary tangles (NFTs) (Steele et al. 1964), recent studies have focused on glial lesions. NFTs in PSP are different from those in Alzheimer's disease by the paucity of ubiquitin (Bancher et al.

---

[*] D.W. Dickson, et al., Mayo Clinic, 4500 San Pablo Road, Jacksonville, FL 32224, USA.

Mapping the Progress of Alzheimer's and Parkinson's Disease
Edited by Mizuno *et al.*, Kluwer Academic/Plenum Publishers, 2002

1987), and the same is true for CBD. In both disorders NFTs have weak or no fluorescence with thioflavin-S and can be missed if this method is used to evaluate neuropathology. The observed differences probably reflect differences in packing, conformation or post-translational modification of tau protein in NFTs.

Tufts of abnormal fibers, also known as tufted astrocytes, are common in PSP in motor cortex and striatum. They are fibrillary lesions within astrocytes based upon double immunolabeling studies (Iwatsubo et al. 1994). The most characteristic astrocytic lesion in CBD, particularly in the neocortex, is an annular cluster of short stubby processes that may be suggestive of a neuritic plaque that are called "astrocytic plaques" (Feany & Dickson 1995). Astrocytic plaques, in contrast to true neuritic plaques, do not contain amyloid, but rather a central astrocyte. Astrocytic plaques of CBD are morphologically distinct from tufted astrocytes of PSP, and the two lesions do not co-exist in the same brain (Komori et al. 1998).

Tau-positive fibers, so-called "neuropil threads," are found in areas affected by PSP and CBD. Tau-immunopositive glial cells in white matter, also referred to as "coiled bodies," (Braak & Braak 1989) are inclusions in oligodendroglial cells found in both PSP and CBD (Iwatsubo et al. 1994; Yamada & McGeer 1995). Thread-like processes in white matter are sparse in PSP, but numerous in CBD. Immunoelectron microscopy has shown white matter threads in the outer mesaxon of myelinated fibers (Arima et al. 1997). No neuronal or glial lesion is pathognomonic for PSP or CBD; the pathologic diagnosis (Hauw et al. 1994) is contingent upon both the morphology and distribution of lesions (Feany et al. 1996).

The cerebral cortex is little affected in PSP, but lesions are increasingly recognized in the peri-Rolandic region (Hof et al. 1992; Bergeron et al. 1997). Furthermore, cases of PSP with dementia have more extensive cortical pathology than typical cases (Bigio et al. 1999; Bergeron et al. 1998). In CBD atrophic frontoparietal cortices shows disruption of cortical lamination, spongiosis and astrocytosis. These findings are dissimilar to those of PSP, where the cortex looks unremarkable with routine histologic methods. The limbic lobe is relatively preserved in both PSP and CBD. When found, NFTs in the hippocampus are consistent with concurrent age-related pathology (Braak et al. 1992).

The basal ganglia have consistent pathology in both PSP and CBD. The nucleus basalis of Meynert has NFT and pre-tangles in both disorders. The red nucleus and subthalamic nucleus show mild neuronal depletion and astrocytic gliosis and may have NFT, especially in PSP. Thalamic nuclei are affected in both PSP and CBD. The subthalamic nucleus is affected in both PSP and CBD, but neuronal loss and gliosis is usually marked in PSP and minimal in CBD. Mild neuronal depletion of the dentate nucleus occurs in CBD, but more marked pathology is very common in PSP.

In CBD swollen and vacuolated cortical neurons, referred to as ballooned neurons, are a characteristic feature, especially in the superior frontal and parietal lobules. Ballooned neurons are also found in cingulate, amygdala and claustrum, especially in argyrophilic grain disease (Braak & Braak 1992), which is not uncommon in PSP and CBD (Martinez-Lage and Munoz 1997). In PSP BN are extremely uncommon outside of the limbic lobe.

Neurons in atrophic cortical areas of CBD have pleomorphic tau-immunoreactive lesions somewhat reminiscent of Pick bodies or small NFTs. In other neurons the filamentous inclusions are more dispersed and disorderly. NFTs are found in brainstem monoaminergic nuclei, such as the locus ceruleus and substantia nigra, in both PSP and

CBD, but NFTs in PSP are typical densely packed globose fibrillary lesions, while those in CBD are ill-defined amorphous or irregular inclusions.

2.1 Grumose Degeneration In PSP

Grumose degeneration (GD) of the cerebellar dentate nucleus is common in PSP (Mizusawa et al. 1989) and it occurs less frequently in CBD, but its clinical significance remains unknown. Grumose degeneration is a characterized by clusters of degenerating presynaptic terminals around dentate neurons. Grumose degeneration is not specific for PSP, but PSP is the most common condition in which it is detected. The dentatorubrothalamic pathway consistently showed fiber loss in cases that were systematically studied with single- and multiple-immunolabeling, as well as stains for myelin and axons (Ishizawa, et al. 2000). Grumose degeneration was associated with demyelination, axonal loss, glial tau pathology and microgliosis in regions juxtaposed to the dentate nucleus. Specifically, demyelination and microgliosis were prominent in the superior cerebellar peduncle, dentate hilus and cerebellar hemispheric white matter. Tau pathology and microgliosis was less prominent in the dentate nucleus itself. The degree of myelin loss correlated with the tau burden in the dentatorubrothalamic pathway. GAP-43, which is a phosphoprotein involved in axonal growth and sprouting, was decreased in grumose degeneration. These results suggest that grumose degeneration may be related to pathology in the dentatorubrothalamic tract and the cerebellar white matter, in other words concurrent degeneration of both output from and input to the dentate nucleus. The results further suggest a possible role for oligodendroglial and myelin pathology in the pathogenesis of neurodegeneration in PSP.

2.2 Microglia in PSP and CBD

The role of microglia in PSP and CBD is unknown, but has been the focus of a recent study of 10 cases of PSP, 5 cases of CBD and 4 normal controls (Ishizawa & Dickson 2001). Microglial and tau burdens were determined with image analysis on brain sections that had been immunostained with monoclonal antibodies to HLA-DR and to phospho-tau. Microglial activation was greater in PSP and CBD than normal controls, and the microglial burden correlated with the tau burden in most areas. There were distinct patterns of microglial activation and tau pathology in PSP and CBD, with PSP showing more pathology in infratentorial structures and CBD showing more pathology in supratentorial structures. These results support the notion that PSP and CBD are distinct clinicopathologic entities. Microglial activation was not well correlated with tau pathology in the brainstem of PSP, which suggests that some of the pathology in PSP is due to neurodegeneration not associated with tau inclusions. While the results do not necessarily support a direct causal link between microglial activation and neurodegeneration in PSP or CBD, they nevertheless suggest that microglia play a role in disease pathogenesis.

3. TAU BIOCHEMISTRY IN PSP AND CBD

Recent biochemical studies have focussed on tau protein abnormalities in PSP and CBD (Buee & Delacourte 1999). Tau protein is a microtubule-associated phosphoprotein that promotes tubulin polymerization and stabilization of microtubules. The tau gene is located on chromosome 17, and it has 16 exons, three of which are alternatively spliced (Andreadis et al. 1992). In the amino-terminal half of the molecule are conserved 30-amio acid tandem repeats that are essential for interaction of tau with microtubules. One of the repeat domains is the product of exon 10 and is alternatively spliced. Depending on whether exon-10 is included or not in the mRNA, the tau protein will have either three (3R) or four (4R) repeats. In PSP and CBD 4R tau predominates in detergent-insoluble tau (Buee & Delacourte 1999). Analysis with antibodies specific to exon 3, another alternatively spliced exon, suggest that exon 3, in contrast to exon 10, is underrepresented in pathological tau protein of PSP and CBD (Feany et al. 1995). The explanation for the increase in 4R tau in PSP and CBD is unknown. The possibility that 4R tau may be overexpressed in PSP has been suggested by a recent study (Chambers et al. 1999).

4. TAU HAPLOTYPE IN PSP AND CBD

A dinucleotide repeat polymorphism in the intron between exons 9 and 10 of the tau gene is associated with PSP (Conrad et al. 1997). The number of dinucleotide repeats varied from 11 (a0) to 14 (a3). In PSP more than 95 percent of the cases are a0. The increased frequency of this tau polymorphism in PSP has been confirmed in several studies among different ethnic groups (Oliva et al. 1998; Hoenicka et al. 1999; Morris et al. 1999). More recent genetic studies of the tau gene revealed a number of polymorphisms throughout its length that were inherited in a non-random manner as extended haplotypes. Two ancestral forms of the tau gene are referred to as H1 and H2 (Baker et al. 1999). Most cases of PSP are H1 and a large proportion are H1/H1 homozygous. This has recently been shown to be true for CBD as well (Di Maria et al. 2000).

5. GENOTYPE-PHENOTYPE CORRELATIONS IN PSP

To determine the influence of the tau haplotype on tau isoform composition and neuropathology, 25 PSP cases and 6 Alzheimer disease (AD) cases matched for age, sex and postmortem delay were studied (Liu et al. 1999). In the basal ganglia, tau and amyloid burdens were determined to see if there was an effect of concurrent Alzheimer type pathology and the ratio of 4R to 3R tau was measured in detergent-insoluble tau fractions. Insoluble tau from PSP was not composed exclusively of 4R tau. All brains had a mixture of 4R and 3R tau, but the ratio was different in AD and PSP. In AD there was less 4R than 3R tau, while the ratio was reversed in PSP. In PSP cases with concurrent Alzheimer type pathology, the ratio of 4R to 3R was intermediate between AD and PSP. The H1 haplotype had no effect on the 4R to 3R ratio or on tau and amyloid burdens. In summary, the H1 haplotype does not have a major influence on the pathologic or biochemical phenotype of PSP.

REFERENCES

Andreadis, A., Brown, W.M., and Kosik, K.S. (1992). Structure and novel exons of the human tau gene. *Biochemistry* **31**, 10626-10633.

Arima, K., Nakamura, M., Sunohara, N., Ogawa, M., Anno, M., Izumiyama, Y., Hirai, S., and Ikeda, K. (1997). Ultrastructural characterization of the tau-immunoreactive tubules in the oligodendroglial perikarya and their inner loop processes in progressive supranuclear palsy. *Acta Neuropathol.* **93**, 558-566.

Baker, M., Litvan, I., Houlden, H., Adamson, J., Dickson, D., Perez-Tur, J., Hardy, J., Lynch, T., Bigio, E., and Hutton, M. (1999). Association of an extended haplotype in the tau gene with progressive supranuclear palsy. *Hum. Molec. Gen.* **8**, 711-715.

Bancher, C., Lassmann, H., Budka, H., Grundke-Iqbal, I., Iqbal, K., Wiche, G., Seitelberger, F., and Wisniewski, H.M. (1987). Neurofibrillary tangles in Alzheimer's disease and progressive supranuclear palsy: antigenic similarities and differences. *Acta Neuropathol.* **74**, 39-46.

Bergeron, C., Davis, A., and Lang, A.E. (1998). Corticobasal ganglionic degeneration and progressive supranuclear palsy presenting with cognitive decline. *Brain Pathol.* **8**, 355-365.

Bergeron, C., Pollanen, M.S., Weyer, L., and Lang, A.E. (1997). Cortical degeneration in progressive supranuclear palsy. A comparison with cortical-basal ganglionic degeneration. *J. Neuropathol. Exp. Neurol.* **56**, 726-734.

Bigio, E.H., Brown, D.F., and White, C.L. 3rd (1999). Progressive supranuclear palsy with dementia: cortical pathology. *J. Neuropathol. Exp. Neurol.* **58**, 359-364.

Braak, H., and Braak, E. (1989). Cortical and subcortical argyrophilic grains characterize a disease associated with adult onset dementia. *Neuropathol. Appl. Neurobiol.* **15**, 13-26.

Braak, H., Jellinger, K., Braak, E., and Bohl, J. (1992). Allocortical neurofibrillary changes in progressive supranuclear palsy. *Acta Neuropathol.* **84**, 478-483.

Buee, L., and Delacourte, A. (1999). Comparative biochemistry of tau in progressive supranuclear palsy, corticobasal degeneration, FTDP-17 and Pick's disease. *Brain Pathol.* **9**, 681-693.

Chambers, C.B., Lee, J.M., Troncoso, J.C., Reich, S., and Muma, N.A. (1999). Overexpression of four-repeat tau mRNA isoforms in progressive supranuclear palsy but not in Alzheimer's disease. *Ann. Neurol.* **46**, 325-332.

Conrad, C., Amano, N., Andreadis, A., Xia, Y., Namekataf, K., Oyama, F., Ikeda, K., Wakabayashi, K., Takahashi, H., Thal, L.J., Katzman, R., Shackelford, D.A., Matsushita, M., Masliah, E., and Sawa, A. (1998). Differences in a dinucleotide repeat polymorphism in the tau gene between Caucasian and Japanese populations: implication for progressive supranuclear palsy. *Neurosci. Lett.* **250**, 135-137.

Di Maria, E., Tabaton, M., Vigo, T., Abbruzzese, G., Bellone, E., Donati, C., Frasson, E., Marchese, R., Montagna, P., Munoz, D.G., Pramstaller, P.P., Zanusso, G., Ajmar, F., Mandich, P. (2000) Corticobasal degeneration shares a common genetic background with progressive supranuclear palsy. Ann. Neurol. **47**, 374-377.

Feany, M.B. and Dickson, D.W. (1995). Widespread cytoskeletal pathology characterizes corticobasal degeneration. *Am. J. Pathol.* **146**, 1388-1396.

Feany, M.B., Ksiezak-Reding, H., Liu, W.-K., Vincent, I., Yen, S.-H., and Dickson, D.W. (1995). Epitope expression and hyperphosphorylation of tau protein in corticobasal degeneration: differentiation from progressive supranuclear palsy. *Acta Neuropathol.* **90**, 37-43.

Feany, M.B., Mattiace, L.A., and Dickson, D.W. (1996). Neuropathologic overlap of progressive supranuclear palsy, Pick's disease and corticobasal degeneration. *J. Neuropathol. Exp. Neurol.* **55**, 53-67.

Hauw, J.-J., Daniel, S.E., Dickson, D., Horoupian, D.S., Jellinger, K., Lantos, P.L., McKee, A., Tabaton, M., and Litvan, I. (1994). Preliminary NINDS neuropathologic criteria for Steele-Richardson-Olszewski syndrome (progressive supranuclear palsy). *Neurology* **44**, 2015-2019.

Hoenicka, J., Perez, M., Perez-Tur, J., Barabash, A., Godoy, M., Vidal, L., Astarloa, R., Avila, J., Nygaard, T., and de Yébenes, J.G. (1999). The tau gene A0 allele and progressive supranuclear palsy. *Neurology* **53**, 1219-1225.

Hof, P.R., Delacourte, A., and Bouras, C. (1992). Distribution of cortical neurofibrillary tangles in progressive supranuclear palsy: a quantitative analysis of six cases. *Acta Neuropathol.* **84**, 45-51.

Ishizawa, K., Dickson, D.W. (2001) Microglial activation parallels system degeneration in progressive supranuclear palsy and corticobasal degeneration. *J. Neuropathol. Exp. Neurol.* 60, (in press).

Ishizawa, K., Lin, W.-L., Tiseo, P., Honer, W.G., Davies, P., and Dickson, D.W. (2000) A qualitative and quantitative study of grumose degeneration in progressive supranuclear palsy. *J. Neuropathol. Exp. Neurol.* 59, 513-524

Iwatsubo, T., Hasegawa, M., and Ihara, Y. (1994). Neuronal and glial tau-positive inclusions in diverse neurologic diseases share common phosphorylation characteristics. *Acta Neuropathol.* **88**, 129-136.

Komori, T., Arai ,N., Oda, M,. Nakayama, H., Mori, H., Yagishita, S., Takahashi, T., Amano, N., Murayama, S., Murakami, S., Shibata, N. Kobayashi, M., Sasaki, S., and Iwata, M. (1998). Astrocytic plaques and tufts of abnormal fibers do not coexist in corticobasal degeneration and progressive supranuclear palsy. *Acta Neuropathol.* **96**, 401-408.

Liu, W.-K., Yen, S.-H., Younkin, L., Hardy, J., Hutton, M., Baker, M., Dickson, D.W. (1999) Relationship of the extended tau haplotype to tau isoform expression and neuropathology in progressive supranuclear palsy. *Soc. Neurosci. Abstr.* **25**, 1095.

Martinez-Lage, P., and Munoz, D.G. (1997). Prevalence and disease associations of argyrophilic grains of Braak. *J. Neuropathol. Exp. Neurol.* **56**, 157-164.

Mizusawa, H., Yen, S.-H., Hirano, A., and Llena, J. (1989). Pathology of the dentate nucleus in progressive supranuclear palsy: a histological, immunohistochemical and ultrastructural study. *Acta Neuropathol.* **78**, 419-428.

Morris, H.R., Janssen, J.C., Bandmann, O., Daniel, S.E., Rossor, M.N., Lees, A.J., and Wood, N.W. (1999). The tau gene A0 polymorphism in progressive supranuclear palsy and related neurodegenerative diseases. *J. Neurol. Neurosurg. Psychiatry* **66**, 665-667.

Steele, J.C., Richardson, J.C., and Olszewski, J. (1964). Progressive supranuclear palsy: a heterogenous degeneration involving the brainstem, basal ganglia and cerebellum with vertical gaze and pseudobulbar palsy, nuchal dystonia, and dementia. *Arch. Neurol.* **10**, 333-339.

Yamada, T., and McGeer, P.L. (1995). Oligodendroglial microtubular massess: an abnormality observed in some human neurodegenerative diseases. *Neurosci. Lett.* **120**, 163-166.

BIOCHEMICAL ANALYSIS OF TAU AND α-SYNUCLEIN IN NEURODEGENERATIVE DISEASES

Masato Hasegawa*, Sayuri Taniguchi, Hirofumi Aoyagi, Hideo Fujiwara, and Takeshi Iwatsubo

1. INTRODUCTION

Filamentous cytoplasmic inclusion bodies made of distinct proteins in neurons and/or glial cells in affected brain regions are defining characteristics of many neurodegenerative diseases. In Alzheimer's disease (AD), hyperphosphorylated tau is assembled into paired helical filaments (PHFs) or related straight filaments (SFs) in neurofibrillary tangles (NFTs) and neuropil threads (NTs) and the presence of NFTs correlates with the presence of dementia. In Parkinson's disease (PD) and dementia with Lewy bodies (DLB), a presynaptic protein, α-synuclein is deposited in a filamentous form in Lewy bodies and Lewy neurites.

The recent discoveries of exonic and intronic mutations in tau gene in frontotemporal dementia and parkinsonism link to chromosome 17 (FTDP-17) and of two missense mutations in α-synuclein gene in familial form of Parkinson's disease have provided genetic links between the abnormality of these proteins and neurodegeneration.

To investigate the molecular mechanisms underlying these protein deposits and neurodegeneration, we have biochemically analyzed tau protein, α-synuclein and other related proteins in AD and DLB. We also have analyzed the effects of tau mutations in FTDP-17.

2. TAU IN AD BRAIN

Tau is one of the microtubule-associated proteins. It promotes microtubule assembly and stabilizes microtubules. Tau exists normally in a soluble form in normal brain and is partially phosphorylated in vivo. In AD brain, hyperphosphorylated tau (PHF-tau), which is functionally incompetent, is deposited in a filamentous form in NFTs and NTs [1].

---

* Masato Hasegawa, Tokyo Institute of Psychiatry, Setagaya-ku, Tokyo, 156-8585, Sayuri Taniguchi, Hirofumi Aoyagi, Hideo Fujiwara, Takeshi Iwatsubo, University of Tokyo, Bunkyo-ku, Tokyo, 113-0033.

Mapping the Progress of Alzheimer's and Parkinson's Disease
Edited by Mizuno *et al.*, Kluwer Academic/Plenum Publishers, 2002

The PHF·tau in a filamentous form has an unusual insolubility to buffers containing Triton-X or Sarkosyl.

In order to investigate the nature of PHF-tau, we purified PHF-tau from AD brain using differential extraction and centrifugation, gel filtration and reverse-phase HPLC, and directly analyzed by mass spectrometry and protein sequencing [2]. The protein chemical analyses of PHF-tau have shown that both three and four-repeat tau isoforms are incorporated in PHFs, where the ratio of three-repeat tau to four-repeat tau isoforms was about three to two, which is similar to that in normal soluble tau [2]. We also identified 19 phosphorylation sites at serine and threonine residues in PHF-tau [2, 3]. Most of the phosphorylation sites are localized to both amino- and carboxy-terminal flanking regions of the microtubule-binding domain.

To confirm a part of these results and to search for kinase(s) responsible for the abnormal phosphorylation, we made an antibody that specifically recognizes phospho-Ser422 (AP422) which has been identified in PHF-tau [4]. Immunoblot analysis of adult tau, fetal tau and PHF-tau with the AP422 antibody have shown that AP422 strongly labeled PHF-tau but only faintly stained fetal and adult tau, showing Ser422 is abnormally phosphorylated in PHF-tau. Interestingly, out of the candidate protein kinases, CDK5, GSK3 and MAP kinase, only MAP kinase could generate the AP422 epitope, suggesting MAP kinase or a MAP kinase-like enzyme may be involved in this abnormal phosphorylation [4]. Additional members of the MAP kinase family, stress-activated protein kinases (SAPKs), which are activated by cellular stresses, generated AP422 epitope, showing that SAPKs also phosphorylate Ser422 [5].

3. p25/p35 CDK5 ACTIVATOR AND CALPAIN IN AD BRAIN

It has recently been shown that p25, a truncated form of CDK5 activator p35, is upregulated and accumulated in neurons in AD brains and that the expression of the p25-CDK5 complex induces cytoskeletal disruption and neurodegenration in cultured neurons [6]. It has also been shown that p35 can be cleaved to p25 by calpain in vitro and in vivo [7, 8]. To investigate whether there is any correlation between the p35 processing and calpain activation we analyzed cleavage of p35 to p25 and activation of calpain in AD and control brains, and in rat brains with various postmortem incubation. Nine control and eight AD brain tissues with relatively short postmortem intervals (~3hrs) which had been preserved in good condition without any freeze-thaw processes, were used for this analysis. PHF-tau was detected in Sarkosyl-insoluble fractions of all AD brains, whereas no immunoreactivities were seen in control brains, showing that all AD patients had tau pathology. Immunoblot analysis with a p25·p35 antibody showed no significant difference in the ratios of p25 to p35 between control and AD brains, although total levels of both p25 and p35 were decreased by about 50 % in AD brains compared with those in normal controls [9]. Because calpain has been implicated in the conversion of p35 to p25, we investigated whether calpain is activated in AD brains using an antibody which specifically recognizes the active form of μ-calpain. The active form of calpain was detected in AD brains and the immunoreactivities were about 7-fold stronger than those in control brains [9].

Immunoblot analysis of human brain tissues with various postmortem delays revealed a marked variation in the levels of p25 or p35. This prompted us to investigate the effect of postmortem delay on p35 protein degradation. We examined the amount of p25 in rat brains with different postmortem intervals by leaving the sacrificed rats at room temperature for 1, 3, 6, 14 and 24 hrs. During incubation at room temperature after death, p35 was rapidly degraded, and the levels of p35 band gradually decreased, whereas those of p25 in turn increased [9]. The intensity of p25 band reached maximum at 14 hrs and then slightly decreased at 24 hrs. These observations suggest that the conversion of p35 to p25 is a postmortem degradation event and may not be upregulated in AD brains. Hyperphosphorylation of tau in AD may not be explained with such a simple mechanism.

4. TAU MUTATIONS IN FTDP-17

Over the past years, 19 different mutations in the tau gene in over 50 families with FTDP-17 have been identified [10]. The mutations are found in both coding region (missense, deletion, or silent) and the intron following exon 10, clustering in the microtubule binding region. All the intronic mutations have been shown to disrupt a predicted stem-loop structure, and lead to an increased production of 4-repeat tau isoforms [10]. Interestingly, tau deposits from one of the families have been shown to contain only 4-repeat tau isoforms [11]. Missense mutations are located in the repeat region or around the microtubule-binding region of the tau molecule. P301L mutation only affects the 4-repeat tau isoforms, because this exon 10 is spliced out of the mRNA that encodes the 3-repeat isoforms. Analysis of affected brain from the family with this mutation reveals that the tau deposits consist mainly of 4-repeat isoforms [12]. On the other hand, V337M mutation found in Seattle family A occurs in all six tau isoforms. Tau deposits from this family have been shown to contain six tau isoforms like those in AD [13].

In order to investigate the mechanisms of neurodegeneration with these mutations, we have produced mutant proteins and analyzed the effect of mutations in the coding region on the ability to promote microtubule assembly. Most missense mutations including K257T, G272V, P301L, P301S, V337M, G389R and R406W have shown a significant reduction in the ability [14, 15], while other missense (N279K and S305N) or silent (L284L, N296N and S305S) mutations were shown to increase the splicing of exon 10 [10, 16]. P301L mutation has the strongest and R406W mutation has the weakest effect on the ability to promote microtubule assembly in these missense mutations [14]. This effect has also been observed in cultured cells transfected with wild or mutant tau on the extension of microtubules [17]. We also confirmed the reduced levels of microtubule or -lar extension with mutations in CHO cells (H. Aoyagi et al, unpublished observation). Additional effects of the missense mutations have been suggested on the self-assembly of tau into filaments[18, 19]. We also observed that tau with some missense mutations formed tau filaments faster than wild-type (H. Aoyagi et al unpublished observation).

Importantly, the tau in abnormal filaments from patients with these mutations is also hyperphosphorylated. The reduction in the ability of tau to promote microtubule assembly (partial loss of function) may lead to the hyperphosphorylation of tau, and assembly into filaments. Similarly, over production of 4-repeat tau isoforms resulting from the intronic mutations may result in the inability of a proportion of the excess tau to bind to microtubules, leading to its hyperphosphorylation and assembly into filaments. Alternatively, functionally defective tau or overproduced unnecessary tau protein may selectively get phosphorylated and gradually deposited in cell bodies and form filaments.

Cells containing tau filaments may die with the abnormal filaments as the consequence of gain of toxic function.

5. α-SYNUCLEIN IN LEWY BODY PATHOLOGY

Lewy body and Lewy neurite are the defining neuropathological features of PD and DLB. Ultrastructurally, Lewy bodies and Lewy neurites consist of abnormal filaments. After the discovery of A53T missense mutation in α-synuclein in some familial cases of PD in 1997 [20], α-synuclein was identified as the major component of Lewy body [21,22]. Human _-synuclein is 140 amino acids in length and is abundantly expressed in brain, where it is located in presynaptic nerve terminals, however its function is unknown. The amino-terminal half of α-synuclein is taken up by imperfect amino acid repeats, with the consensus sequence KTKEGV.

Because α-synuclein depositions in PD and DLB and tau deposition in AD share significant similarities and they may result from common mechanisms, we next investigated α-synuclein in Lewy bodies and Lewy neurites. We took the same approach as that taken in the analysis of PHF-tau, which is direct protein chemical approach. First we extracted the pathological α-synuclein from DLB brain, using differential centrifugations, anion exchange column chromatography and reverse-phase HPLC. Most α-synuclein is recovered in Tris-soluble and Triton-soluble fractions and the levels of α-synuclein in these fractions showed no significant difference between normal and DLB brain. In contrast, strong immunoreactive band at 15 kDa was detected in the Urea-soluble fraction of DLB brain but not in control brain. Since the levels of this Sarkosyl-insoluble α-synuclein band correlated well with the quantity of Lewy bodies and Lewy neurites in the tissue homogenates, this Sarkosyl-insoluble α-synuclein should represent the α-synuclein in Lewy bodies.

Then, we purified the insoluble α-synuclein using anion exchange chromatography and HPLC, and the α-synuclein was chemically digested, and the fragments were separated by HPLC. The HPLC profiles were compared with those of recombinant and soluble normal, α-synuclein. By the analysis of the fragments of soluble and insoluble α-synuclein by protein sequencing and MALIDI-TOF mass spectrometry, a modified peptide fragment was detected in the insoluble α-synuclein from DLB brain, which were never detected in the soluble or recombinant α-synuclein.

To further examine the modification of the α-synuclein in normal and pathological conditions, we raised an antibody against this modification. The polyclonal antibody specifically recognized the insoluble α-synuclein, but never reacted with recombinant α-synuclein. This antibody also strongly and specifically stained Lewy body pathology in DLB brain. These results strongly suggest that the modification of α-synuclein may be involved in the deposition of α-synuclein and pathogenesis of PD and DLB.

REFERENCES

1. M. Goedert, M. G. Spillantini, M. Hasegawa, R. Jakes, R. A. Crowther, and A. Krug, Molecular dissection of the neurofibrillary lesions of Alzheimer's disease, Cold Spring Harbor Symposium on Quantitative Biology, LXI, 565-573 (1996).

2. M. Hasegawa, M. Morishima-Kawashima, K. Takio, M. Suzuki, K. Titani, Y. Ihara, Protein sequence and mass spectrometric analyses of tau in the Alzheimer's disease brain, J. Biol. Chem., 267, 17047-17054 (1992).

3. M. Morishima-Kawashima, M. Hasegawa, K. Takio, M. Suzuki, H. Yoshida, K. Titani, and Y. Ihara, Proline-directed and non-proline-directed phosphorylation of PHF-tau, J. Biol. Chem., 270, 823-829 (1995).

4. M. Hasegawa, M. R. Jakes, R. A. Crowther, V. M. -Y. Lee, Y. Ihara, and M. Goedert, Characterization of mAb AP422 a novel phosphorylation-dependent monoclonal antibody against tau protein, FEBS Lett. 384, 25-30 (1996).

5. M. Goedert, M. Hasegawa, R. Jakes, S. Lawler, A. Cuenda, and P. Cohen, Phosphorylation of microtubule-associated protein tau by stress-activated protein kinases FEBS Lett. 409, 57-62 (1997).

6. G. N. Patrick, L. Zukerberg, M. Nikolic, S. de la Monte, P. Dikkes, and L. -H. Tsai, Conversion of p35 to p25 deregulates Cdk5 activity and promotes neurodegeneration, Nature, 402, 615-621 (1999).

7. G. Kusakawa T. Saito, R. Onuki, K. Ishiguro, T. Kishimoto, and S. Hisanaga, Calpain-dependent proteolitic cleavage of the p35 cyclin-dependent kinase 5 activator to p25, J. Biol. Chem., 275,17166-17172 (2000)

8. M.-S. Lee, Y. T. Kwon M. Li, J. Peng, R. M. Friedlander, and L.-H. Tsai, Neurotoxicity induce cleavage of p35 to p25 by calpain, Nature, 405, 360-364 (2000).

9. S. Taniguchi, Y. Fujita, S. Hayashi, A. Kakita, H. Takahashi, S. Murayama, T. C. Saido, S. Hisanaga, T. Iwatsubo, and M. Hasegawa, Calpain-mediated degradation of p35 to p25 in postmortem human and rat brains, FEBS Lett. 489, 46-50 (2001).

10. M. Goedert, The significance o tau and a-synuclein inclusions in neurodegenerative diseases, Curr. Opin. Genet. Dev. 11, 343-351 (2001).

11. M. G. Spillantini, J. R. Murrell, M. Goedert, M. R. Farlow, A. Krug, B. Ghetti, Mutation in the tau gene in familial multiple system tauopathy with presenile dementia, Proc. Natl. Acad. Sci. USA, 95, 7737-7741 (1998).

12. M. G. Spillantini, R. A. Crowther W. Kamphorst, P. Heutink, J. C. Van Swieten, Tau pathology in two Dutch families with mutations in the microtubules binding region of tau, Am. J. Pathol. 153, 1359-1363 (1998).

13. M. G. Spillantini, T. D. Bird B. Ghetti, Frontotemporal dementia and parkinsonism linked to chromosome 17: a new group of tauopathies, Brain Pathol. 8 387-402 (1998).

14. M. Hasegawa, M.J. Smith, M. Goedert, Tau proteins with FTDP-17 mutations has a reduced ability to promote microtubule assembly, FEBS Lett. 437, 207-210 (1998).

15. S. Pickering-Brown, M. Baker, S. H.Yen, W. K. Liu, M. Hasegawa, N. Cairns, P. Lantos, M. Rossor, T. Iwatsubo, Y. Davies, D. Allsop, R. Furlong, F. Owen, J. Hardy, D. Mann, M. Hutton, Pick's disease is associated with mutations in the tau gene, Ann. Neurol., 48, 859-867 (2000).

16. M. Hasegawa, M.J. Smith, M. Iijima, T. Tabira, M. Goedert, FTDP-17 mutations N279K and S305N in tau produce increased splicing of exon 10, FEBS Lett. 443, 93-96 (1999).

17. R. Dayanandan, M. Van Slegtenhorst, T.G.A. Mack L. Ko, S. -H. Yen K. Leroy, J.-P. Brion, B.H. Anderton, M. Hutton, S. Lovestone, Mutations in tau reduce its microtubule binding properties in intact cells and affect its phosphorylation. FEBS Lettt., 446, 228-232 (1999).

18. M. Goedert, R. Jakes, R. A. Crowther, Effect of frontotemporal dementia FTDP-17 mutations on heparin-induced assembly of tau filaments, FEBS Lett. 450, 306-311 (1999).

19. S. Barghorn, Q. Sheng-Fischhofer, M. Ackmann, J. Biernat, M. von Bergen, E.-M. Mandelkow, and E. Mandelkow, Structure, microtubule interactions, and paired helical filament aggregation by tau mutations of frontotemporal dementias., Biochemistry, 39, 11714-11721 (2000).

20. M.H.Polymeropoulos, C. Lavedan, E. Leroy, S.E. Ide, A.Dehejia, A. Dutra, B.Pike, H.Root, J. Rubenstein, R. Boyer, et al., Mutation in the gene encodeing _-synuclein in Parkinson's disease, Science, 276, 2045-2047 (1997).

21. M.G. Spillantini, M.L. Schmidt, V.M.-Y, Lee, J.Q.Trojanowski, R.Jakes, M. Goedert, _-synuclein in Lewy bodies., Nature, 388, 839-840 (1997).

22. M. Baba, S. Nakajo, P.H.Tu, T. Tomita, K. Nakaya, V.M.-Y. Lee, J.Q. Trojanowski, T. Iwatsubo, Aggregation of α-synuclein in Lewy bodies of sporadic Parkinson's disease and dementia with Lewy bodies, Am J. Pathol., 152 879-884 (1998).

PRODUCTION AND ANALYSIS OF HUMAN MUTANT TAU (V337M) TRANSGENIC MOUSE

Akihiko Takashima* and Kentaro Tanemura*

1. Introduction

Neurofibrillary tangles (NFTs) are neuronal inclusions of microtuble-binding protein tau and are composed of phosphorylated and ubiquitinated tau aggregations that form β-pleated sheet structures. Given this, NFT-bearing neurons are ubiquitin- and phosphorylation dependent tau-immunoreactive, argyrophilic, Congo red birefringent, and thioflavin-S reactive. Tau dysfunction and filamentous tau aggregates are key markers of neurodegenerative pathologies that display NFTs, which is the most common pathway leading to degeneration of neurons in several neurodegenerative diseases, including Alzheimer s disease (AD). In the brains of patients with AD, tau fibrils form abundant paired helical filaments (PHF) and straight tubules. Discovery of the molecular mechanisms of NFT formation may lead to more insight about events occurring during neurodegeneration in AD. In frontotemporal dementia parkinsonism 17 (FTDP-17), genetic studies revealed that the causative factor is a tau gene containing mutations in its exons and introns. Patients possessing this mutation exhibit NFTs and

---

*   Akihiko Takashima and Kentaro Tanemura, Laboratory for Alzheimer s Disease, RIKEN Brain Science

Institute, 2-1 Hirosawa, Wako-shi, Saitama, Japan 351-0198, e-mail: kenneth@brain.riken.go.jp, tel:+81-

48-467-9704, fax:+81-48-467-5916

Mapping the Progress of Alzheimer's and Parkinson's Disease
Edited by Mizuno *et al.*, Kluwer Academic/Plenum Publishers, 2002

loss of neurons (Clark, L., et l., 1998, Goedert, M., 1998, Hutton, M., et al., 1998, Poorkaj, P., et al., 1998, Spillantini, M.G., et al., 1998, D Souza, I. et l., 1999, Iijima, M., et al., 1999). Thus, tau abnormalities cause the formation of NFTs, which in turn leads to neurodegeneration.

In light of the important link between NFT formation and neurodegenerative diseases, a concerted effort has been made to produce tau transgenic mice. Intronic mutation and some missense mutations of the tau gene alter the splicing and increase the expression of the tau transcript within exon 10, which encodes 4 microtubule binding repeat tau (Goedert et al., 1998; Lee & Trojanowski, 1999). Based on this evidence, three lines of transgenic mice expressing human tau have been produced (Brion et al., 1999; Gotz et al., 1995; Ishihara et al.,1999). Neurons from mice of these lines express a 2-10 fold increase in tau, and display tau inclusions. However, none of these exhibit the Congo red birefringent or thioflavin-S reactive-tau aggregations characteristic of NFTs observed in neurodegenerative diseases. More recently, three groups have reported NFT-like structures in Tg mice expressing human P301L (Lewis et al. 2000; Gotz et al 2001) and shortest tau (Ishihara et al. 2001). P301L tau Tg mice express almost the same levels of mutant tau as endogenous tau and, at about 6 months of age, these mice exhibit NFT-like tau inclusions. Shortest human tau Tg mice express 5-7 times more human tau compared to endogenous tau, and at 24 months of age, these mice exhibit congophilic and argyrophilic tau inclusions. These findings suggest that the robust over-expression of tau is not always necessary for the formation of filamentous tau having β-sheet structures, but the expression of mutant tau from FTDP-17 readily induces it.

Taken together, these findings prompted us to develop an animal model that closely mimics the formation of NFTs by using another missense mutation of tau: V337M of FTDP-17.

2. Production of human mutant tau (V337M) transgenic mouse.

A cDNA construct of the V337M human tau with myc and FLAG epitope tags at the N- and C- terminal ends was generated by PCR-based site-directed mutagenesis (Fig.1a). Three sets of PCR primers were used: (1) sense: CTGATCTCCGAGGAGGACCTGATGGCTGAGCCCCGCCAGGAG; antisense:

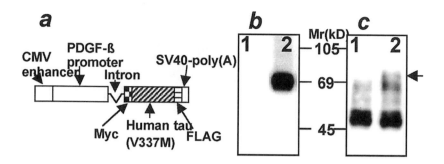

Figure 1. **Analysis of mutant human tau (V337M) expression in brain of Tg (Tg) mice.** **a,** Map of the mutant human tau construct used to generate Tg mice. **b,** Western blot analysis of mutant human tau in brain of Tg mice. The same protein (20 g) of RAB fraction from mice brains was applied to 8% acrylamide gel. Transferred membranes were probed by anti-myc antibody, which specifically recognises the mutant human tau (V337M) at 70 kD. Lane 1, *non*-Tg littermate of Tg214; lane 2, Tg214 **c,** The same samples were probed by the anti-tau antibody, JM, which recognises both endogenous and mutant human tau. Same lane assignments as in **b.**

AGATTTTACTTCCATCTGGCCACCTCC); (2) sense: GGAGGTGGCCAGATGGAAGTAAAATCT; antisense: ATCGTCCTTGTAGTCCAAACCCTGCTTGGCCAGGGA); (3) sense: CCCCTCGAGCCACCATGGAGCAGAAGCTGATCTCCGAGGAGGACCTG; antisense: CCGCGGCCGCTCACTTATCGTCATCGTC).

The entire sequence of cloned cDNA (1.4 kb) was confirmed by an ABI PRISM377 DNA sequencer. This cDNA was inserted into the PDGF-β chain expression vector at the XhoI and NotI sites. A 4.3 kb BglII and NaeI fragment containing PDGF-β promoter, V337M human tau cDNA, and 3 untranslated sequence was used as the transgene to create the mutant tau Tg mice on a B6SJL background. Microinjection of the transgene and generation of Tg mice were performed at DNX Transgenic Sciences (Cranbury, NJ, USA). Southern blot analysis showed that 5-10 copies of the transgene were inserted at a single site in transgenic mice. The expression of mutant tau was confirmed by western blot analysis. Brains taken from transgenic mice and littermates (8-10 months old) were homogenised in

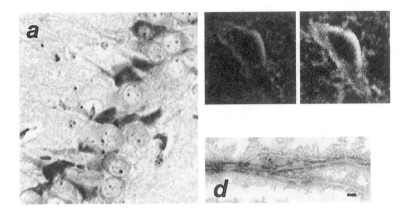

Figure 2. Histochemical characterization of neurons in mutant Tg mice. Sections of the hippocampal region from Tg214 (10 month old) showing: **a**, Gallyas silver staining, Argirophilia is one of the marker for NFTs **b**, Immunostaining with mouse monoclonal antibody Alz50,an antibody specific for a conformational epitope found in PHF. **c**, Thioflavin-S staining, which recognizes the β-sheet structure (**a-e**; scale bars, 10 μm) **d**, Immuno-EM analysis of RIPA-insoluble materials. Before solubilizing the RIPA-insoluble pellet in formic acid, the pellet was dispersed in PBS and investigated at the ultrastructural level after immunostaining with the JM, anti-tau antibodies. Images show bundles of (a) JM immunoreactive (5 nm gold particle) fibrils.

reassembly buffer (RAB), and then tau was recovered as the heat stable fraction. These fractions were separated on SDS-PAGE gel (8% acrylamide). Separated proteins were transferred onto a PVDF membrane (Millipore). Anti-myc antibody recognised the 70kD band in transgenic mice but not in littermates (Fig.1b). Treatment with JM, an anti-tau antibody, recognised both 70kD human tau from the transgene and 50kD endogenous tau (Fig. 1c). Although the amount of mutant tau varied among mice, even within the same transgenic line, the levels of transgenic tau expressed were generally less than one tenth of the endogenous tau levels. The regional localisation of mutant tau in the brains of transgenic mice was demonstrated by immunostaining with anti-myc antibody and anti-tau antibody JM. Although PDGF- β chain promoter activity is found throughout the mouse brain, anti-myc immunoreactivity localised to some neurones in the cerebral cortex and hippocampal region. Myc-immunoreactive neurones were also double-labelled by anti-tau antibody JM (Fig.2c-f), suggesting that myc-positive neurons do expressed human mutant tau.

3. Pathological determination of human tau transgenic mouse

To determine whether these Tg mice form NFT, mice brain was histologically investigated by Gallyas silver staining, (Fig.2a) phosphorylation or conformational change of PHF-tau dependent anti-tau antibodies (Alz50, PS199 and AT8) (Fig.2b), and Thioflavin-S (Fig.2c). Neurons were immunoreactive with tau antibodies recognizing conformational (Alz50), and phosphorylated epitopes (PS199 and AT8) in CA1 and CA2 regions of hippocampus. In the same region, Gallyas silver staining positive neurons were detected (Fig. 2a), suggesting that Tg mice form NFTs in hippocampus region. This was confirmed by Thioflavin-S. We stained by Thioflavin-S, and observed under polarizing filter. Thioflavin-S positive neurons were detected in hippocampus region (Fig. 2c). Ultrastructually, hippocampal neurons showed filamentous aggregates in cytoplasm, and most filaments were straight with 15 nm diameter (Fig.2d). Thus, the expression of mutant tau V337M in mice brain induced the fibrilar tau inclusion in neurons of hippocampus, similar to human NFTs observed in neurodegenerative disorders. Since the expression level of mutant tau is less than 10 % of endogenous tau in our Tg mice, the nature of tau, rather than the intracellular concentration of tau, may be the determining factor in the formation of NFTs. The exact biochemical events underlying the formation of NFTs from tau in these Tg mice is yet to be determined.

REFERENCES

Brion, J. P., Tremp, G., and Octave, J. N. (1999). Transgenic expression of the shortest human tau affects its compartmentalization and its phosphorylation as in the pretangle stage of Alzheimer's disease. Am J Pathol 154, 255-70.

Clark, L. N., Poorkaj, P., Wszolek, Z., Geschwind, D. H., Nasreddine, Z. S., Miller, B., Li, D., Payami, H., Awert, F., Markopoulou, K., Andreadis, A., D'Souza, I., Lee, V. M., Reed, L., Trojanowski, J. Q., Zhukareva, V., Bird, T., Schellenberg, G., and Wilhelmsen, K. C. (1998). Pathogenic implications of mutations in the tau gene in pallido-ponto- nigral degeneration and related neurodegenerative disorders linked to chromosome 17. Proc Natl Acad Sci U S A 95, 13103-7.

D'Souza, I., Poorkaj, P., Hong, M., Nochlin, D., Lee, V. M., Bird, T. D., and Schellenberg, G. D. (1999).

Missense and silent tau gene mutations cause frontotemporal dementia with parkinsonism-chromosome 17 type, by affecting multiple alternative RNA splicing regulatory elements. Proc Natl Acad Sci U S A *96*, 5598-603.

Goedert, M., Crowther, R. A., and Spillantini, M. G. (1998). Tau mutations cause frontotemporal dementias. Neuron *21*, 955-8.

Gotz, J., Probst, A., Spillantini, M. G., Schafer, T., Jakes, R., Burki, K., and Goedert, M. (1995). Somatodendritic localization and hyperphosphorylation of tau protein in transgenic mice expressing the longest human brain tau isoform. Embo J *14*, 1304-13.

Gotz, J., Chen, F., Barmettler, R., Nitsch, R.M. (2001). Tau Filament Formation in Transgenic Mice Expressing P301L Tau. J. Biol. Chem. *276*, 529-534.

Hutton, M., Lendon, C. L., Rizzu, P., Baker, M., Froelich, S., Houlden, H., Pickering-Brown, S., Chakraverty, S., Isaacs, A., Grover, A., Hackett, J., Adamson, J., Lincoln, S., Dickson, D., Davies, P., Petersen, R. C., Stevens, M., de Graaff, E., Wauters, E., van Baren, J., Hillebrand, M., Joosse, M., Kwon, J. M., Nowotny, P., Heutink, P., and et al. (1998). Association of missense and 5'-splice-site mutations in tau with the inherited dementia FTDP-17. Nature *393*, 702-5.

Iijima, M., Tabira, T., Poorkaj, P., Schellenberg, G. D., Trojanowski, J. Q., Lee, V. M., Schmidt, M. L., Takahashi, K., Nabika, T., Matsumoto, T., Yamashita, Y., Yoshioka, S., and Ishino, H. (1999). A distinct familial presenile dementia with a novel missense mutation in the tau gene. Neuroreport *10*, 497-501.

Ishihara, T., Hong, M., Zhang, B., Nakagawa, Y., Lee, M. K., Trojanowski, J. Q., and Lee, V. M. (1999). Age-dependent emergence and progression of a tauopathy in transgenic mice overexpressing the shortest human tau isoform. Neuron *24*, 751-62.

Lee, V. M., and Trojanowski, J. Q. (1999). Neurodegenerative tauopathies: human disease and transgenic mouse models. Neuron *24*, 507-10.

Lewis,J., McGowan, E, Rockwood, J., Melrose, H., Nacharaju, P., Van Slegenhorst, M., Gwinn-Hardy, K., Murphy, M.P., Baker, M., Yu, X., and et al. (2000). Neurofibrillary tangles, amyotrophy and progressive motor disturbance in mice expressing mutant (P301L) tau protein. Nature Genetics *25*, 402-405.

Poorkaj, P., Bird, T. D., Wijsman, E., Nemens, E., Garruto, R. M., Anderson, L., Andreadis, A., Wiederholt, W. C., Raskind, M., and Schellenberg, G. D. (1998). Tau is a candidate gene for chromosome 17 frontotemporal dementia. Ann Neurol *43*, 815-25.

Spillantini, M. G., Murrell, J. R., Goedert, M., Farlow, M. R., Klug, A., and Ghetti, B. (1998). Mutation in the tau gene in familial multiple system tauopathy with presenile dementia. Proc Natl Acad Sci U S A *95*, 7737-41.

INDEX